ACTUATORS

Other SAE books in this series:

Sensors and Transducers
by Ronald K. Jurgen
(Order No. PT-68)

Object Detection, Collision Warning and Avoidance Systems
by Ronald K. Jurgen
(Order No. PT-70)

Electronic Instrument Panel Displays
by Ronald K. Jurgen
(Order No. PT-71)

Navigation and Intelligent Transportation Systems
by Ronald K. Jurgen
(Order No. PT-72)

Electronic Engine Control Technologies
by Ronald K. Jurgen
(Order No. PT-73)

For information on these or other related books, contact SAE by phone at (724) 776-4970, fax (724) 776-0790, e-mail: publications@sae.org, or the SAE website at www.sae.org

ACTUATORS

Automotive Electronics Series

PT-74

Edited by

Ronald K. Jurgen

Published by
Society of Automotive Engineers, Inc.
400 Commonwealth Drive
Warrendale, PA 15096-0001
U.S.A.
Phone (724) 776-4841
Fax: (724) 776-5760
www.sae.org

ISBN 0-7680-0395-4
Library of Congress Catalog Card Number: 98-88834

SAE Order No. PT-74

Introduction

The Workhorses of Electronic Control Systems

No electronic control system can function without sensors at the input and actuators at the output. Both are as important to the operation of the system as the electronic control unit, which acts on the sensor inputs and commands the actuators to regulate energy, mass, and volume flows.[1] In a real sense, actuators are the workhorses of electronic control systems.

Actuators are available in a variety of sizes, shapes, and basic operating principles. There are, for example, magnetic actuators, electrical actuators and thermal actuators. Magnetic actuators include dc solenoids, torque motors, electromagnetic step motors, moving coils, dc motors, and magnetostrictive actuators.[1] Electrical actuators include piezoelectric and electrostatic devices and electrorheological fluids. Thermal actuators include temperature-sensitive bimetallic elements, memory alloys, and expansion elements.[1] These, however, are only a few examples; there are many other types of actuators.

This book contains a collection of papers on actuators that have been selected to give a broad overview of the versatility of actuators in a variety of different uses. In general, the papers are grouped by type of application; however, within each application group, a number of different types of actuators are represented. For example, in the fuel metering systems group, the papers describe various fuel injectors including those based on piezo-fluidic combinations, pulse-width-modulated sonic flow, in-cylinder injection, metering valves, and monostable fluids. In the suspension systems group, the papers describe actuators that are based on electrorheological fluids, electromagnetics, and piezoceramics.

In addition to the papers grouped by type of application, there is a group of papers on multipurpose actuators; and another on mechatronics—the use of mechanical, electrical, electronic, and computer concepts in a single device or system.

In a look to the future, Randy Frank of Motorola discusses the increasing use of power electronics for actuators, the growing trend in mechatronics, and the need for higher operating voltages in vehicles to handle the ever increasing power consumption.

Reference

1. Klaus Muller, Chapter 10, "Actuators," in the *Automotive Electronics Handbook*, McGraw-Hill, Inc., New York, N.Y., 1995.

* * * * *

This book and the entire automotive electronics series is dedicated to my friend Larry Givens, a former editor of the SAE monthly publication, *Automotive Engineering.*

Ronald K. Jurgen, Editor

Table of Contents

Introduction

Actuators in Fuel Metering Systems

Engine Valve Control

Actuators in Brake and Cruise Control Systems

Actuators in Steering Systems

Actuators in Suspension Systems

Actuators in Vibration and Noise Control

Actuators in Restraint Systems

Multipurpose Actuators

Mechatronics

A Look to the Future

ACTUATORS IN FUEL METERING SYSTEMS

980493

CFD-Aided Development of Spray for an Outwardly Opening Direct Injection Gasoline Injector

Min Xu and Lee E. Markle
Delphi Automotive Systems

ABSTRACT

A high pressure outwardly opening fuel injector has been developed to produce sprays that meet the stringent requirements of gasoline direct injection (DI) combustion systems. Predictions of spray characteristics have been made using KIVA-3 in conjunction with Star-CD injector flow modeling. After some modeling iterations, the nozzle design has been optimized for the required flow, injector performance, and spray characteristics. The hardware test results of flow and spray have confirmed the numerical modeling accuracy and the spray quality.

The spray's average Sauter mean diameter (SMD) is less than 15 microns at 30 mm distance from the nozzle. The DV90, defined as the drop diameter such that 90% of the total liquid volume is in drops of smaller diameter, is less than 40 microns. The maximum penetration is about 70 mm into air at atmospheric pressure. An initial spray slug is not created due to the absence of a sac volume.

INTRODUCTION

Improved fuel economy, reduced engine-out emissions, quick starting and fast transient response have been considered as major advantages of gasoline DI spark-ignition (SI) engines over port fuel injection (PFI) SI engines. Although DI's great theoretical potential has been recognized for a long time, the technology realization and implementation has been delayed due to many technical difficulties [1-3]. The current development has demonstrated some of the DI advantages, but it has also unveiled many problems and new challenges. In particular, excessive unburned hydrocarbon (UHC) emissions, injector fouling, low conversion efficiency and durability of the lean NOx catalysts remain as the major concerns [1]. New technology breakthroughs and in-depth understanding of the fundamentals are necessary for further improvement and development of gasoline DI engines.

In gasoline DI engines, fuel is directly injected into the cylinder, therefore the spray characteristics and mixture formation are of primary importance. Depending on the engine operating condition, the injection time window and the available cylinder volume for the injected fuel are often very constrained. Small droplet size, rapid evaporation, stable ignition, appropriate mixing (or charge stratification), and minimal wall impingement are generally required for DI engines [4].

For high engine load conditions, injection during the intake stroke (early injection) requires a large spray angle and sufficient penetration for maximum air utilization. However, excessive penetration aided by intake air flow and low cylinder pressure can result in significant cylinder wall and/or piston wetting [5]. A well atomized spray is required for homogeneous mixture formation prior to ignition.

For part load conditions, injection occurs during the compression stroke (late injection) to achieve adequate charge stratification for lean burn combustion, which is desired for high thermal efficiency. During the compression stroke, the cylinder pressure and temperature increase with crank angle. The charge turbulence intensity increases substantially near TDC due to the collapse of the large scale flow structures. This phenomenon tends to improve the fuel droplet evaporation and air-fuel mixing. However, the increased air drag causes rapid deceleration of fuel droplets and increases the probability of droplet coalescence and the formation of large droplets. The prompt atomization of a spray with a short intact liquid length must take place in such severe environments. The excessive UHC emissions experienced in many gasoline DI engines have been attributed to the flame quenching in the overly lean region and to wall wetting. To avoid piston impingement and to create an appropriate degree of spatial charge stratification for stable ignition and combustion, the spray should be compact with low penetration [6].

Consequently, to ensure a high combustion efficiency and low emissions levels, a well-atomized spray is always required over the entire engine operating range. An SMD of 15 microns or smaller with Rosin-Rammler N values greater than 1.9 (equivalent to DV90 smaller than 43 microns) was set as the target for the DI sprays [7]. DI combustion systems are very sensitive to spray quality. A few large droplets could significantly degrade the emissions and combustion efficiency. Therefore, a spray with a maximum droplet size larger than 50 microns cannot be accepted. For typical automotive engines, the spray penetration distance should be less than 80 mm in the early injection mode and less than 30 mm in the late injection stratified charge mode. Fortunately, the cylinder pressure variation as a function of the crank angle affects the penetration and angle of the spray favorably for the DI requirements.

In order to meet the severe DI requirements, significant effort has been placed on fuel injector development. Among many design concepts, the inwardly opening high pressure swirl type injector is predominant [1]. High liquid velocity resulting from high fuel pressure is needed for rapid breakup of the bulk liquid and secondary atomization. Swirl motion imparted on the fuel droplets has been effectively utilized for spray penetration reduction while maintaining adequate atomization. In high cylinder pressure environments, angular momentum was found to be preserved for a longer time than the momentum in other directions [8]. Thus, smaller droplet sizes can be expected for a swirling spray compared to a non-swirling spray in the late injection regime. A hollow cone spray can be generated from a swirl type injector. Swirl also affects the spray angle due to the centrifugal forces exerted on the individual droplets. Various spray angles can be obtained by varying the swirl intensity.

Although the inwardly opening high pressure swirl injector has many advantages, there are also drawbacks associated with the design. Because the spray characteristics such as drop size, spray angle, and penetration are all closely related to the swirl intensity, the customized tailoring of a spray for a specific combustion system requirement is difficult. Trade-offs and compromises have to be made case by case, depending on which characteristics are most critical to the application. For example, combustion chamber geometry constraints may require a small spray angle. A small spray angle may be accompanied by relatively large droplets and penetration. Therefore, a lower fuel supply pressure may be required for penetration control, but a lower pressure may increase the droplet size.

In general, excessive spray penetration and droplet size are still problems associated with most DI sprays. Wall-controlled charge stratification combustion systems have been developed and realized in production engines, partially because of their insensitivity to spray quality [2,3]. These spray-wall impingement systems, however, risk producing UHC emissions. An initial fuel slug, created by most inwardly opening DI injectors, travels ahead of the main spray body [9]. Compared to the main spray body, the slug has a larger speed, and a longer penetration distance. The slug possesses large droplets which contain a notable percentage of the total fuel injected per pulse. Also, combustion product deposits on the inwardly opening injector tip may degrade spray quality and reduce fuel flow.

A new design concept which has the potential to produce an improved spray meeting the critical gasoline DI requirements, has been implemented into a Delphi DI gasoline injector, by successfully utilizing computational fluid dynamics (CFD). The Delphi DI injector has an outwardly opening nozzle. A hollow cone spray is formed from the nozzle. Different from the inwardly opening pressure-swirl injectors, the Delphi DI injector has design flexibility which may allow the spray angle, penetration and droplet size to be controlled separately. Experimental spray analysis has shown that the Delphi DI injector's spray meets the stringent DI combustion system requirements. The nozzle design procedure, numerical modeling and test results will be discussed in this paper.

NOZZLE DESIGN

Of the hundreds of different nozzle concepts for spray generation, two main classifications exist: round jet and liquid sheet disintegration. For a fixed flow rate, an annular liquid sheet has a much larger surface area than a cylindrical liquid jet. Therefore, liquid sheets are more easily disintegrated into small droplets. Annular liquid sheets can be generated by either pressure-swirl inwardly opening nozzles or outwardly opening nozzles. However, outwardly opening nozzles have the following advantages over inwardly opening nozzles. A liquid sheet is formed by extrusion of the liquid through the conical seat annulus, rather than by centrifugalization of swirling liquid as in inwardly opening nozzles. As the sheet moves outwardly from the nozzle axis, the thickness of the sheet will decrease because of mass conservation. The initial sheet thickness is directly controlled by the pintle valve stroke regardless of swirl intensity. In general, the spray droplet size, characterized by SMD, is affected by the combination of the sheet thickness, the sheet velocity (i.e., the fuel supply pressure), and the initial disturbances. Outwardly opening nozzles, therefore, allow for additional atomization control by adjusting the initial liquid sheet thickness. The spray angle is also predetermined geometrically by the cone angle of the seat passage. Although swirl is not necessary for the liquid sheet generation and spray angle determination, it can be used for reducing penetration, and also for extending the spray angle and evenly distributing the fuel. The control of the droplet size, spray angle and penetration can be decoupled, to some extent, by varying the initial sheet thickness, the valve and seat angle, and the swirl intensity, respectively. Therefore, it is possible to create a well-atomized and compact spray for late injection stratified charge combustion, even though small drop size, small penetration and small spray angle in a single spray have been considered as contradictory requirements [6].

The outwardly opening nozzle has other benefits as well. During the valve opening and closing, the initial sheet thickness is small. The upstream pressure of the discharge orifice is greater than the fully opened condition due to decreased pressure drop at the swirler. Therefore, the initial spray and the spray tail have even smaller droplets than the main spray. There is no sac volume between the valve seal surface and the discharge orifice. The potentially detrimental initial fuel slug is not created. The outwardly opening valve continuously cleans the conical valve seat. There are no nozzle holes exposed to the combustion environment. Therefore, this design should be robust to combustion product deposits, and nozzle clogging should not be a problem.

Another problem associated with the inwardly opening swirl nozzle is flow control. The minimum flow area varies from the valve seat to final nozzle orifice during opening. The discharge coefficient also varies depending on the instantaneous flow velocity. However, in outwardly opening nozzles, the flow area is a simple function of the valve stroke, and the discharge coefficient is nearly constant and almost independent of the flow velocity. Hence, precise flow control is easily realized. Based on the above-mentioned potentials, the design concept of the outwardly opening nozzle with an internal swirl generating feature was selected for the Delphi DI injector.

Recently, computational fluid dynamics techniques have been used increasingly in industry and research facilities for improved accuracy, fast turn-around and relatively low cost, as compared to experiments. In the Delphi DI injector development process, CFD modeling was used extensively prior to the hardware design and build. Star-CD (Adapco Co.) was chosen for the injector internal flow modeling. The angles of the valve and seat were determined from the spray angle target. The valve stroke was derived by considering the required spray drop size, the internal contamination sensitivity and the actuator constraints. The critical dimensions of the injector were optimized through the flow modeling until the specific requirements, such as the steady-state flow rate, swirl strength, pressure distribution inside the injector, and force balance for reliable seal and fast actuation, were met. Once the flow modeling was accomplished, its output was input to the multidimensional CFD program, KIVA-3, for spray simulation and assessment [10,11]. Ambient pressures of 100 kPa absolute and 1.5 MPa absolute were chosen for representing the cylinder pressures of the early injection and late injection modes, respectively. Special treatments and modifications were incorporated into the program for simulation of the hollow cone spray emanating from the annular orifice of the outwardly opening Delphi DI injector. The predicted spray characteristics were used to modify the injector spray controlling features as needed. This procedure was repeated several times until the resulting spray met the targets. During the design process, the limitations and constraints of actual injector design and manufacturability were taken into consideration. Finally, the nozzle design was completed and implemented into the Delphi DI injector by efforts of the injector design and manufacturing teams. Figure 1 shows a sketch of the Delphi DI injector.

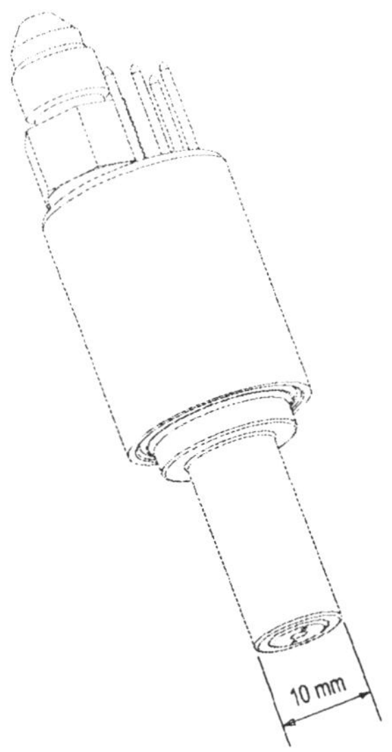

Figure 1: Sketch of Delphi DI gasoline injector.

CFD MODELING RESULTS

The flow passages inside the Delphi DI injector are complicated. However, the most restriction is in the lower part of the injector including the nozzle. The pressure drops occurring in all upstream passages are assumed to be negligible. Therefore, only the flow domain in the lower part of the injector was meshed and modeled with Star-CD. Pressure boundary conditions were imposed on the inlet and outlet of the model. The standard k-ε turbulence model was used for the turbulence simulation. Flow was assumed to be incompressible. Typical gasoline fluid properties (e.g., density of 736 kg/m^3 and viscosity of 4.25×10^{-4} Pa·s) were used in the model. A steady-state flow solution was obtained for the fully opened injector during the main injection period.

Figure 2 is a contour plot of the pressure distribution in the nozzle model. The main pressure drops occur at the swirler holes and the annular discharge orifice. Negative pressure in a portion of the valve seat passage indicates the presence of cavitation, but the actual cavitation phenomena is not simulated because Star-CD does not have a cavitation model. However, the flow model is still valid. Figure 3 shows the flow velocity vectors of the flow model. Large rotational velocities are generated in the swirl chamber. They are sustained downstream until exiting the injector. The modeling results at the injector exit for the two ambient pressures are summarized in Table 1. A slightly reduced flow is predicted in the elevated ambient pressure case due to decreased pressure differential across the nozzle.

Table 1: Summary of the Star-CD flow modeling of the Delphi DI injector

Characteristic	P_a=100 kPa	P_a=1.5 MPa
Initial swirl angle*	17.8°	17.7°
Initial cone angle	60°	60°
Steady-state flow rate	11.59 g/s	10.74 g/s
Turbulence intensity	7.1%	7.1%

* Initial swirl angle is defined as the arctangent of the ratio of swirl velocity to metering velocity.

KIVA-3 is a multidimensional modeling code developed for spray dynamics and combustion simulation particularly in internal combustion engines. In this paper, only the work of modeling the DI sprays injected into a constant volume, initially still chamber is reported. To accommodate the annular opening orifice of the Delphi DI injector, the KIVA-3 code was appropriately modified. The Star-CD modeled injector flow was directly input into the KIVA-3 model. The initial swirling velocity component and the turbulence were incorporated into the KIVA-3 model with some modifications [12,13]. A square wave form was assumed for the injection pulse. A reference flow rate of 14 mg per injection was chosen for the spray analysis. The injected gasoline was divided into 5000 droplet parcels. The initial drop size was set as the initial liquid sheet thickness. Chamber pressure was varied from 100 kPa absolute to 1.5 MPa absolute at room temperature (20°C).

Many of the complex processes related to spray including: liquid breakup, secondary atomization of the droplets, liquid droplet vaporization, air entrainment into the spray, mixture formation, gas and liquid droplet dynamics, droplet collision and coalescence, as well as air and liquid droplet interactions are modeled in the KIVA simulation [10]. By analyzing the spatially and temporally resolved spray simulation, characteristics of interest, such as the spray angle, penetration, SMD, DV90 and the Rosin-Rammler N value, are obtained. In addition, the spray-induced air flow, vapor concentration distribution, pressure and temperature distribution, and turbulence intensity, etc., are also available.

The snapshot of the spray modeled at 1 ms after the start of injection (SOI) into air at 100 kPa absolute pressure is illustrated in Figure 4 (a), and its sectional view is shown in Figure 4 (b). The spray induced air flow field is illustrated in Figure 4 (c). The penetration of the spray leading edge is 34 mm from the injector exit. In the model, instead of an intact liquid sheet, discrete but closely packed liquid droplet parcels are assumed. A very narrow high flux region is predicted along the hollow cone sheet path. The initial spray has an angle of approximately 60°, following the injector conical seat. Beyond the distance of 15 mm, the spray region becomes much thicker. Many small droplets are produced as a result of the secondary atomization and the continuous evaporation of the droplets. The small droplets lose their momentum quickly against the air drag force, and therefore are easily carried by the air flow. Figure 4 (c) depicts a large vortex in the air flow near the outer edge of the spray's main flux cone. This was confirmed numerically and experimentally in both swirling sprays and non-swirling sprays. Therefore, this vortex is not induced by the swirl motion imparted on the spray. It is more likely that the velocity profile relaxation of the liquid spray increases the thickness of the air entraining layer. This layer consists of a large coherent vortex structure with low velocity. The spray is also widened by the small droplets laden in this vortex. Accordingly, the spray angle is increased to 70°. This large scale vortex enhances the momentum transfer between the liquid droplets and surrounding gaseous phase, accelerating the spray dispersion process. As a result, a short penetration and a slightly increased spray angle are obtained.

In Figure 4 (a) and (b), larger droplets are shown at the spray leading edge than elsewhere. This result could be attributed to droplet coalescence, vanishing of the small droplets due to rapid vaporization, and large droplets with high initial momentum. Figure 5 shows the temporal variation of the SMD, spatially integrated at the plane 30 mm from the injector. At 0.6 ms after the SOI, the spray leading edge arrives at the sampling plane. The SMD of the spray leading edge is about 23 microns, becoming smaller with time as the main spray body passes the 30 mm plane. The average SMD of the main spray body is 12 microns. The SMD as a function of time and radial position in the same plane is shown in Figure 6. The large droplets reside on the spray outer edge as expected for a swirling spray. However, the number density of these large droplets is very low. Figure 7 shows the radially-integrated SMD as a function of the axial location and time. It is evident that the drop size of the spray leading edge is larger than the main spray. It increases with distance and time. DV90 is an appropriate term describing the large droplet characteristics of the spray.

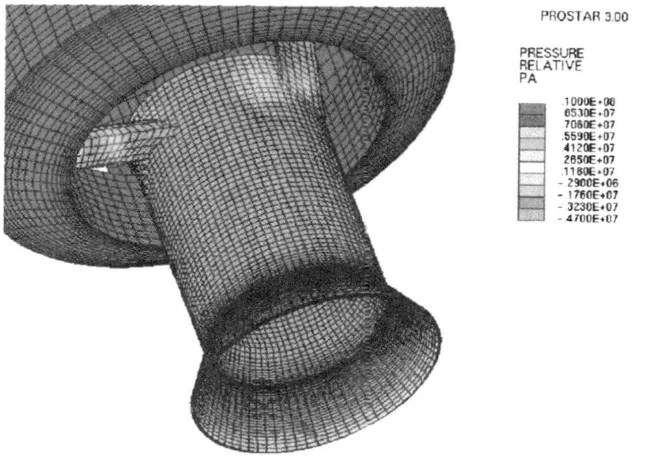

Figure 2: Star-CD modeled pressure distribution inside the Delphi DI injector.

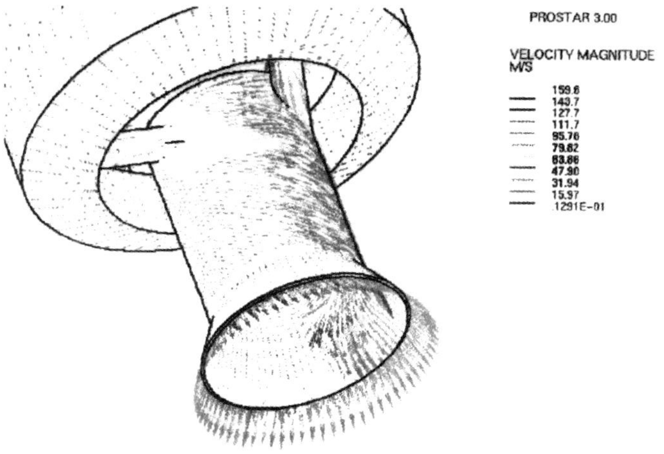

Figure 3: Velocity vectors of the injector internal flow predicted by the Star-CD injector flow model.

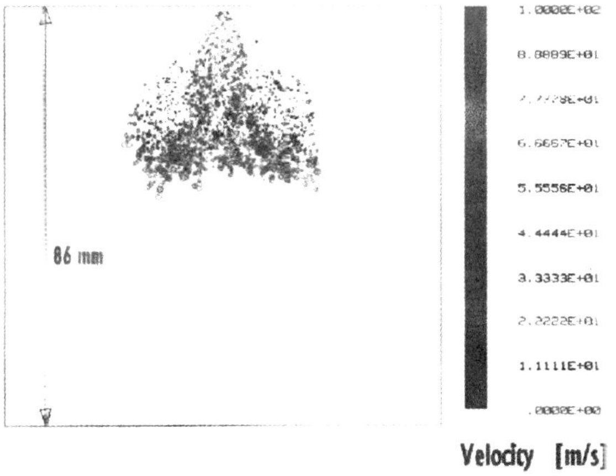

(a)

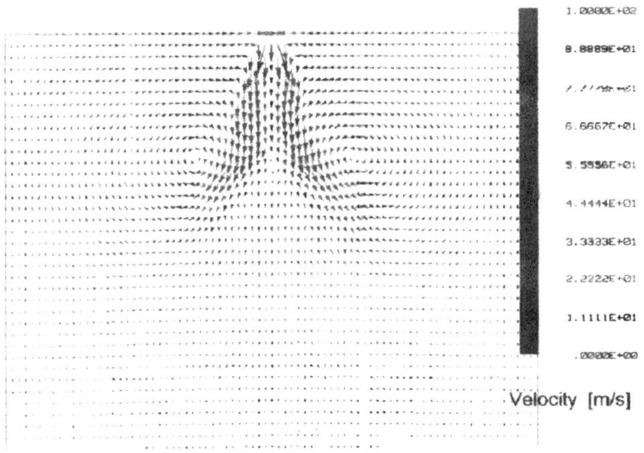

(c)

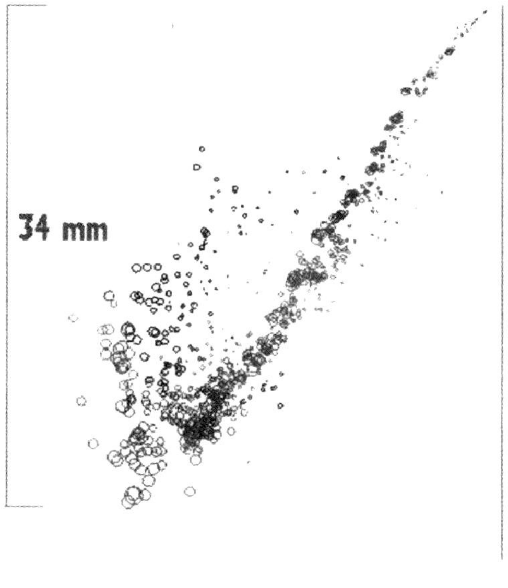

(b)

Figure 4: KIVA-3 spray modeling results for the Delphi DI injector spray into 100 kPa absolute ambient pressure at 1 ms after the start of injection. (a) Spray droplet location, relative size and velocity; (b) Enlarged 2-D sectional view of the half spray; (c) Spray induced air flow velocities.

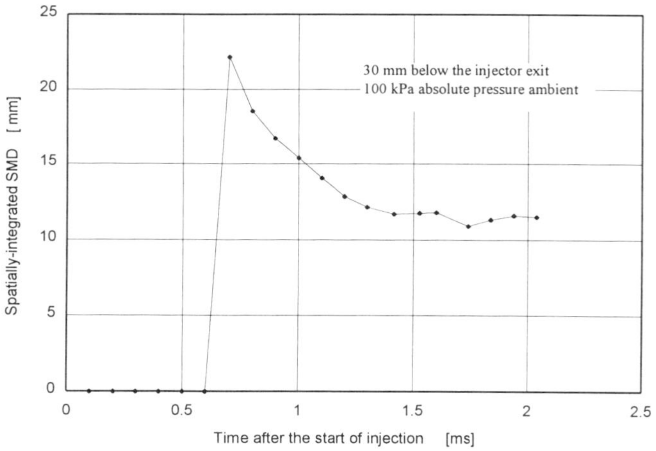

Figure 5: Spatially-integrated SMD as a function of time predicted by KIVA-3.

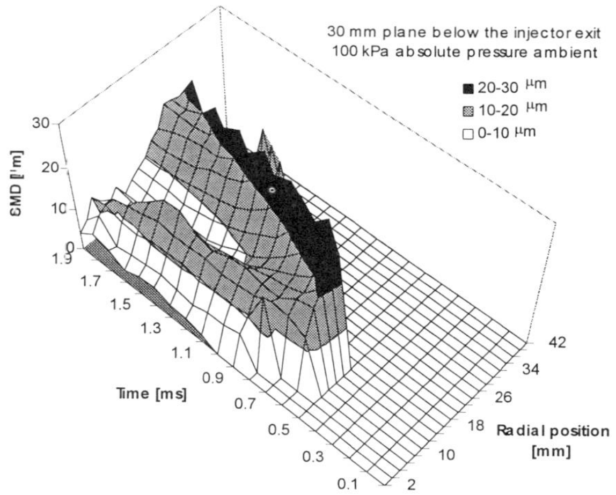

Figure 6: Spatially-resolved SMD as a function of time after the SOI predicted by KIVA-3.

Table 2: The Delphi DI injector spray characteristics predicted by KIVA-3

Characteristic	P_a=100 kPa	P_a=1.5 MPa	Requirement/ Design target (100 kPa)
Penetration at 1 ms	34 mm	16 mm	30 mm
Spray angle at 1 ms	70°	76°	60°
SMD at 30 mm	13.4 μm	42 μm	15 μm
DV90 at 1 ms	28 μm	58 μm	40 μm
Rosin-Rammler N value	2.138	1.646	>1.9

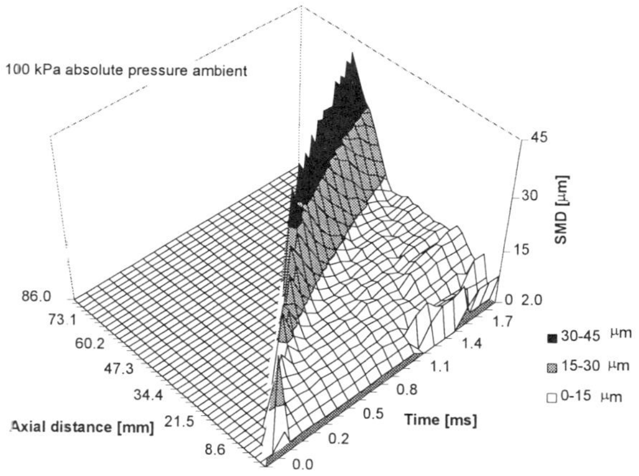

Figure 7: Spatially-integrated SMD as a function of axial distance and time after the SOI computed by KIVA-3.

The modeled spray characteristics are summarized in Table 2 with the DI spray requirements. Since the requirements are specified only for the spray injected into 100 kPa absolute pressure, the Delphi DI injector spray is verified to be sufficient. The spray injected into 1.5 MPa absolute pressure is very compact, even though the spray angle increased slightly. Both the axial penetration and the radial width of the spray are much smaller compared to the 100 kPa absolute pressure condition. The largely increased droplet size of the spray may be due to the increased probability of droplet collision and coalescence. However, in a real engine, high cylinder pressure is usually accompanied by high temperature during the compression stroke. The enhanced heat transfer and rapid evaporation rate could cancel some of the possible coalescence effects.

EXPERIMENTAL SPRAY ASSESSMENT

The Delphi DI fuel injector was tested per a standard protocol developed for evaluation of DI injectors. This protocol involves identifying the injector flow as a function of pulse width at the operating fuel pressure of 10 MPa and spray tests. Experimental assessments of sprays were conducted so as to identify the macroscopic spray parameters first, then to perform more detailed analyses by using the results obtained from the initial spray studies. An injector flow of 14 mg per pulse was selected for the spray testing. Spray axial symmetry was investigated to provide indication of mass distribution of the spray. The results of the spray symmetry evaluation guided the testing to characterize the spray structure by time-resolved still imaging techniques. The injector was oriented in a manner to expose the maximum maldistribution of the spray in the images. All subsequent spray testing was performed with the injector in this same orientation. Instantaneous images were sequentially recorded to visualize the spray structure and the development of the spray from the injector exit. Spray angle and penetration were determined from the images of the spray captured as a function of time after injector logic.

The spray droplet size distribution was evaluated in two ways: spatially-integrated and spatially-resolved. The spatially integrated droplet size measurement was performed along a line passing through the spray center and the spray's maximum flux region at an axial distance of 30 mm from the injector exit. The breakup of the liquid sheet occurred before the measuring plane. This was verified by the imaging spray structure evaluation. Using the results of the spatially-integrated droplet sizing, the appropriate optics and instrument settings were chosen for performing a spatially-resolved droplet size and axial velocity interrogation of the spray.

FLOW TESTING

Prior to spray evaluation, the flow performance of the injector was verified. Stoddard solvent at 10 MPa was supplied to the injector, and the injector steady-state flow rate of 11.5 g/s was verified with the pintle held open. Then, the dynamic response of the injector was determined. The injector logic pulse width corresponding to 14 mg of liquid per pulse was determined with the injector cycling at 100 Hz. An accelerometer was used to check the opening and closing response of the pulsing injector. All of the spray experiments were conducted at the 14 mg per injection pulse width, 1.30 ms, as determined by this procedure. The fluid used for all of the spray tests was California Phase II gasoline supplied to the injector at 10 MPa. The frequency of injections was varied for the spray tests to accommodate limitations in the measurement techniques and equipment, while allowing the spray data to be collected most efficiently. The operating characteristics of the Delphi DI injector are summarized in Table 3.

Table 3: Delphi DI injector operating characteristics

Fuel supply pressure	10 MPa
Operating voltage	13.5 volts
Steady-state flow rate	11.5 g/s
Opening and closing response	0.3 ms
Minimum flow	1.5 mg/pulse

SPRAY SYMMETRY ASSESSMENT

The injector spray symmetry was determined by use of a spray patternator. The patternator is approximately 90 mm in diameter and consists of 253, 4.5 mm diameter cells arranged in a hexagonal pattern to minimize the space between the cells. Each cell is shaped to allow the spray droplets which hit the patternator between the cells to be guided into the adjoining cells. The center-to-center distance between two adjacent cells is 5.5 mm. Testing was performed by centering the injector 30 mm above the patternator. The injector was cycled at a frequency of 25 Hz for 1000 pulses, so as to fill the cells as much as possible without overflowing any cell. Analysis of a typical Delphi DI injector spray is shown in Table 4. Some spray skew was observed. The orientation of the region of maximum flux was noted and used for subsequent experiments.

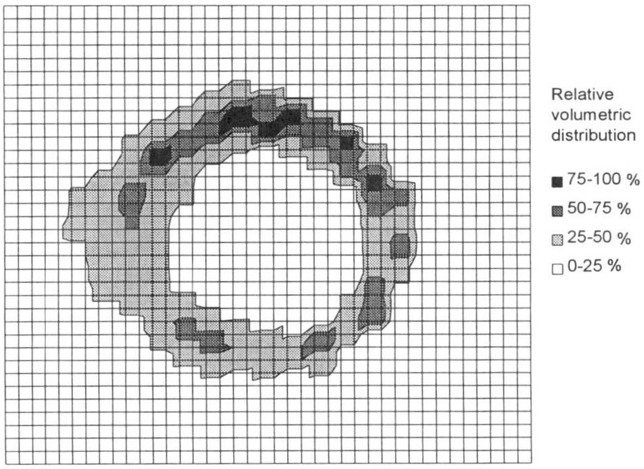

Figure 8: Patternator spray volume distribution for the Delphi DI injector.

Table 4: Spray patternator test results

Characteristic	Test result
Offset of spray centroid point	5.5 mm
10 % included volume radius	18.3 mm
50 % included volume radius	24.4 mm
90 % included volume radius	32.2 mm

The individual cell volume data was transformed to Cartesian coordinates and is shown in Figure 8.

SPRAY STRUCTURE AND DEVELOPMENT CHARACTERIZATION

The structure and development of the Delphi DI injector spray were assessed with imaging. The spray angle, penetration and breakup of the liquid sheet can be observed in the collected images. By capturing multiple images of the spray at various times after injector logic, it was possible to measure spray penetration versus time. The injector was housed in a nitrogen-purged windowed chamber capable of simulating engine cylinder pressures. Images of the spray at 100 kPa absolute and 1.5 MPa absolute chamber pressures were captured with a Kodak 8-bit, 1000 by 1000 pixel digital camera and a Bitflow frame-grabber board. These images were processed with Optimas image analysis software. The illumination source for the images was a 10 ns pulse, Nd:YAG laser. The laser beam was transformed optically to a vertical sheet of light approximately 0.5 mm thick. The light sheet was centered on the injector axis and the injector was rotated to allow the light sheet to intersect the spray's maximum flux region. The injection frequency was set at 4 Hz because of the required time for the laser and camera synchronization. Figure 9 shows an image of the Delphi DI injector spray at 1 ms after injector logic at 100 kPa absolute pressure. The large vortices predicted by the modeling (Figure 4) are apparent in the image. A comparison of Figure 4 and

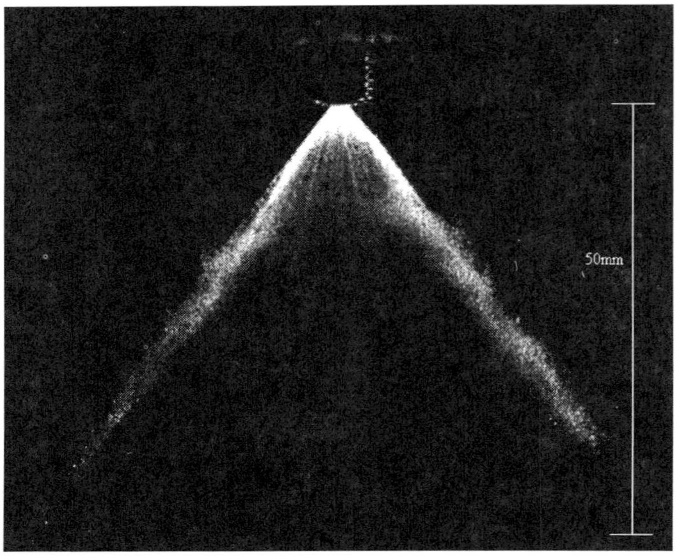

Figure 9: Laser sheet spray image at 1 ms after the SOI at 100 kPa absolute pressure ambient.

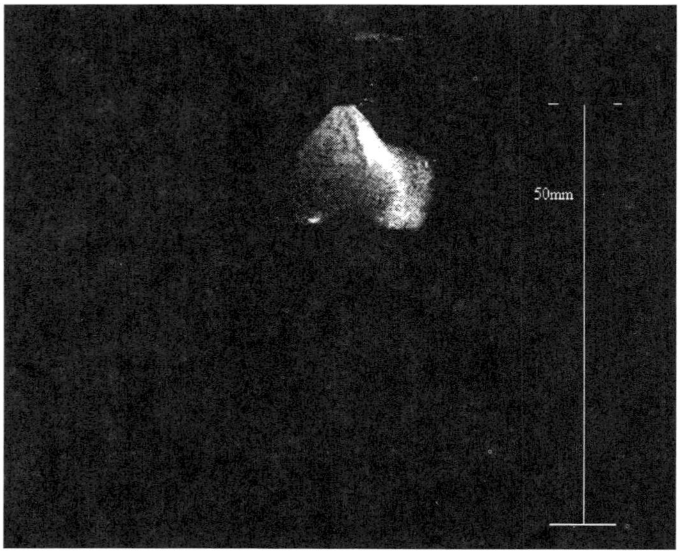

Figure 10: Laser sheet spray image at 1 ms after the SOI at 1.5 MPa absolute pressure ambient.

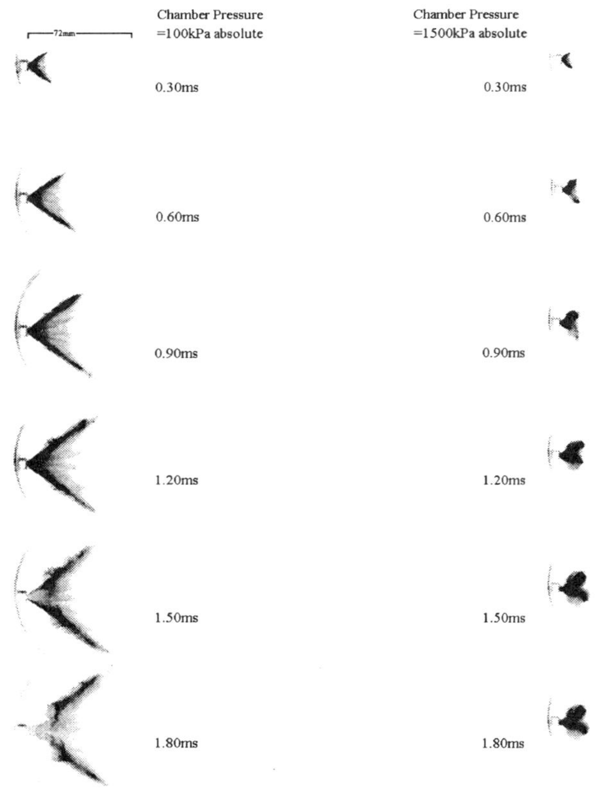

Figure 11: Sequential laser sheet spray images as a function of time at 100 kPa absolute and 1.5 MPa absolute ambient pressures.

Table 5: Spray characteristics as a function of ambient pressure as determined from imaging

Characteristic	P_a = 100 kPa	P_a = 1.5 MPa
Spray angle at 1 ms	74 °	75 °
Penetration at 1 ms	51.5 mm	14.4 mm

Figure 9 shows that the modeled spray is very similar to the spray observed in the physical testing. Figure 10 shows an image of the spray at 1 ms after injector logic at 1.5 MPa absolute chamber pressure. Table 5 shows the spray angle and penetration determined from these images.

Spray penetration as a function of time was determined from 30 sequential images of the spray captured beginning at 0.1 ms after injector logic and ending at 3 ms after injector logic. The penetration as a function of ambient pressure was obtained from images of the spray at 100 kPa absolute and 1.5 MPa absolute chamber pressures. Figure 11 shows two series of laser sheet illuminated digital images of the spray depicting the spray development as a function of time for the two chamber pressures. The axial penetration distance of the spray leading edge was measured from each image of the sequence. Figure 12 shows the axial spray penetration as a function of time. For comparison, the modeled spray penetration is included in this plot. There is good agreement between the modeled and measured penetrations for the two chamber pressures.

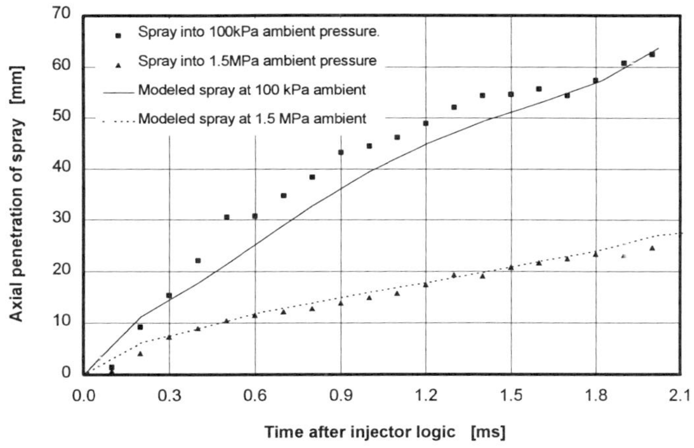

Figure 12: Measured and predicted spray penetration as a function of time at 100 kPa absolute and 1.5 MPa absolute ambient pressures.

SPATIALLY-INTEGRATED DROPLET SIZE DISTRIBUTION CHARACTERIZATION

The spatially-integrated droplet size distribution tests were conducted using a laser diffraction instrument. All of the droplet size measurements were performed with the injector spraying into air at 100 kPa absolute pressure, with the line of measurement centered 30 mm below the injector exit. The laser diffraction instrument used was a Sympatec model HELOS/Vario-KF fitted with a 5 mW, 632.8 nm Helium Neon laser. The receiver of this instrument has 31 semicircular detector rings and 3 centering elements. The receiver was fitted with a lens of 100 mm focal length, so that particle sizes from 0.9 to 175 microns could be measured. Testing was performed with laser beam diameters of 2.2 mm and 13 mm, with similar results. The injection frequency was 50 Hz. The laser diffraction instrument was set to sample the diffracted light signal for 10 seconds, thereby temporally averaging the spray droplet distributions of 500 injection cycles. The light intensities from the 31 detector rings were processed by the HELOS software to derive a spray droplet size distribution and characteristic sizes that are reported in Table 6.

Table 6: Laser diffraction spray sizing results

Characteristic	Test result
Arithmetic mean diameter	8 microns
SMD	15 microns
DV90	28 microns
DV50	12 microns
DV10	7 microns

SPATIALLY-RESOLVED DROPLET SIZE DISTRIBUTION CHARACTERIZATION

The spatially-resolved spray droplet size measurements were conducted using phase Doppler interferometry. Using the results of the spatially-integrated spray droplet size measurements, the appropriate optics and instrument settings were chosen for performing a spatially-resolved droplet size and axial velocity interrogation of the injector spray. The test conditions of the laser diffraction testing were reproduced for the phase Doppler testing. Pointwise measurements were performed along the same line of measurement as in the laser diffraction tests. The phase Doppler interferometry instrument used was an Aerometrics Phase Doppler Particle Analyzer (PDPA) operated so as to measure the size and axial velocity of the spray droplets. The instrument was used in forward scatter mode with the receiver offset 30 degrees from the axis of the transmitter. Focal lengths of the transmitting and receiving lenses were 250 mm and 500 mm, respectively. The droplet diameter range was set at 2 to 100 microns. The argon ion laser power was set so that the transmitted light intensities were 5 mW for the frequency shifted and unshifted beams. The PDPA was set to automatically adjust the photomultiplier tube high voltage. For the Delphi DI injector spray in the region of peak flux, the photomultiplier high voltage was 400 volts. Phase calibration of the instrument was performed at this setting. The PDPA's intensity validation function was enabled to remove false measurements due to light reflected from droplets. The injector was axially centered 30 mm above the PDPA probe volume. PDPA measurements were performed along a radius of the spray by moving the injector in 1 mm increments from the axial center point, through the region of peak flux, to the edge of the spray where the spray flux was less than 1% of the maximum measured spray flux. The number of droplet samples collected at each point was 10,000. Figure 13 shows the droplet size variation with radial

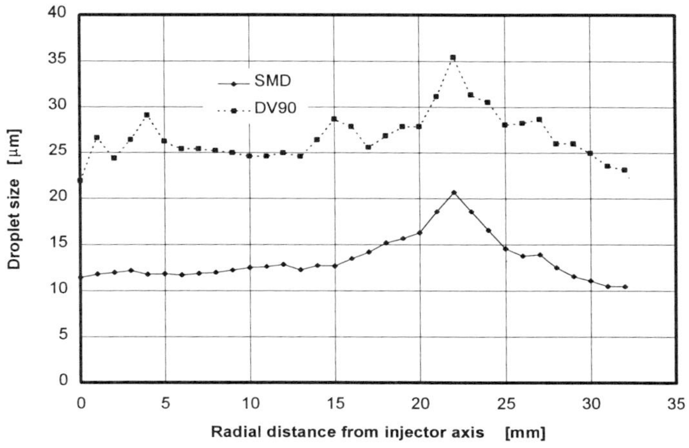

Figure 13: Radially-resolved SMD and DV90 for the Delphi DI injector spray measured by PDPA.

Table 7: Phase Doppler interferometry measured spray characteristics

Characteristic	Test result
Radially-integrated SMD	13.6 microns
Minimum SMD	10 microns
Maximum SMD	21 microns
Minimum DV90	22 microns
Maximum DV90	35 microns
Rosin-Rammler N value	2.2 - 3.3
Minimum axial velocity	0.2 m/s
Maximum axial velocity	23 m/s

distance. Table 7 shows the PDPA measured spray droplet size and axial velocity. The averaged SMD value was radially-integrated from the PDPA measurements.

The physical assessment of the Delphi DI injector spray confirmed the CFD modeled results. The injector flow rate, spray angle, penetration, SMD and DV90 were accurately predicted by modeling. Table 8 shows a summary of the modeling results, the testing results and the DI spray requirements. The injector spray satisfies the stringent DI spray requirements.

Table 8: Comparison of modeling and testing results with DI requirements

Characteristic	Modeling result	Testing result	Requirement
Steady-state flow rate [g/s]	11.6	11.5	11.5
Spray maximum penetration [mm]	63	63	<70
Spray angle [°]	70	65	60~70
SMD [μm]	13.4	13.6	<15
DV90 [μm]	28	35	<40

DISCUSSION

Despite the many advantages of DI gasoline engines, one major obstacle is the increased cost of the fuel system compared to that of the PFI engines. This cost is assumed to be proportional to the system fuel pressure. If the flow and spray requirements can be met, the fuel pressure should be reduced as much as possible. Figure 14 shows the relation of SMD of the DI spray versus fuel supply pressure at 100 kPa absolute pressure and room temperature. The thick solid line represents the KIVA-3 modeling results. The discrete diamond symbols denote the values integrated from the PDPA measured radial SMD profile at 30 mm distance from the Delphi DI injector. For comparison, similar data from some inwardly opening DI injectors are also included [8, 14, 15]. In general, the SMD decreases with

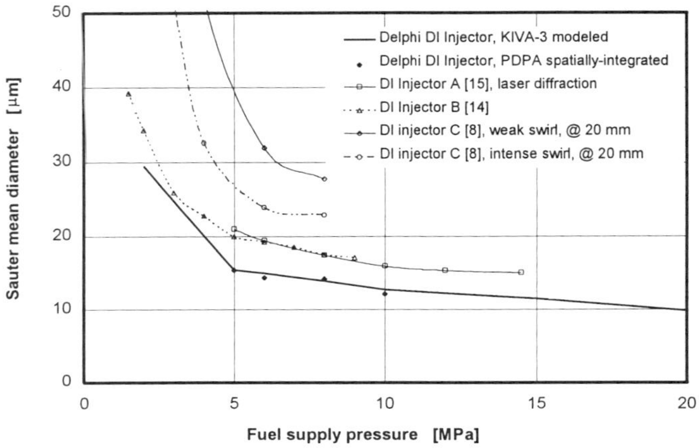

Figure 14: Effects of fuel pressure on SMD of DI sprays at 100 kPa absolute pressure ambient.

increasing fuel pressure. In the low pressure range, the SMD is very sensitive to the fuel pressure. For the pressures greater than 5 MPa, further pressure increase does not lead to significant drop size reduction. The Delphi DI injector sprays show the same trend, however, the SMD values of these sprays are at least 4 microns smaller than from the inwardly opening swirl injector sprays. This spray drop size difference is considered to be a result of the smaller initial liquid sheet thickness of the Delphi DI injector sprays.

In Figure 14, the fuel pressure of 5 MPa seems to be high enough for producing an acceptable DI spray. Nonetheless, a higher fuel pressure is recommended for the following reasons. As the fuel pressure was increased from 5 MPa to 10 MPa, the SMD of the Delphi DI spray decreased from 15.4 microns to 13.6 microns. This difference does not seem significant, but the total surface area for 14 mg of fuel increased from 7411 mm^2 to 8392 mm^2. This surface area increase of 13% should lead to direct enhancements in heat transfer, vaporization and combustion. More importantly, high fuel pressure is required to reduce the maximum droplet size in the spray, expressed by DV90. Laser diffraction measurements have confirmed that the DV90 of the Delphi DI injector spray at 30 mm from the injector exit decreased from 40 microns to 28 microns, as the fuel pressure increased from 5 MPa to 10 MPa. Therefore, the DV90 is more sensitive to fuel pressure than the SMD.

The DI sprays, however, behave very differently in the high pressure, high temperature combustion chamber than they do in an atmospheric pressure, room temperature environment. The maximum cylinder pressure could be above 2 MPa at the end of the compression stroke. Since the pressure differential across the nozzle orifice dictates the spray quality, the spray may degrade significantly if the fuel pressure is not sufficiently high. According to KIVA-3 predictions, even though the Delphi DI injector is operated at 10 MPa

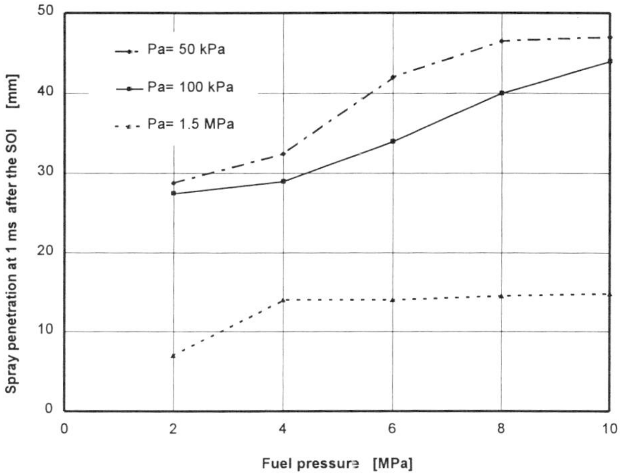

Figure 15: Effects of fuel pressure and ambient pressure on Delphi DI injector spray penetration measured at 1 ms after the SOI.

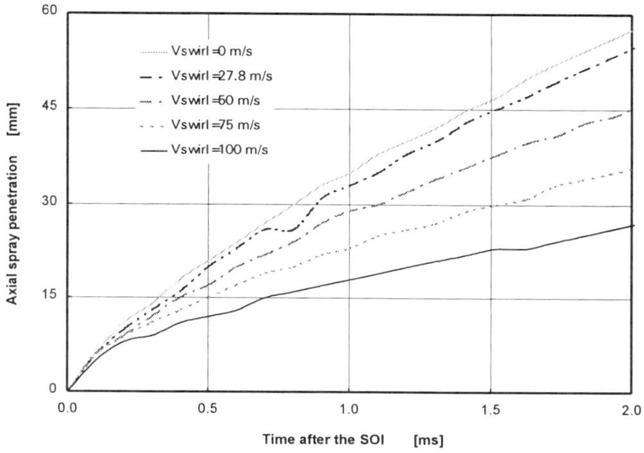

Figure 16: KIVA-3 predicted effects of initial swirl velocity at the injector exit on spray penetration as a function of time at 100 kPa absolute pressure ambient.

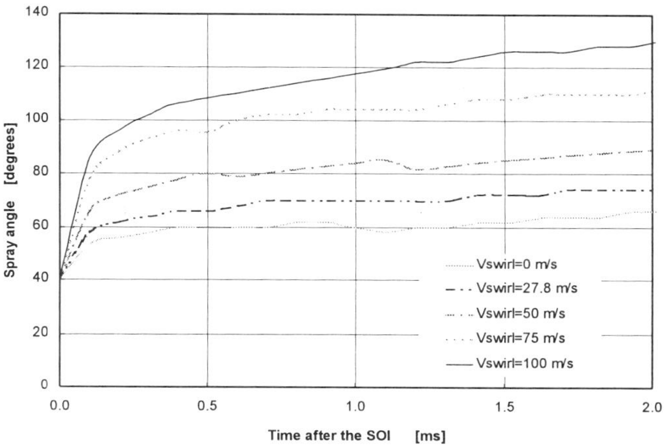

Figure 17: KIVA-3 predicted effects of injector initial swirl velocity on spray angle as a function of time at 100 kPa absolute pressure ambient.

fuel pressure, the drop size will increase substantially with increased chamber pressure (Table 2). Essentially, the high fuel pressure is needed for fine sprays at the late injection stratified charge operation rather than the atmospheric air pressure or vacuum early injection conditions. In addition to its effects on the spray, the higher fuel pressure is also required for reducing the injector flow rate sensitivity to the injector valve stroke variations. Because of anticipated fuel pressure pulsation in the fuel system and unpredictable variations of the injector valve stroke, an adequate pressure drop must occur elsewhere in the fuel system besides the discharge orifice. In the Delphi DI injector, this pressure drop was used for swirl generation.

Fuel pressure and ambient pressure also have pronounced effects on spray penetration. Figure 15 shows both pressures' effects on the Delphi DI injector spray penetration at 1 ms after the start of injection. The spray penetration was measured from the laser sheet spray images. As fuel pressure was reduced and ambient pressure was increased, spray penetration was diminished. Penetrations of the sprays into a vacuum environment of 50 kPa absolute pressure were about 5 to 10 mm longer than those into 100 kPa absolute pressure. Sprays injected into 1.5 MPa absolute pressure had much shorter penetrations. At this ambient pressure, fuel pressures above 4 MPa did not increase the penetrations of the sprays at 1 ms.

In a DI gasoline engine, cylinder pressure varies with crank angle. The cylinder pressure's effect on the spray penetration depends on the injection timing and duration. In the early injection mode, the cylinder pressure is low, resulting in a long spray penetration. However, the distance from the injector exit to the piston is large. In the late injection mode, the increased cylinder pressure causes a shorter spray penetration. Short penetration is favorable because the distance from the injector exit to the piston is small. Some DI injection systems require reduced fuel pressure to limit the in-cylinder spray penetration. It is important to note, however, that the use of reduced fuel pressure may compromise spray droplet size.

Swirl imparted on the spray droplets is known to be effective in reducing the spray penetration, with simultaneous enhancements to atomization. Figure 16 shows the relation between the spray penetration and the tangential velocity component at the Delphi DI injector exit, predicted by the KIVA-3 spray modeling. For the range of the swirl velocities less than 28 m/s, only a slight change in the penetration was noted. However, it was verified experimentally that the spray mass distribution of a swirl type Delphi DI injector was much more symmetric than one without swirl. The swirl motion in the azimuthal direction helped to evenly distribute the spray. As the tangential velocity was increased further, the spray penetration reduction became significant. For instance, when the tangential velocity component was varied from 28 m/s to 50 m/s, the spray penetration at 1 ms after the SOI was reduced from 34 mm to 29 mm. However, the spray angle also was increased from 70° to 84°, as shown in Figure 17. If the increased spray angle is not desirable due to potential wall impingement in the spray radial direction, the injector pintle and seat angle could be reduced appropriately to compensate for this change. Since the penetration reduction by using swirl is always accompanied by increased spray angle, careful spray tailoring is required to avoid the wall impingement in radial and axial directions. Ideally, a promptly atomized spray with very small drop size and penetration, the "puff" spray, is required to achieve the maximum potential of the gasoline DI engine concept. The spray droplets are so small that their momentum are very quickly consumed by air drag. The KIVA-3 simulation also predicted a slight effect of the swirl intensity on the spray drop size. Comparing a spray with a tangential velocity component of 50 m/s and a non-swirling spray, only a 1 micron decrease in SMD was found in the swirling spray.

CONCLUSIONS

The outwardly opening Delphi DI injector has been developed to meet the stringent spray requirements of the direct injection gasoline engines. CFD modeling software has been effectively utilized to optimize the injector nozzle design towards specific performance and spray requirements. The experimental spray evaluation has verified the modeling accuracy. Confidence has been developed for the numerical modeling tools because of acceptable agreement between the modeling and test results. Regarding the nozzle concept selection, advantages of the outwardly opening nozzles and problems associated with the inwardly opening pressure-swirl nozzles were discussed. Strategies for separate control of the three major spray characteristics: droplet size, angle and penetration, have been described. The influences of the fuel pressure, ambient pressure and tangential swirl velocity on these characteristics have been investigated. In summary, the following main conclusions can be extracted from this work:

- The Delphi DI injector produces a fine spray which meets the stringent DI requirements. This design has flexibility for custom tailored sprays to satisfy various combustion systems.

- For the spray injected into 100 kPa absolute pressure, the radially-integrated SMD measured by PDPA, was 13.6 microns. The maximum DV90 along the radial scanning was 35 microns. The spatially-integrated DV90 measured by the laser diffraction technique was 28 microns. The spray angle was 74°. The spray penetration at 1 ms after the start of injection was 51.5 mm. The maximum penetration was about 70 mm.

- The spray became very compact with increased ambient pressures. The spray penetration did not exceed 16 mm at 1.5 MPa absolute ambient pressure at 1 ms after SOI. The hollow cone spray collapsed into a solid cone spray. KIVA-3 simulation showed the tendency of increased droplet size in the high ambient pressure conditions.

- Fuel pressure has significant influence on the spray droplet size. SMD decreases with increasing fuel pressure. As fuel pressure increased above 5 MPa, the SMD was reduced only slightly. However, the DV90 reduction due to the increased pressure was still notable in this range.

- Effects of the fuel pressure on the spray penetration are dependent on the ambient pressure. Generally, penetration increases with the fuel pressure, and decreases with the ambient pressure. The penetration in a vacuum of 50 kPa absolute was 5 to 10 mm longer than at 100 kPa absolute pressure. In 1.5 MPa absolute ambient pressure, fuel pressures greater than 4 MPa have negligible effects on the penetration at 1 ms.

- Increased swirl intensity reduces spray penetration, increases spray angle, and slightly reduces spray droplet size.

ACKNOWLEDGMENTS

The authors gratefully acknowledge the significant contributions and support from those who assisted in this work. Their help was invaluable in discussion, hardware preparation and testing. Particular thanks are extended to: R. Cooper, Y. Kazour, K. Keegan, E. Kobos, S. Maczynski, H. Mieney, D. Moran, E. Schneider, D. Varble and J. Zizelman.

REFERENCES

1. Zhao F.-Q., Lai M.-C., and Harrington D. L. "A Review of Mixture Preparation and Combustion Control Strategies for Spark-Ignited Direct-Injection Gasoline Engines," SAE Paper 970627.
2. Harada J., Tomita T., Mizuno H., Mashiki Z., and Ito Y. "Development of a Direct Injection Gasoline Engine," SAE Paper 970540.
3. Tomoda T., Sasaki S., Sawada D., Saito A., and Sami H. "Development of Direct Injection Gasoline Engine - Study of Stratified Mixture Formation," SAE Paper 970539.
4. Fraidl G. K., Piock W. F., and Wirth M. "Gasoline Direct Injection: Actual Trends and Future Strategies for Injection and Combustion Systems," SAE Paper 960465.
5. Han Z., Fan L., and Reitz R. D. " Multidimensional Modeling of Spray Atomization and Air/Fuel Mixing in a Direct-Injection Spark-Ignition Engine," SAE Paper 970884.
6. Solomon A. Panel Discussion on Direct Injection Gasoline Engines, ILASS-Americas '97.
7. Dodge L. G. "Fuel Preparation Requirements for Direct-Injected Spark Ignition Engines," SAE Paper 962015.
8. Kume T., Iwamoto Y., Lida K., Murakami M., Akishino K.,

and Ando H. “Combustion Control Technologies for Direct Injection SI Engine,” SAE Paper 960600.
9. Parrish S., and Farrel P. V. “Transient Spray Characteristics of a Direct-Injection Spark-Ignited Fuel Injector,” SAE Paper 970629.
10. Amsden A. A., O’Rourke P. J., and Butler T. D. “KIVA II - A Computer Program for Chemically Reactive Flows with Spray,” Los Alamos Labs, LS 11560 MS, 1989.
11. Amsden A. A. “KIVA-3: A KIVA Program with Block-Structured Mesh for Complex Geometries,” Los Alamos Labs, LS 12503 MS, 1993.
12. O’Rourke P. J. and Amsden A. A. “The Tab Method for Numerical Calculation of Spray Droplet Breakup,” SAE Paper 872089.
13. Lai M.-C., Zhao F.-Q., Amer A. A., and Chue T.-H. ”An Experimental and Analytical Investigation of the Spray Structure from Automotive Port Injectors,” SAE Paper 941873.
14. Croissant K. and Kendlbacher C. “Requirements for the Engine Management System of Gasoline Direct Injection Engines,” Presentation Technische Akademie Esslingen, December 1996.
15. Toyota Press Information ‘96, “Direct-Injection 4-Stroke Gasoline Engine,” 1996.

980494

Improvements of GDI-Injector Optimization Tools for Enhanced SI-Engine Combustion Chamber Layout

M. Pontoppidan and G. Gaviani
Magneti Marelli - Engine Control Division

G. Bella and V. Rocco
Tor Vergata University, Rome - Department of Mechanical Engineering

ABSTRACT

The suggestions for upcoming Euro 2000 clean air act puts an increasing legislative pressure for lower specific fuel consumption in order to reduce the emission of CO_2. and thereby decrease the impact of the "green house" effect. One of the possible suggestions to meet these requirements for SI-engines is the gasoline direct injected (GDI) power unit.
One of the key points of the success of a layout of a GDI system is the optimization of the fuel injector and combustion chamber charge formation parameters. A brief description of the basic GDI-system used during the study is given. Hereafter are outlined the computational and experimental optimization tools which have been used to produce, on a reasonable industrial time scale, the main indications to optimize the design of a given injector/chamber configuration.
The paper discusses in detail the results produced by the latest enhancements introduced into the 3D multi-phase computational approach, NCF-3D. These are a fully implemented combustion sub-model as well as a sub-model for an auto-ignition mechanism.
Three different injector nozzle geometries are evaluated in the same combustion chamber. The computional and experimental results, obtained with different injector positions, are compared for the given combustion chamber of a 4 cylinder 16 valves 2 liter engine.
An alternative ignition strategy, the multi-spark concept, is discussed.

INTRODUCTION

The current European exhaust emissions legislation for SI-engines is limiting only the contents of carbon monoxide (CO), hydrocarbons (HC) and nitrogen oxides (No_x). However with the suggestions for the upcoming Euro-2000 legislation, the situation is very likely to change. In order to decrease the impact of the "green house effect", political pressure is very strong for a reduction of carbon dioxide (CO_2) emissions. The most effective means to obtain this reduction is to reduce the specific fuel consumption of the SI-engine.
The strategies used up to now to obtain fuel consumption reductions by operating the SI-engine in lean burn conditions with sequential port injection combined with air-assisted injectors, high EGR-rate and eventually variable valve timing, provide only an approximate 10 % reduction of the specific consumption. The main reason for this is that the port injection system is limited to a homogeneous lean burn charge concept in part load engine operation, due to driveability requirements. On the other hand, it is now well known [3, 14, 17, 18] that a highly stratified combustion chamber charge preparation achieved by a very late injection (about 40 degree crank angle before TDC) allows for much higher air/fuel ratios, and hereby a substantial further reduction of the specific fuel consumption.
One possible solution to obtain a cycle to cycle well controlled stratified charge preparation is the use of direct fuel injection into the combustion chamber. This has led to an important activity by almost all the European car manufacturers to try to introduce on the market, from the start of next century, gasoline direct injected powered vehicles.

GDI-SYSTEM MAIN FEATURES

One of the key points of a successful layout of a GDI-system is the optimization of the fuel injector parameters. This is due to the fact that the entire combustion process (initiation of the fire ball, the flame front propagation nature and velocity and the combustion duration) is heavily influenced, particularly in the case of stratified lean burn operation, by the shape of the combustion chamber (generated gas motion), by the crank angle (CA) for start of injection and its duration as well as the spray atomization pattern produced by the injector. The latter can cause eventual fuel droplet splashing on the piston crown or on the chamber walls as well as fuel accumulation in the chamber crevices [3, 9, 10, 18]. But

*Numbers in brackets designate references at the end of the paper

before going into details with this optimization process a brief description of the system within which the fuel injector operates will be given.

ENGINE DATA - The reference engine used in test rig and vehicle tests has the following main characteristics:

Type	**4 Stroke SI**
Displacement	**1970 cc**
No of Cyl.	**4**
No of Valves/Cyl.	**4**
Ignition	**1 - 3 - 4 -2**
Bore	**83 mm**
Stroke	**91 mm**
Compression ratio	**10:1**

Table 1

SYSTEM LAYOUT - As illustrated in figure 1 the system implemented on the engine test rig as well as in a test vehicle is of the common rail type (3) with an integrated pressure sensor. The fuel supply is delivered by a variable pressure (4 MPa $<P_{Rail}<$ 12 MPa) mechanical piston pump (2). The pump concept is of the radial type with 3 pistons separated by an angle of 120 degrees, which enables a maximum equalization of the internal forces for long duration as well as a very compact layout. On the pump housing are mounted a fast acting electromagnetic pressure regulator (9) and an electronically controlled by-pass unit (10) enabling the engine start at a 0.3 MPa pressure delivered by the tank fuel supply pump (5).
The concept of a variable fuel pressure strategy has emerged from engine test rig and vehicle tests. Test rig combustion analysis has shown that the natural mechanical dynamic range of the fuel injector (which is approximately 14:1) is insufficient to account for both super lean stratified charge preparation in the engine part load area (late injection - approximately 40 degree BTDC), and the homogeneous richer charge preparation, in the high load area (injection during intake stroke). To overcome this problem the natural dynamic range must be multiplied with a factor of approximately 1.5, thus requiring a regulated working pressure range from 4 MPa to 12 MPa. Vehicle tests show the convenience of the low pressure (0.3 MPa) engine cranking strategy for mainly 2 reasons:

* The electrical delivery pump gives a steady pressure at "key on" before cranking, whereas the mechanical pump supplies a steady pressure only when running at a speed beyond 120 rev/min.
* Security: when the engine is cut off the system pressure goes down to near atmospheric pressure.

The fact that the fuel injector nozzle/atomizer must produce an imposed spray pattern, not only in the main pressure range between 4 and 12 MPa, but also at the low pressure value of 0.3 MPa means that the nozzle layout as well as the test measuring equipment must be very carefully designed.
Engine test rig experimentation has shown that theoretically the ultra lean stratified charge preparation allows for air/fuel ratios sufficiently high to run the engine unthrottled (A/F 80 to 90). Nevertheless, as the raw No_x emission in this working area is too important to comply with European emission requirements, and as so far no industrial solution for a real DeNox catalyst is available, a drive by wire electronically controlled throttle body (1) is present in the system. This is due to the fact that the management of the mixture enrichment peak (frequency 1 to 2 Hz) necessary for the recovery cycle of a trap/storage and reduction 3-way catalyst requires a highly sophisticated torque stabilization strategy, controlling simultaneously the air flow, the instantaneous fuel flow and the ignition.
All engine experiments were made with a general control strategy based on the speed-density concept without use of a specific air flow meter.
The fuel injector (4) is of the electromagnetic type. Great care has been taken to ensure that the design of the coil unit and the magnetic path of the injector guarantees a small but fast acting device (pintle displacement time < 500 μ-seconds) with very high cycle stability over the full pressure range (figure 2).
As mentioned above the quality of combustion for the stratified GDI concept is closely related to the spray shape produced by the fuel injector. To achieve maximum tuning flexibility, the injector design has been made to allow the same electro-mechanical housing to receive several types of nozzle/atomizer seats. Three different versions have been tested:

* Hollow cone spray with a cone angle between 70 and 90 degree produced by a single cylindrical metering hole in which is axially placed a deviation pintle attached to the metering needle.
* Rotating spray, ideally hollow cone at maximum needle lift, with an outer border angle between 80 and 115 degree and an inner border angle equal to 0.55 times the outer border angle (12 MPa). The spray is produced by a single cylindrical metering hole with n (2=< n =<5) upstream swirl bores.
* n (3=< n =<5) Solid cone sprays produced by a single cylindrical metering hole with a smooth pintle attached to needle tip and a closed break up cap with the n non metering bores (slits) machined to guide the sprays in the wanted direction with respect to the injector main symmetry axis.

INVESTIGATION AND DEVELOPMENT TOOLS

With the today's "time to market" philosophy imposed even in the early development stages of a project in which time consuming fundamental work, like the fluid dynamic optimization of the fuel spray temporal and spatial formation with respect to the combustion chamber geometry must be done, it is absolutely necessary to follow a very precise systematic work scheme. To reduce the experimental part, which is the most time consuming, to a minimum, the chosen approach a combination of real scale fired engine test rig experiments and a powerful 3D numerical simulation tool together with a detailed optical analysis of the dynamic fuel injector behavior within very well defined repetitive boundary conditions.

By the simultaneous use of the numerical simulation tool and the experimental study much time can be gained. This is particularly important in the phase of the choice of the fuel injector atomizer layout and positioning in the combustion chamber as well as for suggestions for design modifications to the chamber itself , while only promising combinations will be transformed into physical experiments.

A brief description of the special GDI-system optimization tools is given below:

HIGH-PRESSURE FLOW RIG - To obtain dynamic and steady state visualizations and measurements of different spray patterns produced by the different nozzle atomizers, in equal test conditions , a specific high pressure flow rig is used. On the rig is installed a pressure vessel in which the injector is mounted and can be submitted to a pre-heated ambient gas condition at a counter pressure between - 0.07 MPa < P_C < 0.8 MPa with respect to atmospheric conditions. The rig can produce a steady state regulated fuel rail pressure between 0.3 MPa< P_{FR} <15 MPa. The pressure vessel contains transparent windows for optical access to the spray area. The rig is equally equipped with the necessary software and hardware to dynamically control the injector pulse width and the duty-cycle. An example of a Schlieren scattering recording in the pressure vessel is given in figure 3. The rail pressure is at 8 MPa, the vessel counter pressure at 0.8 MPa and the gas temperature at 500 K. The injector works at a pulse width of 1,5 millisecond. The dense center zone represents the non-evaporated part of the spray, whereas the illuminated border areas correspond to the fuel vapor products.

The different optical tools used to visualize and monitor the spray formation process are based on either continuous or phase controlled flashed illumination techniques of the spray in 3D or 2D (laser sheet) conditions. Continuous illumination is mostly used for high speed photographic recording, whereas phase illuminated techniques apply to CCD camera recording. Measurements of velocity vector distributions in the sprays are made either by Particle Image Velocimetry (PIV) used on laser sheet recordings or by Phase Doppler Anemometry (PDA-Dantec particle size analyzer). The latter is equally used for droplet diameter analysis. Detailed examples of these recording techniques on this flow rig were given in [16, 17].

NUMERICAL COMPUTATION TOOL - The developed numerical computation tool "NCF 3D" is based on the well known KIVA III code originally developed by the Los Alamos Laboratory [4, 5]. The original version employs a finite volume approximation of the governing 3D Navier-Stokes (N-S) equations in a Cartesian or cylindrical reference system in which a multi-block grid structure is generated. A modified κ-ε turbulence model, accounting for compressibility effects, is used during the arbitrary Lagrangean Eulerian (ALE) integration of the averaged N-S equations. The original version has been enhanced by a certain number of specific submodels necessary for the computation of the dynamic fuel spray behavior, including spray atomization, internal droplet ballistics and gas phase motion as well as the combustion The GDI-specific enhancements of the NCF 3D code can be listed as follows:

1. Introduction of the experimentally obtained initial spray formation characteristics (spray angle and main axis direction, initial droplet diameter and position in the spray, initial droplet velocity) [17].
2. Improvements of the formulation of the gas/wall heat transfer model in order to deal with transient effects and compressibility. The re-writing of the submodel follows the approach suggested in [11, 19].
3. Implementation of a droplet wall interaction submodel [2].
4. Implementation of a spark initialization submodel [18].
5. Introduction of a law for the turbulent characteristic time scale [3, 18].
6. Implementation of a flamelet combustion submodel
7. Introduction of a "shell" submodel to simulate the auto-ignition phenomena of hydrocarbons.

As the first five items have been discussed in detail in previous papers [17, 18] only the last two items will be subject to a more detailed description in the following.

<u>Flamelet combustion model</u> - The combustion process is initiated by the flame kernel produced by the energy release in the spark plug gap. In the early stage of the flame kernel its dimension is small compared to the eddy turbulence length scale. It can therefore be assumed that the turbulence has no influence on the early kernel growth, which can then be considered as a perfectly spherical growth of which the time rate change of the radius can be easily estimated. In this first phase the flame growth remains typically laminar. Once the expansion radius becomes larger than the eddy turbulent length scale the flame front propagation becomes heavily influenced by the turbulence [9].

The flamelet submodel allows the de-coupling of the chemistry from the physics of turbulence. The modeling approach to the flame structure estimates the surface area of a thin, wrinkled reaction-surface flame. In this case the problem of modeling the turbulence can be reduced to a description of time dependent flame surface behavior. The mass burning rate can be computed as:

$$\varpi = \rho_u S_{LO} I_o \sum \quad (1)$$

where ρ_u is the density of unburned gas, S_{LO} is the laminar flame speed, I_O is a stretching factor and $\sum$ is the surface area per unit volume.

Assuming a local equilibrium in the wrinkled, thin reaction-surface, it is possible to identify a mean flame contour to which can be attached a mean turbulent flame speed. In that case the mass burning rate can be computed from the flame surface density given by:

$$\sum = \frac{1}{L}\left(\frac{S_T}{S_{LO}}\right)\bar{c}(1-\bar{c}) \quad (2)$$

where L is a characteristic length of the order of the turbulence integral scale, $\bar{c}$ is a reaction progress variable which gives the local mean degree of reaction and S_T/S_{LO} is the turbulent to laminar velocity ratio.

To estimate the laminar reaction-surface area, a fractal geometric model is used for the description of the wrinkled flame surface. The fractal theory assumes self similarity between scales of different size, producing a power-law scaling between surface area and wrinkling scales.. This can be applied to derive an expression for the velocity ratio S_T/S_{LO} [13, 15]. Gouldin uses fractal geometry to compute the area per unit volume of wrinkled surface [7]. In particular he suggested that, at high Reynolds numbers, the flame front can be considered a "passive scalar". Hence the ratio S_T/S_{LO} can be expressed as:

$$\frac{S_T}{S_{LO}} = \frac{A_T}{A_L} = \left(\frac{\varepsilon_o}{\varepsilon_1}\right)^{d_3-2}$$

where A_T is the area of a wrinkled surface, A_L that of a reference surface. The parameter d_3 is the fractal dimension and ε_0 and ε_1 representing the maximum and the minimum wrinkling scales. Using this expression Zhao [20, 21] has suggested a Fractal Flame Model (FFM) evaluating the fuel burning rate by means of the equations:

$$\varpi = \frac{c_f y_{fu}}{b_1 - b_2}\frac{\rho_u S_{LO}}{MW_f}\left(\frac{\delta_L}{L_T}\right)^{0.5}\left(1 - A'\left(\frac{\delta_L}{L_T}\right)^2 \mathrm{Re}^{3/2}{}_L\right)\cdot$$

$$\cdot\frac{\mathrm{Re}_l^{\frac{3d_3-4}{4}}}{L_t}\bar{c}(1-\bar{c})$$

where:

$$\delta_L = \frac{\nu_l}{S_{LO}}$$

represent the flame thickness, while L_T is the turbulence integral scale computed as :

$$L_T = 0.39\frac{k^{2/3}}{\varepsilon}$$

The Reynolds number is given by:

$$\mathrm{Re}_L = \frac{u'}{S_{LO}}\left(\frac{\delta_L}{L_T}\right)^{-1} \qquad u' = \left(\frac{2}{3}k\right)^{0.5}$$

A' and c_f are model constants, The fractal dimension d_3 in the FFM can expressed as a function of the Reynolds number using the expression suggested in [15]:

$$d_3 = 2.0\frac{200}{\mathrm{Re}_L + 200} + 2.35\frac{\mathrm{Re}_L}{\mathrm{Re}_L + 200}$$

The progress variable can be expressed as:

$$\bar{c} = \frac{\bar{y}_{fu} - \bar{y}_f}{\bar{y}_{fu} - \bar{y}_{fb}}$$

where $\bar{y}_f, \bar{y}_{fu}$ and $\bar{y}_{fb}$ are respectively the local mass fractions of unburned and burned species. *MW* represents the molecular weight of the fuel species and b_1 and b_2 its stoichiometric coefficients.

The local values of the turbulent parameters are provided by the κ - ε model included in the original version of the code. As will be seen later, the implemented combustion model produces results with a quite reasonable correlation with experimental data.

<u>Auto-ignition hydrocarbon model</u> - Halsead has suggested [8] that the "shell" model, to predict knock, can be used for the computation of hydrocarbon fuel auto-ignition phenomena. Based on the coupling stability of hydrocarbon the auto-ignition mechanism is generated in eight generic reactions containing five generic species. The reaction rate constant for each equation is of the Arrhenius type. The mathematical model for this generalized reaction set contains 26 kinetic parameters [6]. The generic species and reactions involved in the model are listed below:

Auto-ignition Chemical Mechanism	Reaction rate
$RH + O_2 \Rightarrow 2R$	k_q
$R \Rightarrow R + P + Q$	k_p
$R \Rightarrow R + B$	$f_1 k_p$
$R + Q \Rightarrow R + B$	$f_2 k_p$
$R \Rightarrow out$	$f_3 k_p$
$R \Rightarrow R + Q$	$f_4 k_p$
$2R \Rightarrow out$	k_t
$B \Rightarrow 2R$	k_b

Table 2. - Auto-ignition Chemical Mechanism

The mathematical auto-ignition model for the above described mechanism can be expressed in the following five equations:

$$\frac{1}{V}\frac{dn_r}{dt} = 2\{k_q[RH][O_2] + k_b[B] - k_t[R]^2$$
$$-f_3 k_p[R]$$

$$\frac{1}{V}\frac{dn_B}{dt} = f_1 k_p[R] + f_2 k_p[Q][R] - k_b[B]$$

$$\frac{1}{V}\frac{dn_Q}{dt} = f_4 k_p[R] - f_2 k_p[Q][R$$

$$\frac{1}{V}\frac{dn_{O2}}{dt} = pk_p[R$$

$$n_{RH} = \frac{\left(n_{o_{21}} - n_{o_2}(t=0)\right)}{pm} + n_{RH}(t=0)$$

Equations (1)-(5) represent the reaction rate for each constituent in the auto-ignition mechanism.
As the above written equations represent a "stiff" differential system, it is necessary to use a smaller time step to obtain a convergent solution compared with conventional solvers. A particular technique has to be applied to guarantee a solution within the time step used by the KIVA solver. The approach chosen is the use of the VODE routine.

NUMERICAL ANALYSIS

As the grid generation sometimes can be a rather long and tedious operation a specific approach involving several industrial codes has been made. For this kind of fluid-mechanic computation, a very detailed grid is absolutely necessary. This is why the grid generation is performed directly within the main frame of the original design in CAD-CATIA , while the final multi-block mesh for the KIVA III solver is made by the IBM created interface, the ENGAGE code. As stated above, the presence of a stiff system of differential equations, imposes the necessity of an externally connected solver, in order to deal with this kind of time step problem. A partial differential equation solver "VODE" that solves at each time step and in all the computational cells the stiff equation system (1)-(5), has been directly coupled to the NCF3D code. The routine VODE, developed at Sandia National Laboratories, is a general code and a very efficient solver for stiff differential systems. All the obtained results are analyzed using appropriate nets of the Data Explorer visualization code

COMBUSTION CHAMBER - For the primary analysis, no change was made to the original layout of the combustion chamber. As in an earlier analysis, reported in [16, 17], this chamber enables either a center (top) mounting (point 1 on figure 5) or a side mounting position (point 2 on figure 5) for the injector and another side mounting position for the spark plug (point 3 on figure 5).

The axis of the number 1 position is on the cylinder symmetry axis, whereas the number 2 position has a clockwise off-set of 73 degrees with respect to the cylinder axis. Figure 5 shows equally the generated grid of the chamber with its pent-roof profile and a near-to-flat piston top. The latter feature is a major difference compared to the previous investigation in which the same cylinder head and unitary displacement cylinder was mounted with a bowl type piston.
From a mixture preparation optimization point of view the compression stroke late injection is the most interesting to deal with, as it is in this phase the major combustion difficulties appear. During this period the inlet and exhaust valves are closed. This is the reason for which the valve movement is eliminated from the computation. The initial charge movement is introduced by tumble and swirl conditions in the cylinder at inlet valve closure. These data are known from measurements done on this engine in an earlier phase.
As mentioned earlier 3 different injector types were tested with the NCF3D numerical analysis and on the corresponding test engine. The raw spray data of these injectors were sampled on the high pressure rig. The values corresponding to a 12 MPa rail pressure in free jet conditions can be summarized as follows:

\# Injector type 1 - The hollow cone spray
Single hole nozzle.
Initial primary spray angle 80 degree.
Initial mean penetration velocity 90 m/s.
SMD at 15 mm from injector tip 20 μm.

\# Injector type 2 - Rotating spray - swirl
5 hole nozzle.
Initial primary spray angle 90 degree.
Initial mean penetration velocity 65 m/s
SMD at 15 mm from injector tip 25 μm.

\# Injector type 3 - Closed cap injector
3 hole nozzle
Initial primary angle 15 degree for each spray and an axis off-set of 70 degree with respect to the injector main axis.
Initial mean penetration velocity 40 m/s.
SMD at 15 mm from injector tip 28 μm.

As the data was recorded with different counter pressures (up to 0.8 MPa) in the pressure vessel, it was observed that over this pressure range, the spray angle is subject to a mean decrease of 15 to 17 % and the mean penetration velocity to a decrease of 30 to 35 %. These values were in good agreement with the droplet velocities computed by the NCF3D code for late injection during the compression stroke.
Basically the injector types 1 and 2 were tested in location 1 and 2 whereas the injector type 3 was tested in position 2 only.
As the most critical situation for the charge preparation is generally found in the part load area with late injection the illustrative examples given below are chosen at a throttle valve opening of approximately 40 % for the following engine working conditions:

Engine speed	**2500 rev/min**
Swirl/Tumble ratio	**2**
Tumble	**0.5**
Start of injection	**80 degree BTDC**
Duration	**39 degree CA**
Start of ignition	**41 degree BTDC**
Duration	**10 degree CA**
Injector rail pressure	**12 MPa**

Table 3.

Figure 6 shows, for the type 2 injector in the center position, the gas and droplet flow field as well as the plane containing a stoichiometric mixture in the combustion chamber at 30 degree BTDC. The computation was made without ignition. It can be noted that this type of result is equally interesting as the image corresponds to what can be observed in a single cylinder transparent motored test engine.

It can be observed that the plane containing the stoichiometric air/fuel vapor ratio is at this CA-value rotating and expanding in the direction of the spark plug location (3), but will probably arrive too late to produce a correct combustion. As engine experience shows, as well as reports made by others [10, 12], the precision of a ignitable mixture in the vicinity of the spark plug gap at the moment of discharge is rather critical.

As a matter of fact, in this type of GDI-engine, the ignitable vapor cloud should not be further than approximately 2 mm from the spark gap. If not, high hydrocarbon emissions and eventual misfiring occur. This of course is an obvious handicap for several reasons. It imposes the requirement of a very high cycle to cycle temporal and spatial repetitivity of the spray. In other words, very small tolerances are allowed in the manufacturing process of the injector, and in particular its atomization nozzle, as well as on the electronic control strategy.

In order to decrease the spatial sensitivity it is often suggested to locate the spark plug near to the injector. But this solution has the major inconvenience that the injector tip, at least for the type 1 and 2 (the latter although less than type 1) can be rather rapidly (within 20 hours of functioning in the engine) contaminated by an important hydrocarbon deposit. As this deposit is made near to the nozzle outlet it causes change in flow characteristics and spray geometry. In the worst case reported on this engine with the type 1 injectors, located in center position, a mean flow decrease was observed, with respect to new clean injector data, between 25 and 35 % as a function of the injector pulse width as well as a spray angle distortion. As the secondary cracking process of the liquid hydrocarbon present on the injector tip is a function of the injector tip material temperature and the surface roughness tests are made at present with a vitreous ceramic coating of the vicinity of the nozzle outlet and the lower part of the injector tip. Although not an inexpensive solution it seems that it could apply to cases where the injector has to be placed very near to the spark plug. For the moment the experimental data gathered on this solution are statistically insufficient to draw a final conclusion. Another suggestion to decrease the spatial sensitivity of the ignitable part of the spray is to change the ignition philosophy. In current systems one single energy discharge of about 30 to 60 mJ with a duration of about 1.5 to 2.5 milliseconds is made. Some quite promising engine tests have shown that with an adequate coil position on the top of the spark plug, it is quite possible to divide the ignition process in a sequence of short duration sparks (multi-spark concept) each having a discharge energy of about 10 to 15 mJ. An acceptable duration of each spark seems to be about 200 μs. By this concept it is possible to gather up to five or six discharges instead of an exponentially decaying energy tail, which in most cases is unfit to generate a kernel initiation. In other words, the temporal window in which an energy burst is present has been increased and by the same means the sensitivity to variation in time of arrival of the ignitable vapor has been decreased.

<u>Injector type 1 side mounted</u> - As seen above, the center mounting of the injector type 1 in this combustion chamber does not provide an excellent mixture preparation for late injection. This is why the side mounted position was suggested. Figure 7 shows the situation at 20 ATDC indicating on the left of the figure the instantaneous position of remaining unburned stoichiometric vapor and on the right three different iso-surfaces representing the 1500, 1750 and 2000 degree K levels. Although far from being perfect, as the burning process has gone much faster on the left side of the chamber compared to the right, a substantial improvement was achieved with respect to the center mounting. Engine tests confirm this prediction as the HC-emission, still above the acceptable limit, is divided by a factor 2. At this stage it is worthwhile remembering the rather high penetration velocity of the injector type 1, which in the later part of the injection stroke (between 50 and 40 degree CA BTDC) introduces an important fuel splashing on the flat piston top surface. As stated above the injector type 3, the closed cap injector, produces spray with very much lower penetration velocities. Equally the spray axes are with an off-set with respect to the main injector axis ejecting thus less fuel in the direction of the piston.

<u>Injector type 3 side mounted</u> - Figure 8 shows the behaviour of the injector type 3 in identical conditions at 20 ATDC. The figure shows a more symmetrical expansion, centered around the cylinder main axis, of the high temperature areas. The smaller amount of fuel impingement on the piston and the wall enables a better vaporization in the center part of the cylinder. On the other hand, it can be detected from the air/fuel ratio representation that the border areas are in a leaner state compared to the previous configuration. As the equivalence ratio in this region is still around 0.25 to 0.35, a retarded flame propagation still takes place. But the computation is carried out with iso-octane, which is mostly correctly consumed if the temperature is higher than 1150 K. For real fuels (Euro 95), however, a content of larger, complex molecules, which do not vaporize as easily as iso-octane, is present. This must be taken into account, when comparing computed results to experimental engine data. A relative improvement from the previous configuration, as predicted by the

simulation, can be observed on the engine but the combustion quality is still unable to fulfill the required absolute level.

Injector type 2 center mounted - A side location of the type 2 injector was tested, but due to the greater amount of fuel present at the outer spray borders of the large angle spray the piston/wall impingement problem observed with the type 1 injector was amplified. For this reason the side mounted configuration was eliminated.
On the contrary, as shown in figure 9, the center mounted position seems interesting as a result is obtained, similar to the injector type 3 in the side mounted position although with a slightly larger 1500 K surface.
The data concerning the burned fuel mass fraction and the developed cylinder pressure as a function of the CA, for this solution, is given in figure 10 and 11.
As would be expected, the slope of the burned fuel mass fraction is smoother than that obtained with a fully stoichiometric homogeneous charge As the simulations were made with the flat piston top and the original tumble and swirl conditions of the physical engine, the obtained slope was not a surprise. The corresponding cylinder pressure curve is in good agreement with the measured engine data.
However, it was clear that for the two interesting solutions, the injector type 2 in center position and the injector type 3 in side position, a further optimization of the fuel spray movement was necessary. A 30 % increase in tumble (modification of inlet ports) and a piston top profile modification provided an improvement of approximately 18 to 20 % in burned fuel mass fraction at 20 degrees CA ATDC.
For the center mounted injector type 2, the piston top was modified adding at the center a bulge of the form of a symmetrically rotated planer tractrix curve with the parametric equations:

$$x = a(\ln(\cot(0.5\phi)) - \cos(\phi)) \text{ and } y = a(\sin(\phi)).$$

For the side mounted injector type 3 a longitudinal rim bulge was added perpendicular to the injector axis and displaced a small amount away from the cylinder axis towards the injector location. The profile is of the "witch of Agnesi" type with the parametric equations:

$$x = 2a(\cot(\phi)) \text{ and } y = a(1-\cos(2\phi)).$$

On each side of the rim two concave bowls were grooved in order to form a continuous wave-form across the piston surface.
As the purpose of the study was to look for an optimization of the coupling between the injector type, nozzle geometry, location and chamber shape, the influence of a high rate EGR was not considered in this phase.
Comparison with the earlier reported studies [17] on this same engine equipped with a piston with one simple diesel-like central bowl shows that the more sophisticated profiles introduce a higher level of improvement on the burned fuel mass fraction at the 20 degree CA value ATDC.
Figure 12 shows, in the angular window of +/- 60 degrees CA around the TDC, an example of the improvement in percentage of burned fuel mass fraction as a function of the piston crown shape.
The experimental data presented in figure 12 are extracted in engine test rig part load conditions at 2500 rev/min. The central bowl piston (continuous line) and the above mentioned "witch of Agnesi"-profiled piston (broken line) were tested. It can be observed that the rim-profiled piston improves the combustion process with respect to the bowl-profiled condition. For the rim-profile, compared to the bowl-profile, the start of the combustion process takes place closer to the TDC and the slope of the burned mass fraction is steeper, thus producing an improvement of 13.3 % at 20 degrees CA ATDC.

CONCLUSION

The consumption gain of contemporary stratified charge GDI engines can be an effective way to respond to probable future CO_2.-legislation. But the quality of the result depends entirely of how well the engine implementation is accomplished. Especially the raw HC and NO_x emissions are problematic. The latter must be dealt with principally through catalytic after-treatment. At present, this operation is performed by a thick-coated 3-way trap/storage catalyst reducing at regular high equivalence ratio peaks the stored NO_x while waiting for an industrial solution for a real $DeNO_x$ catalyst. However, these catalytic converters require, to guarantee an acceptable conversion ration over the entire life-time, a very low sulfur content in the gasoline (less than 25 to 30 ppm). For the trap/storage version the recovery cycle imposes very sophisticated control strategies to avoid abrupt engine torque behavior.
The HC emission is mostly due to incomplete flame propagation and post combustion at the boundaries of the stratified charge area and in the crevices as well as to wall and piston wetting. These phenomena are very dependent of the complex spray-geometry-flow interactions in the combustion chamber.
To meet the "time to market" delay requirements to find an optimized compromise between a nearly "frozen" chamber geometry and 3 possible injector atomizers a development methodology was used, based on the simultaneous use of highly specialized rig equipment and visualization techniques, for the initial description of the injector spray, and the 3D multi-phase NCF-3D numerical simulation code. The latter has been enhanced with combustion and auto-ignition sub-models.
The analysis made on the original chamber shape, with a flat piston profile, points out the following two combinations: a center mounted swirl atomizer and a side mounted closed cap atomizer.
A further analysis showed that an active piston bowl shape will improve the burned fuel mass fraction for each of the two solutions.
Test rig experiments were made on the corresponding 2 litre 4 cylinder real engine, in conjunction with the numerical analysis. A reasonable good correlation was

found between the numerically predicted results and the measured engine data.
As the cycle to cycle stability of the charge preparation is a critical factor for the fuel injector and its control strategy, through very small tolerance requirements, efforts are made to decrease the injector dependent sensitivity. Engine test experiments indicate that the multi-spark strategy can be a possible way to widen the temporal window in which kernel initiation can take place, and thus allow for an increase in cycle to cycle spray instability.
Future works will focus on further enhancements of the NCF-3D modeling tool by introducing more complex fuel molecules as well as on the correlation of the computed later stages of the lean stratified combustion with experimental visualizations made in a transparent single cylinder test engine.

REFERENCES

[1] Abraham J., Bracco F.V., Reitz R.D. "Comparison of Computed and Measured Premixed Charge Engine Combustion" Combustion and Flame 60: pp 309-322, 1985

[2] Allocca L., Corcione F.E., Golini S., Papetti F. "Measurement and Modeling of spray/wall Dynamics and Heat Transfer" 3rd Int. Kiva Users Meeting SAE Congress Detroit MI, 1993

[3] Ando H. "Mitsubishi GDI Engine - Strategies to meet the European Requirement.", Proceedings from Engine and Environment, Graz, September 1997

[4] Amsden, A.A., O'Rourke, P.J, and Butler, T.D., "KIVA II A Computer Program for Chemical Reactive Flows with Sprays," Los Alamos National Labs. LA -11560-MS-1989.

[5] Amsden, A.A., O'Rourke, P.J, and Butler, "KIVA III: A KIVA Program with Block Structured Mesh for Complex Geometries" Los Alamos National Labs. LA-12503-MS.

[6] De Petris C., Diana S. Giglio V., Police G. "Numerical Simulation of Combustion in Premixed S.I. Engine Using Fractal Models" SAE Paper 952383

[7] Gouldin F.G. Bray K.N.C., Chen I,J., "Chemical Closure Model For Fractal Flamelet" Combustion and Flames 77, pp. 241-259, 1989

[8] Halsted M., Kirsh L. and Quinn C. "The Auto-ignition of Hydrocarbon Fuels at High Temperature and Pressure - Fitting of a Mathematical Model," Combustion and Flames, Vol. 30, 1977, pp. 45,60

[9] Heywood J. B. "Internal Combustion Engine Fundamentals." - Mc-Graw Hill Book Co, New York, 1995

[10] Herweg R., Maly R. R., "A fundamental Model for Flame Kernel Formation in SI-Engines.", SAE Paper 922243

[11] Kang, Y. Huh, I-Ping ,Chang , Jay. K. Martin, "A Comparison of Boundary Layer Treatments for Heat Transfer in IC Engines," SAE Paper 900252, 1990.

[12] Kuma t., Iwamoto I., Lida K., Murakami M., Akishina K. and Ando H., "Combustion control Technologies for Direct Injection SI-Engine.", SAE Paper 960600

[13] Mantzares J., Felton P.G., Bracco f.V. "Fractals and Turbulent Premixed Engine Flames" Combustion" Combustion and Flame 77: pp. 295-310, 1989

[14] Nohira H., "Development of Toyota's Direct Injection Gasoline Engine.", Proceedings from Engine and Environment, Graz, September 1997

[15] North G.L., Santavicca D.A. " The Fractal Nature of Premixed Turbulent Flames", Combustion" Combustion Science and Technology, vol 72. pp. 215-253, 1990

[16] Pontoppidan M., Bella G. and Rocco V. "Mixture formation in SI-Engine intake Systems - A simplified theoretical Model with Experimental Verification." - ASME 1996 Spring Congress

[17] Pontoppidan M, Gaviani G., Bella G. and Rocco V. "Direct Fuel Injection - A Study of Injector Requirements for Different Mixture Preparation Concepts" SAE Paper 970628

[18] Pontoppidan M, Gaviani G., Bella G. and Rocco V. "Experimental and Numerical Study of Fuel Spray Behavior to Optimize Charge Motion in Direct Injected SI-Engine", 3rd International ICE Conference on Internal Combustion Engines, Capri September 1997

[19] Reitz, R.D., "Assessment of Wall Heat Transfer Models for Premixed-Charge Engine Combustion Computations" SAE Paper 910267

[20] Zhao X., Matthews R.D., Elizey J.L., "Three Dimensional Numerical Simulation of Combustion and Flame Propagation in Spark Ignition Engines" SAE 932713

[21] Zhao X., Matthews R.D., Elizey J.L., "Numerical Simulation of Combustion in S.I. Engines: Comparison of the Fractal Flame Model to the Coherent Flame Model" COMODIA 94

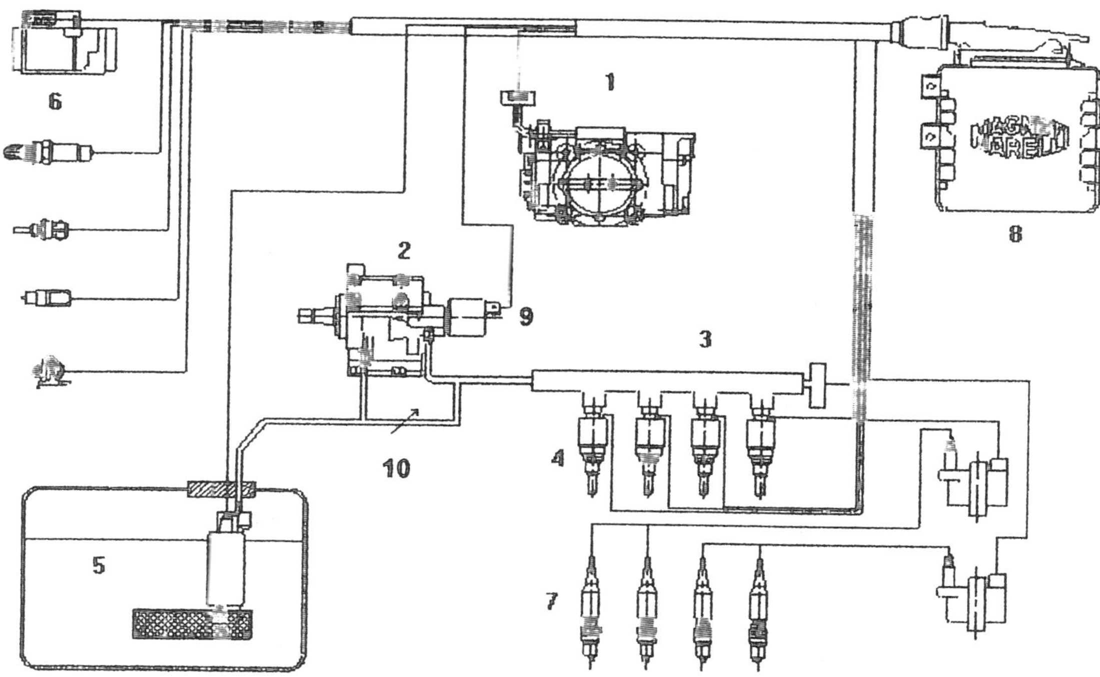

1. Electronic Throttle Body - 2. High Pressure Pump - 3.Common Fuel Rail - 4. Fuel Injectors
5. Fuel Tank - 6. Engine Sensors - 7. Spark Plugs - 8. Control Unit - 9. Pressure regulator - 10. By-pass

GDI-System Layout
Figure 1.

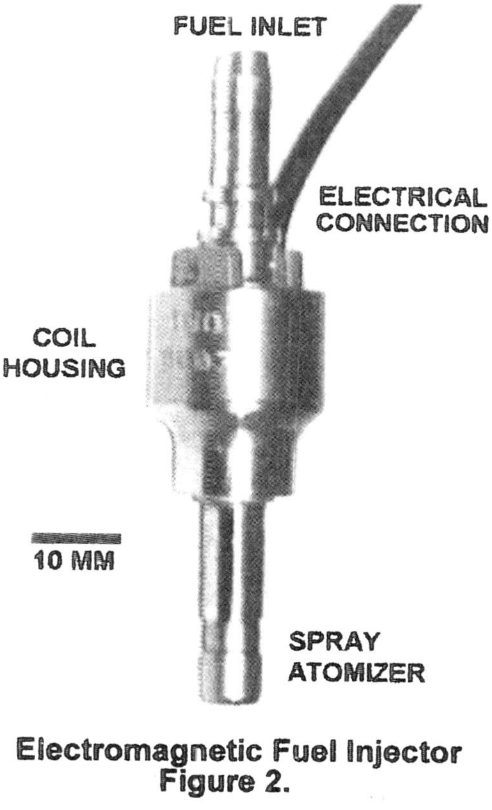

Electromagnetic Fuel Injector
Figure 2.

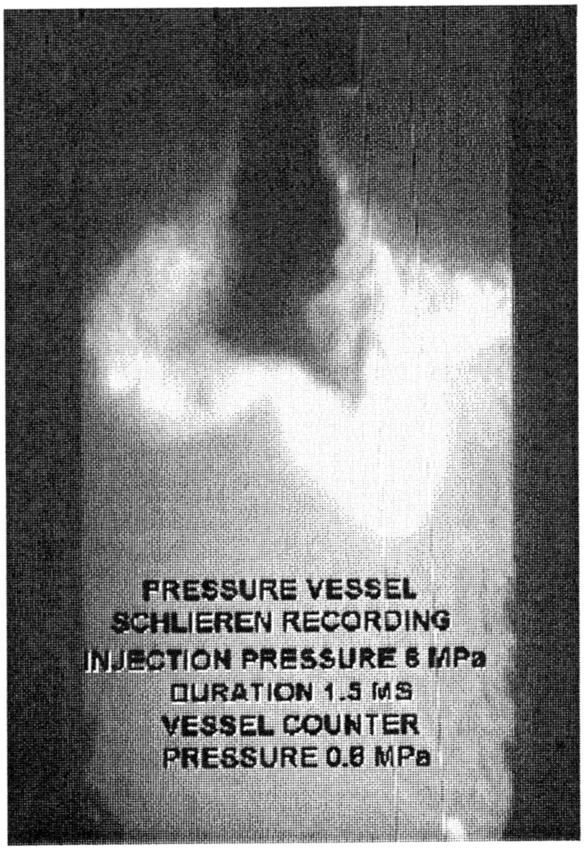

Figure 3

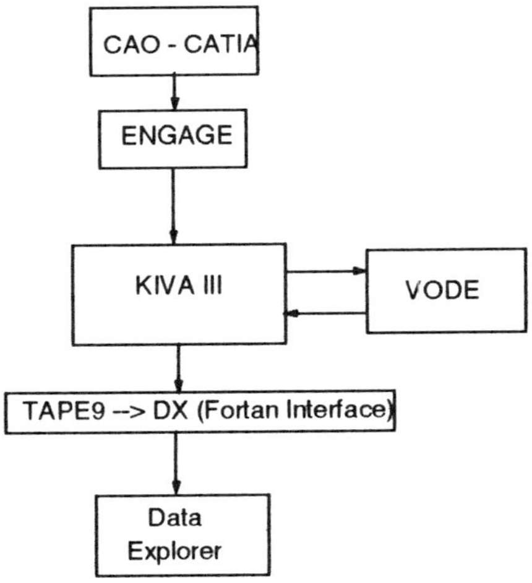

NCF3D Software Flow Chart
Figure 4

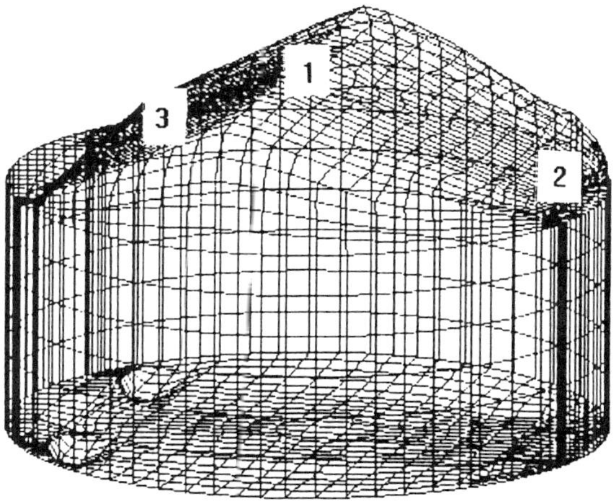

Combustion Chamber - Dynamic computation grid
Figure 5.

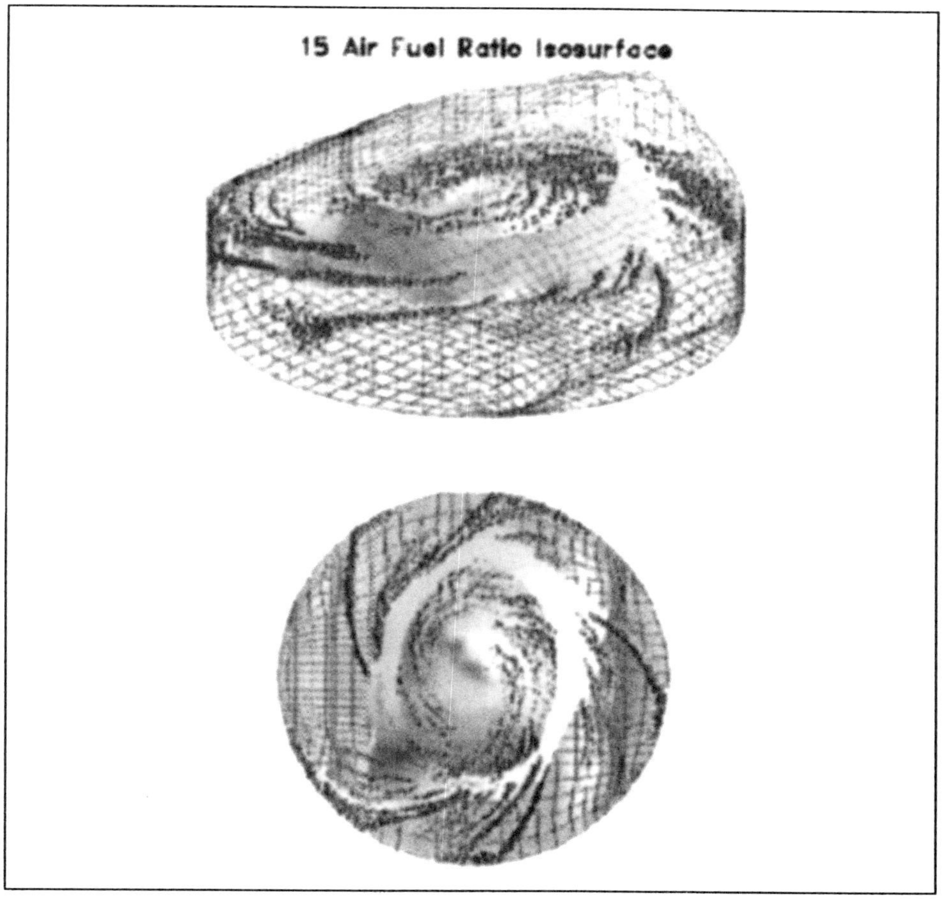

Injector type 2 - swirl - center mounted - 30 degree BTDC
Figure 6.

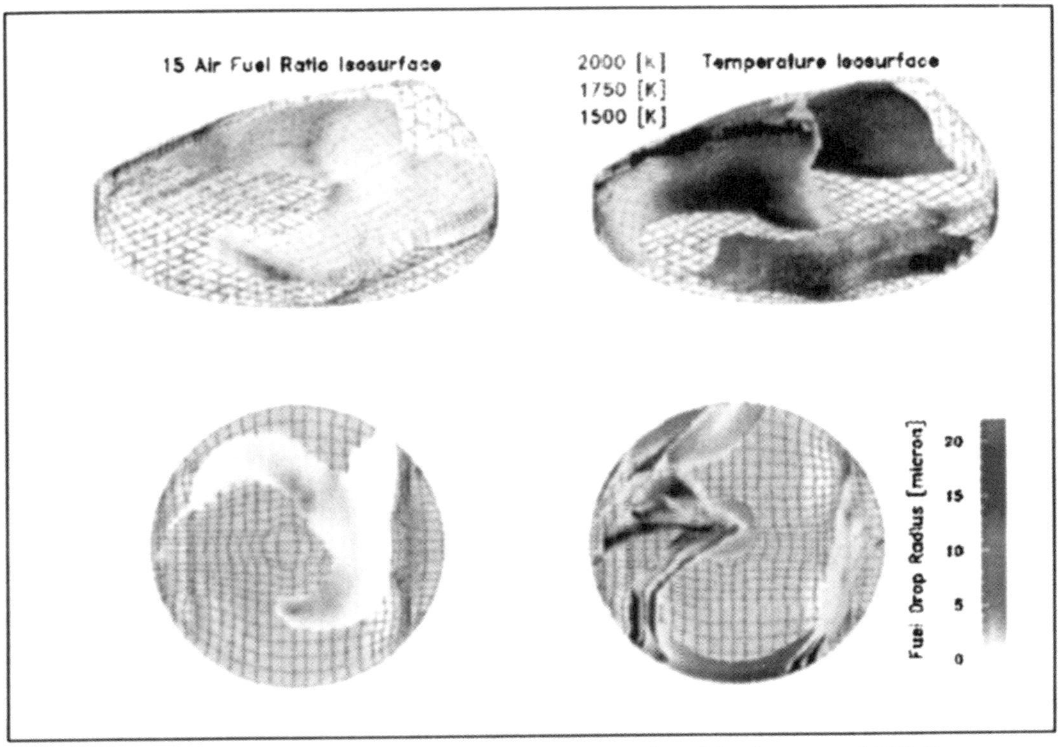

Injector type 1 - Pintle - Side mounted - 20 degree ATDC
Figure 7.

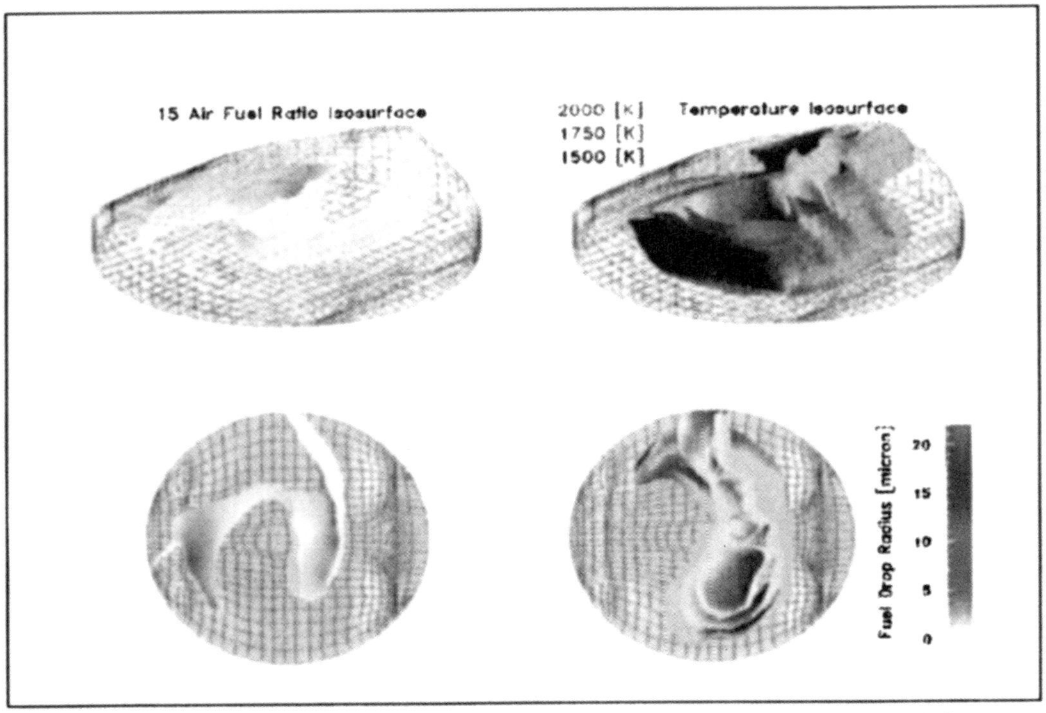

Injector type 3 - Closed cap - side mounted - 20 degree ATDC
Figure 8.

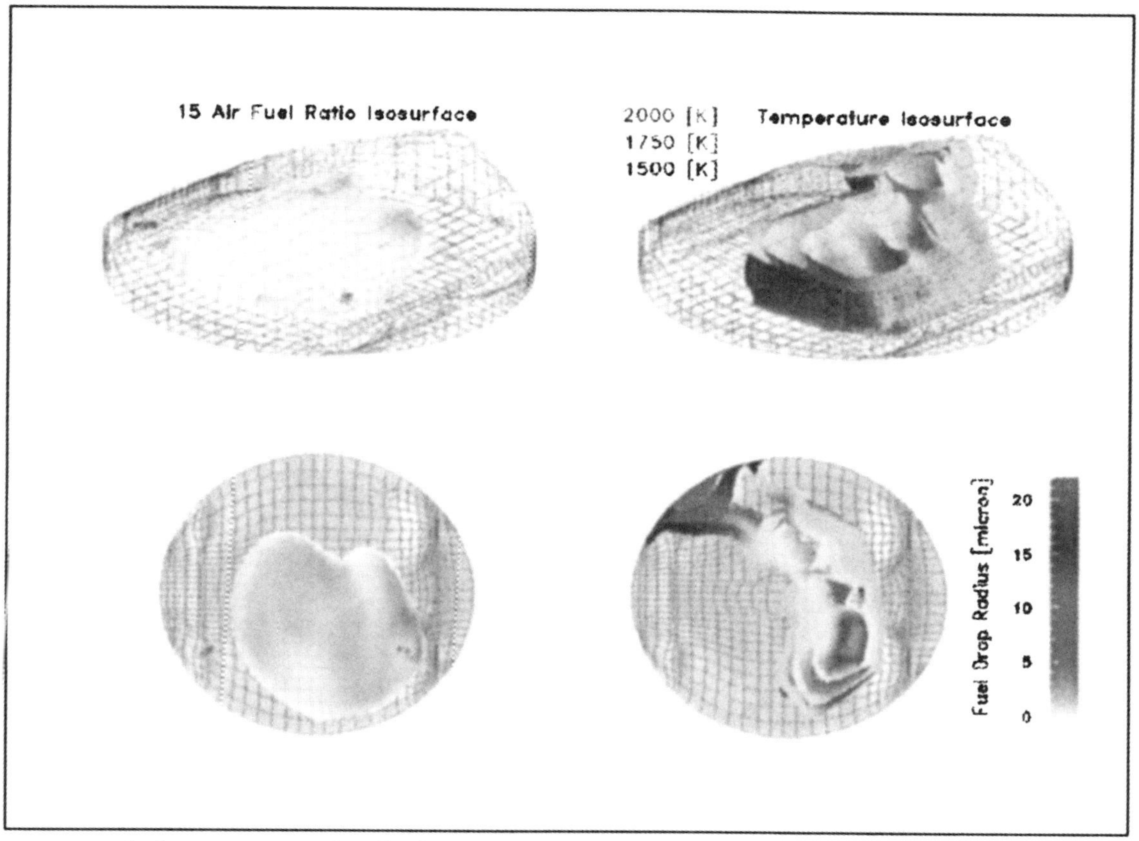

Injector type 2 - Swirl - center mounted - 20 degree ATDC
Figure 9.

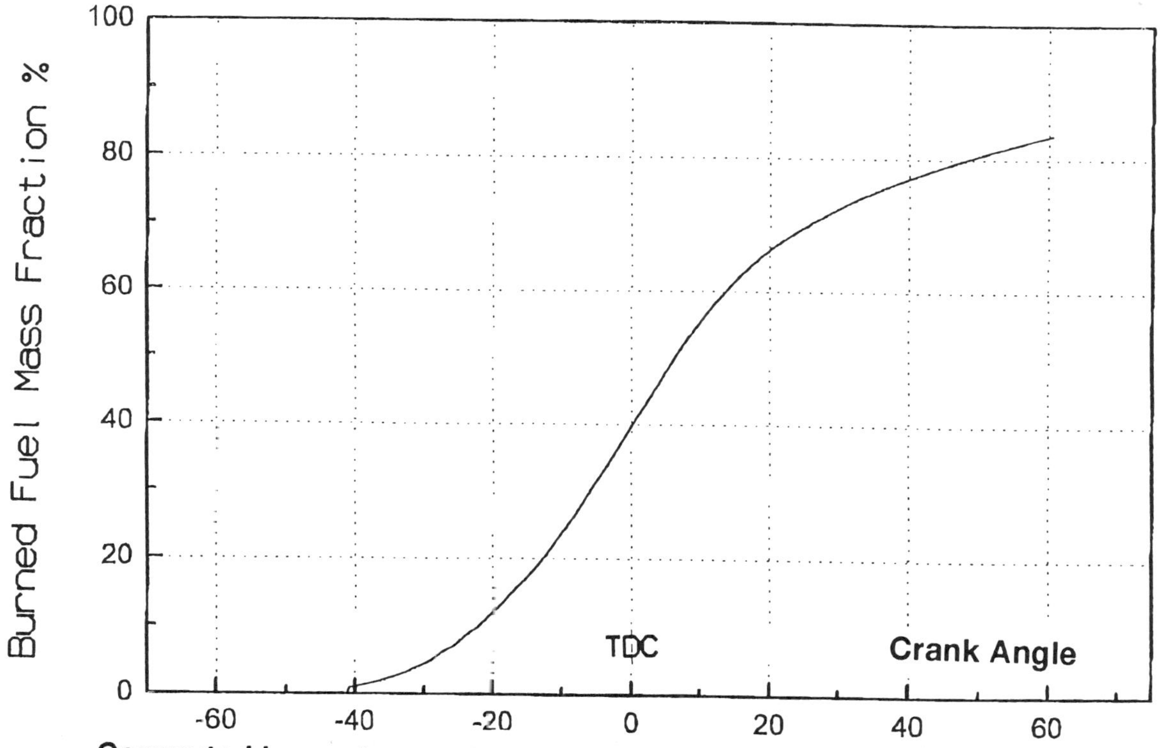

Computed burned mass fraction for center mounted type 2 injector
Figure 10.

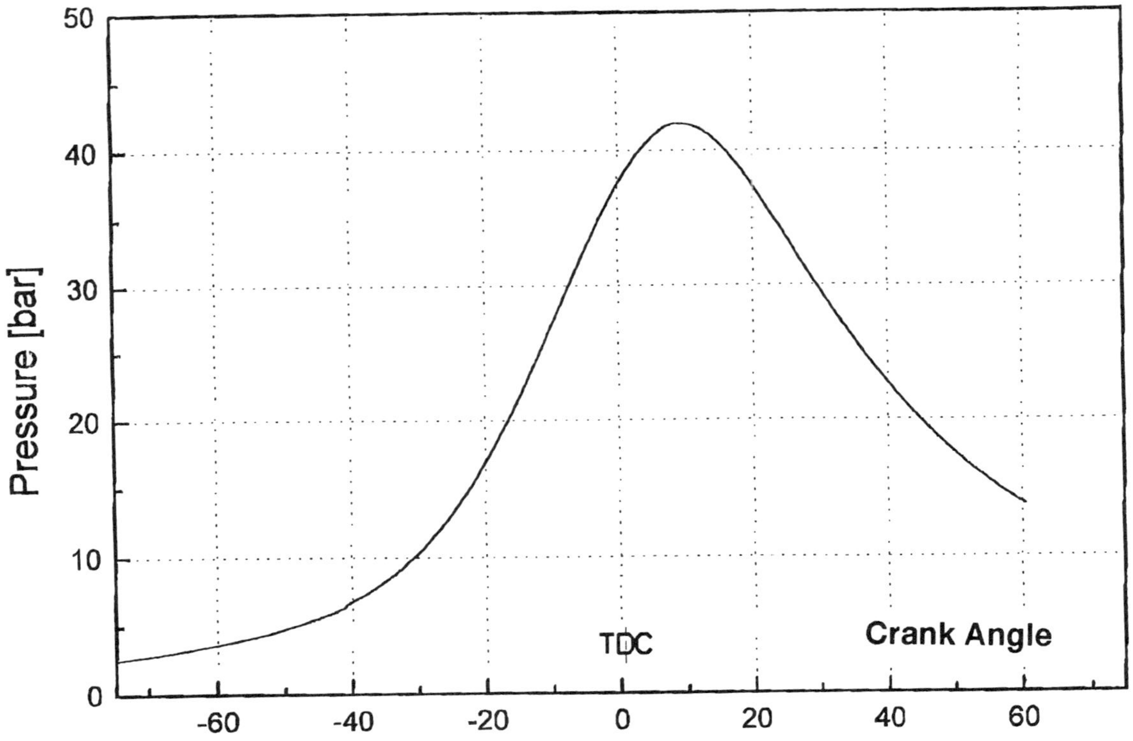

Computed cylinder pressure for center mounted type 2 injector
Figure 11.

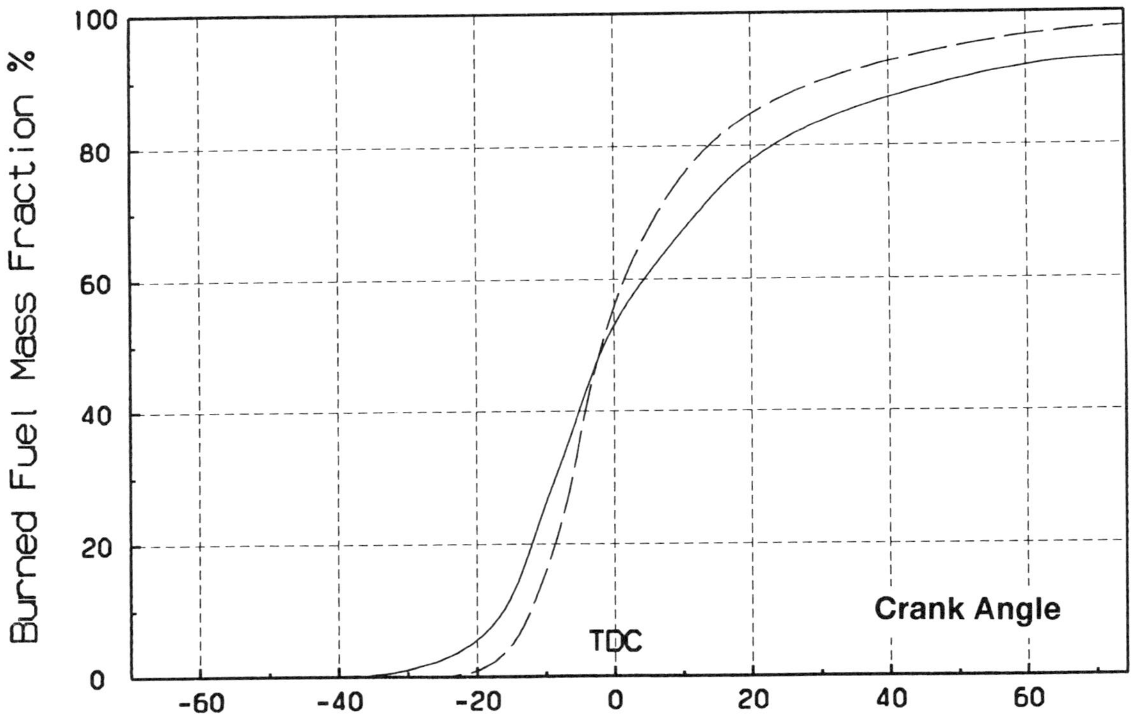

Measured burned mass fraction for bowl- and rim-profiled pistons
Figure 12.

940483

Preliminary Investigation of Solenoid Activated In-cylinder Injection in Stoichiometric S.I. Engines

T. H. Lake, M. J. Christie, and J. Stokes
Ricardo Consulting Engineers, Ltd.

O. Horada and K. Shimotani
Isuzu Motors, Ltd.

ABSTRACT

An investigation into the application of electronically controlled solenoid activated high pressure in-cylinder gasoline injection systems has been carried out in both conventional and novel four-valve four-stroke pent-roof chamber single-cylinder engines. Air motion requirements were studied and their effects on port design and layout were assessed. Alternative injector types, locations and spray characteristics were investigated. Transient and steady-state comparisons of the engines were made under both normal and cold running conditions. The outlook for the use of in-cylinder injection technology compatible with ULEV emissions requirements is discussed in the light of the results obtained.

INTRODUCTION

The advent of low cost, highly controllable direct injection equipment for gasoline, initially in air-assisted form and latterly using fuel only, has created the recent resurgence of interest in two stroke engines. This technology has been applied to both simple and complex two-stroke engines and development experience of these has been widely reported (1,2). However, recently there has been a waning of interest due to fears over durability and ability to meet US ULEV emissions limits after high mileages with these engines. Attention has now turned to the use of direct injection equipment in four stroke engines as a possible route to the lowering of emissions and fuel consumption and improving driveability and refinement.

Development work conducted at Ricardo and other establishments (3,4) has shown that increasingly stringent emissions legislation, for example ULEV, requires a major reduction in engine out hydrocarbon emissions in the first 30 seconds of operation after a cold start. In addition, transient air-fuel ratio control must be improved throughout any drive cycle to maintain and improve upon current catalyst conversion efficiencies.

Furthermore there is mounting pressure to make the drive cycles reflect more accurately driving conditions actually encountered to improve air quality. 90% of air quality failures in US metropolitan areas occurred when the ambient temperature was below the 20°C to 30°C temperature window used in the FTP test and 20% of the failures occurred at temperatures below -7°C (5). This issue has already been partly addressed with the introduction of a carbon monoxide emissions standard over the FTP drive cycle conducted at -7°C. Gradually it is assumed that there will be increased pressure to achieve low levels of all emissions at low temperatures.

Fuel economy legislation such as CAFE in the USA is being continually toughened and in Europe and Japan there is a strong emphasis on the need to reduce CO_2 emissions.

The evolution of gasoline fuelling systems is described in figure 1. The development of single, multiple and air-assisted injectors has been driven by the need for improved transient and cold start air/fuel ratio control to minimise emissions. Although significant advances in air/fuel ratio control have been achieved, this has always been at the expense of hot steady state mixture quality. The development of direct injection technology is a logical step in the process to further improve cold and transient operation. The major challenge facing direct injection technology is the ability to achieve competitive mixture quality during stabilised engine operation.

The direct injection spark ignition engine can operate in two alternative regimes; early injection with a stoichiometric air/fuel ratio, i.e. homogeneous charge, or late injection with stratified charge.

An early injection strategy with essentially homogeneous charge promises a number of benefits aimed directly at cold start and transient emissions

Year	Fuel System	Transient Control/Emissions	Cold Engine Control/Emissions	Hot Steady-State Mixture Quality	Cost
1900 - 1985	Carburettor	*	*	*****	*
1980 - 1985	SPI	**	*	***	**
1980 - 1991	MPI	***	**	**	***
1991-2020?	MPI (Air-Assisted / Elec. Fuel Heater)	****	***	**	***
2000?	DI (Medium Pressure / Solenoid Injector)	*****	*****	*?	*****

SPI Single point injection
MPI Multi-point injection
DI Direct Injection

Figure 1 - Evolution of gasoline fuel systems

reductions. A direct injection combustion system should be capable of stoichiometric cold start combustion, eliminating the need for a secondary air pump. Direct injection will also remove the need for port heaters and may also reduce the demands on ULEV aftertreatment systems such as electrically heated catalysts and hydrocarbon traps. Stoichiometric cold start operation will also improve urban fuel consumption.

A late injection strategy to provide stratified charge offers the potential to reduce fuel consumption and NOx emissions. One of the advantages of a late injection strategy is to reduce octane requirement at high load operation. This enables an increase in compression ratio with the attendant thermodynamic efficiency gains. However, it is exceptionally difficult to achieve full air utilisation with a late injection strategy due to incomplete mixing leading to unacceptable smoke emissions. Concepts using homogeneous charge at full load to eliminate smoke require a reduction in compression ratio reducing some of the fuel economy benefit at part load. It is widely reported that control of hydrocarbon emissions at part load has also proved exceptionally difficult. Additionally no stratified charge engine has demonstrated low enough engine out NOx emissions for achievement of ULEV emissions levels in the absence of lean NOx catalysis. However, the fuel economy potential of a late injection strategy has attracted extensive research studies. A high compression ratio stratified charge engine operating with a lean NOx catalyst continues to be an attractive concept.

The work reported in this paper is based on an early injection strategy aimed at improvements in cold start emissions and driveability which could use an aftertreatment system based on a three-way catalyst to meet ULEV emissions standards.

OBJECTIVES

The aim of the work described in this paper was to investigate the combustion system and charge motion requirements of a homogeneous direct injection four stroke engine. Initially the objectives were as follows:-

- Investigate alternative packaging arrangements for direct injection in four-valve pent-roof combustion chambers.
- Develop the speed/load operating envelope to be suitable for a passenger car engine, using pump gasoline of 95 RON.
- Achieve peak power, torque, part load emissions and fuel consumption equivalent to the engine on which the conversion was based.
- Achieve high EGR tolerance for stoichiometric fuel economy improvement potential and high NOx emission reduction potential.
- Demonstrate improvements relative to MPI engines under transient and cold conditions.

INJECTOR CHARACTERISTICS/SYSTEMS

The fuel injectors used throughout the programme were supplied by Siemens Automotive. Initially these were the Phase 2 type and laterly the improved Phase 4 injectors, both of which are fuel only types. Detailed information about the Phase 4 system is shown in table 1. These systems are solenoid operated with variable injection timing and period metering. The injectors themselves do not generate the high pressure, but

TABLE 1

PHASE 4 SIEMENS AUTOMOTIVE FIE DETAILS

Designation	:	HP4 Mid Feed
Cone angle	:	15 to 75 degrees available
Fuel Injection Pressures	:	3 - 12 MPa
Fuel particle Size (SMD)	:	5 to 10 microns
Driver Unit	:	Siemens High Voltage
- Max Voltage	:	70.5 V
- Max Current (Peak)	:	9.9A
- Max Current (Hold)	:	3.0A
Maximum Delivery	:	50mg/inj
Minimum Delivery	:	1mg/inj
Dynamic Range	:	> 10
Leak	:	< 0.5cc/min (Nitrogen at 380kPa)
Durability	:	> 500 million cycles

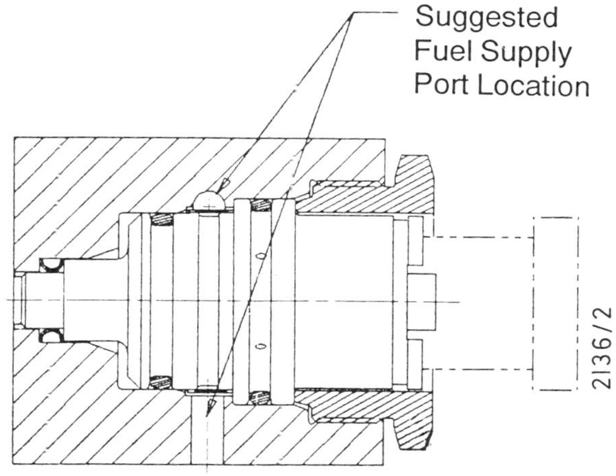

Figure 2 - Injector Fuel Supply Location

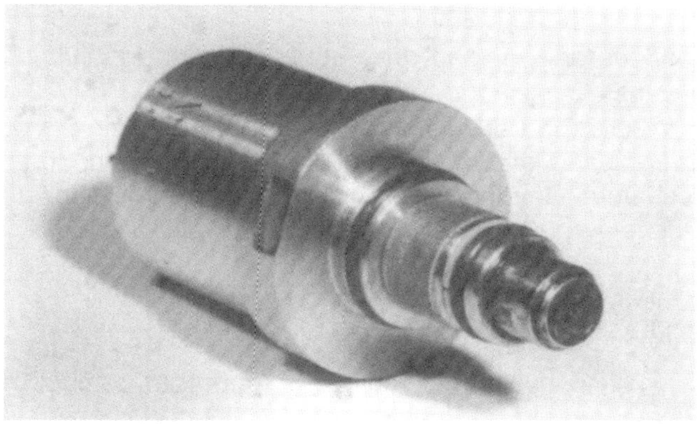

Figure 3a - Phase 2 Injector

instead rely upon an externally driven fuel pump to supply fuel to a gallery around the base of the injector, figure 2. Fuel pressure was supplied by a test shop system and was variable up to 15 MPa, independent of engine condition; the typical operating range being 3 to 10 MPa.

The fuel injector variables are thus, spray cone angle and fuel pressure. Reduction of fuel pressure tends to lower penetration of the fuel spray. Fuel spray droplet size also increases as pressure drops below 5 MPa.

Typically with the Phase 4 injectors, the fuel particle size is between 5 and 10 microns SMD, with cone angles available from 15 to 75 degrees. Typical injection

Figure 3b - Phase 4 Injector

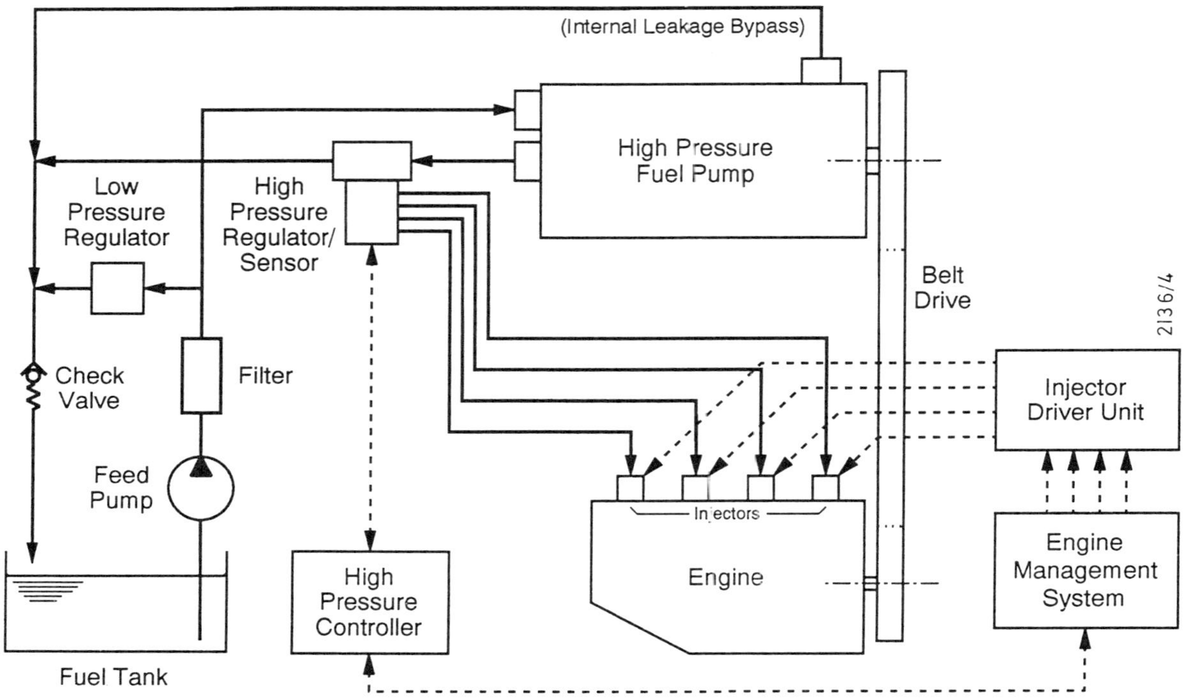

Figure 4 - Schematic Layout of a Vehicle Fuel System

periods at part load are about 2.0ms and at full load 4.5ms.

Both the earlier Phase 2 and later Phase 4 types were used and are shown in figure 3a and 3b. A schematic layout of a vehicle fuel system is shown in figure 4.

COMBUSTION SYSTEM, PORT DESIGN AND INJECTOR LAYOUT.

The conventional four valve pent-roof combustion chamber was used in all cases as it has demonstrated very good breathing and power potential combined with low emissions and fuel consumption characteristics (6). It is considered the most likely candidate for the initial application of injector systems in the near term.

Within the constraints of the combustion chamber configuration, a number of different port and injector layouts were investigated. These are shown in figures 5a,b,c.

Initially the injector was installed in the side of the cylinder head under the inlet valves. This was considered the most attractive packaging possibility for a multi-cylinder engine and is thought to provide an opportunity for spray entrainment with the air flowing into the cylinder.

The larger Phase 4 injector could not be fitted in this way and had to be fitted in the front of the cylinder head between the inlet and exhaust valves.

The third layout used a top entry inlet cylinder head because of its previous success in achieving good performance and part load emissions and fuel consumption during a poppet-valve direct injection two-stroke engine development programme carried out by Ricardo (7).

DEVELOPMENT EXPERIENCE

The experimental programme was based on a Hydra single cylinder research engine (8). A schematic cross section of the engine in the top entry configuration is shown in figure 6. The specifications of the engines and builds are described in table 2.

The experience described below is an accumulation of the typical responses encountered during the work.

The flow performance and tumble generating properties of the inlet ports were measured and developed using the Ricardo steady flow bench described in (6). Steady flow data is integrated over the valve event to produce a "Tumble Ratio", defined as the ratio of the angular velocity of the charge at the end of induction to the crankshaft speed, assuming a forced vortex exists in the cylinder.

INJECTION TIMING -- Injection timing response characteristics were examined at part load and typical results are shown in figure 7.

The best emissions and fuel consumption were achieved with injection timings between 380 and 300 deg BTDCF. The early injection timing necessary is taken as evidence of a long mixing and evaporation time.

Head Number One

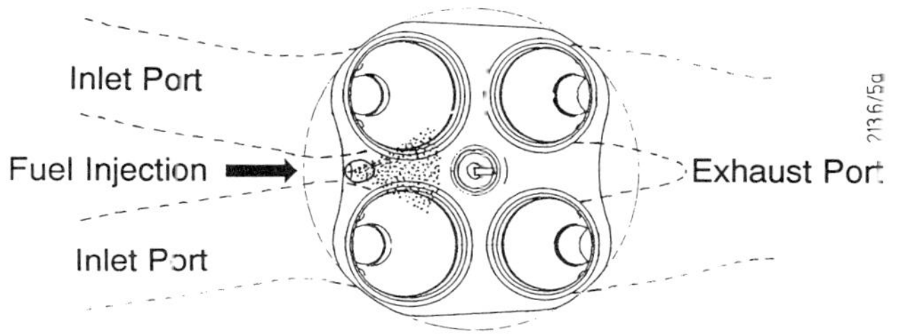

Figure 5a

Head Number Two

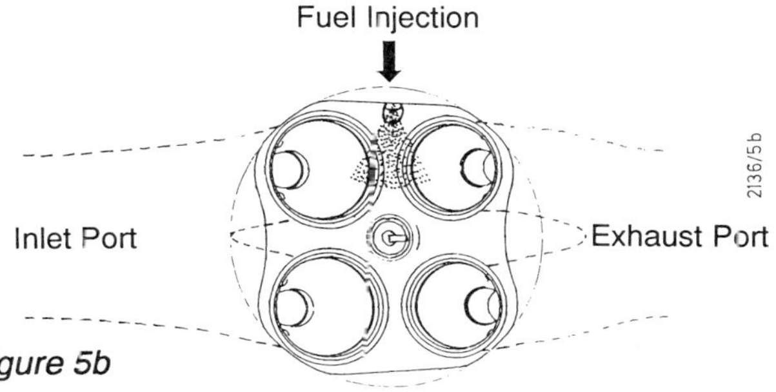

Figure 5b

Head Number Three

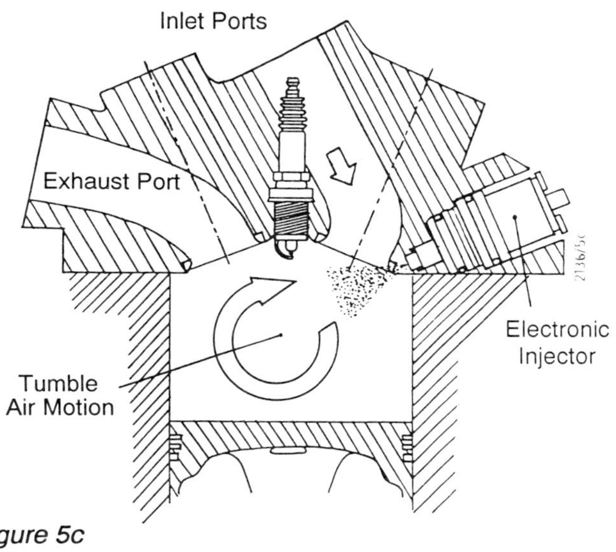

Figure 5c

The Different Port and Injector Configurations Tested

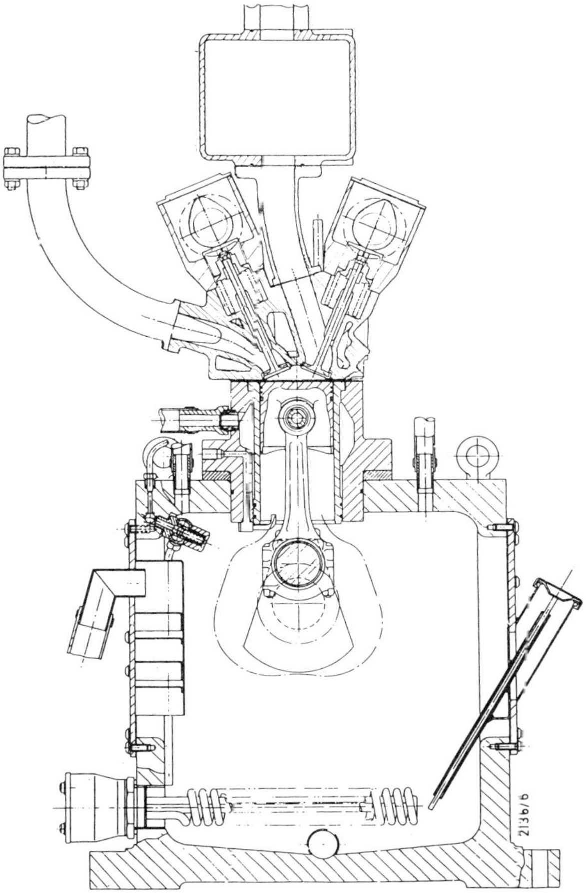

Figure 6 - *Cross Section of the Ricardo Hydra Engine in Configuration 3*

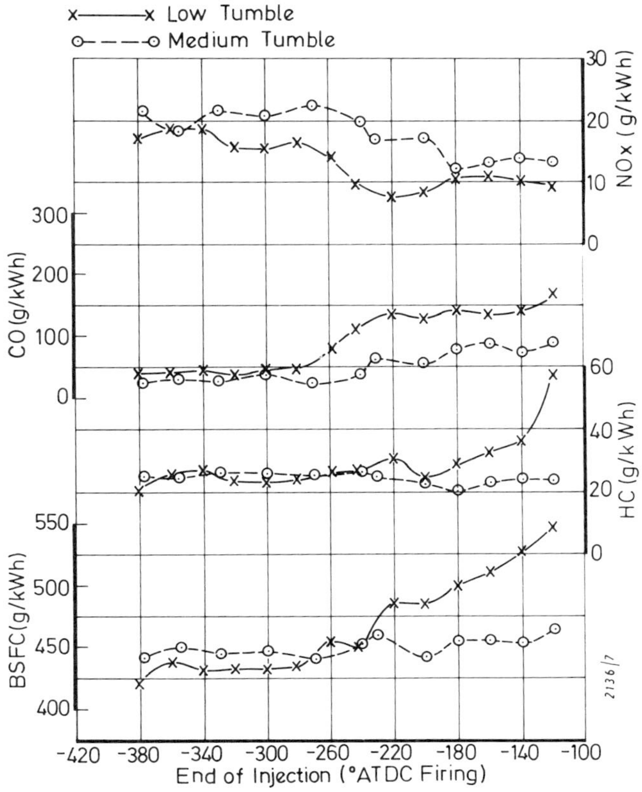

Figure 7 - *Typical Injection Timing Response Characteristics at Part Load*

TABLE 2

ENGINE DETAILS

Engine 1		Engine 2	
Bore	93.7mm	Bore	80.26mm
Stroke	77.0mm	Stroke	72.0mm
Cylinder Head Type	Head 1 and head 2	Cylinder Head Type	Head 3 (Top entry)
Compression Ratio	9.6:1	Compression Ratio	9.5:1
Valve Timings	IVO - 12 BTDC IVC - 48 ABDC EVO - 54 BBDC EVC - 6 ATDC	Valve Timings	IVO - 12 BTDC IVC - 48 ABDC EVO - 63 BBDC EVC - 12 ATDC

At the lower tumble level, difficulty was experienced in setting ignition timing at minimum advance for best torque (MBT) at injection timings 180 BTDC and later; the engine would eventually stop. It is possible the advance is limited because the in-cylinder conditions earlier in the cycle are poorer for either propagation or generation of the flame kernel. As a result the engine ran more retarded ignition timing than optimum. This has resulted in lower NOx, and higher fuel consumption than would have been observed at MBT timings.

AIR MOTION LEVEL -- Applying higher levels of tumbling air motion meant that the mixing and evaporation can be improved and later injection is achievable at the same emission and fuel consumption levels as well as achievement of MBT.

INJECTOR CONE ANGLE -- At the higher air motion level, the injector cone angle and fuel pressure response was assessed. This test aimed to optimise the injector cone angle and penetration characteristics with the level of tumbling air motion present. These tests were performed at different injection timings, ranging from 180 to 140 deg BTDCF. These timings give a long spray path length which emphasises the influence of cone angle and pressure. The results are shown in figures 8-11.

At the earlier injection timing condition, a clear optimum cone angle of 25 degrees is established. At the later injection timing similar trends but generally poorer characteristics can be achieved by the 18 and 25 degree injectors. The more difficult later injection timing regime now shows larger differences in the response to fuel pressure.

FUEL PRESSURE -- The engine will run with a very wide range of fuel pressures, from 13.8 MPa to below 0.7 MPa. Reducing the fuel pressure from 7 MPa bar shows signs of deteriorating combustion. This might be expected as fuel particle size will be increasing and mixing becomes more difficult. It is noted that at the later injection timing the combustion stability improves below 3.5 MPa fuel pressure. This may be a function of the decreasing injection rate and longer injection periods. These factors may combine to improve the mixing process.

VARIABLE AIR MOTION -- Different levels of tumble air motion were required at different loads and speeds for the best emissions and fuel consumption, suggesting the requirement for a variable air motion system. Such devices are now becoming an integral part of some combustion systems (9).

The optimum injection timing proved insensitive to air fuel ratio as well as engine coolant temperature.

DEVELOPED BUILD PERFORMANCE

FULL LOAD PERFORMANCE -- Figure 12 shows a comparison between the baseline MPI build and a developed DI build with a variable tumble inlet system giving lower tumble at wide open throttle.

The performance of the DI engine is competitive with the MPI engine. The torque of the DI engine at high speed is greater because the tumble ratio of the DI engine at full load is lower and the inlet port is therefore somewhat less restrictive. Full load tests demonstrated that the injection system is capable of achieving a typical automotive speed range and can deliver and mix the fuel in sufficient time for competitive performance to be obtained.

In the case of early injection timing, tests at 1800 rev/min showed there to be no significant difference in octane requirement between the DI and MPI systems. This indicates that a substantially homogeneous charge

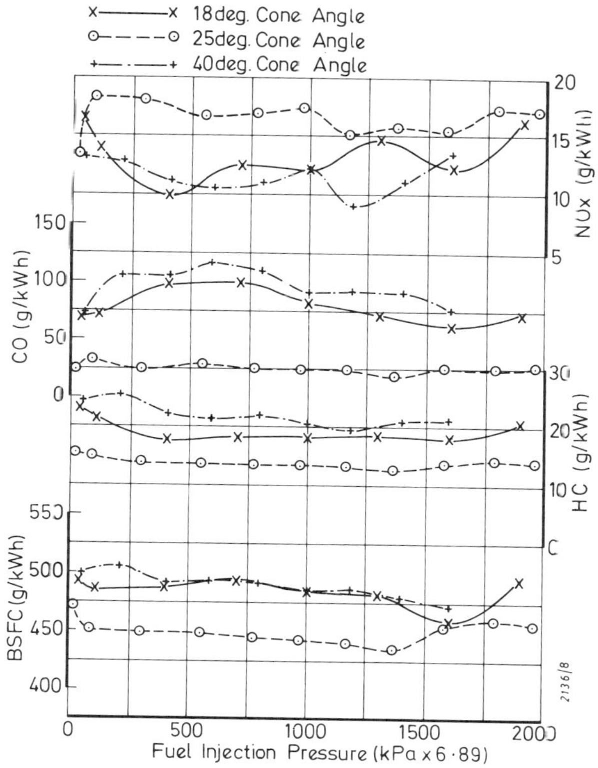

Figure 8

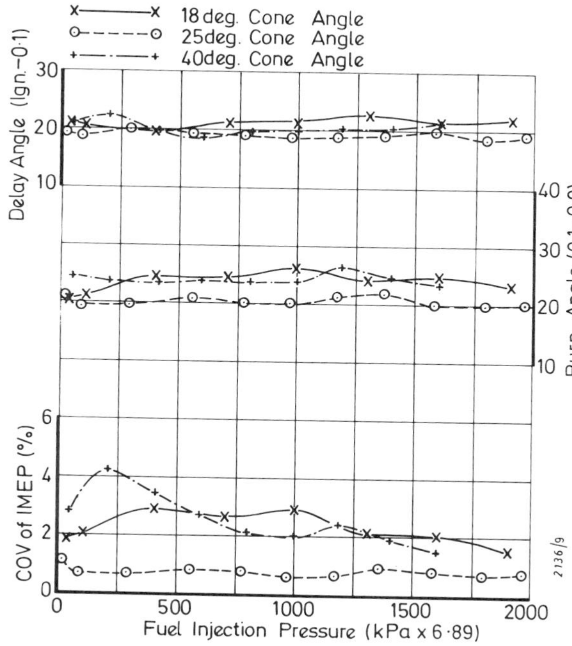

Figure 9

Fuel Pressure Response at Stoichiometric Part Load Conditions. End of Injection Timing, BDC. Medium Tumble Ratio

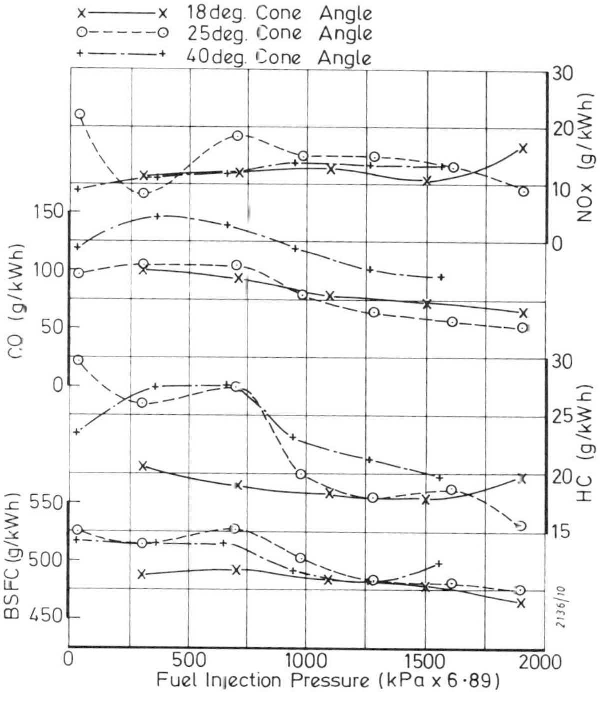

Figure 10

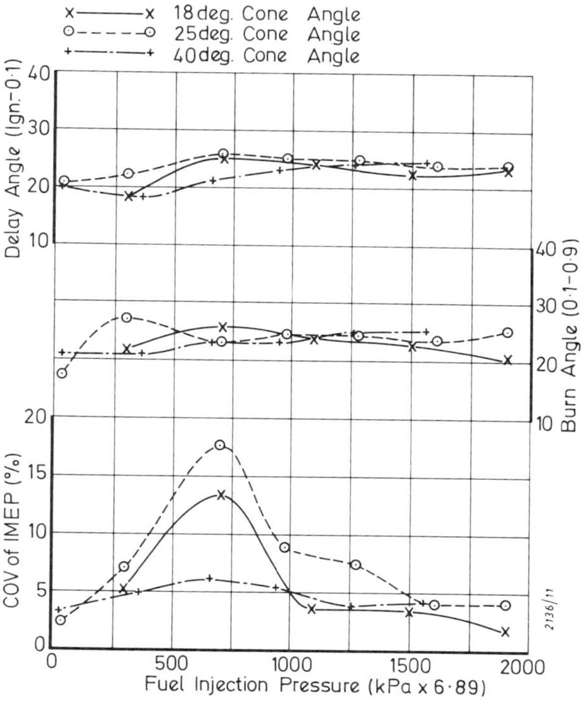

Figure 11

Fuel Pressure Response at Stoichiometric Part Load Conditions. End of Injection Timing 140 BTDC. Medium Tumble Ratio

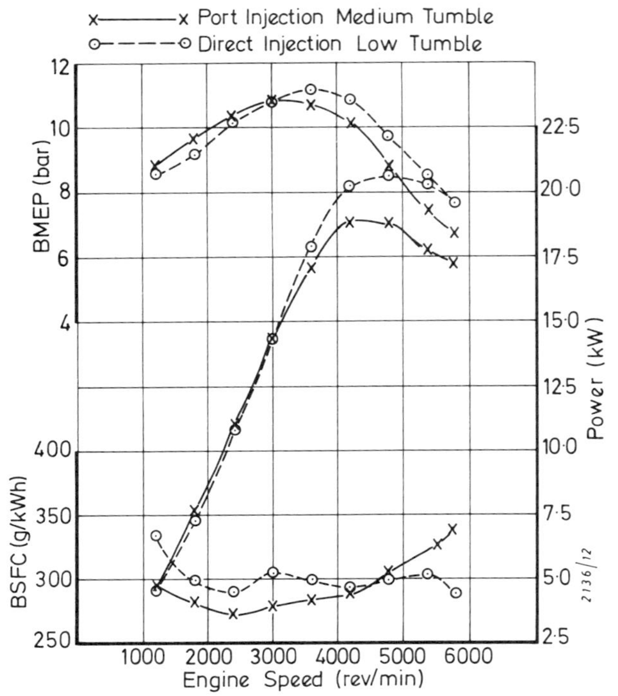

Figure 12 - Comparison of Power Curves for the Conventional Engine and a Direct Injection Version

has been achieved without lean regions that may incite detonation.

PART LOAD PERFORMANCE -- A comparison between a developed DI build with the baseline MPI build at part load is shown in figure 13a and b.

The developed DI produces almost identical hydrocarbon levels to those from the MPI baseline at stoichiometric air fuel ratio. The NOx emission levels are identical indicating similar air fuel ratio at the spark plug for both the MPI and DI engines. The CO response indicates that a substantially homogeneous charge has been achieved by the early injection strategy. The fuel consumption at stoichiometric air fuel ratio is 6% worse. The DI engine had slightly higher mechanical friction, but the difference in fuel consumption is mainly due to the higher heat losses associated with the higher levels of air motion required for good mixing and fuel preparation.

The combustion characteristics of the DI engine are markedly different to the MPI engine. The higher levels of air motion have contributed to the large reduction in delay angle and burn angle at the same air fuel ratio with the DI engine. The resulting combustion stability is maintained at high levels of dilution. At the dilution limit the rapid deterioration of stability is typical of high air motion regimes where flame initiation becomes difficult or excessive flame stretching causes extinguishment of

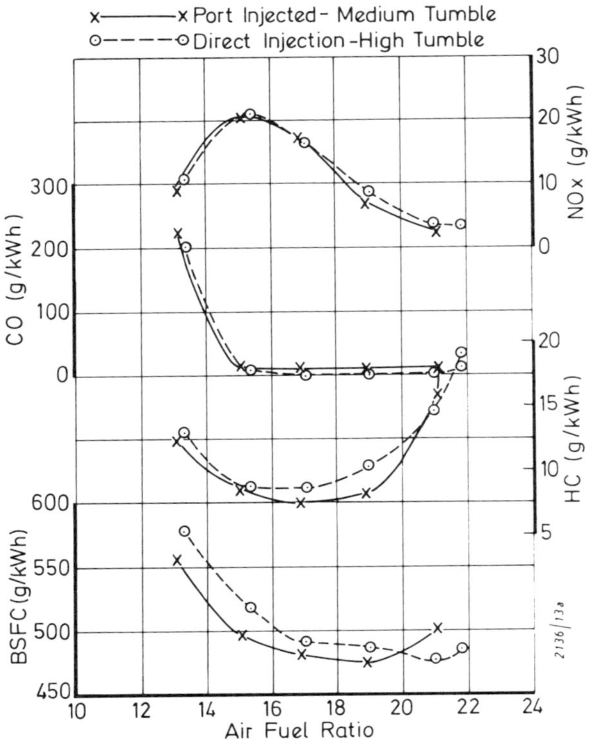

Figure 13a

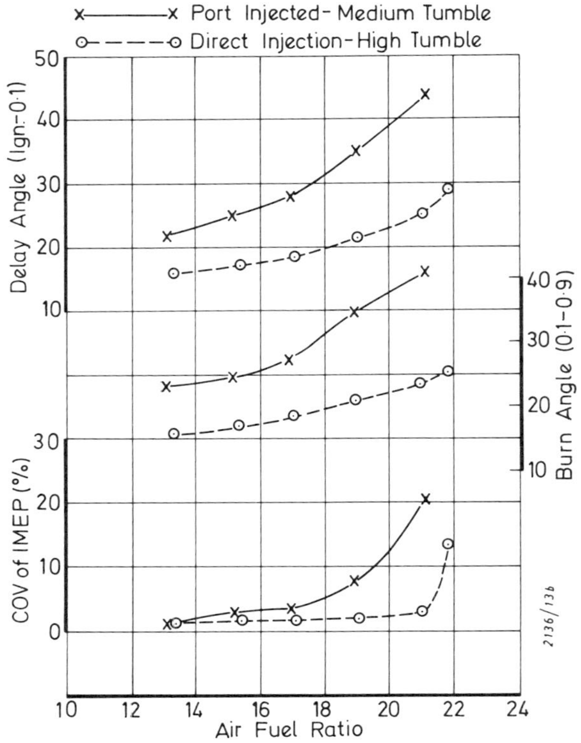

Figure 13b

Air Fuel Ratio Response at 1500 rev/min 1.5 bar BMEP Baseline MPI Engine versus a Developed DI Engine

the flame kernel.

The burn angle at stoichiometric is approximately 16 degrees. The high rate of pressure rise associated with such a short burn angle is incompatible with high refinement. However, the very short delay and combustion periods are advantageous when considering requirements for dilution tolerance. The delay and combustion periods lengthen with exhaust gas dilution due to a reduction in the laminar flame speed and thus the pressure rise is reduced to an acceptable level for refinement.

EGR TOLERANCE -- Figure 14 shows the high EGR tolerance that has been achieved with the DI combustion system. In this case the burn angle at zero EGR is about the ideal of 20 degrees and there is less increase in the burn angle with the application of EGR than would be expected of a typical MPI engine. The fuel consumption improvement in this case was 4% which is about the maximum achievable with a conventional homogeneous EGR strategy (10).

TRANSIENT CAPABILITIES

The response of an engine during cold start conditions was briefly investigated. It was not possible to simulate an actual cranking start as the single cylinder engine is not fitted with a starter motor nor a fully mapped engine management system.

A simulated cold start test procedure was established as follows:

> Throttle settings and fuelling levels were established at 1500 rev/min, 1.5 bar BMEP during hot, steady state operation for each build.
>
> The engine was stopped and allowed to cool to ambient temperatures. (20 - 25 °C)
>
> Whilst cold, the engine was then rapidly motored to 1500 rev/min at which point the fuelling was switched on.
>
> Combustion and emissions data were then logged for 500 cycles following the start of fuelling.

Examples of the results from this test procedure are shown in figures 15-16.

The results show that the DI engine clearly responds more quickly than the MPI engine when cold and with lower hydrocarbon emissions. The initial cycle IMEP data from the MPI engine shows a considerable amount of variation compared to the DI engine. The MPI engine hydrocarbon emissions are initially very high because of the failure of the engine to fire for a number of cycles as there is a film of fuel built up on the manifold and port surfaces. Whilst a reasonably sophisticated engine management system would be capable of controlling the first few cranking cycles for the MPI build, the intrinsic benefit of direct injection is demonstrated. The DI engine could run with little or no fuel enrichment from start up with obvious benefits to fuel consumption and emissions when cold.

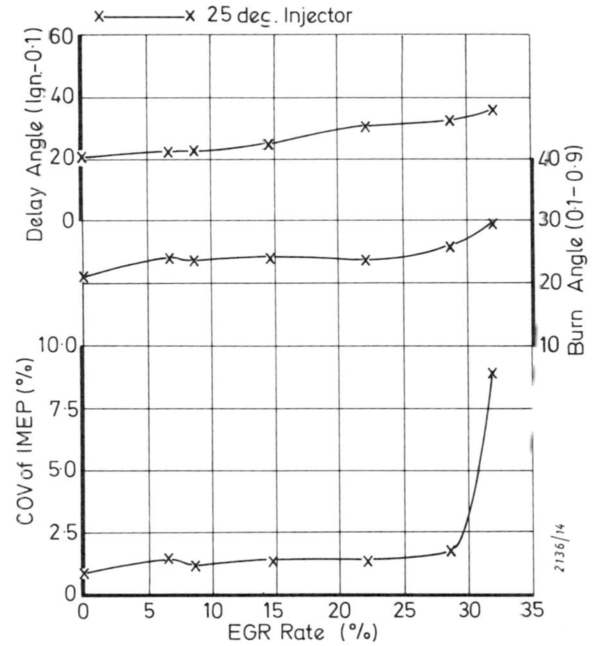

Figure 14 - Part Load EGR Response at Stoichiometric. End of Injection 300 BTDC. Medium Tumble Ratio

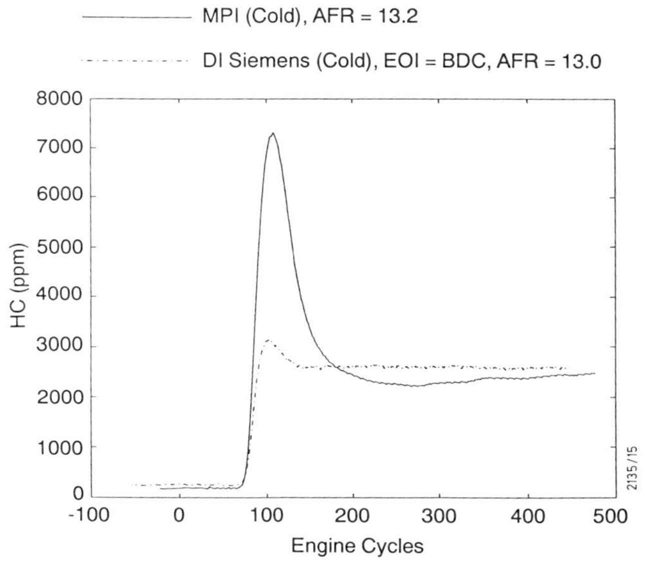

Figure 15 - Raw Engine out Hydrocarbons During Cold Start

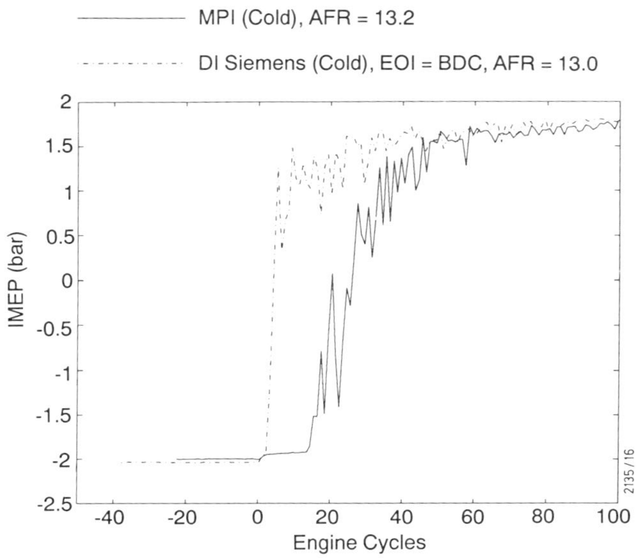

Figure 16 - Engine IMEP During Cold Start

OUTLOOK

The application of DI technology has traditionally been aimed towards unthrottled late injection stratified charge engines, where the intention is to obtain significant fuel consumption benefits (11). The late direct injection stratified charge engines have yet to demonstrate satisfactorily low enough NOx emissions or light load hydrocarbons for ULEV though they still remain highly attractive.

However, there has been little or no published data demonstrating the benefits to be obtained from the modern solenoid systems when implemented in a homogeneous charge strategy. Homogeneous operation allows the use of EGR and a three-way catalyst (TWC) for the control of NOx emissions whilst hydrocarbon emissions at low load are comparable to a port injected engine.

The results demonstrate that a direct injection homogeneous charge combustion system is compatible with ULEV emissions legislation when combined with conventional aftertreatment technologies.

CONCLUSIONS

Homogeneous charge direct injection and combustion has been achieved at early injection timings of between 380 and 300 deg BTDCF. The optimum timing is affected by injector spray characteristics, in-cylinder charge motion and fuel pressure.

Part load performance has been developed to a level where NOx, CO, HC emissions and fuel consumption are competitive with an MPI version of the same engine.

Comparable wide open throttle performance was obtained with no increase in octane requirement.

The combination of high in-cylinder air motion and direct injection has demonstrated exceptional EGR tolerance.

Cold transient tests have demonstrated that a DI combustion system has the potential to operate without fuel enrichment at current test cycle start temperatures.

ACKNOWLEDGEMENTS

The authors would like to thank the Directors of Ricardo Consulting Engineers Ltd for permission to publish this paper and Isuzu Motors Ltd for their support. In particular the authors would like to thank Siemens Automotive for supply of the fuel injection equipment.

REFERENCES

1. Sato K., Ukawa H., Nakano M. "A Two-Stroke Cycle Engine with Poppet Valves in the Cylinder Head - Part 2." SAE 920780
2. Thornhill D., Fleck R. "Design of a Blower-Scavenged, Piston-Ported, V6, Two-Stroke Automotive Engine." SAE 930980
3. Hadded O., Stokes J., Cartmell A.J., Brisley R.J., Bruck R. "The Achievement of ULEV Emission Standards for Large High Performance Vehicles." SAE 930389
4. Hadded O., Stokes J., Grigg D.W. "Low Emission Vehicle Technology for ULEV and European Stage 3 Emissions Standards." Autotech 1993
5. Larson R.E. "Vehicle Emission Characteristics Under Cold Ambient Conditions." SAE 890021
6. de Boer C.D., Johns R.J.R., Grigg D.W., Train B.M., Denbratt I., Linna J-R. "Refinement with Performance and Economy for Four-Valve Automotive Engines." I.Mech.E. C394/053
7. Stokes J., Hundleby G.E., Lake T.H., Christie M.J. "Development Experience of a Poppet-Valved Two-Stroke Flagship Engine" SAE 920778
8. French C.C.J. "A Universal Test Engine for Combustion Research." SAE 830453
9. Stokes J., Lake T.H., Christie M.J., Denbratt I. "Improving the Fuel Economy Trade-Off for Gasoline Engines with the CCVS System." SAE 940482
10. de Boer C.D., Stokes J., Lake T.H. "Advanced Gasoline Combustion Systems for Fuel Economy and Emissions." IMechE Conference, Worldwide Emissions Standards and How to Meet Them, May 1993
11. Schäpertöns H. et al. "VW's Gasoline Direct Injection (GDI) Research engine." SAE 910054

961987

An Investigation into the Use of Piezo-Fluidic Combined Units as Fuel Injectors for Natural Gas Engines

Rui Chen and Gordon G. Lucas
Loughborough University of Technology

ABSTRACT

A novel piezo-fluidic gaseous fuel injector system designed for natural gas engines is described in this paper. The system consists mainly of no-moving-part fluidic devices and piezo electro-fluidic interfaces. The steady state and dynamic characteristics of the system were tested on a laboratory experimental rig. The results show that the system can handle the large gas volume flow rate required by natural gas engines and is capable of operating via pulse width modulation. A few typical commercial solenoid type gas injectors were also tested and the results were compared with those from the piezo-fluidic injector system. It was found that the piezo-fluidic gaseous fuel injector system has faster switching responses and smaller injection cycle-to-cycle variations.

INTRODUCTION

Natural gas consists primarily of methane. As an alternative automotive fuel, natural gas can produce significantly less harmful emissions. However, to meet the increasingly stringent vehicle emission legislation, the fuel metering system of natural gas engines must be capable of accurate control and have a fast switching response, especially when stoichiometric air/fuel ratio combustion strategy incorporated with a three-way catalyst exhaust system is employed [1].

Gaseous fuel injection, combined with a sophisticated engine management system, is able to provide good air/fuel ratio control [2]. However, since natural gas has a much lower energy density and less lubrification than conventional liquid fuels, the gaseous fuel injectors must, therefore, be capable of handling the large gas volume flow rate. Thus, for conventional solenoid type gas injectors, not only will the components be larger, with higher pintle lift than liquid fuel injectors to achieve a larger orifice for the large gas volume flow rate, but they also have to work in an environment with poor lubrication.

In order to partially overcome the low energy density drawback, a conventional solution for solenoid type gas injectors is to use a high injection pressure (over 700 kPa) which is obtained directly from natural gas onboard storage tank. As a consequence, when tank pressure drops lower than injection pressure, the tank must be refuelled. This leaves a part of onboard fuel unused. If the onboard natural gas is stored as compressed natural gas (CNG), this part of unused natural gas may not affect vehicle travel range obviously since the storage pressure is up to 20 MPa. However, if adsorbed natural gas (ANG) is utilised, the high injection pressure will certainly seriously limit vehicle travel range since the storage pressure is only up to 3.5 MPa [3].

It is to meet these difficulties that the development of a novel fluidic gaseous fuel injector system based on the use of mono-stable fluidic devices has been undertaken. This system has the potential of providing a very accurate and fast response gas control as well as being able to handle large gas volume flow rates. Also, this system can be arranged as a multi-point gas injection system and be operated with pulse width modulation mode [4,5].

One of the important features of the fluidic device is the amplification characteristic (ratio of the operating pressure to the control pressure). It offers the possibility of using a very low energy consumption, and very fast response piezo unit to control the fluidic device to inject gaseous fuel into the engine manifold. In this paper, this piezo-fluidic gaseous fuel injector system is proposed and evaluated. Both steady state flow rates and dynamic switching properties of the system were tested on a laboratory experimental rig. The test results were compared with those from typical commercial gas injectors. It was found that the piezo-fluidic gaseous fuel injector system has faster switching responses and smaller injection cycle to cycle variations.

PIEZO-FLUIDIC GASEOUS FUEL INJECTOR SYSTEM

MONO-STABLE FLUIDIC DEVICES - The fluidic device which is being used in the piezo-fluidic gaseous fuel injector system is a mono-stable fluidic amplification device based on the wall re-attachment Coanda effect as shown in Figure 1. It has two control ports, namely the positive control port and the vacuum control port. These are located on each side of the input jet nozzle. The output flow channel which connects with the positive control port is called the stable side output. The one which connects with the vacuum control port is called the unstable side output.

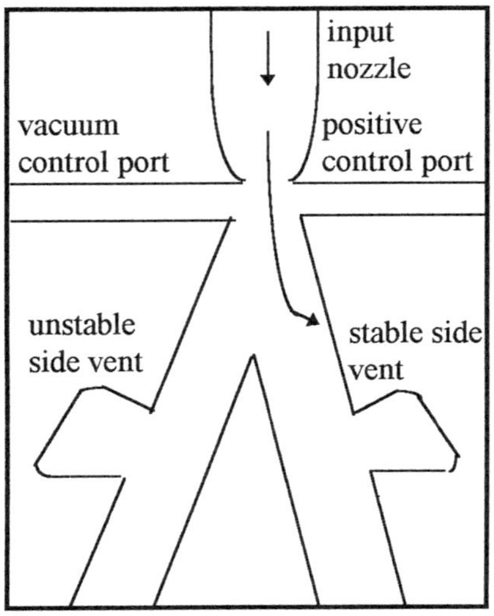

Figure 1 Mono-stable fluidic device

If the pressure in the two control ports is atmospheric, and the pressure of the gas stream entering the input nozzle is higher, the jet flow issuing from the input nozzle will attach to the stable side output. This is because the geometry design is such that this bias exists. When a gas flow at a sufficiently high pressure is initiated in the positive control port through an electro-fluidic interface, the jet flow will switch to the unstable side output and remain there. When this control flow is removed, the jet flow will shift back to re-attach to the stable side output. This switching control method is called the "positive control" method.

Alternatively, the electro-fluidic interface can be used to close off the vacuum control port. Because of the jet flow entrainment, a "vacuum" is created in the vacuum control port. This too, will cause the jet flow to switch to the unstable side output from the stable side output till the vacuum control port is re-opened to atmospheric. This switching method is called the "vacuum control" method.

Opening the electro-fluidic interface and applying a sufficient high pressure to the positive control port will result in a rapid switching of the jet flow. Conversely, when the electro-fluidic interface is closed, that is to cut off the high pressure from the positive control port, the pressure in the port will take some time to decay. The switching from the unstable side output to the stable side output, therefore, will be relatively slow by positive control.

While by vacuum control, the switching of the jet flow from the stable side output to the unstable side output is relatively slow, since it will take some time for the vacuum to be established. Conversely, when the vacuum control port is opened suddenly, the jet flow will switch from the unstable side output to the stable side output rapidly. This is because the necessary pressure change in the vacuum control port is relatively small and the switching is helped by the bias design.

The overall view is that, to switch the jet flow from the stable side output to the unstable side output, the fast method is the positive control method. Conversely, to switch the jet flow from the unstable side output to the stable side output, the fast method is the vacuum control method. However, there is an advantage in using the vacuum control method because the amplification is higher. Hence, the power requirement for the electro-fluidic controlling interface with the vacuum control method is lower.

The speed of switching the jet flow from the stable side output to the unstable side output by vacuum control can be greatly improved by connecting the stable side vent to the positive control port as illustrated in Figure 2. This will create a pressure slightly above the atmospheric in the positive control port when the jet flow attaches to the stable

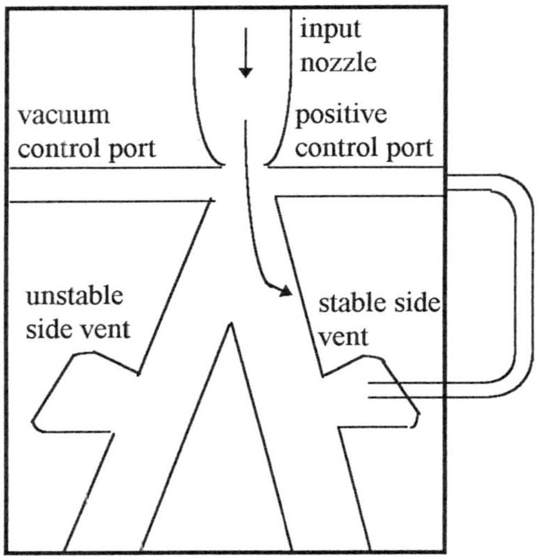

Figure 2 Improvement on the mono-stable fluidic device

side output. This, in itself, will not be high enough to cause the jet flow to be switched. However, it is a help when the vacuum control port is closed. This improvement has been experimentally proved in the previous research [5] and was adopted in this work.

PIEZO ELECTRO-FLUIDIC INTERFACE - The fluidic device needs an electro-fluidic interface to switch the jet flow to the opposite side output. Therefore, the total switching response of the fluidic injector is largely affected by the interface response. In the previous research [5] which was focused on the concept of the fluidic gaseous fuel injector system, a conventional central fuel injector (CFI) used for Ford engines was selected. The fluidic gas injector system driven by this CFI interface has shown its faster response and higher injection stability over the tested commercial gas injectors. However, the gas injection problems associated with the solenoid were still not solved properly.

Piezo-electrical material can produce an almost instantaneous deformation when an electric field is applied. This physical effect has been known for a long time, but it is rarely used for engine fuel injection systems. The major reason is that the deformation is much too small and the required driving voltage is much too high.

There are two basic methods to increase the deformation. One is to stack the multi-layer piezo materials together to generate an accumulated deformation. This requires a high strength electric field and incurs high manufacturing costs. Therefore, it is inefficient to use it in an engine fuel injector.

The second solution is to bond two-layer piezo-electrical material together by polarising in the opposite direction to generate a large bending deformation. This principle is well known in bimetallic strips. The piezo bending strip based on this principle can generate a relatively large deformation at a much lower driving voltage with a low manufacturing cost, but exerts a little force. Also, the strip bending response time is proportional to its geometric size and much slower than the multi-layer piezo stack. Therefore, if the piezo bending strip is used as a fast responding gas flow switching valve, it must be made small, work with a relatively low operating pressure and control a very small gas flow. Hence, it cannot be used as the gas injector directly. However, the small gas flow controlled by the piezo bending strip can be used as the control flow of the fluidic device. The large jet flow of the fluidic device can then be controlled by this small flow and switched between the two outputs.

Figure 3 shows a schematic set-up of the piezo-fluidic combined unit where a piezo bending strip is used as the controlling interface and fitted on the vacuum control port. When the piezo strip is energised, it will bend off the fluidic device, and let the vacuum control port open. The jet flow will then switch to the stable side output from the unstable side output. When the piezo strip is de-energised, it will bend to the fluidic device and close the control port. The jet flow will then switch back to the unstable side output.

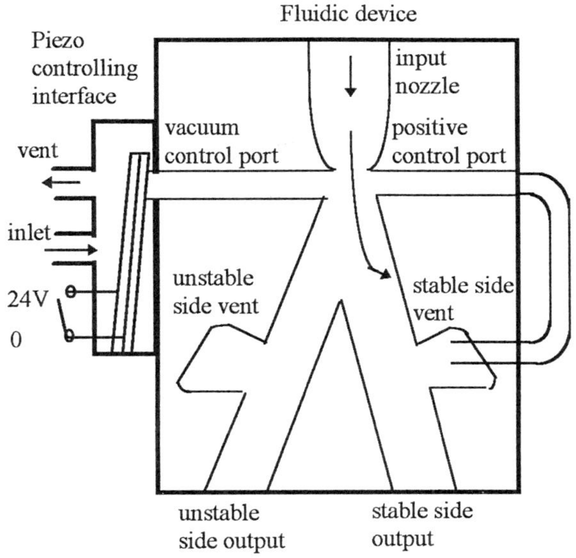

Figure 3 Piezo-fluidic combined unit

The potential advantages of the piezo-fluidic combined unit in comparison with the previous solenoid controlled fluidic unit [5] are lower energy requirement, lower wear and faster switching response. Unlike a solenoid interface which is controlled by electric current, the piezo bending strip exhibits an electric capacitor behaviour and is driven by electric voltage due to its two-layer construction. Therefore, no further energy input is required after the bending work has been done. The piezo bending strip unit being used in this project can be driven by 24 V. Its energy consumption is only about 0.001 W. In contrast, the solenoid fuel injector needs current to hold it open. Its energy consumption is over 100 W. Since the switching action of the piezo-fluidic combined unit is achieved by simply bending the piezo strip, and the fluidic device itself is a no-moving-part unit, there is no direct contact movement within the piezo-fluidic combined unit, therefore, the wear can be expected to be reduced to an extremely low level. The fast switching response of the piezo bending strip unit will be discussed in later experiments.

PIEZO-FLUIDIC GAS INJECTOR SYSTEM - In the piezo-fluidic gas injector system, the piezo-fluidic combined unit is arranged to be "submerged" in an encapsulated gas reservoir as shown in Figure 4. A conventional three-stage pressure regulator is used to reduce the gas pressure from 20 MPa CNG storage pressure to atmospheric. It is by connecting the output from the regulator to the reservoir that the "submerged" pressure of atmospheric can be maintained.

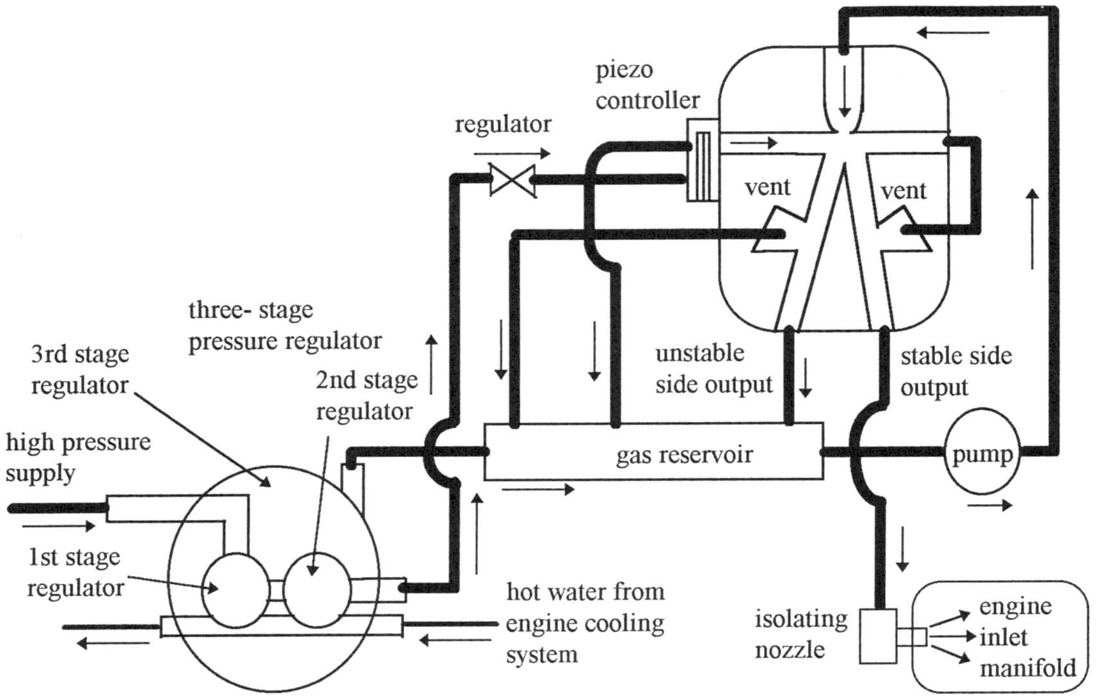

Figure 4 A schematic set-up of the piezo-fluidic gaseous fuel injector system

A low pressure gas pump is connected to the gas reservoir and supplies a gas flow to the input nozzle of the fluidic device. When the piezo unit closes off the vacuum control port, the jet flow will attach to the unstable side output and be returned to the reservoir. In this state, the system is switched off. When the piezo unit is energised, the vacuum control port is opened. The jet flow will then switch to the stable side output and so provide fuel to the engine inlet manifold. In this state, the system is switched on.

The requirement of the pump is a drawback of the system due to its extra power consumption. However, since the current research is in the early stage to investigate the feasibility of using the piezo-fluidic combined unit as a gas injector, the optimisation of the pump requirement is, therefore, left for future research.

A diaphragm type isolating nozzle is used to connect the stable side output of the fluidic device to the engine inlet manifold. The nozzle only contains a free moving diaphragm as illustrated in Figure 5. At the bottom of the diaphragm, around the outlet, is the gas pressure. At the top side, a hole is opened to the atmosphere. Therefore, when the jet flow is switched to the stable side output, the gas pressure becomes higher than atmospheric. The diaphragm opens and the gas is injected into the engine manifold. When the jet flow is switched off, the gas pressure drops to atmospheric. The diaphragm closes. The vacuum in the manifold, then, cannot suck the gas from the fluidic injector.

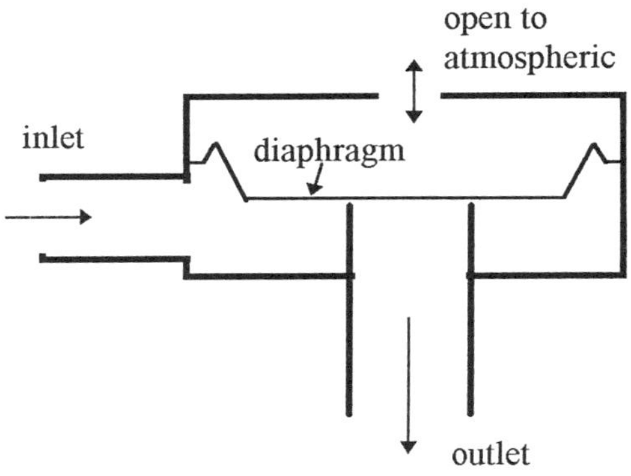

Figure 5 Diaphragm type isolating nozzle

EXPERIMENTAL STUDY

The experimental work was focused on the dynamic switching characteristics of the piezo electro-fluidic interface, steady state flow delivery and the dynamic switching response of the piezo-fluidic gas injector system. Several commercial solenoid type gas injectors were also tested and the results were compared with those from the piezo-fluidic gas injector system.

DYNAMIC TEST OF THE PIEZO ELECTRO-FLUIDIC INTERFACE - The dynamic switching

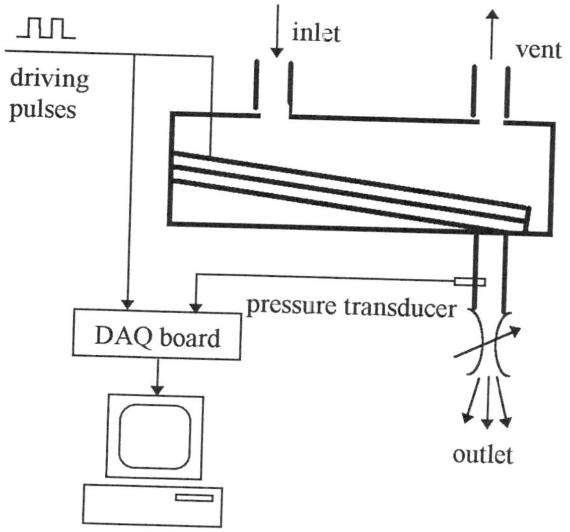

Figure 6 Dynamic test rig for the piezo-fluidic interface

response of the piezo electro-fluidic interface plays an important part of the total switching response of the piezo-fluidic gas injector system. It was measured by a fast responding Entran pressure transducer (type EPI-M4) on the laboratory test rig as shown in Figure 6.

A PC based data acquisition system was used to collect the test results and analyse the switching responses and the cycle-to-cycle switching variations. The software used in the PC was written in "LabVIEW"; a language developed by National Instrument. An Amplicon data acquisition (DAQ) board was chosen for its high frequency and low cost. Advanced linking driver software had to be developed in order to operate the Amplicon DAQ board in the LabVIEW language satisfactorily.

Figure 7 to Figure 9 show the measured pressure trace of the piezo bending strip unit and the response results from 500 consecutive cycles. The supply pressure at the inlet of the piezo unit was 550 kPa at which the piezo unit can switch enough gas flow to control the fluidic device. It can

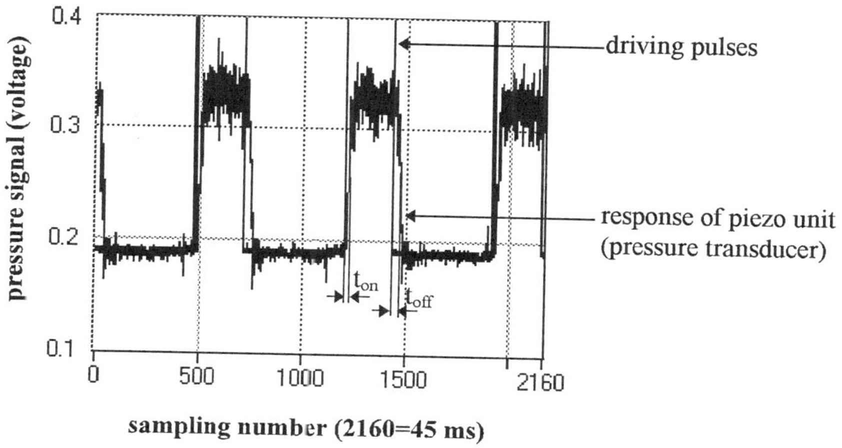

Figure 7 Measured output pressure of the piezo unit

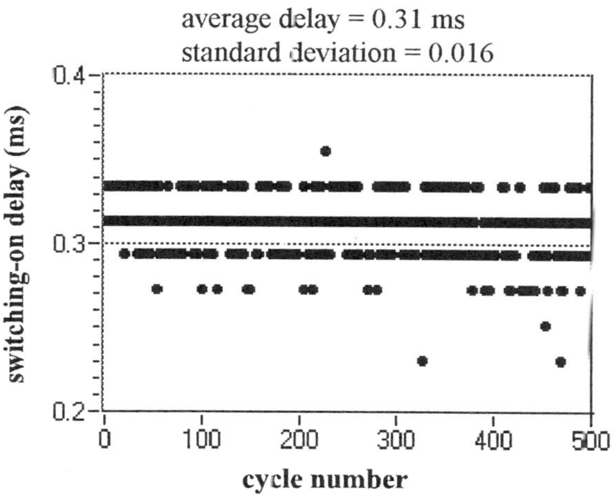

Figure 8 Switching-on response time of the piezo unit from 500 cycles

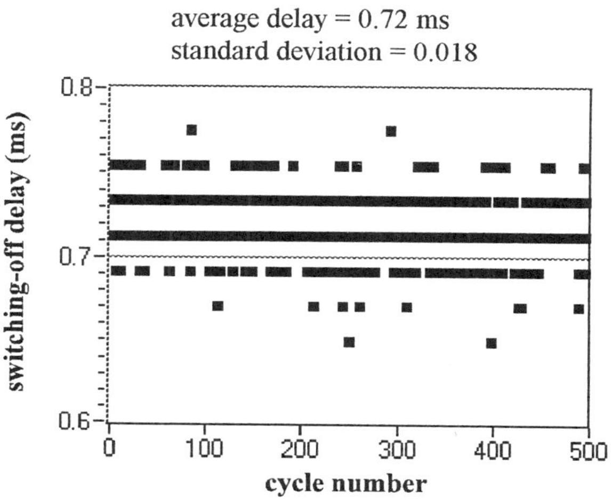

Figure 9 Switching-off response time of the piezo unit from 500 cycles

be seen that the average response time of switching-on t_{on} is 0.31 ms with a standard deviation (SD) of 0.016, and switching-off t_{off} is 0.72 ms with a SD of 0.018.

Figure 10 shows the switching-on and -off responses of the piezo unit under varying supply pressure. It can be seen that the switching-on response delay is very consistent while switching-off response delay increases slightly as supply pressure increases. The slight increase of the switching-off may due to the high restriction at the outlet of the piezo unit to obtain a clear pressure trace. Therefore, it is reasonable to predict that both switching-on and -off of the piezo unit will be more or less independent of its supply pressure when outlet is opened to ambient.

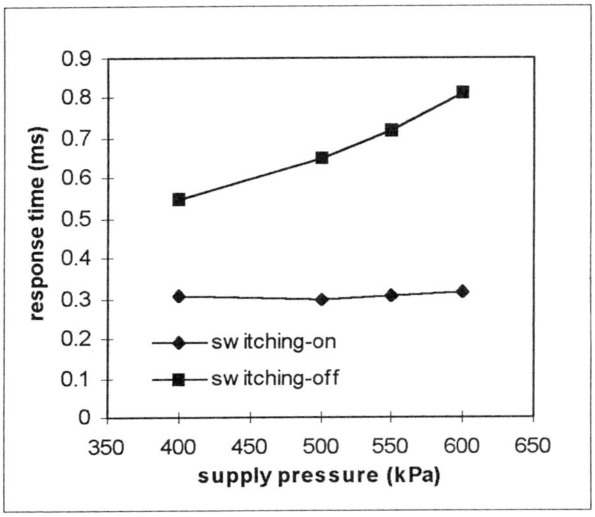

Figure 10 Switching response times of the piezo unit

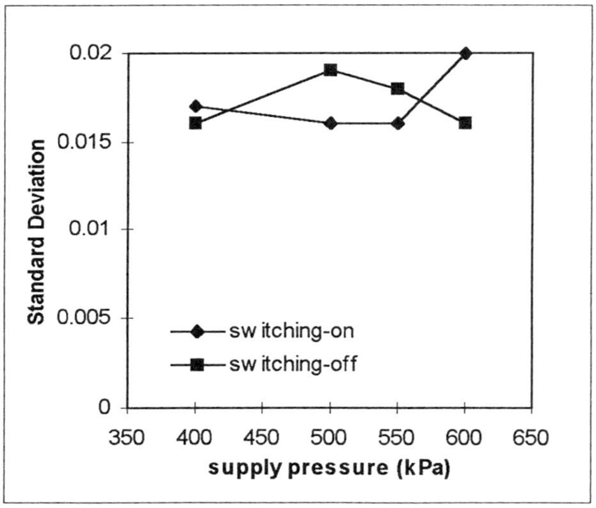

Figure 11 Switching standard deviations of the piezo unit

Figure 11 shows the switching-on and -off switching stability of the piezo unit under varying supply pressure. It can be seen that the standard deviations of both switching on and -off of the piezo unit are fairly consistent as supply pressure increases.

In comparison, Figure 12 and Figure 13 show the switching-on and -off responses of the previously used CFI solenoid interface. The average response time of switching-on is 1.08 ms with SD of 0.017 and switching-off is 0.93 ms with a SD of 0.015. The supply pressure to the solenoid was 100 kPa which was the optimal operation pressure as the electro-fluidic interface. The previous experimental results [5] showed that both switching-on and -off response times of the CFI solenoid interface were not affected by the control pressure.

Comparing the switching response times of the pizeo unit at 550 kPa supply pressure with those of the solenoid, the switching-on response of the piezo unit is over 70% faster and the switching-off is about 23% faster.

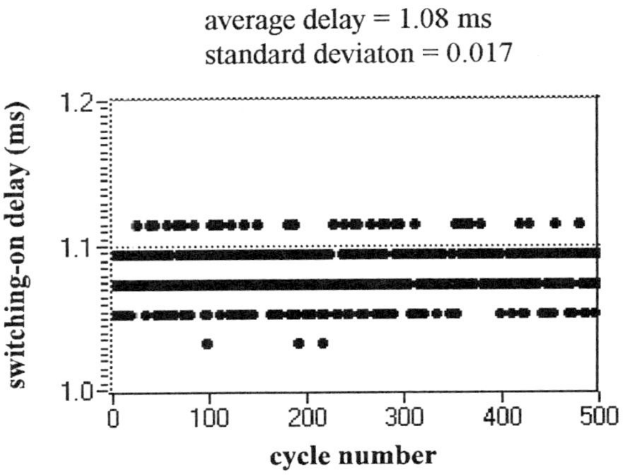

Figure 12 Switching-on response time of the CFI solenoid unit from 500 cycles

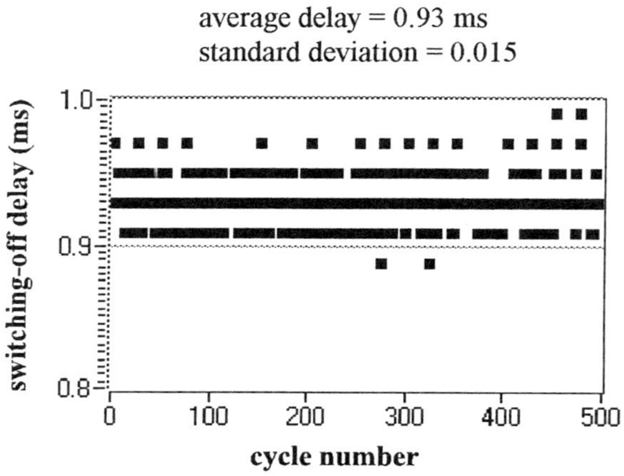

Figure 13 Switching-off response time of the CFI solenoid unit from 500 cycles

STEADY STATE TEST OF THE PIEZO-FLUIDIC INJECTOR SYSTEM - The gas flow handling capability of the piezo-fluidic gas injector system was estimated by the steady state flow test operating on air. Figure 14 shows a

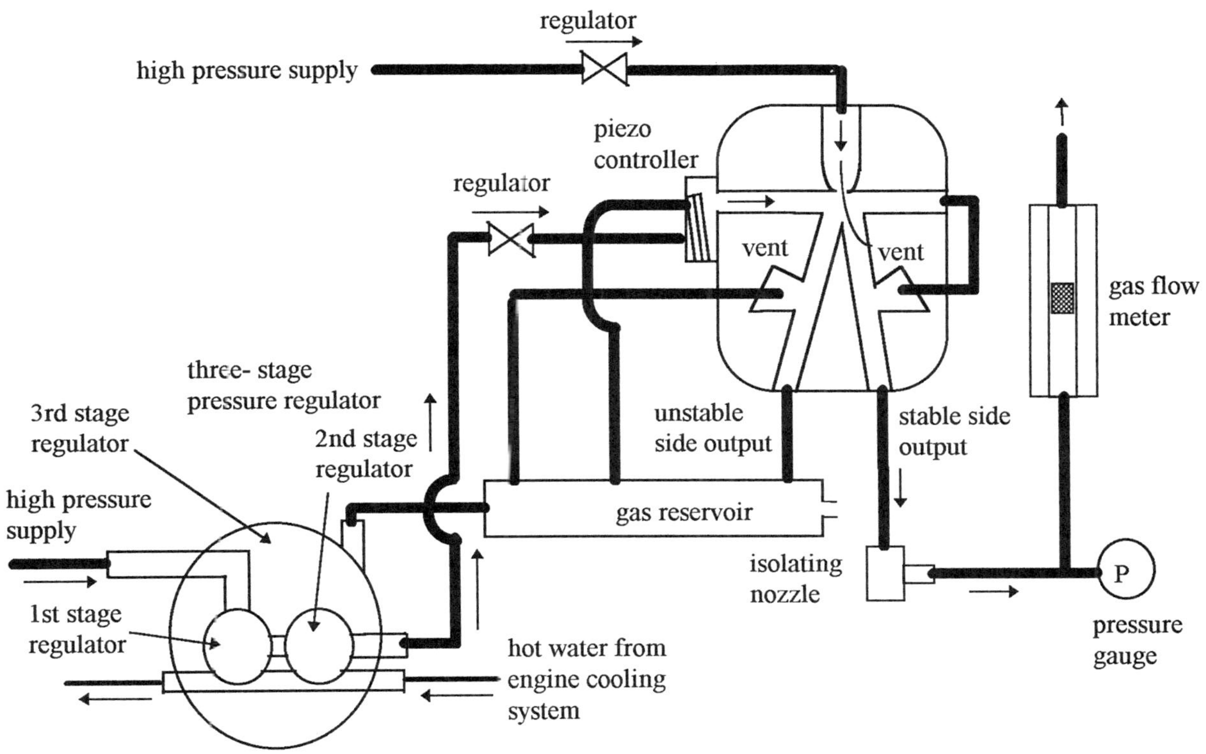

Figure 14 Steady state test rig of the piezo-fluidic gaseous fuel injector system

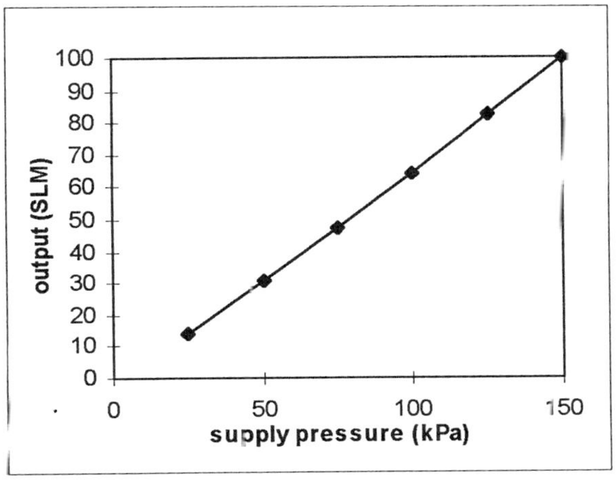

Figure 15 Steady state flow rate of the piezo-fluidic gas injector system

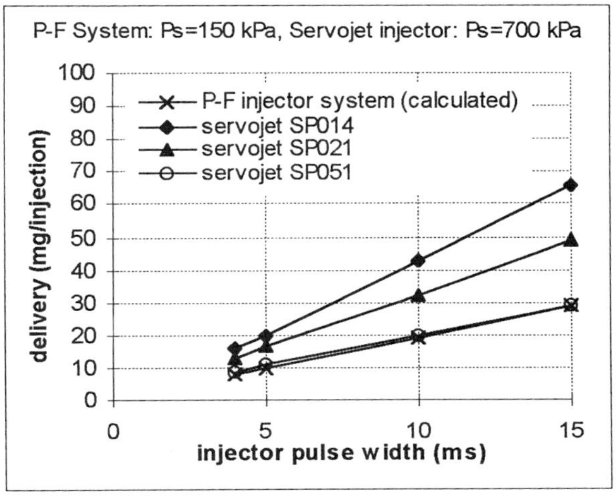

Figure 16 Dynamic flow delivery of Servojet injectors and the piezo-fluidic (P-F) gas injector system

schematic set-up of the laboratory test rig. The piezo interface was energised during the test. The jet flow attached to the stable side output and was routed to a gas flow meter through the isolating nozzle.

Figure 15 shows the measured steady state flow rate under varying supply pressure of the system. It can be seen that the steady state flow rate of the piezo-fluidic gas injector system is linearly proportional to the system supply pressure. At 150 kPa supply pressure, the gas flow rate of the system is about 100 SLM (standard litre per minute). This flow rate is equivalent to the fuel requirement of a 2-litre natural gas engine.

Figure 16 shows the comparison of the results between commercial Servojet solenoid gas injectors and the dynamic flow delivery of the fluidic device. The Servojet data were provided by the manufacturer. The dynamic flow delivery of the piezo-fluidic gas injector system was calculated from

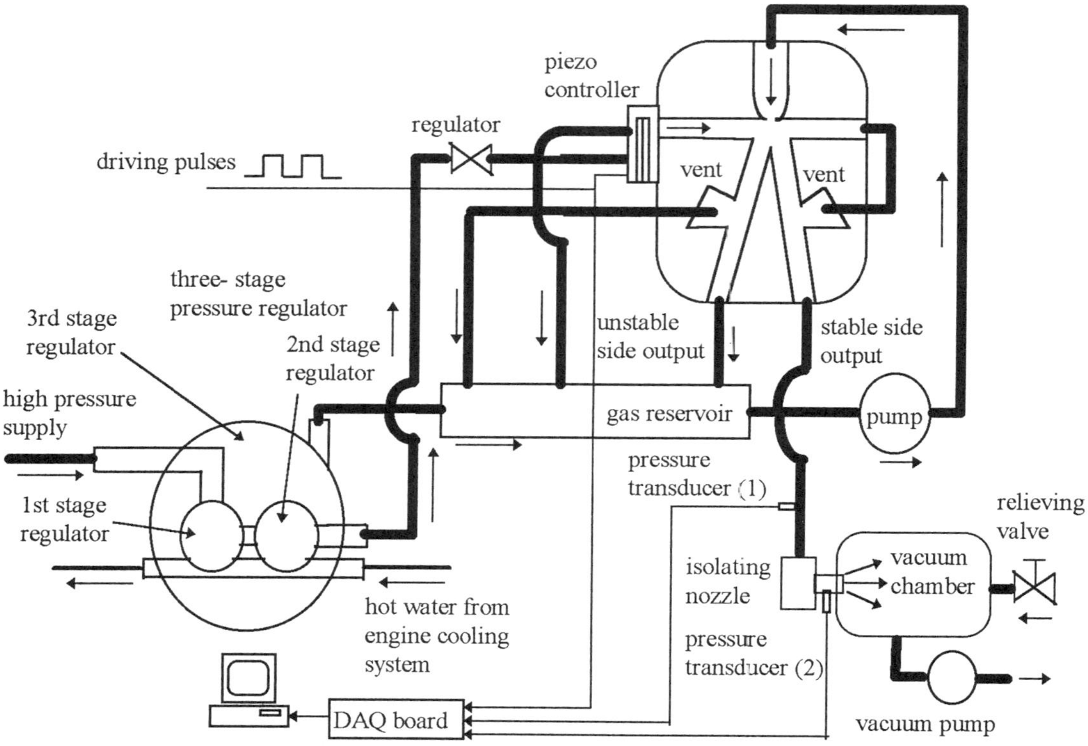

Figure 17 Dynamic test rig of the piezo-fluidic gas injector system

the measured steady flow rate at 150 kPa supply pressure because the dynamic gas flow rate cannot be measured by the current test facility in the laboratory.

Three Servojet units were used, namely SP014, SP021 and SP051. The supply pressure for these injectors was 700 kPa. It can be seen that the dynamic flow delivery of the piezo-fluidic gas injector system at the supply pressure of 150 kPa is same as the SP051 which is the smallest Servojet unit. In order to achieve the flow delivery of the SP014, which is approximately twice that of the SP051, two fluidic devices can be stacked together to give the required flow delivery for such larger engine applications. The advantages of the stacked arrangement are the flexibility of adapting the fluidic unit for varies engine sizes and the avoidance of a slow response time.

DYNAMIC TEST OF THE PIEZO-FLUIDIC GAS INJECTOR SYSTEM - In order to evaluate the dynamic switching characteristics of the piezo-fluidic gas injector system and to compare them with those from commercial gas injectors, a laboratory dynamic test rig was built as shown in Figure 17.

Two Entran pressure transducers were used in the rig. One was fitted on the stable side output of the fluidic device to detect the response of the piezo-fluidic combined unit to the driving signal. The other one was fitted downstream of the isolating nozzle to monitor the response of the whole piezo-fluidic gas injector system.

A vacuum chamber equipped with a relieving valve and driven by a vacuum pump was used in the test rig to simulate the engine inlet manifold. Different manifold conditions were obtained by adjusting the relieving valve. The output of the piezo-fluidic injector system was delivered to the vacuum chamber by connecting the outlet of the isolating nozzle to the vacuum chamber.

Figure 18 shows the pressure traces from the two transducers with the vacuum chamber at atmospheric pressure. The supply pressure of the system was 150 kPa. The control pressure to the piezo interface was 550 kPa.

Figure 19 and Figure 20 show the switching response results of the piezo-fluidic combined unit measured by the pressure transducer (1) from 500 consecutive cycles. It can be seen that the average response time of switching-on is 1.15 ms with a SD of 0.059. The average response time of switching-off is 0.98 ms with a SD of 0.038.

Figure 21 and Figure 22 show the switching response of the whole piezo-fluidic gas injector system to the driving signal measured by pressure transducer (2). The average response time of switching-on is 1.46 ms with a SD of 0.055. The average response time of switching-off is 1.31

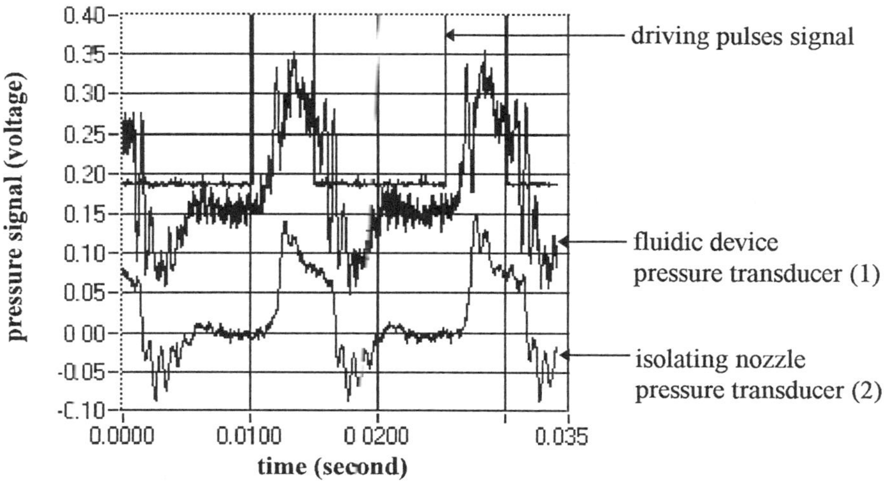

Figure 18 Measured pressure traces of the piezo-fluidic gas injector system at 150 kPa supply pressure and 550 kPa control pressure

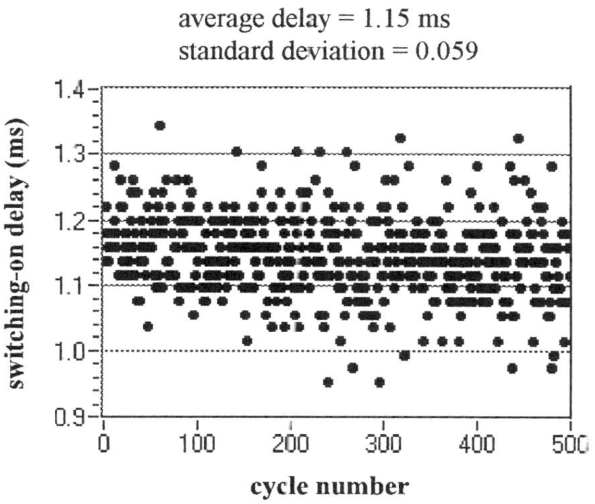

Figure 19 Switching-on response time of the piezo-fluidic combined unit from 500 cycles

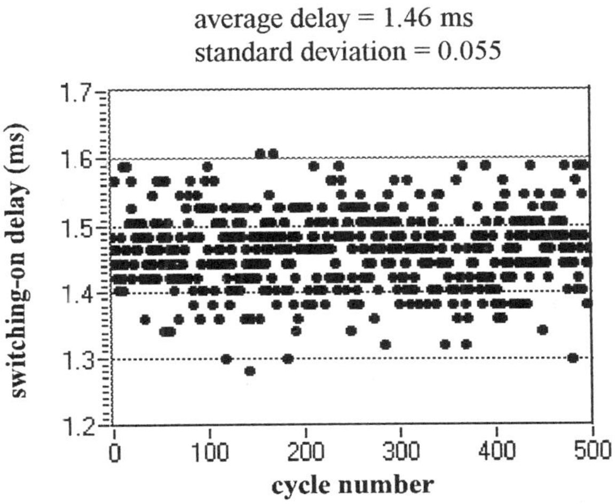

Figure 21 Switching-on response time of the piezo-fluidic gas injector system from 500 cycles

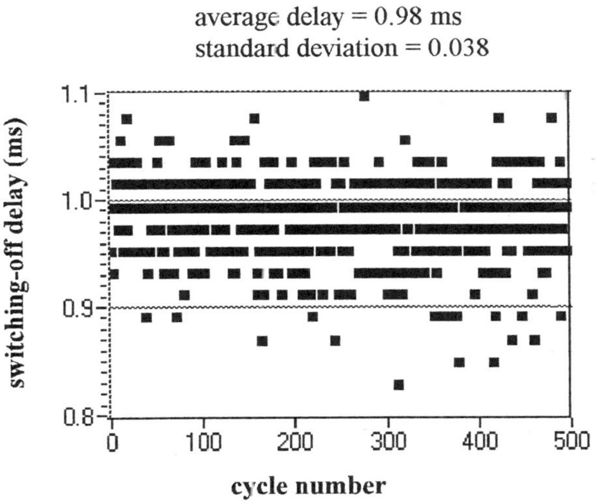

Figure 20 Switching-off response time of the piezo-fluidic combined unit from 500 cycles

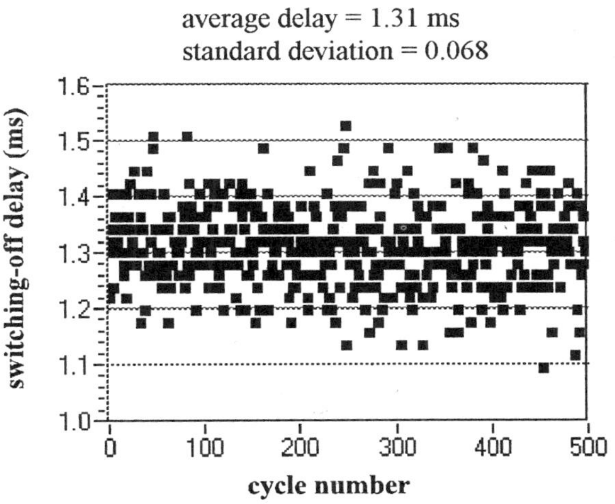

Figure 22 Switching-off response time of the piezo-fluidic gas injector system from 500 cycles

ms with a SD of 0.068.

Deducting the response times of the piezo-fluidic combined unit from those of the system, the net response times of the isolating nozzle can be obtained as 0.32 ms of switching-on and 0.33 ms of switching-off.

Figure 23 and Figure 24 show the piezo-fluidic gas injector system switching response times and injection cycle-to-cycle variations under varying supply pressure. The control pressure to the piezo interface was 550 kPa, and the pressure of the vacuum chamber was atmospheric. It can be seen that both switching-on and -off response times are fairly stable as the system supply pressure increases. This feature permits the gas flow rate through the piezo-fluidic gas injector system also to be adjusted by varying the supply pressure of the system.

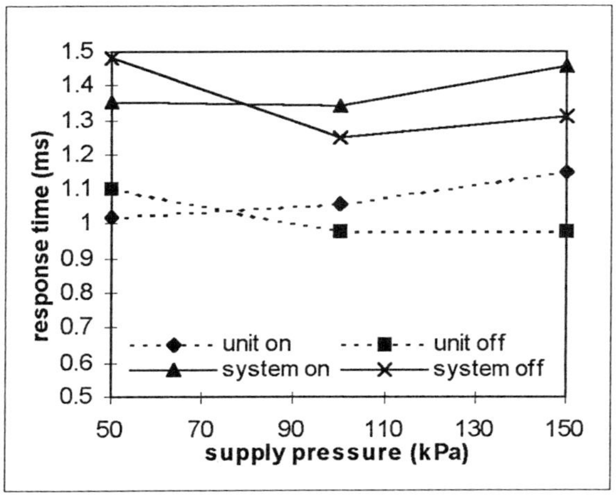

Figure 23 Switching response times of the piezo-fluidic gas injector system at varying supply pressure

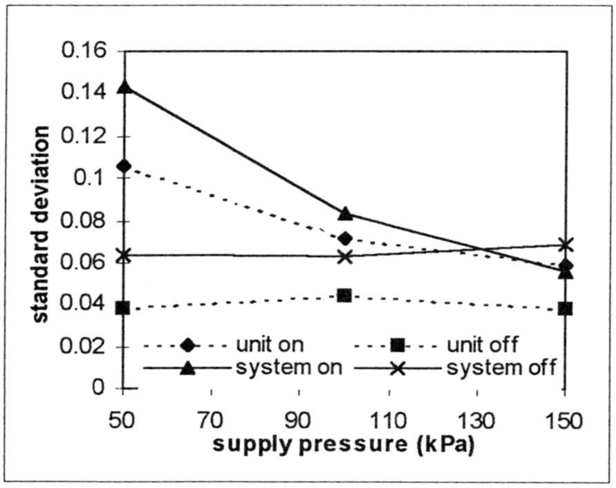

Figure 24 Standard deviations of the piezo-fluidic gas injector system at varying supply pressure

Figure 25 shows the system switching response times under different vacuum chamber pressures (absolute pressure). It can be seen, when the chamber pressure drops to vacuum, both switching-on and -off switching response times of the system will slightly increase. This is because when the pressure in the chamber drops to vacuum, the pressure of the gas surrounding the outlet of the isolating nozzle has to take some time to build up to overcome the force acting on the diaphragm due to the suction of vacuum. However, since the size of the outlet hole is small, the suction force is limited. Therefore, as vacuum increases, the system switching response delay will not increase very much.

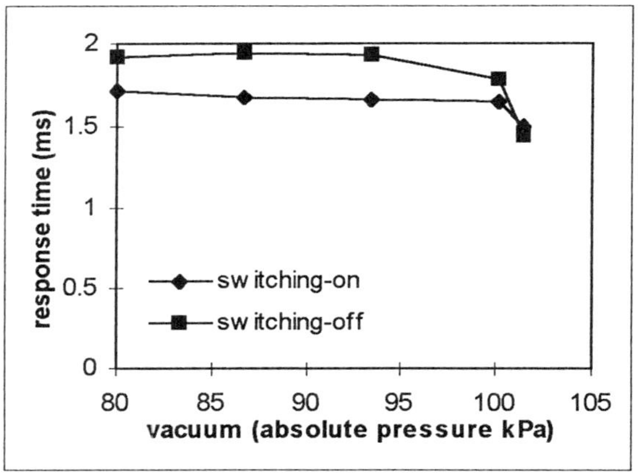

Figure 25 Switching response times of the piezo-fluidic gas injector system at varying outlet pressure

Figure 26 shows the switching stability of the system under varying chamber pressure. It can be seen that the system switching stability is slightly increased when the pressure of the chamber drops to vacuum. This, too, is due to the contribution of the suction force acting on the diaphragm.

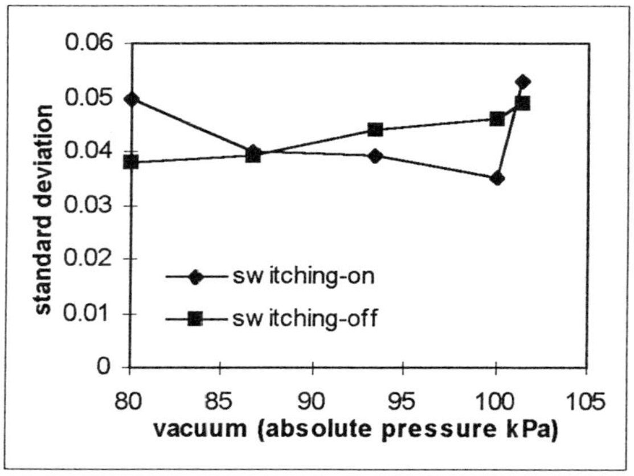

Figure 26 Standard deviations of the piezo-fluidic gas injector system at varying outlet pressure

DYNAMIC TEST OF THE COMMERCIAL GAS INJECTORS - The commercial gas injectors available for comparison tests were three Servojet gas injectors and one Bosch gas injector. Since no dynamic characteristics were available for these proprietary injectors, it was decided to carry out similar tests on the laboratory test rig so that their switching performances could be compared with those from the piezo-fluidic gas injection system.

Figure 27 and Figure 28 show the switching-on and -off responses of these solenoid gas injectors under varying supply pressure. It is known that the operating pressures of these injectors are over 700 kPa on gas engines. However, the maximum pressure used in the test was 350 kPa due to limited peak pressure availability in the laboratory. It can be seen that the switching-on response time increases as the supply pressure increases for all four injectors. As mentioned earlier, the switching-on response time achieved by the piezo-fluidic injection system at the required injection pressure of 150 kPa is in the region of 1.46 ms. Therefore, it is reasonable to predict that the fluidic injector has a faster switching-on speed than these commercial gas injectors, when equivalent quantities of gaseous fuel are injected.

For the switching-off response the Bosch unit showed the best response time (1.4 ms) among all the commercial units. However, the switching-off response time achieved by the piezo-fluidic injector at 150 kPa injection pressure is 1.3 ms. It should be noted that the Bosch gas injector used here is a relatively small unit designed for gas engines possibly below 2 litres capacity. Therefore, it has the advantage of a light armature and a small passage area resulting in a faster response than the Servojet injectors.

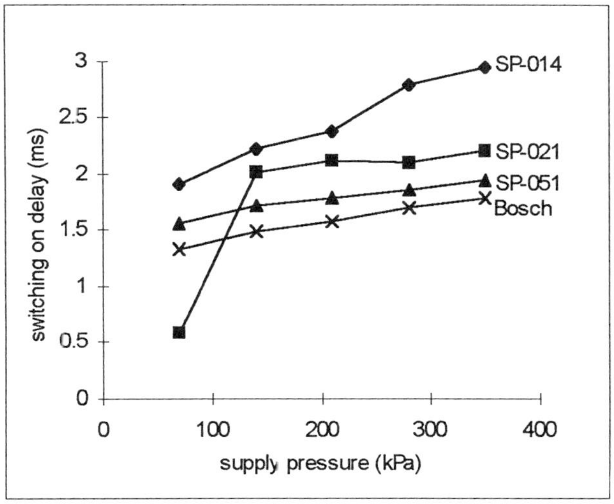

Figure 27 Switching-on response times of the commercial gas injectors

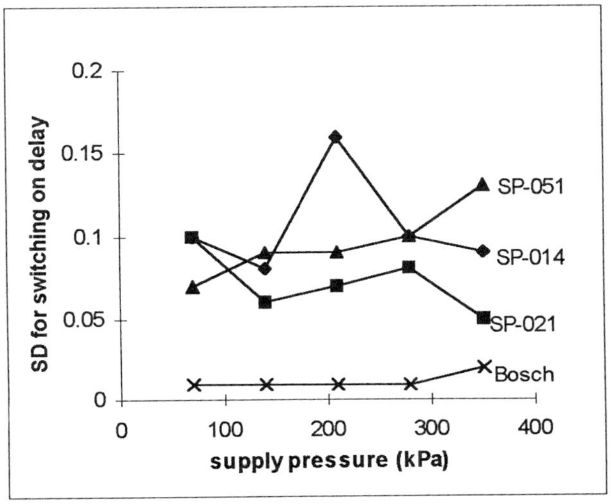

Figure 29 Switching-on standard deviations of the commercial gas injectors

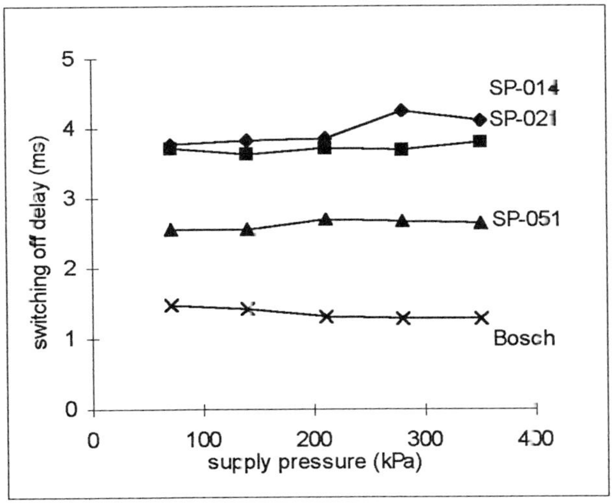

Figure 28 Switching-off response times of the commercial gas injectors

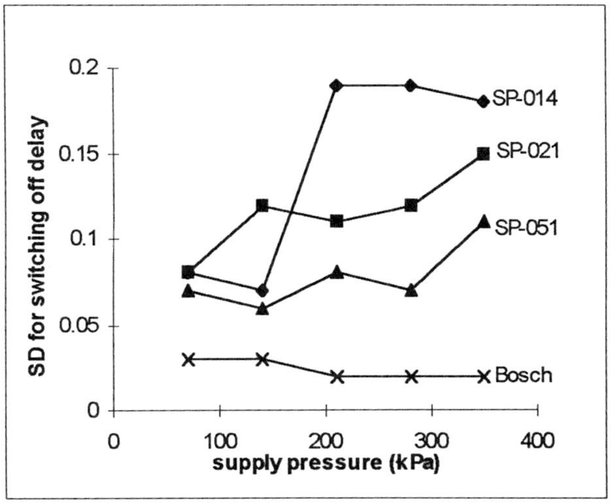

Figure 30 Switching-off standard deviations of the commercial gas injectors

Figure 29 and Figure 30 show the injection stability of these solenoid injectors. It can be seen that the Bosch unit showed extremely consistent cycle-to-cycle variations with a SD of 0.02 for both switching-on and -off delays. This is largely due to its compactness and light weight design. In comparison, the switching cycle-to-cycle variations of the piezo-fluidic injection system are capable of switching-on with a SD of 0.055 and switching-off with a SD of 0.068.

FUTURE WORK

The research reported in this paper has proved the possibility of using the piezo-fluidic combined unit as a gaseous fuel injector for natural gas engines. Future work will concentrate on the evaluation of the piezo-fluidic gaseous fuel injection system on a research natural gas engine, and the optimisation of the circulation pump.

CONCLUSIONS

(1) A novel piezo-fluidic gaseous fuel injection system was proposed. The system consists mainly of the no-moving-part fluidic device and the bending strip type piezo electro-fluidic interface. The operating principle of the system was described and verified by experiments on a laboratory test rig.

(2) The volume flow rate of the system at 150 kPa supply pressure is equivalent to the Servojet SP-051 unit. Higher flow rates can be obtained by stacking the multi-layer fluidic injector.

(3) The vacuum control method of the mono-stable fluidic device is ideal for gaseous fuel injection due to the higher amplification. The switching properties can be improved by connecting the positive control port to the stable side vent.

(4) The piezo electro-fluidic interface was tested and compared with the CFI solenoid interface. It was found that the piezo unit has faster switching response speeds and much lower energy consumption.

(5) The switching responses of the system at the supply pressure of 150 kPa are: 1.46 ms of switching-on with a SD of 0.055, and 1.31 ms of switching-off with a SD of 0.068.

(6) Inlet manifold vacuum will slightly increase the system switching response delay, but decrease injection cycle to cycle variations.

(7) Three Servojet and one Bosch commercial solenoid gas injectors were tested. It was found that the piezo-fluidic gas injector system at 550 kPa control pressure and 150 kPa supply pressure has faster switching responses than all four commercial units at their normal operation pressure (over 700 kPa). Also, the switching cycle to cycle variations of the system are smaller than all Servojet units. However, the Bosch unit showed a slightly better cycle to cycle switching repeatability than the piezo-fluidic gas injection system.

ACKNOWLEDGEMENT

The financial support from British Gas plc. for this project is gratefully acknowledged. The authors also wish to record their appreciation of technical assistance provided by Ms. R. Palmer and Mr. A. Brightwell, British Gas plc., Mr. J. Elliott, Preci-Spark Ltd. UK and Technician Mr. D. Harkis and Mr. G. Knowles, Loughborough University.

REFERENCES

1. J. Klimstra et al. "Catalytic Converters for Natural Gas Fuelled Engines - A Measurement and Control Problem", SAE paper 872165, 1987.
2. M. Perrault et al. "An Alternative to the Conventional Natural Gas Conversion Kit for Vehicles - Computer Controlled Upstream Injection", SAE paper 881659, 1988.
3. C. Komodromos, S Pearson and A. Grint, "The Potential of Adsorbed Natural Gas (ANG) for Advanced On-board Storage in Natural Gas-fuelled Vehicles", International Gas Research Conference (IGRC), 16-19th November 1992, Orlando, Florida.
4. R. Chen, Q. Huang, "Prospects for a Fluidic Device Based Electronic Gas Fuel Injection System for Natural Gas Engines", XXV FISITA Congress, 17-21 October 1994, Beijing.
5. R. Chen, Q. Huang, "An Investigation into the Use of Fluidic Devices as Gas Fuel Injectors for Natural Gas Engines", SAE paper 960768, 1996.

951912

Application of Digital, Pulse - Width - Modulated, Sonic Flow Injectors for Gaseous Fuels

Robert L. Barkhimer and Hoi-Ching Wong
BKM, Inc.

ABSTRACT

Sonic flow, pulse-width-modulated electronic fuel injectors for gaseous fuels provide precise, stable and reliable service for over 1 billion cycles. Techniques for precision flow calibration are described along with dynamic response characteristics. Application techniques including pressure regulation, filtration and procedures for adjusting flow calibration for changes in gas pressure, temperature and composition are presented. Applications include single point (throttle body or mixer), multi-point constant flow and multi-point sequential strategies. Durability testing in parallel with a utility pipeline regulator fully conserves power and gaseous fuel in a simple, low cost, multiple injector test bank.

INTRODUCTION

The Servojet pulse-width-modulated electronic gas injector is a two way, normally closed, solenoid operated valve. Designed with corrosion resistant materials, it meters gaseous fuels into the intake system of internal combustion engines. The injector is a cartridge design for easy installation. The injector is typically pulse-width-modulated to meter fuel quantity at any operating frequency even above 100 Hz. With sonic flow across the valve inlet when fully open, mass flow rate is approximately proportional to the supply pressure.

PERFORMANCE REQUIREMENTS

Very high demands are placed on engines equipped with electronically controlled fuel injection (EFI) systems, with regard to starting performance, driveability, fuel consumption and exhaust emissions. Gaseous fuel injectors have a decisive influence upon these characteristics, and therefore must comply with stringent quality requirements. The objectives of the Servojet gas injector are to satisfy the following requirements:

- Precise fuel metering at all operating conditions
- Flow vs pulse width linearity
- Adequate dynamic flow range
- Low cost
- Durable
- Functional reliability
- Minimal supply voltage effect on operating range

BASIC DESIGN AND FUNCTION

The basic design of the gas injector was derived from the fast responding solenoid valve which is described in detail in Reference (1). The injectors are composed of a solenoid coil and valve assembly. The coil assembly consists of solenoid winding and electrical connection. The valve assembly is comprised of the valve body which holds the solenoid armature, ball poppet and seat. Figure 1 shows the operating concept of the injector. The cross-section of the actual injector is shown in Figure 2. With the solenoid de-energized, the supply pressure, assisted by a spring, forces the ball poppet on its seat, prohibiting gas flow. When the solenoid is energized, the ball poppet is lifted off the seat and held against the stop. Gas then passes through the valve seat and outlet port of injector. The solenoid is a low resistance coil designed for rapid response, and is typically actuated by a current regulated driver which limits the initial opening current to 4 amperes, then reduces and holds at 1 ampere to conserve energy for the duration of the energized time (pulse width).

GAS FLOW CHARACTERISTICS

In electronically controlled fuel-injection systems, solenoid operated injectors are used for fuel metering. Gaseous fueling consistency between injectors and cycles is just as important as with liquid injectors. The static gaseous flow rate is defined by the following formula:

$$Q_{static} = C_d \cdot A \cdot V_t \ ,$$

where C_d = Coeff. of discharge
A = Area of throat at valve seat

V_t = Velocity at throat

For sonic flow, the velocity of gas through the orifice is given by:

$$V_t = \sqrt{\frac{\gamma P_t}{\rho_t}} ,$$

where γ = Ratio of specific heats
P_t = Pressure at throat
ρ_t = Density at throat

The static mass flow rate becomes:

$$Q_{static} = \frac{KAP_1}{\sqrt{T_1}}$$

where P_1 = Absolute inlet pressure
T_1 = Absolute inlet temperature
K = Constant

provided that,

$$\frac{P_2}{P_1} \leq r ,$$

where P_2 = Absolute outlet pressure
r = Critical pressure ratio (0.546 for methane)

and,

$$r = \left(\frac{2}{\gamma+1}\right)^{\frac{\gamma}{\gamma-1}}$$

Therefore, it is recommended the ratio of manifold absolute pressure to absolute supply pressure not exceed the critical pressure ratio to maintain the sonic flow characteristic. The static flow of injectors at various inlet pressures are plotted against the throat area as shown in Figure 3, illustrating that static flow is directly proportional at constant pressure and temperature.

The mass quantity per injection depends on the response during opening and closing events, which is influenced by the biasing spring, armature mass, and the magnetic and electronic driver circuits.

The dynamic flow at a constant gas supply pressure is linear and directly proportional to pulse width over the full range of the injector if the total effective opening duration is equal to or proportional to pulse width. Figure 4 shows the typical flow curve of Servojet gas injectors. In most applications, the operating pressure and injection pulse width can be kept in the linear range. If it becomes necessary to operate at pulse widths below the linear range, an appropriate adjustment in the flow calibration table can be made. Figure 5 shows the flow characteristics at low pulse widths which are quite repeatable from injector to injector.

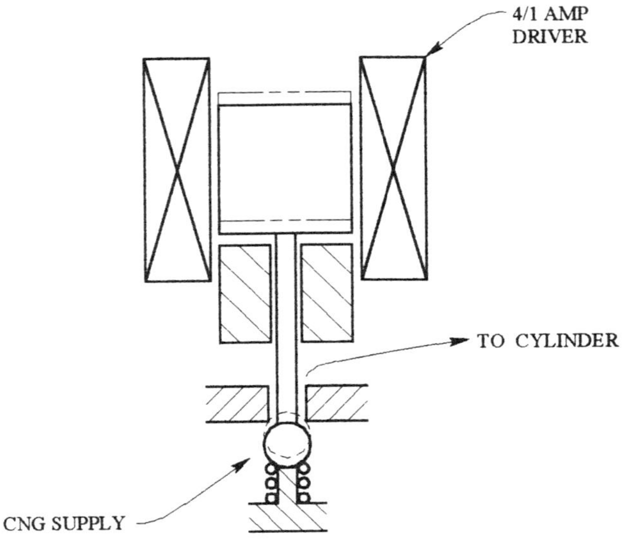

Figure 1: CNG Injector Schematic

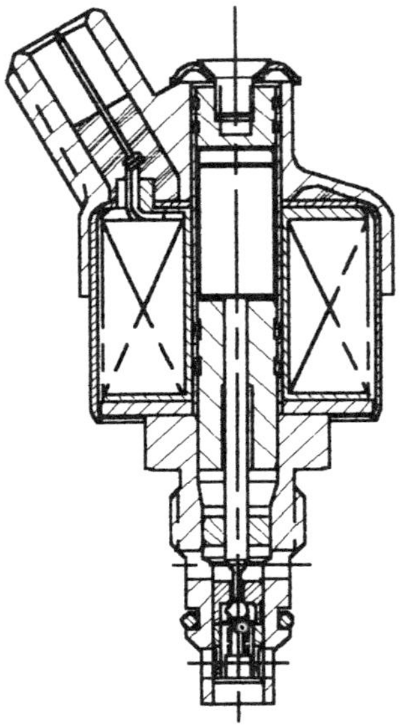

Figure 2: Injector Cross-Section

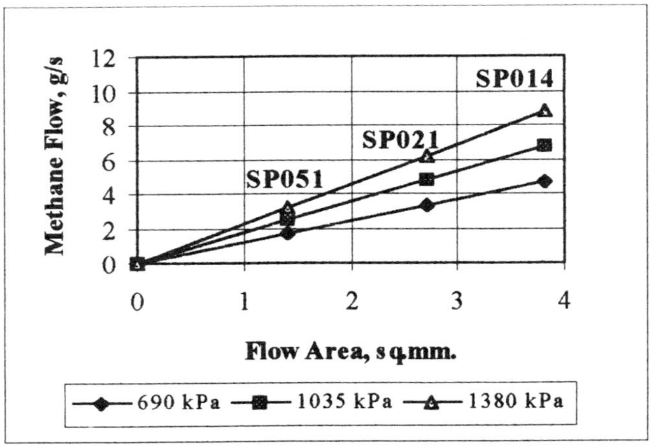

Figure 3: Steady Flow vs Flow Area

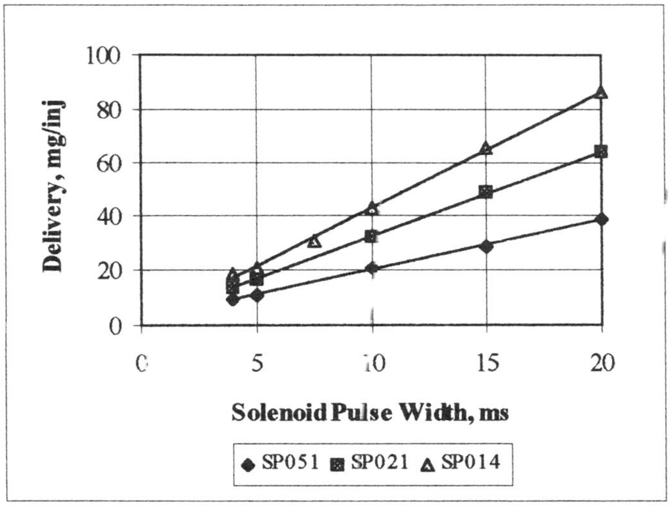

Figure 4: Methane Delivery at 690 kPa (100 psig)

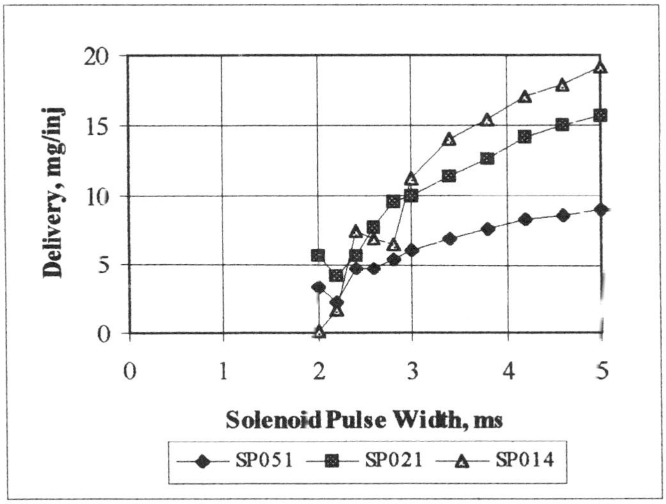

Figure 5: Methane Delivery at 690 kPa (100 psig) and Low Pulse Widths

DYNAMIC RESPONSE CHARACTERISTICS

Response time of a solenoid depends upon the physical size of the solenoid, power input, driver circuit, mass and travel of armature, and load. The opening time of the injector, which is the time required for the armature to reach its fully opened position after initiation of the driver circuit pulse input, depends upon the gas supply pressure. It takes less than 3 milliseconds at supply pressures up to 2070 kPa (300 psig). The closing time, which also depends on supply pressure, is less than 3 milliseconds at 690 kPa (100 psig) supply pressure, after de-energizing. The travel time of the armature is less than 1 millisecond, as it takes 2 milliseconds for the coil current to rise to a required energy level for opening and for the residual magnetism to dissipate during closing.

A piezoelectric pressure transducer mounted at the outlet of injector is typically used to measure the opening and closing time intervals. Typical time histories of coil voltage and outlet pressure are shown in Figure 6. The inflection of voltage trace during opening event clearly indicates the armature has reached its fully opened position.

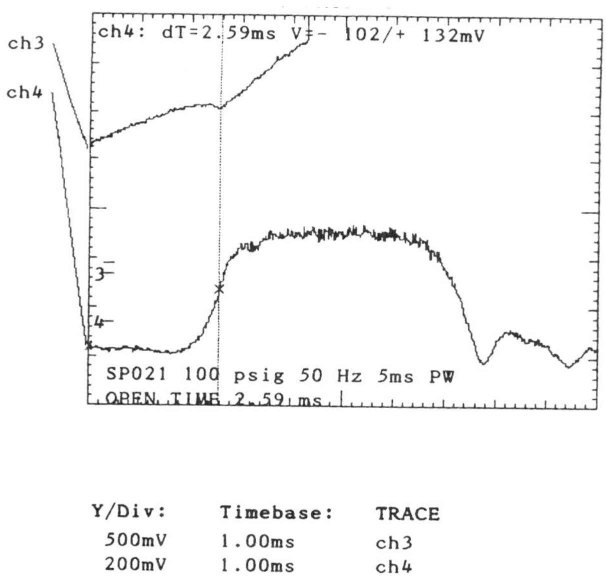

Figure 6: Pressure and Voltage Traces

EFFECT OF VOLTAGE VARIATION

Under certain circumstances, such as starting and low battery operation, normal voltage is not available. Having a low resistance coil winding, the minimum opening voltage of the Servojet gas injector is less than 4 VDC. The minimum current required to actuate the injector from the closed to open position is approximately 2.5 amperes at 690 kPa (100 psig) inlet pressure. Due to the low resistance coil and the current limiting driver circuit characteristics, the injector is not as sensitive to low battery voltage as are the high resistance units. Furthermore, the required flow rate of injectors at low voltage is to provide a satisfactory start, after which full battery charging voltage is normally available.

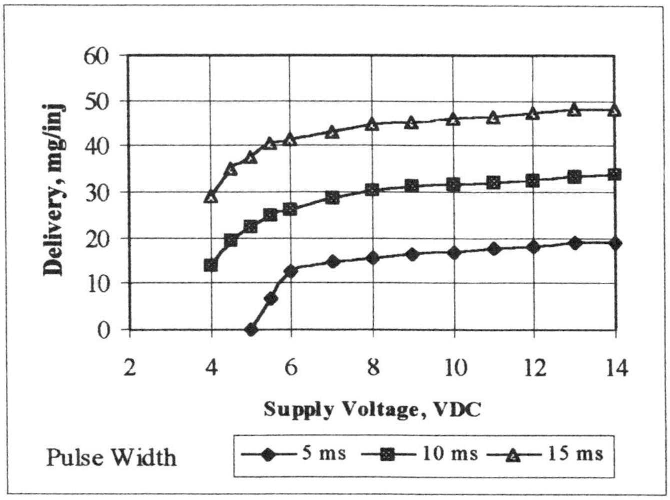

Figure 7: Effect of Supply Voltage

Experiments have shown the dynamic minimum opening voltage (DMOV) at 690 kPa (100 psig) supply pressure is below 4.5 VDC for model SP021 injectors. The DMOV is defined as the voltage at which the flow rate at 10 ms pulse width exceeds 50% of the flow at 14.0 VDC. Figure 7 shows the deviation of injected CNG quantity at 690 kPa (100 psig) supply pressure and various pulse widths over a supply voltage range of 4.0 to 14.0 VDC.

DURABILITY

The CNG injector endurance test was conducted with the cooperation of San Diego Gas & Electric at a gas regulating station. Two multiple injector test banks were installed in parallel between a high pressure supply and a low pressure outgoing gas pipelines to conserve gaseous fuel. The test program commenced in September 1991 using prototype and pre-production injectors. Injectors randomly selected from production lots were put on test in February 1993, continuing to June 1993.

Since the downstream gas pipeline is regulated at 414 kPa (60 psig), the inlet pressure to the injectors was set at 1310 kPa (190 psig) and 1724 kPa (250 psig) at each bank, respectively, maintaining a pressure ratio below 0.55 for sonic choked flow. All injectors were tested at a frequency of 100 Hz, pulse-width sweeping between 2.0 to 5.0 milliseconds. A total of 43.5 billion cycles were accumulated on 100 injectors during a test period of almost two years. The injector has been proven capable of a minimum of one billion cycles. Although the allowable dynamic flow shift limit has not been established, the results showed that after one billion cycles the dynamic flow of injectors shifted less than 10% from their original calibrations at a 7 ms pulse width. Test samples were also evaluated for static leakage. The leakage rate was found reduced.

BENCH CALIBRATION

For simplification of controls, it is desirable to operate all gas injectors at the same pulse width, dependent on the commanded fueling. Therefore, the controller produces one single pulse width signal shared by all gas injector driver circuits. Calibration accuracy of gas injectors is therefore just as important as for liquid injectors.

Each gas injector is dynamically flow tested using clean and dry compressed air. The measured air flow is then converted to methane mass flow as described below:

$$\frac{Q_{methane}}{Q_{air}} = \frac{C_{methane}}{C_{air}} \cdot \frac{\rho_{methane}}{\rho_{air}}$$

where Q = Mass flow
C = Gas correction factor
ρ = Density

The gas correction factor is calculated by:

$$C = k \cdot \sqrt{R \cdot \gamma \cdot \left(\frac{2}{\gamma+1}\right)^{\frac{(\gamma+1)}{(\gamma-1)}}}$$

where k = unit conversion factor
R = Gas constant
γ = Ratio of specific heats

Since the dynamic flow of the gas injector is influenced by absolute inlet pressure and temperature, frequency and effective opening duration, the following parameters are set and held constant during bench testing:

- Temperature of air entering the injector is maintained at 20° ± 1° C.
- Pressure at the injector inlet is set at a value determined by the application, and is held to within ± 0.7 kPa (0.1 psi).
- Frequency is 50 ± 0.05 Hz to maintain a stable and steady inlet pressure.
- Pulse width is set at a value determined by the application, and held to within ± 0.005 ms.
- Voltage supplied to the injector driver is 13.8 ± 0.05 VDC.

Based on the allowable variations, the dynamic flow measured from the same injector tested at 690 kPa (100 psig) inlet pressure and 10 ms pulse width, is expected to have a tolerance within ± 0.5% providing that the accuracy and repeatability of the mass flow meter are neglected.

OPERATION ON LNG & PROPANE

The SP series of gas injectors can also operate on vaporized liquid natural gas or propane. For liquid natural gas (LNG), it is desirable to utilize a low gas supply pressure to minimize LNG storage pressure and maximize the storage time before venting. The preferred operating supply pressure is 586 kPa (85 psig) which is compatible with the recommendations of most LNG tank suppliers and the manifold absolute pressure of usual turbocharged engine.

Because of the different heating values and physical properties between natural gas and propane, the required gas absolute supply pressure for gaseous fueled engines fitted with the same fuel injection hardware and software, operating on propane, would be approximately 60% of that when natural gas is used. The gas supply pressure should range from 207 ~ 414 kPa (30 ~ 60 psig), which is compatible with the critical pressure ratio (0.576) and the vapor pressure of propane. Figure 8 delineates vapor pressures at various temperatures for propane.

The natural gas and propane flow curves for the SP014 injector, which was designed for use at lower supply pressures, are shown in Figures 9 and 10 respectively.

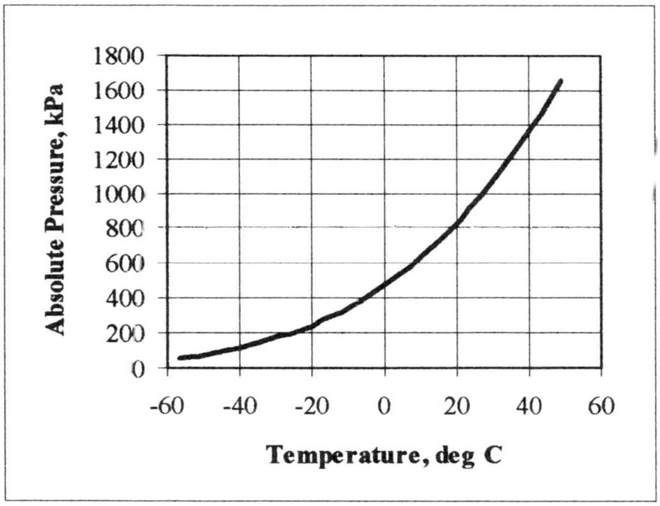

Figure 8: Propane Vapor Pressure

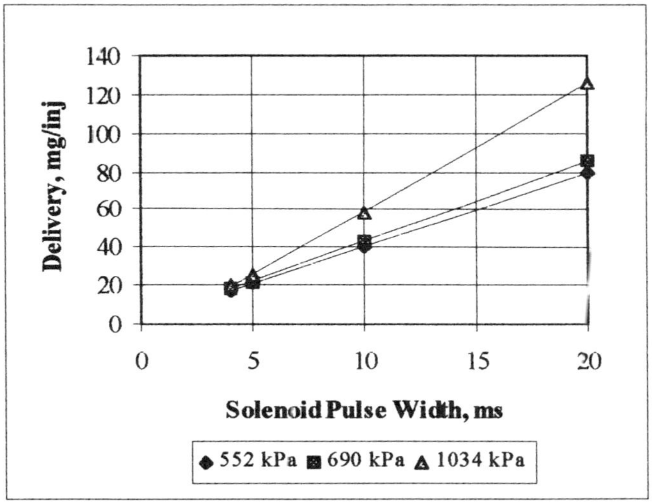

Figure 9: SP014 Methane Delivery

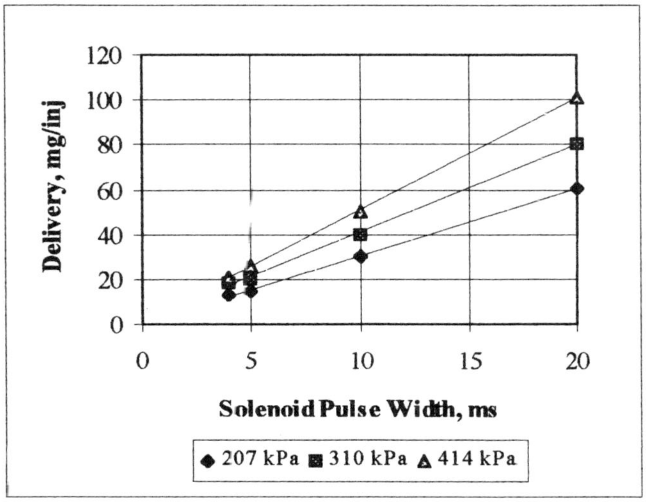

Figure 10: SP014 Propane Delivery

INJECTOR APPLICATIONS

The injector application can take various forms, but those which have proven successful are characterized below:

SINGLE POINT - This application which is shown in Figure 11, is similar to the gasoline throttle body configuration locating the injector outlets below the throttle plates. The injectors are typically operated sequentially at engine cycle frequency. The advantage of this configuration is a minimal number of injectors used, however, an efficient mixing ring will usually be required to control distribution and avoid excess hydrocarbon emissions.

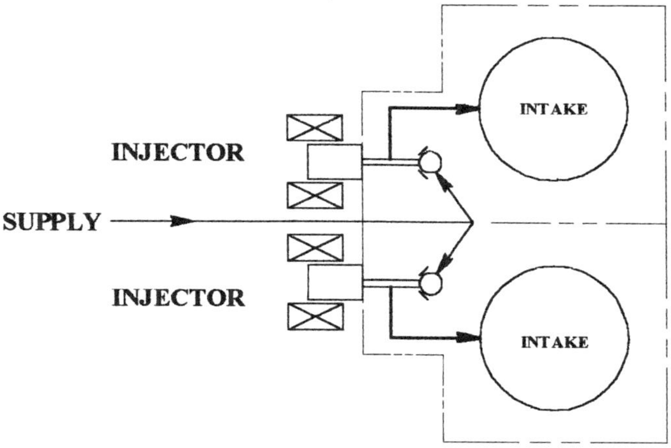

Figure 11: Single Point Injection System

MULTI-POINT CONSTANT FLOW - As shown in Figure 12, a plenum, common to the required injectors is used with the fuel supplied to each intake port by a tube assembly. This places the point of injection closer to the intake valve improving transient response and reducing the unburned methane content of hydrocarbon compared to the single point system, particularly, in throttle body having a two step throttle with a large venturi used only at high power levels.

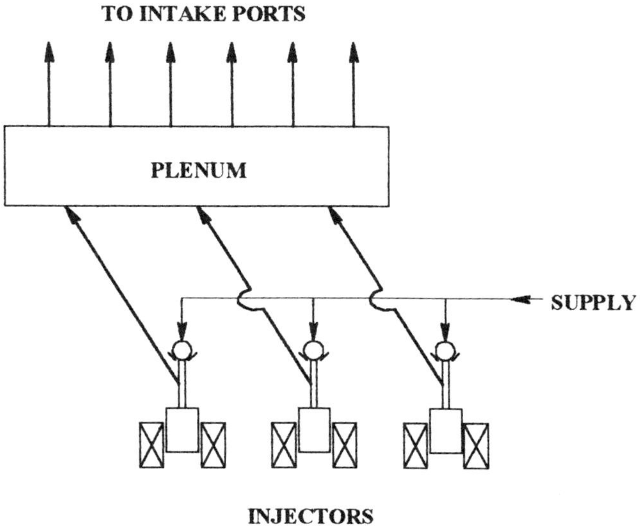

Figure 12: Multi-Point Constant Flow Injection

MULTI-POINT - Figure 13 represents the best configuration for transient performance and the lowest total hydrocarbon emissions. Where engines, typically diesel, have common intake port opening, one injector per cylinder is used with a variable operating strategy. For low to medium power, one injector per cylinder is typically employed. At higher power levels, two injectors are utilized. This system offers the best fuel distribution in addition to the above performance benefits. The fuel system used on the 1992 Dodge B250/B350 CNG Ram Van/Wagon described in Reference (2) is a typical example.

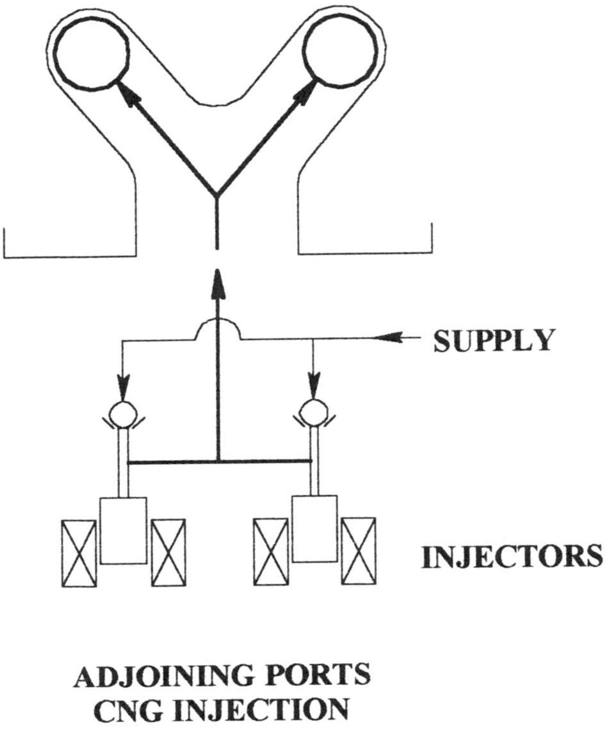

Figure 13: Timed Multi-Point Shared Port Sequential Injection

FLOW COMPENSATIONS

Discussed previously were the separate effects of the following:

- Gas supply pressure
- Battery voltage
- Solenoid pulse width

To precisely calculate the amount of gas delivered through the injector, continuous and instantaneous adjustments must be included in the software strategy. It is necessary to include gas pressure and temperature sensors in the engine fuel system. Their outputs are used to compensate injector pulse width for fuel density changes caused by transient pressure and temperature variation. In addition to gas pressure and temperature, compensation for supply voltage variation is employed in Servojet's electronic engine control module. A gas composition sensor is also needed for flow adjustment due to variations of methane content.

SUMMARY & CONCLUSIONS

The Servojet electronic gas injector was developed from the high speed solenoid valve typically used for hydraulic applications. These ball poppet solenoid valves have been used as gas injection valves since 1987, in the Servojet's electronic dual fuel injection systems described in References (3) and (4). With more than 50,000 units manufactured to date, the electronic gas injectors have provided precise, stable and reliable service for over one billion cycles. The performance and durability of the injector have also been validated in the field, where at least 25 fuel systems have been in service for more than 50,000 miles each. With the application techniques including flow compensations for variations of voltage, pressure, temperature and gas composition, optimum and precise mixture of air and gaseous fuel can be obtained using a closed-loop electronic fuel injection system for achieving current and future emission goals.

REFERENCES

(1) Barkhimer, R.L.; Beck, N.J.; Weseloh, W.E.: "Development of a Durable, Reliable and Fast Responding Solenoid Valve", SAE Paper 831326, 1983.

(2) Geiss, R.O.; Burkmyre, W.M.; Lanigan, J.W.: "Technical Highlights of the Dodge Compressed Natural Gas Ram Van/Wagon", SAE Paper 921551, 1992.

(3) Beck, N.J.; Johnson, W.P.; George, A.F.; Petersen, P.W.; van der Lee, B.; Klopp, G.: "Electronic Fuel Injection for Dual Fuel Diesel Methane"' SAE Paper 891652, 1989.

(4) Johnson, W.P.; Beck, N.J.; Lovkov, O.; van der Lee, A.; Koshkin, V.K.; Piatov, I.S.: "All Electronic Dual Fuel Injection System for the Belarus D-144 Diesel Engine", SAE Paper 901502, 1990.

980814

New Concept of a Low Cost, Fast Acting Electronically Controlled Fuel Metering System With Back-Up Capability

S. S. Mohtasebi, H. Hong and T. Krepec
Centre for Industrial Control, Concordia University, Montreal

ABSTRACT

Typical fuel metering systems for small gas turbine engines consist of a metering valve and a bypass valve which is maintaining a constant differential pressure across the metering valve orifice. The metering valve is operated by the compressor pressure signal and the bypass valve acts automatically.

It is proposed to bring both the metering and the bypass valves under the control of two digital linear actuators. The first one would move the metering plunger according to the compressor pressure and the second would move the bypass plunger in such a way, that the differential pressure across the metering orifice would be maintained constant only during steady state engine operation. During the engine transient processes, however, it would be increased or decreased as required, to speed up the change in the nozzle flow rate and consequently the engine dynamic response.

Independent actuation of the metering and bypass valves gives the opportunity to meter the fuel flow to the nozzles only by one of the valves, in case if the second valve would fail. When the metering valve fails, the nozzle flow is controlled by the change of the differential pressure across a fixed metering orifice. When the bypass valve fails, the nozzle flow is controlled by the variation of the metering orifice flow area at constant fuel supply pressure. The change from the first mode of operation to the second one is provided by the rotation of both plungers which open or close the particular orifices.

The new fuel metering system design is explained at all three modes of operation. It is computer simulated and its dynamic response is investigated on an engine model to determine the most effective control strategy.

INTRODUCTION

The present electronic fuel control units (EFCU's) for small gas turbine engines do not differ much from their older hydromechanical predecessors (FCU's) regarding their principle of operation. The main modifications are in the use of electronic actuators instead of bellows for the fuel flow metering and in replacing the mechanical speed governor with an electronic one based on the measurement of engine rotational speed by a simple magnetic pickup sensing the Hall-effect. These two main electronic components are improving the engine operational performance without changing much the mode of operation of the fuel control system. To reduce more the cost of the modern fuel control system, further simplification of its main components would be required, however, without compromising the engine performance and the aircraft safety [1].

SYSTEM OPERATION AND DESCRIPTION

The hydraulic schematic of the proposed electronic fuel metering system (Single Barrel, Double Plunger Fuel Metering System) is shown in figure 1. This system employs two pressure balanced plungers which are located in one single barrel at opposite directions and are operated by two digital linear actuators simultaneously but independently. This means that this system is able to keep a constant differential pressure across the metering valve at steady state condition and vary the differential pressure during the transient engine operation, in order to speed up the dynamic response of the engine. This is due to the fact that both, the metering valve and the bypass valve plungers are operated independently, but under the same electronic control management.

During the normal operation of the system, the metering valve controls the nozzle fuel flow and the bypass valve spills the excess of fuel back to the tank. However, during the back-up operation in case of a failure of the bypass valve, an on-off solenoid, which actuates a radial lever connected to the plungers, is used to rotate both plungers. When the bypass valve fails, the plungers are rotated clockwise (figure 2) so that, the bypass port

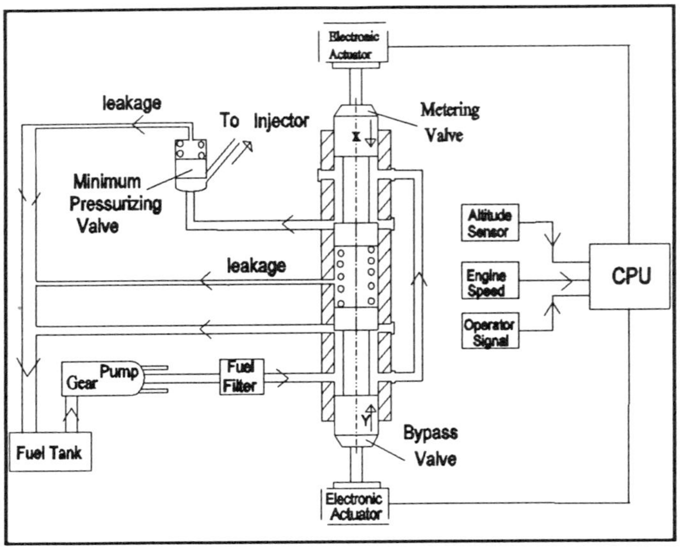

Figure 1 Hydraulic Schematic of the Fuel Metering System

becomes closed; however, the metering valve still operates. As the result of bypass valve closure, the fuel pressure increases and opens a relief valve which is originally used to protect the system from excessive pressure.

When the metering valve fails, both valves are rotated counterclockwise by the solenoid (figure 3). In this case, the metering valve is closed, however, at the same time, a fixed orifice in the barrel is opened to allow the fuel flow to the nozzle and the bypass valve controls the differential pressure across the fixed orifice by spilling the excess fuel back to the tank.

MATHEMATICAL MODEL DESCRIPTION

A schematic diagram of the fuel metering system is shown in figure 4 and the equations describing the different parts of the system are presented as follows [2]:

Fuel Flow Rate Equations

Pump Delivery (Fuel Pump Driven by the Engine)

$$Q_P = C_1 N - C_2 (P_1 - P_0) \tag{1}$$

Metering Valve Flow

$$Q_{mv} = C_{dmv} A_{mv} \sqrt{2 (P_1 - P_2)/ \rho} \tag{2}$$

where $A_{mv} = n_{smv} X_{mv} W$, n_{smv} is the number of slits in the metering valve barrel and *W* is the width of the slits.

Bypass Valve Flow

$$Q_{bv} = C_{dbv} A_{bv} \sqrt{2 (P_1 - P_0)/ \rho} \tag{3}$$

The flow area A_{bv}, is the combined area of the slits in the bypass valve barrel and is a function of the plunger position X_{bv} i.e. $A_{bv} = X_{bv} n_{sbv} w_{bv}$.

Minimum Pressurizing Valve Flow

$$Q_{mp} = C_{dmp} A_{mp} \sqrt{2 (P_2 - P_n) / \rho} \tag{4}$$

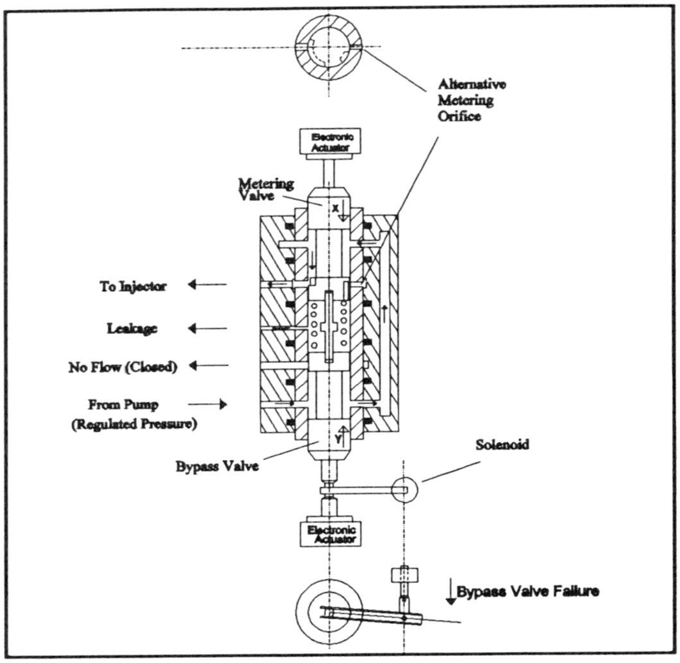

Figure 2 Schematic of the Fuel Metering System During Bypass Valve Failure

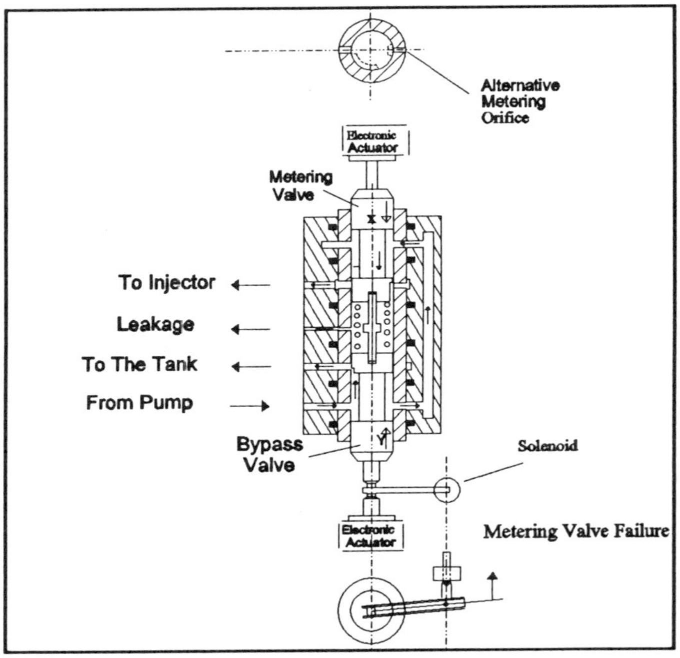

Figure 3 Schematic of the Fuel Metering System During Metering Valve Failure

Injector Nozzle Flow

$$Q_n = C_{dn} A_n \sqrt{2 (P_n - P_0) / \rho} \tag{5}$$

Continuity Equations

Metering Valve Volume

$$\frac{dP_1}{dt} = \frac{\beta}{V_1}(Q_P - Q_{bv} - Q_{mv}) \tag{6}$$

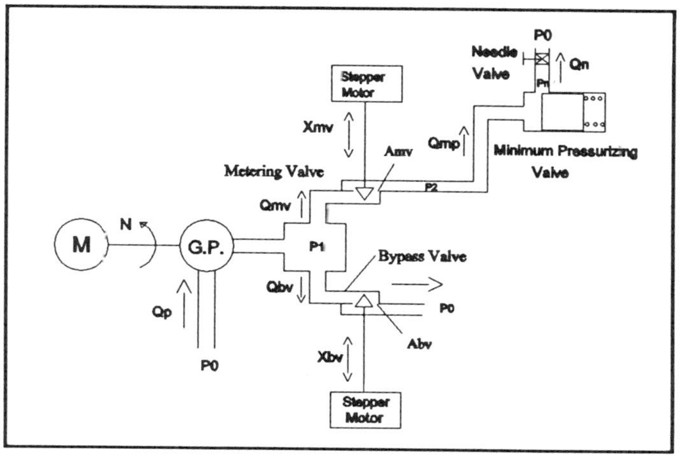

Figure 4 Schematic of the Fuel Metering System Model

Metering Valve Down-Stream Volume

$$\frac{dP_2}{dt} = \frac{\beta}{V_2}(Q_{mv} - Q_{mp}) \tag{7}$$

Nozzle Manifold Volume

$$\frac{dP_n}{dt} = \frac{\beta}{V_n}(Q_{mp} - Q_n) \tag{8}$$

Valve Motion Equations

Metering and Bypass Valves Plungers Motion

$$X_{mv} = \Delta X_{mv} \ INTEGER \ (f_{mv} \ t) \tag{9}$$

$$X_{bv} = \Delta X_{bv} \ INTEGER \ (f_{bv} \ t) \tag{10}$$

where f is actuator triggering frequency and Δx is the step size of the actuator.

STEADY STATE SIMULATION RESULTS

The model for steady state operation is obtained by setting the transient terms (d/dt , d^2/dt^2) equal to zero in the above equations. This results in a set of nonlinear equations which are solved for the full travel of the metering valve. The step sizes and the maximum stepping rate of the stepper motors for both, the metering and the bypass plungers, are considered equal and are set at 0.025 mm and 500 steps / s respectively.

Figure 5 shows the simulation result for the nozzle fuel flow versus the metering valve position which varies linearly with respect to the metering valve movement. This is due to the fact that the differential pressure across the metering valve is kept constant by the bypass valve. Figure 5 also shows the simulation result for the bypass valve failure. The relief valve was set at 55 bar pressure. As it is seen, the nozzle flow is no longer varying linearly and this is due to the fact that the differential pressure across the metering valve increases with the higher nozzle flow rate. However, even in this case, the metering valve is still able to maintain the nozzle flow required for the engine and can return the aircraft home safely.

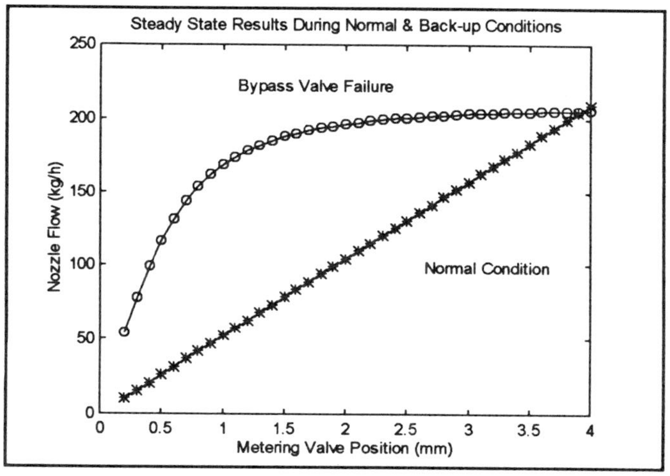

Figure 5 Steady State Simulation Results for the Nozzle Flow During Normal Operation and Bypass Valve Failure

Figure 6 shows the simulation results of the nozzle flow versus the full travel of the bypass valve plunger for the metering valve failure. As seen, by movement of the bypass valve plunger, the nozzle flow changes accordingly.

It has to be mentioned that, at both back-up operation modes, the fuel metering system is not operating as per the best performance of the engine. However, in both cases, the objective is to supply the engine with an adequate fuel flow rate, allowing the aircraft to return back home.

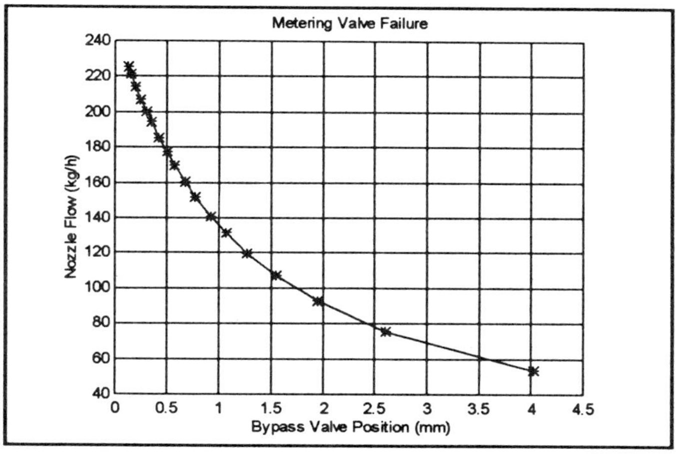

Figure 6 Simulation Results During the Metering Valve Failure

TRANSIENT SIMULATION RESULTS

Due to the flexibility of the proposed fuel metering system, different control strategies can be used. This is due to the fact that both, the metering valve and the bypass valve are under independent electronic control and, therefore, it is possible to provide a constant or variable differential pressure across the metering valve. Since the different control strategies have various impacts on the dynamic response of the system, therefore, the dynamic response of the system with each strategy is first presented and the results of all the control strategies are next compared to determine the most advantageous one for the proposed fuel metering system.

Figure 7 shows the block diagram of the first control strategy. The error e_1, which is the difference between the demanded and the actual fuel flow, acts on the first digital controller to produce a voltage signal. The second error e_2, which is the difference between the desired and the actual differential pressure, acts on the second digital controller. In both cases, the voltage signals enter two voltage to frequency converters to produce the rectangular waveforms. These rectangular waveforms are used to trigger the stepper motor drivers which send a sequence of pulses to drive the actuators of the first and the second stepper motors in proper directions and thus, increase or decrease the fuel flow and the differential pressure across the metering valve, respectively. Note that the controllers output voltages are limited to the maximum values which are provided by the stepper motors manufacturers to ensure that the actuators produce a linear force high enough to drive the valve when the controller outputs are very high.

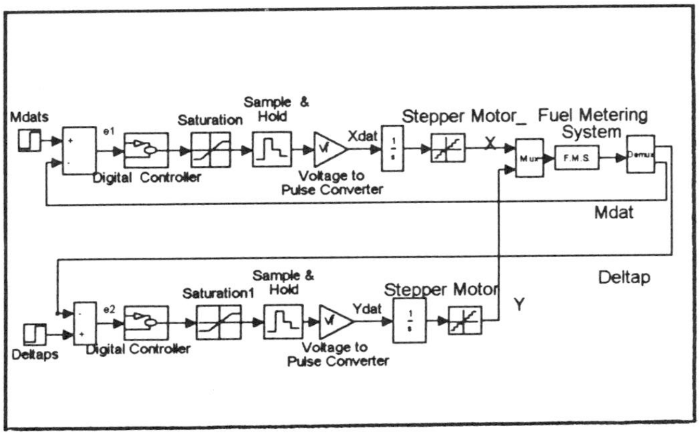

Figure 7 The Block Diagram of the First Control Strategy

In general, the purpose of this option is to keep a constant differential pressure across the metering valve, even during the transient period. This ensures that the relation between the metering valve position and the fuel flow is linear and unique and is not changing during the operation.

Figure 8 shows the typical dynamic responses of the system to a request for an abrupt change in the fuel flow rate from the initial value of 83.5 kg/h to the final value of 183.5 kg/h. Next, after 2 seconds, the fuel flow rate is commanded to return to its initial value of 83.5 kg/h. Meanwhile, the differential pressure must be kept constant by the second actuator.

At the start of the transient process, the error e_1 is very large, therefore, the output of the first controller is saturated and the metering valve moves at its maximum stepping rate. With the opening of the metering valve, the differential pressure across the metering valve drops suddenly and therefore, the bypass valve moves back with its maximum rate trying to keep a constant differential pressure.

It has to be mentioned that at steady state operation, the positions of both valves may fluctuate about their final values, as shown in the figures. This is due to the fact that the metered fuel flow is a function of the differential pressure and of the opening of the metering valve. This means that around steady state, the deviation of the first actuator from its final value results in the deviation of the second actuator and, therefore, both actuators never stop at their final values and fluctuate around them. These fluctuations depend on the gains of the controllers and the step size of the stepper motors and, to some extend, can be eliminated by deactivation of the controllers when the errors are smaller than a minimum value.

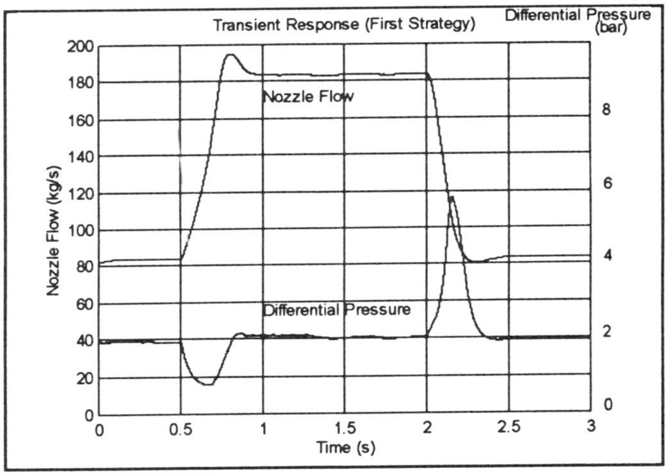

Figure 8 The Transient Response of the Differential Pressure and Nozzle Flow For the First Control Strategy

To improve the transient processes over those obtained with the first control strategy, the second control strategy was investigated. In this strategy, the main advantage is that the differential pressure across the metering valve changes only during the transient part of the response and then, it returns to its initial value during the steady state part. According to this strategy, when higher fuel flow rate is demanded by the engine, the differential pressure is set to a higher value and therefore, the response is faster. After the nozzle flow reaches 85% of its final value, the differential pressure returns to its steady state value and from this moment the response is slower. This ensures that there is just one positioning of the actuators. When a nozzle flow decline is required, the differential pressure is set to its lower value as long as the nozzle flow is more than 15% of the final value and then the differential pressure returns to its steady state value.

Figure 9 shows the transient response of the metered fuel flow for the second control strategy. As seen, the response is very fast, as long as the flow is less than 85% of its final value. This is due to the fact that, as the metering valve moves forward, the differential pressure is also increasing. However, when the nozzle flow reaches 85% of the final value, the differential pressure starts to return to its steady state value and thus the response is slower.

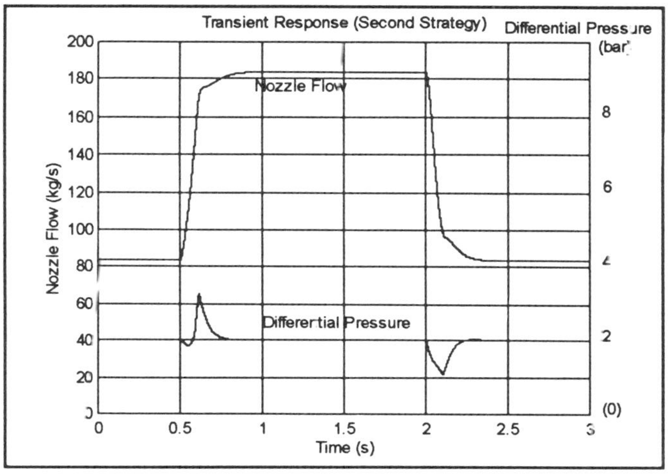

Figure 9 The Transient Response of the Differential Pressure and Nozzle Flow For the Second Control Strategy

During the simulation, the maximum and the minimum of the differential pressure was 5.5 and 0.34 bar respectively. This means that the gain of the second controller is not constant and varies during the transient part of the response.

Figure 10 shows the block diagram of the control system which was used during the bypass valve controller failure. As it was mentioned before, during the bypass valve failure, the bypass valve orifice is closed and a relief valve protects the system from excessive pressure. In this case, the relief valve is set at 55 bar.

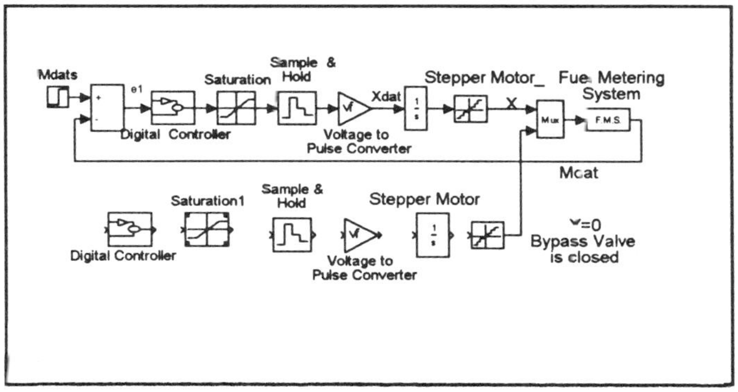

Figure 10 The Block Diagram of Control System During Bypass valve Failure

Figure 11 shows the simulation results for different fuel flow rate demands. It illustrates that the transient fuel response of the system is very fast, and that the system operation, even during the failure, is satisfactory. However, as shown in the figure, the relief valve was set for a higher pump delivery pressure, and the differential pressure across the metering valve is also high. This means that the pump is operating with a higher load which might not be acceptable for a long time extensive operation.

During the metering valve failure, the metering valve controller is not operating and the bypass valve is reacting solely to the error which is the difference between the fuel flow demand and the actual fuel flow. Figure 12 shows the dynamic response of the system for different nozzle flow demands. As seen, the response is very fast despite the fact that the metering valve is not operating.

In general, both back-up modes of operation ensure that the proposed fuel metering system is acting quite well during the back-up conditions and therefore, it can be recommended to be used in small gas turbine engines to improve their reliability and safety. In the next section, a simple engine model is presented to be used as a tool for comparison of different control strategies discussed earlier.

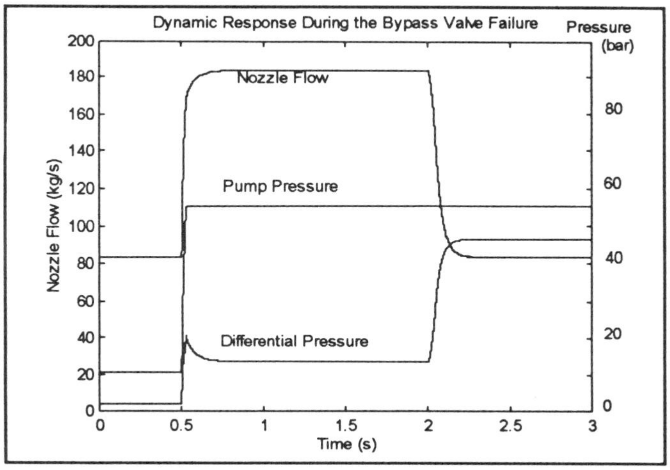

Figure 11 Dynamic Response of the System During Bypass Valve Failure

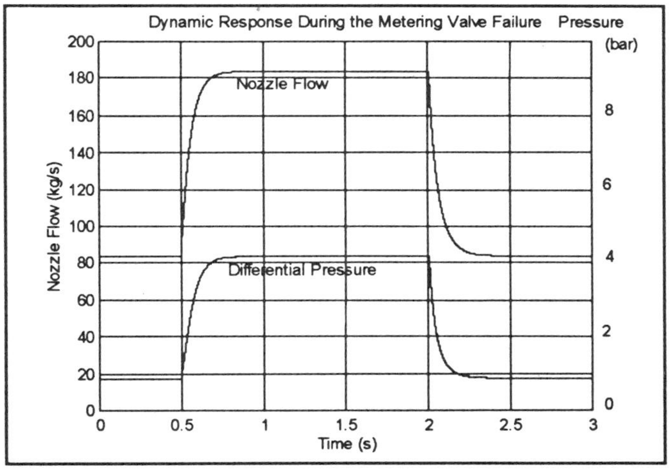

Figure 12 Dynamic Response of the System During Metering Valve Failure

ENGINE MODEL

The investigation of the dynamic response of the system, including the gas turbine engine would be recommended, in order to properly evaluate the behaviour of the complete system. However, to achieve the real time control, the engine model has to be simplified. Therefore, a dynamic analysis is used to obtain the transfer function describing the response of the engine to the fuel flow changes. It is supposed that the fuel metering system is mounted on the gas generator shaft of a gas turbine engine with the following specifications:

P_{max}=533.2 kw, N_{nom}= 33000 rpm, $\dot{m}_{max}$= 200 kg/ h and SFC=0.3410 kg/ kw/ h.

For the purpose of simplifying the dynamic simulation, the ratio of fuel flow to the engine torque at maximum speed is assumed constant and is not affected by the speed change in the narrow speed range of the maximum governing engine operation; then:

$$SFC = \frac{\dot{m}_f}{P} = \frac{\dot{m}_f}{T\omega} \quad and \quad T = \frac{1}{SFC}\frac{\dot{m}_f}{\omega} \qquad (11)$$

and finally, the dynamic balance of the torques is found as follows:

$$T_t - T_c - T_d - T_l = J\,\dot{\omega}$$
$$and \qquad (12)$$
$$T_d = B\,\omega$$

where subscripts t, c, d and l stand for turbine, compressor, damping and load. Taking the Laplace transform and simplifying the equation:

$$\frac{\omega}{(T_t - T_c) - T_l} = \frac{1}{JS + B}$$
$$and \qquad (13)$$
$$T_t - T_c = \frac{1}{SFC}\frac{\dot{m}_f}{\omega}$$

Figure 13 shows the dynamic model of the engine. This model is linked with the fuel metering system model in order to investigate the impact of different control options on the complete system model.

COMPARISON OF DIFFERENT CONTROL STRATEGIES

In this section, the dynamic response of the whole system with the engine speed feedback is investigated in order to evaluate the behaviour of the complete system when different control strategies are applied. Figure 14 shows the block diagram of the complete system model. As seen, the demand and the actual feedback engine speed are compared and the error is generated. Note that the controllers gains were obtained based on the fuel flow feedback, therefore, for the current system, the error should be evaluated based on the fuel flow. To do this, the engine speeds are converted to their equivalent fuel flow based on the fuel flow schedule for a PT6-A27 engine. Then, the error is processed in the fuel metering system for different control options and thus the new fuel flow is evaluated based on the error and the current engine speed. The output fuel flow is applied to accelerate or decelerate the engine.

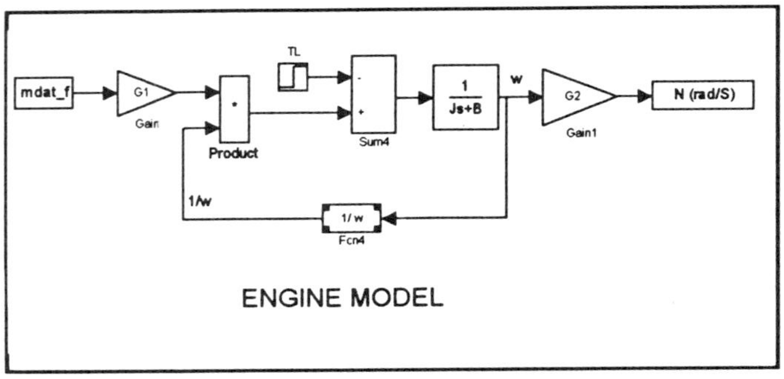

Figure 13 The Block Diagram of the Engine Model

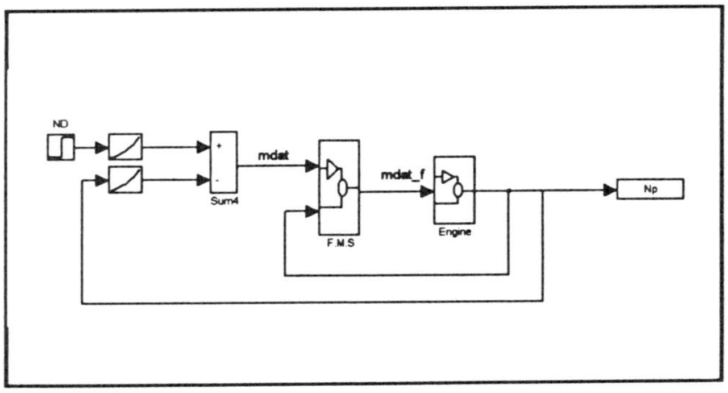

Figure 14 The Block Diagram of the Complete System Model

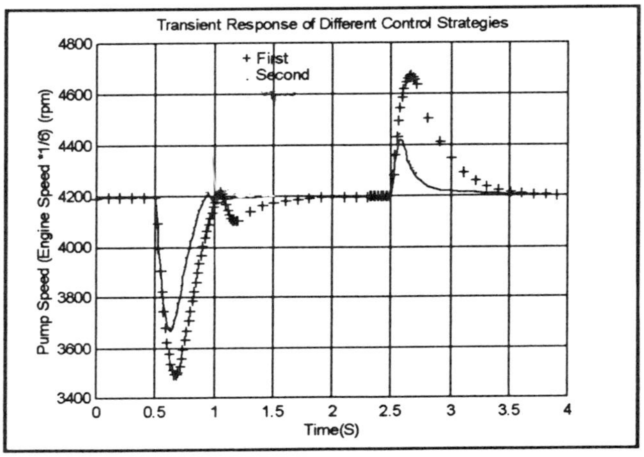

Figure 15 The Transient Response of the Engine Speed For Different Control Strategies

Results for the transient response of the engine speed using different control strategies for an abrupt change in the load are shown in figure 15. It is assumed that the load increases abruptly by 250% (from 80 N.m to 200 N.m) and then after 2 seconds, it reduces by 50%.

According to figure 15, the response of the second control strategy is faster. The reason is that in the second control strategy, both valves move in appropriate direction to speed up the fuel flow changes. The first control strategy which operates as a conventional fuel metering system, has the slower response and its settling time is 0.5 second longer than the other one.

CONCLUSION

The feasibility of a new fuel metering system with a single barrel fitted with two plungers, which are operated by two independently acting linear digital actuators was investigated. This system has the ability to speed up the dynamic response of the engine by varying the differential pressure across the metering valve. Due to the flexibility offered by this fuel metering system, different control options were introduced and to compare them, a dynamic model for a gas turbine engine was presented and used. Based on the results obtained from the simulation of the engine with different control strategies, it was concluded that the best operation is provided by the second control strategy. In this control strategy, the differential pressure across the metering valve varied only in the initial part of the transient process and next, it is returned to its steady state value. Moreover, an additional advantage of this fuel metering system is its back-up capability. It was shown that during the failure of either the bypass valve or the metering valve actuator, the system was able to maintain the required fuel flow for the engine in order to return the aircraft home safely.

NOMENCLATURE

A_{bv}	Bypass Valve Flow Area (m^2)
A_{mv}	Metering Valve Flow Area (m^2)
A_{mp}	Minimum Pressurizing Valve Flow Area (m^2)
A_n	Nozzle Manifold Inlet Cross Section Area (m^2)
C_1	Pump Displacement (m^3/ s/ rpm)
C_2	Pump Leakage Coefficient (m^5/ s/ N)
C_{dbv}	Flow Coefficient for Bypass Valve
C_{dmv}	Flow Coefficient for Metering Valve
C_{dmp}	Flow Coefficient for Minimum Pressurizing Valve
C_{dn}	Flow Coefficient for Nozzle Orifice
f_{mv}	Metering Valve Actuator Triggering Frequency (Hz)
f_{bv}	Bypass Valve Actuator Triggering Frequency (Hz)
J_M	Engine Moment of Inertia ($N.m.s^2$)
n_{sbv}	Number of Bypass Valve Slits
n_{smv}	Number of Metering Valve Slits
N	Engine Speed (rpm)
N_{nom}	Engine Nominal Speed (rpm)
N_P	Pump Speed (rpm)
P_0	Ambient Pressure (N/ m^2)
P_1	Pump Pressure (N/ m^2)
P_2	Metering Valve Down Stream Pressure (N/ m^2)
P_C	Compressor Pressure (N/ m^2)
P_{max}	Engine Maximum Power (W)
P_n	Nozzle Pressure (N/ m^2)
Q_{bv}	Bypass Valve Flow (m^3/ s)
Q_{mv}	Metering Valve Flow (m^3/ s)
Q_{mp}	Minimum Pressurizing Valve Flow (m^3/ s)
Q_n	Nozzle Flow (m^3/ s)
Q_P	Pump Flow (m^3/ s)
T_C	Compressor Torque (N.m)
T_d	Damping Torque (N.m)
T_l	Load Torque (N.m)
T_{max}	Engine Maximum Torque (N.m)
T_t	Turbine Torque (N.m)
V_1	Volume of the High Pressure Chamber (m^3)
V_2	Volume of the Metering Valve Down Stream (m^3)
V_n	Volume of Nozzle Manifold (m^3)
w_{mv}	Width of Metering Valve Slits (m)
w_{bv}	Width of Bypass Valve Slits (m)
X_{mv}	Metering Valve Plunger Position (mm)
X_{bv}	Bypass Valve Plunger Position (mm)
α	Angular Acceleration (rad / s^2)
β	Bulk Modules of Fuel (N/ m^2)
ΔX_{mv}	Metering Valve Actuator Step Size (m)
ΔX_{bv}	Bypass Valve Actuator Step Size (m)
ρ	Fuel Density (kg/ m^3)
ω	Angular Velocity (rad/ s)

ACKNOWLEDGEMENTS

The first author would like to thank the Ministry of Culture and Higher Education of Iran for providing scholarship and financial support.

REFERENCES

[1] T. Krepec and A.I. Georgants, "New Family of Low Cost Electronic Fuel Control Units for Small Gas Turbine Engines", SAE paper 901039, April 1990.

[2] S. S. Mohtasebi, " Investigation on New Low Cost Electronically Controlled Fuel Metering Systems for Small Gas Turbine Engines " Ph.D. Thesis, Concordia University, Montreal, Canada, February 1997.

960768

An Investigation into the Use of Fluidic Devices as Gas Fuel Injectors for Natural Gas Engines

Q. Huang and R. Chen
Loughborough Univ.

ABSTRACT

A novel gas fuel injector system based on the use of monostable fluidic devices is described in this paper. The proposed system consists of non-moving-part fluidic devices which are capable of operating in a Pulse Width Modulated (PWM) control mode and of handling a large amount of gas flow for engine operations. The system also includes an electro-fluidic interface for fluidic switching and air-gas mixing nozzles for better mixing quality.

Two prototype fluidic injector units were produced and their steady-state and dynamic characteristics were evaluated on a laboratory test rig. The results were compared with those from several commercial gas injectors and it was found that the fluidic injector has a faster dynamic response and a smaller cycle-cycle variations.

INTRODUCTION

Newly legislated exhaust emission standards have revived interest in the use of natural gas as an alternative fuel for automotive engines mainly due to the fact that much less harmful emissions are produced from gas engines than those from gasoline engines. Natural gas consists primarily of methane and has high knock resistance characteristic (typical octane number of 120) which will permit engines to be operated at higher compression ratios resulting in improved fuel economy and reduced emissions. Since natural gas is in a gaseous state, the use of heavy fuel enrichment during engine cold start and transient conditions can be eliminated and as a result, lower exhaust emissions and lower fuel consumption levels can be obtained.

In order to meet increasingly stringent emission legislations, it is necessary to employ a similar air-fuel ratio (AFR) control strategy for gas engines as spark ignition (SI) engines, i.e. stoichiometric AFR control with a Three-Way Catalytic converter (TWC) **[1,2]**. This requires that the fuel metering device must be able to provide an accurate amount of gas fuel under various engine operating conditions and also be able to provide fuel compensations for various ambient and gas storage conditions.

The most commonly used fuel metering device for gas engines is the venturi type carburettor or mixer. The mixer operates on the venturi principle and is designed to admit a stoichiometric air-fuel ratio of 9.5 on the basis of volume flow under one specified ambient condition. However, the required stoichiometric AFR for the TWC operation is based upon a mass flow ratio (16.5:1). Unfortunately, gas carburettors are not capable of providing accurate stoichiometric AFR control on the basis of mass flow under various engine and fuel tank conditions and no compensations can be provided for various gas fuel compositions. Therefore, the use of gas mixers will result in considerable errors in stoichiometric AFR control and result in unsatisfactory emission control **[2,3]**. It was concluded **[2,3]** that it is essential to utilise a timed gas fuel injection system similar to the one used in SI engines in order to ensure accurate stoichiometric AFR control.

Numbers in brackets designate references at the end of the paper

Gas fuel injection combined with a modern engine management system not only provides more accurate AFR control but also provides opportunities for utilising positive upstream pressure from the tank to improve the air-fuel mixture quality. The accuracy of stoichiometric AFR control can be ensured for various engine and gas tank conditions through fuel compensations by the engine management system and gas fuel injectors. However, reliable and satisfactory gas fuel injectors with a low manufacturing cost are not yet available **[2,3,4]**.

It is known that the volumetric energy density of natural gas is approximately one order of magnitude lower than that of typical liquid fuels when the gas is injected at 10 bar by the solenoid injector (three orders of magnitude lower at atmospheric conditions). Therefore gas fuel metering devices must be capable of handling correspondingly larger gas volumes to obtain equivalent energy flows. This need for large volume flow per engine cycle requires substantially larger injector components and relatively high pintle lift for the required flow passage area. As a result, high power drive stages are required in the drive circuitry and a relatively slow response will be expected.

Methane is not regarded as an effective lubricant for solenoid injectors, therefore fast wearing and aging could also be expected. It is not unusual to hear complaints about the noise level produced by solenoid gas injectors. This is due to the impact of the armature of large mass on the injector seat which will also contribute to the fast wear and aging of the injectors and relatively high injection pressures used. It is due to the above problems associated with solenoid gas injectors that a novel fluidic device based gas injector is proposed.

The purpose of this project is to investigate the use of fluidic techniques for the gas fuel injector stages for natural gas engines. The proposed fluidic gas injectors are potentially capable of providing more accurate and faster response gas fuel control while meeting the the flow requirements. Considerable cost benefits are also envisaged from the fluidic system due to it's mechanical simplicity. The fluidic injectors can be incorporated into a multipoint gas fuel injection system together with a modern engine management control unit and by doing so both reduced exhaust emissions and improved fuel economy for gas engines can be obtained.

FLUIDIC GAS FUEL INJECTORS

MONOSTABLE FLUIDIC DEVICES - The fluidic devices being considered for natural gas injectors are wall reattachment type monostable fluidic amplifiers. Due to their inherent amplification characteristics, large volumes of gas flow for engine operation can be easily controlled by relatively small flow signals (with a typical gain of 10 plus) through an electro-fluidic interface in a PWM mode. This offers potentials for utilising a low power, fast response interface for fluidic stage switching and for reducing the size and power requirements of the ECU's (Electronic Control Unit) drive circuitry in the engine management system.

Figure 1 shows a monostable device with a biased configuration. The jet flow is initially attached to the right wall (output 1) due to the biased geometry ("Coanda effects"). When sufficient control flow is applied on the positive control port, the main jet flow will be switched to the output 2 and remains on that output so long as the control flow is present. When the control flow is removed the jet flow will be automatically shifted to the output 1 again due to the unbalanced flow entrainment (differential pressure) on both sides of the jet caused by the biased configuration. This method is so-called "Positive control method".

Alternatively, the switching of flow jet to the output 2 can be caused by closing the the vacuum control port cutting off the flow entrainment through this port. The flow attachment on the output 2 will be maintained so long as the vacuum control port is closed off. As the vacuum control port is opened to the medium, the flow jet will be shifted to the output 1 again due to its monostable nature. This switching control method is so-called "Vacuum control method". It can be seen that output 1 represents "stable" side of the fluidic device which the main jet flow always tends to attach to.

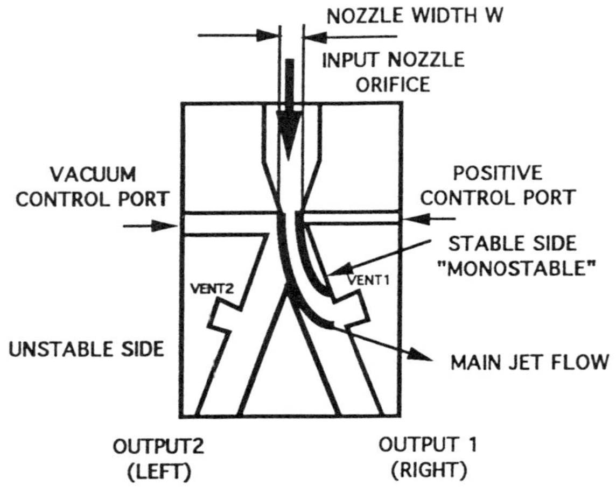

Figure 1 A monostable fluidic device

SYSTEM OPERATION - **Figure 2** shows a schematic set-up of the fluidic gas injector system with the vacuum control switching method employed. The fluidic device is arranged to be "submerged" in an encapsulated gas reservoir under atmospheric pressure P_{atm}. This is to maintain normal wall reattachment and switching operation of the flow jet in the device. A conventional three-stage pressure regulator is used to reduce gas pressure from 200 bar nominal tank pressure to atmospheric one. The output of the regulator is connected to the gas reservoir and therefore an atmospheric pressure condition in the reservoir can be maintained.

A low pressure gas pump or compressor is connected to the other side of the gas reservoir and is used to provide a supply of gas flow to the fluidic device main supply nozzle. When the solenoid actuator is in the closed position cutting off the flow entrainments into the vacuum control port of the device, the main jet flow will be attached to the output 2 (unstable side), which is open to the reservoir itself (unused spill flow). Under such conditions, the gas flow will simply be circulated around the pump, fluidic device and the reservoir.

When the solenoid is energised by drive pulses from ECU, the valve opens allowing the gas flow in the reservoir to be entrained into the device through its vacuum control port. This will cause the main flow jet to be switched to the output 1 (stable side) due to its monostable nature. The gas flow will then be introduced into the engine manifold through an air-gas mixing nozzle. The fuel flow jet will be maintained on the stable side output 1 so long as the solenoid valve is kept open.

As the solenoid is de-energised, the valve closes and no more flow will be entrained into the device from the vacuum control port. As a result, the main jet flow switches back to output 2 (i.e. reservoir) due to unbalanced pressure across the main jet flow. It can be seen that fuel flow delivered into the engine is determined by the steady-state flow rate at the output 1 and pulse period applied on the solenoid interface by the ECU.

Due to the amplification characteristics of fluidic devices, the mass flow of gas through the vacuum control port causing jet switching is only a fraction (less than 10%) of the main jet flow delivered into the engine. Therefore, the demand for the solenoid interface has been greatly reduced and a low power, low flow rate and fast response miniature solenoid or equivalent device can be used. In this arrangement the gas pump is required to provide the whole amount of gas flow needed for the fluidic device. Constant gas flow into the the reservoir at atmospheric pressure from the regulator will ensure that the reservoir will not "dry out" and atmospheric pressure P_{atm} within the reservoir can always be maintained.

The purpose of the air-gas mixing nozzles is two fold; one is to introduce an auxiliary air flow from the upstream of the air throttle plate (not shown in this figure) to assist gas mixing; the other, probably more significant, is to maintain atmospheric pressure at the output of the fuel discharge tube inside the nozzle (not shown) thus preventing the fuel jet flow inside the fluidic stages from premature switching to the output 1 under engine manifold depression and pressure fluctuations. Such techniques were used in the previous fluidic gasoline injector system

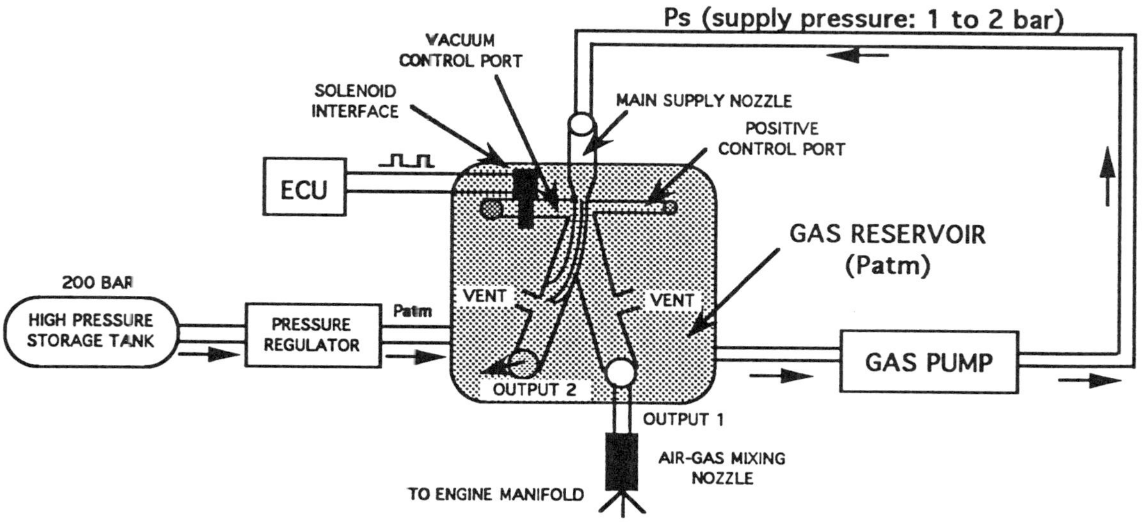

Figure 2 A schematic arrangement for the fluidic injector system (vacuum switching method)

developed by the author **[5,6,7]** and it is believed that it is equally applicable to the gas injector arrangement.

An alternative solution is also under consideration by fitting a small one-way valve between the fluidic stage and the engine manifold. The valve will be fully open when the fuel flow is present in the fluidic output allowing the fuel to be delivered into the manifold; and the valve will be closed by a light spring when the fuel is in the other output directing the fuel back to the main reservoir. By doing so, the effects of the manifold vacuum conditions on the fluidic stage can be removed.

A positive switching control method can also be used. This can be achieved by placing the solenoid interface at the positive control port allowing the main jet flow to be attached on the output 1 (stable side). The output 2 will be connected to the mixing nozzle and engine manifold. For this arrangement a minimum control pressure for fluidic switching is required depending on the value of supply pressure.

BENEFITS OF FLUIDIC INJECTORS - The fluidic devices are very reliable due to having no moving parts and they can be produced at low manufacturing costs due to their mechanical simplicity. The fluidic injector offers capability for handling large amount of gas flow and fast and accurate fuel metering for stoichiometric AFR control. It overcomes the weakness of both gas carburettors and solenoid injectors and it potentially offers considerable benefits of improved air-fuel mixture preparation and reduced exhaust emissions for natural gas engines.

The improved switching performance and the reduced switching-off delay in particular will result in wider dynamic flow range and improved turn-down ratio for the gas injectors. This will ensure a highly accurate stoichiometric AFR control for a wide range of engine operating conditions and will reduce exhaust emissions from gas engines.

Due to amplification characteristics of the fluidic devices, the demand for the solenoid interface has been greatly reduced and a low power, low flow rate and fast response miniature solenoid or equivalent device can be used. This will reduce the size and power requirements of the drive stages in the ECU.

It is believed that the fluidic gas injector provides an unique opportunity for using a multipoint gas fuel injection system for natural gas engines and will bring gas engine fuel control technology to the same level as that of the well established SI engine Multi-Point Injection (MPI) systems. It's simplicity and low manufacturing cost can also allow such systems to be used not only in Natural Gas Vehicles (NGV's) applications but also in stationary gas engines.

EXPERIMENTAL WORK

The initial experimental work is focused on two prototype fluidic devices developed in Loughborough University and manufactured by Preci-spark Ltd. Their steady-state flow delivery and dynamic switching characteristics together with cycle to cycle variability were evaluated on a laboratory test rig. Several commercial solenoid injectors were also tested and the results from the fluidic system were compared with those from commercial injectors. A separate test rig including a reservoir enclosure together with a three-way regulator and a recirculating pump was also set up and it is used to investigate the pressure stability in the enclosure under spill flow and vent flow conditions and to prove that the operation of the fluidic injector will be unaffected by the added reservoir enclosure.

STEADY-STATE TESTS - The potential of flow handling capability by the fluidic device was estimated by the steady-state flow tests operating on air. **Figure 3 & Figure 4** show the output pressure-flow characteristics of two prototype fluidic injectors with main input orifice size of 1 mm (width) by 2 mm (depth) and 1.5 mm by 3 mm respectively. These results were obtained by varying input supply pressure and output restricting nozzle size. It can be seen that a larger output flow (i.e. steady-state flow in Standard Litre per Minute or SLM) can be obtained for a less restricted output (i.e. larger nozzle size) and for the same supply pressure conditions.

It is known that a typical maximum gas flow requirement for a two litre, four cylinder gas engine is about 400 litre/min., or 100 litre/min. per cylinder (i.e. per injector) at atmospheric pressure (SLM). It can be seen from **Figure 3** that for the smaller device (1 mm by 2 mm) an output of 55 SLM can be obtained at the supply pressure of 140 kPa or 20 pounds per square inches (psi) and at the output nozzle size of 3 mm in diameter. For the larger device (1.5 mm by 3 mm), however, an output of 100 SLM can be obtained at the supply pressure of 140 kPa and the output nozzle size of 3 mm in diameter, as shown in **Figure 4**. Therefore, a single device of the large version would be suitable for a 2 litre engine flow requirement and two stacked devices of the smaller version would be required for the same sized engine application.

For larger gas engine applications, several more basic fluidic devices can be stacked together with appropriate channels connected internally providing required larger steady-state flow rates. The advantages of stacked arrangement are the flexibility of adapting the fluidic unit for various engine sizes and avoidance of possible slow response time associated with excessively large fluidic devices.

Figure 5 shows the calculated dynamic flow delivery by the fluidic injector (the larger version) based on the steady state flow characteristics obtained in **Figure 4**. It can be seen that the higher the supply pressure, the higher the flow delivery per injection. The proposed supply pressure for the fluidic unit is between 140 kPa (20 psi) and 175 kPa (25 psi) as adequate steady-state flow can be obtained from these conditions.

The dynamic flow delivery by the fluidic unit was further compared with those from commercially available Servojet solenoid injectors (data provided by the manufacturer) and the result is shown in **Figure 6**. Three types of Servojet units are quoted here, namely SP014, SP021 and SP051 and the pressure for these injectors is 7 bar (700 kPa or 100 psi). It can be seen that the dynamic flow delivery by the larger version of the fluidic device with input nozzle size of 1 mm by 3 mm is slightly larger than that from the SP051, which is the smallest Servojet unit here. In order to achieve the flow delivery by the SP014, which is approximately twice as much as the one by the SP051 and the larger fluidic unit, two fluidic devices can be stacked together as described above to give the required flow delivery for larger engine applications.

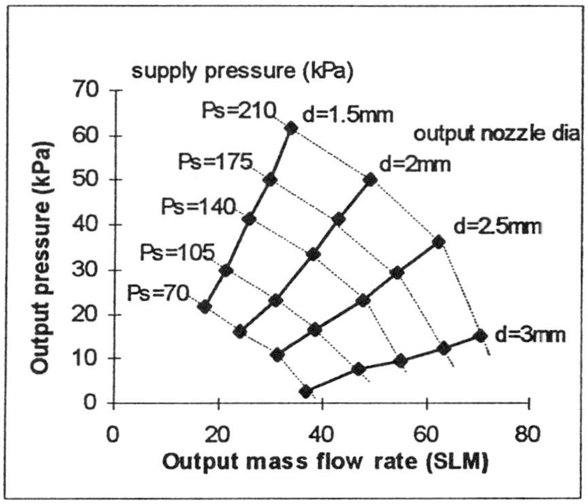

Figure 3 Steady-state output pressure-flow characteristics for the 1st prototype fluidic device

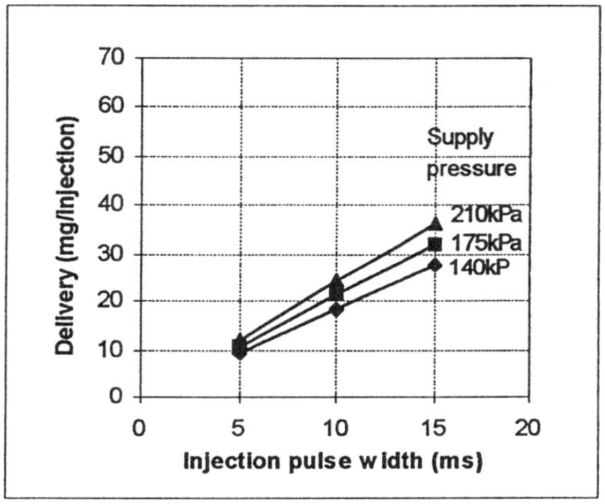

Figure 5 Dynamic flow delivery by the second fluidic injector

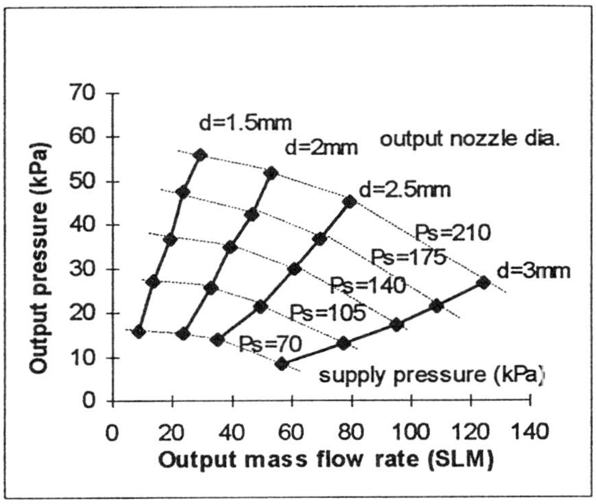

Figure 4 Steady-state output pressure-flow characteristics for the 2nd prototype fluidic device

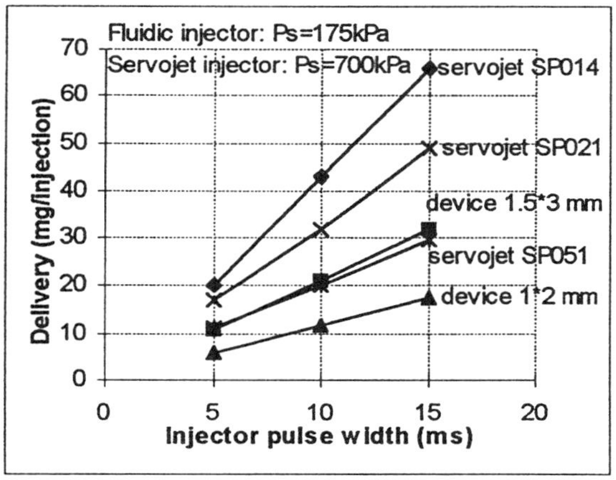

Figure 6 Comparison of the dynamic flow delivery between the fluidic and Servojet injectors

DYNAMIC TEST RIG AND INSTRUMENTATION - In order to evaluate dynamic switching characteristics of the fluidic injector units and to compare them with those from some commercial gas injectors, a laboratory test rig with a compressed air supply and control board together with appropriate instrumentation was set up. **Figure 7** shows a schematic arrangement for the dynamic tests of the fluidic injector. Two fast response Entran pressure transducers (type EPI-M4) were used. One was used to detect the response of the solenoid interface (i.e. the control flow signal) to the drive pulses; the other one is used to detect the response of the fluidic stage to the control flow.

Figure 7 shows the arrangement for the vacuum switching control method with the solenoid interface being placed at the vacuum port of the fluidic device. For dynamic testing using positive switching method, the solenoid interface should be placed at the positive control port of the device. The solenoid interface chosen is a conventional central fuel injector (CFI) used in Ford 1.4 litre engines. The solenoid drive circuitry operates in a "peak and hold" mode(type CS-453) and it was specially designed for fuel injector use for rapid opening and closing.

A PC based data acquisition system was used to collect and analyse switching responses of the fluidic devices and their cycle-cycle switching variability. The software used in the PC is called "LabView" developed by National Instrument. An Amplicon data acquisition board (DAQ board) was chosen instead of the DAQ board from National Instrument after considering its low costs. An advanced linking driver had to be developed by the authors in order to operate the Amplicon DAQ board in LabView language satisfactorily.

DYNAMIC TESTS ON THE FIRST PROTOTYPE DEVICE - The first prototype fluidic device was produced based on the scaled-up version of the existing device which was used for the liquid fuel injector in a previous work **[5,6,7]**. The input nozzle orifice size of this prototype device is 1 mm width by 2 mm depth. The objective of the dynamic tests on this device is to establish its dynamic capability and the solenoid interface response and to compare the results from

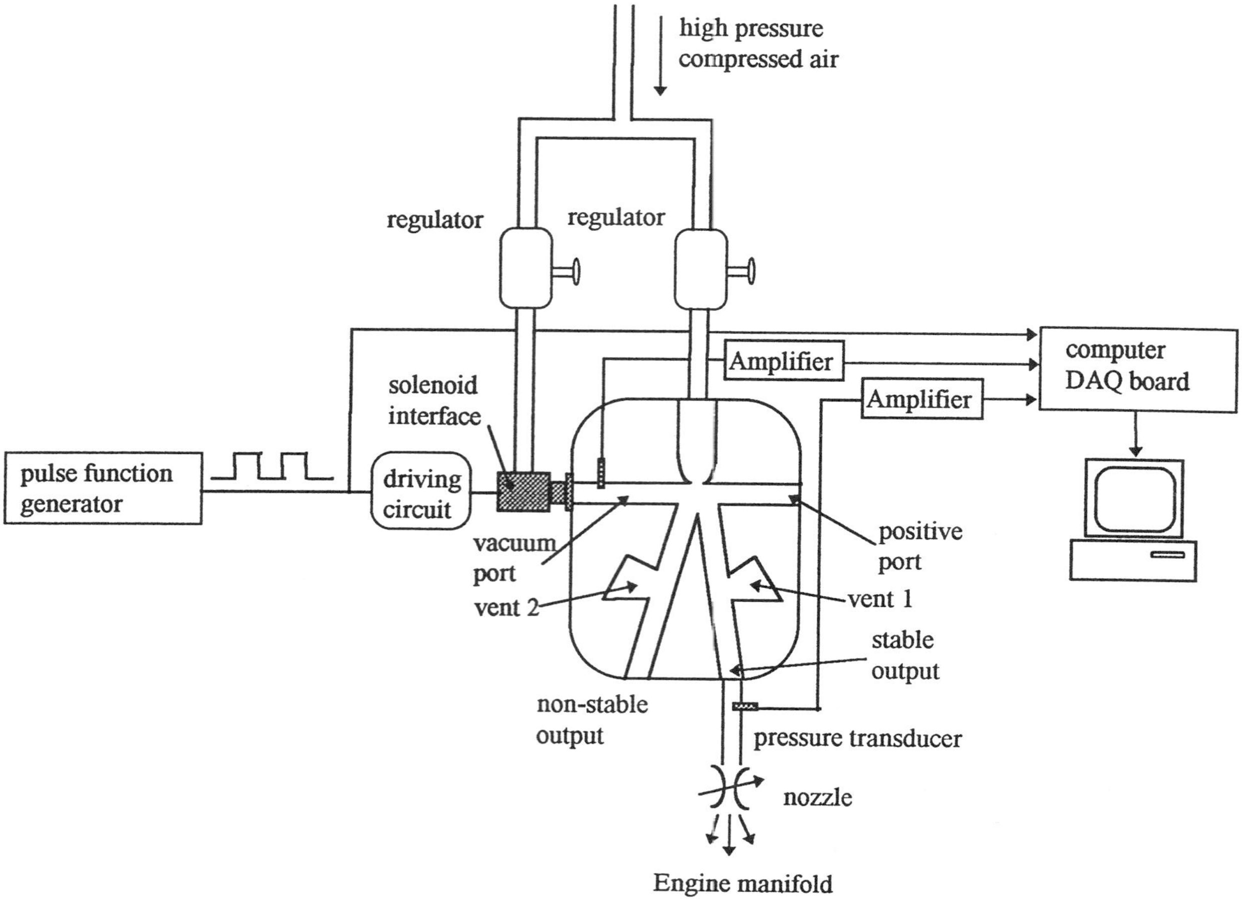

Figure 7 A schematic arrangement of dynamic tests of the fluidic injector

two switching control methods, namely positive and vacuum switching.

Figure 8 and **Figure 9** show the results of the response times of the fluidic stage to the control flow signals from the solenoid interface when positive switching method is used. It can be seen that as control flow pressure increases, the switching-on delay reduces for all supply pressure conditions. This is due to the increased control flow momentum by a higher control pressure resulting in a more rapid jet switching action within the fluidic stage. However, under the same control flow pressure condition, a higher supply pressure will result in a slower switching. This is because a high supply pressure flow possesses a higher flow inertia.

The switching-off delay of the fluidic stage does not seem to be affected considerably by the supply pressure or control pressure. This is because the switching of the jet flow from the unstable output to the stable one, when the drive pulse (i.e. control flow) is removed from the solenoid, largely depends on the biased geometry of the monostable design and therefore the delay time is irrelevant to the external parameters **[8]**.

Figure 10 & **Figure 11** show the response of the solenoid interface (i.e. control flow signals) to the drive pulses. It can be seen that both switching-on and switching-off delays of the solenoid are more or less independent of the control pressure or supply pressure used and they are in the region of 1.1 ms and 0.9 ms respectively. It was found later that the response of the solenoid is not affected by the switching method used, and both switching-on and -off delays of the solenoid remains unchanged when a

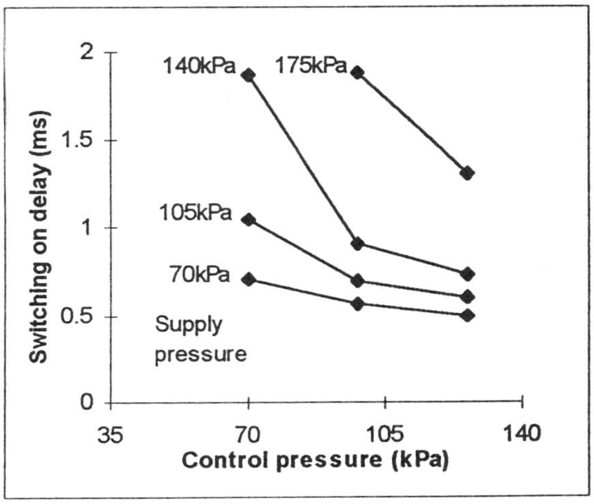

Figure 8 Switching-on delays of the fluidic stage to the control flow signals

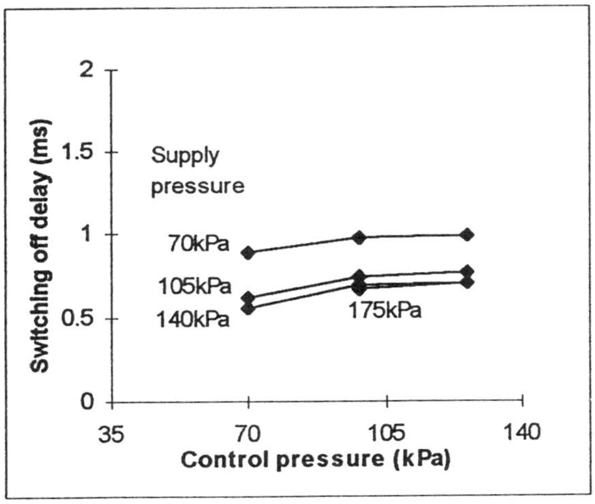

Figure 9 Switching-off delays of the fluidic stage to the control flow signals

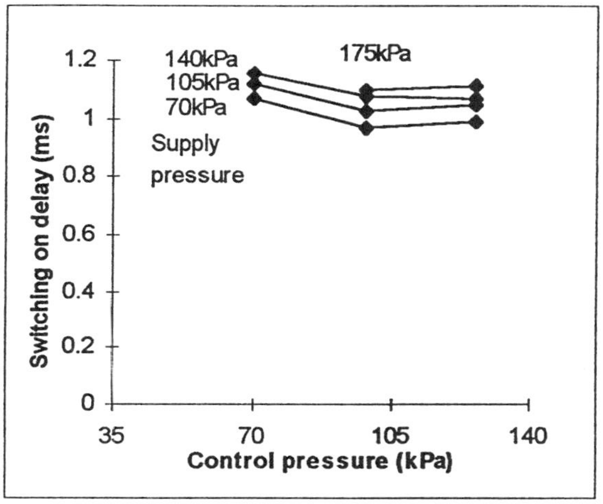

Figure 10 Response of the solenoid interface (i.e. control flow signals) to the drive pulses

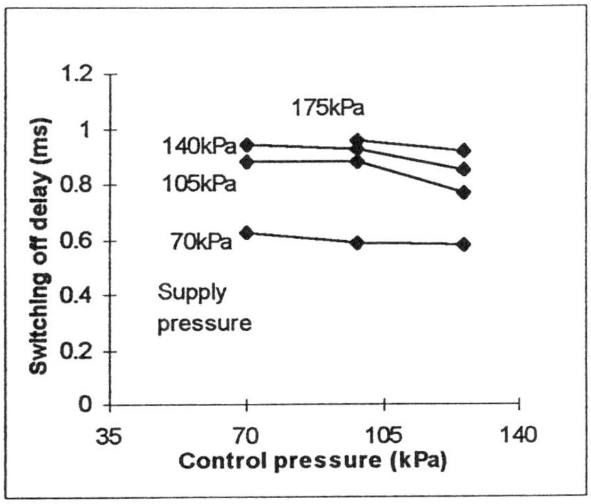

Figure 11 Response of the solenoid interface (i.e. control flow signals) to the drive pulses

vacuum switching method is employed. **Figure 12 & Figure 13** show the total switching-on and -off delays by the fluidic injector unit with solenoid and fluidic delays added together. It is the total delays which will be used to compare the performance of the fluidic injector against commercial injectors.

In order to compare the fluidic response characteristics with those from vacuum switching control, the solenoid interface was placed at the vacuum port as described in **Figure 7.** The total switching-on and -off delays were obtained and the results are shown in **Figure 14 & Figure 15**. It can be seen from **Figure 14** that the total switching-on delay was reduced by up to 2 ms in the vacuum switching arrangement. Since the solenoid delay remains unchanged (1.1 ms for -on delay, as mentioned above) for the two switching arrangements, any reduction in total switching-on delay must be due to the improvement in the response of the fluidic stage.

The improvement in the fluidic response by the vacuum switching method can be explained as follows. When the drive pulse is applied on the solenoid, the valve opens allowing the fuel flowing through the vacuum port into the fluidic device. The main jet flow will switch to the stable side spontaneously due to the biased monostable nature. Additional control flow pressure will provide extra momentum resulting in a even faster switching action. In the positive switching arrangement, however, the main flow jet is initially attached to the stable output and in order to cause the jet to switch to the unstable side, a higher positive control pressure is required. The overall effect is that the switching-on response of the fluidic stage by the positive switching method is considerably slower than that by the vacuum

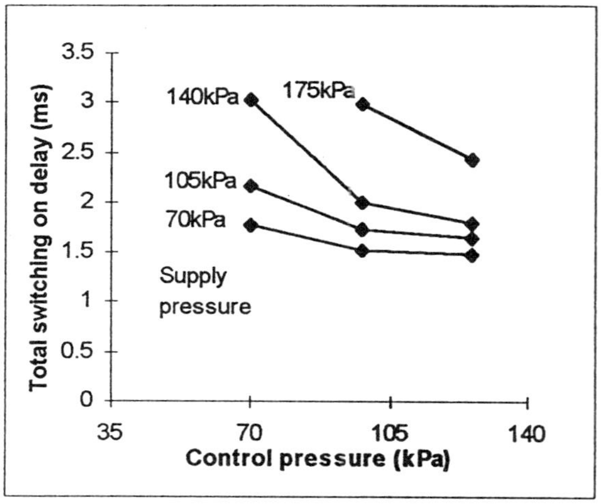

Figure 12 Total switching-on delays by the fluidic injector

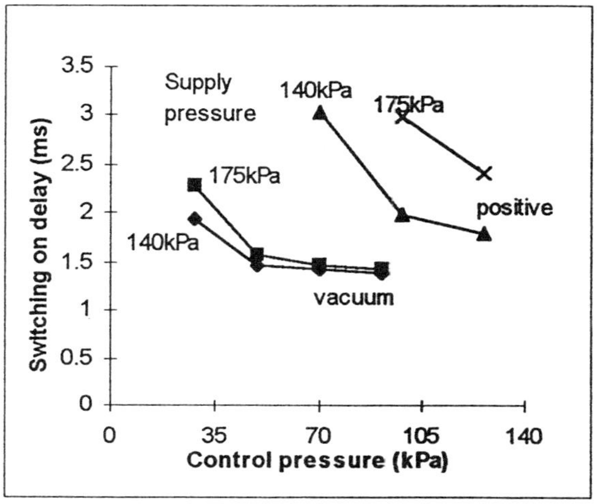

Figure 14 Comparison of switching-on time by the positive and vacuum switching methods

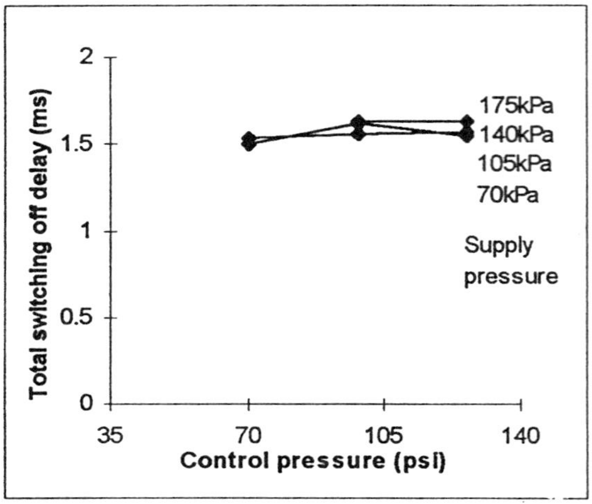

Figure 13 Total switching-off delays by the fluidic injector

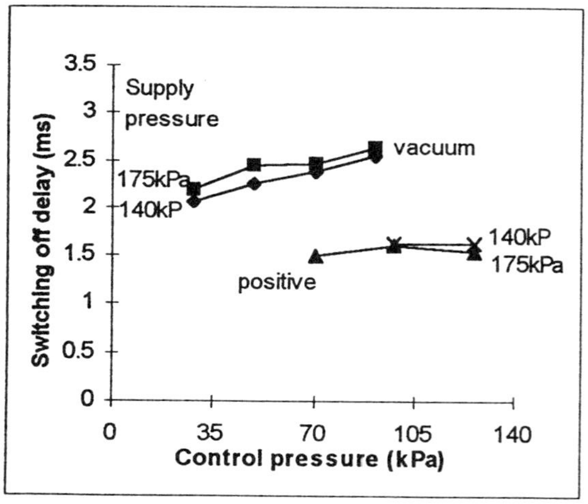

Figure 15 Comparison of switching-off time by the positive and vacuum switching methods

switching method.

On the other hand, the switching-off response of the fluidic stage by the vacuum switching method is more sluggish than that by the positive method and the difference is in the region of 1 ms. as shown in **Figure 15**. It is considered that the main reason for this result is due to the fact that a relative long time is required to to establish a sufficient pressure differential across the main flow jet causing it to switch from the stable side to the unstable one. A solution to this problem will be discussed in a later section.

It can be concluded from the above discussions that for the fluidic stage, switching from the stable side to the unstable side is always more difficult regardless which switching method is used. There is no clear favourable choice between the two switching method for the basic fluidic device arrangement. However, when a small modification is made on the fluidic port and vent arrangement, the vacuum switching will show a clear advantage. This will be discussed later.

The repeatability tests for the cycle-cycle switching variability can further prove the above conclusion. **Figure 16 & Figure 17** show the standard deviation (SD) for both -on and -off delays by the two methods. It can be seen that the SD for the switching-on delay by the positive method (i.e. from stable output to the unstable one) is consistently worse than that by the vacuum method (i.e. from unstable output to the stable one). Similarly, the SD for the switching-off delay by the vacuum method is consistently worse than that by the positive method. To conclude, the switching from unstable output to the stable one always gives a better switching repeatability and consistency.

DYNAMIC TESTS ON THE SECOND PROTOTYPE DEVICE - Subsequent to the first prototype, a second prototype device with an input orifice size of 1.5 mm (width) by 3 mm (depth) was produced. This is to increase the flow handling capability by a single device. Although for large engine applications, several basic devices can be stacked together to provide required flow, it is more convenient to have a few basic devices so that different combinations can be made for different engine requirements. It was found that the 1.5 mm by 3 mm device is capable of providing sufficient gas flow for a 1.6 litre to a 2 litre engine, when a supply pressure of 1.5 bar (22 psi) and a 3 mm output nozzle are used. The steady-state flow results were discussed already in an early section.

Initial evaluations on this large device

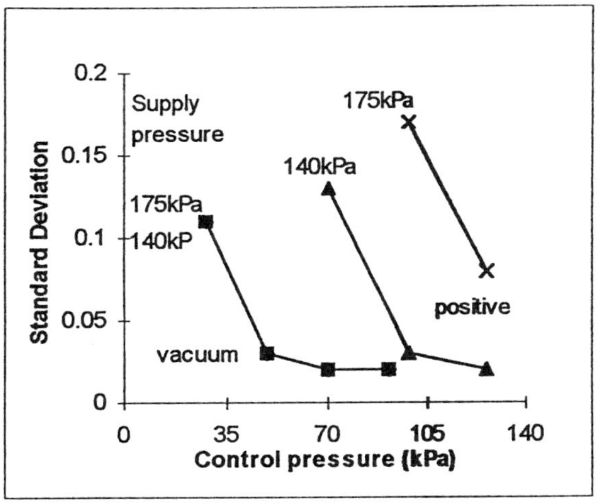

Figure 16 Standard deviation (SD) of the switching-on response by the two methods

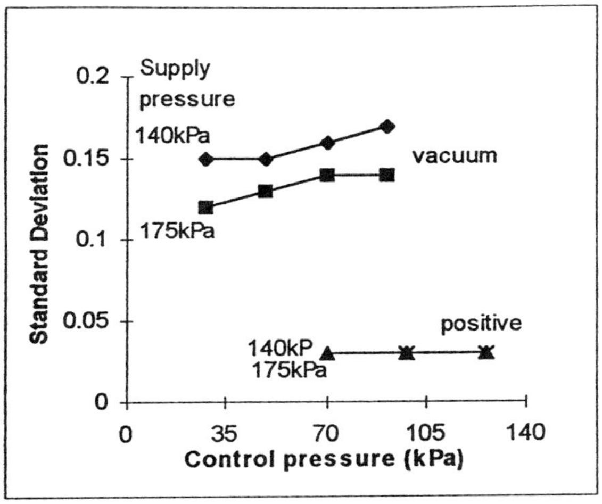

Figure 17 Standard deviation (SD) of the switching-off response by the two methods

showed that it was unable to operate under a positive switching control due to insufficient control flow provided by the same solenoid interface. Therefore only vacuum switching method was used for the second device. Similar response characteristics to those from the first device were obtained and the switching-off performance (i.e. from the stable output to the unstable one) is again unsatisfactory.

In order to reduce the switching-off delay and to improve the SD of the fluidic stage, a small modification was made to the basic device. **Figure 18** shows the modified version with positive control port connected with the vent on the same side (stable side). Such a connection can be easily made internally for production units and no interference with the device operation is envisaged. **Figure 19 & Figure 20** show the comparison of response times from the basic device and the modified version when a vacuum switching control is used.

It can be seen that there is a slight increase (4 to 5%) in switching-on delay when the modified version is used. However, in the gas reservoir enclosure arrangement where a recirculating pump is used, there is actually an improvement in this switching-on time. This will be discussed later. For the switching-off delay, the modified device showed a considerable improvement in the response time and it is reduced by about 20% as shown in **Figure 20.** Also the cycle-cycle repeatability (i.e. SD) of the switching-off response is improved by up to 38% in the modified device as shown in **Figure 21**.

The reason for the improvement in the switching-off performance of the modified version can be explained as follows. During the fuel delivery period, the main flow is attached to the stable output. Due to a slight restriction at the output by the nozzle orifice, some fuel may flow out of the device through the vent on the stable side. When this vent is connected to the positive control port, a pressure which is slightly higher than an atmospheric one may be established at that port. However, this pressure is not sufficiently high to cause the main flow to switch back to the unstable side. This is the case that is required, otherwise the flow switching would have become uncontrollable by the solenoid interface. When the drive pulse is removed, the control flow at the vacuum port is cut off causing the pressure at that port to collapse. The presence of a slight high pressure at the positive port will accelerate the process of establishing a sufficient high pressure differential cross the main jet, which determines the switching-off speed of the fluidic device.

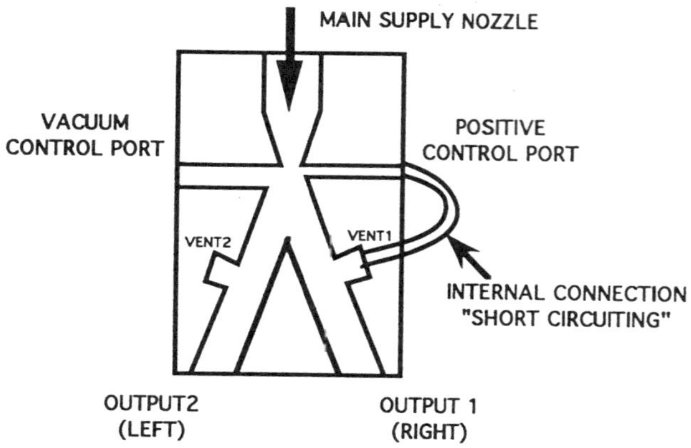

Figure 18 A modified version of the fluidic device with connected control port and the vent

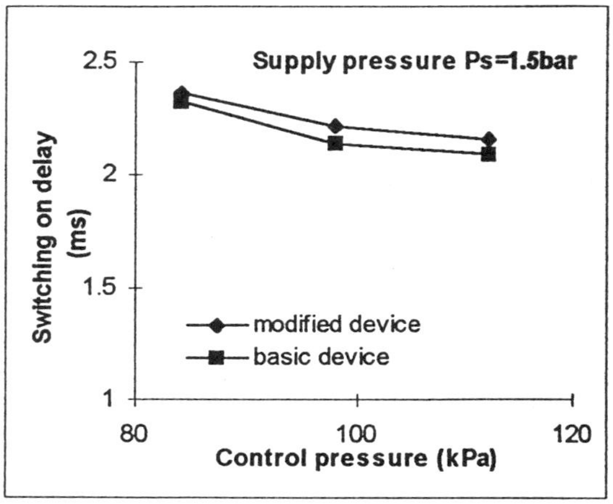

Figure 19 Response times (-on) from the basic device and the modified one (vacuum switching)

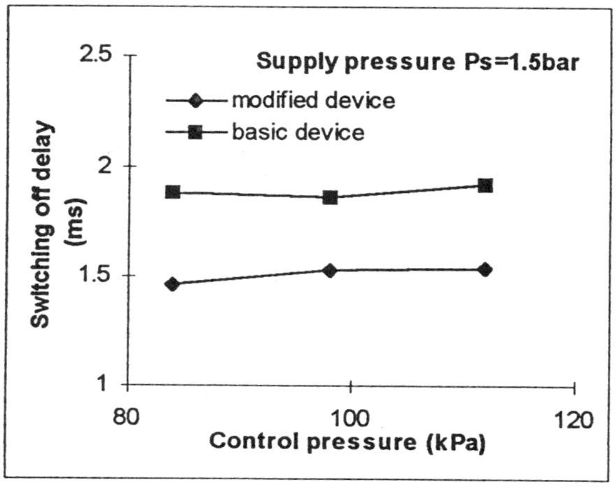

Figure 20 Response times (-off) from the basic device and the modified one (vacuum switching)

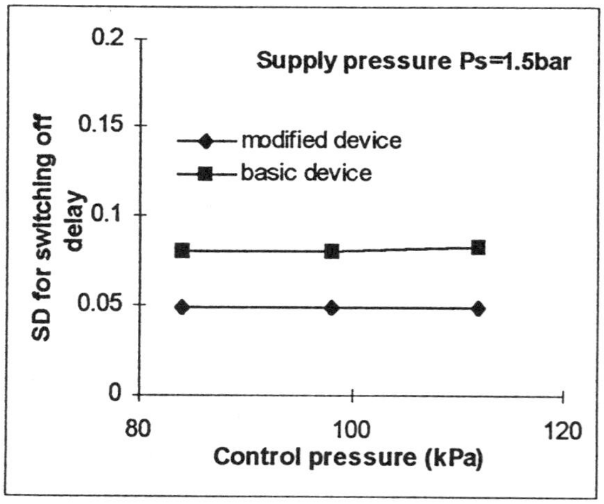

Figure 21 Standard deviations of the switching times from the basic device and the modified one

If the response time of the solenoid (on delay is 1.1 ms and off delay is 0.9 ms) is deducted from the above total response time (2.2 ms for on delay and 1.5 ms for off delay for a control pressure of 1 bar or 100 kPa), the switching-on and -off delays by the fluidic stage itself are 1.1 ms and 0.6 ms respectively. It will be seen later that both these delay times are further improved in the gas reservoir arrangement.

RESERVOIR ENCLOSURE AND DYNAMIC TESTS - The above tests on the fluidic device were carried out on an "open" test rig, in which supply flow is provided by a compressed air mains and the vents and outputs of the device were both open to atmosphere. In this section the proposed reservoir enclosure system (shown in **Figure 2**) was set up and the second fluidic device was tested in the reservoir environment.

Figure 22 shows the schematic set-up of the proposed system. A conventional three stage pressure regulator was used to bring a high pressure down to required levels. The output of the second stage pressure reduction was connected to the solenoid interface providing a control pressure of 1 bar for the fluidic switching. The output of the third stage, which is at an atmospheric pressure, is connected to a gas reservoir. The spill flow from the unstable output and the vent are also connected to this reservoir.

Both basic device and the modified version (with an internal connection) were tested. In **Figure 22,** a modified version of the device is shown. In the case of the basic device, the positive port and its vent are also connected to the gas reservoir. A recirculating pump is used to provide all the required supply flow to the device input nozzle. **Figure 22** shows a practical implementation for the schematic arrangement shown in **Figure 2.**

It should be noted that a single operating point was identified with a supply pressure of 1.5 bar provided by the pump. An additional pressure transducer was fitted at the reservoir to detect its pressure variations. The operating characteristics of the basic device with the reservoir was compared with those in an "open" rig test, and evaluations on the modified device was also carried out on this reservoir test rig.

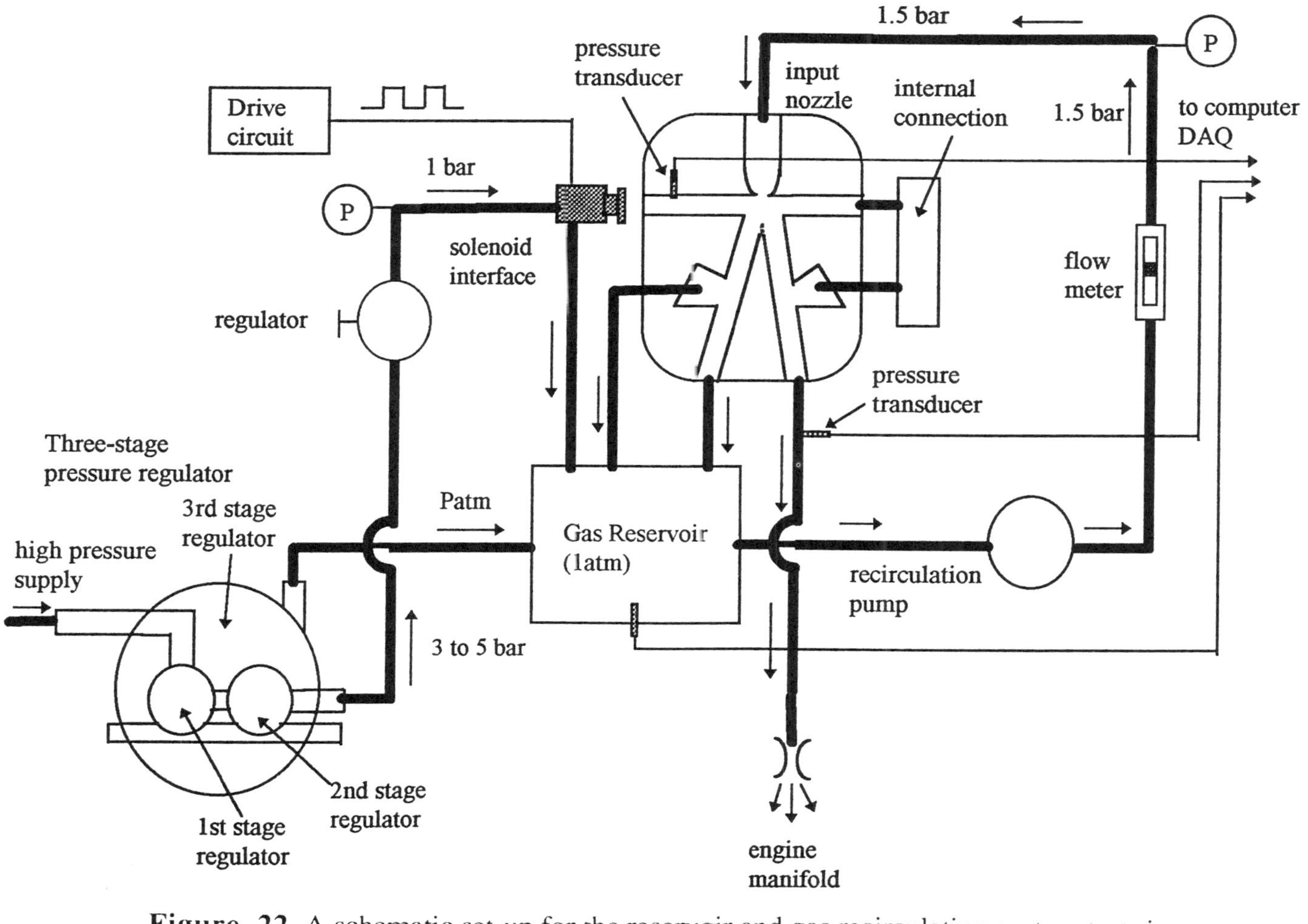

Figure 22 A schematic set-up for the reservoir and gas recirculating system test rig

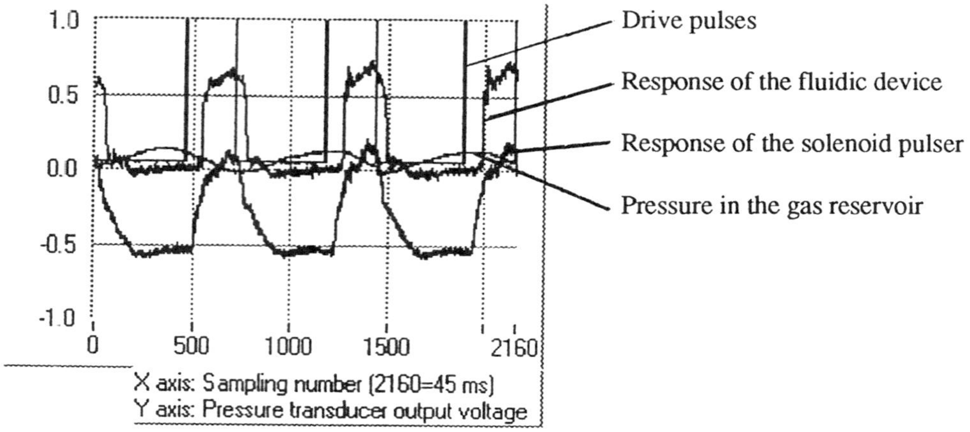

(Supply pressure = 20 psi or 140 kPa, Control flow pressure = 14 psi or 98 kPa,
Average reservoir pressure = 0.85 psi or 6 kPa)

Figure 23 A plot of three pressure traces: control flow, fluidic output and the reservoir pressure

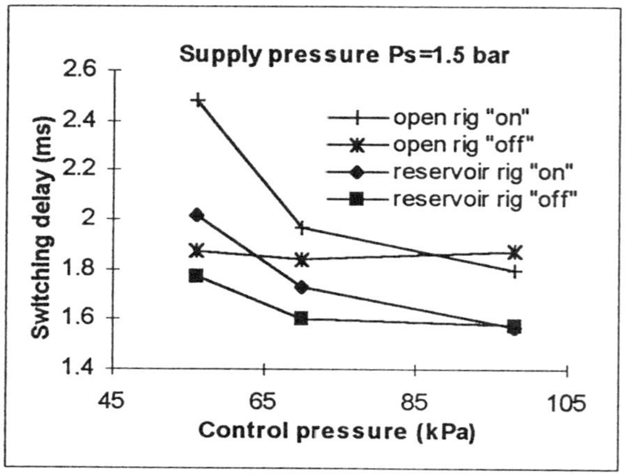

Figure 24 Switching responses of the basic device on both "open" and the reservoir test rig

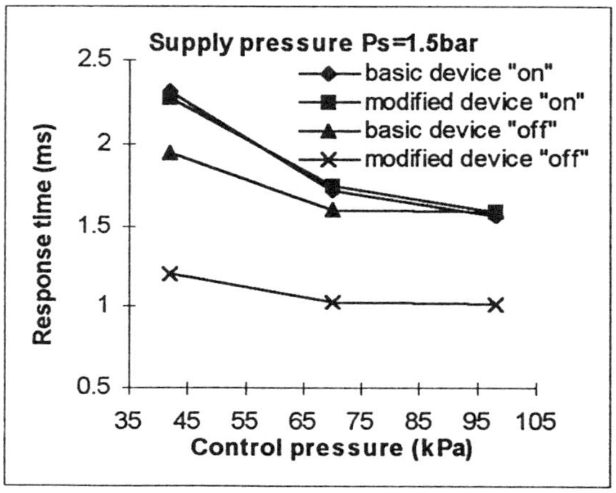

Figure 25 Switching responses of the basic device and the modified version on reservoir rig

Figure 23 shows a results from three pressure transducers. The supply pressure and control pressure used for the fluidic device were 140 kPa (1.5 bar) and 98 kPa (about 1 bar) respectively. The reservoir did show some pressure fluctuations as expected. However, the average value under this condition is only 6 kPa and no significant effects on the switching characteristics on the fluidic stage were observed. The frequency of the pressure fluctuation in the reservoir depends on the pulse width and the Pulse Repeat Period (PRP) applied on the solenoid.

Figure 24 shows the comparison of the switching response of the basic device on both "open" and reservoir rig test. It can be seen that some small improvements were obtained for both switching-on and -off delays in the reservoir environment. This may be attributed to the presence of a slight positive pressure in the reservoir, which has benefited both switching responses.

The use of internal connection to link the positive port and its vent together would further reduce the switching-off delay, as demonstrated earlier in the "open" rig test. **Figure 25** shows comparison of the switching response of the basic device and the modified device on the reservoir test rig. It can be seen that there is a significant decrease (up to 42%) in switching-off response time compared to the basic device arrangement.

For a control pressure of 100 kPa (1 bar), the total switching-on and -off delays are 1.6 ms and 1 ms respectively. If the response of the solenoid (on delay of 1.1 ms and off delay of 0.9 ms) is deducted from the total delay, the response of the fluidic device itself are 0.5 ms and 0.1 ms for switching-on and -off delay respectively. In other words, the total on and off response of the fluidic stage is merely 0.6 ms. It is believed that

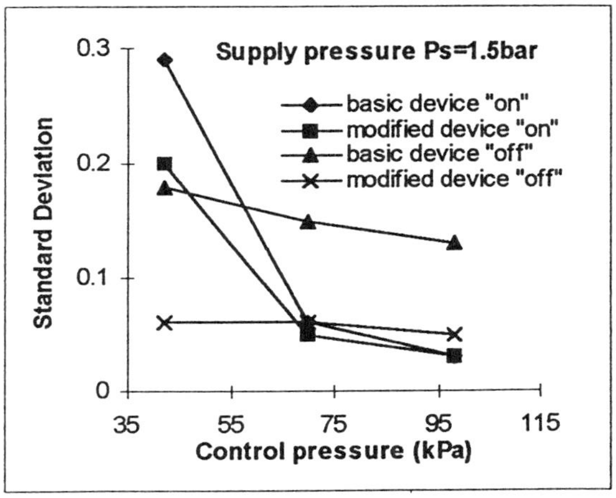

Figure 26 Standard deviations of the basic device and the modified one on a reservoir test rig

there is a considerable potential of using these devices for fast response fuel injectors, and that when a faster, dedicated fluidic interface is developed, the performance of the fluidic injector unit will be comparable to that in current gasoline injectors.

Figure 26 shows SD's of the basic device and the modified device on the reservoir test rig. It can be seen that the switching-off repeatability has improved by 53% in the modified device and a very good cycle-cycle consistency with a SD of 0.06 was achieved for a control pressure of 100 kPa. **Figure 27** to **Figure 29** show the pressure trace and response results from 500 consecutive cycles in the modified device. **Figure 30** to **Figure 32** show the pressure trace and response results in the basic device for comparison. A supply pressure of 1.5 bar and a

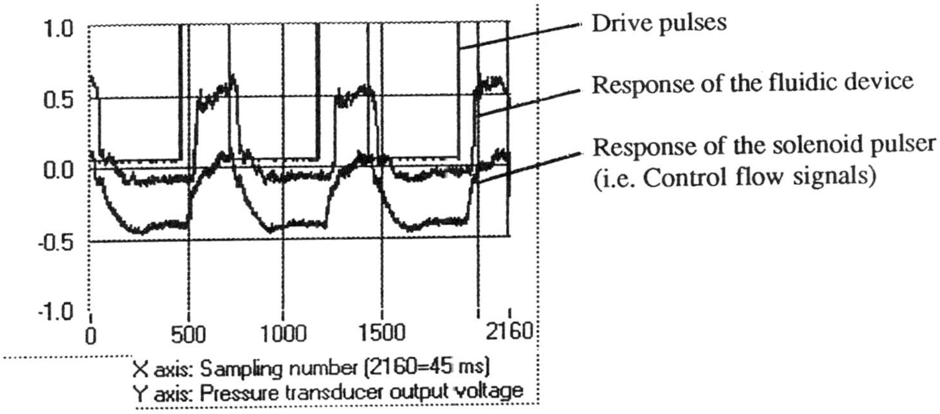

Figure 27 A plot of the pressure traces in the modified device

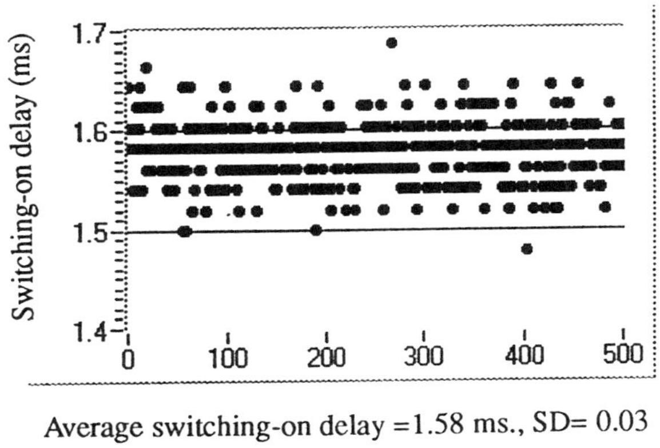

Figure 28 Total switching-on time from 500 cycles by the modified device

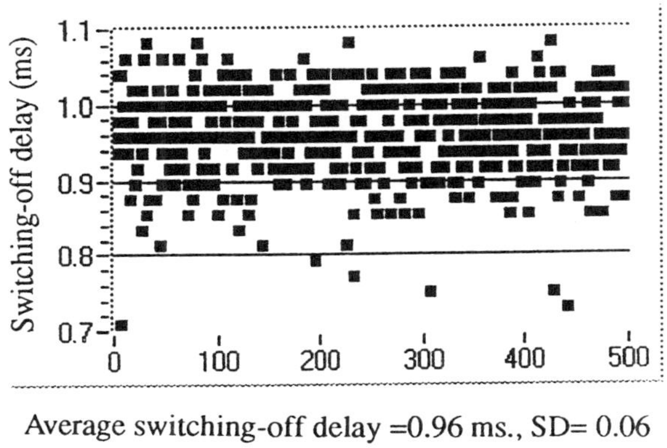

Figure 29 Total switching-off time from 500 cycles by the modified device

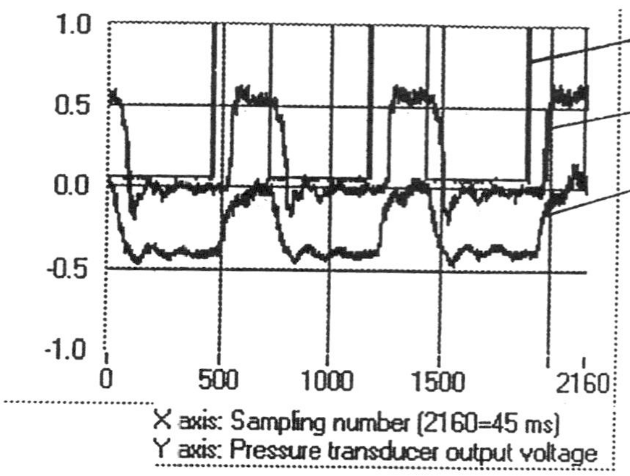

Figure 30 A plot of the pressure traces in the basic device

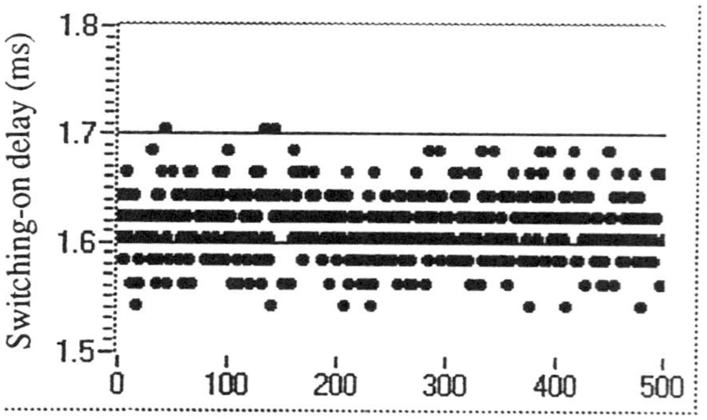

Average switching-on delay =1.62 ms., SD= 0.03

Figure 31 Total switching-on response time from 500 cycles by the basic device

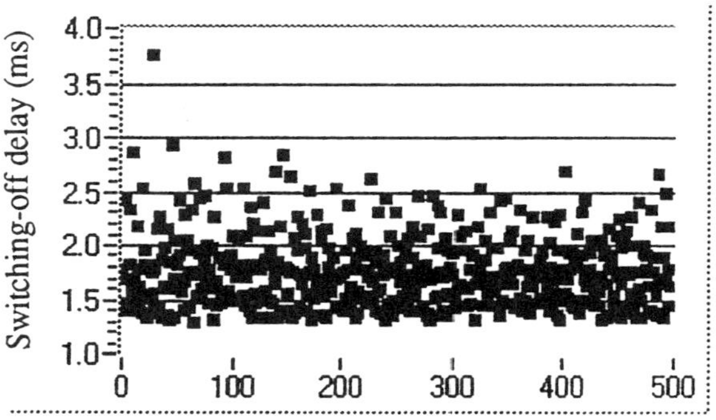

Average switching-off delay =1.76 ms., SD= 0.34

Figure 32 Total switching-off response time from 500 cycles by the basic device

control pressure of 1 bar were used to produce these results.

DYNAMIC TESTS ON THE COMMERCIAL GAS INJECTORS - Three Servojet solenoid gas injectors (type SP-014, SP-021 and SP-051) and one Bosch gas injector were obtained. No dynamic characteristics were provided by the manufacturers. It was therefore decided to carry out similar dynamic tests on these injectors in the laboratory test rig so that their switching performance can be compared with those from the fluidic injector unit.

Figure 33 shows the switching-on response of these injectors under varying pressures. It is known that these injectors normally operate between 7 to 10 bars on gas engines. However, the maximum pressure used in the tests was 3.5 bar (51 psi) due to limited peak pressure available in the laboratory. It can be seen from the result in **Figure 33** that the switching-on response time increases as the supply pressure increases for all four injectors. As mentioned earlier, the best switching-on time achieved by the fluidic unit (fluidic plus solenoid interface) is in the region of 1.6 ms, it is therefore reasonable to predict that the fluidic unit has a faster switching speed than all these four injectors, when they operate at their normal pressure range.

Figure 34 shows the switching-off response of these injectors under varying pressures. The Bosch unit again showed the best response time (1.4 ms). However, the best switching-off time achieved by the fluidic unit (fluidic plus solenoid interface) is in the region of 1 ms (for a supply pressure of 1.5 bar and a control pressure of 1 bar shown in **Figure 25**). It should be noted that the Bosch gas injector used here is a relatively small unit designed for gas engines possibly below a 2 litre capacity.

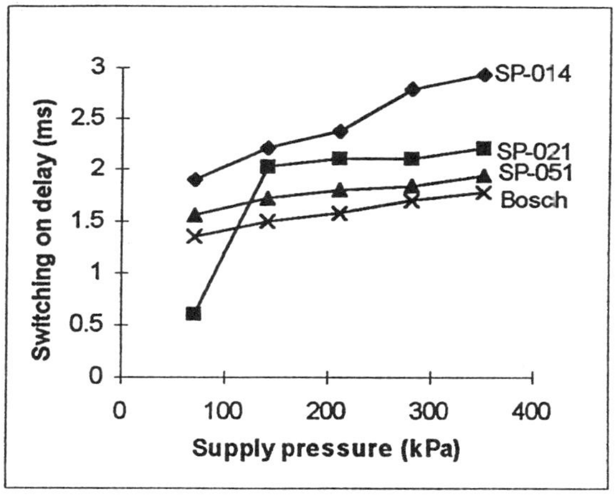

Figure 33 Switching-on response of the commercial gas injectors

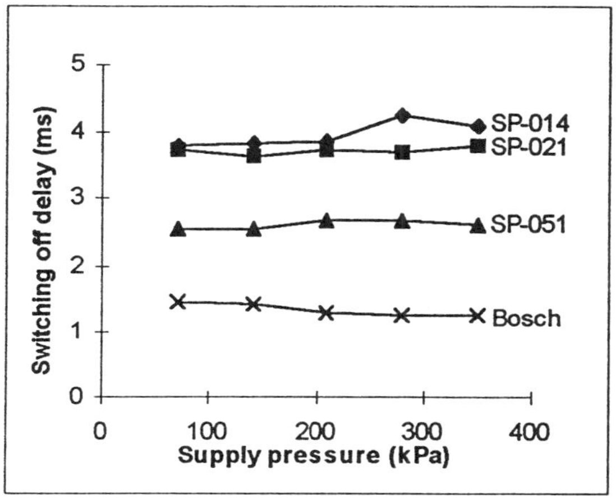

Figure 34 Switching-off response of the commercial gas injectors

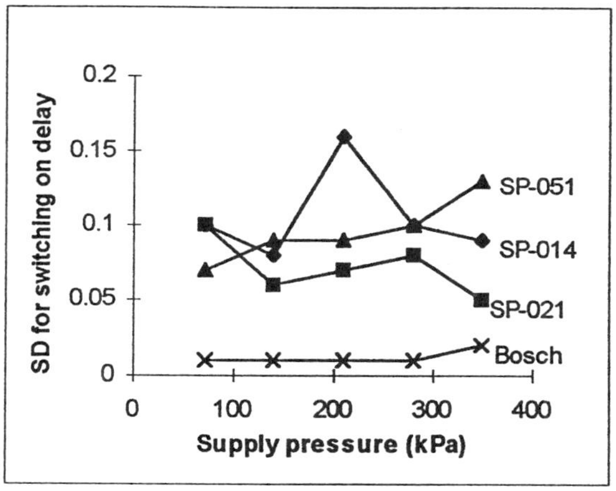

Figure 35 Standard deviations of the switching-on times from the commercial gas injectors

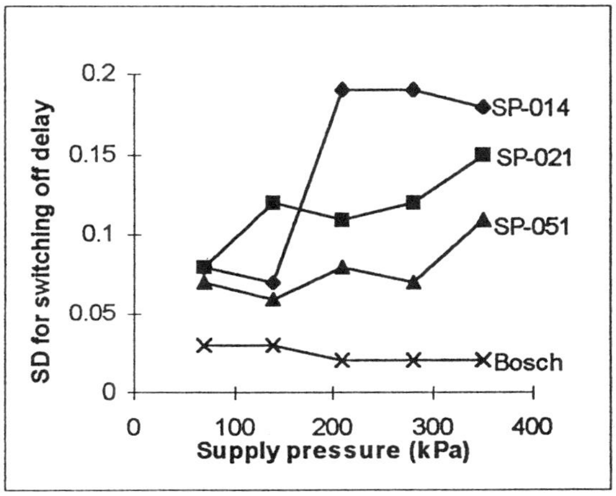

Figure 36 Standard deviations of the switching-off times from the commercial gas injectors

Therefore it has advantages of light armature and small passage area, which results in its faster response than Servojet injectors.

Figure 35 & **Figure 36** show the standard deviation of the response time of these four injectors. It can be seen that the Bosch unit showed extremely consistent cycle-cycle variations with a SD of 0.02 for both switching-on and -off delays. This is largely due to its compact and light design for small flow capacity requirements. In comparison, the fluidic unit is capable of switching with SD of 0.05 and 0.06 for on and off response respectively.

FUTURE WORK

Future work will concentrate on evaluating the fluidic injector under a simulated engine vacuum condition in a laboratory test bench and carrying out comparative engine tests on an engine test bed. The purpose of the laboratory investigation is to ensure that the injector system operates satisfactorily before it is implemented on a research engine. The second purpose is to establish a suitable air-gas mixing nozzle or some sort of one-way diaphragm (or valve) to eliminate the effect of vacuum conditions on the fluidic stage.

When the laboratory test is complete, the fluidic injector will be implemented on a research engine so that its engine operating characteristics of the fluidic injector system can be compared with those from other gas injection systems and basic gas carburettor systems.

Efforts will also be made on developing a dedicated fluidic interface for the fluidic gas injector, as it was identified that the existing interface used in the fluidic unit has limited dynamic response. A novel piezo interface is under investigation for the fluidic gas injector.

CONCLUSIONS

[1] A novel gas fuel injector system based on the use of monostable fluidic devices is proposed in this paper. The system consists of non-moving-part fluidic devices, a solenoid interface and air-gas mixing nozzles. The operating principle of the proposed system was described and verified in experiments.

[2] Two prototype fluidic devices were produced and tested in the laboratory. Suitable operating points of the fluidic stage with required steady-state flow characteristics for gas injectors were identified.

[3] Dynamic tests on the first prototype device showed that switching from the stable output to the unstable output is sluggish with poor response characteristics and switching repeatability, regardless any switching method is used.

[4] A modified version of the basic fluidic device (second prototype) was proposed and tested on both an "open" test rig and a gas reservoir enclosure test rig. It was found that the modified version showed a considerably improved switching-off response for the vacuum method used while maintaining a fast switching-on response.

[5] The fluidic injector with a modified fluidic device can achieve switching-on response (fluidic plus solenoid) of 1.6 ms with a Standard Deviation (SD) of 0.03, and switching-off response of 1 ms with a SD of 0.06. If the response of the solenoid is deducted from the above values, the response of the fluidic device itself are 0.5 ms and 0.1 ms for switching-on and -off respectively.

[6] Comparing with commercial gas injectors, the fluidic injector has a faster dynamic response (on and off response) than all the commercial units tested and it has a smaller cycle-cycle variations than the three Servojet injectors. However, the Bosch injector showed a slight better cycle-cycle switching repeatability than the fluidic unit.

[7] Future work will concentrate on engine tests of the fluidic injector and on developing a dedicated piezo interface for the fluidic device.

ACKNOWLEDGEMENT

The financial support from British Gas plc. for this project is gratefully acknowledged. The authors also wish to record their appreciation of technical assistance provided by Mr. Alan Brightwell, British Gas plc, Mr. John Elliott, Preci-Spark Ltd. U.K. and Technician Mr. D.M. Harkis, Loughborough University.

REFERENCES

1. J. Klimstra et al. "Catalytic Converters for Natural Gas Fuelled Engines - A Measurement and Control Problem". SAE paper 872165.
2. J. Klimstra et al. "Carburettors for Gaseous Fuel - On Air-fuel Ratio, Homogeneity and Flow Restriction". SAE paper 892141.
3. M. Perrault, G.Allard et al, "An Alternative to the Conventional Natural Gas Conversion Kit for Vehicales - Computer Controlled Upstream Injection" SAE paper 881659.
4. C. J. Green et al. "Electronically Actuated Injectors for Gaseous Fuels". SAE paper 892143.
5. Q. Huang et al. "Electro-fluidic Devices as Fuel Injectors for Spark-ignition Engine Fuel Injection Systems". SAE paper 910378.
6. Q. Huang, B.Jones. "Evaluation of Engine Operating Characteristics of a Fluidic Device Based Fuel Injection Systems". 1993 SAE Transaction, No. 930324.
7. Q. Huang, D. Sansum, "The Investigation into a PC-based Fluidic Fuel Injection System for Passenger Cars", SAE Paper 950070.
8. R.A. Comparin, et al. " Fluid Switching Elements and Amplifiers". IBM Research Report RZ-94, July 23 1962.

980813

Study of the Injection Control Valve in a New Electronic Diesel Fuel System

Minggao Yang, Qilong Lu, Jianqiu Li, Xiaodong Lu and Chuanjin Du
Tsinghua University

ABSTRACT

At first, the dynamic electromagnetic characteristics of a pulsed solenoid valve is analyzed by experiments. The fast valve response is obtained by material modifications. Then, the intelligent solenoid driving method is discussed. The new techniques of the "active" PWM and the "d^2i/dt^2" detection are developed for feedback control of the solenoid holding current and the valve closure timing. Finally, the control and diagnosis method for the valve closure duration is investigated. A sensing mechanism utilizing momentary camshaft speed fluctuations of fuel injection pump is presented, which provides the basis for feedback control and diagnosis of the valve closure duration and diesel fuel injection process.

INTRODUCTION

Aimed at the most commonly used pump-pipe-injector system, especially the in-line pump-pipe-injector system, a new type of electronic diesel injection system has been presented by one of the authors. It is called electronically controlled in-line Pump-Pipe-Valve-Injector system[1][2]. In this system, conventional in-line pump is simplified, a solenoid-controlled spill valve is employed to trim injection process under the management of an electronic control unit. In order to change the system concept into product, lots of research and development work have been done. The more practical system structure is developed which is called electronic in-line Pump-Valve-Pipe-Injector injection system or PVPI system. Its basic system structure is shown in Figure 1. The solenoid-controlled spill valve is put on the side of the pump. A new injection rate shaping method is developed, which takes advantage of the inlet port in pump to maintain a small variable spill passage after control valve is closed. In the process of the development, further study on the injection system reveals that the technology of injection control valve is the crucial technology to improve the system performance after the structure and function of the system are determined.

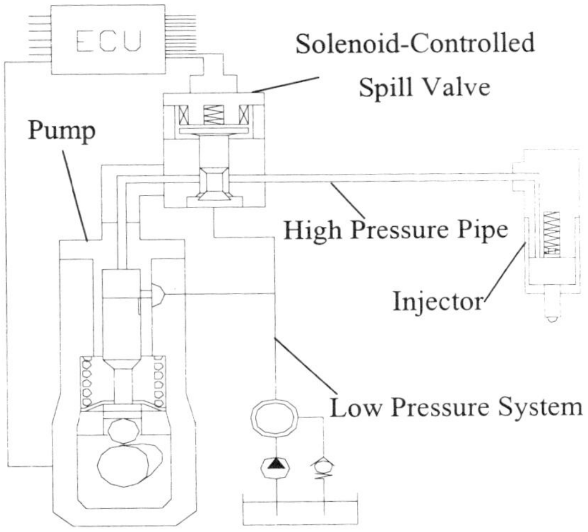

Figure 1 System structure

Figure 2 shows the block diagram of the injection control valve technology. It is an integration of solenoid valve, solenoid power driver and injection control algorithms. Solenoid valve is connected with the hydromechanical system and regulates the injection process directly. Its closure timing and duration determine the injection timing and fuel quantity. The power driver controls energy into the solenoid. Smart power driving can ensure fast response of solenoid valve and compensate for different variations. The injection control algorithm is used to process signals from different sensors, especially the angular position sensor of camshaft and generate injection command pulses to solenoid valve through power driver, which determines the control accuracy of injection process. Some introductions about the technology of injection control valve have been made by several authors in their papers which describe the whole diesel control systems[3][4][5][6]. In this paper, we will focus on the subject of injection control valve of PVPI system for further mechanism analysis and new technique development.

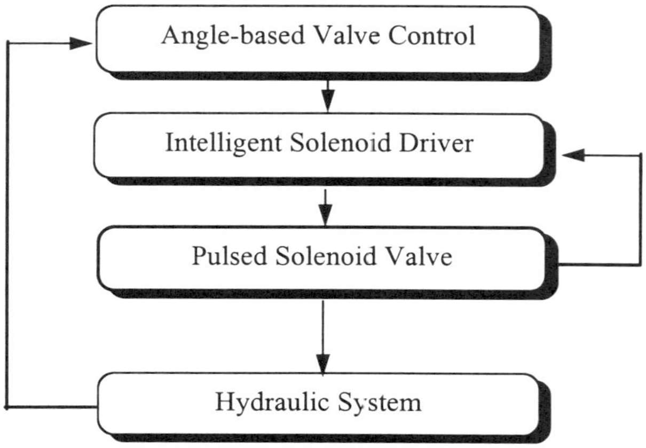

Figure 2 Block diagram of the technology of injection control valve

In the following sections, the technology of injection control valve will be discussed in three parts. In part one, the pulsed solenoid valve is studied. The dynamic electromagnetic characteristics of a multipole solenoid is analyzed by a special experimental system. The magnetic response time is reduced by material modifications. Testing results show a significant improvement in injection control performance. In part two, the intelligent solenoid driving method is discussed. In particular, the active modulation of holding current based on microprocessor is designed to provide more accurate control upon holding current than conventional hardware hysterisis comparator. Predictive control of valve closure timing based on "d^2i/dt^2" valve closure detection and momentary camshaft speed estimation is proposed to compensate for variations more effectively. In part three, the control and diagnosis of control valve closure duration is investigated. Based on results of open-loop fuel metering and valve closure duration control, the closed-loop control and diagnosis of fuel metering and valve closure duration is evaluated. A sensing method utilizing momentary camshaft speed fluctuations of fuel injection pump is presented and applied to valve closure duration diagnosis. Since the main component of the momentary camshaft speed fluctuations results from the high pressure fuel actuation which reflects the characteristics of injection process in multicylinder diesel engines, this signal also provides the basis for fuel injection feedback control and diagnosis of all diesel injection systems based on jerk pumping principle.

PULSED SOLENOID VALVE

CONCEPT OF PULSED SOLENOID VALVE

The structure of the solenoid valve used in the PVPI diesel injection system is shown in Figure 3. It consists of a spill valve and a high speed solenoid. The main function of the spill valve is to seal and spill high pressure fuel in the injection system. The high speed solenoid actuates the spill valve to regulate the fuel injection process. To meet the fast high pressure formation and spill process of the injection system, it is essential for the solenoid valve to close and open quickly.

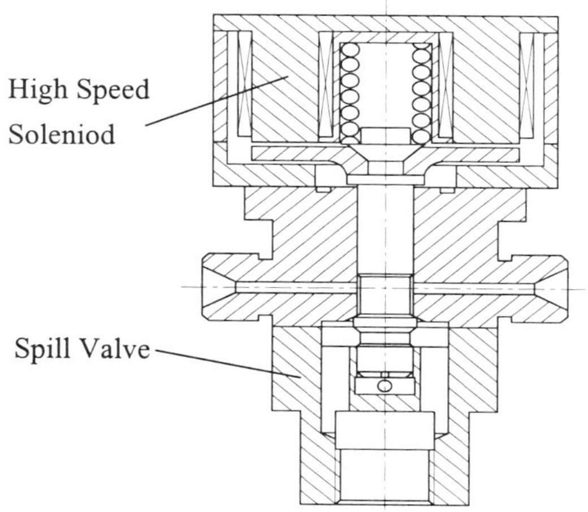

Figure 3 Solenoid valve structure

Different from conventional DC solenoid and AC solenoid, the solenoid valve in the injection system works in the condition of high frequency pulse voltage drive and possesses special electromagnetic and mechanical dynamic characteristics. That is why the solenoid valve is named the pulsed solenoid valve. In this study, a multipole solenoid is adopted. The concept of multipole solenoids was first proposed by Schechter[7]. Its prominent feature is that the force to the moving mass ratio is independent of the magnitude of the force and the size of the solenoid. The multipole solenoids were made of low carbon steel. Based on the original concept, solenoid valves are developed and tested. It is found that this kind of solenoid is suitable to be used as pulsed solenoid valve in electronic diesel injection system, at the same time, there is further potential of response time reduction. Therefore, a study on dynamic electromagnetic characteristics is made.

DYNAMIC ELECTROMAGNETIC CHARACTERISTIC

In order to estimate the dynamic characteristics of the pulsed solenoid under the operation of pulse voltage, a special measuring system is developed. The coil current is examined by a hall effect current sensor and the electromagnetic force is measured by a piezoelectric force transducer. Multipole solenoids in which the coils are connected in series are used in the experiment. The coil currents and forces versus time are examined in order to analyze dynamic electromagnetic characteristics of the pulsed solenoid.

Figure 4 shows the forces versus coil currents for various airgaps. The solenoid is made of low carbon steel. Comparing point A with point B in the curve at a constant airgap of 0.1mm, it can be seen that forces are equal at both points, while the current values differ greatly. The coil current value of point A is three times that of point B. This phenomenon indicates that to obtain the same force, much larger coil current is required during the dynamic process than in the static state. On the other hand, after the pulse voltage is cut off, the coil current falls to zero instantly, however, the electromagnetic force cannot change to zero abruptly. These experiment results show that because of the magnetic response time, the electromagnetic forces lag behind the driving pulse during the dynamic process, i.e. the start and end of the pulse voltage drive. Because the actual switch velocity of the valve is dependent on the electromagnetic force generated during the dynamic process, it is very important to reduce the magnetic response time from the point of view of the fast action of the pulsed solenoid valve.

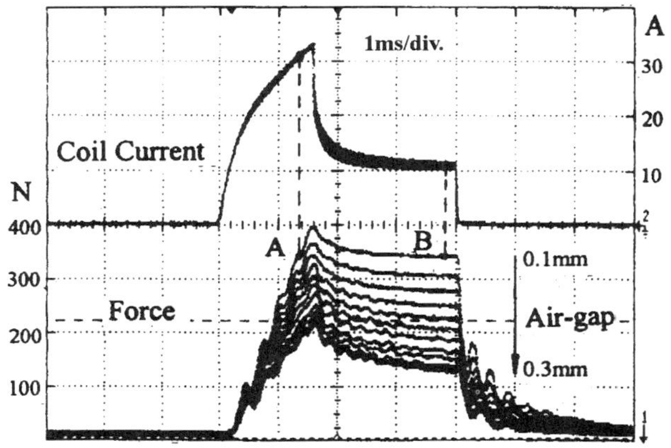

Figure 4 Oscilloscope trace of currents and forces at various constant airgaps (from 0.1mm to 0.3mm, a curve per 0.02mm)

It is well known that the magnetic response time is determined by the eddy currents to a large extent. To reduce the eddy currents in solenoids, it is essential that the material is kept as thin as possible. In general, a laminated structure is chosen. Although the structure design of mutipole solenoids is a very efficient method to decouple the electromagnetic force and acceleration, it is difficult to make it in a laminated structure. So, to decrease the magnetic response time of the solenoid, the soft magnetic material of high specific electrical resistance should be applied to make the solenoid.

MODIFICATION AND TEST

Based on the analysis of dynamic electromagnetic characteristics of the pulsed solenoid valve, the soft magnetic material which provides high specific resistance is chosen to accelerate the closing and opening process. At present, in PVPI injection system, the pulsed solenoid is made of Fe-Cr alloy. Its specific resistance is about $100\mu\Omega\cdot cm$, which is about ten times that of low carbon steel.

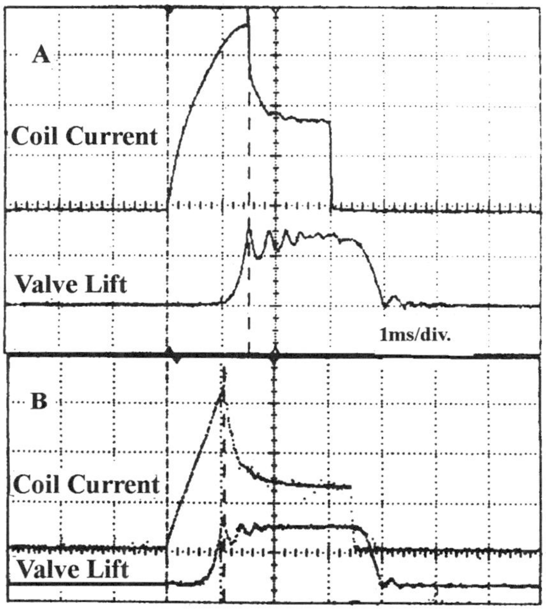

Figure 5 Comparison of the closing and opening process of dry valve testing

A low carbon steel pulsed solenoid

B Fe-Cr alloy pulsed solenoid

Figure 5 A and B respectively show the experiment results of the pulsed solenoid valves made of low carbon steel and Fe-Cr alloy. The structure parameters of two solenoids are the same except the different magnetic materials. Figure 5A shows the closing and opening process of the solenoid valve made of low carbon steel. The closing time is 1.5ms and the opening time is 1.0ms. In Figure 5B, the closing time is 1.0ms and the opening time is 0.6ms. The switch time is reduced to two thirds of the low carbon steel solenoid valve. The effect of accelerating the closing and opening process is prominent. However, in the above results under the condition of dry valve testing, large bouncings occur during the period of the closing and opening. The bouncings will influence the formation of high pressure badly and result in pressure fluctuation and unsteady injection characteristics in the injection system. To eliminate the bouncings, the method of adding fluid damping is adopted. Figure 6 shows injection performance experiment results on the injection system test rig. The closing time is 1.50ms and the opening time is 0.6ms. The bouncings are completely eliminated. The method of adding fluid damping to eliminate the bouncings is a good compromise at the expense of the closing and opening time of the pulsed solenoid valve.

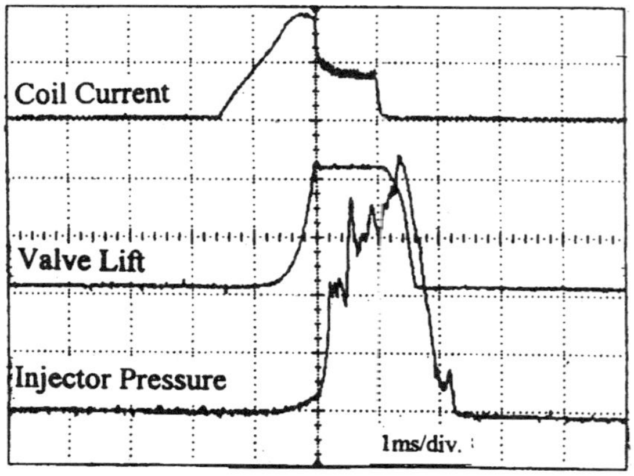

Figure 6 Injection performance experiment results

INTELLIGENT INJECTION CONTROL VALVE DRIVING

DRIVING ENERGY CONTROL PRINCIPLE

With the consideration of precision and reliability of injection control, the objectives pursued in the design of power driver can be summarized as follows:

1. Fast energizing before closure. The power driver should infuse driving current into solenoids as quick as possible to minimize closure delay of control valve, without complicating the circuit configuration.

2. Efficient holding during injection. Once the valve shuts off the spillport, the reluctance in the solenoid is reduced as the air gap decreases. So a holding current, much less than peak value but large enough to seal up high pressure fuel , goes through the coil in order to bring down heat dissipation as well as to favor the open response of control valve.

3. Instantaneous cutting off driving current at the end of injection. The sharper the current falls off, the more rapid the valve opens.

To meet those requirements, an intelligent solenoid power driver that involves the least number of power MOSFETs for multi-cylinder engine control has been designed, as shown in figure 7. The PWM FET regulates the driving current, while the CS FETs determine which cylinder will be driven. The driving current is cut off instantaneously by taking advantage of the high avalanche energy capability of parasite diodes in CSFETs.

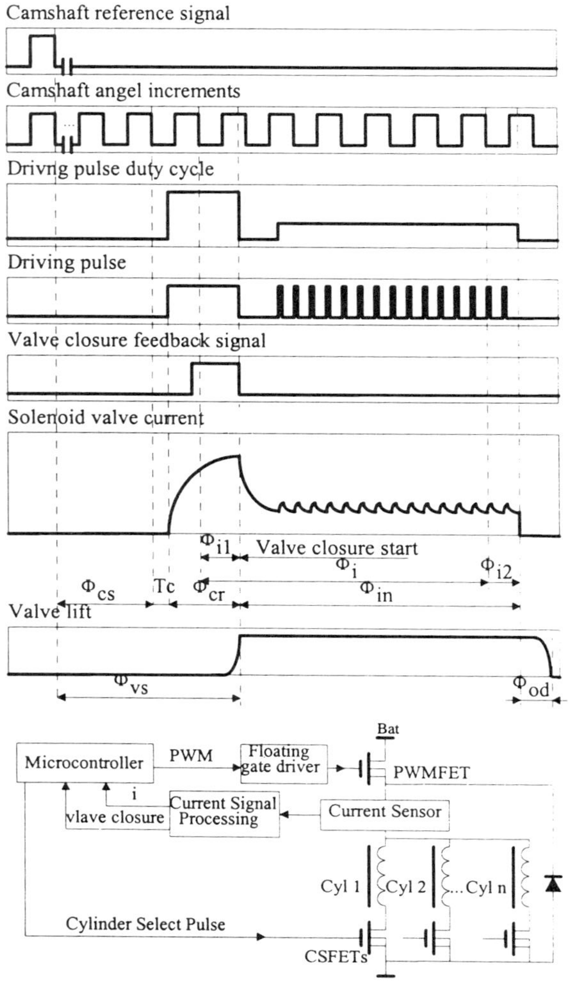

Figure 7 Injection control valve driving timing (up) and the intelligent power driver (down).

ACTIVE MODULATION OF HOLDING CURRENT

This power driver abandons the conventional current regulation method that depends wholly on the hardware hysterisis comparator to pulse the current. The microcontrollor directly controls the powerful driving current, not only the timing and duration but also the amplitude at any time with very low fluctuation by 25kHz PWM pulse. The resistor of multi-pole solenoid used in the PVPI injection system is so small that the driving current is very sensitive to the variation of conducting line resistor and battery voltage. Therefore, a regulation method is brought forward to guarantee a constant holding current(Id) in spite of the factors outside. As shown in figure 8. It is the set point of holding current stored in

EEPROM of the electronic control unit, and If is the feedback value from current sensor. A variable parameter PI controller is adopted to regulate the duty cycle of PWM pulse(P). During engine start, a fixed duty cycle(Ps) is sent to power driver to ensure the reliable operation of the control valve under the condition of sharp reduction in battery voltage. The experiment results in different voltage are compared in figure 9. (Control valve testing without fuel damping)

CLOSED-LOOP VALVE CLOSURE TIMING CONTROL

As shown in figure 7, in the angle-based electronic control of diesel injection, the optimum injection start point calculated according to such work condition as accelerator pedal, engine speed and so on, can be expressed as the valve closure start angle Φvs relative to the reference signal from an angular position sensor installed on camshaft. To ensure the precision of timing control, Φvs is subdivided into three part: Φcs, Tc and Φcr. Among them, Φcr-angles the camshaft revolves from current start to valve closure must be predicted according to the delay time (Tcr) and the average camshaft speed (ϖcr_{n+1}) during Φcr. Considering the camshaft speed fluctuation of cycle by cycle and position to position, ϖcr_{n+1} can be predicted from the following formula:

$$\varpi cr_{n+1} = \varpi cr_n \times \varpi_{n+1} / \varpi_n$$

where ϖcr_{n+1} and ϖcr_n are average speed during valve motion of present cycle and last cycle, ϖ_{n+1} and ϖ_n are average speed during certain degrees before reference signal of present cycle and last cycle. Delay time Tcr varies with the stiffness, pre-load of return spring,

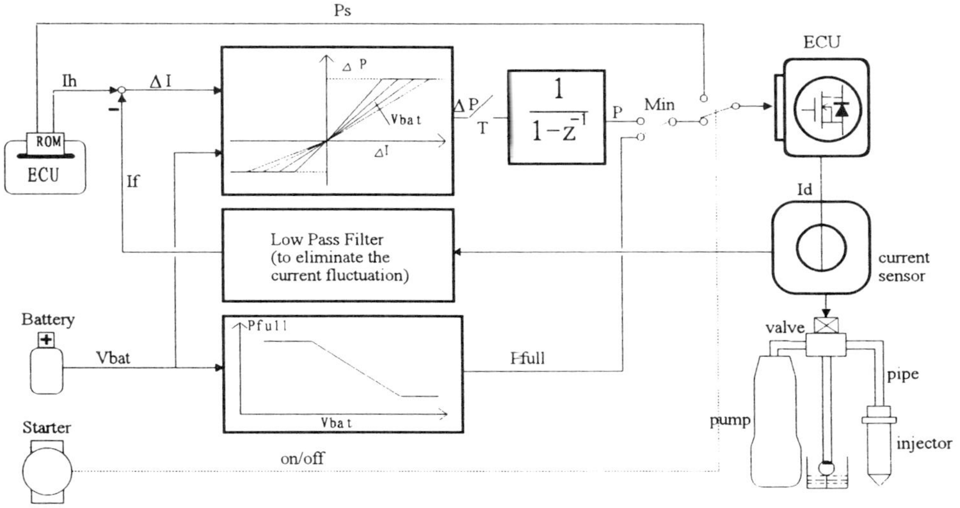

Figure 8 Holding current control logic

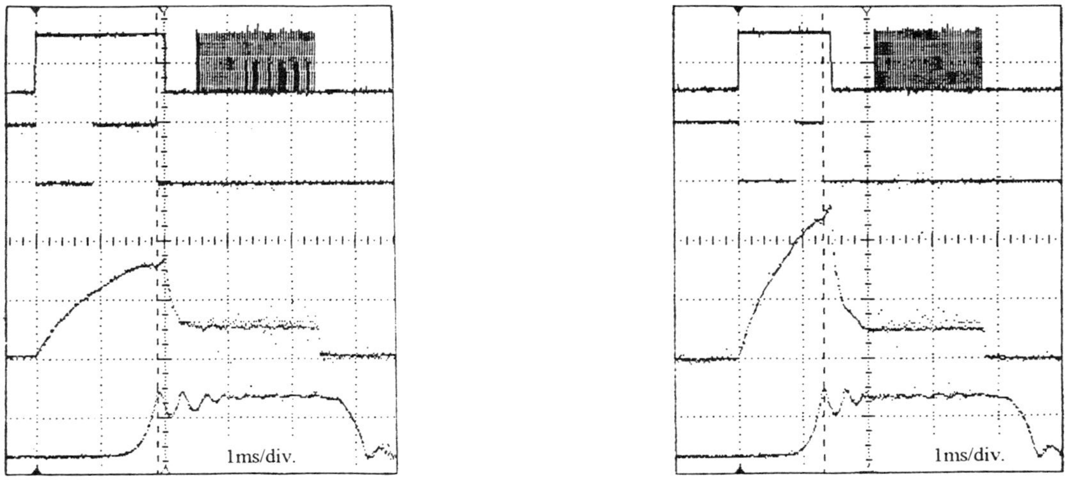

Figure 9 Constant holding current control results under different battery voltages, the left is under the voltage of 16V and the right is under the voltage of 24V, (testing without fuel damping).

resistance in valve motion, fuel pressure, voltage of battery, resistor in the coil and so on. So the feedback of the closure start of control valve or valve closure delay with regard to the current start is necessary to valve closure timing control. The first-order differential signal of the driving current (di/dt)reflects the valve closure start in the condition that the valve moves very fast. However, that signal will become indistinctive when the resistance which damps the valve increases. To avoid this defect, a second-order differential current signal(d^2i/dt^2), which reveals the characteristic of driving current more essentially, is chosen to feedback the closure start of the solenoid valve as shown in figure 10. Compared with the di/dt method, d^2i/dt^2 signal is insensitive to variation of the rule of valve motion. Figure 11 describes the control logic of valve closure timing.

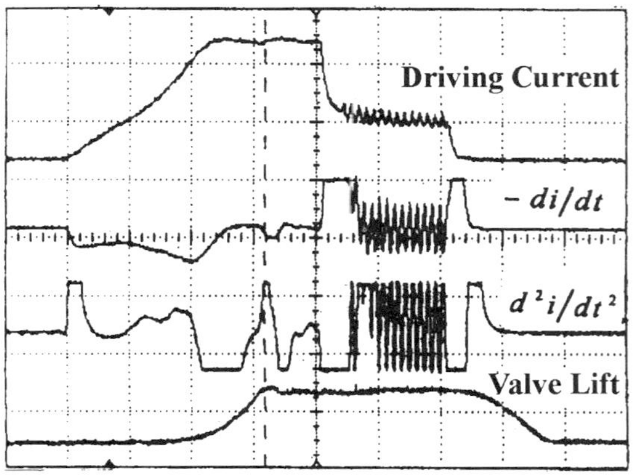

Figure 10 Current signal processing result

CONTROL AND DIAGNOSIS OF THE VALVE CLOSURE DURATION

THE OPEN-LOOP FUEL METERING AND VALVE CLOSURE DURATION CONTROL

As shown in figure 7, the injection fuel quantity is regulated by command pulse duration from valve closure start. As a result, the angle-based control of valve closure duration(Φin) becomes the key technology in the injection control. To be decoupled from timing control, the fuel quantity is metered from the point of valve closure start. And Φin is expressed as follows:

$$\Phi in = -\Phi i1 + \Phi i + \Phi i2$$

where Φi1 is the duration from the latest angle pulse before valve closure to the valve closure start, Φi and Φi2 are the integrate and decimal fraction of Φin plus Φi1 respectively. As Φi is counted by the accumulator of MCU and Φi2 is counted by the Timer, the accuracy of Φin is ensured. But this is an open-loop control. As a matter of fact, there is a valve open delay time that corresponds to a variable open delay angle Φod, which is important to fuel quantity control accuracy because in that period the injection pressure is very high and it becomes larger and larger when engine speed increases. The conventional method is to resort to so-called pump map, which is not ideal and accurate because operating time, tolerance of solenoid valve and hydraulic parts cause variation in open delay time. Furthermore, in the valve failure mode, the open delay time will become so long that fuel quantity is out of control. Therefore, some sensing technique must be developed for valve closure duration diagnosis and closed-loop fuel metering.

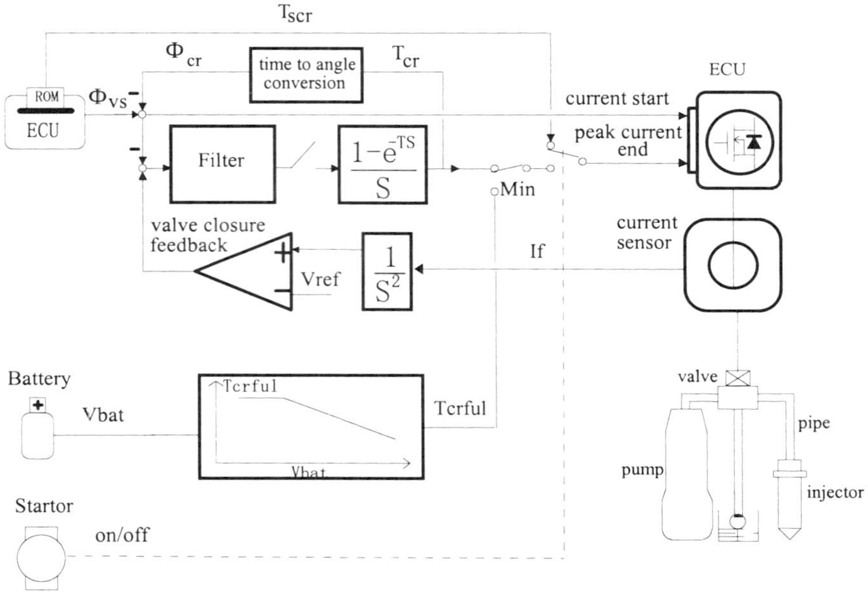

Figure 11 The solenoid valve closure timing control logic

SENSING MECHANISM UTILIZING MOMENTARY CAMSHAFT SPEED FLUCTUATIONS

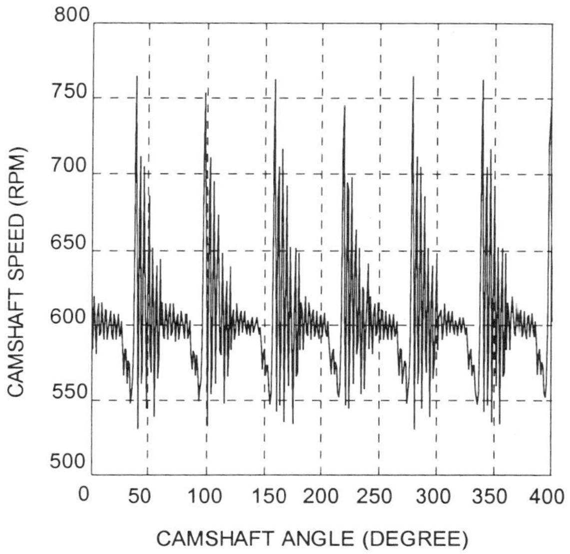

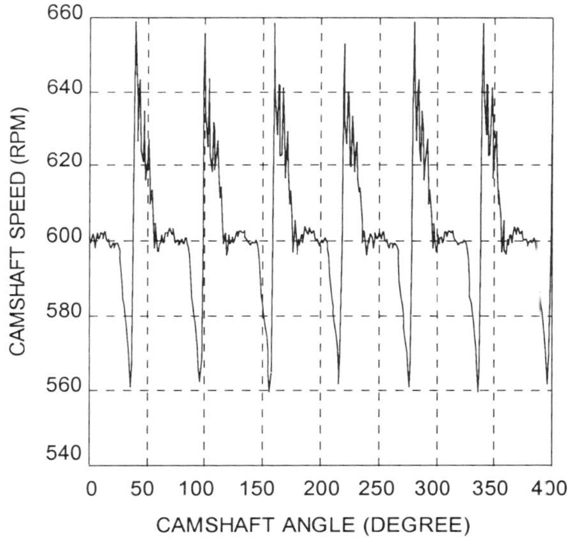

Figure 12 A typical measurement result of camshaft momentary speed fluctuation during one cycle of a six cylinder diesel engine (raw measurement data and filtered signal).

In respect that the motion of control valve influences the fuel pressure inside plunger chamber that brings about fluctuation of camshaft speed, a deep investigation of momentary camshaft speed has been done. A typical measurement result and filtered signal of camshaft speed during one cycle of a six cylinder diesel engine is shown in figure 12. The waveform of camshaft speed results mainly from the change of fuel pressure, though it is influenced by crankshaft speed too. When the plunger moving up closes inlets, the fuel is spilled through the valve to low pressure system as the control valve is still open, and the fuel pressure inside plunger chamber increases slowly, which results a slow decrease in the camshaft speed. Once control valve is closed, the plunger chamber pressure increases suddenly, thus a corresponding sharp decrease in camshaft speed is observed. When the valve is fully open, fuel is spilled again and the fuel pressure decreases violently, which makes a sudden download from camshaft that excites a damping vibration at inherent frequency in camshaft. At last, the camshaft speed converges to a average value. It can be noticed that momentary camshaft speed fluctuations reflect characteristics of diesel fuel injection process and can be used for a closed-loop control of the valve closure duration and fuel metering process.

BASIS FOR CLOSED-LOOP CONTROL OF VALVE DURATION AND DIESEL INJECTION PROCESS

The first application of camshaft speed fluctuation is to feedback the valve closure duration including open delay. From figure 13, it is obvious that the fluctuation varies according to different command pulse durations. So, the actual valve closure duration can be extracted from that fluctuation through special algorithms. Of course this method can be used in PVPI system to diagnose control valve of its two extreme failures: the valve is broken or the valve is kept constant close as the result that the valve is clamped in the control valve assembly, which is a very important safety measure for PVPI system. Another potential application is to feedback the fuel quantity supplied during valve closure which can be induced from the fuel pressure in plunger chamber, since the pressure can be calculated from momentary camshaft speed fluctuation. This method is most suitable for electronic unit injector, for the supplied fuel quantity is just the fuel injected into the cylinder. Even to other second generation of electronic diesel fuel injection system, such as rotary pump, unit pump and in-line pump, with the consideration of propagation of fuel pressure wave, the feedback of injected fuel quantity is also possible. Compared with other feedback methods of fuel quantity already discussed [8][9][10], this method is more convenient. Those methods with the aid of needle lift sensor, no matter digital or analog, complicate the structure of injector and have to install one sensor for each cylinder, while the camshaft speed sensor is only one for multi-cylinder diesel engine without changing the conventional injectors. Furthermore, when the nozzle is fully open, the needle lift sensor will no longer reflect the injection pressure, while the camshaft fluctuation at this time is still applicable. Finally, it should be mentioned that the momentary camshaft speed fluctuation is a universal signal for all diesel injection systems based on jerk-pumping principle, so, it can be applied to all systems of this type for control and diagnosis purpose.

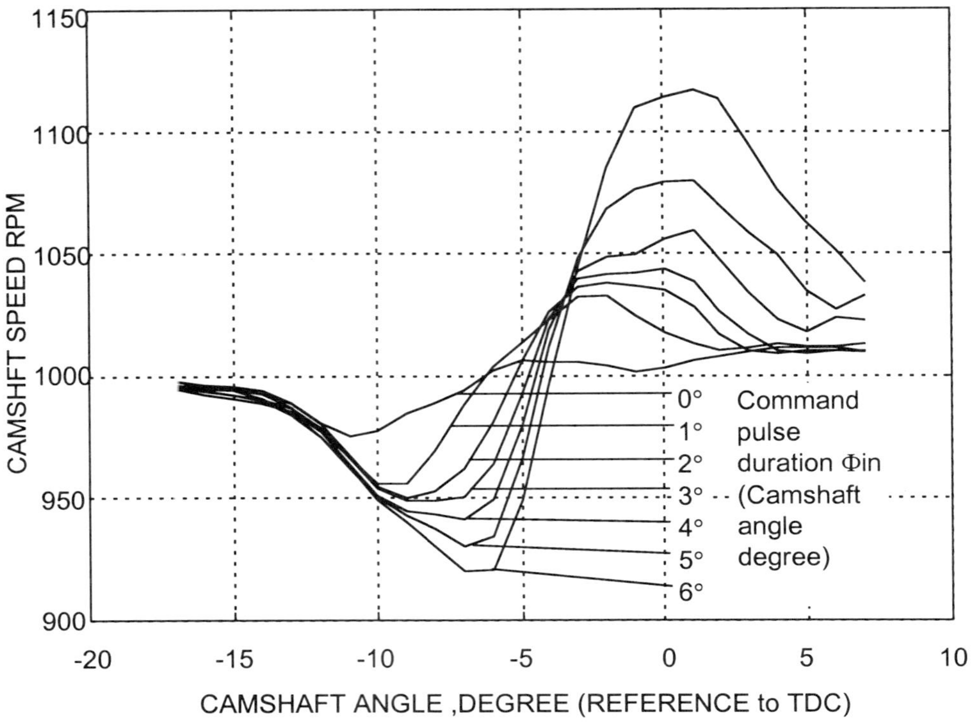

Figure 13 Momentary camshaft speed fluctuations regarding to different injection angles

CONCLUSION

1. The pulsed solenoid valve is studied by experiments. Analysis on the dynamic electromagnetic characteristics of a multipole solenoid reveals that response speed of this solenoid is mainly limited by magnetic lag. Therefore, soft magnetic material of high specific resistance, Fe-Cr alloy is used to replace original low carbon steel. Testing results verify that a significant improvement on control response has been achieved.

2. The intelligent solenoid valve driving method is discussed. In particular, the active modulation on holding current based on microprocessor can provide more accurate control over driving current. The predictive control of valve closure timing based on "d^2i/dt^2" valve closure detection and momentary camshaft speed estimation is proposed to compensate for variations more effectively.

3. The control and diagnosis especially, the closed-loop control and diagnosis of the valve closure duration and fuel metering process are investigated. A sensing method utilizing momentary camshaft-speed fluctuations of fuel injection pump is presented and applied to diagnosis and control of the valve in PVPI system. Since the main component of the momentary camshaft speed fluctuation results from the high pressure fuel actuation which reflects the characteristics of injection process in multi-cylinder engines, this signal also provides the basis for fuel injection feed back control and diagnosis of all diesel injection system based on jerk pumping principle.

REFERENCES

1. Minggao Yang, S.C.Sorenson, "Direct Digital Control of Diesel Fuel Injection Process", SAE Paper 920626
2. Minggao Yang, S.C.Sorenson, "Direct Digital Control of Diesel Engines", SAE Paper 940372
3. Werner Fischer et al, "Solenoid-Valve Controlled Diesel Distributor Injection Pump", SAE Paper 930327
4. R.J.Hames et al, "DDECII-Advanced Electronic Diesel Control", SAE Paper 861049
5. Pierre Lauvin et al, "Electronically Controlled High Pressure Unit Injector System For Diesel Engines". SAE Paper 911819
6. R.J.Hames et al, "DDEC-Detroit Diesel Electronic Control", SAE Paper 850542
7. Michael M. Schechter, "Fast Response Multipole Solenoid", SAE Paper 820203
8. C.A.MacCarley et al, "An Indirect Sensing Technique for Closed-Loop Diesel Fuel Quantity Control", SAE Paper 900494
9. Keiichi Yamada et al, "The Second Generation of Electronic Diesel Fuel Injection Systems-Investigation with a Rotary Pump", SAE Paper 860145
10. C.A.MacCarley et al, "An Auxillary Sensor Processor to Provide Real Time Fuel Delivery Feedback for a Microprocessor Based Diesel Engine Controller", SAE Paper 870435

960869

Shaping of Fuel Delivery Characteristics for Solenoid Operated Diesel Engine Gaseous Injectors

Henry Hong, Tadeusz Krepec, and Harry Kekedjian
Concordia Univ.

ABSTRACT

Solenoid operated gaseous injectors, when compared to conventional liquid fuel diesel injectors, differ in the way the fuel dose and its discharge rate are controlled. While in conventional diesel systems, the fuel dose and its injection rate depends on the fuel injection pump effective stroke and on the plunger diameter and velocity, the solenoid injectors operate in an "on-off" manner which limits the ability to control the gas discharge rate, resulting in its profile to be basically rectangular in shape. To reduce the gas injection rate at the beginning of the injection process in order to suppress the "diesel-knock" phenomenon, similar procedures as used in diesel engines could be implemented. One such approach is to use a throttling type pintle nozzle, and another method is to use a double-spring injector with a hole nozzle. The rationale for using such nozzle configurations is that gaseous fuels do not require atomization, and therefore, can be injected at lower discharge velocities than with liquid fuels.

The gas delivery characteristics from a solenoid injector has been computer-simulated in order to assess the impact of the investigated three modes of fuel discharge rate control strategies. The simulation results confirmed that the gas dose and its discharge rate can be shaped as required. An experimental set-up is described to measure the gas discharge rate using a special gas injection mass flow rate indicator with a strain-gage sensor installed at the entry to a long tube, similar to that proposed by Bosch for liquid fuel volumetric flow rate measurements.

INTRODUCTION

The use of gaseous fuels, such as natural gas, in the Otto cycle for internal combustion engines, does not involve any significant complications to the fuel injection process. Solenoid operated injectors, as produced for gasoline injection, can be used with minor modifications; the changes are primarily in the nozzle flow area. On the contrary, when diesel engines are considered, an entirely new fuel delivery system has to be developed to ensure the right timing of the gas injection process. The diesel type injection pump cannot be used due to the high compressibility of the gas, and a common rail approach similar to gasoline injection but with much higher gas pressure, would be the logical solution.

The recent research and development work in the field of high pressure gas injection is heading in two directions. The first approach is to use a conventional diesel injection system with a hydraulic actuator installed at the upper side of the injector which is operated by the injection pump fuel pressure. The actuator lifts the needle of the gas injector, providing the gas flow from a common rail fuel delivery system to the combustion chamber of the engine, with the injection timing and the gas dose corresponding to the original diesel fuel injection pattern. As described in the literature [1][2], although this approach is practical for engine development work, the method does not take advantage of the substantial cost reduction offered by the common rail type of injection system; the diesel injection pump must still be used.

The second approach is to use solenoid operated injectors to open the gas flow from the common rail supply system. However, the gas is not supplied to the intake manifold as in the Otto engines, but directly into the engine combustion chamber. This method introduces serious difficulties in that the injector must be resistant to the high temperature gases and that the accuracy of the injected gas dose and timing must be provided by the injector fast dynamic response. This approach [3][4] allows for a reduction of cost and size of the injection system, but creates some new problems, one of which is the shaping of the injected gas flow profile to conform to the diesel engine requirements.

There are no special problems associated with the control of the gas dose injected from a common rail system. Regulation can be achieved by altering the command voltage pulse width or by varying the needle position of the metering valve that throttles the gas flow from the common rail to the injector, as is reported in reference [3]. The subject under discussion in this paper is rather related to the shape of the gas discharge rate characteristic versus time which has a significant impact on the combustion process in the diesel engine.

POSSIBILITIES OF SHAPING THE GAS DISCHARGE RATE CHARACTERISTIC

The solenoid operated injector under consideration is shown in Fig. 1. The needle is partly pressure balanced and acts as a valve upon the injector nozzle; it is connected to the electric solenoid along the pulling rod. When the solenoid is energized, the needle opens the valve and the gas is injected into the cylinder of the engine.

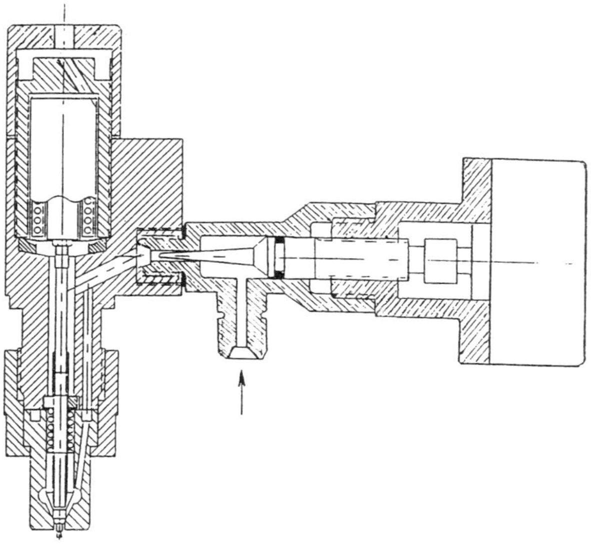

Fig. 1 Solenoid operated gas injector with metering valve

The supply of gas to the injector is controlled by a metering valve operated by a stepper motor. During the gas injection period, gas flow is throttled which reduces the pressure in the injector that limits the injected gas dose. Together with the pulse width control, the basic shape of the gas discharge rate characteristic can be varied as shown in Fig. 2a and 2b. The experiments performed with such an injection system, as described in reference [5], are confirming the expected operation of this concept, and the optimization procedure presented in reference [6] shows that it is possible to achieve a fast dynamic response of the injector as required by a high speed diesel engine. What is not assured by this system is the ability to provide a shape of the gas discharge characteristic which would be particularly suitable for the diesel combustion process, as shown in Fig. 2c.

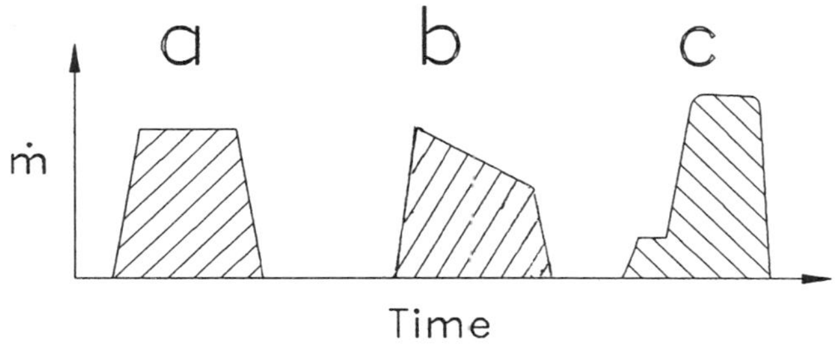

Fig. 2 Different basic gas discharge rate characteristics obtained from a solenoid operated injector:
a. Without metering valve throttling
b. With throttling by metering valve
c. More suitable for diesel engine combustion

The last shape (2c) which is more suitable to the typical diesel combustion characteristics, is recommended specifically for higher speed diesel engines, in order to reduce the amount of fuel injected during the ignition delay in the combustion process and to limit the following pressure increase rate in the cylinder after ignition.

There are design features used in conventional diesel injectors which could also be applied to solenoid gas injectors in order to obtain a more suitable gas discharge rate characteristic. Such modifications include: the throttling type pintle nozzles (Fig. 3) and the double spring hole injectors (Fig. 4). Both of these concepts have been computer simulated in order to investigate the possibility of shaping the gas discharge rate characteristic.

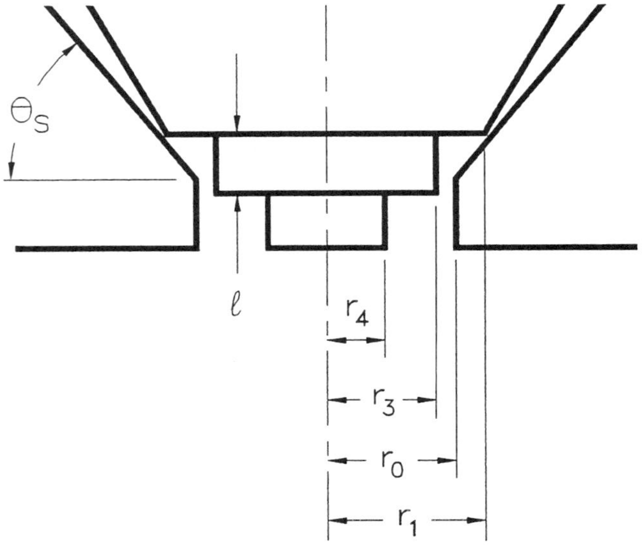

Fig. 3 a) Schematic of throttling pintle nozzle geometry

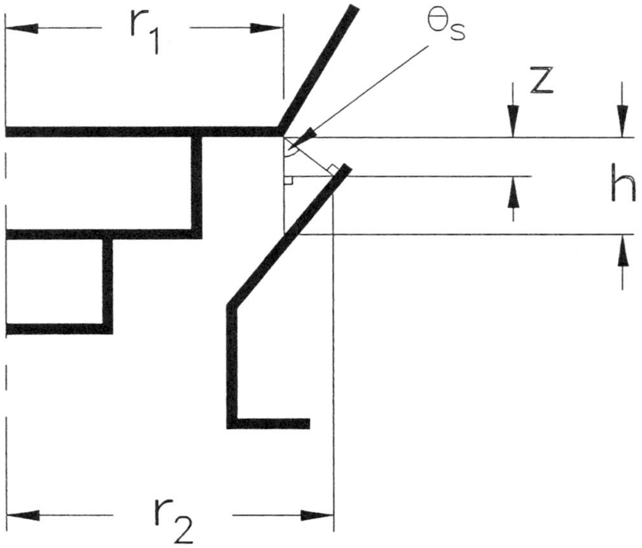

Fig. 3 b) Injector minimum geometric flow area

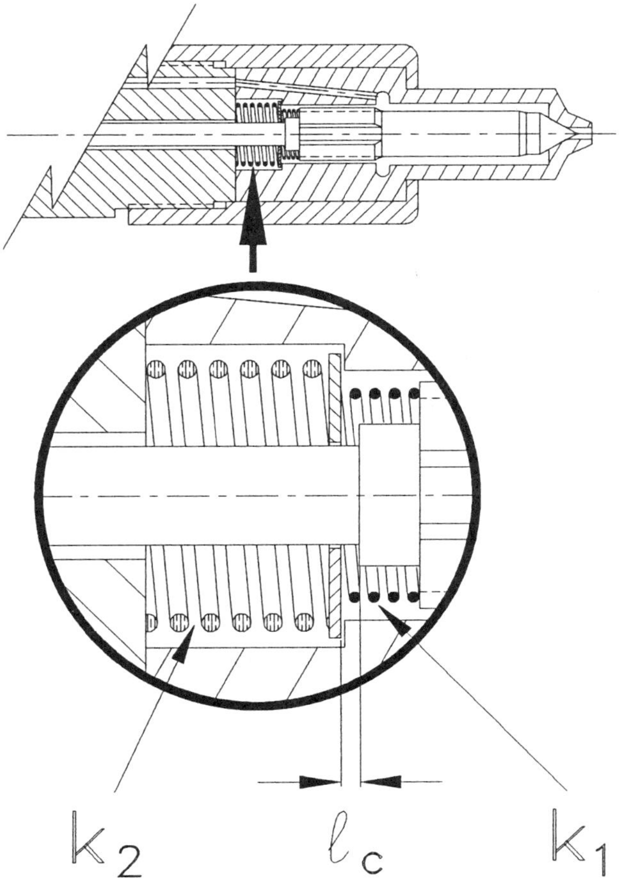

Fig. 4 a) Double spring injector design

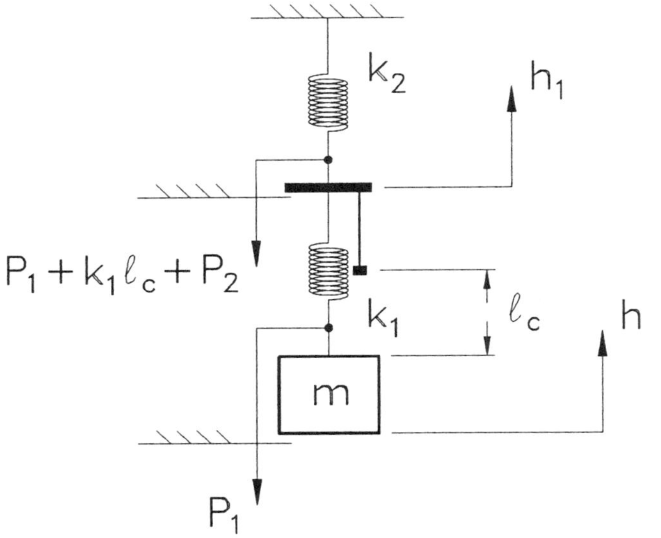

Fig. 4 b) Double spring-mass model

MODELLING AND COMPUTER SIMULATION OF GAS DISCHARGE RATE CHARACTERISTICS

To investigate the possibility of adapting the gas discharge characteristic to the diesel engine needs, computer simulation is used based on the modelling of the solenoid [7] and on the programs developed for injector optimization [6]. While the optimized injectors are equipped with conventional hole type nozzles, the control of the gas discharge rate also requires throttling type pintle nozzles and double spring injector modelling. Figs. 3 and 4 show schematically the design of these devices and the mathematical formulation of their discharge characteristics are described in the following.

For the pintle nozzle injector shown in Fig. 3, once the cone part of the pintle lifts off its seat, the minimum geometric flow area is given by the surface of the frustum of a right cone:

$$A_m = \pi \cdot (r_1 + r_2) \cdot \sqrt{z^2 + (r_1 - r_2)^2} \tag{1}$$

where
h = pintle lift
θ_s = angle of the seat (60°)
z = altitude of the right cone = $h \cdot \cos^2\theta_s$
r_0 = radius of the orifice
r_1 = top radius of the right cone
r_2 = base radius of the right cone
= $r_1 + h \cdot \sin\theta_s \cos\theta_s$

Next, up to a certain needle lift, the minimum geometric flow area changes to the flow area given by the pintle of radius r_3, as expressed by:

$$A_m = \pi \cdot (r_0^2 - r_3^2) \tag{2}$$

After the pintle has lifted by l, the minimum geometric flow area is given again by EQ (1), where the base radius of the right cone is the orifice radius r_0, the top radius of the right cone is the pintle radius r_3, and the altitude of the right cone is the pintle lift minus the pintle length, $h-l$. Then, the flow area increases to its final area given by the pintle tip of radius r_4:

$$A_m = \pi \cdot (r_0^2 - r_4^2) \tag{3}$$

For the double spring injector shown in Fig. 4a, the minimum geometric flow area is described by the needle tip cone angle 2θ (59°40'), by the orifice diameter $2r_0$, and by the needle lift h, given by:

$$A_m = \pi \cdot (h \cdot \sin\theta)(2r_0 - h \cdot \sin\theta \cdot \cos\theta) \tag{4}$$

With reference to the schematic diagram of Fig. 4b for the spring-mass system configuration of the double spring injector, the spring force is given by:

$$F_s = k_1 \cdot h + P_1 \qquad \text{if} \quad 0 \le h < l_c$$
$$F_s = k_2 \cdot h_1 + P_1 + k_1 \cdot l_c + P_2 \qquad \text{if} \quad h \ge l_c \tag{5}$$
$$\text{where} \quad h_1 = h - l_c$$

where l_c = compression limit of the first spring
k_1 = spring rate of the first spring
P_1 = preload of the first spring
k_2 = spring rate of the second spring
$P_1 + k_1 l_c + P_2$ = preload of the second spring

Fig. 5 simulation results are showing the dynamic response of a solenoid operated injector obtained with a special driving circuit equipped with a capacitor. The system parameters have been optimized for best dynamic performance [6]. It can be seen that the injector is opening and closing very fast, however the gas discharge rate $\dot{m}$ is remaining almost constant throughout the full injection process, as shown in Fig. 2a.

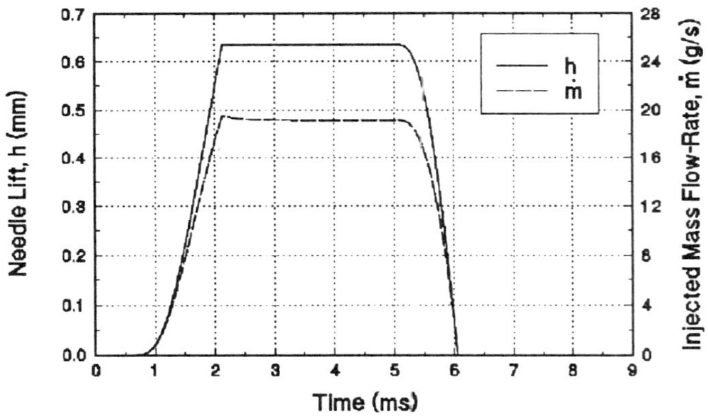

Fig. 5 The gas discharge process from an unmodified solenoid injector operating at optimal dynamic performance

Fig. 6 shows the dynamic response of the same injector, however with the gas flow to the injector throttled by the metering valve. The record of the gas discharge rate shows a negative slope, according to the pattern shown in Fig. 2b.

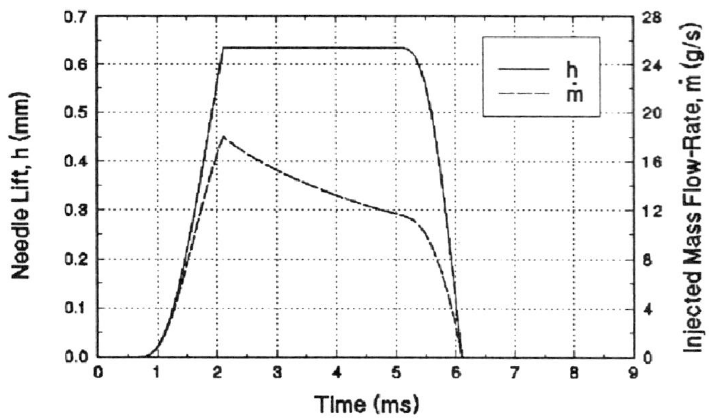

Fig. 6 The gas discharge process affected by the gas throttling at the inlet to the injector

To obtain the shape of the gas discharge characteristic, as proposed in Fig. 2c, the two injector design modifications, described respectively in Figs. 3 & 4, have been simulated. The results are shown in Figs. 7 & 8, confirming the ability to shape the gas discharge rate process in order to adapt a solenoid injector to the diesel engine combustion process.

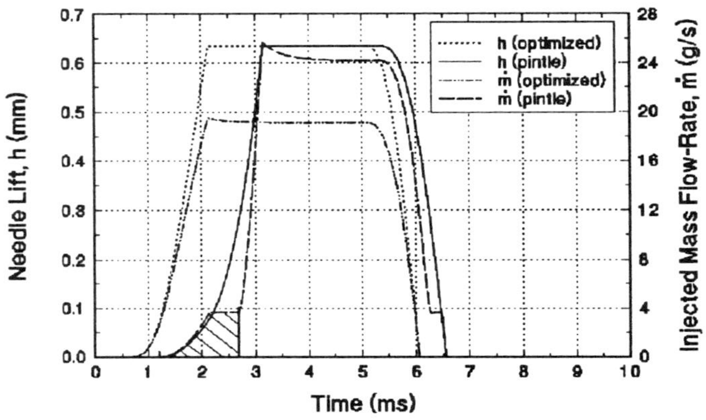

Fig. 7 Gas discharge rate with throttling pintle nozzle injector

Both modifications allowed to obtain the same gas dose and gas injection period as for the basic injection process shown in Fig. 5, which are re-plotted in the two figures for comparison. The initial injection period of 1.5 ms, shown with crosshatching in Figs. 7 & 8, is arbitrarily chosen for comparison purposes; it could be altered, depending on particular engine needs. However, some compromises had to be made to obtain the required gas discharge shape.

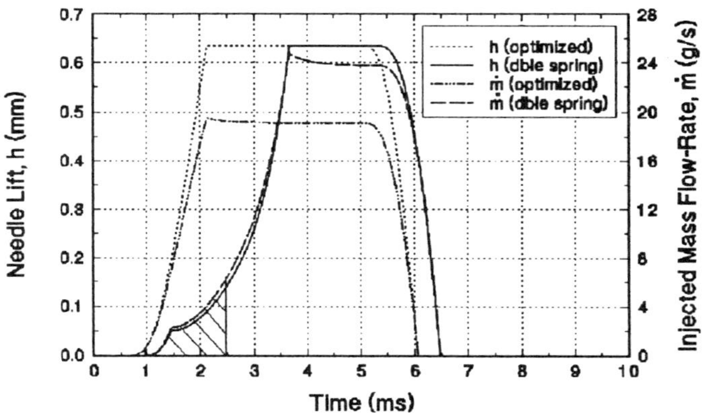

Fig. 8 Gas discharge rate for two spring hole nozzle injector

For the injector with the pintle nozzle, the spring preload is increased and the spring rate is decreased, as compared to the unmodified injection system. The spring preload is increased so that the initial needle rise is slowed down to effectively throttle the initial flow. The decrease in the spring rate allows the injector to rapidly accelerate to its fully open position after the initial flow throttling. Because of the geometric design of the pintle, the needle seat diameter had to be increased which causes the seat sealing diameter to be larger than the orifice diameter. This increases the unbalanced gas pressure force acting on the needle, and together with the increase in the spring preload, prolongs the injector opening delay period, as compared with the original process shown in Fig. 5.

To accomplish similar task in the hole type injector, the two spring design requires also an increase in the nozzle orifice flow area to obtain the higher gas flow rate after the initial throttling. As in the pintle nozzle design, the injector opening delay is also prolonged because of the increase in the unbalanced gas force on the needle. For both designs, this increase in opening delay can be compensated by advancing the injection timing on the engine.

MEASUREMENT OF THE GAS DISCHARGE RATE

To ensure the validity of the computer simulation results, experiments have been conducted using a specially made gas discharge rate indicator, which is shown in Fig. 9. The device consists of a thin aluminium pipe with strain gages installed on the pipe periphery to measure circumferential expansion. The pipe is installed at the exit of the injector nozzle and is very long so that the pressure wave reflected from the pipe end does not interfere with the measurements before completion of the injection process. This requires that the pipe length is greater than the velocity of sound in the compressed gas multiplied by the time of measurement.

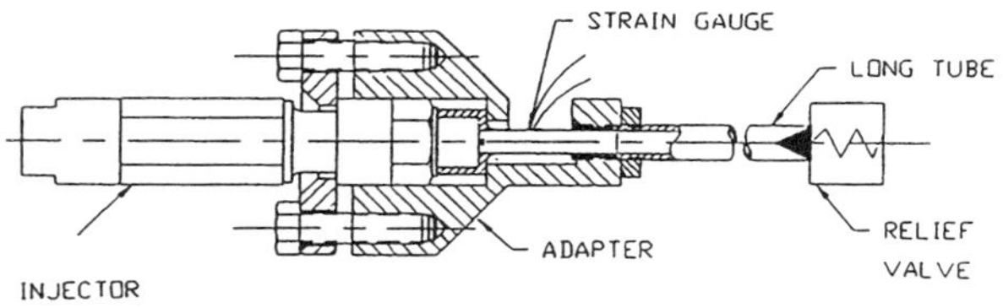

Fig. 9 Gas discharge rate indicator schematic

The flow measurement concept used is based on the diesel fuel injection indicator developed by R. Bosch in 1964 and described in reference [8]. That indicator did measure the volumetric discharge rate of liquid fuel, while the proposed device is measuring the mass flow rate of gaseous fuel. The theoretical background used is somewhat different and it can be described as follows [9].

Fox and McDonald [10] utilized a control volume approach with a propagating pressure wave to obtain the pressure-velocity relationship for a compressible flow through a constant cross sectional area. The momentum equation in integral form can be reduced to:

$$dV = \frac{1}{\rho a} dP \tag{6}$$

where, dV = velocity of flow medium relative to undisturbed medium
dP = pressure of flow medium relative to undisturbed medium
ρ = density of flow medium
a = velocity sound in flow medium

It is assumed that the flow is steady, inviscid and body forces are neglected. Since the undisturbed medium is at rest and the amplitude of the pressure wave in the experiment will be measured relative to that of the undisturbed medium, the above equation can be written in terms of final values as:

$$V = \frac{P}{\rho a} \tag{7}$$

The density term can be eliminated from the equation knowing the mass flow rate:

$$\rho = \frac{\dot{m}}{AV} \tag{8}$$

where A is the cross sectional area of the pipe. Substituting EQ (8) into EQ (7) and rearranging results in the equation for the mass flow rate as:

$$\dot{m} = \frac{A}{a} P \tag{9}$$

Assuming that the velocity of sound does not change significantly throughout the flow field, then:

$$\dot{m} = KP \tag{10}$$

where $K = A/a$ can be considered as a constant.

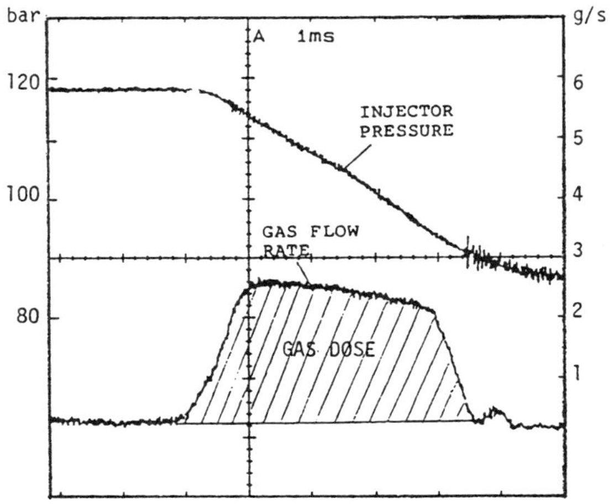

Fig. 10 Plots of gas pressure in the injector and the gas discharge rate

Fig. 10 is showing the natural gas discharge characteristic obtained with reduced opening of the metering valve. The declining discharge rate results from the pressure drop inside the injector due to the inlet gas throttling. At the end of the discharge characteristic, a small gas flow increase can be noted which results from bouncing of the injector needle in the seat after injector closing.

CONCLUSIONS AND RECOMMENDATIONS

Further development work on solenoid operated gas injectors has led to the modifications of the injector design in order to better adapt the gas discharge characteristic to the needs of a high speed diesel engine. The results are showing that there are two different means available to alter the injected gas dose: by varying the voltage pulse width and/or the metering valve opening at the inlet to the injector. There are also two possible ways to change the shape of the gas discharge rate characteristic of the injectors through the use of a throttling pintle nozzle or a double-spring configuration. Finally, a measuring instrument is presented to plot the mass flow rate of the gas discharged from the injector nozzle. All these means might contribute to better adaptation of the solenoid operated injectors to the high speed diesel engines.

ACKNOWLEDGEMENT

The authors are grateful to the Natural Sciences and Engineering Research Council of Canada for supporting this research.

REFERENCES

1. Furuhama, S. and Kobayashi, Y., "A Liquid Hydrogen Car with a Two-Stroke Direct Injection Engine and LH-Pump", *Int. J. Hydrogen Energy*, Vol. 7, No. 11, 1982.
2. J.K. Wakenell, G.G. O'Neal, Q.A. Baker and C.M. Urban, "An Investigation of High Pressure/Late Cycle Injection of CNG as a Fuel for Rail Applications", Report, Southwest Research institute, 1988.
3. Miele, D., Krepec, T. and Giannacopoulos, T., "Electronic Injection System for Natural gas in a Diesel Engine, Development and Testing", SAE Paper No. 890852, 1989.
4. Green, C.J. and Wallace, J.S., "Electronically Actuated Injector for Gaseous Fuels", SAE Paper No. 892147.
5. Krepec, T., Carrese, G. and Miele, D., "Further Investigations of Electronically Controlled Hydrogen Storage and Direct Injection System for Automotive Applications", *Int. J. Hydrogen Energy*, Vol. 17, No. 12, 1992.
6. Hong, H., Krepec, T. and Cheng, R.M.H., "Computer Aided Design Optimization of Fast Operating Electronically Controlled Fuel Injectors", 2nd International Conference on Road Vehicle Automation ROVA '95 INTERNATIONAL, Bolton Institute, Bolton, England, September 11-13, 1995.
7. Hong, H., Krepec, T. and Cheng, R.M.H., "Transient Response of Fast Acting Solenoids in Automotive Applications", *Journal of Circuits, Systems, and Computers*, World Scientific Publishing Co., Vol.4, No.4, 1994.
8. Bosch, W., "Der Einspritzgesetz-Indikator, Ein neues Messgerät zur directen Bestimmung des Einspritzgesetzes von Einzeleinspritzungen", M.T.Z., No. 7, 1965.
9. Kekedjian, H., "Further Development in Solenoid Operated Injectors for High Pressure Direct Fuel Injection", A Thesis in the Department of Mechanical Engineering, Concordia University, Montreal, Quebec, May 1994.
10. Fox, R.W. and McDonald, A.T., Introduction to Fluid Mechanics, 3rd Ed., John Wiley & Sons, U.S.A., 1985, pp. 569-570.

ENGINE VALVE CONTROL

960590

An Initial Study of Variable Valve Timing Implemented with a Secondary Valve in the Intake Runner

Olivier Vogel, Kimon Roussopoulos, and Lino Guzzella
Swiss Federal Institute of Technology

James Czekaj
Ford Motor Co.

ABSTRACT

This paper describes an investigation of implementing variable valve timing with the use of a secondary valve in the intake tract. The secondary valve acts in series with the conventional poppet valve of the engine, which is actuated in the conventional manner and with fixed timing. It is significantly easier to adjust the timing of the secondary valve than that of the main valve. There are two possible modes of operation of this device. One is to control valve overlap, improving low speed low load performance. The other is to use it as a load control mechanism, replacing the conventional throttle and therefore reducing the 'pumping losses' associated with conventional throttling. We have implemented such a system on a motored single-cylinder research engine using pneumatic actuation. Our experiments have demonstrated successful load control, significant pumping loss reduction, and control of backflow from the exhaust to the intake manifold.

INTRODUCTION

Conventional automotive engines suffer from the drawback that the timing of the intake event is invariant with engine speed and load. It has long been known that improvements in gasoline engine performance would be obtainable if the valve timing were variable so that it could be optimised for any combination of load and speed in the engine's operating range. A great deal of work is currently under way on schemes for variable valve timing (VVT), some of which are starting to come to market. It remains true, however, that the difficulties of providing a variable valve timing mechanism with acceptable performance, cost and reliability for general usage have yet to be satisfactorily solved.

This paper describes an investigation of a method of implementing VVT with the use of a secondary valve in the intake tract. The principle is that the secondary valve acts in series with the conventional poppet valve of the engine, which is actuated in the conventional manner and with fixed timing. The secondary valve is not required to withstand the temperatures and pressures of combustion, and therefore can be of relatively lightweight design. It is therefore significantly easier to adjust the timing of the secondary valve than that of the main valve.

There are two possible modes of operation of this device. One is to use it as a load control mechanism, replacing the conventional throttle and therefore reducing the 'pumping losses' associated with conventional throttling. The other is to control valve overlap, improving low speed low load (idling) performance. In the load control mode, the secondary valve closes early in the intake stroke of the engine, limiting the flow of air and fuel into the cylinder. The timing of the valve closure governs the load of the engine. In this manner some losses resulting from the irreversible nature of flow past a throttle plate are avoided, and a fuel economy benefit should accrue. This is very similar to load control by use of early intake valve closure. In the overlap control mode, the secondary valve remains closed as the main valve opens, preventing the 'backflow' of air and exhaust residual from the exhaust to the inlet port that can deteriorate engine performance at idle. The secondary valve opens once the exhaust valve has been effectively closed. This of course allows the main valve timings to be set to a high degree of overlap, resulting in higher possible torque and power at high load high speed operating points (at which the secondary valve would be permanently open).

The purpose of this study was not to develop actuators and algorithms suitable for deployment on a prototype or production vehicle, but to investigate in a laboratory what magnitude of benefits could be expected from such a deployment. We have implemented such a secondary valve and tested its performance on a motored single cylinder research engine. We wish to emphasise that we have made no effort yet to develop an actuator suitable for in-car use, and also that we have made no effort to develop an energy-efficient actuator: these tasks will be considered if it becomes clear that doing so is worthwhile.

SURVEY OF PREVIOUS WORK

There is a large body of technical literature on the subject of variable valve timing: the reader is referred to, for example, the review articles by Gray (1)* and Stone and Kwan (2). Several authors have considered, and some tested, the use of early intake valve closure (EIVC) to reduce pumping losses: see for example (3-9). These papers all investigated making the timing of a conventional poppet intake valve variable. They are in general agreement on two important aspects of our work: that the use of variable valve timing to control overlap at low engine speeds is desirable, and that in general EIVC is advantageous as a means of load control. The benefits derived from EIVC vary between the various papers, fuel economy savings figures between 4% and 15% being mentioned (this is of course a function of load and engine speed). There is widespread agreement on the emissions implications of EIVC; basically, NO_X emissions are significantly reduced (due to lower combustion temperatures), and HC and CO emissions are slightly increased (due to lower combustion temperatures, hence less complete combustion). It is also suggested in several papers that EIVC results in worse combustion due to lower in-cylinder turbulence (although Tuttle (3) suggests it is higher), lower in-cylinder temperatures, and/or less mixing through the inlet valve.

Gray in his review (1) makes explicit mention of the possibility of using a supplemental valve in the intake runner, and states the possible advantages of this concept as identified in our study. He makes no mention however of any investigation of this concept. The closest study to our own known to us is that of Low (10). A sliding shutter device was used as a secondary valve in series with the intake valve of a conventionally-valved engine. The shutter device was sandwiched between the inlet manifold and the cylinder head block. The shutter was arranged so that it blocked the intake runner at a variable time before the inlet valve closure, thus implementing EIVC type load control. Experiments on a 4-cylinder 1.6 litre Ford engine were performed at low engine speeds (between 360 and 1500 rpm). Meager detail is presented of the engine instrumentation or experimental technique, but the presented results imply improvements of brake thermal efficiency of between 30% and 40%. In the light of the other results discussed above we find these claims astonishing. No analysis is presented in his paper.

Several other authors have presented studies using secondary valves in the intake, usually implemented with rotary valves (11-15). There has been a lot of interest in implementing Miller cycles on large engines using this technique: e.g. the paper from Azechi et al. (11), reporting on experiments on a 23-litre gas engine. Ichimaru et al. (12) carried out simulations and experiments with such a valve on a production engine with an increased compression ratio, with mixed results. Abdalla (13) suggested the use of rotary secondary valves, and considered several design parameters.

THEORETICAL STUDIES

A brief investigation of the possibilities for load control was made using simple models of the charging process and the ETHZ Engine Simulation Program 'CIRCE'. This is a thermodynamic process simulation incorporating heat release and empirical wall heat transfer into a 2-zone (burned and unburned gases) model with no fluid dynamics. The outline results of these simple investigations are listed below.

- Use of a secondary valve to implement EIVC load control should lead to an increase in overall engine efficiency and therefore a fuel saving at part and idle loads, the saving being as much as 15% at idle.
- The model indicated that the cooling during the charge expansion after valve closure would lead to burn times increasing, reducing or reversing the efficiency gain. Pre-heating the air charge would offset this effect. It is noted however that other researchers have not to our knowledge used pre-heating.
- It is important both to minimise the 'dead volume' between the secondary and primary inlet valves, and to close the secondary valve quickly, to optimise the efficiency gains. These are equally significant causes of loss of efficiency.

IMPLEMENTATION OF SECONDARY VALVE

The main technical challenge of this study was the implementation of the secondary valve. The design targets established at an early stage of the project were: fast closing (of the order of 1 ms) in the load control phase, to reduce flow losses; sufficiently fast opening (of the order of 5 ms) in the valve overlap control phase; and minimum 'dead' volume between this valve and the poppet valve. These design targets, notably the fast closing, are severe requirements for a mechanical system. A number of possible technologies for achieving this were considered; these included spring driven systems, electromagnetically driven systems, rotating slotted disks, cam driven systems, hydraulically driven systems and pneumatically driven systems.

It was decided that the most suitable for our

* Numbers in parentheses designate references at the end of the paper

purposes was to use pneumatic actuation. We based our device on a double-acting electro-pneumatic actuator, model number RZ1540B, manufactured by Eugen Seitz AG of Wetzikon, Switzerland. In this device the prime mover is a double acting pneumatic piston. The flow of the high pressure air to one side or the other of the piston is controlled by incorporated electromagnetic valves. The performance of the device was improved by three measures: operating at a higher air pressure than specified (12 bar, up from 8 bar), by driving the device with a higher voltage, and by removing the acoustic absorption material. With these measures the primary actuator was able to fully open and close in a few milliseconds. Pneumatic power for the actuator was taken from a suitably regulated compressed air cylinder.

The valve implementation chosen is illustrated in Figure 1. When the device is open, flow into the runner is like a 'sink flow' from a small plenum-like space adjoining the cylinder head. The valve, when open, fits into the wall of the plenum opposing the inlet hole to the runner. The valve closes by being driven across the plenum and sealing the intake hole to the runner. The total motion required is 9 mm, and the valve is in motion for about 3.5 ms. The plenum was fitted to the rest of the intake manifold, which consisted of a conventional throttle and, for the backflow control experiments only, airflow measurement apparatus. Great efforts were made to ensure a smooth flow of air into and through the plenum, in order to reduce the fluid mechanical losses.

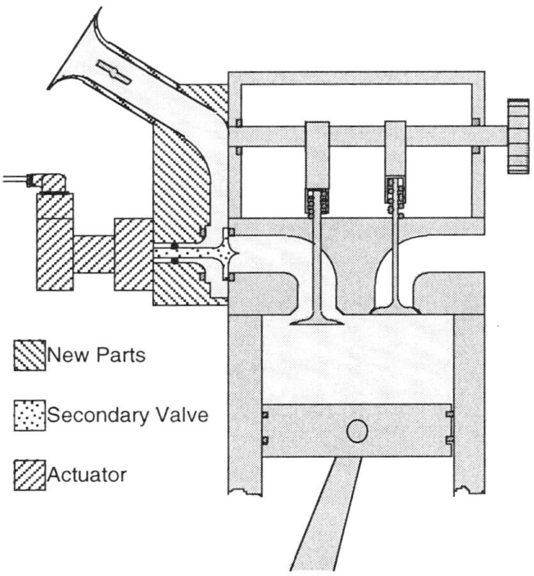

Figure 1: Diagram of cylinder head and manifold, showing secondary valve (closed)

Our secondary valve inevitably constitutes an additional flow constriction in the intake of the engine, and would therefore be expected to adversely affect the peak power obtainable at high engine speeds. The valve was designed so that the minimum area through which the flow must pass is never less than that of the intake port entrance, and there are no sudden flow expansions; these measures were an attempt to reduce this constriction effect. At the engine speed of our experiments (1200 rpm and lower) the flow constriction was too small to be measurable, and we were unable to run the engine at higher engine speeds because it was fitted with a piston extension and quartz piston insert (for LDA measurements). We are therefore not able to make a definitive statement on the peak power implications of the flow restriction due to our actuator.

Our implementation does have the significant advantage over other schemes proposed in the literature that it need not operate at high engine speeds, as implementations mechanically linked to the crankshaft must do -- it can be simply left open.

We re-iterate that we made no effort to implement an energy-efficient actuator: this was beyond the remit of this project. Our actuator consumed a maximum of 1.6 J of electrical energy and about 2.6 J of pneumatic energy per cycle (which does compare favorably with our pumping work savings of about 8 J per cycle, see below).

ENGINE AND EXPERIMENTAL SETUP

The engine used for these experiments was an AVL model, 500 cm^3, single cylinder research engine. For the load control/pumping loss experiments, the engine was operated with its standard camshaft. This has conservative timing and effectively no overlap.

For the backflow control experiments a modified camshaft was used . This was made by changing the phasing of the cam lobes on a spare camshaft, to generate a large overlap of about 110 degrees of crankshaft timing. Since the total lift time was not altered, this inevitably resulted in either later exhaust valve opening or earlier intake valve closing (or both). We accept that this results in an unrealistic gas exchange process: but we found that it was the only method of obtaining a significant backflow in this motored engine. More realistic measurements would require a fired engine, an extra order of complexity.

All our experiments were undertaken with the engine motored, and all the results presented in this paper were taken at an engine speed of 1200 rpm. The exhaust manifold was removed to reduce distracting dynamic effects, so the exhaust port vented directly to the atmosphere. During all the experiments the engine coolant was kept at a constant 50° C; this helped ensure repeatability.

The secondary valve described above was bolted to the engine intake port (as illustrated in Figure 1). The intake to the valve was connected to a throttle via a short pipe. For the load control/pumping loss reduction experiments, the engine was operated with the throttle open to the atmosphere. For the valve overlap/backflow control experiments, the throttle was connected to an airflow metering system. This consisted of a constant-pressure 'lung' unit, a large (2000 litre) surge tank, and a venturi-type air flow meter through which the surge tank was replenished from a compressed air line. The 'lung' was a sealed unit with flexible rubber side walls; it was monitored visually to determine when the mean steady airflow in was equal to the (unsteady) airflow to the manifold. The steady airflow was then measured with the air flow meter. The air supply to the surge tank was adjusted manually while watching the lung.

Instrumentation consisted of an in-cylinder pressure sensor, a pressure sensor located just outside the secondary valve in the manifold, and the air flow meter. In addition, during all experiments, engine position/timing data and the secondary valve control signal were recorded. All experimental data was recorded on a PC via an A-D card for subsequent post-processing.

During all experiments the engine was motored with air only. The peak cylinder pressure, at the end of the compression stroke, was used as a measure of equivalent engine 'load' -- it is approximately an indication of the mass of air in the cylinder (16). The accuracy of this load estimation technique is slightly reduced by the possibility of the charge temperature varying depending on the load control technique used, but it was determined that this small effect could be reasonably neglected.

The manifold pressure at part load conditions was not as constant as would be desired; the manifold tended to fill almost to atmospheric pressure between intake strokes, and fall during the stroke. This phenomenon, typical of port-throttled engines, tended to reduce the pumping losses. Some improvement was made at lower loads by using a second supplementary throttle in series.

LOAD CONTROL EXPERIMENTS

Throttleless load control was attained by closing the secondary valve during the intake stroke while the piston was still moving downwards in the cylinder, thereby limiting the charge induced to the cylinder. In this mode of operation, the secondary valve was set to close abruptly at some variable point during the intake stroke, and then to re-open during the compression stroke well after the main intake valve had closed.

During experiments the 'load' was set to a suitable value using the throttle, with the secondary valve fully open. The pressure-time traces and other data were recorded for later analysis. The throttle was then opened fully, and the secondary valve activated. The time of opening of the secondary valve was then adjusted to obtain the same load (estimated from the peak cylinder pressure). The instrumentation data was then re-recorded.

Figure 2 shows typical pressure-volume traces, for the same load, with conventional throttling and with the secondary valve operating. Only the 'pumping loop' part of the trace is shown. In the throttled case, it is noted that the pressure in the cylinder falls significantly during the intake stroke. (This effect is actually more obvious in Figure 4, which is for a rather higher load.) This is because of the throttle oversizing problem mentioned earlier. Note that this tends to reduce the pumping losses as well for a given load, therefore reducing the apparent effectiveness of the early intake valve closure as a pumping loss mechanism. If the throttle were 'ideal', so the manifold pressure was constant during the intake stroke, the throttled case would follow a path closer to the dotted line also shown: this would be more nearly the case with a multicylinder engine with a throttle upstream of an intake plenum.

The term 'pumping work' is somewhat arbitrary. For the purposes of obtaining quantitative values we have integrated the pressure-volume diagram area between the in-cylinder pressure and atmospheric pressure, from top dead centre to bottom dead centre of the intake stroke. This is illustrated in Figure 2. While this definition neglects the effects of the actual valve timing, it does provide an intuitively reasonable basis for comparing the two load control mechanisms.

Figures 2 - 6 show the pressure-volume 'pumping loop' diagram for a number of different loads. In each case the data with throttle load control and secondary valve load control are shown (they coincide at full load). In Figure 7 the magnitude of the pumping losses with the conventional throttle, using the secondary valve for load control, and the savings therefrom, are given for all the loads tested. The greatest absolute savings in pumping work is at mid load, but the savings are almost constant at all loads, and at loads near full load the pumping work is reduced to almost nothing.

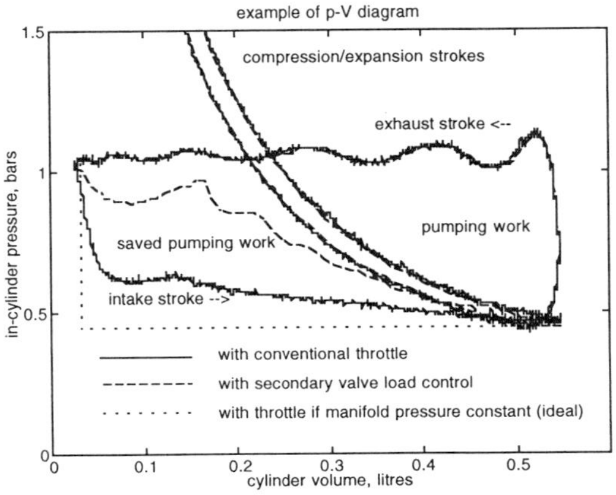

Figure 2: Typical pressure-volume traces at part load condition, with the load controlled both by throttle and by the secondary valve. This was at 34% of full load.

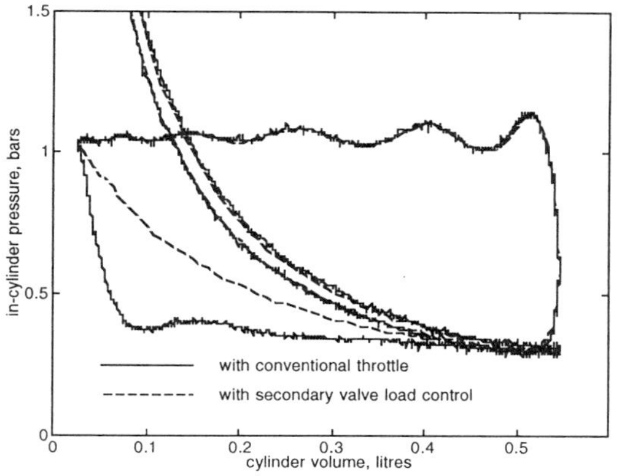

Figure 3: Pressure-volume traces at 26% of full load.

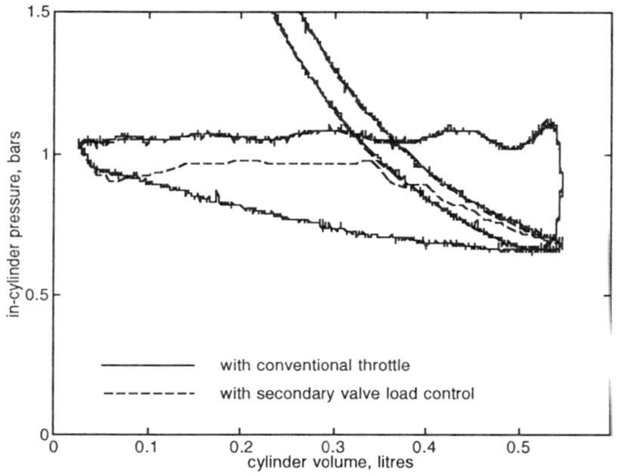

Figure 4: Pressure-volume traces at 70% of full load.

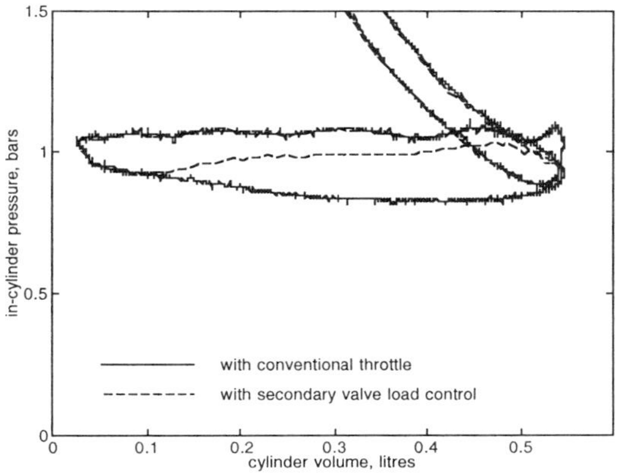

Figure 5: Pressure-volume traces at 95% of full load.

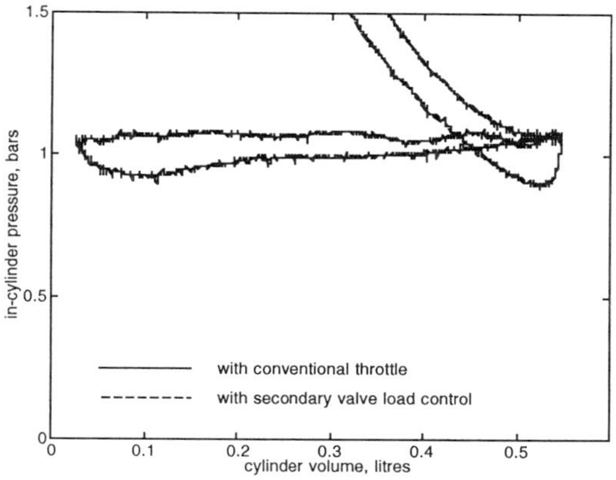

Figure 6: Pressure-volume traces at full load. Note that the traces coincide, since at full load the secondary valve is always open.

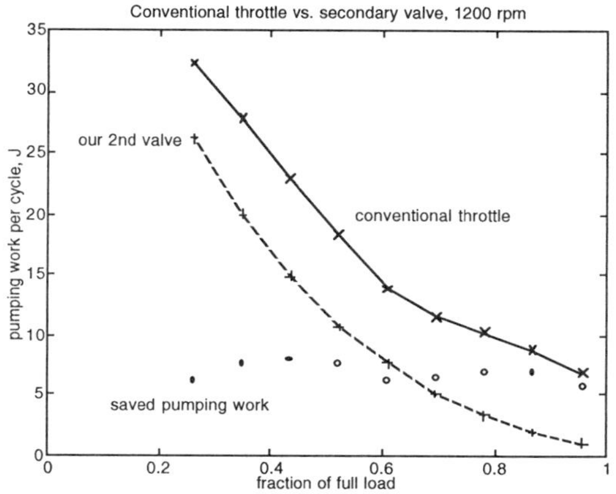

Figure 7: Summary of 'pumping work' with throttle and with secondary valve load control.

That the saved pumping losses are almost constant with load is initially surprising, since the total pumping work of course decreases rapidly with increasing load. From considering the form of the plots in Figures 2 to 6, though, it is clear that at lower load there is a large apparent pumping component which cannot be recovered. It should be noted that this is slightly artificial, since this 'pumping work' component (which corresponds to the section labelled 'pumping work' on the right hand side of the compression stroke region in Figure 2) is in fact recovered during the compression stroke.

We have computed the minimum possible pumping work attainable with a variable valve system, by computing what the pumping work would be if the valve closed instantly and there was no trapped volume between the secondary valve and the poppet valve (or, indeed, if load control was achieved by varying the timing of the poppet valve). This is an ideal that could never of course be attained in practice. The performance of this 'ideal' secondary valve is shown in Figure 8. It is seen that our valve is not far off this, particularly at higher loads. Given the significant size of our trapped volume, and that our valve is relatively slow acting, this was rather gratifying. At the lowest load tested, our valve has about 55% of the performance of such an 'ideal' valve; at mid loads, it was about 80% as efficient as the 'ideal'.

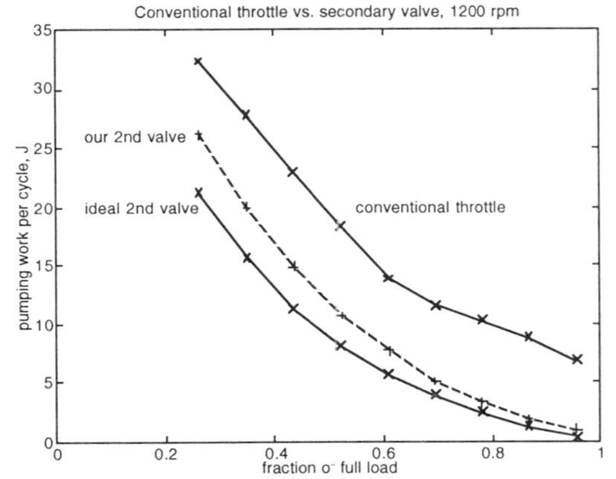

Figure 8: Comparison of experimental performance of our system with theoretical best performance of ideal early closing intake valve system. The work below the 'ideal 2nd valve' curve cannot be saved, and is in fact fully recovered at the start of the intake stroke.

We did attempt to evaluate the effect of the trapped volume by inserting a 1cm diameter tube into the intake port, between the secondary valve and the poppet valve, and filling the rest of the port with a sealant. We found however that the potential savings of pumping work were more than nullified by the introduction of significant pressure losses due to flow friction in this small tube. We concluded that reducing the trapped volume was only reasonably achievable by locating the secondary valve physically closer to the primary valve.

It is not obvious how to relate these savings to actual fuel savings in use, but it must be remarked that the absolute savings in power are significant. If we take a typical saving of 8 J per cycle at 1200 rpm, this corresponds to 80 W of power. If we interpolate that to a 4-cylinder 2-litre displacement engine at twice the engine speed, to obtain a reasonable operating condition for a modern automobile engine, and assume the same savings can be achieved per firing cycle, this corresponds to 640 W, slightly less than one horsepower. Since such an engine may be producing of the order of 20 horsepower or 15 kW the fuel savings are likely to be of the order of 5%.

Another way of interpolating these results to a full engine is to consider the work (power) saved per cycle as a fraction of the power produced by the engine at the output, making reasonable assumptions about the engine's performance. We show this in Figure 9. Saving the pumping losses would appear as more power available to the user for the same fuel, as marked on the axis; this can be taken roughly as a measure of the fractional saving in fuel possible for the same power. The lower curve is what we have demonstrated with our engine. It will be recalled that our pumping work with the throttle is less than expected for a multicylinder engine, because the manifold pressure falls so much during the intake stroke rather than being constant at the lowest value (see e.g. Figure 4). The upper curve shows an upper bound for the savings possible, presuming the pumping losses were greater due to this mechanism. The real possibilities for a multicylinder engine are expected to lie between these two bounds.

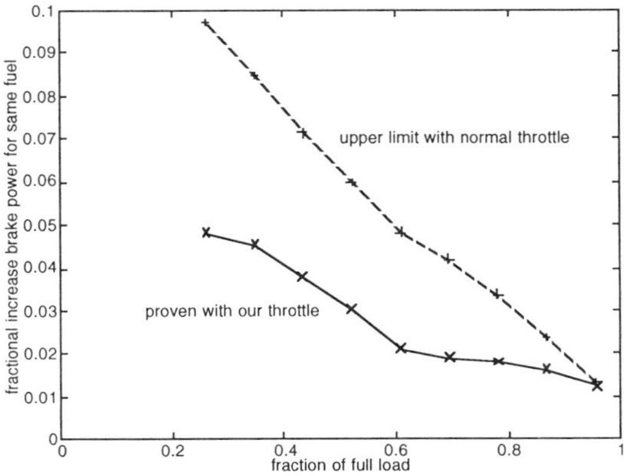

Figure 9: Extrapolation to fractional power savings possible in a firing engine. The upper curve is made with the assumption that the manifold pressure is kept constant during the induction stroke, as is more nearly the case for a multicylinder engine.

VALVE OVERLAP CONTROL EXPERIMENTS

An important possible application of variable valve timing, implemented with a secondary valve or otherwise, is to enable some control over backflow from the exhaust to the inlet at low loads, while still permitting significant valve overlap at higher loads and higher rpm's to enhance volumetric efficiency and peak power. In this mode of operation the secondary valve is opened after the conventional poppet valve. The purpose is to reduce the flow of gas from the exhaust to the intake manifold. In these experiments the load was regulated with the use of a throttle, so the manifold pressure is below the exhaust pressure. In a real engine, the benefits expected to accrue from such backflow control would be better burning and better controllability at light load conditions. These unfortunately could not be demonstrated on a motored engine. In addition, the significant effect of the higher temperature and lower density of the exhaust gases in a fired engine is absent.

The engine was fitted with the modified camshaft with large overlap of approx. 110 degrees crank angle. We shifted the camshaft timing so that the exhaust valve closed at 40° ATDC, with the intake valve opening at 70° BTDC. This resulted in a very late exhaust valve opening (at 20° ABDC) and early inlet valve closing (at 40° BBDC), a highly impractical state for an operating engine. With these changes we were able to clearly observe backflow with this engine, and to control it: but we accept that the manner we obtained it was unrealistic.

The experiments were carried out with the airflow measurement apparatus previously described. We ran the engine at a load slightly above the minimum load possible with our throttle (to ensure some net flow through the engine) and with the secondary valve deactivated. We recorded the same data as before and the airflow measurement, repeating the experiments 3 times. We then activated the secondary valve, timed to open after the exhaust valve closed. We readjusted the throttle to ensure the engine load was the same as in the earlier case (the backflow control had the effect of slightly reducing the load, probably as a result of the increased flow past the throttle). We then recorded all the data as usual, and again repeated the experiment 3 times.

Figure 10 shows the measured in-cylinder and manifold pressure during exhaust and intake strokes when the device was not operating. It can be clearly seen that the intake manifold pressure rises abruptly to atmospheric when the intake valve opens, indicating that the manifold is being filled from the exhaust manifold through the cylinder. This is evidence for the presence of significant backflow. The intake manifold and cylinder pressure then drop during the intake stroke, after the exhaust valve closes, as the throttle takes effect. Figure 11 shows the same data recorded with the secondary valve acting. In this instance it is clear that the backflow into the bulk of the intake manifold is eliminated. (It should be noted that the pressure sensor is located just upstream of the secondary valve, and there would still be some backflow into the trapped volume between the secondary valve and the conventional poppet valve.) A slight rise in manifold pressure can be discerned as the secondary valve opens; this is to be expected as the cylinder and trapped volume pressure is greater than the intake manifold pressure. It is seen that this is accompanied by an abrupt drop in cylinder pressure, as the pressure equalises across the secondary valve. Thereafter the data follows the same curve as in the case without the secondary valve acting.

The airflow measurements were difficult because of the very low levels of airflow at these low loads and high backflows, and by the inevitable unsteadiness in the airflow. The best estimate of the airflow in the case when the secondary valve was not operated was 0.75 ± 0.25 l/s, and with the secondary valve operated at the same load 1.32 ± 0.16 /s. There is therefore no doubt that the total mass airflow through the engine was increased by the action of the secondary valve, and since the cylinder charge was the same in both cases, the backflow from the exhaust to the intake manifold must have been reduced.

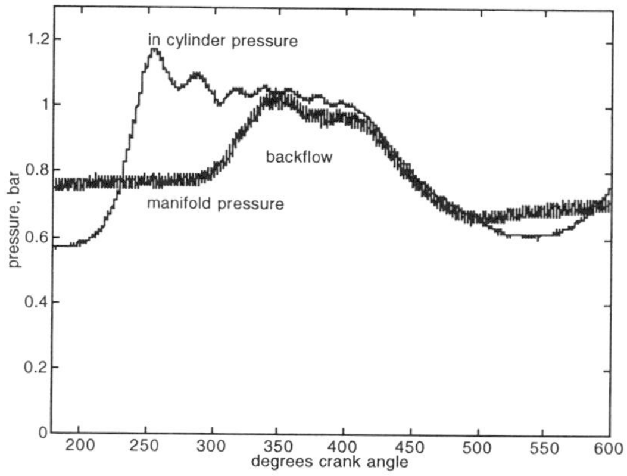

Figure 10: In-cylinder and manifold pressure during exhaust and intake stroke. The valve overlap is approx. 110 degrees crank angle. It is seen that the manifold pressure rises to atmospheric during overlap because of backflow through the exhaust valve and cylinder.

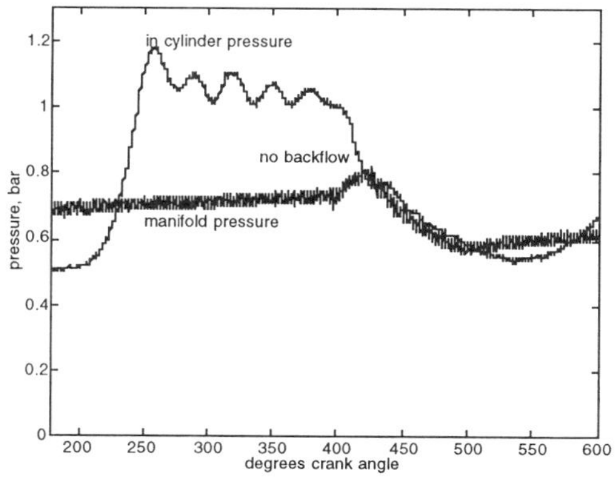

Figure 11: The same configuration as Figure 10, and the same load, but with the secondary valve operating to prevent backflow.

These measurements are not indicative of the reductions in backflow to be expected in a firing engine because the valve timings necessary to clearly observe valve flow in this motored engine were unrealistic. We can only assert that having obtained unrealistically significant backflow, we were able to control it with our device.

DISCUSSION

We have implemented an actuator able to test the secondary valve concept. This actuator is able to open or close the intake port upstream of the poppet valve in a few milliseconds. With this actuator we were able to use the secondary valve concept to demonstrate the main identified benefits of variable valve timing: throttleless load control, with accompanying savings of pumping work, and backflow control at low loads (the latter obtained with extreme valve timings). The results obtained are in line with the predictions made on the basis of theoretical work: i.e., fuel savings at part load of around 5% and at idle of about 15% (the idle figure being based purely on analytical work alone). These predictions, along with the expected benefits of the backflow control possibilities, await validation on a firing engine.

It is worth noting that when a secondary valve is used to regulate the engine load it becomes possible to vary the engine load from one cycle to the next -- indeed, in a multicylinder engine, from one cylinder firing to the next. This might have considerable value in enabling better control of an engine in transient conditions, as well as in controlling cylinder to cylinder variations in torque, air-fuel ratio, etc.

The next stage of our study will be to implement the secondary valve concept on a production engine. We currently intend to implement a very similar actuator to that described in this paper on such an engine. This device has proved itself to be remarkably robust, reliable and repeatable. If the benefits deduced from the experiments described in this paper are indeed demonstrated on a production engine, the next task would be to investigate practical implementations of this concept for automotive application.

ACKNOWLEDGMENTS

We acknowledge assistance from colleagues and technicians at ETHZ in making this work possible. We also wish to thank those who reviewed this paper on behalf of the SAE for valuable comments and for bringing further references to our attention.

REFERENCES

(1) A review of variable engine valve timing. C. Gray, SAE paper 880386, 1988.
(2) Variable valve actuation methods and the potential for their application. R. Stone and E. Kwan, SAE paper 890673, 1989.
(3) Controlling engine load by means of early intake valve closure. J.H. Tuttle, SAE paper 820408, 1982.
(4) Effects of intake-valve closure timing on spark-ignition engine combustion. S. Hara, Y. Nakajima, S. Nagumo, SAE paper 850074, 1985.
(5) Variable valve timing - a possibility to control engine load without throttle. H.P. Lenz, K. Wichart, SAE paper 880388, 1988.
(6) Variable valve actuation systems for the optimisation of engine torque. W. Demmelbauer-Ebner, A. Dachs, H.P. Lenz, SAE paper 910447, 1991.
(7) Performance evaluation of a camless engine using valve actuators with programmable timing. L.A. Gould, W.E. Richeson and F.L. Erickson, SAE paper 910450, 1991.
(8) Variable gas exchange systems for SI engines - layout and experimental data. L. Mikulic, J. Schommers, B. Geringer, K. Wolf and Ch. Enderle, SAE paper 920296, 1992.
(9) Strategies to improve SI-engine performance by means of variable valve intake lift, timing and duration. P. Kreuter, P. Heuser and M. Schebitz, SAE paper 920449, 1992.
(10) A non-throttled gasoline engine using a variable shutter valve. S.C. Low, SAE paper 911229, 1991.
(11) The development of new throttle valve actuation system for natural gas engine. M. Azechi, H. Kanesaka, A, Nishida, Y. Terauchi and T. Yoshimitsu, SAE Paper 920454, 1992.
(12) A high expansion ratio gasoline engine with intake control rotary valve installation. K. Ichimaru, H. Sakai, H. Kanesaka, D.E. Winterbone, JSAE Review Vol. 14, No. 4, p. 12 -- 15, October 1993.
(13) Cut-off control: a promising method of load regulation in spark ignition engine. M.M.Y. Abdalla, SAE Paper 940202, 1994.
(14) Suppression of pumping loss at partial load by throttling with a rotating distributor. Martin and Tazerout, International Conference on New Developments in Powertrain and Chassis Engineering, Strasbourg, June 1987.
(15) Variable Steuerzeiten und variable Verdichtung beim Ottomotor. P. Walzer, P. Adamis, H. Heinrich and V. Schumacher, MTZ, Vol. 47, No. 1, p. 15 - 20, 1986.
(16) A convenient technique for determining comparative volumetric efficiency. K. Roussopoulos, SAE paper 900352, 1990.

970248

Dynamic Model of a Springless Electrohydraulic Valvetrain

Dean H. Kim, Mark D. Anderson, and Tsu-Chin Tsao
University of Illinois at Urbana-Champaign

Michael B. Levin
Ford Motor Co.

ABSTRACT

A dynamic model for the springless electrohydraulic valvetrain has been developed. The model speeds up the valvetrain development process by simulating effects of parameter changes, thus minimizing the number of hardware variations. It includes dynamic characteristics of check valves that enable energy recovery, hydraulic snubbers that limit seating velocity of the engine valves, and leakage in the control solenoids. A good match of the experimental data has been obtained for a single valve system, and the model calibration and validation have been completed. The known parameters are used together with some unknown calibration constants which have been tuned to match the experimental data. The simulation results for a twin valve system are also presented. The model applications for system performance analysis and for the closed-loop control of the engine valve lift are described. The cyclic variability of the experimental data is also discussed.

INTRODUCTION

The increasing need to improve fuel economy and to reduce exhaust emissions has forced engine designers to consider alternative methods of valve operation. An experimental springless electrohydraulic valvetrain has been developed at Ford Research Laboratory. Hydraulic pressure controlled by a pair of solenoid valves and a pair of check valves is the only means to operate engine valves. The valvetrain system enables continuously variable control of engine valve timing, lift, and event duration as described by Schechter and Levin [1]. The hydraulic pendulum concept permits energy recovery for both the opening and the closing of engine valves. It is done by converting the potential energy of the high pressure fluid to accelerate the engine valve, and then by converting the kinetic energy of the moving engine valve to the potential energy of the high pressure fluid.

A dynamic model for the camless valvetrain has been developed using the Matlab-Simulink software. The model aids in the design process by demonstrating effects of parameter changes for a given design, and by showing the effects of system design alterations. The model uses known physical parameters and a few unknown calibration constants that have been tuned to match the simulation results to the experimental data.

This paper is organized as follows. The basics of the springless electrohydraulic actuation are described briefly. The model for a single valve system is presented with its simplifying assumptions. The submodels for the snubber chamber, the solenoid leakage, and the check valve are described. A calibration procedure has been developed to tune the listed calibration constants in order to match the given experimental data for three lifts of the engine valve at a fixed engine speed. The model is then used to study and improve the system performance with various design changes. The model application for a closed-loop lift control is shown as a response to changes in lift and speed commands. The single valve model is further expanded to include a twin valve system consisting of two identical and hydraulically coupled engine valves. The twin valves use the same four control components, i.e. a pair of solenoids and a pair of check valves. An unequal lift modifier enables lift adjustments for either of the two engine valves. The cyclic variability study of the experimental system includes the variations of the engine valve lift and the cycle time.

OPERATION OF A SINGLE VALVE SYSTEM

The schematic of a single valve system is depicted in Figure 1. Hydraulic oil is supplied to the high pressure reservoir by a pressure-compensated variable displacement pump, while the low pressure reservoir is held at a constant pressure by a back pressure regulator. The high pressure solenoid valve (HPSV) and the low pressure solenoid valve (LPSV) are opened and closed in order to allow the fluid flow into and out of the control chamber located above the valve piston. This flow determines the control chamber pressure, and ultimately, the engine valve lift. Just before the valve seating, a smaller chamber called snubber chamber is isolated from the rest of the control chamber, and its rising pressure acts as a brake. The snubber chamber becomes effective when the engine valve lift is less than the snubber overlap, typically within 0.5 mm of the seated position. In Figure 1, the length of the snubber overlap is exaggerated for the sake of clarity, and the engine valve is shown in the seated position (x_v=0). During engine valve operation, the high pressure check valve (HPCV) and the low pressure check valve (LPCV) are opened when the control chamber pressure exceeds the high pressure or falls below the low pressure, respectively.

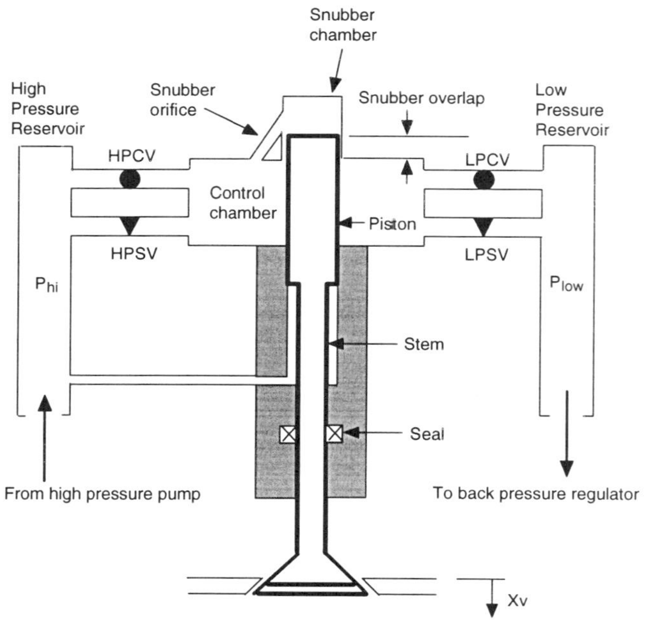

Figure 1: Schematic of single-valve system.

MODEL FOR A SINGLE VALVE SYSTEM

The initial model development focuses on a single valve system, with the eventual goal to simulate the operation of a 16-valve 4-cylinder valvetrain. The fundamental governing equations are provided in Appendix A. Several simplifying assumptions are made. The discharge coefficient for each turbulent flow is assumed constant. The respective pressure distributions within the control chamber and snubber chamber are neglected. The reservoir pressures are assumed constant, as are the oil properties which include the viscosity and the bulk modulus. The cylinder pressure acting on the engine valve is neglected. The model inputs are the solenoid valve lifts that control most of the fluid flow into and out of the control chamber.

The control chamber pressure is described as follows, based on EQ (A.2):

$$p_r = p_{cc},\ V_o = V_{cc},\ \text{and}$$
$$\Sigma Q = Q_{hpsv} - Q_{lpsv} - Q_{hpcv} + Q_{lpcv} + Q_{vel} + Q_{lkhpsv} - Q_{lklpsv} + Q_{sn} + Q_{bet} + Q_{ex} \quad (1)$$

where

Q_{hpsv} is flow across HPSV, based on EQ (A.3):

$$\text{flow area} = W_{hpsv}\, x_{hpsv},\ \Delta p_r = P_{hi} - p_{cc} \quad (2)$$

Q_{lpsv} is flow across LPSV, based on EQ (A.3):

$$\text{flow area} = W_{lpsv}\, x_{lpsv},\ \Delta p_r = p_{cc} - P_{low} \quad (3)$$

Q_{hpcv} is flow across HPCV, based on EQ (A.3):

$$\text{flow area is from EQ (A.6)},\ \Delta p_r = p_{cc} - P_{hi} \quad (4)$$

Q_{lpcv} is flow across LPCV, based on EQ (A.3):

$$\text{flow area is from EQ (A.6)},\ \Delta p_r = P_{low} - p_{cc} \quad (5)$$

Q_{vel} is rate of control chamber volumetric change due to the valve motion:

$$Q_{vel} = -A_{up}\,\dot{x}_v \quad (6)$$

Q_{khpsv} is leakage flow through HPSV, based on EQ (A.4):

$$\Delta p_r = P_{hi} - p_{cc},\ K_{lam} = K_{lkhpsv} \quad (7)$$

Q_{lklpsv} is leakage flow through LPSV, based on EQ (A.4):

$$\Delta p_r = p_{cc} - P_{low},\ K_{lam} = K_{lklpsv} \quad (8)$$

Q_{sn} is flow through snubber orifice, based on EQ (A.3):

$$\text{flow area} = \frac{\pi}{4} D_{snubber}^2,\ \Delta p_r = p_{sn} - p_{cc} \quad (9)$$

Q_{bet} is flow between snubber chamber and control chamber in the overlap region, based on EQ (A.4):

$$\Delta p_r = p_{sn} - p_{cc},\ K_{lam} = K_{Qbet} \frac{\pi D_{piston} C_{sn}^3}{12 \mu L(x_v)} \quad (10)$$

Q_{ex} is flow between snubber chamber and control chamber when piston is not in the overlap region, from EQ (A.3):

$$\text{flow area} = K_{ex} x_v,\ \Delta p_r = p_{sn} - p_{cc} \quad (11)$$

The control chamber pressure is affected by several flows. The flows from the control valves are assumed to be turbulent because of the small orifices, as shown in EQ (2,3,4,5). The volume flow from the engine valve displacement is represented by EQ (6). Its effect is most apparent when all control valves are closed, as the engine valve motion affects the control chamber pressure due to the oil compressibility. Solenoid leakage flows, which have been observed in the experimental system, are included in EQ (7,8). These leakage flows typically result from production imperfections and natural wear. Leakages are difficult to model because of the variations in solenoid valve surface, which cause unpredictable flow areas. Here, a generic laminar flow is assumed, with the coefficients to be estimated from experimental data.

Accurate modeling of the flow between the control chamber and the snubber chamber EQ (9,10,11) is important because of the effect on valve unseating and seating. The flow through the fixed snubber orifice EQ (9) is assumed turbulent. This snubber orifice is always open. The flow between the two chambers along the piston surface area is represented by either EQ (10) or EQ (11), depending on the engine valve lift. When the valve lift is smaller than the snubber overlap, i.e. when the snubber chamber is isolated from the control chamber, laminar flow EQ (10) is assumed because the involved geometry resembles a journal inside a sleeve. This expression has a flow gain (K_{Qbet} > 1) which is tuned to ensure that the snubber chamber pressure and control chamber pressure converge rapidly during valve opening. This flow gain also ensures that the valve is decelerated properly during seating. When the valve lift exceeds the snubber overlap, i.e. when snubber chamber and the control chamber are joined, turbulent flow EQ (11) is assumed because the flow area here is similar to the flow areas for the solenoid valves and check valves.

Snubber chamber pressure is described by EQ (A.2):

$$p_r = p_{sn},\ V_o = V_{sn}\ \text{, and}$$
$$\Sigma Q = Q_{vel} - Q_{sn} - Q_{ex} - Q_{bet} \qquad (12)$$

EQ (12) shows that the snubber chamber pressure is affected primarily by the flows between the snubber chamber and the control chamber EQ (9,10,11). Typically, the enclosed volume in the snubber chamber is considerably smaller than the control chamber volume, which results in more dramatic pressure changes in the snubber chamber.

The engine valve is described by EQ (A.1):

$$x = x_v, M = M_v, \Sigma F = A_{up}\, p_{sn} - A_{dn} P_{hi} - B_v \dot{x}_v \qquad (13)$$

where

$$B_v = \text{viscous damping coeff.} = \frac{K_{calib}\, \mu\, A_{side}}{c} \qquad (14)$$

The snubber chamber pressure is used in EQ (13), instead of the control chamber pressure, in order to ensure that the correct pressure acts on the piston when the snubber action begins. After the valve clears the snubber chamber, the two pressures converge quickly because of the large flow area between their respective chambers. The friction between the engine valve piston and its sleeve, caused by shear stress, is modeled as translational viscous damping. The viscous damping coefficient of EQ (14) contains a calibration constant that is tuned using experimental data to account for unmodeled damping effects. These include fluid damping and seal friction on the valve stem shown in Figure 1.

High pressure check valve is described by EQ (A.1):

$$x = x_{hpcv}, M = M_{hpcv},$$
$$\Sigma F = (p_{cc} - P_{hi})A_{seat} - F_{preload} - K_{hpcv}\, x_{hpcv} \qquad (15)$$

Low pressure check valve is described by EQ (A.1):

$$x = x_{lpcv}, M = M_{lpcv},$$
$$\Sigma F = (P_{low} - p_{cc})A_{seat} - F_{preload} - K_{lpcv}\, x_{lpcv} \qquad (16)$$

For EQ (15,16), the net force must exceed the preload, typically 5 N, in order to open each check valve. Unlike the engine valve EQ (13,14), the check valve equations do not consider viscous damping forces because there is little contact area between the check valve and its seat.

CALIBRATION OF THE SINGLE VALVE MODEL USING EXPERIMENTAL DATA

Simulations were run on Matlab-Simulink using Euler's method with a fixed step size of 2 μsec. Other simulation algorithms such as Runge-Kutta and Adams were tried, however, they increased the simulation time by nearly twenty times with no differences in results. The model for the single valve system includes several calibration constants which must be tuned properly to ensure that the simulation results match the experimental data: the solenoid leakage coefficients K_{lkhpsv}, K_{lklpsv} from EQ (7,8), the flow gain K_{Qbet} from EQ (10), the engine valve damping constant K_{calib} from EQ (14), the amount of entrapped air in the oil (α) from EQ (A.5), and the snubber overlap x_{vcrit}. Figure 2 shows the top level of the Matlab-Simulink model that includes icons for the system components and indicates the computational sequence.

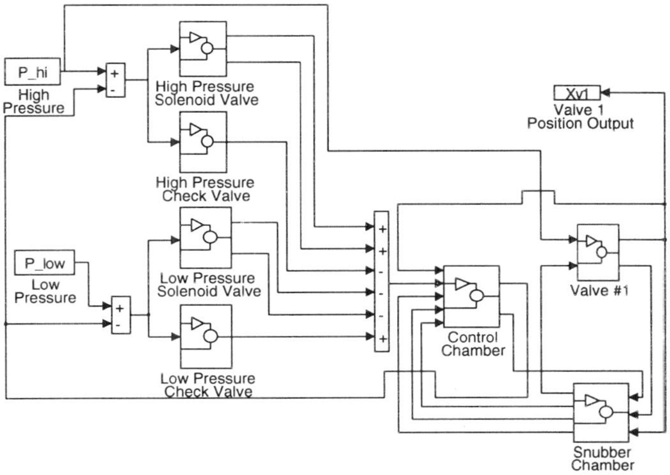

Figure 2. Simplified Matlab-Simulink model.

The experimental system consisted of a 16-valve cylinder head with only one pair of solenoids operating and the lift modifier adjusted to shut off one of the twin valves. The experimental data was available for the engine valve lift, solenoid valve lifts, check valve lifts, the control chamber pressure, and the snubber chamber pressure. The data was taken at an engine speed of 1500 RPM for event durations typical for that speed. The system parameters were: P_{hi} = 4.14 MPa (600 psi), P_{low} = 0.69 MPa (100 psi), D_{piston} = 8.5 mm, and D_{stem} = 6.0 mm. The data were obtained for maximum lift values of 2 mm, 4 mm, and 6 mm, which cover a wide range of engine operating conditions.

The range of typical lifts is suggested in Figure 3, where the valve lift and event duration for both the exhaust and the intake valves were adjusted to minimize fuel consumption and NOx emission for an unthrottled operation using an engine combustion model. (The unthrottled operation in this case means that there is no throttling by either a throttle plate or an intake valve, which admits the required amount of air without a detrimental pumping loss. The exhaust valve closing timing is used to control the residual gas content for NOx control). The lifts and events are compared for a 2.0 liter, spark-ignited engine and two valvetrains: a conventional camshaft-type and the camless electrohydraulic. Two different loads are shown for the engine speed of 1500 RPM. Here the near idle load of 1 bar BMEP (Brake Mean Effective Pressure) is obtained with an intake valve lift set to 1 mm and a 7 bar BMEP is reached with an approximately 4 mm lift. The exhaust valve lift was reduced to 2 mm to minimize the valvetrain energy consumption without any negative effects on the engine performance and emissions.

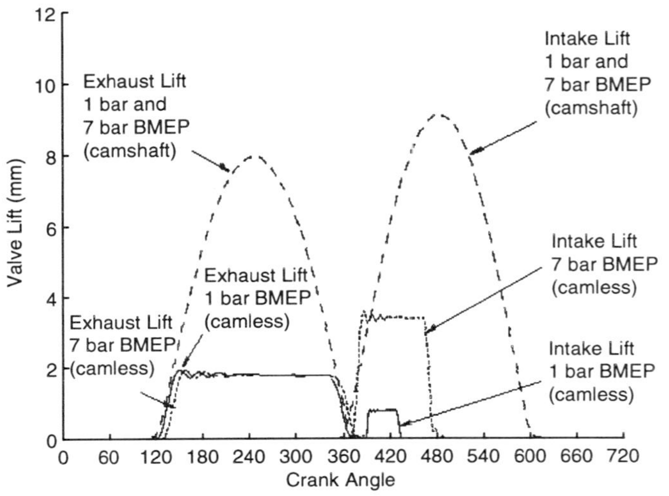

Figure 3. Valve lifts for throttled (camshaft) and unthrottled (camless) operation.

For an accurate comparison of the valvetrain model output with the experimental data, the experimental solenoid lifts were used as the model inputs (Figure 4).

Here the HPSV pulse is applied first for the valve opening, and then the two LPSV pulses are applied for the valve closing and seating, respectively. As the desired lift decreases, so do the opening HPSV pulse and the first (closing) LPSV pulse. For each of the three lifts the duration of the closing LPSV pulse is chosen such that the valve loses its kinetic energy at a lift (called lift undershoot) just prior to snubber engagement. This valve closing strategy is chosen to maximize the energy recovery, since the valve's kinetic energy during the seating process (snubber engagement) is wasted. The greater need for energy recovery exists at larger lifts because more energy is required to propel the valve. The second (seating) LPSV pulse does not usually require an adjustment with the lift and speed changes.

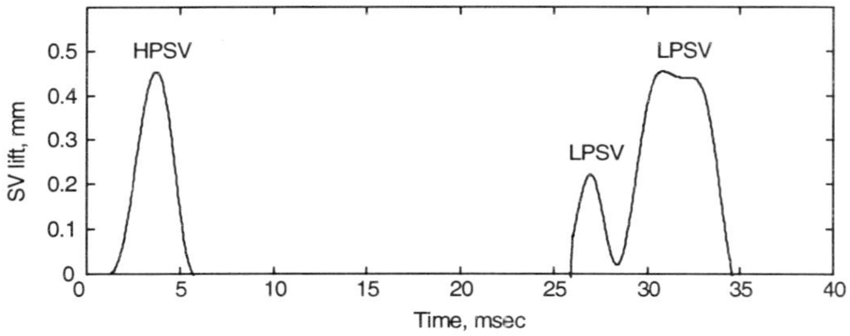

Figure 4. Input solenoid valve lifts for 6 mm EV lift.

The availability of the experimental pressure data for both the control chamber and snubber chamber is important for tuning the model because the flow calibration constant "K_{Qbet}" from EQ (10) can be set to ensure that both pressures converge properly (Figure 5).

The simulations are matched with the experimental data for the 6 mm lift (Figures 6, 7, 8) using the following calibration constants: $K_{lkhpsv} = 1.86 \times 10^{-13}$, $K_{lklpsv} = 2.06 \times 10^{-13}$, $K_{Qbet} = 2$, $K_{calib} = 6.3$, $\alpha = 0.0017$ (0.17 % air by volume), and $x_{vcrit} = 0.6$ mm. Appendix B describes how the experimental data are used to tune the calibration constants. Using the same calibration constants, the simulations for 4 mm lift and 2 mm lift are compared with the corresponding experimental data.

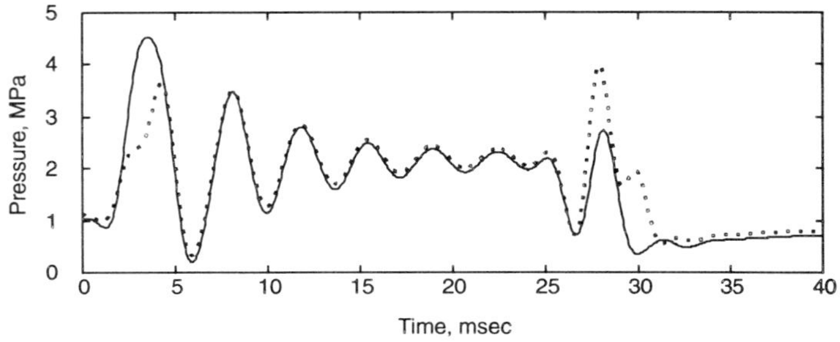

Figure 5. Experimental snubber pressure (dot) and control chamber pressure (solid) for 6 mm EV lift.

At the start of engine valve opening (Figures 6, 7, 8), the HPSV flow causes the control chamber pressure to approach the system high pressure. The snubber chamber pressure is also increased, although not as quickly. After the snubber action ends, the control chamber pressure and the snubber chamber pressure are the same. After the HPSV closes, the pressures initially drop due to oil expansion. At this point, the low pressure check valve (LPCV) should open, if the control chamber pressure falls below the low pressure. This would improve energy consumption because less oil is needed from the high pressure reservoir. Due to the relative lack of system damping, the pressures oscillate as the engine valve approaches its settled-down lift in the middle of the cycle.

At the start of engine valve closing, the flow from the closing LPSV pulse causes the pressures to approach the system low pressure. After this LPSV pulse closes, the pressures initially rise due to oil compression. At this point, the high pressure check valve (HPCV) should open if the control chamber pressure exceeds the high pressure. This would improve energy recovery because oil is returned to the high pressure reservoir. In order to open the check valves, and thereby improve system efficiency, the solenoid valves must close quickly. Due to the previously described seating strategy, the control chamber pressure and the snubber chamber pressure are the same until the seating LPSV pulse. Afterwards, the snubber chamber pressure increases, which reduces the seating velocity below 0.1 meters/sec. A better tuning of the seating keeps the snubber overlap in the 0.2 - 0.3 mm range.

The effects of the solenoid leakage can be seen at the end of the cycle, as the pressures are starting to increase. The solenoid leakage does not seem to affect pressures in the middle of the cycle, when the engine valve is reaching the settled-down lift. This suggests that the leakage flows past the high pressure solenoid and past the low pressure solenoid are similar at this point.

The engine valve unseating is delayed by 1 msec because the snubber chamber pressure initially is less than the control chamber pressure. During opening the valve lift has an overshoot that is not expected to affect engine performance, and the lift continues to oscillate similarly to the pressures.

During closing the valve approaches zero velocity just before the snubber action begins, and then is seated by the second LPSV pulse.

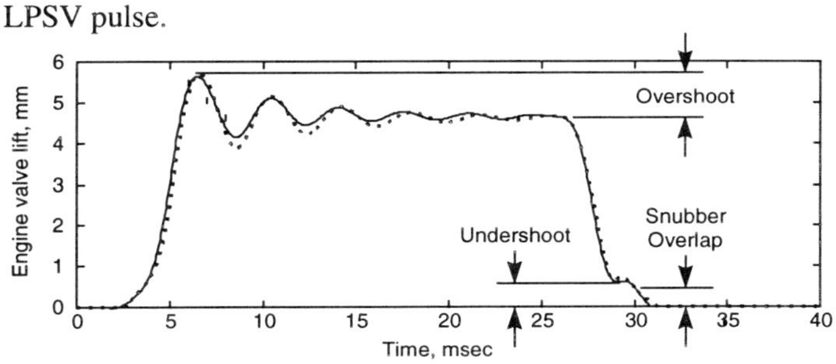

Figure 6. Engine valve lifts for 6 mm: experiment (solid) and simulation (dot).

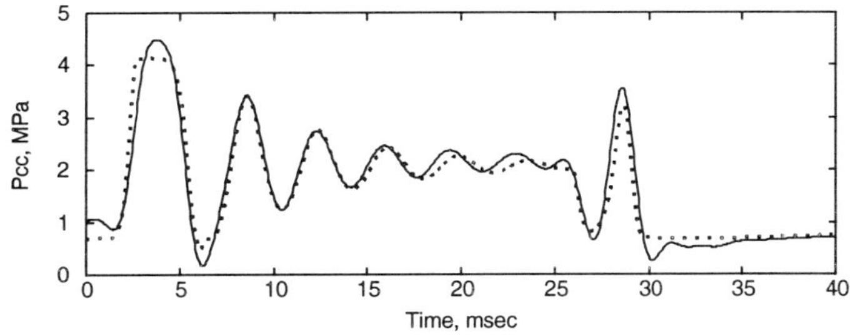

Figure 7. Control chamber pressure for 6 mm EV lift: experiment (solid) and simulation (dot).

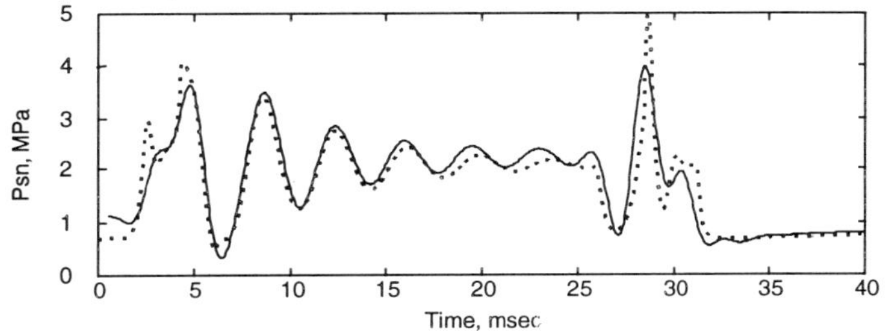

Figure 8. Snubber chamber pressure for 6 mm EV lift: experiment (solid) and simulation (dot).

Figure 7 illustrates that the simulated and the experimental responses for the control chamber pressure are matched. Figure 8 does likewise for the snubber chamber pressure, although the simulated values are initially more oscillatory than the experimental ones. This discrepancy can be attributed to the inexact model of flow between the snubber chamber and control chamber. Where the snubber is not engaged, i.e. for $x_v > x_{vcrit}$, the control chamber and the snubber chamber pressures are equivalent.

The results for the 4 mm lift are shown in Figures 9-11, as the simulations mirror the experimental data, using the calibration constants previously computed from the 6 mm data. The responses are less oscillatory when compared to the 6 mm case. Figure 10 suggests that for the smaller lifts the check valves are less likely to open because the solenoid valves close more slowly. This occurs because the solenoid valves available for these experiments do not open fully at small flows. On the other hand the overall energy consumption will be lower, hence less need for recovery. Throttling losses also increase since the solenoid valves spend less time in the fully open position. Figure 11 shows characteristics similar to Figure 8.

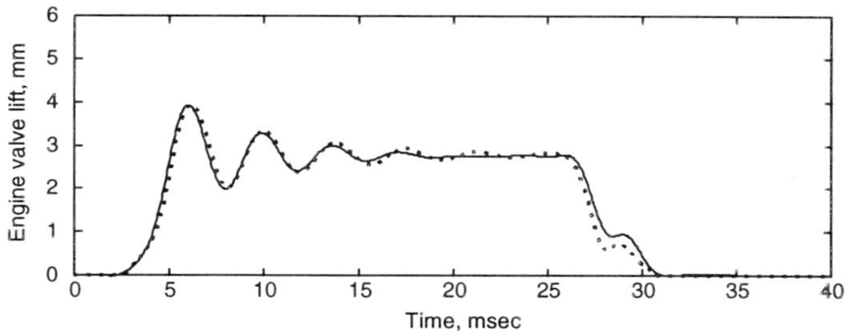

Figure 9. Engine valve lifts for 4 mm: experiment (solid) and simulation (dot).

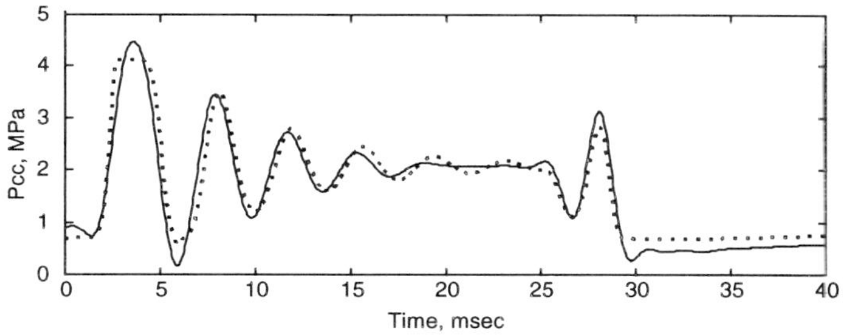

Figure 10. Control chamber pressure for 4 mm EV lift: experiment (solid) and simulation (dot).

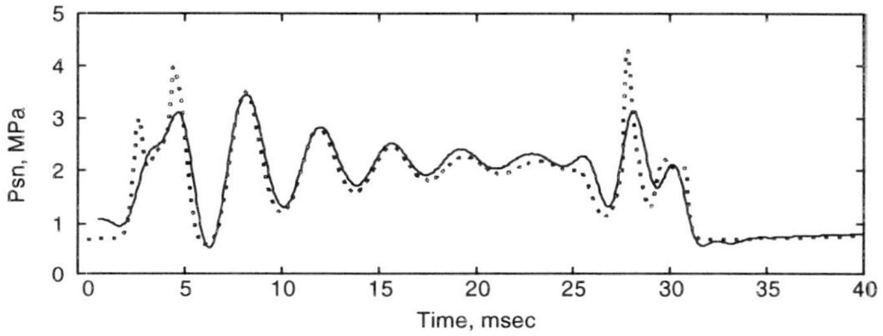

Figure 11. Snubber chamber pressure for 4 mm EV lift: experiment (solid) and simulation (dot).

The results for the 2 mm lift are shown in Figures 12-14, where the simulations again mirror the experimental data using the same calibration constants. The pressure and lift oscillations have been reduced even further.

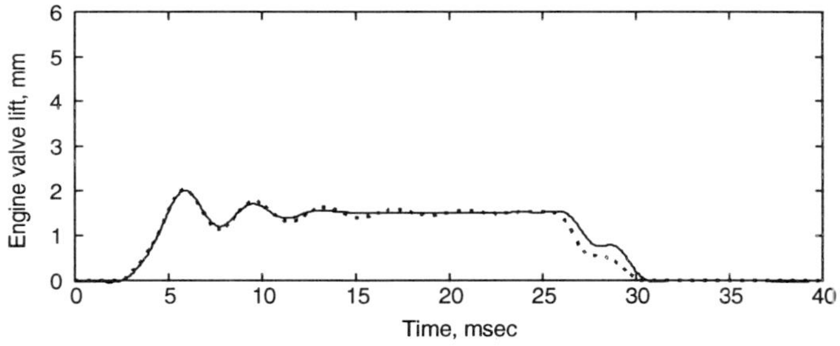

Figure 12. Engine valve lifts for 2 mm: experiment (solid) and simulation (dot).

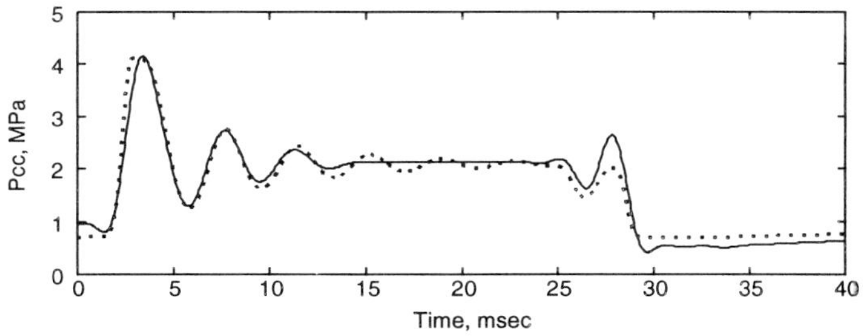

Figure 13. Control chamber pressure for 2 mm EV lift: experiment (solid) and simulation (dot).

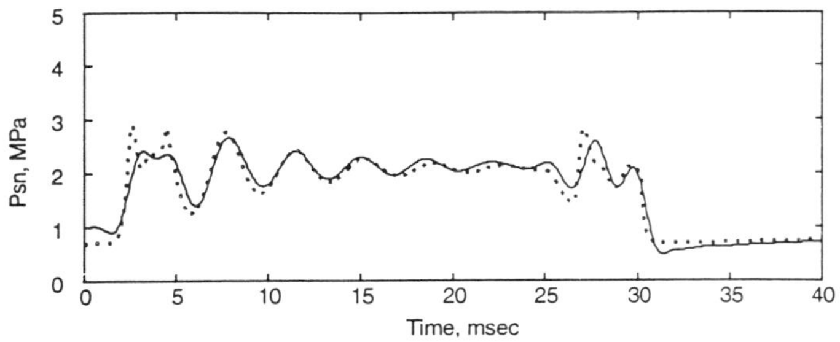

Figure 14. Snubber chamber pressure for 2 mm EV lift: experiment (solid) and simulation (dot).

MODEL FOR A TWIN VALVE SYSTEM

The model for the single engine valve system is expanded to simulate a twin valve system. Each valve pair shares the same control chamber, and utilizes the same control valves (i.e. HPSV, LPSV, HPCV, and LPCV). Figure 15 shows the schematic for a twin valve system.

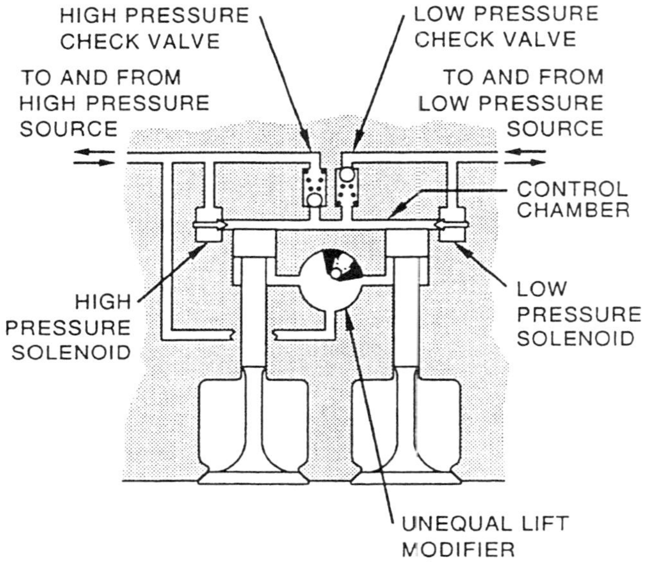

Figure 15. Schematic of a twin-valve system.

Control chamber pressure is described by EQ (A.2):

$$p_r = p_{cc},\ V_o = V_{cc},\ \text{and}$$
$$\Sigma Q = Q_{hpsv} - Q_{lpsv} - Q_{hpcv} + Q_{lpcv} + Q_{vel1} + Q_{vel2} + Q_{lkhpsv} - Q_{lklpsv} + Q_{sn1} + Q_{sn2} + Q_{bet1} + Q_{bet2} + Q_{ex1} + Q_{ex2} \qquad (17)$$

In EQ (17) the previous flows from EQ (6,9,10,11) now have a component for each of the twin valves.

Snubber chamber pressures are described by EQ (A.2): For the first valve,

$$p_r = p_{sn1},\ V_o = V_{sn1},\ \text{and}$$
$$\Sigma Q = Q_{vel1} - Q_{sn1} - Q_{bet1} - Q_{ex1} \qquad (18)$$

and for the second valve,

$$p_r = p_{sn2},\ V_o = V_{sn2},\ \text{and}$$
$$\Sigma Q = Q_{vel2} - Q_{sn2} - Q_{bet2} - Q_{ex2} \qquad (19)$$

The engine valves are described by EQ (A.1): For the first valve,

$$x = x_{v1},\ M = M_{v1},\ \text{and}$$
$$\Sigma F = A_{up1} p_{sn1} - A_{dn1} p_{hi1} - B_{v1} \dot{x}_{v1} \qquad (20)$$

and for the second valve,

$$x = x_{v2},\ M = M_{v2},\ \text{and}$$
$$\Sigma F = A_{up2} p_{sn2} - A_{dn2} p_{hi2} - B_{v2} \dot{x}_{v2} \qquad (21)$$

Initially, the chamber below each valve piston is assumed to have the same pressure:

$$P_{hi1} = P_{hi2} = P_{hi}. \qquad (22)$$

UNEQUAL LIFT MODIFIER - In a twin valve system, the unequal lift modifier allows flexibility in setting the lift of each valve, as shown in Figure 15. A detailed description is given in [1], while a brief description is given here. The unequal lift modifier is an adjustable throttle which restricts the flow into and out of the chamber below a particular valve piston. As a result, during the valve opening the pressure below that piston is increased, thereby reducing its lift. The pressure is assumed to change instantaneously in response to the volume of oil being moved by their corresponding valves. The flow areas A_1 and A_2 can be adjusted separately.

$$P_{hi1} = P_{hi} + \text{sign}(\dot{x}_{v1}) \frac{\rho}{2} \left(\frac{A_{dn1} \dot{x}_{v1}}{C_d A_1} \right)^2 \qquad (23)$$

and

$$P_{hi2} = P_{hi} + \text{sign}(\dot{x}_{v2}) \frac{\rho}{2} \left(\frac{A_{dn2} \dot{x}_{v2}}{C_d A2} \right)^2 \qquad (24)$$

The simulated response of a twin valve system when one valve (i.e. "x_{v1}") has a slightly larger area A_{dn} (by approximately 0.5%) is shown in Figure 16. The lift traces cross over, so that the valve with the greater maximum lift ends up with the smaller settled-down value. Another case is shown in Figure 17, where one valve (i.e. "x_{v1}") initially has greater lift because it has less viscous damping (by approximately 10 %). Both of these variations have been observed in the physical system, although the unequal lift modifier is more effective in the second case. The first case would require a tradeoff between the maximum and the settled-down lift values. For the second case (Figure 18), the unequal lift modifier reduces the lift of x_{v1}. The two lifts are now the same. Figure 19 illustrates how the pressures below the valves have changed after the lifts have been adjusted.

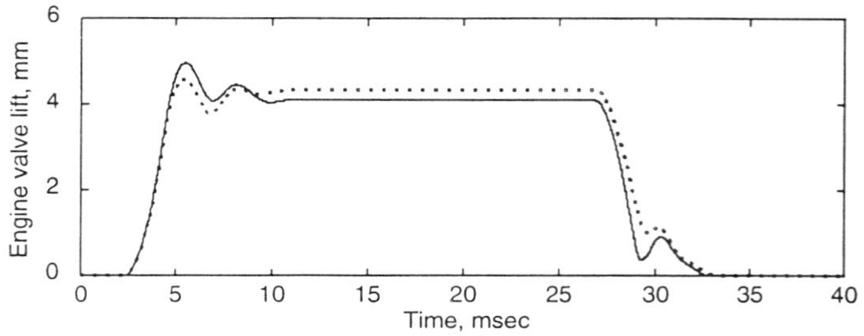

Figure 16. Crossover effect illustrated for non-identical valves: Xv1(dot) and Xv2(solid).

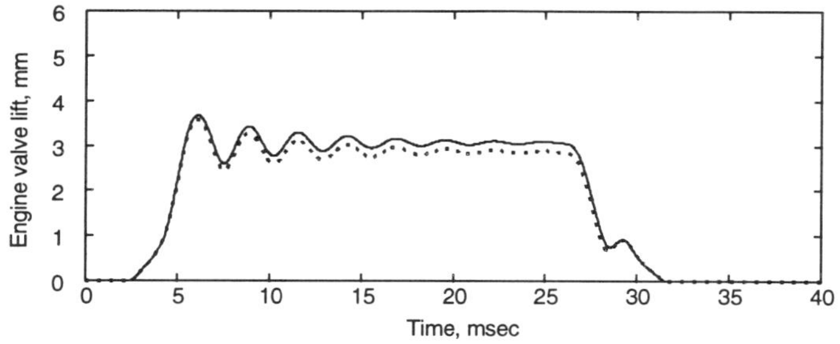

Figure 17. Uncompensated non-identical valve operation: Xv1 (solid) and Xv2 (dot)

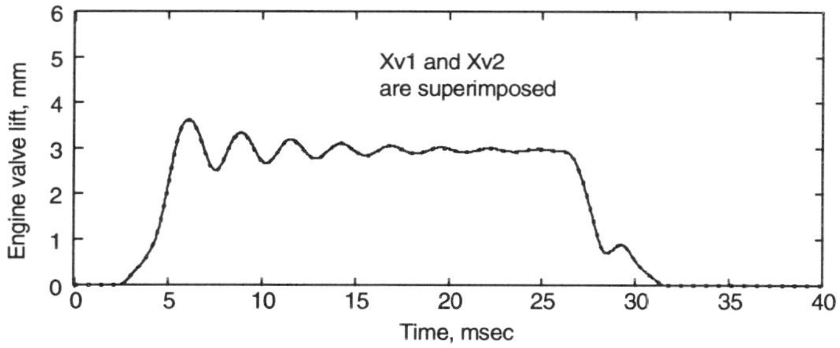

Figure 18. Compensated non-identical valve operation: Xv1 (solid) and Xv2 (dot).

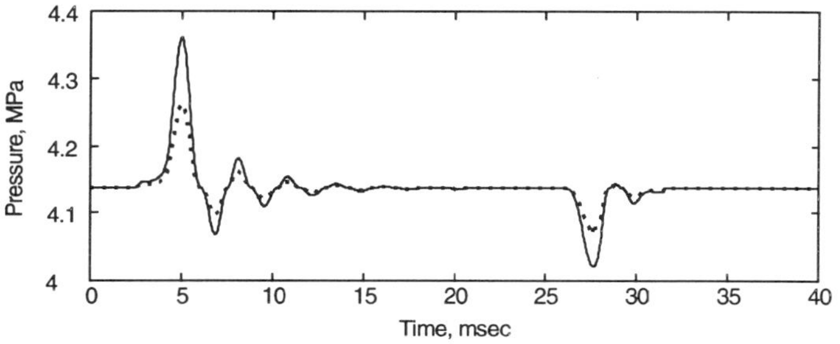

Figure 19. Pressures below pistons during unequal lift modifier operation: Phi1(solid) and Phi2 (dot).

MODEL APPLICATION FOR PERFORMANCE ANALYSIS AND ENHANCEMENT

VISCOSITY EFFECT - The validity of the model has been confirmed by the fact that the simulations and experimental responses are matched well. One of the first practical applications was the need to investigate how the system reacts to changes in fluid temperature. The baseline viscosity of μ=11.9 centistokes for the MIL5606 fluid was used for all of the previous simulations, which corresponds to a temperature of 40 degrees C. To simulate the low temperature operation the viscosity has been changed to μ=390 centistokes corresponding to a fluid temperature of -40 degrees C. The high and the low pressure solenoid lifts have been adjusted to obtain the same settled-down value of the engine valve lift (Figure 20). Increases in solenoid valve opening intervals of approximately 120% were needed, which would increase the electrical energy consumption. As expected, the motion of the valve is slower at low temperature, but proper seating is maintained.

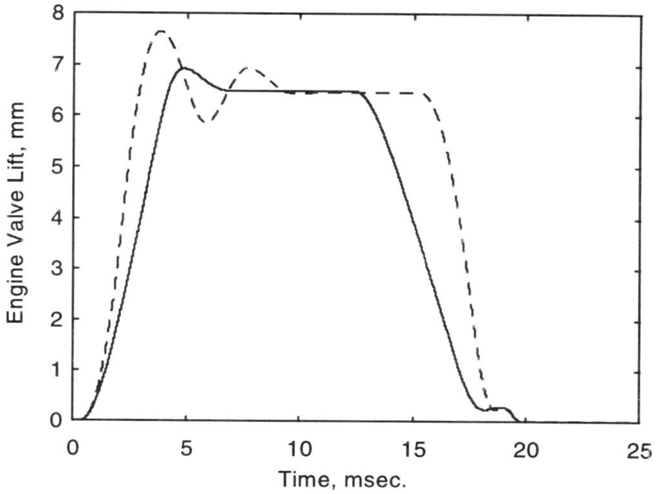

Figure 20. Simulated low and normal temperature operation (-40°C, solid; 40°C, dash).

UNSEATING DELAY - The model provides insight for design changes which can enhance system performance. One such example is the addition of a snubber chamber check valve. This device increases the flow into the snubber chamber during engine valve opening. Figure 21 shows the simulated lift response with and without the snubber chamber check valve, and the simulated pressure responses with and without the check valve are shown in Figures 22 and 23, respectively. The unseating delay has been eliminated, and the required HPSV pulse width has been reduced. On the basis of this analysis, the hardware modifications were made and tested. The experimental results of Figure 24 confirm the simulation observations.

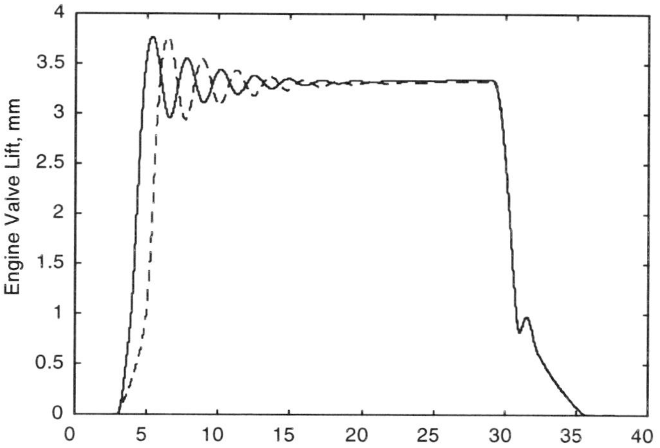

Figure 21. Simulated lift response with (solid) and without (dash) snubber chamber check valve.

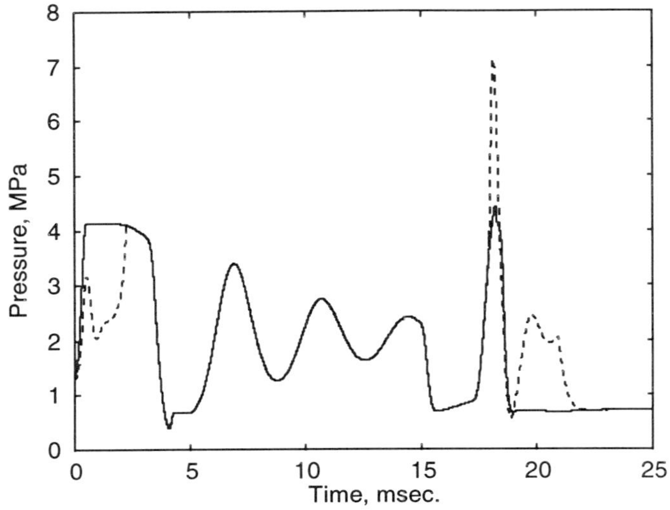

Figure 22. Simulated snubber (dash) and control (solid) chamber pressure response *without* snubber chamber check valve.

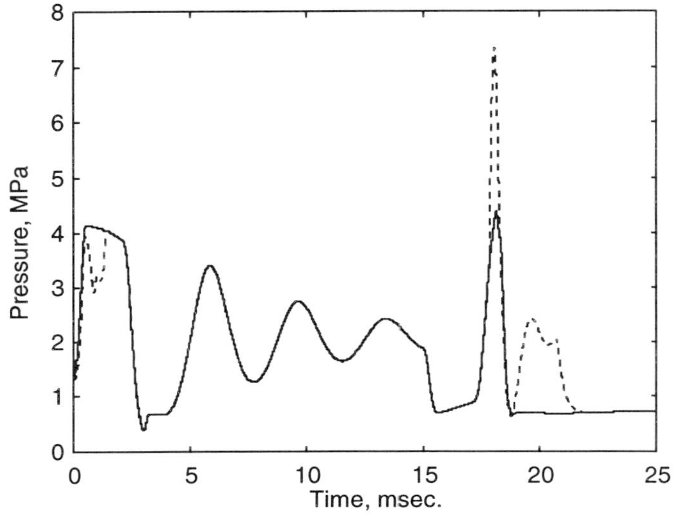

Figure 23. Simulated snubber (dash) and control (solid) chamber pressure response *with* snubber chamber check valve.

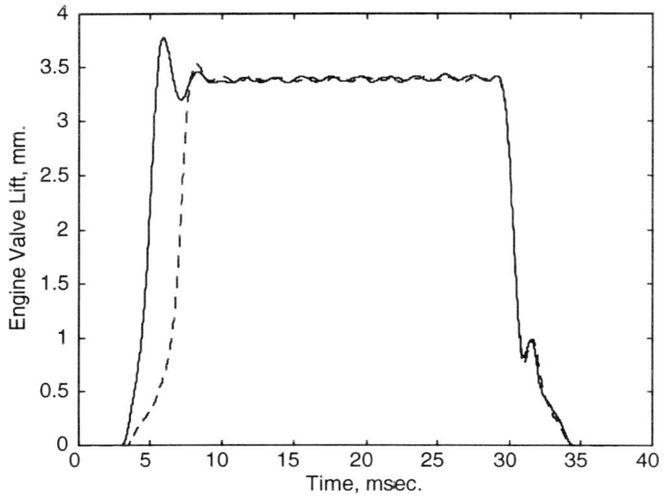

Figure 24. Experimental response with (solid) and without (dash) snubber chamber check valve.

ENERGY RECOVERY - As described previously, the opening of the LPCV is an indicator of the energy recovery during the opening of the engine valve. The recovery is defined as:

$$ER_{opening} = \frac{\int_0^{t_{cycle}} Q_{lpcv}(t)\,dt}{\left(A_{up} - A_{dn}\right) X_{lift}} \qquad (25)$$

The model has been used to determine the extent of the energy recovery improvement, if the solenoid closing time is reduced. Table 1 illustrates the energy recovery in terms of the volume passing through the check valve for a 5 mm lift during the valve opening. The greater flow means more recovery and more efficient valvetrain. As the solenoid closing time becomes shorter the total volume through the low pressure check valve (LPCV) increases, hence, more energy recovery. As a direct consequence of this experiment the design of control solenoids has been modified to substantially reduce the closing times.

Table 1: Effect of solenoid closing time on energy recovery

Solenoid closing time (msec)	Total volume through LPCV (mm^3)	Energy Recovery
2.00	35	25%
1.00	51	36%
0.30	58	40%
0.05	64	46%

The experimental and the simulated data presented in this paper indicate little or no possibility for recovery during closing of the engine valve. Two factors account for somewhat lower recovery during the closing part of the lift compared to the opening part for a symmetric pendulum (a pendulum with the net opening force equal to the net closing force) [1]. First, the closing part of the lift is always less due to the overshoot during the opening, which results in a difference between the maximum and the settled-down values of lift. Second, depending on how well the tuning of the quiet seating of the engine valve is done, some part of the lift is always used for the seating. However, with the shorter closing times and much greater recovery in both directions, this system's energy consumption (including the electric part) is expected to be equal to or less than losses in most camshaft-based valvetrains.

CLOSED-LOOP LIFT CONTROL - The model can be modified to provide a platform for control strategy development. The first issue to be resolved in the control of the valvetrain is to provide a method to insure proper tracking of the valve commands. An "inner-loop" control must be developed to take valve lift and timing commands from the global control system and guarantee that those commands are followed in the physical system, in the presence of system disturbances (such as engine speed changes, pressure and temperature fluctuations, and dimensional changes). Since the hydraulic valvetrain model provides an accurate nonlinear representation of the system, it is used to develop some basic strategies for the control of the maximum valve lift. Artificially simulated high pressure solenoid lifts are used as the inputs, because a mathematical model for the solenoid motion has not been developed at this time. It is assumed that there will be some method to determine the actual maximum valve lift, to be used as the feedback to the control. In Figure 25, it is seen that this initial controller design responds to instantaneous changes in lift command fairly well, changing from 6 mm to 3 mm in approximately 3 engine cycles. The controller's ability to maintain constant valve lift during engine speed transients is depicted in Figure 26, where the engine undergoes a speed change of 1500 RPM to 3500 RPM in approximately 0.5 seconds. A closed-loop control to obtain the best tuning of the valve seating while maximizing the energy recovery during the valve closing is also necessary, and it is currently being developed.

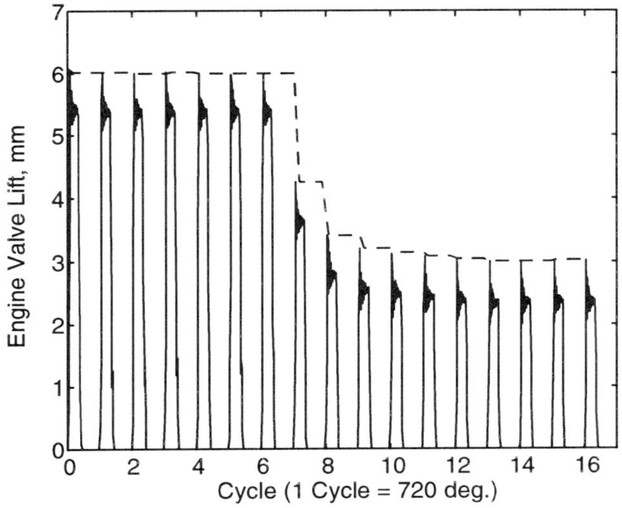

Figure 25. Simulated closed-loop response for change in lift command from 6.0 to 3.0 mm.

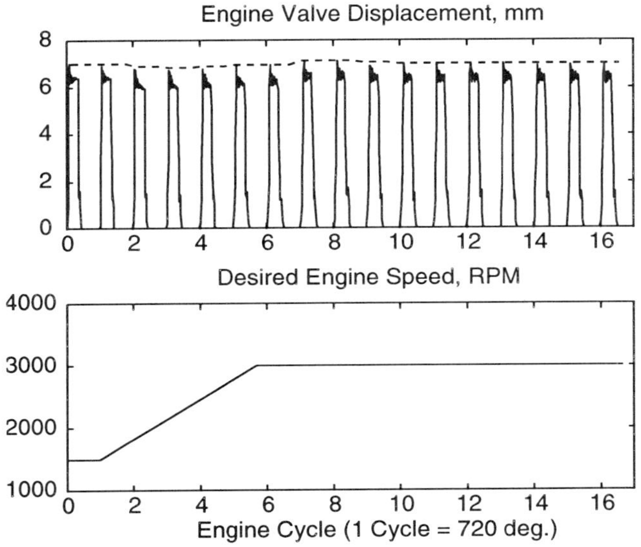

Figure 26. Simulated closed-loop response to change in engine speed: constant 7.0 mm lift desired.

OPEN-LOOP CYCLIC VARIABILITY - EXPERIMENTAL RESULTS

A major concern regarding the operation of the camless valvetrain is its ability not to exceed the cyclic variability of a typical camshaft. A study was performed to assess the contributions of the control pulse generator, the power electronics, and the hydraulics to the cyclic variability of the engine valve events. For a set engine speed, it is expected that a given point on the rising trace of the high pressure solenoid opening pulse occurs once every 720 crankshaft degrees, exactly. Likewise, a given point on the rising trace of the high pressure solenoid current occurs once every 720°, and a given point on the rising trace of the engine valve motion occurs exactly every 720°. Any internal variations in the system would cause any of these points to occur later or sooner relative to the last cycle, indicating mis-timed valve events. By observing the system operation for a number of cycles and recording the high pressure solenoid control pulse voltage, the high pressure solenoid current, and the engine valve lift, the trigger points can be picked off and the cycle times can be determined. By comparing the distributions of the cycle times from the average cycle times for each of the three metrics, the relative contributions to system variability can be identified.

The system was operated for a long period of time (without closed-loop control), and the needed data were recorded using a digitizing oscilloscope operating at 100 kHz and an effective 16-bits of vertical resolution. The data were gathered for approximately 50 engine cycles. The times when the control pulse reached 4 volts (out of 5 volts maximum) on the rising edge were recorded, as well as the times when the current reached 4 amps (out of 10 amps maximum) on the rising edge and the times when the engine valve lift reached 1 mm (out of approximately 5 mm of maximum lift) during the valve opening. The cycle times based on each of these three metrics were then computed by measuring the elapsed time between two adjacent reference points (see Figure 27 for a graphical representation of the process).

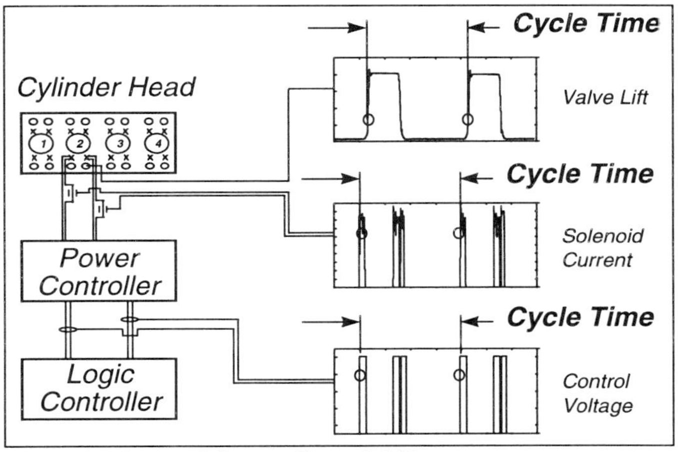

Figure 27. Determining cyclic variability.

The distributions of the actual cycle times minus the average cycle times for each approach are plotted in Figure 28. As can be seen, the control pulse has a relatively high variability, with 66% of the points (2-sigma) falling within 10 microseconds. It is expected that improved control electronics will decrease this to less than 1.0 microsecond. Next, the solenoid current has a slightly higher variability of 20 microseconds; however, half of this comes from the variability of the control pulse itself. Finally, the cycle time distribution of the engine valve lift has a range of 40 microseconds. Again, the contribution by the system hydraulics is half of this, as 20 microseconds come from the solenoid current variation alone. Therefore, 66% of the cycle times based on the hydraulics of the camless system operating with a somewhat inferior power controller deviate less than 20 microseconds, or approximately 0.175 degrees at 1500 RPM. This variability is rather small; however, it would have to be verified at a variety of speeds and loads.

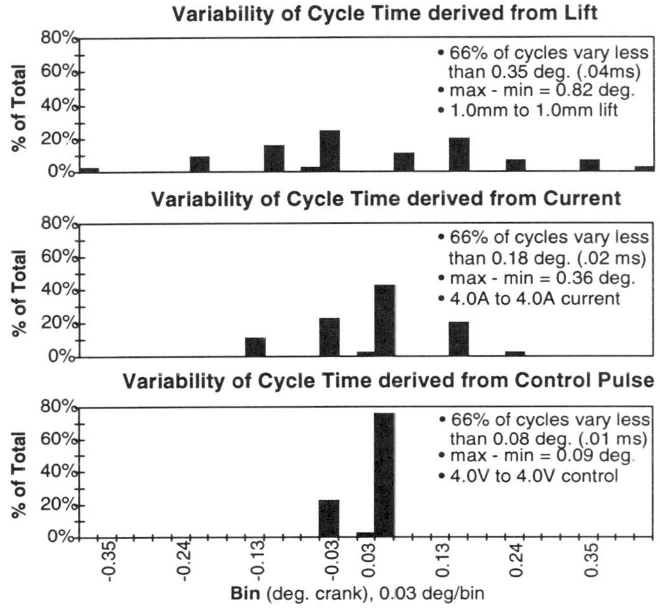

Figure 28. Experimental cyclic variability.

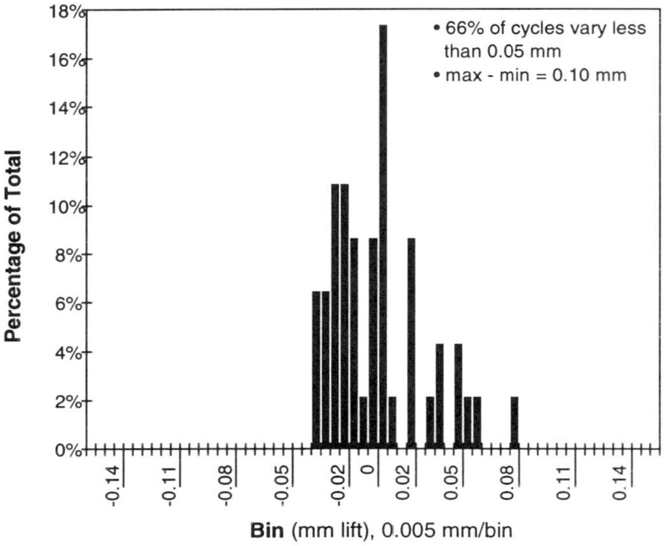

Figure 29. Experimental open-loop variability of maximum engine valve lift.

The cyclic variability of engine valve lift also was studied for a range of desired maximum lift values. The distribution of the actual lift errors for 4 mm is presented in Figure 29, where 66% of the cycles fell within a range of 0.05 mm. The distributions for other nominal lifts are similar.

The cyclic variabilities of timing and lift should be viewed in light of the influence they exert on the engine performance and emissions. As predicted by the engine combustion model for a 2.0 liter, spark-ignited engine operated unthrottled at 1500 RPM and 4 bar IMEP (Indicated Mean Effective Pressure), a variation of 1 degree of crankshaft angle in the intake valve closing timing will produce 1.8% variability in the engine IMEP. The same 1 degree variation in the exhaust valve closing timing will vary the residual gas mass fraction, which affects NOx emission, by 2.0%. A 0.1 mm cyclic variability of lift will affect the IMEP by 1.6%. These percentages for engine performance and emissions variability are generally acceptable for most engines.

CONCLUSIONS

A dynamic model has been developed for the Ford springless electrohydraulic camless valvetrain. The initial efforts have focused on the single engine valve system, with the solenoid lifts being the system inputs. The fundamental laws of displacement dynamics, pressure dynamics, and orifice flows have been applied to the system components. The main points of this work are summarized as follows.

1) Model elements such as the snubber chamber pressure, the solenoid valve leakage, and the check valve have been included, so that the system model more accurately represents the experimental system.

2) A procedure, described in Appendix B, has been developed for evaluating the calibration constants from experimental data. These constants include the engine valve damping, the solenoid leakage coefficients, and the amount of entrapped air.

3) For a given set of system parameters at an engine speed of 1500 RPM, the simulations matched the experimental data provided by Ford for lifts of 2 mm, 4 mm, and 6 mm, using the given parameter values and the tuned calibration constants. These observations confirm the accuracy of the model.

4) The single valve model has been expanded to include the twin valve model with the unequal lift modifier to allow flexibility in setting the lift of each valve. The simulation results demonstrate the effect of the unequal lift modifier.

5) The usefulness of the model for system design and development, performance enhancement and implementation of control system strategies has been demonstrated.

6) The preliminary experimental data for the open-loop cyclic variability of valve timing and lift indicate that a hydraulic pendulum-based system should be capable of satisfying emissions and performance requirements for an unthrottled operation.

ACKNOWLEDGEMENTS

The authors express their sincere appreciation to Mike Schechter for valuable consulting, to Sam Asaro and Dennis Roper for helping with the experimental work, and to George Davis, Chuck Newman, and Ron Miller for running the engine combustion model and optimizing valve lift traces for the unthrottled operation.

REFERENCES

1. Schechter, M. and Levin, M. "Camless Engine", SAE paper 960581.

2. Merritt, H. E., Hydraulic Control Systems, John Wiley and Sons, New York, NY, 1967.

3. Robertson, J. and Crowe, C., Engineering Fluid Mechanics, Houghton Mifflin Company, New York, NY, 1975.

NOMENCLATURE

Fluid properties:

β_m:	manifold compressibility (assumed infinite)
β_l:	oil compressibility
β_a:	air compressibility = (1.4)*(Pressure in given chamber)
α :	relative volume fraction of entrapped air in oil
ρ:	oil density
μ:	oil viscosity

Engine valve parameters:

M_v:	engine valve mass
D_{piston}:	piston diameter
D_{stem}:	stem diameter
A_{up}:	area on piston top = $(\pi/4) * D_{piston}^2$
A_{dn}:	area on piston bottom = $(\pi/4) * (D_{piston}^2 - D_{stem}^2)$
K_{calib}:	damping calibration factor
Ev_{length}:	contact length for viscous friction
V_{cc}:	control chamber volume
c:	radial clearance between piston and sleeve
x_{vcrit}:	snubber overlap
$D_{snubber}$:	snubber diameter
$A_{snubber}$:	snubber area $= \frac{\pi}{4} D_{snubber}^2$
V_{sn}:	snubber chamber volume
Aside:	friction rubbing area of piston = $\pi D_{piston} * Ev_{length}$
A_1:	flow area for chamber under first valve piston (twin valve system)
A_2:	flow area for chamber under second valve piston (twin valve system)

Volume flow parameters:

P_{hi}:	high pressure
P_{low}:	low pressure
C_d:	flow discharge coefficient
W_{hpsv}:	flow area gradient for HPSV
W_{lpsv}:	flow area gradient for LPSV
W_{hpcv}:	flow area gradient for HPCV
W_{lpcv}:	flow area gradient for LPCV
θ:	valve angle = 45 deg
γ:	flow angle = 45 deg
K_{ex}:	flow calibration constant
K_{Qbet}:	flow calibration constant
c_{sn}:	radial clearance between snubber chamber and control chamber
$L(x_v)$:	effective length of flow chamber for Q_{bet}.
K_{lkhpsv}:	HPSV leakage coefficient
K_{lklpsv}:	LPSV leakage coefficient

Check valve parameters:

M_{hpcv}:	HPCV mass
M_{lpcv}:	LPCV mass
K_{hpcv}:	HPCV spring constant
K_{lpcv}:	LPCV spring constant
$F_{preload}$:	Spring preload force
D_{seat}:	Seat diameter
A_{seat}:	Seat area $= \frac{\pi D_{seat}^2}{4}$

APPENDIX A
FUNDAMENTAL GOVERNING EQUATIONS

DISPLACEMENT DYNAMICS - The fundamental equation for displacement dynamics is Newton's second law of motion:

$$\ddot{x} = \frac{1}{M}(\Sigma F) \qquad (A.1)$$

where M is the mass and ΣF represents the forces acting on the mass.

PRESSURE DYNAMICS - The fundamental equation for pressure dynamics is:

$$\dot{p}_r = \frac{\beta_e}{V_o}(\Sigma Q) \qquad (A.2)$$

where β_e is the fluid compressibility, V_o is the chamber volume, and ΣQ represents the net volume flow into the chamber.

TURBULENT FLOW THROUGH AN ORIFICE:

$$Q = [\text{sign}(\Delta p_r)]C_d(\text{Flow Area})\sqrt{\frac{2}{\rho}\text{abs}(\Delta p_r)} \qquad (A.3)$$

LAMINAR FLOW THROUGH AN ORIFICE:

$$Q = K_{lam}(\Delta p_r) \qquad (A.4)$$

OIL COMPRESSIBILITY - The effective bulk modulus of the fluid mixture is as explained in Merritt [2]:

$$\beta_e = \left(\frac{1}{\beta_m} + \frac{1}{\beta_l} + \frac{\alpha}{\beta_a}\right)^{-1} \qquad \text{(A.5)}$$

FLOW AREA THROUGH A POPPET VALVE - Figure 30 illustrates the flow area across a check valve, from Robertson and Crowe[3],

$$\text{Flow area} = [\pi D_{seat} \sin(\theta)] x_{cv} - [\pi \cos(\gamma) \sin^2(\theta)](x_{cv})^2 \qquad \text{(A.6)}$$

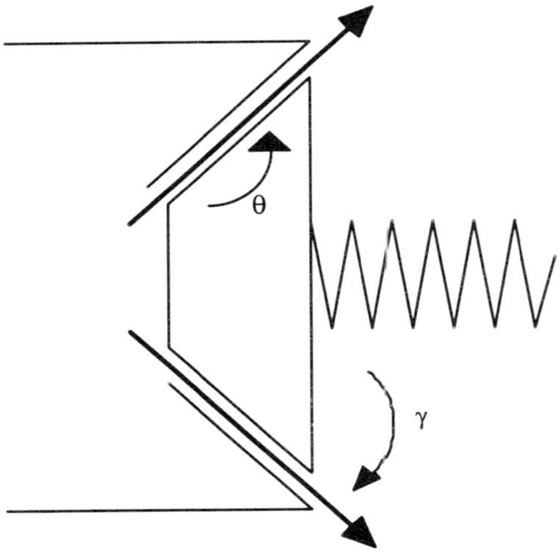

Figure 30: Volume flow through check valve.

APPENDIX B
DETERMINATION OF CALIBRATION CONSTANTS

The experimental data from the 6 mm lift response is used to determine the calibration constants. When all of the control valves are closed (i.e. solenoid valves, check valves), the state equations are, with $v_v = \dot{x}_v$,

$$\dot{p}_{cc} = \beta_e \frac{K_{lkhpsv}(P_{hi} - p_{cc}) - K_{lklpsv}(p_{cc} - P_{low}) - A_{up} v_v}{V_{cc}} \qquad \text{(B.1)}$$

$$\dot{v}_v = \frac{A_{up} p_{cc} - A_{bot} P_{hi} - B_v v_v}{M_v} \qquad \text{(B.2)}$$

This leads to the second-order characteristic equation:

$$s^2 + \left[\frac{\beta_e}{V_{cc}} K_L + \frac{B_v}{M_v}\right] s + \left[\frac{\beta_e B_v}{V_{cc} M_v} K_L + \frac{\beta_e A_{up}^2}{V_{cc} M_v}\right] =$$

$$s^2 + [2\zeta\omega_n] s + \omega_n^2 = 0 \qquad \text{(B.3)}$$

where $K_L = K_{lkhpsv} + K_{lklpsv}$.

1) The leakage coefficients (K_{lkhpsv}, K_{lklpsv}) are determined using the previous observation that the leakage flows past the high pressure solenoid and past the low pressure solenoid are equivalent, at the middle of the cycle. This leads leads to the relationship:

$$K_{lkhpsv} = K_{lklpsv} \frac{P_{ccss} - P_{low}}{P_{hi} - P_{ccss}}, \qquad \text{(B.4)}$$

where $P_{ccss} = P_{hi} \dfrac{A_{dn}}{A_{up}}$, which is used in conjunction with the measured derivative of the control chamber pressure at the end of the cycle, shown in Figure 7.

2) The snubber overlap (x_{vcrit}) is determined from the time where the control chamber pressure and snubber chamber pressure converge (Figure 5), and then by using Figure 6 to determine the lift corresponding to that time.

3) The flow gain K_{Qbet} from EQ (10) is adjusted in order to ensure that during simulation, these two pressures converge properly.

4) The expected values of ω_n and ζ are computed from the portion of the control chamber pressure response during which all control valves are closed. One such portion is the response of Figure 7, during the interval from 4 msec to 20 msec.

5) K_{calib} and α are adjusted so that the simulations match the experimental responses, affecting the engine valve damping of EQ (14) and the oil compressibility of EQ (A.5), respectively. The calculated values of ω_n and ζ from the simulated response match the measured values from the experimental data.

981029

Adaptive Lift Control for a Camless Electrohydraulic Valvetrain

Mark D. Anderson and Tsu-Chin Tsao
University of Illinois at Urbana-Champaign

Michael B. Levin
Ford Motor Co.

ABSTRACT

Camless actuation offers programmable flexibility in controlling engine valve events. However, a full range of engine benefits will only be available, if the actuation system can control lift profile characteristics within a particular lift event. Control of the peak value of valve lift is a first step in controlling the profile. The paper presents an adaptive feedback control of valve lift for a springless electrohydraulic valvetrain. The adaptive control maintains peak value of lift in presence of variations in engine speed, hydraulic fluid temperature and manufacturing variability of valve assemblies. The control design includes a reduced-order model of the system dynamics. Experimental results show dynamic behavior under various operating and environmental conditions and demonstrate advantages of adaptive control over the non-adaptive type.

INTRODUCTION

Current production internal combustion engine systems provide a wide range of actuation methods to control operating characteristics, such as emissions and power output. One of the few engine sub-systems that are still restricted via mechanical linkage to the crankshaft is the actuation mechanism for intake and exhaust valves. Replacing the mechanical valvetrain with a fully variable camless actuation system will allow for complete control of mass transfer into and out of the engine cylinders, and allow improvements in emissions and power.

Many approaches at providing this actuation have been developed, each allowing varying degrees of system control. At the most basic, systems have been developed which allow the exhaust camshaft to be phased relative to the crankshaft, which allows for camshaft control of exhaust gas recirculation, improving emissions (Bassi et al. [1]). Another approach is to use two or more completely different sets of cams for different operating conditions (Inoue, [2]). These systems use a series of actuators to switch the valvetrain between different cam lobes, one for low speed operation and one for high speed, with two sets of lift and timing parameters. While improving design freedom, it is still limited to two distinct lift and timing schemes, unless coupled with other mechanical systems. A system proposed by Kreuter el al. [3] and Aura Systems Inc. [4], which operates only with electromagnets with fixed lift, allows a further degree of control. In this case, the electromagnets are operated independently so that variable timing and duration are achieved. Finally, a system presented by Schechter and Levin [5] combines electronically controlled solenoids with hydraulic fluid power to provide control over engine valve lift and timing. This system is the basis of the investigation of this paper.

SYSTEM DESCRIPTION

The system as shown in Figure 1 operates by controlling two solenoid valves and two check valves for

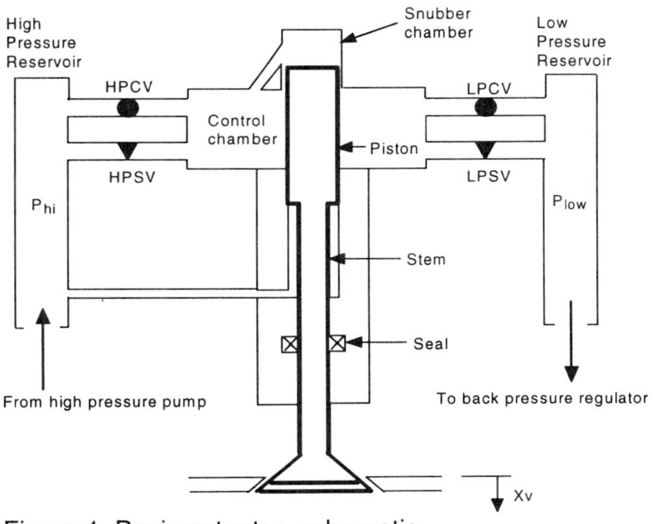

Figure 1. Basic actuator schematic.

each set of paired engine valves (with four valves per cylinder). The solenoids control the flow of high pressure fluid into and out of a control chamber which is connected to the top of the engine valve piston.

At the same time, a constant force is applied to the bottom of the valve piston, such that when the high pressure is removed from the valve top, the valve returns to the seated position. By controlling the timing and duration of the solenoid valve openings, the pressure in the chamber is precisely controlled for continuously variable lift with independent control over engine valve opening and closing timing. This system has been designed so as to allow compact packaging. Without camshafts, substantial engine height and weight reduction have been achieved. Also, the design provides high energy efficiency through use of check valves for maximum energy recovery and minimum parasitic drain on engine power.

For high energy efficiency, low cyclic variability, and low valve seat wear, the camless valvetrain system will require the use of a closed-loop electronic control, which makes the valve timing and lift follow commands from an outer loop engine control system, as shown in Figure 2. The inner loop valve lift and timing should have much faster response than the outer loop engine control, which uses slower feedback sensors like MAP and MAF. Fast and robust inner loop dynamic performance also facilitates the outer loop control design and implementation, where the inner loop dynamics may be neglected or substantially simplified. To aid in the design of a camless valvetrain, as well as control strategy development, a dynamic mathematical model of the camless valvetrain system has been developed (Kim, et. al., [6]). The model allows simulation of various control approaches to determine their effects on the lift tracking and variability.

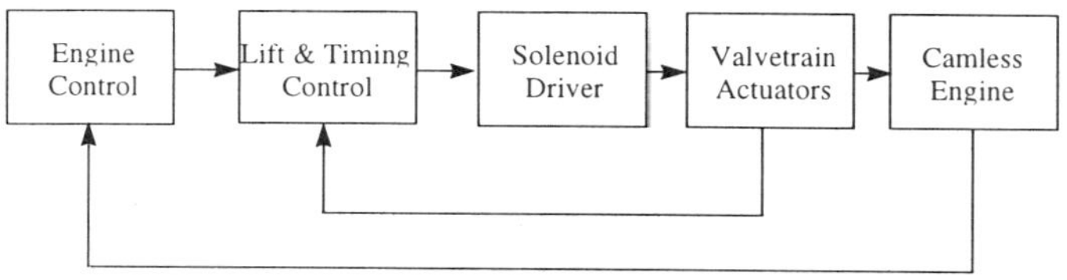

Figure 2. System integration overview.

OPERATION OF SYSTEM – It is necessary to precisely control the solenoid activation sequence to insure that the given lift and speed trajectories are achieved. In all, there are four parameters that determine the timing sequence (see Figure 3). First, the occurrence (T_{open}) of the opening pulse determines the valve opening timing. The opening pulse width (*int1*) determines the amount of high pressure fluid admitted into the control chamber, which ultimately determines the peak lift value (see Figure 4). The control of this parameter is referred to as *peak lift control.* Next, the occurrence (T_{close}) of the closing pulse determines the valve closing timing. The

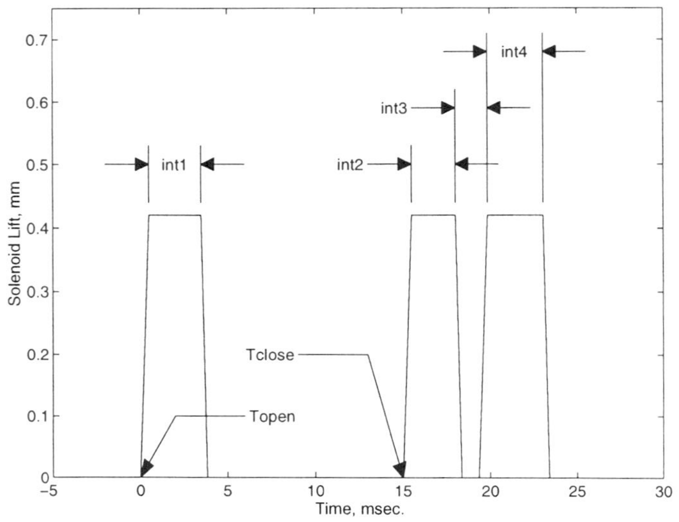

Figure 3. Solenoid pulse sequence parameters.

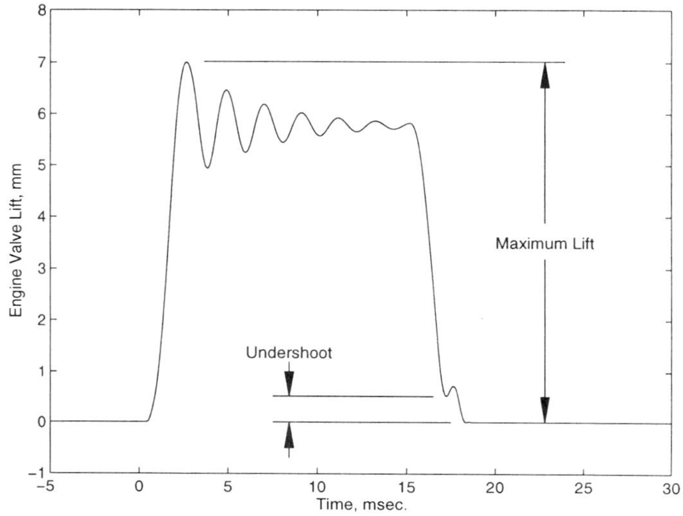

Figure 4. Diagram of maximum valve lift and undershoot.

closing pulse width (*int2*) determines how far the valve moves in its closing descent; this determines the valve 'undershoot'. As the undershoot determines the end of cylinder gas exchange and partially determines valve seating velocity, its control is also of importance and is referred to as *closing control.* Next, the seating delay time (*int3*) determines how far the valve will rebound after the undershoot. This rebound will also affect the final seating velocity. As the seating velocity determines both noise and seat wear, it is also of control interest and is referred to as *seating control.* Finally, the seating pulse (*int4*) finishes the closing of the valve, and insures that the valve is completely seated with a proper closing force. It is desired to minimize the electrical energy consumption of the system; therefore this seating pulse should be no longer than necessary. In this paper, only the peak lift control is considered as the first part of the investigation of various valve control problems. A fixed ratio of the closing pulse *int2* to the opening pulse *int1*, and fixed seating pulses *int3* and *int4* were used throughout this study.

HARDWARE/SOFTWARE INTERFACE

The experimental system consists of the electrohydraulic engine valve actuators, solenoid valve driver power electronics, an embedded controller, and a host personal computer. The embedded controller generates pulses to trigger the solenoid drivers and process the lift sensor feedback signals for feedback control. The host computer communicates with the embedded controller through a serial port. The valve pulse timing and duration commands in the embedded controller can be changed in real-time by the host personal computer through this communication. In a standard operation, the interface sends a series of commands to the controller to set up the operating parameters, such as lift sensor calibrations, sensor sampling parameters (rising or falling edge of control pulse starts window for sample, length of sample window, solenoid driver parameters (hold and peak current, etc.), cylinder firing order, etc.). The interface then sends a command to start operation, at which point the controller enters its real-time control loop. The interface software may then send changes in certain parameters to the controller, which are implemented on the next cycle. For instance, a valve may be commanded to go from a seated position to operating, at which point the normal operation occurs using existing solenoid pulse width settings. These pulse widths may also be changed during operation, allowing for open-loop control of the opening and closing. Also, since this experiment does not involve an actual engine operation, the change of engine speed is simulated either through a shaft encoder simulator, or through commands generated by the interface software.

In order to allow closed-loop control with the computer interface, a series of commands are provided by the controller that return samples of the engine valve peak lift. However, as the controller is operating in real-time and the interface is operating on a non-real-time operating system (Microsoft Windows 95), a polling method is used to indicate when new lift samples are available. This method does introduce communications latencies that must be taken into account in closed loop control design.

Another hardware limitation is the quantization of the sampled lift and the commanded pulse width. The resolution of the latter is 1.0 μsec, which may limit the amount of fine lift control available. This is of lesser concern in this case, though, as the resolution on the sampled lift is only 0.1 mm, due to internal limits of the A/D converter of the controller hardware. Thus, the best that one may hope to achieve in terms of lift variability in steady-state operation is 0.1 mm. As indicated in [6], this is the maximum allowable variability; it is used here as the design goal for steady-state operation. A primary focus of future controller development should be the refinement of this sampling resolution, to allow for better overall performance.

To ensure acceptable performance in the wide range of operating conditions expected in vehicle operation, an adaptive feedback control algorithm is developed.

MODELING AND ADAPTIVE LIFT CONTROL ALGORITHM

SYSTEM IDENTIFICATION - The first step in designing the feedback controller is to identify a simplified model of the plant using the previously developed mathematical model [5]. The input to the plant is the high pressure solenoid valve pulse width; the output is chosen to be the engine valve peak lift. These are both good choices as the pulse width is easily and precisely controlled, and the peak lift is both easily sampled and critical for engine combustion and emissions.

Nominally, the system was run at a 5.52/2.28 MPa (800/330 PSI) high/low pressure combination. Pulse widths from approximately 3.95 to 4.60 msec were used, and the peak lift was measured in each case. The data were plotted comparing peak lift with high pressure solenoid pulse width, as shown in Figure 5. The nonlinearity in the curve is due to the transient dynamic relationship between the solenoid pulse width and the solenoid opening area. A first order linear fit was made however. As there is a delay in the plant between command input and system output, this was also incorporated into the model. The resulting model is then:

$$y(k) = a \cdot u(k-d) + b \qquad (1)$$

where the sample time k is defined as one complete engine cycle, or 720 crank degrees, d is the combination of sensing and controller-host computer communication delay, y is the peak lift output, u is the solenoid valve pulse width, a is the gain and b is an offset parameter.

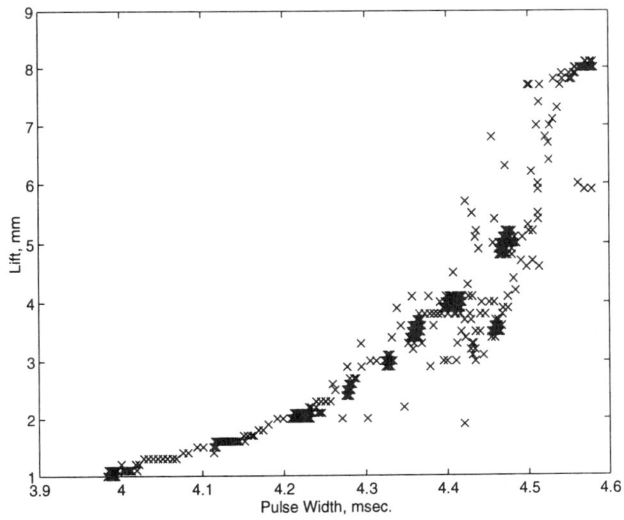

Figure 5. Typical lift vs. solenoid pulse width for the experimental system.

In an ideal case, there should only be sensing delay but no communication delay and hence d = 1. In this study however, the serial communications make the delay d = 1 or 2 in most cases and maybe more than 2 in some cases where the serial communications may not be fast enough for the high engine speed regime.

POLE - PLACEMENT DESIGN - This reduced-order model was then used to perform a pole-placement design for a feedback controller. The design was performed such that integral control is provided, and a good level of damping is achieved (no more than 5% overshoot) for either case of delay, i.e. d = 1 or d = 2. The resulting control law is:

$$G_c(z) = \frac{K_c(z-0.05)}{(z-1)(z+0.45)} \tag{2}$$

where G_c is the controller transfer function. Figure 6 shows the compensated closed-loop root locus plot in the z-domain for both cases. Therefore, a relationship structured as:

$$K_c = m/a \tag{3}$$

, where m = 0.287, gives similar dominant pole location as indicated by the '+' symbol in Figure 6. This results in similar step responses for both cases, as shown in Figure 7. A step change of 1 mm lift takes about 10 steps to reach within 0.1 mm, which is the sensor signal resolution. To further reduce the settling time to 2 to 5 engine firing events, which is desirable for camless operation, the communication delay must be eliminated so that d = 1 is always the case. A fully embedded control implementation would eliminate such communication delay problem.

PLANT PERTURBATIONS - The system nonlinear input-output relation illustrated by Figure 5 has been modeled as a variable gain linear system by Eq. (1). Other system changes such as varying high/low pressure combinations, component size variations, oil viscosity, or air content in the hydraulic fluid could be modeled as variations in the first-order gain factor. As only the gain in Eq. (1) is altered, the only modification that must be made in the controller to compensate is to increase or decrease the controller gain. Doing so makes sure that the closed loop poles and zeros remain at the same location, providing consistent system performance in varying operating conditions. Adaptive control is used to account for such plant perturbation in real-time.

ADAPTIVE POLE-PLACEMENT FEEDBACK CONTROL - The plant gain is determined in real-time through a recursive projection parameter adaptation scheme, and this gain is used to calculate a new feedback control gain. It was expected that using the adaptive method would result in improved transient response and improved steady-state noise rejection in conditions

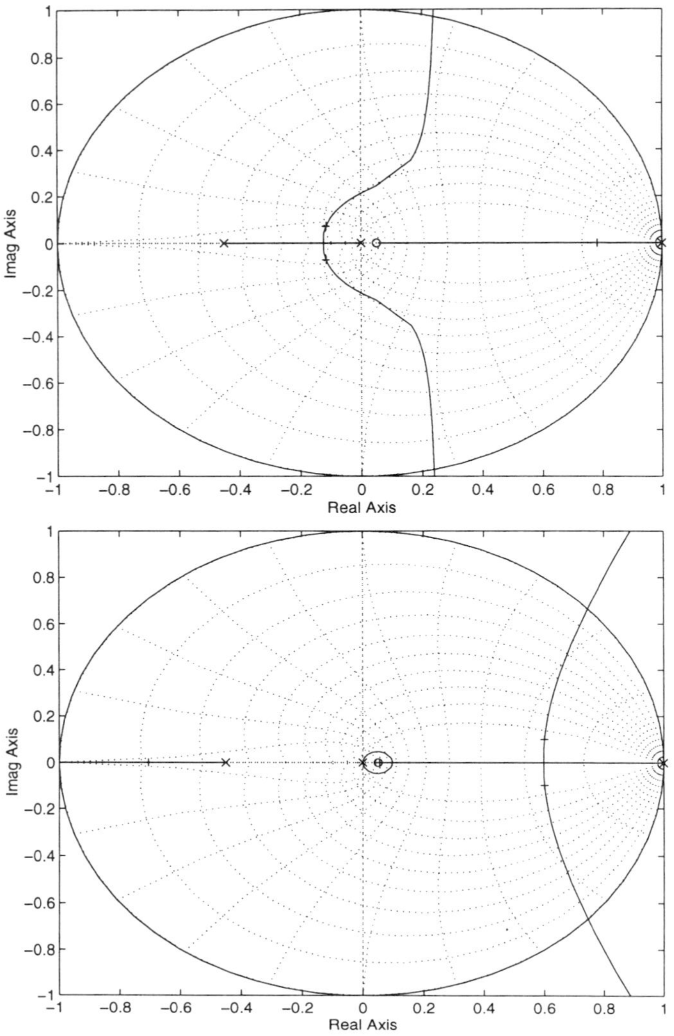

Figure 6. Pole-zero diagram of compensated closed-loop system (upper plot for d =1, lower plot for d =2).

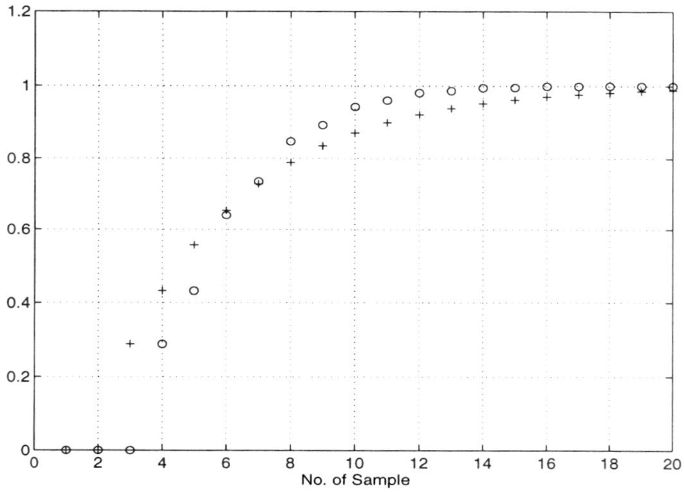

Figure 7. Nominal step response of compensated closed-loop system. ('+' for system delay d = 1, and 'o' for d = 2)

deviating from the nominal condition. The estimated parameter is used directly in determining the new control gain. As the plant gain, *a*, increases, the controller gain, K_c (eq. 2), should decrease to maintain closed loop pole placement, and vice versa.

To correlate with the assumed plant model, the general model for the estimation was

$$y(k) = \phi^T (k-1)\hat{\theta}(k-1) \tag{4}$$

where

$$\hat{\theta}(k) = \begin{bmatrix}\hat{a} & \hat{b}\end{bmatrix}^T \tag{5}$$

$$\phi(k-1) = \begin{bmatrix}u(k-d) & p\end{bmatrix}^T \tag{6}$$

This case assumes two parameters to be estimated; *a* corresponds to the plant gain, and *b* corresponds to the plant offset. The *b* parameter may be left out if it is believed to be small, or may also be fixed while leaving the *a* parameter to adapt. The *p* parameter may be set to unity to match the model of Eq. (1), or, it may be scaled to more closely match the magnitude of *u(k-d)*. This scaling may be necessary to insure equal adaptation of both parameters.

The general form of the projection adaptation algorithm follows (Åström and Wittenmark, [7]):

$$\hat{\theta}(k) = \hat{\theta}(k-1) + \frac{\alpha \cdot \phi(k-1)\,e(k)}{c + \phi^T (k-1)},$$

$$e(k) = y(k) - \phi^T (k-1)\hat{\theta}(k-1) \tag{7}$$

where the parameter α is provided as a design parameter to adjust the gain rate (with $0 < \alpha < 2$), and the parameter *c* also provided as a design parameter; for all cases here, it is set to 1. However, in order to insure a somewhat stable parameter estimation under bounded noise, such as the quantization, it is necessary to place a minimum limit on changes in this parameter. Therefore, a dead zone is also incorporated into the estimation methods as follows:

$$e = y(k) - \phi^T (k-1)\hat{\theta}(k-1),$$

$$if\ |e| \le \delta, e = 0, else$$

$$e = e - \delta \cdot sign(e) \tag{8}$$

where the dead zone magnitude δ is a design parameter which may be chosen to achieve one of a variety of performance objectives.

EXPERIMENTAL RESULTS

NOMINAL CONTROL PERFORMANCE - With the parameters for the adaptation method set (projection method, scaling factor *p* of 4.5, error gain α of 1.0, and dead zone δ of 0.075), the baseline performance of the system while undergoing set point changes was investigated. The main objective of the adaptive control is to maintain model matching performance regardless of the variable system parameters.

Due to the limitation of the hydraulic pump capacity used in the experiment, the system could not run in a consistent and stable manner at speeds higher than 3000 rpm. Also lift over 5 mm could not be controlled consistently due to the lack of an accumulator in the hydraulic system to supply transient flow demand. To test the step response, the commanded lifts were set to change between 1.0 mm and 3.5 mm as shown in Figure 8, all with a constant speed of 1500 RPM. The system was tested both with the adaptive control and with a fixed control. The responses of the fixed and adaptive controls are shown in Figure 8 and 9, respectively. Both fixed and adaptive control were designed based on the pole placement method mentioned earlier, where the valve gain for the fixed control is *a* = 7.0. The response of θ for the adaptive control is shown along with the set points in Figure 10. In order to aid in comparing the lift responses, the data have been filtered by a second order low pass filter, whose break frequency is 0.2 of the Nyquist frequency. The filtering eliminates the quantization noise without distorting the transient response data.

For both the adaptive and the fixed control, there does not appear to be much dependence on starting lift for transient performance. The presence of unusual lift excursion is mainly due to air bubbles introduced into the pump flow and causing a variation in valve lift. When this occurs, the feedback control generally appears to compensate quickly and return the lift to the reference value. The results show that the adaptive control has less oscillatory response, especially at the highest lift points. This is because the nominal *a* and *b* values used for the fixed control design result in a 25% higher controller gain than that obtained with the adaptive control.

PERFORMANCE WITH VARYING ENGINE SPEED - To test the control system response to changing engine speed, two tests were used: one in which only the engine speed is changed while the lift is held constant, and another which follows both lift and speed transients. For the speed test, the lift was set to 2.0 mm while the speed was changed from 1000 RPM to 1500, then 2000, 2500, 3000, 2500, 2000, 1500, then back to 1000 RPM, each at 125 cycle intervals. The control should adequately reject disturbances at any of the speed conditions, and the adaptive control should make the transients behave in the same manner regardless of starting and ending speeds. The lift response for the fixed and adaptive controls are shown in Figure 11. The lift data has been low-pass filtered in both cases to make it easier to follow the responses. The parameter adaptation along with the

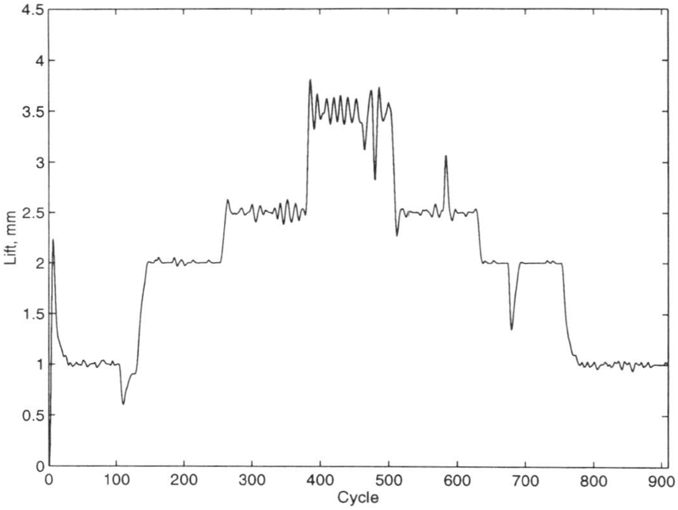

Figure 8. Fixed control response to set point changes

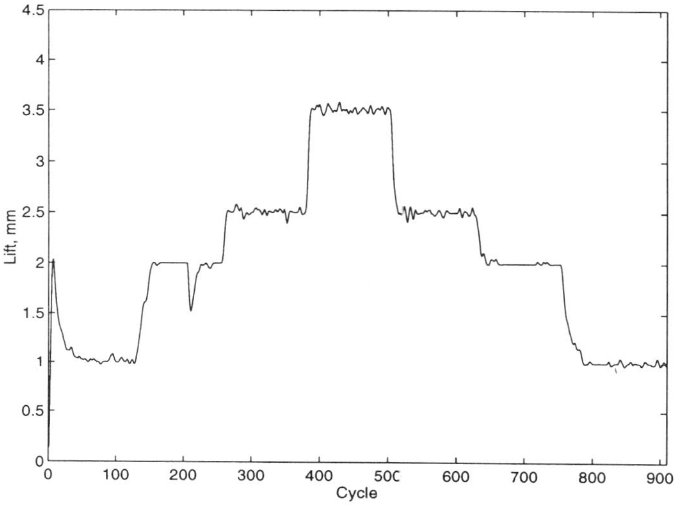

Figure 9. Adaptive control response to set point changes.

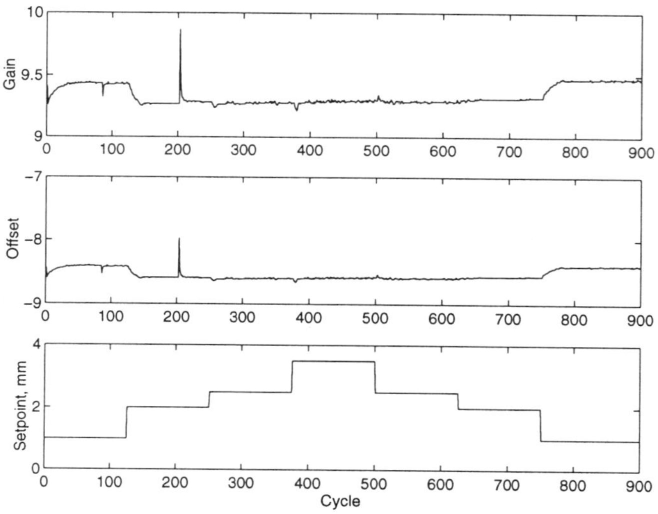

Figure 10. Response of adaptation parameters to set point changes (with set points).

speed settings are shown in Figure 12. The adaptive control maintains a consistent transient response over all speeds. Overall, the change in parameters between the high and low speeds was approximately 12%.

COMBINED LIFT SET POINT AND ENGINE SPEED VARIATIONS – The adaptive control was then used in a test where both the speed and the set point were changed. The transients specified for this test represent a combination of the set point and engine speed changes, and is called the 'combined test'. The lift response is shown in Figure 13, and the parameter and pulse width responses are shown in Figure 14. The adaptive control is able to hold the desired lifts in both conditions; however, the transient response, especially

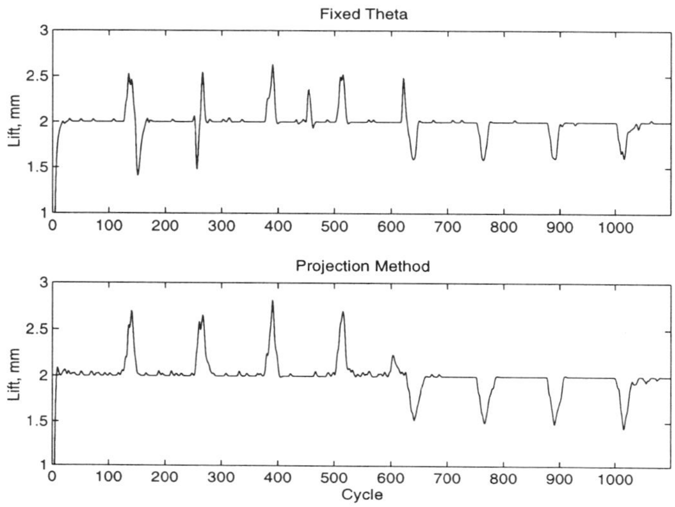

Figure 11. Fixed and adaptive control response to speed changes, fixed lift.

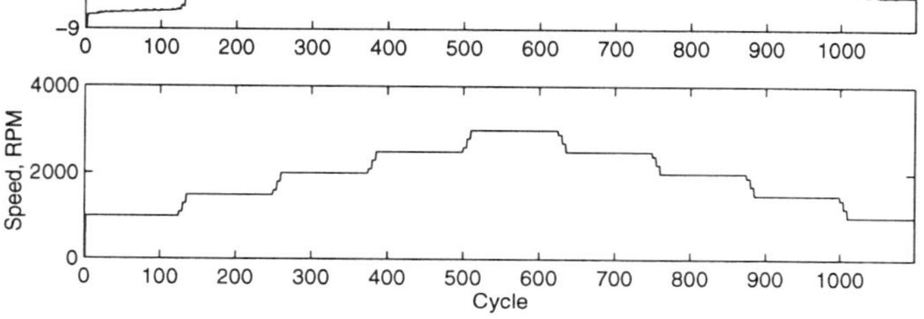

Figure 12. Response of adaptation parameters to speed changes (with speeds).

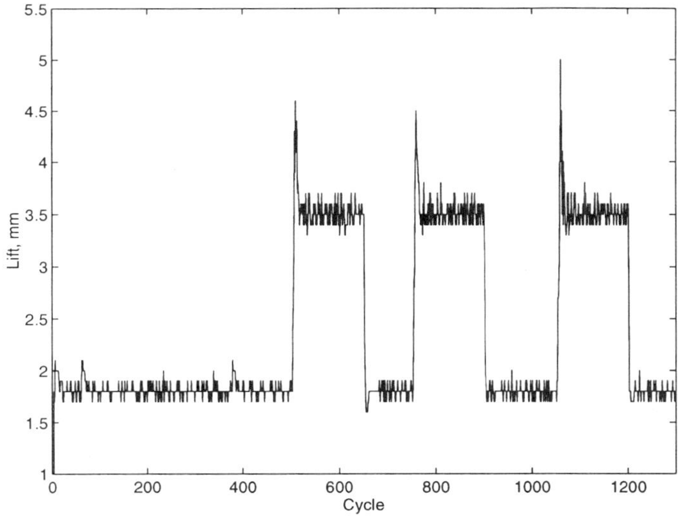

Figure 13. Combined test of speed and lift set point changes: lift response.

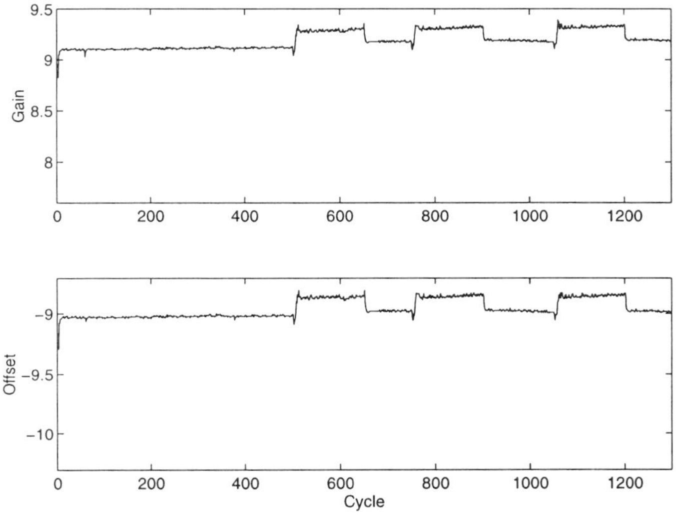

Figure 14. Combined test: adaptation parameter responses.

during the initial speed change, has a large degree of overshoot. This is due to the parameter adaptation not occurring rapidly enough. As this overshoot is not likely to be acceptable in a production engine, a separate table that stores the parameters identified at different speed and lift conditions may be required.

CHANGE IN VALVE PAIRS - A major source of variation in plant parameters can arise from changes in the physical components. Each valve has a slightly different characteristics due to manufacturing variability. From variations within dimensional tolerance to different levels of wear, factors such as solenoid leakage, valve friction, and flow efficiency may all be different for different pairs of engine valves. These cases cause differences in the estimated plant parameters. Therefore, the previous test was performed on another valve pair, and the results in Figures 15 and 16 compared with those for the baseline

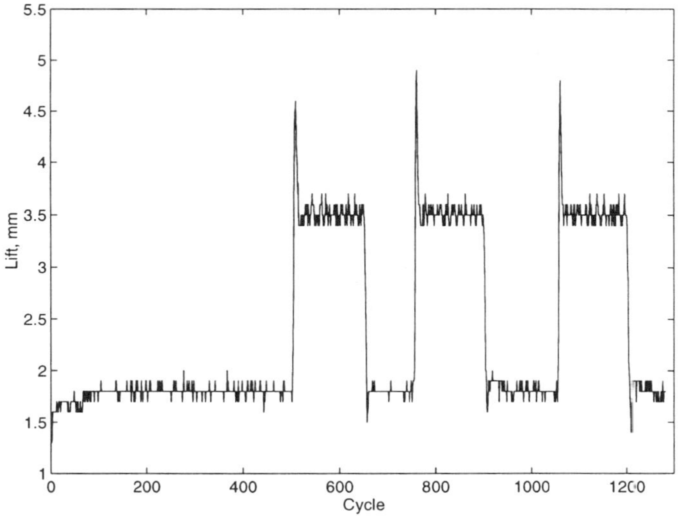

Figure 15. Combined test lift response for a different valve pair.

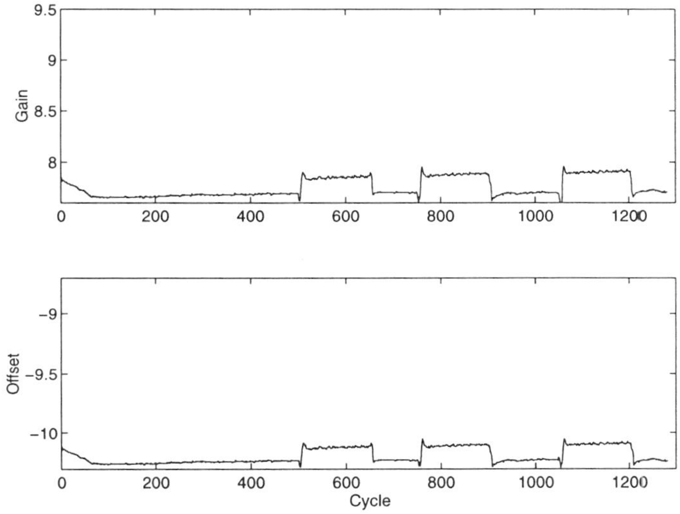

Figure 16. Estimation parameter response for a different valve pair.

system in Figures 13 and 14, where only one valve lift is shown for each valve pair. Clearly, the average values for the parameter estimates are substantially lower than the baseline system (7.8/-10.2 *a*/*b*; a decrease of about 16%). The transient lift response has been maintained consistently as compared to the baseline, indicating proper functioning of the adaptive control.

PERFORMANCE WITH VARYING TEMPERATURE - The next test performed investigates the ability of the control to compensate for the effects of fluid temperature. As the fluid temperature is decreased, the oil viscosity rises. This then increases friction on the valve stem, requiring longer solenoid pulse widths to achieve a given peak lift. In these experiments, the fluid was heating up quite substantially on its own, reaching 55°C in the lab where no cooling apparatus is used and the system remains stationary. To achieve lower temperature, the fluid was cooled by immersing dry ice in the reservoir tank. To measure control response, the standard 'combined' test was performed multiple times while the fluid warmed up, with the temperature being recorded for each test.

Unfortunately, because the temperature of the complete system was not being maintained, the minimum temperature achieved was 16 °C. However, the fluid did eventually reach 58° C, providing a total range of 42° C, and a viscosity range of about 25.0 centistokes (approximately 35.0 centistokes at 16° C, 10.0 centistokes at 58° C for MIL5606 fluid). The steady-state parameters and pulse width values at the 1.8 mm/1000 RPM state are shown versus temperature in Figure 17. As expected, as the temperature of the system increases, the system gain increases, and the required pulse width for 1.8 mm lift decreases. Although the temperature differential achieved only results in a change in θ of approximately 6% (for either gain or bias), it is known that the system will be required to operate at lower temperatures, and likely at higher temperatures.

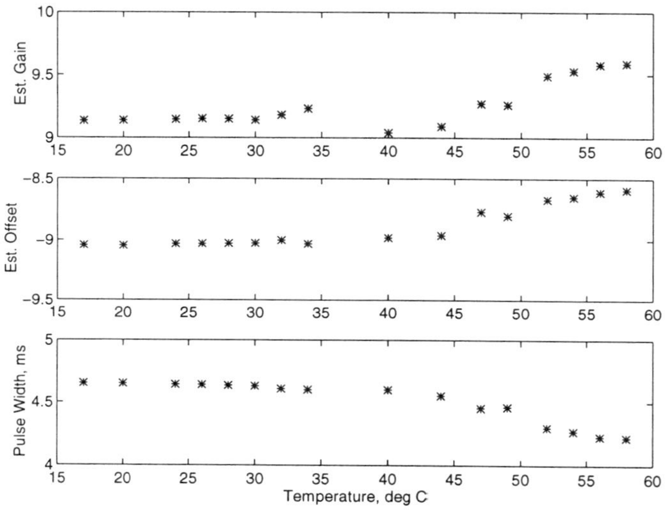

Figure 17. Steady-state pulse width and θ vs. temperature at 1.8mm/1000 RPM.

CONCLUSIONS

A closed loop adaptive lift control has been developed for an electrohydraulic camless valvetrain. A reduced order model has been used to perform a pole placement design, which achieves good level of damping. The experimental results show that the adaptive lift control can

1) maintain consistent performance in following set point changes and rejecting disturbances under a range of operating conditions that change the system parameters;

2) compensate for manufacturing variability in valve assembly;

3) Compensate for changes in hydraulic fluid viscosity as affected by temperature variations.

Future improvements for the experimental hardware set-up have also been identified through the control Implementation. The adaptive control transient performance can be further enhanced by incorporating a parameterized nonlinear model. Engine speed could also be included in the model to make the model parameter insensitive to speed changes, and hence achieve faster transient response. Future work with the valvetrain closed loop control will be directed toward the design and integration of closing (*int2*) and seating control (*int3* and *int4*), and opening (T_{open}) and closing (T_{close}) timing control, as well as improvement over current experimental hardware set-up.

REFERENCES

1. Bassi A., Arcari F., and Perrone F. "CEM-The Alfa Romeo Engine Management System - Design Concepts - Trends for the Future," SAE paper 850290.

2. Inoue, K., Nagahiro K., Ajiki Y., and Kishi N. "A High Power, Wide Torque Range, Efficient Engine with a Newly Developed Variable-Valve-Lift and -Timing Mechanism," SAE paper 890675.

3. Kreuter P., Heuser, P., and Scheeibitz, M., "Strategies to Improve SI-Engine Performance by Means of Variable Intake Lift, Timing and Duration," SAE Paper 920449.

4. Aura Systems, sales literature. El Segundo, California, 1995.

5. Schechter M. and Levin M. "Camless Engine," SAE paper 960581.

6. Kim D., Anderson M., Tsao T. C., and Levin M. "A Dynamic Model of a Springless Electrohydraulic Camless Valvetrain System," SAE paper 970248.

7. Åström K. and Wittenmark B. *Adaptive Control.* Addison-Wesley Publishing Company, Reading, Massachusetts, 1995.

940816

Control of Engine Load via Electromagnetic Valve Actuators

Mark A. Theobald, Bruno Lequesne, and Rassem Henry
General Motors North American Operations Research and Development Center

ABSTRACT

The purposes of this research were to demonstrate new electromagnetic valve actuators on a running engine and to explore the efficiency and emissions advantages of variable valve actuation. A single-cylinder research engine was equipped with programmable electromagnetic valve actuators and operated with early intake-valve closing in addition to conventional throttling, both methods with and without exhaust gas recirculation. The engine was run at a fixed speed of 1500 r/min over varying load at stoichiometric air-fuel mixture, with and without swirl generated by a shrouded intake valve. Engine performance and emissions were measured with spark timing adjusted to minimum advance for best torque. Early intake-valve closing showed a net-efficiency gain over conventional throttling without EGR, but essentially no gain over conventional throttling with heavy EGR dilution in the test engine. Electrical input energy for the actuators was less than the equivalent friction loss for a conventional non-roller valvetrain at this speed.

INTRODUCTION

"In fact, a practised observer (with a carburettor well known to him...)...should be able to estimate the optima of the valve opening and closing points more closely than they could be calculated by any other means. While they are unknown they are by no means unknowable."

L. Mantell in Automobile Engineer, March 1940 [1][1]

Optimizing valve timing before emissions regulations, fuel-economy rules, and variable valve actuation (VVA) came on the scene was simple but very tedious indeed. A multitude of mechanical cams had to be ground and tested to provide guidelines for each new engine design. The valve timings, once chosen, were then fixed by the valve gear for all speeds and loads. The possibility of non-throttling load control by VVA introduces more variables into the selection of optimum valve timings. To make matters even more complex, several widely differing valve-actuation strategies have been proposed, and an advanced VVA engine might switch between strategies over the driving schedule. Missing in 1940 were actuators capable of quickly selecting timing points, as well as the computational powers now provided by engine simulation programs and vehicle computers. The actuators used in this study originate from a design patented in the 1970's [2] but include a permanent magnet that eliminates the need for a holding current while the valve is fully open or closed. Thus the required electrical power should be reduced, but at an increase in actuator cost.

BACKGROUND ON LOAD-CONTROL STRATEGIES

The main motivation for exploring VVA for efficiency improvements can be understood from a breakdown of energy sinks for an auto on a combined highway/urban driving cycle ([3], see Fig. 1). Pumping air to supply the throttled engine consumes about 6% of the input fuel energy. If this pumping loss could be cut in half using non-throttled load control then, assuming no increase in engine friction or degradation in indicated efficiency, the brake efficiency and hence the fuel economy of the car would increase by the significant increment of 12%. Completely eliminating pumping loss (an impossibility because of pressure drops in the ports and ducting) would boost fuel economy for the driving schedule by 24% under these assumptions. Additional benefits potentially available with VVA include improved idle stability, power enhancement, and refined emissions control [4].

Previous data available for VVA were reviewed in [4] and elsewhere, so only the most relevant prior data are

[1]Numbers in brackets refer to References listed at the end of the text.

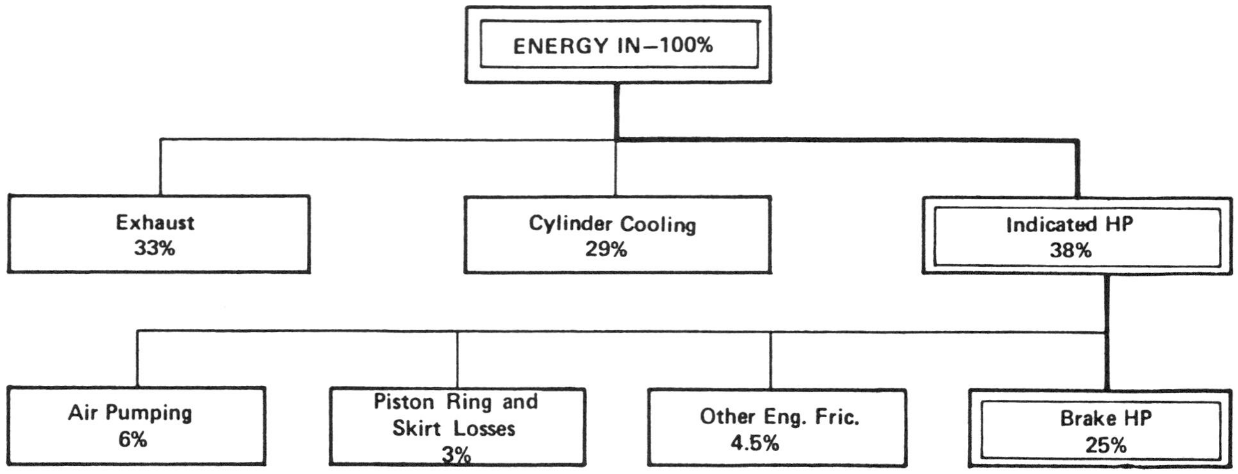

Fig. 1 - Energy breakdown for drive cycle (from [3]).

Table 1 - Comparison of Measured Efficiency Improvements via EIVC re Throttling

Increase in Thermal Efficiency (%)†	Net or Brake Efficiency	Diluent*	Test Conditions	Reference
18	net	none	1000 r/min, 200 kPa net	Kreuter [5]
10	"	"	1000 r/min, 300 kPa net	
6	"	air	"	
4	brake	none	2000 r/min, 200 kPa brake	Wichart [7]
4	"	air	"	
5	"	EGR	"	
10	net	none	1500 r/min, 270 kPa net	present study
2	"	EGR	1500 r/min, 300 kPa net	

† rounded to nearest 1%.
* pairwise if used, i.e. for both EIVC and throttling.

mentioned here (see Table 1). Most VVA experiments in the literature have not included the effects of dilution by EGR although this is the method now most used in conventional engines to decrease NO_x emissions and boost efficiency. The use of EGR with load control by early intake-valve closing (EIVC) could be advantageous for further reducing NO_x and pumping loss. More information is available for lean mixtures [5-6]. Wichart [7] found that adding EGR with EIVC operation did not improve brake efficiency over stoichiometric EIVC operation, whereas dilution with air improved brake efficiency significantly, and these same trends also held for throttled operation. These results are apparently explained by the poor dilution tolerance of Wichart's unusual engine. Kreuter [5] compared EIVC against throttling for lean mixtures and explored the effect of intake valve opening (IVO) on residual fraction for these methods. After an extensive study, Kreuter concluded that there was an optimal combination of residual fraction and air-fuel ratio for each speed/load combination and for each method of load control examined. In his experiments, EIVC yielded a 6-10% improvement in net thermal efficiency over throttling for the best-efficiency lean mixtures at 1500 r/min, part load. Südhaus [6] explored the lean-mixture/residual-fraction combinations needed to minimize engine-out emissions and maximize efficiency for EIVC. Hara, et. al. [8] tested the effect of EGR on EIVC operation but with some throttling retained for fine load control. Improvement in net efficiency relative to throttling ranged from 7% without dilution to nearly zero with EGR dilution to the stability limit. The authors ascribed the lack of a net-efficiency gain to slower combustion caused by 1) a decrease in effective compression ratio, resulting in 2) reduced cylinder temperature and pressure, and 3) reduced flowrate of hot EGR (necessary to retain stability with EIVC operation), all relative to conventional throttled operation with EGR. Measurements and results from cycle simulation supported these conclusions. Turbulence intensity was little changed between EIVC with some throttling and throttling alone.

To summarize, it is known that using EGR with throttling decreases the relative efficiency advantage of non-throttled load control because pumping losses are reduced. Adding EGR to EIVC should further decrease pumping losses but also slow combustion, so the relative advantages of the methods are not clear a priori. Combustion simulation codes are not presently able to describe combustion near the dilute limit or to predict emissions accurately, so experiments are required to make the final determination. Because the turbulence fields affecting combustion depend upon engine geometry as well as valve timing, the relative ranking of load-control methods may vary with engine design. Since major engine modifications are usually required to generate different valve-lift profiles mechanically, experimental data often have been compromised to the extent that comparing results from disparate engines is difficult. The use of programmable valve actuators eases considerably the strain of performing such experiments and increases the reliability of the data obtained.

In the experiments reported here, one engine with an electromagnetic valvetrain required no changes other than to software tables in order to vary the valve-lift profiles radically. The paper points out differences between swirling and quiescent combustion in the same engine with trapezoidal valve-lift profiles. EIVC was compared against conventional throttling, both cases with and without heavy EGR dilution. Characteristic valve-lift profiles for the conventional throttling are sketched in Fig. 2, in which a throttle plate modulates the charge density and the lift profiles do not change. Load control by EIVC (Fig. 3) uses an atmospheric-pressure intake to reduce throttling loss and closes the intake valve early to restrict cylinder charging [9]. Other known effects of VVA were not tested such as the improvement in idle stability due solely to phasing intake and exhaust events as well as the ability to increase maximum high-speed power or low-speed torque. Said another way, variations in IVO timing made solely to change residual fraction were not studied. Although this effect was shown by Kreuter [5] to be important for a lean charge, residual fraction control by IVO variation should be of less importance for the cases studied here with heavy EGR dilution for use with existing 3-way catalysts as required to meet foreseeable emission requirements.

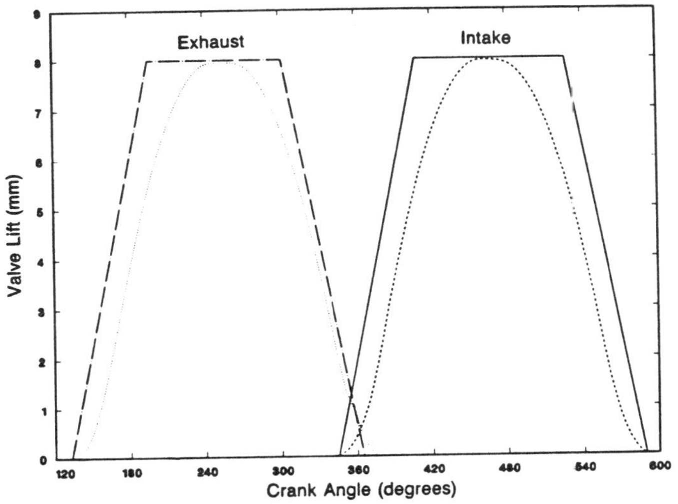

Fig. 2 - Baseline electromagnetic and standard mechanical valve-lift profiles.

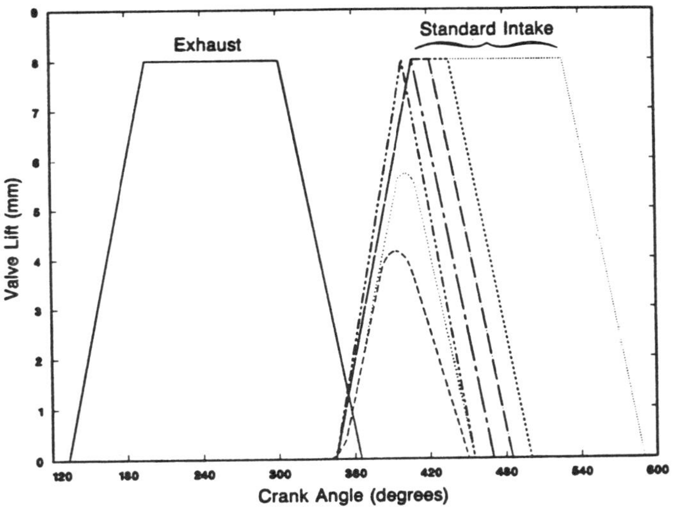

Fig. 3 - Typical EIVC valve-lift profiles.

VALVE ACTUATORS

A cross-section drawing of the actuator [10-11] is shown in Fig. 4 and a force diagram is given in Fig. 5. In the valve-closed position shown, spring A is compressed and spring B is released. The radially poled permanent magnet holds the valve in position by creating the necessary latching magnetic force on the plunger in airgap A. The magnet's hold is released by exciting the coils to reduce the magnetic flux in the airgap, and the springs then provide motive force. As the valve reaches the other end of its travel, the permanent magnet latches the compressed spring and holds the valve. The travel time T for the valve to move from full-open to closed or vice-versa is approximately one-half the period of the spring - mass system $T = \pi \sqrt{(m/2k)}$, where m is the total effective moving mass and k is the spring stiffness (see [11] for a more complete formula including the main aspects of the magnetic circuit design). Because the intake actuator need not open against elevated cylinder pressure at exhaust blowdown, an approximate 20% reduction in spring stiffness and a consequent downsizing of the actuator is possible relative to the design for the exhaust actuator. But identical actuators were used for simplicity in these experiments, see Table 2 and a photograph of the components in Fig. 6. Valve lubrication was

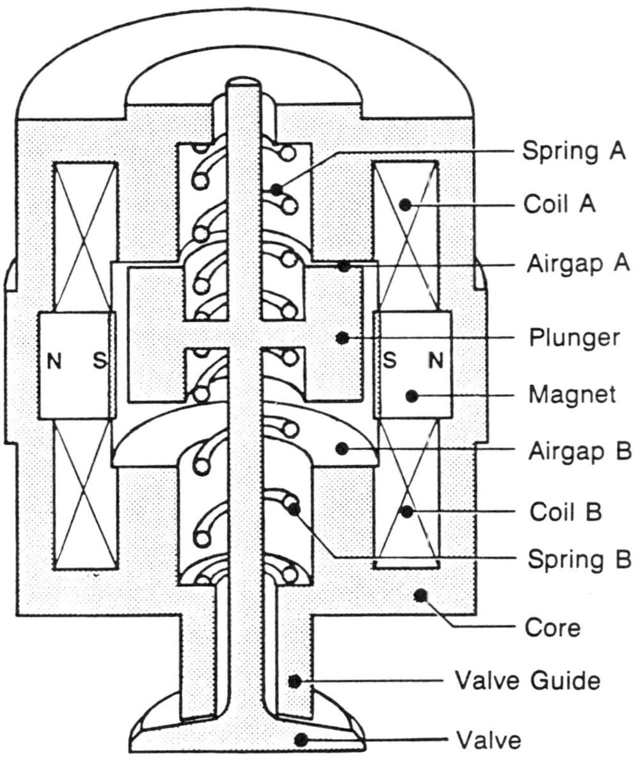

Fig. 4 - Cross-section drawing of the valve actuator.

performed manually during the tests as is commonly done for the laboratory engine. Actuator cooling was provided by external coolant jackets circulating engine coolant. The coolant's main function was to prevent the magnet from overheating (exceeding the Curie

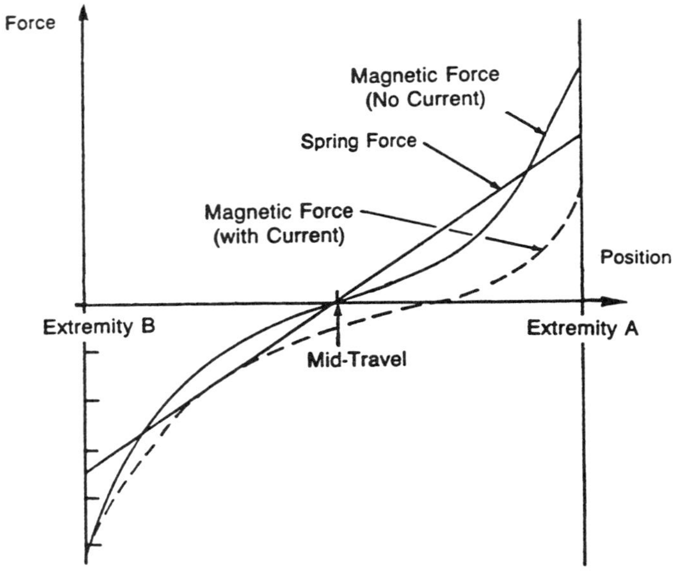

Fig. 5 - Actuator force diagram. (Coil A is energized to begin motion of plunger toward Extremity B.)

temperature), but in this experimental set-up it also minimized needed adjustments of the coil currents: A warmer magnet exerts a weaker latching force and thus requires less energy to release the spring, with travel time left unaffected. External cooling jackets would be unnecessary in a practical engine design in which the actuators were mounted directly in the cylinder head, as opposed to atop the head as in this experimental set-up.

Each actuator coil was supplied at 45 Vdc by a

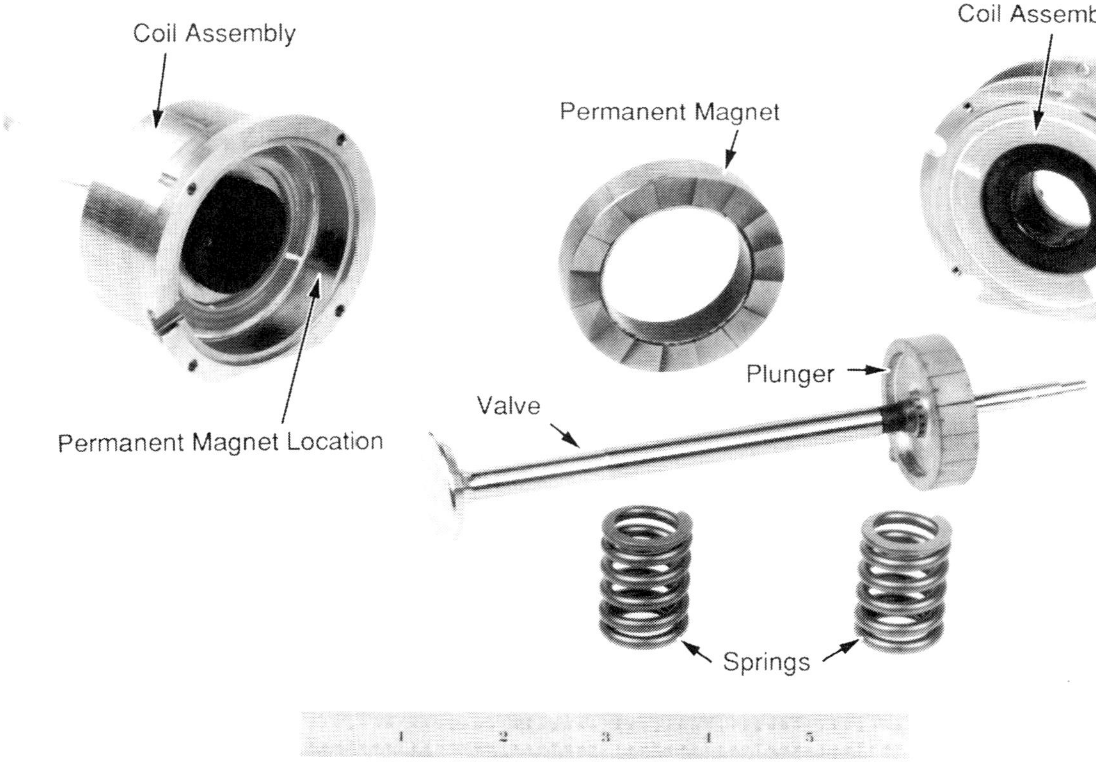

Fig. 6 - Components of the exhaust-valve actuator.

half H-bridge inverter controlled by a microprocessor, and a sample oscillogram showing current and displacement is reproduced as Fig. 7. Timing and current settings were entered manually and monitored continuously during the testing. Occasional adjustments were necessary to maintain the specified valve timings

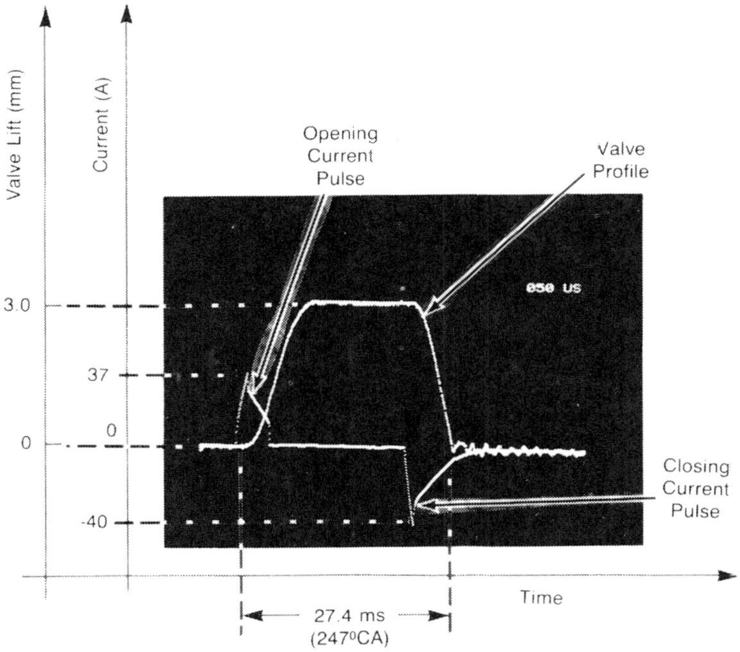

Fig. 7 - Oscillogram of current and displacement.

and ensure reliable and relatively smooth valve seating. Because these adjustments were infrequent, it is anticipated that an adaptive control scheme will allow the actuators to run unattended. Partial valve lift was produced either by stopping the valve at mid-lift where the actuator has an equilibrium point or by countermanding a lift command before the valve reached the opposite latching pole [12].

Table 2 - Actuator Design for CFR Engine

Valve Lift	8 mm
Moving mass:	
Valve+extensions+plunger	217 g
Springs	33 g each
Magnet: Material	NdFeB alloy
Field strength	35 MGOe
@ temperature	20° C
Spring stiffness	50 N/mm
Outside diameter (excludes coolant jacket)	76 mm
Height	70 mm

TEST ENGINE AND CONDITIONS

The Cooperative Fuels Research (CFR) engine used here is a single-cylinder research engine having one intake valve and one exhaust valve mounted vertically in a disk-shaped combustion chamber with variable clearance height. This engine and a compression ratio of 7.6 were chosen to prevent piston/valve contact in case of accidental loss of synchronization between the electronic valve system and crankshaft (i.e. "free-spin" design). The combustion chamber and valve arrangement are shown in Figs. 8 and 9, and the engine specifications are listed in Table 3. Dual spark plugs and a dished piston were employed to increase the burn rate. Whereas a 180°-shrouded intake valve is usually used in the CFR engine to produce swirl, an unshrouded intake valve was also used here for minimum throttling loss with a quiescent chamber. Swirl numbers computed from steady flow measurements under throttled operation were 8.9 and 0.0 with the shrouded and unshrouded valves, respectively. The disk combustion chamber of the CFR has a geometry less representative of modern engines having pentroof or hemispherical combustion chambers with central ignition and the valve/bore-size ratio is smaller than for conventional engines because of the vertical valve placement. The under-square bore/stroke ratio is also smaller than for most modern engines. Because the bore/stroke ratio as well as the valve size and cam profile affect the fraction of charge flow/unit crank angle, the mean flow and turbulence formation for a given valve-lift profile are also affected [13-14]. For this reason, the results reported here are likely to be correct as trends but not in absolute value as applied to all engines. That is, the results are engine specific to a degree that remains to be investigated.

Table 3 - Engine Specifications

Bore		82.5 mm
Stroke		114.3 mm
Displacement		0.61 L
Compression ratio	(adjustable)	7.56
Clearance height	(adjustable)	12.0 mm
Valves:	-head diameter	34.3 mm
	-inner seat diameter	30.4 mm
Spark plugs: (2)	-large tapping	**AC** C86
	-small tapping	**AC** 42XLS

The engine's load range was swept at a fixed speed of 1500 r/min at stoichiometric air-fuel ratio with a premixed charge at the conditions listed in Table 4. Load was controlled in the throttled case by adjusting the manifold pressure using a critical air system, which also measured airflow. In the non-throttled cases, intake manifold pressure was held constant and only the valve-lift profiles determined the load. Premixed charge was prepared by spraying liquid fuel onto a heated plate in a baffled mixing tank upstream of the engine.[2] For

[2]By design, the use of a premixed charge and a critical flow meter meant that possible improvements in fuel-air mixing associated with low valve lifts were not included. Similarly, problems with hot-film air meters and degradations in mixture formation by loss of intake

external EGR, exhaust gas was routed from the exhaust pipe, cooled to room temperature in a long line, and then introduced into the fuel-air mixing tank where the mixture was heated to 45° C.

Table 4 - Operating Conditions

Speed	1500 r/min
Fuel	Indolene clear (97 Research Octane)
Fuel-air equivalence ratio	0.98
Spark timing	MBT, except as noted
Intake-charge temperature	45° C
EGR temperature	25° C
Coolant and oil temperatures	85° C
Non-throttled intake pressure	99.5 kPa absolute
Exhaust manifold pressure	102.9 kPa absolute

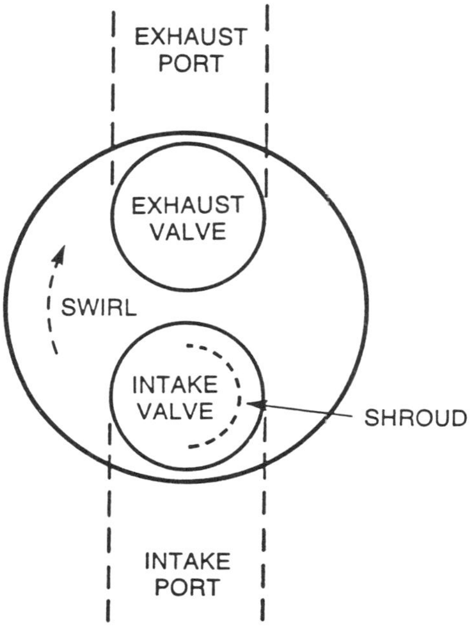

Fig. 8 - Schematic drawing of port and valve arrangement.

Cylinder-pressure data were acquired for 300 consecutive engine cycles using a flush-mounted Kistler 6061 water-cooled pressure transducer. Approximate minimum-advance-for-best-torque (MBT) timing was set by placing the peak cylinder pressure at 15° CA ATDC [15-16], except when trace knock limited the spark advance at some test conditions. External EGR, when used, was added in an amount to increase the coefficient of variation (COV) in indicated load to a target value of 3%.[3] All information regarding engine

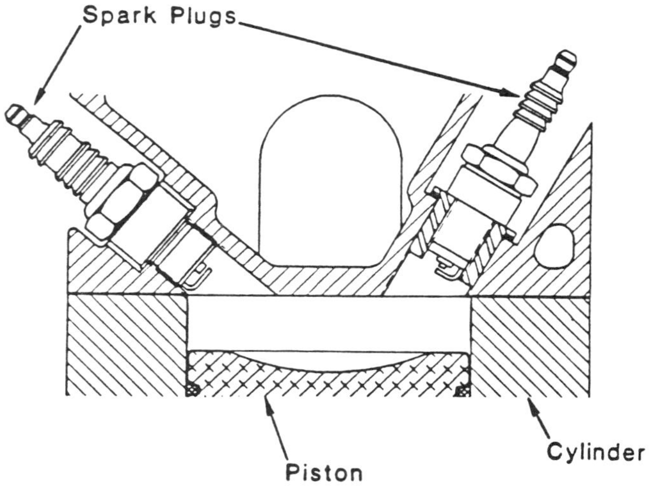

Fig. 9 - Schematic drawing of the combustion chamber.

load was determined from the cylinder-pressure data rather than the dynamometer torque. The cylinder-pressure data were processed with a single-zone heat-release program to yield statistics on the combustion events. In order to recover burn-rate data, the cylinder-pressure signals were pegged to force agreement with the measured time-averaged exhaust pressure [17]. This practice was necessary because the more common method of pegging cylinder pressure to the intake pressure at bottom dead center of intake cannot be used with EIVC profiles at low loads, where the cylinder charge is expanded during the intake stroke following IVC. The pressure data were also integrated with respect to volume over an engine cycle in accordance with the definitions of [18] to yield the ensemble-averaged mean effective pressures: indicated (IMEP), pumping (PMEP), and net (NMEP).[4]

---

manifold vacuum as reported by other researchers using port fuel injection [7] were not seen. This means that the maximum efficiency improvement without constraint from mixture problems was found for this engine.

[3]Because EGR flow was set while the COV was measured in the test cell using 100 cycles of nonsequential pressure data, the reported COV based on 300 sequential cycles did not always match the 3% target value, but the data are reported nonetheless. It is noted that other target values appear in the literature, where often 10% COV has been used. Due to time constraints, EGR dilution was carried out at 3% COV of IMEP at MBT timing, not as EGR hooks to find the dilution for maximum efficiency. This simplification may have biased the results slightly in an unknown direction.

[4]A brief digression on the manner of quantifying pumping loss is warranted. As Kreuter [5] and others contend, the PMEP definition of [18] is not thermodynamically consistent for variable-geometry engines. That is, by defining PMEP as the integral of pressure with respect to volume over a revolution from bottom center to bottom center for the gas-exchange process (the shaded area plus the cross-hatched area in the following diagram), extra work is included in the calculation that would not go to zero if all flow losses were to vanish. Because the same extra work is also included in the IMEP calculation, the results for net quantities are still completely equivalent. However, a second, more meaningful definition for pumping work (shaded area only) has been suggested that requires

ACTUATOR PERFORMANCE

TIMING/LIFT COMBINATIONS - A map of measured combinations of timing and lift obtained with the actuator is shown in Fig. 10. The line-shaped region **1** corresponds to maintaining full peak lift. The shortest full-lift duration measured to be 7.3 ms. In order to obtain such short durations, the energy input must be increased to assist the spring motion. This additional energy is not recovered by the spring but is dissipated as high seating velocity of up to 3 m/s. With a 1 m/s velocity limit, the shortest duration was 13.5 ms. For partial lifts in region **2** in Fig. 10, just enough energy is given to the actuator for the valve to settle at the mid-lift equilibrium point, which occurred here in a band between 35% and 60% of peak lift because of friction and asymmetries. Beyond that range, in regions **4** and **5**, there is no equilibrium and the valve coasts either towards the center or the open or closed positions, depending on whether the magnetic or spring forces are dominant. The valve follows a usable pseudo-ballistic profile for the combinations of part-lifts and durations comprising region **3**.

ENERGY REQUIREMENTS - Actuator performance was characterized by the valve-travel time as a function of electrical energy input, as shown in Fig. 11, with magnet temperature as a parameter. Differences between opening and closing strokes were due to asymmetries in the actuator construction. The average energy usage for the intake actuator was about 2.20 J per engine cycle while the exhaust actuator used about 3.45 J at one-half load because of the need to open against cylinder pressure. These values can be used to estimate the equivalent valvetrain-friction torque for an actuator-equipped, 2-valve multi-cylinder engine. An energy usage of 5.65 J per cylinder per engine cycle at 1500 r/min corresponds to 70.6 W. In a vehicle, this energy would come from the engine via the alternator. Assuming an alternator efficiency of 70%, which is realistic for an alternator of the size required to accommodate the extra electrical load of the valve actuators, the mechanical power required from the engine is 100 W per cylinder at 1500 r/min or 400 W for a 4-cylinder engine, equivalent to a load torque on the crankshaft of 2.55 Nm. This value falls below the range of friction torque (3.2 - 5.2 Nm at 1500 r/min) measured for various (non-variable) production valvetrains excluding roller-follower designs [19]. However, the use of roller elements can decrease friction torque to only about 20% of the value for sliding elements [20], thereby substantially undercutting the values measured here.

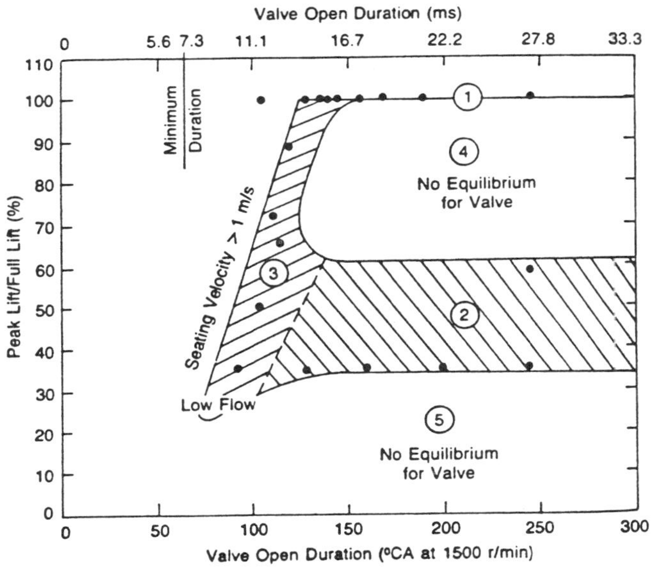

Fig. 10 - Map of possible lift/timing combinations.

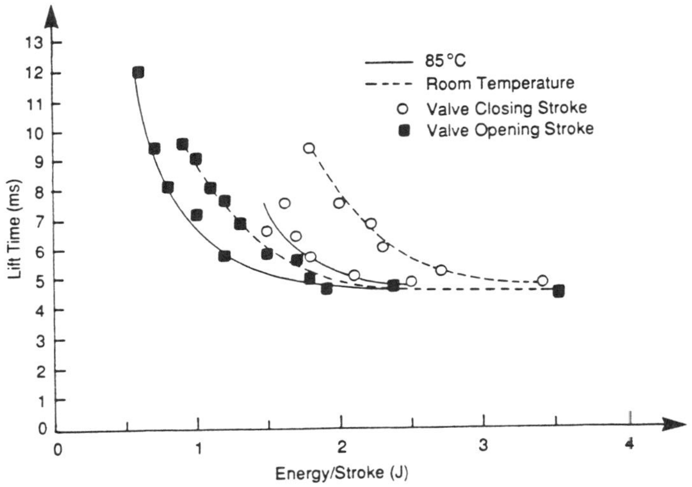

Fig. 11 - Performance characteristics of the exhaust actuator.

SEATING VELOCITY - The prototype actuators mounted two polymeric rings on the plunger, one to stop the valve at maximum lift and the other to absorb some of the kinetic energy just before the valve seated. Valve seating velocity was measured to be between 0.35 m/s and 1 m/s during the tests when operated under open-loop control. Although these values at 1500 r/min were unacceptable for a production engine, the validity of the following performance and emissions data was not affected.

---

determining the intersection point between closed high-pressure and pumping loops in the pressure-volume diagram. The first definition offers the convenience of simple on-line calculation in the test cell while the second definition is more useful for thermodynamic calculations. The first definition of pumping loss was used in reporting results here both for convenience and in order to maintain consistency with the majority of past work. The definitions diverge most strongly at low load.

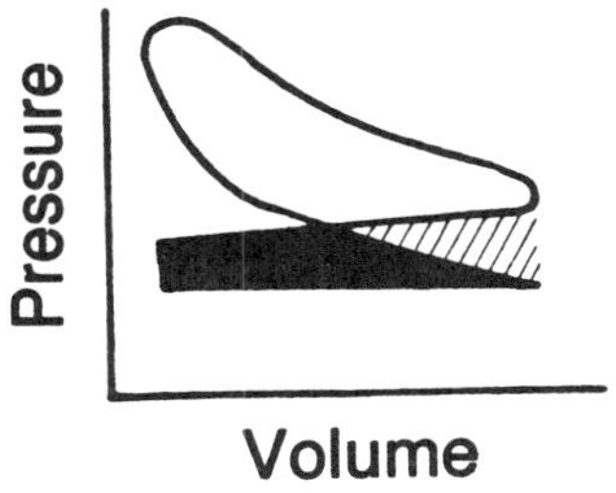

COMPARISON OF LOAD-CONTROL METHODS

EFFICIENCY AND COMBUSTION - Pumping loss measured for the different control methods is plotted in

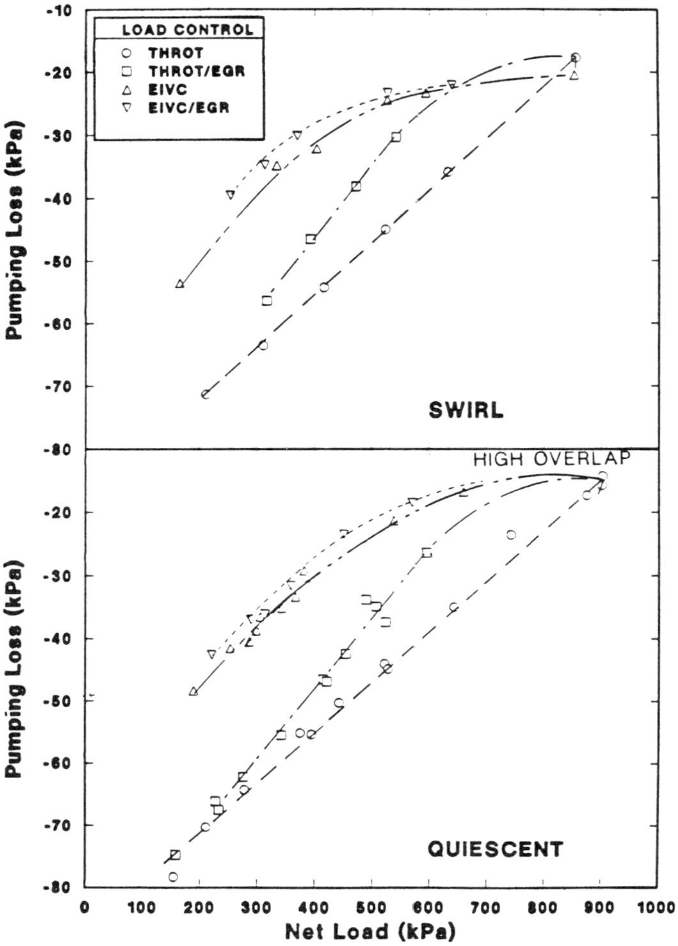

Fig. 12 - Comparison of pumping work.

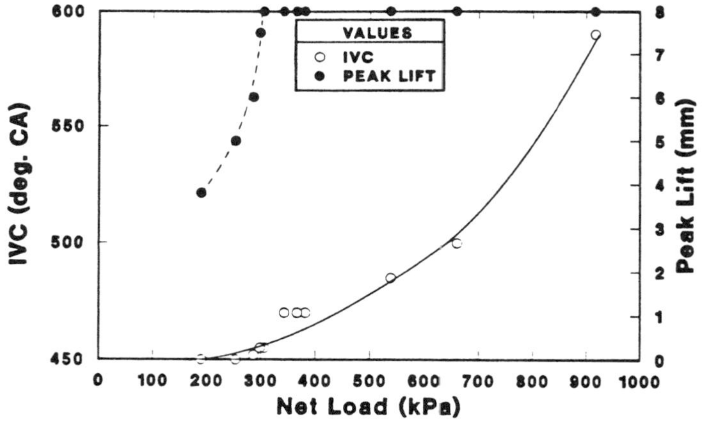

Fig. 13 - Valve-timing diagram for EIVC for the quiescent chamber.

Fig. 13, where large negative values are least desirable. Compared to throttling without EGR, pumping loss was reduced at a given net load by all of the other methods shown, and both EIVC conditions also reduced pumping

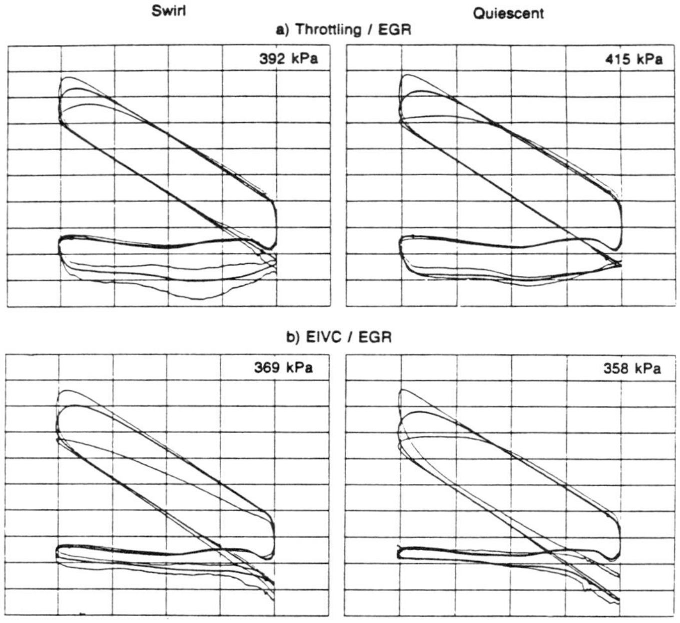

Fig. 14 - Logarithmic indicator diagrams for methods with charge dilution. (Maximum, minimum, and mean pressures are plotted in each case for 300 consecutive cycles.)

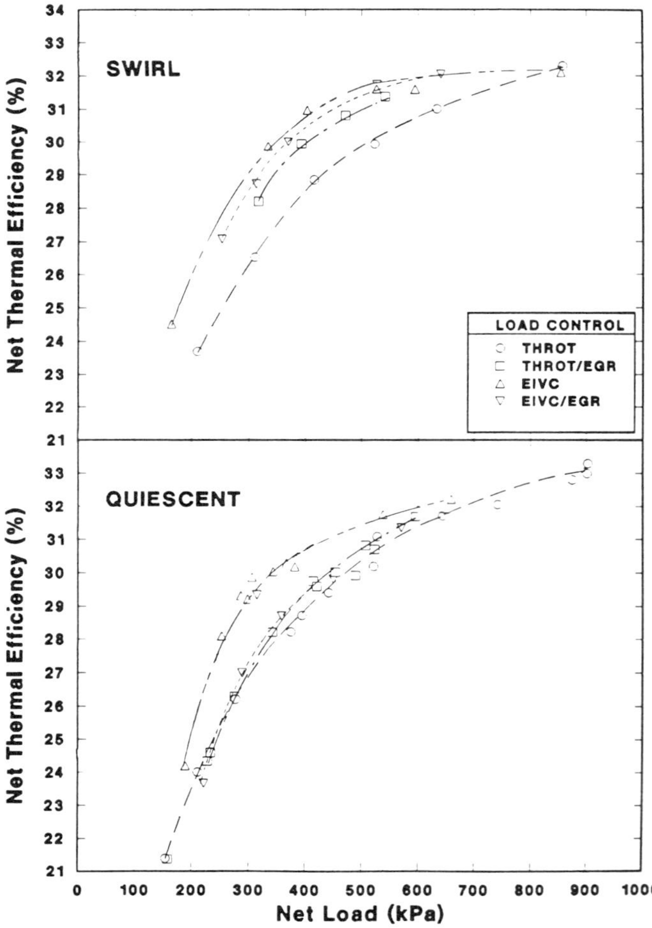

Fig. 15 - Comparison of net thermal efficiency.

loss compared to throttling with EGR. The peak intake lifts and timings used for EIVC control are plotted in Fig. 14. Compared in Fig. 14 are logarithmic indicator diagrams on the same scales for both cases with dilution, with and without swirl at nearly the same net load. The reduction in pumping work for EIVC, although not represented quantitatively by the apparent area, is still striking.

Net thermal efficiency is plotted in Fig. 15. Both EIVC and throttling without EGR were tested down to the minimum loads where 3% COV of IMEP could still be maintained. The relative net efficiency ranking from best to worst at part load was EIVC, EIVC/EGR, throttling with EGR, and finally throttling. The improvement in net efficiency relative to throttling with EGR is shown in Fig. 16. Note that the efficiencies of all

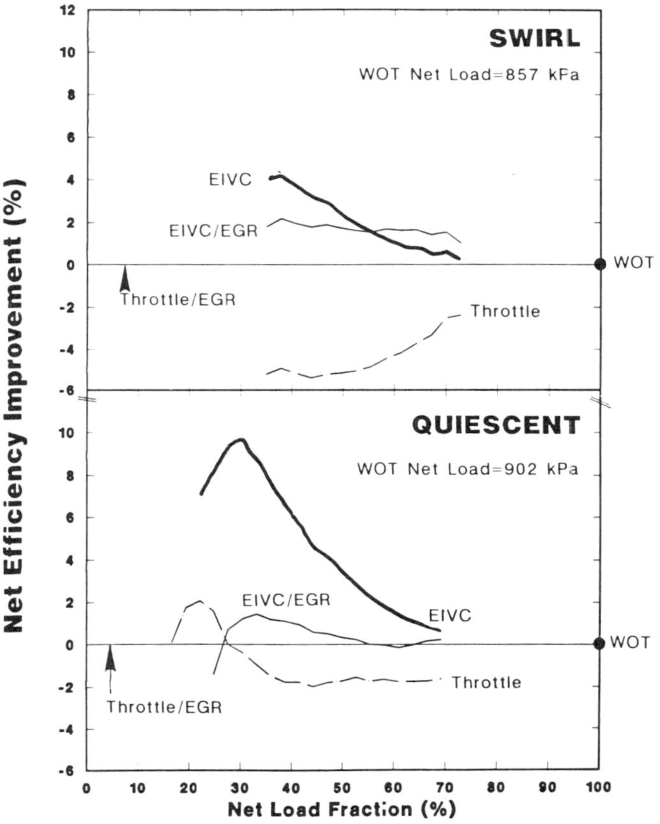

Fig. 16 - Comparison of improvement in net efficiency re throttling with EGR.

methods must converge at WOT and, similarly, the net efficiency of all methods decreased radically at load fractions below about 20%, which made the calculation of relative improvements between the methods unreliable. The present data for efficiency improvement with EIVC operation are compared in Table 1 with other data from the literature. These data are from different engines with different valve-lift profiles at light load and low speed. In general, there is a spread in the results for efficiency improvement. Why? First, the data for improvements in brake efficiency may reflect in part a reduction of valvetrain friction at low valve lift. The experiments by Kreuter and by Südhaus used optimized residual fractions via changing EVC and IVO timings from stock values, in contrast to this study. The present results for the CFR engine show greater improvement without dilution by using trapezoidal lift profiles (as did Kreuter's engine) that could reduce pumping loss for EIVC relative to smoother lift profiles. As discussed below, the CFR engine showed no improvement for EIVC with EGR dilution, apparently due to a poorly burning combustion chamber, so differences in the combustion chambers and stock valve-lift profiles between engines probably also affected the results.

Spark timing is plotted in Fig. 17, where one sees that the spark was retarded from MBT at WOT to suppress knock. (Arrowheads are used in the plots of experimental data to stress the fact that results for all load-control methods must converge at WOT.) MBT spark advance was increased by 5 to 20° by eliminating swirl flow. Combustion durations estimated from pressure data for 10-90% mass-fraction-burned, Fig. 18, showed that EIVC/EGR burned much slower than throttling with EGR, but spark advance was limited by misfire at low loads. Without dilution, EIVC increased combustion duration relative to throttling by about 10° for 0-10% and by about 20° for 0-90% at the same net load. Dilution rates tolerated by the methods are plotted in Fig. 19. As is well known, swirl combustion tolerated considerably more dilution than could quiescent combustion, and EIVC was found to tolerate less EGR than could throttling. Indicated efficiency did not vary

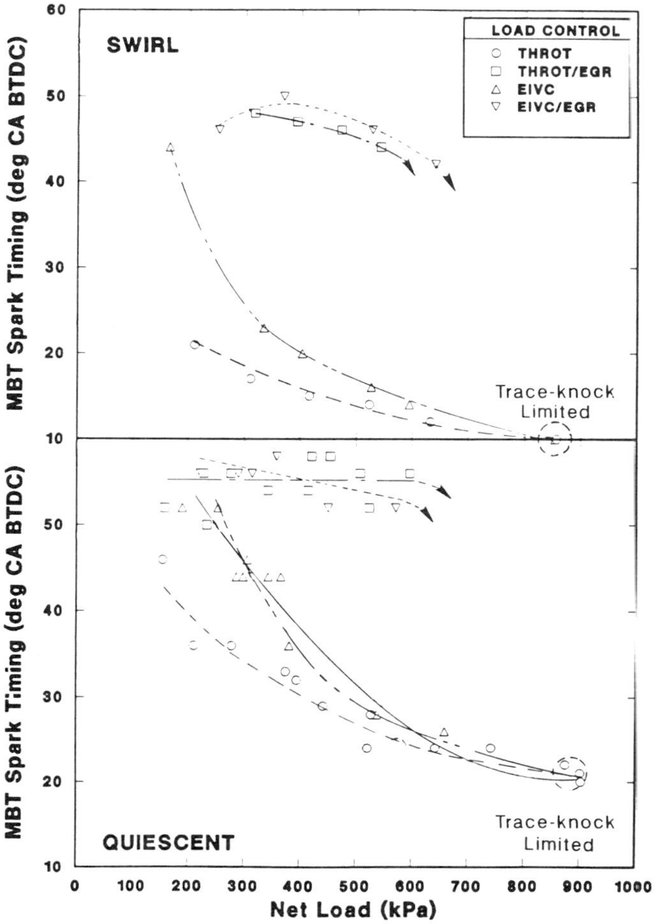

Fig. 17 - Comparison of MBT spark timings.

greatly between the methods except for the EIVC/EGR quiescent case, which had the lowest efficiency and deteriorated with decreasing load as combustion slowed.

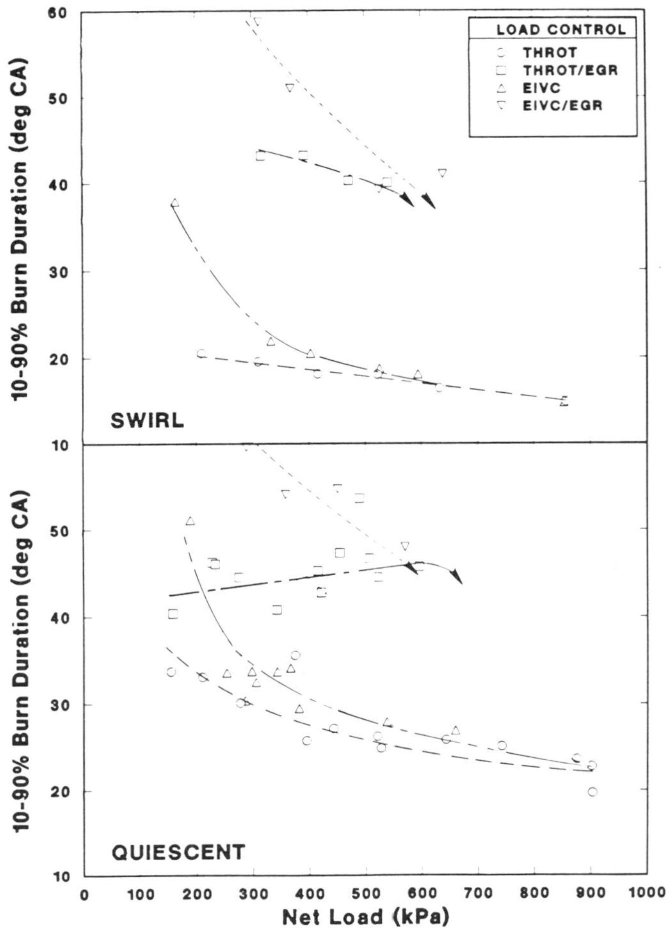

Fig. 18 - Comparison of 10-90% burn durations.

EMISSIONS - Measured NO_x emissions are plotted in Fig. 20. The expected increase in throttled NO_x output for the swirl case relative to the quiescent case was confirmed. Also as expected, EIVC operation decreased NO_x relative to throttling by a significant margin. However, reducing NO_x far enough to be comparable to throttling with EGR required the addition of EGR for EIVC load control, and the minimum NO_x values attained were effectively the same for fixed combustion stability. It is worth noting that in the swirl case, NO_x for throttling was reduced from a value higher than without swirl to a value lower than without swirl by adding EGR. Apparently the much-improved dilution tolerance with swirl, reflected in a higher EGR rate, (Fig. 19), allowed more NO_x reduction.

The results for hydrocarbon (HC) emissions are shown in Fig. 21. Absolute levels of HC emissions were high compared to a modern engine, probably due to the presence of substantial crevice volume, the lack of valve-guide seals, and the use of manual drip oiling. The HC data for EIVC with EGR spanned the values between EIVC alone to throttling with EGR as the dilution rate increased. Adding EGR to throttling showed a 20-30% increase in HC, similar to other conventional engines. Whether these HC results are repeated in a production-like engine with conventional valve seals and much lower overall HC levels would be of interest.

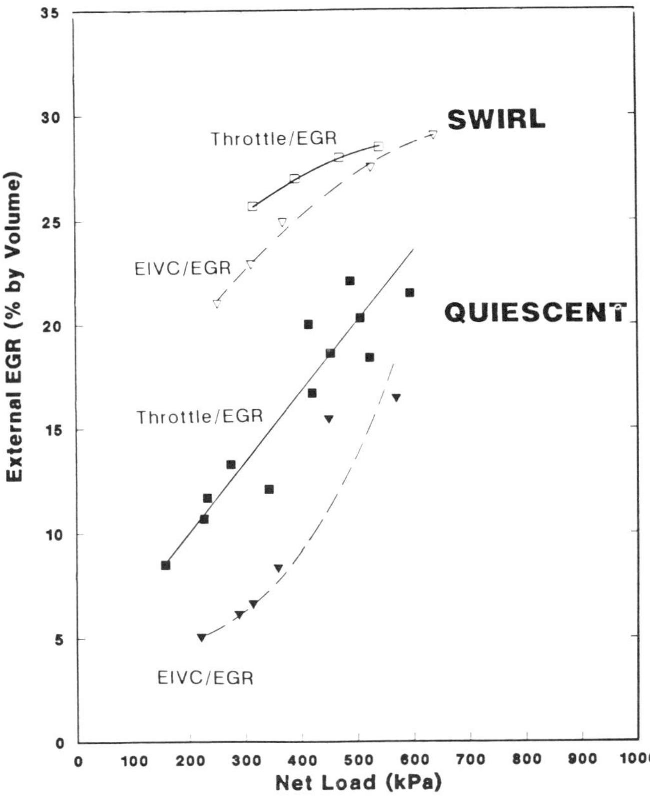

Fig. 19 - Comparison of EGR rates.

CONCLUSIONS

1. The use of heavy EGR dilution reduced to near zero the net efficiency advantage of early intake-valve closing relative to conventional throttling in this engine. The cause of this reduction was slower burning believed to be caused by lower peak cycle temperatures and lower turbulence velocities. This conclusion should be verified on a more compact combustion chamber with central ignition representative of modern production engines.
2. The average electrical energy requirement was 2.2 J/engine cycle for the intake and 3.5 J/engine cycle for the exhaust for the conditions tested. These values are lower than the equivalent friction losses for conventional non-roller production valvetrains at this engine speed. Valve-seating velocity, measured to be 0.35 m/s and above at 1500 rpm, was excessive for production use.
3. Performance and emissions data from operation with intake swirl showed the same trends as data at quiescent conditions, but the numerical values were

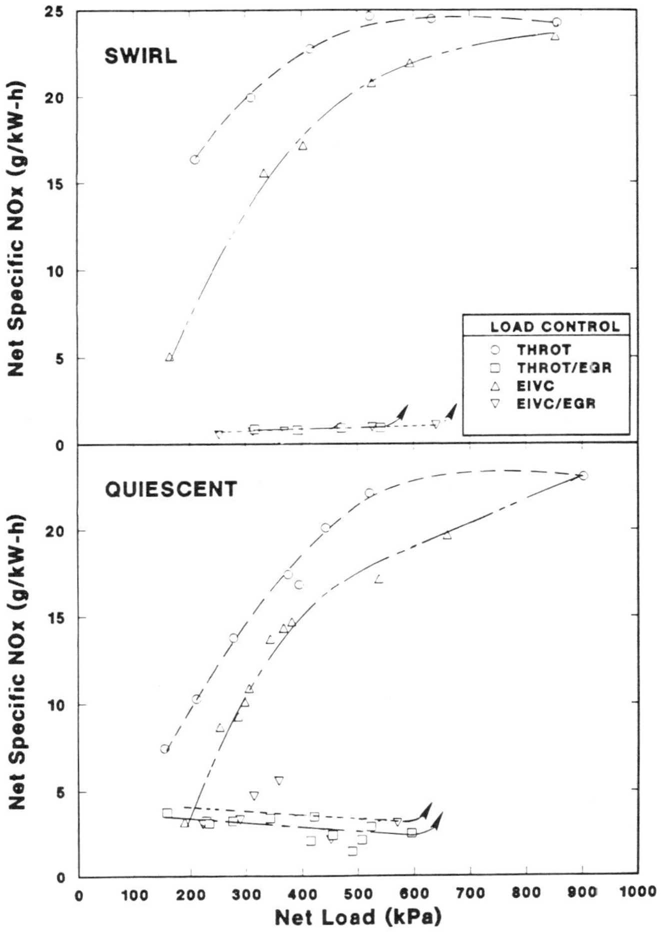

Fig. 20 - Comparison of NO_X emissions.

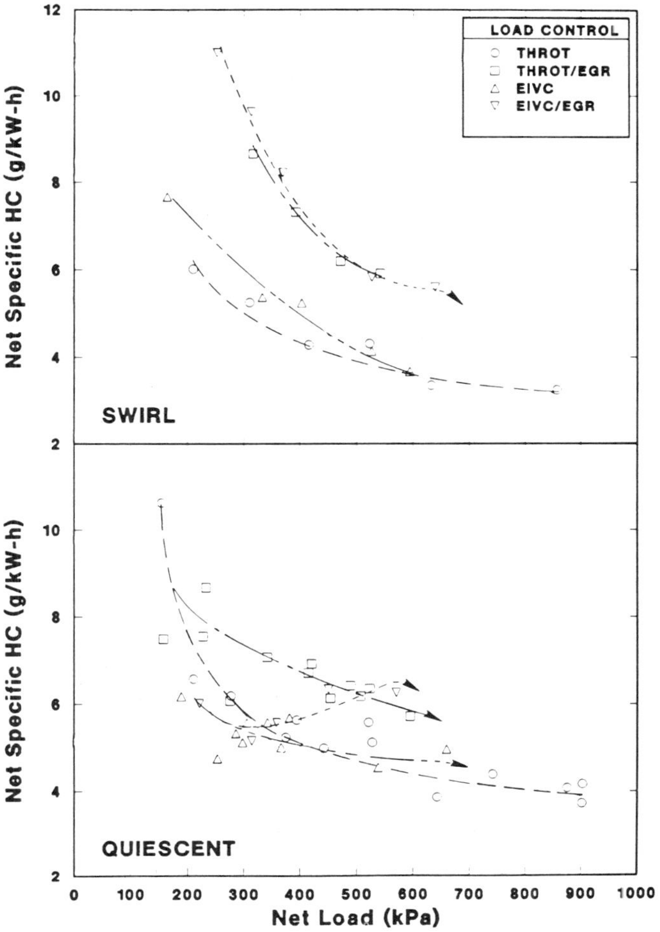

Fig. 21 - Comparison of HC emissions.

different. Use of a programmable valve system made it relatively easy to compare drastically different methods of load control in the same combustion-chamber geometry.

ACKNOWLEDGMENTS

It is a pleasure to acknowledge the contributions of colleagues. Kalyan Gokhale and David Sidlosky engineered the microprocessor-based actuator drive. Stewart McEwen, Jerry Silvas, Bill Omell, and Lonnie Frost operated the engine and valve system and recorded the data. Jim Gallagher provided design assistance.

REFERENCES

1. L. Mantell, "Valve Timing," Automobile Engineer, Vol. 30, No. 395, pp. 85-87, March 1940.

2. K. Longstaff and S. Holmes, "Internal Combustion Engines," U.S. Patent No. 3,882,833, May 13, 1975.

3. O. Pinkus and D. Wilcock, Strategy for Energy Conservation Through Tribology, American Society of Mechanical Engineers, New York, 1977, p. 26.

4. T. Ahmad and M. A. Theobald, "A Survey of Variable-Valve-Actuation Technology," SAE Paper 891674, SAE Transactions, Section 3, 1989.

5. P. Kreuter, "Untersuchungen zur Leistungssteuerung für Ottomotoren," (Studies of Power Control for Spark-Ignition Engines"), Doctoral Thesis, Technische Hochschule Aachen, 1983.

6. N. Südhaus, "Möglichkeiten und Grenzen der Inertgassteuerung für Ottomotoren mit Variablen Ventilsteuerzeiten," ("Possibilities and Limits of Inert-Gas Control for Spark-Ignition Engines with Variable Valve Timing"), Doctoral Thesis, Technische Hochschule Aachen, 1988.

7. K. Wichart, "Möglichkeiten und Massnahmen zur Verminderung der Ladungswechselverluste beim Ottomotor," ("Possibilities and Means for Reducing Pumping Losses of the Spark-Ignition Engine"), Fortschritt-Berichte VDI, Category 12, Number 91, VDI Verlag, Düsseldorf, 1987.

8. S. Hara, Y. Nakajima, and S. Nagumo, "Effects of Intake-Valve Closing Timing on Spark-Ignition Engine Combustion," SAE Transactions, Vol. 94, Section 1, pp. 1.396-1.405, 1985.

9. J. H. Tuttle, "Controlling Engine Load by Means of Early Intake-Valve Closing," SAE Paper 820408, SAE Transactions, Vol. 91, 1982.

10. B. Lequesne, "Bistable Electromechanical Actuator," U.S. Patent No. 4,779,582, October 25, 1988.

11. B. Lequesne, "Fast-Acting, Long Stroke Solenoids with Two Springs," IEEE Trans. on Industry Applications, Vol. 26, No. 5, Sept./Oct. 1990.

12. B. Lequesne, "Variable Lift Operation of Bistable Electromechanical Poppet Valve Actuator," U.S. Patent No. 4,829,947, May 16, 1989.

13. T. Asmus, "Effects of Valve Events on Engine Operation," in Fuel Economy in Road Vehicles Powered by Spark Ignition Engines, Ed. by J.C. Hilliard and G. W. Springer, Plenum Press, New York, 1984.

14. G. C. Davis, R. J. Tabaczynski, and R. C. Belaire, "The Effect of Intake Valve Lift on Turbulence Intensity and Burnrate in S.I. Engines - Model Versus Experiment," SAE Transactions, Vol. 93, pp. 1.186-1.197, 1984.

15. F. A. Matekunas, "Spark Ignition Engines - Combustion Characteristics, Thermo-dynamics, and the Cylinder-Pressure Card," GM Research Publication GMR-4660, March 1984.

16. J. Klimstra, "The Optimum Combustion Phasing Angle - A Convenient Engine Tuning Criterion," SAE Transactions, Vol. 94, Sec. 7, pp. 602-610, 1985.

17. A. L. Randolph, "Methods of Processing Cylinder-Pressure Transducer Signals to Maximize Data Accuracy," SAE Paper 900170, 1990.

18. D. R. Lancaster, R. B. Krieger, and J. H. Lienesch, "Measurement and Analysis of Engine Pressure Data," SAE Transactions, Vol. 84, Sec. 1, pp. 155-172, 1975.

19. W. B. Armstrong and B. A. Buuck, "Valve Gear Energy Consumption: Effect of Design and Operation Parameters," SAE Paper 810787, 1981.

20. S. Bair, J. A. Griffionen, and W. O. Winer, "The Tribological Behaviour of an Automotive Cam and Flat Lifter System," ASME Journal of Tribology, Vol. 108, July 1986, p. 478.

970249

A Novel, Fully Flexible, Electro-Mechanical Engine Valve Actuation System

Rassem R. Henry and Bruno Lequesne
General Motors Research and Development Center

ABSTRACT

A novel, fully flexible engine valve actuation mechanism was built and tested for the first time. It consists of a permanent magnet brushless dc motor driving a cam mechanism to actuate each engine poppet valve. This mechanism has the advantages of low friction, low seating velocity and speed range comparable to that of production valve trains. The electromechanical system has also regeneration capabilities which result in an energy requirement that is equivalent to or lower than the valve-train friction of current production engines. The valve event duration is changed by increasing or decreasing the cam/motor angular velocity during valve opening in order to shorten or lengthen the valve event, respectively. Part-lift operation is also possible by oscillating the mechanism around the valve opening or closing points. The prototype mechanism was run on the bench on an actual engine cylinder head at speeds of up to 3225 r/min, equivalent to 6450 engine r/min. It demonstrated good flexibility in variable timing and part-lift operation. In addition, valve deactivation should be possible at all engine speeds.

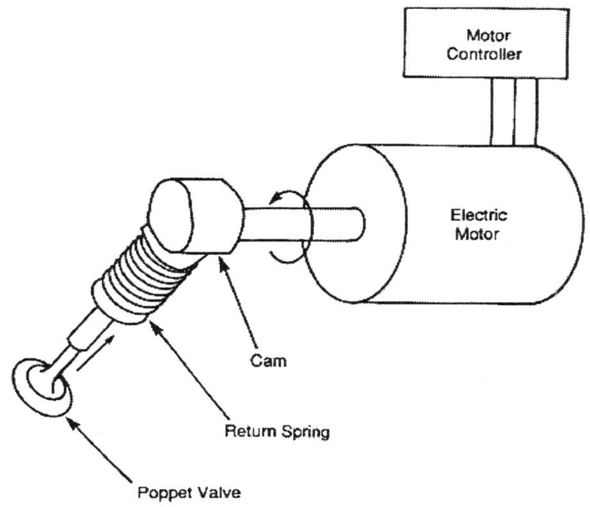

Figure 1. Basic Concept

I. INTRODUCTION

Varying the engine valve event duration, lift and phasing is a known way to improve engine performance, particularly increase fuel efficiency and lower emissions, and has therefore been the object of considerable work. The most flexible actuators usually entail the separate actuation of each valve. The actuator, then, uses hydraulic [1], pneumatic [2] or electromagnetic [3,4] means to move the valve. Some of these actuators are capable of fast valve motion. In many cases, however, the energy requirement is high, with the notable exception of those regenerative schemes that, for example, trade kinetic energy for potential energy at the end of travel. An example of this is the two-spring concept [3-6]. All of these devices, however, suffer to various degrees from high seating velocities, i.e. the valve at the end of the closing motion hits its metallic seat at velocities that produce unacceptable levels of stress and audible noise.

In this paper, section II describes the general concept of our system. Section III explains the mechanism design details. A high torque/inertia ratio is the basis for a large valve-event flexibility, and assessing just how large this flexibility could be and how much energy would be required to achieve it was the primary purpose of the bench tests described in section IV.

II. GENERAL CONCEPT DESCRIPTION

II. 1. Overall Description and Valve-Timing Variations

In its simplest form, this new concept consists of an electric motor with a rotating cam and a reciprocating (poppet) valve [7], see Fig. 1. There is one mechanism per engine valve. If the motor runs all the time at half the engine speed, the valve will behave exactly as it does in present production engines, and the valve-event duration is then equal to the time taken to span the cam lobe, see Fig. 2-a. However, if the motor runs faster than one half of the engine speed while the valve is open, the valve event duration will be shorter, proportional by the ratio of actual motor speed to half the engine speed. This scenario is shown in Fig. 2-b. To maintain synchronism between the valve and the engine, the motor must then be slowed down after the valve is closed so that, overall, the motor makes one revolution while the engine makes two. If a longer valve-event duration is desired, the motor is slowed down and run at a lower speed while the valve is open, see Fig. 2-c. Then, the motor must be accelerated while the valve is closed in order to maintain the valve/engine synchronism. In short, the average motor speed over one engine cycle has to equal half the engine speed at all times.

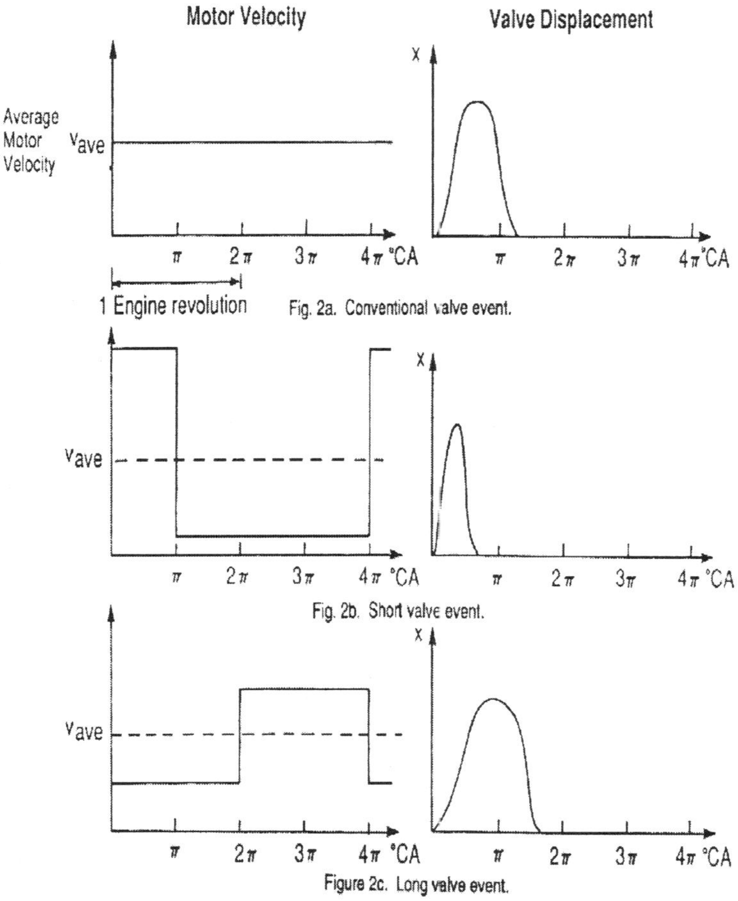

Figure 2. Principles of Variable Valve Timing Operation

II. 2. Valve Phasing

Phasing is achieved by changing the instant at which the valve opens by allowing the motor to slip by until the desired crankshaft angle is reached. Therefore, since both the phasing with respect to the engine of the valve start-of-lift and the valve-event duration can be changed, there is complete valve timing freedom. Furthermore, it is possible to stop the valve between two valve events, thus making cylinder deactivation possible.

This system flexibility is influenced by the motor/mechanism bandwidth, i.e. by how fast they can react to commands, and also by the maximum available motor torque. The fast response of the motor is measured by its torque/inertia ratio. High torque/inertia ratio has been made practical by the invention of high energy magnets which have been utilized in our system.

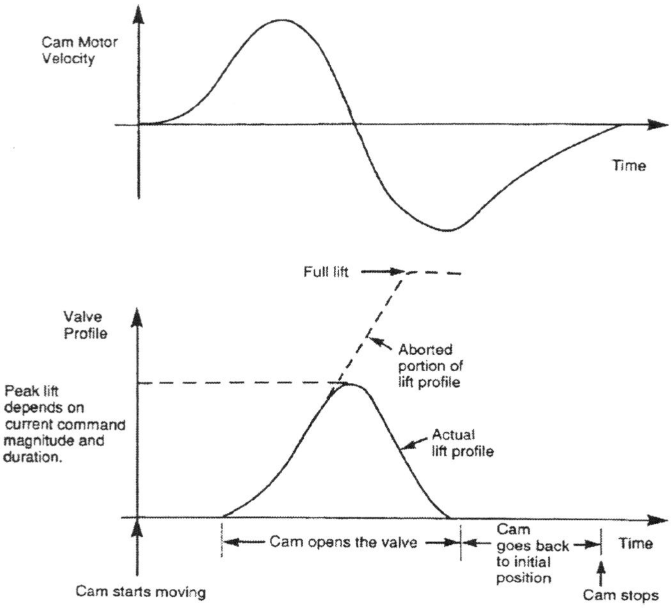

Figure 3. Principle of Part Lift Operation

II. 3. Part-Lift Operation

The electromechanical system can also be operated in part lift. The cam, then, is run in an oscillatory mode around either the valve opening or valve closing position. Assuming the valve opening position is chosen, part-lift operation is done by setting the motor first at standstill with the cam positioned ahead of the start of lift. At some desired time during the engine cycle, a command is sent to the motor which starts running and opens the valve, see Fig. 3. At some point while the valve is partially opened, another command is sent to the motor in order to make the motor stop and reverse direction. The valve then closes and the motor stops. For regular operation on an engine, an appropriate closed-loop position control must then take the cam mechanism back to its initial position, waiting for the next engine cycle to repeat the motion. How high the lift is and how long the valve-event duration is will be a function of the timing, duration and magnitude of the various control commands. They are also a function of the initial angular position, i.e. where the motor was stopped prior to being cycled.

III. MECHANISM DESCRIPTION

III. 1. Main Components

As can be seen from fig. 4, the mechanism consists of the following main components:

- A cylindrical cam, with an external lobe, that is rotating with the motor rotor.
- Two conical roller followers connected to the reciprocating valve through a housing.

- A guide mechanism to absorb the reaction torque on the reciprocating housing.
- A lash adjustment mechanism which is comprised of a compliant roller follower thrust support, and a light spring helping the valve during closing.
- A conventional valve.
- Bearings that allow cam rotation as well as housing translational motion relative to the cam.

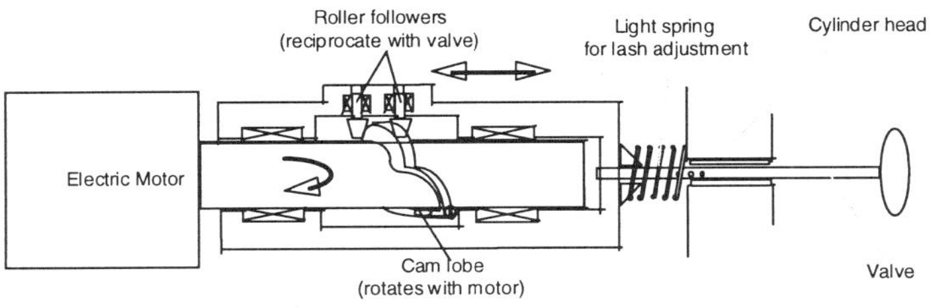

Figure 4. The Mechanism (not to scale)

III. 2. Cam Design

The approach taken here to design the cylindrical cam is similar to designing a 2-D translating cam with a reciprocating roller follower. The base distance is taken as the unwrapped circumference at that particular radius. Undercut is checked graphically at each radius, and the internal and external envelopes are created (fig. 5). In order to construct the cam motion curves, we use known standard cam motions segments and build it up. For the sake of continuity, we match the lift, first derivative, and second derivative at points of intersection. The resultant jerk (third derivative) is only checked for being finite. The chosen lift is symmetrical between rise and fall with a half-cycloidal rise that leads to a half-harmonic rise up to the maximum lift. The cycloidal motion has the lowest jerk among basic cam curves and the harmonic one has the lowest acceleration and jerk when matched with the cycloidal [8,9]. Figure 6 shows the resultant motion curves for the following parameters (chosen to match the actual test engine parameters):
Total angle of valve opening = 148 deg
Maximum lift = 11 mm

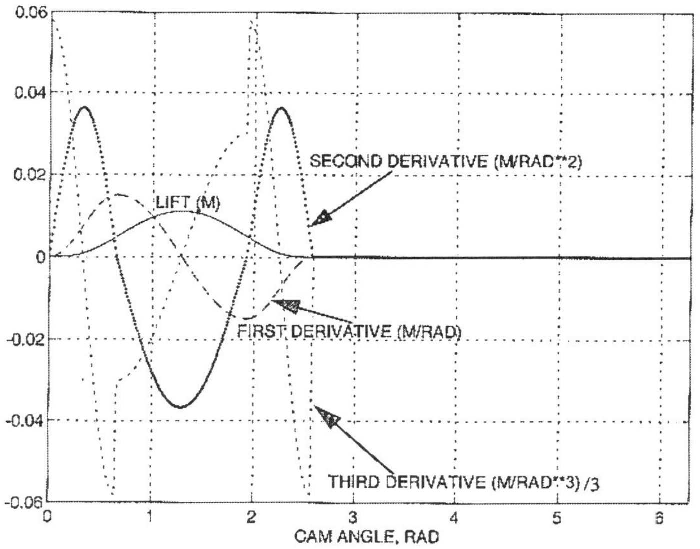

Figure 6. Cam Motion Curves

III. 3. Lash Adjustment Mechanism

Lash adjustment is achieved by adding a spring behind each of the two roller followers and one spring between the cylinder head and the reciprocating part with the valve as shown in fig. 4. The latter spring, "lash spring", is not used as the valve spring and can, therefore, be very light. When the valve lengthens during thermal cycling, it tends to be pushed up more towards the motor by the lash spring. The roller follower closest to the motor is accordingly pushed back against its spring, and the roller follower farthest from the motor is pushed forward by its spring against the cam lobe. The opposite happens as the valve shortens when it is cold e.g. during cold starting. The same principle may apply to compensate for manufacturing and assembly tolerances. The lash spring is also suitable for valve preloading against unseating that may occur during engine operation due to in-cylinder vacuum.

III. 4. The Motor and Electrical Drive

Permanent magnet brushless dc motors were selected because they are high performance electric motors which are particularly suited for this low-power application.

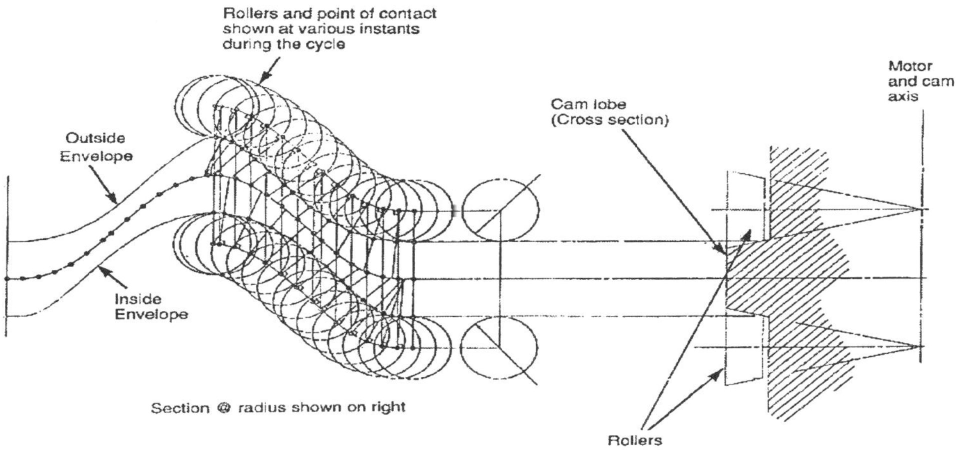

Figure 5. Cam Design

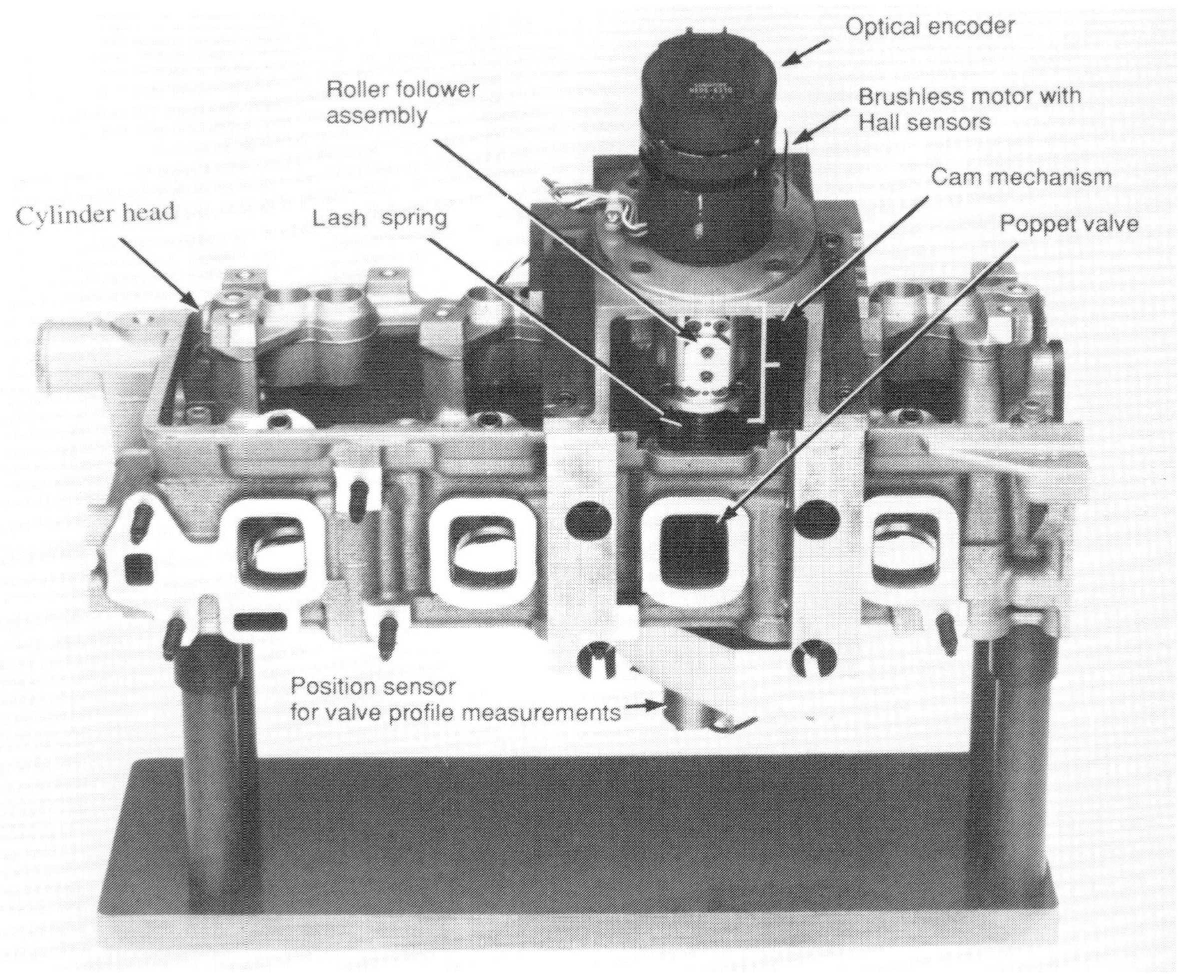

Figure 7. Electro-mechanical system test set-up on actual cylinder head.

Those motors require no maintenance and the recently developed high-energy magnets (such as Nd-Fe-B magnets) provide high torque-to-inertia ratio. With the use of a 4-quadrant inverter to drive the motor, the electrical system can become regenerative when the machine acts as a generator during braking (negative torque). The surplus energy is then stored in capacitors or fed back to the battery, thus providing full regeneration capability to the system.

IV. EXPERIMENTAL EVALUATION

IV. 1. Prototype Description and Test Set-Up

The prototype was mounted on an actual cylinder head (2L, 2-valve/cylinder engine). Photograph of the prototype is shown in fig. 7. The mechanism has an outside diameter of 51mm and a height equal to 94.5mm from the motor mounting flange to the bottom of the end bell that holds the valve. The motor is an 8-pole machine with surface-mounted rare-earth magnets on the rotor, 68 mm long and 57 mm in diameter. The inverter generates rectangular 120° current waveforms according to the input from three Hall effect position sensors mounted inside the motor. The magnitude of motor current is proportional to a user-defined input signal (in brushless motors, the torque is approximately proportional to current).

In order to test the valve flexibility of the concept, a digital incremental optical encoder was mounted on the motor. Its output train of pulses was averaged to yield the instantaneous motor velocity. This velocity was compared to a velocity command, and a current (therefore torque) command proportional to the velocity error was sent to the inverter (P control). Timing variations to test variable valve timing operation were then obtained as follows: The Hall sensors on the motor were used to obtain a pulse once per revolution. This pulse initiated a velocity command of magnitude V_1 and duration τ, after which the velocity command changed to some other level V_2 until the end of the revolution, at which point the cycle was repeated, see fig. 8-a. V_1, V_2 and τ were defined by the user to obtain a variety of valve-event durations at all speeds. A different control pattern was used to assess part-lift operation: The cam was manually positioned a few degrees away from the valve opening position. Then, two consecutive pulses of current/torque command, shown in fig. 8-b, one positive and one negative for clockwise and then counterclockwise rotation, were sent directly to the inverter, with no velocity control. Phasing of the valve event with a running engine was beyond the scope of this initial study.

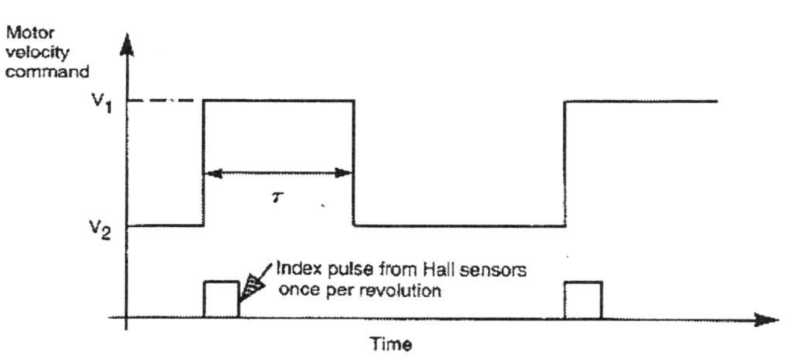

Fig. 8a. Controller velocity command for continuous full-lift operation.

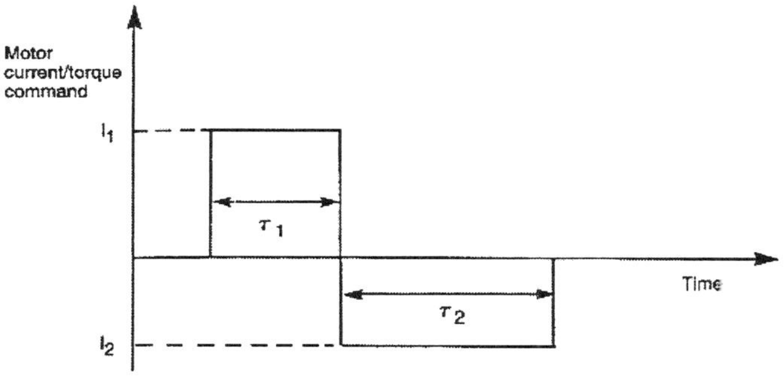

Fig. 8b. Controller current / torque command for part-lift operation.

Figure 8. Controller Command Profile

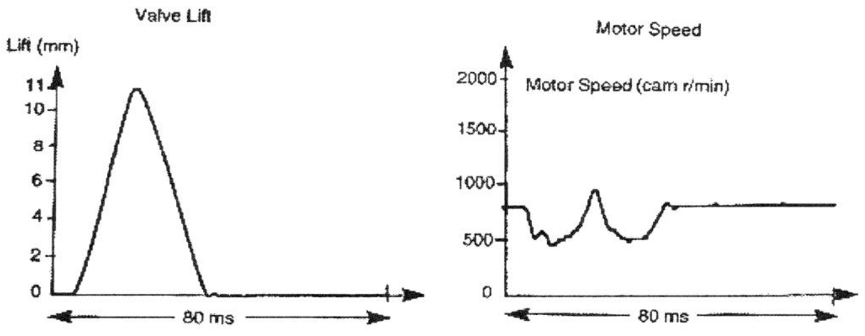

Figure 9a. Waveforms at 1500 r/min.

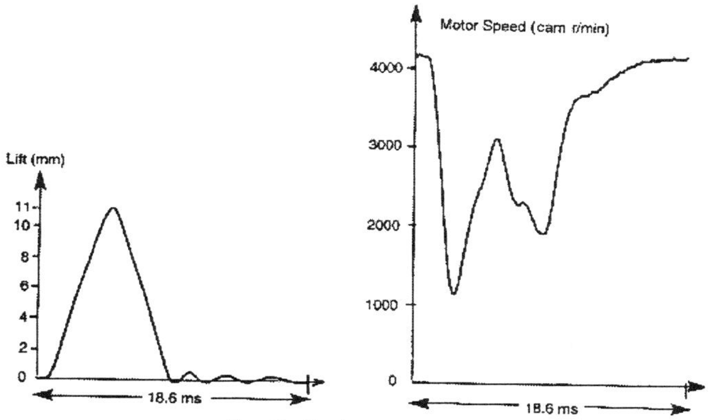

Figure 9b. Waveforms at 6450 r/min.

Figure 9. Examples of Valve Profiles and Motor Speeds

IV. 2. High Speed Capability

The mechanism was run at all speeds from standstill up to 3225 cam r/min, which corresponds to 6450 engine r/min. It has no fundamental limit in high speed capability, and in particular in the absence of a stiff valve spring there is no spring-surge problem as in conventional valve trains. Two valve profiles (1500 and 6450 engine r/min) are shown in fig. 9, along with the corresponding cam-motor velocity. The motor velocity waveforms show two dips which are due to the non-constant nature of the load torque: The first dip corresponds to the valve opening, where the cam inertia load is positive (i.e. opposing motion), and the second dip corresponds to the beginning of the valve closing, for the same reason. The magnitude and duration of these velocity dips depend also on the available motor torque and could be reduced by using more sophisticated control techniques.

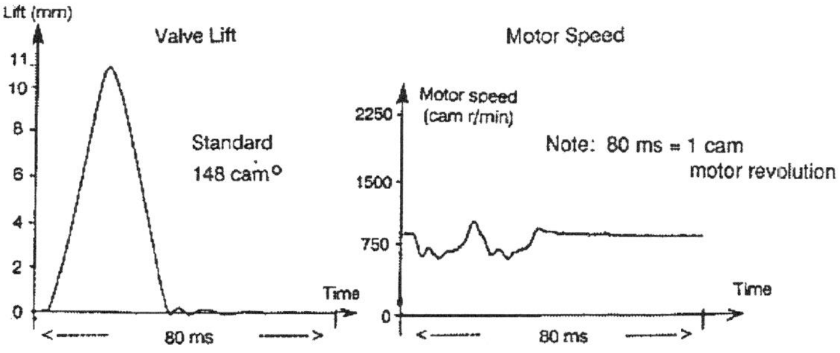

Figure 10a. Standard profile.

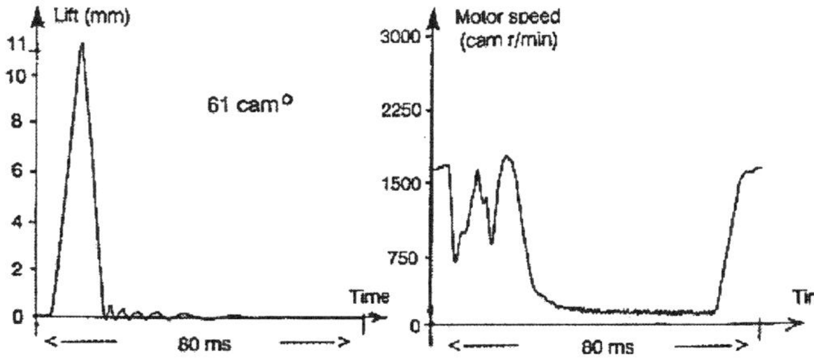

Figure 10b. Short valve event (61 cam°)

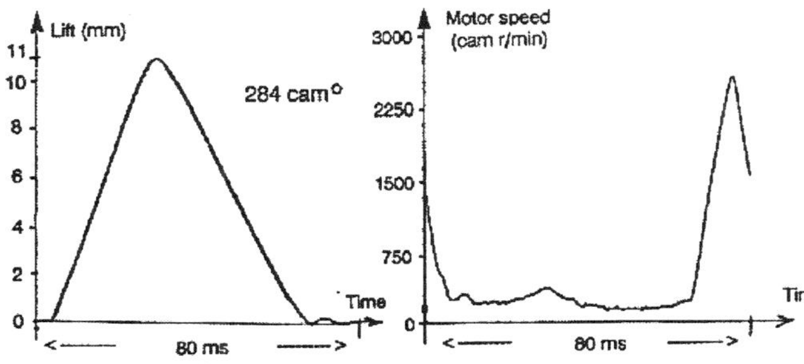

Figure 10c. Long valve event (284 cam°)

Figure 10. Variable Valve Timing at 1500 Engine rpm

IV. 3. Valve-Event Duration Flexibility

Examples of various valve-event durations are shown in fig. 10 at what would be 1500 r/min engine speed (i.e. 750 cam-motor r/min). The standard displacement shown in fig. 9 is repeated here as fig. 10-a for comparison. Its duration, 148 cam°, corresponds to the standard valve opening for this specific two-valve per cylinder engine. Figure 10-b shows a short valve event, 61 cam°. The corresponding motor velocity shows a pattern of high velocity when the valve is open, then low velocity when the valve is closed. When the valve is open, there are again two velocity dips due to the cam positive acceleration, i.e. inertia load, when the valve starts to open and when it starts to close.

Because the motor goes faster when the valve is open, the seating velocity is generally higher. At the time of the valve closing, the cam-motor velocities are ca. 1850 r/min and 750 r/min for the 61 cam° and 148 cam° durations, respectively. As a result, more valve bounce can be observed on the valve displacement waveform, see fig. 10-b. Future control strategies may mitigate this effect. Finally, fig. 10-c shows a very long valve event, 284 cam degrees, along with the corresponding cam-motor velocity profile. The velocity pattern is the reverse of that of fig. 10-b: low velocity while the valve is open, and high velocity when the valve is closed. The system was run with various speed patterns throughout the engine speed range. A map of the valve-event durations observed during the tests is shown in fig. 11 versus engine speed.

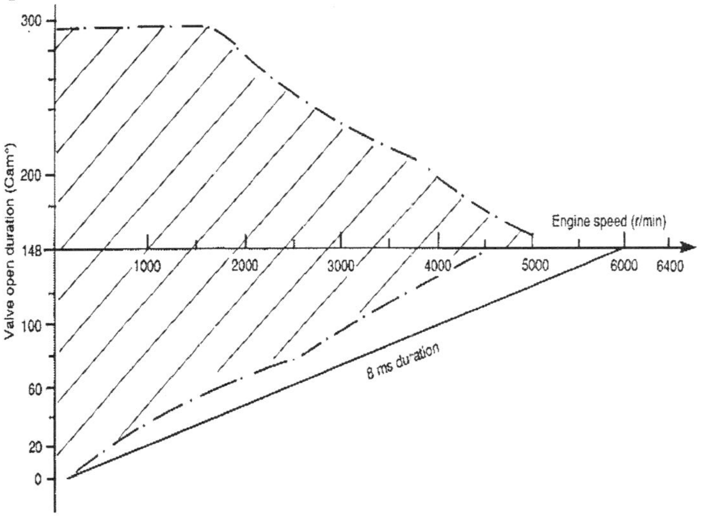

Figure 11. Map of Observed Valve Event Durations vs. Speeds

IV. 4. Part-Lift Capability

Part lift is obtained by reversing the motor velocity, and operating the mechanism in an oscillatory manner around the valve-open or valve-closed position. Figures 12-b and 12-c show two part-lift valve event examples, shown along with the standard valve profile (fig. 12-a) for comparison. The motor velocity is shown in all three cases. A projected map based on all observed part-lift valve events is shown in fig. 13, as peak lift versus duration in milliseconds. Also shown on the abscissa of the figure is the valve-event duration in cam degrees at 1500 engine r/min. The figure indicates that most combinations of peak lift and durations are possible given the following limitations: Very short durations are limited by the system dynamics, with higher peak lifts being more limited than smaller ones because of the time necessary to go up the higher lift. Although the flexibility is higher at lower speeds than at high speeds, there is still room at high speeds for part lift with long durations: For instance, a 2-mm lift lasting 180 cam° is still possible at 6000 r/min. No peak lift below 1 mm was attempted. Such low lifts should be possible, however their repeatability from cycle to cycle will be an issue even with a closed loop position control.

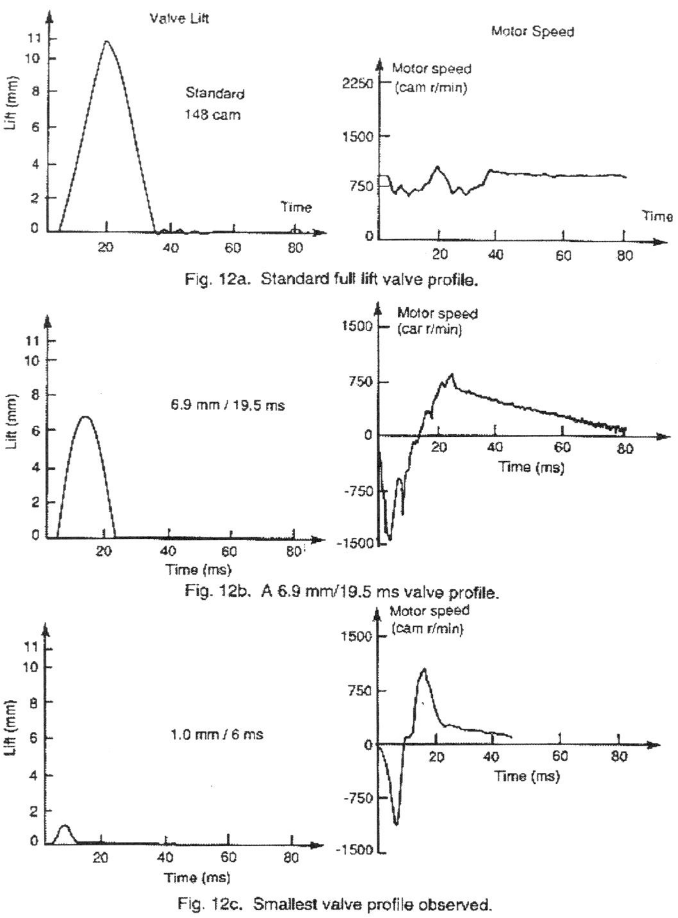

Figure 12. Part Lift Valve Profiles

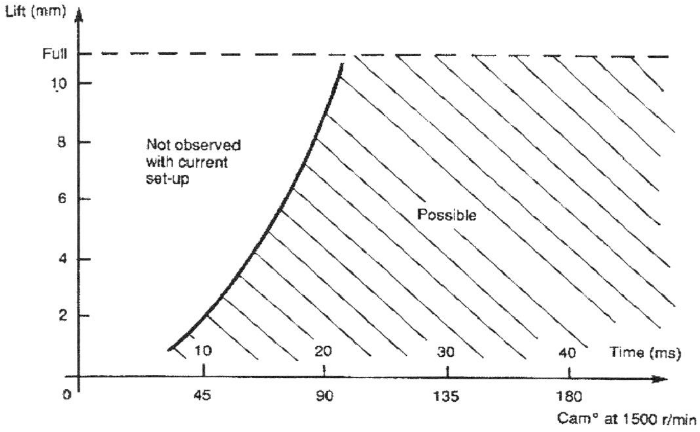

Figure 13. Part Lift Map vs. Duration at 1500 Engine rpm

IV. 5. Energy Requirement

IV. 5. 1. Measurement Method

The energy required by the electromechanical system was measured as electric energy on the dc bus, i.e. dc supply voltage multiplied by current. It, therefore, includes all losses (motor, inverter and cam mechanism). It was transformed into the mechanical energy that would be needed from the engine crankshaft by dividing by the alternator efficiency. An alternator efficiency of 70% was assumed, based on the alternator technology that would be required to provide the power to such a system. The numbers were then multiplied by 8 to emulate a 4-cylinder engine with two valves per cylinder, divided by the engine speed and expressed as torque T in Nm to follow the convention used for engine friction. Compared to valve operation on an engine, the numbers so calculated are optimistic because they do not include the energy the exhaust valves need to overcome the pressure difference across the valve at the time of opening. However, comparable production valve-train friction data do not include the exhaust valve energy requirement either, because valve-train friction is measured by motoring the valve-train by itself without the combustion chamber pressure. Therefore, overlooking the added energy for the exhaust valves was necessary to make a meaningful comparison of the proposed system with conventional valve trains.

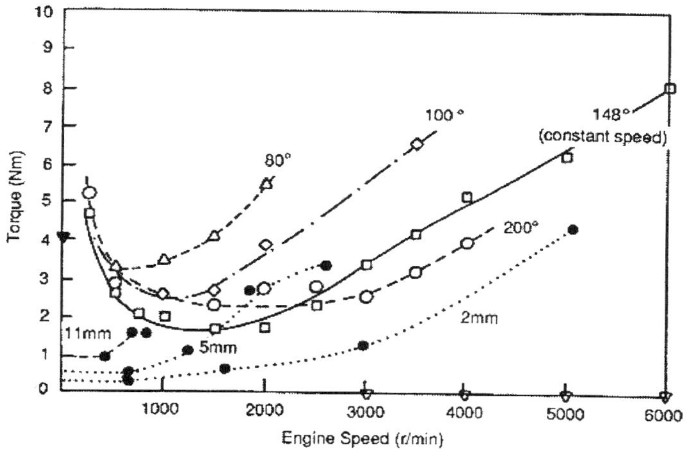

Solid -> Full Lift, Dashed -> Variable Timing, Dotted -> Part Lift

Figure 14. Crankshaft Load Torque for Eight EM Valves Including Alternator Efficiency

IV. 5. 2. General Results

Figure 14 compares the various operating modes of the system. A solid line in Fig. 14, with a caption '148°", plots the results for the standard valve-event duration. The torque is high at low speeds because the brushless motor has some constant losses which are divided by a small speed. Actual power values, though, are low (11 W for one intake valve). The total torque is low, around 1.8 Nm, in the 800 to 2500 engine r/min range, then increases above 2500 r/min, for the reasons explained below.

IV. 5. 3. High Engine Speed Energy Requirement

The higher energy requirement at high speed can be understood by studying the following relations. Since the lift "y" is a function of the cam angle of rotation "θ", the time derivatives of the lift are given by:

$$\dot{y} = \frac{dy}{dt} = \frac{dy}{d\theta}\frac{d\theta}{dt} = y'\omega \qquad (1)$$

$$\ddot{y} = \frac{d^2y}{dt^2} = y''\omega^2 + y'\alpha$$

The first term of the lift time second derivative in this equation is proportional to ω^2 which is the square of the cam angular velocity, and the second term is proportional to α which is the cam angular acceleration. This means that inertia load increases with increasing cam velocity and acceleration, which leads to increasing friction force/torque, and this is the first cause for the increasing power required at high speeds. The second major reason for increased friction torque at speeds higher than 2500 engine r/min (as displayed in the bench test results shown in fig. 14) is the skidding of the roller followers which yields sliding instead of rolling friction, with a possible order of magnitude higher coefficient of friction. Ways were later identified to eliminate or reduce roller follower skidding at high speeds, and they were a focus point in subsequent designs. However, these solutions may be constrained by packaging.

Many factors can affect that rolling without sliding condition:

- the roller followers and cam surface finish that affects directly the coefficient of friction,
- the roller followers preload that affects the normal force between them and the cam,
- the roller followers diameter at the contact surface,
- the squareness of the roller followers with the cam surface that is affected also by elastic deformation of the rollers and bearings,
- the follower bearings friction especially at high speeds,
- the roller followers' mass moment of inertia,
- the increased angular acceleration of rollers which is proportional to that of the cam.

IV. 5. 4. Variable Timing Energy

Varying the duration, since it involves varying the cam speed, needs additional energy to overcome the additional friction losses due to the increased inertia load. Shorter valve event duration is accomplished by increasing the motor speed during valve opening and decreasing it when valve is closed. This magnifies the friction torque due to increased inertia load during opening which is a major part of total load. In addition, when the command speed during opening reaches ca. 2500 r/min the friction torque increases at higher rate than below 2000 r/min. Therefore, it is reasonable to expect a higher energy requirement for shorter durations. This can be observed on fig. 14, where the captions, in °, correspond to the valve-event duration. Shorter durations resemble the 148 cam° data except that the energy requirement rises faster with speed as explained above. The high friction seen at high speed for a standard valve-event duration is, therefore, reflected at low speeds when short durations are used. Conversely, as valve event duration increases, friction torque increases due to higher inertia load but friction decreases because of lower speed. The break even point for 200 cam° seems to be around 2500 r/min (same speed as point of higher rate of friction increase). All curves converge at very low speeds because the motor no-load losses are then dominant.

IV. 5. 5. Part Lift Energy

The energy needed for part-lift operation is also shown in fig. 14 for 3 different lifts (2-mm, 5-mm and 11-mm) on the basis of 148 cam° durations. Part lift is accomplished by reversing the direction of the cam speed which means increased energy required to overcome inertia. It still requires energy to overcome the friction over the no-lift part of the cam. Therefore, we should expect the part lift energy requirement to be more than the energy required for full lift prorated by the ratio of lift to full lift. At the same time the motor is idle for part of the cycle during which obviously no energy is consumed. There is also another factor in this operation which is: since we need to accelerate then decelerate in one direction, and do the same in the opposite direction, if y" is positive during that time (see eqn. 1), that part of the torque (hence the energy) will be negative; i.e. the total energy required will be smaller. As seen in fig. 14, this is the case for the 2 mm lift, where the calculated equivalent friction torque is lower than the standard 148° friction by almost the same ratio over the speed range. The 5 mm and the 11 mm lifts do not quite benefit from that condition since their energy requirements are lower than standard at low speeds but then become higher as the speed increases.

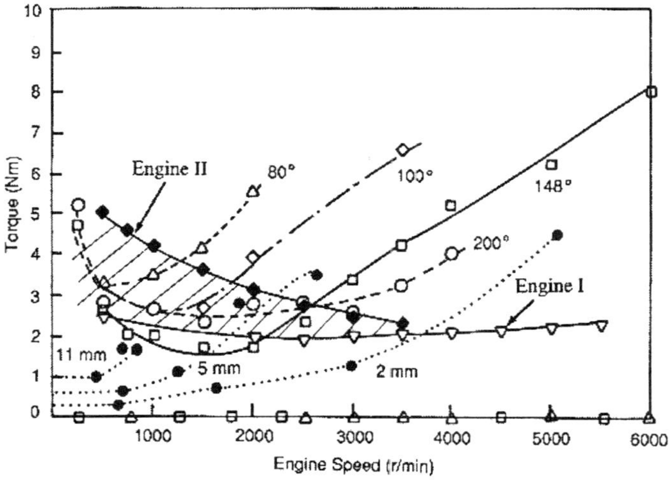

Fig. 15. Torque Requirement Comparison between EM and 4-Cyl. Production Valves

IV. 5. 6. Comparison with Production Engines

Figure 15 repeats the data of fig. 14 and superposes over it the valve-train friction measured on a certain number of production engines. Two engines are shown specifically: One engine, engine I, which has an overhead camshaft with center pivot rockers and lifters, and importantly, roller followers and caged needle bearings, which explains its low friction. The other engine, engine II, is a conventional pushrod engine. The valve-train friction of other engines is shown to fall between these two engines, and is represented here as a shaded area. This data was measured by motoring the valve trains. The conclusion inferred from Fig. 15 is that the proposed electromechanical system data falls below the valve-train friction up to 2000 r/min, and is higher above 2500 r/min. The reason for the lower friction at low speeds is mainly the regeneration effect, as for high speeds see explanation in section IV.5.3.

Altogether, it appears that an electromechanical system would not affect the engine friction in a large manner. However, its impact on the electric power generation system would be very significant. In that respect, Fig. 16 plots the electric power needed to run one electromechanical valve system at standard valve-event duration. This power is low and approximately constant (11 W to 14 W) up to 800 engine r/min. This power corresponds mostly to the motor no-load losses. As the speed increases, the power increases more than linearly, particularly above 2500 engine r/min. The measured power was 315 W and 443 W at 5230 and 6450 r/min, respectively, for one valve. Assuming the exhaust valves consume an extra 56% energy requirement, then for the 4 exhaust valves this translates into 4.5 kW at 6450 r/min for a 4-cylinder, 2-valve/cylinder engine. Of course, since high speed operation is intermittent, at least in the North American driving pattern, some of that power can be drawn from the battery during peak demand as is done already for other electric loads. However, a significant increase in alternator and battery capacities will be needed.

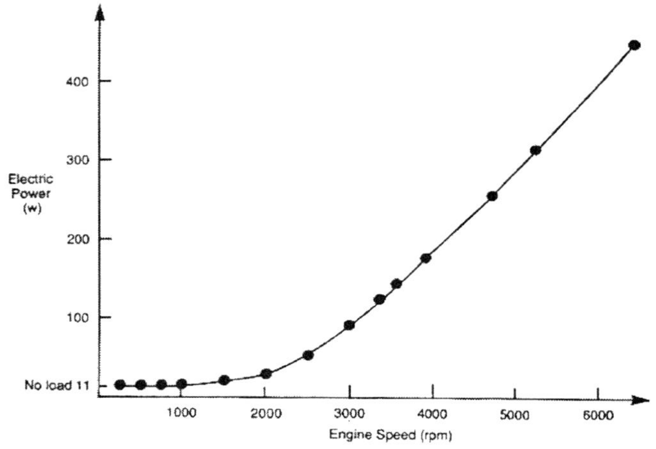

Fig 16. Electric Power for One Standard Valve

IV. 6. Lash Adjustment

The proposed system includes a mechanism to adjust for lash (see also section III. 5.), thereby guaranteeing that the valve always closes despite manufacturing variations, thermal cycling and wear. The prototype was designed for a 0.018" peak-to-peak lash (similar to the usual production requirement). The prototype lash adjustment capacity was tested by adding or subtracting shims between the valve and the reciprocating mechanism, thereby emulating valve lengthening and shortening. The valve was observed to operate normally, without a clearance between the valve and the valve seat, for valve-length variations of up to 0.018", thereby demonstrating the validity of the original design concept.

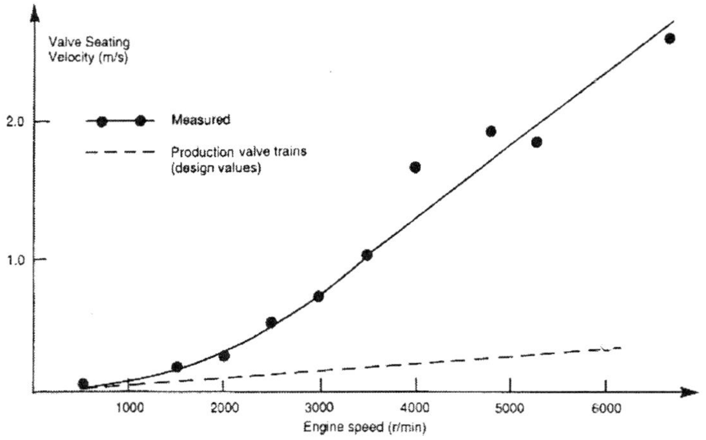

Figure 17. Valve Seating Velocity

IV. 7. Valve Seating Velocity

The valve seating velocity must be kept low in order to reduce valve fatigue and audible noise. This problem has been resolved in conventional systems by designing the cam with a closing ramp (which is at valve closing by definition). The proposed mechanism could adopt the same strategy and attain similar results, but this was not done in this preliminary study since the emphasis was on VVA flexibility assessment and component sizing. The seating velocities were therefore higher than on production engines, between 2 and 3 times higher at speeds up to 2500 engine r/min, and from 4 to 8 times at higher speeds. This is by comparison with typical production valve-train design values of valve seating velocities, i.e. 0.03 m/s and 0.30 m/s at 600 and 6000 engine r/min, respectively. Note, though, that during normal engine operation, actual seating velocities are higher due to the dynamics and compliance effects of the valve train. Figure 17 plots measured velocities at various engine speeds for the standard 148 cam° duration. The 2-3 times higher seating velocities at low speeds are due to the lack of closing ramp. At high speeds, in addition to the above mentioned factor, increased friction forces the motor effort to increase just before closing, hence, increases angular velocity at time of closing. The increased friction and loading beyond 2500 r/min (obvious from increased energy requirement at same speeds) also leads to elastic deformation in the reciprocating valve assembly which is similar in its effect to having axial lash. Those effects result in 4-8 times higher seating velocities at high engine speed. In turn, these high seating velocities result in valve bounce, as seen on the valve profiles of figs. 9-b and 10-b, as well as less durability. Future efforts to reduce energy at high speed should also have a positive impact on seating velocity and the other effects.

IV. 8. Valve Deactivation, Engine Transients and Cam Phasing

One way to control engine load consists of deactivating one or more cylinders. The motor, then, is stopped between two valve events and restarted later on. The proposed system's potential in that regard was evaluated by measuring the cam angular displacement necessary to accelerate the mechanism from standstill to 3000 cam rpm: 120 cam°, covered in 12.9 ms, which is equivalent to ca. 25,000 rad/s^2 angular acceleration for motor and mechanism combined. Since the portion of the cam lobe corresponding to the valve-closed position spans 212°, much more than 120°, the valve can be synchronized from standstill to 6000 engine r/min in less than one engine cycle. Stopping the valve is done by using the electromagnetic braking torque of the motor. Therefore, valve deactivation and re-synchronization are possible within a couple of engine cycles (an engine cycle lasts 20 ms at 6000 engine r/min). This mode of operation could be important to control engine load at high engine speeds where valve-event duration flexibility is reduced. Another consequence of these observations is that cam phasing, i.e. advancing or retarding the valve event with respect to crankshaft TDC (top dead center), is feasible at all engine speeds. Overall, on the basis of these observations, throttleless engine-load control should be possible at those engine speeds most commonly used on the road. Part-lift operation and valve deactivation will be important complements of valve timing flexibility in that regard.

IV. 9. Packaging

This first electromechanical mechanism prototype had active diameters of 51 mm (mechanism) and 57 mm (motor), and was 204-mm high from the surface of the cylinder head to the top of the motor position encoder. On going engine tests show that more torque may be needed from the exhaust valve motor. This would mean that the motor would be longer accordingly. In all cases, the total height is likely to be above the present valve cover.

IV. 10. Durability and Timing Repeatability

There was no intent to conduct an in-depth investigation of durability and timing repeatability at this early stage of the project. However, the prototype mechanism was run for 8 hours continuous at an equivalent 1000 engine r/min to detect unforeseen flaws, in preparation for future designs. The system ran with no problems throughout the 8 hours of the test, and the mechanism was disassembled after the test for inspection. Inspection revealed no fatigue signs on cam or rollers surfaces (e.g. pitting), no measurable wear of plastic bearings, and no signs of excessive vibration like loose screws, etc. The operational stability of the system was assessed by measuring the motor speed and the valve-open duration at regular intervals during the test. The motor average speed varied by less than 3% overall, and the valve open duration averaged 146.1 cam degree ±9%. Engine operation, though, will obviously require a closed-loop control of valve timing.

V. CONCLUSIONS

The proposed system demonstrated full flexibility in variable timing and variable lift operations. The device (motor/mechanism) can be still packaged under hood even though it is higher than conventional valve train. Bench tests showed that energy requirements are comparable to the existing production valve trains up to 2500 engine rpm, but higher at higher engine speeds. The mechanism provides for lash adjustment which is important for proper functioning. Since the prototype did not include a closing ramp, high seating velocities were observed which negatively impact the mechanism durability. This issue should be studied further and the effect of a closing ramp on the seating velocity should be investigated. Also the repeatability of valve-event-duration in case of variable timing and lift is an issue which will be a material for future work.

REFERENCES

1. H.P. Lenz, B. Geringer, G. Smetana and A. Dachs, "Initial Test Results of an Electro-Hydraulic Variable-Valve Actuation System on a Firing Engine," SAE Paper 890678, 1989.

2. W.E. Richeson and F.L. Erickson, "Pneumatic Actuator with Permanent Magnet Control Valve Latching," U.S. Patent No. 4,852,528, Aug. 1, 1989.

3. Mark A. Theobald, Rassem R. Henry, and Bruno Lequesne, "Control of Engine Load via Electromagnetic Valve Actuators," SAE Paper # 940816, 1994.

4. F. Pischinger and P. Kreuter, "Electromagnetically Operating Actuator," U.S. Patent No. 4,455,543, June 19, 1984.

5. F. Pischinger and P. Kreuter, "Arrangement for Electromagnetically Operated Actuators," U.S. Patent No. 4,515,343, May 7, 1985.

6. M. Gottschalk, "Electromagnetic Valve Actuator Drives Variable Valve Train," Design News, Nov. 1, 1993, pg. 123-125.

7. T. Schroeder, R. Henry, B. Lequesne and B.V. Murty, "Method and Apparatus for Electrically Driving Engine Valves," US Patent No. 5,327,856, July 12, 1994.

8. Chen, Fan Y., "Mechanics and Design of Cam Mechanisms," Pergamon Press, 1982.

9. Tesar, "The Dynamics, Synthesis, Analysis, and Design of Modeled Cam Systems," Lexington Books, 1976.

950976

A Study on Microcomputer-Controlled Variable Valve-Timing System

Jian-hong Li and Kangnian Zhang
Changchun Automotive Research Institute

ABSTRACT

This paper deals with the development of a system to control the inlet—valve—timing of a single—cam gasoline engine. With this scheme the camshaft can be phased continuously to any position within 60 crank degree range by an electronic control unit to reflect changes in engine operating conditions. A control strategy was established which would optimize power at wide—open throttle and minimize emissions at small throttle openings. For this system, a mechanical self—protection device and a fail—safe system have been developed. Development testing has shown that with the proper control strategies, the power, emissions, and idle quality of the engine can all be improved.

INTRODUCTION

Since the nineteen seventies, engine designers have been concentrating most of their effort on reducing emissions levels and improving specific fuel economy while improving the power output of the engines. Advanced technology such as turbochargers, intercoolers, multivalve heads, and variable intake systems are being used more and more frequently. Some of the new engines, especially gasoline engines, have broad dynamic operating ranges, operating smoothly at both 1000 and 7500rpm. For all automotive engines having fixed valve—timing, unless these engines are run at speeds near the design point, their overall performance is less than ideal.

One means for achieving improved performance is through the use of inlet — valve — timing control. Variable valve—timing allows for controlled scheduling of valve—timing events. When applied to spark—ignition engines, it yields improvements in emissions, economy, and performance. The benefits for diesel engines further include: improved starting, noise reduction, and the ability to vary compression ratios. These improvements can be more significant when VVT is applied to twin—cam engines, for which the valve overlap can be changed to achieve high speed volumetric efficiency with large valve overlap and smooth idle condition and low speed hydrocarbons with small overlap. The timing of valve actuation is accomplished with one or more camshafts and is determined by the angular relationship between the camshaft(s) and crankshaft. Because of the inertia effects of the gases being induced into and expelled from the cylinders, the optimum valve—timing varies with engine speed.

Variable valve—timing is not new to automotive engineers. For decades, many mechanisms have been proposed to vary the lift, duration, and timing of the valve events of high—speed engines. Many of these mechanisms are expensive and complicated, often adding noise, vibration, and durability concerns. Others have had crude control strategies or were limited in the range or degree of variability. Variable valve — timing mechanisms of the past and present have been summarized into three types by actuation means: Mechanical, Hydraulic, and Magnetoelectric mechanisms[1][3][4].

In January 1991, Changchun Automotive Research Institute began to take this topic as its important research task. The authors have so far developed a microcomputer controlled variable valve—timing system(MCVVT) for the light—duty truck engine CA488. In this research, the authors have solved two key problems, making it possible for valve—timing to be electronically controlled:

(1) Cam — Phasing Measurement—the practical valve — timing can be measured by using two engine speed sensors;

(2) Mechanical Self — Protection Device—when errors occur in electronic control unit or the drive unit loses power, the mechanical self—protection device—arrow washer will function to limit the cam—phasing floating to a permissible range.

The MCVVT system yields accurate cam — phasing adjustment with rapid system responses. The variable valve — timing mechanism of the MCVVT system is simple and relatively inexpensive to manufacture.

This paper describes the design and development of the MCVVT system controlling the inlet — valve — timing of a single — cam gasoline engine CA488.

SYSTEM CONFIGURATION

Figures 1 and 2 show a schematic representation of the microcomputer controlled variable valve — timing system. The actuator drives the cam to change the inlet — valve — close (IVC) phasing. Since the engine to which the MCVVT system applied is a single cam engine, the exhaust timing also varies together with the intake timing by the same amount. The control unit takes over the task of reflecting changes in engine operating conditions and continuously updating the phasing angle of the cam. A control strategy was established which would optimize power at large throttle angles and minimize emissions at part throttle openings as well as provide a smooth idle condition.

THE CA488 ENGINE

The engine chosen for the study is a CA488 engine produced by Chrysler (USA). The engine's specifications are listed below in Table 1. The MCVVTsystem adds no modifications to the engine ignition and fuel systems, and requires no engine casting modifications.

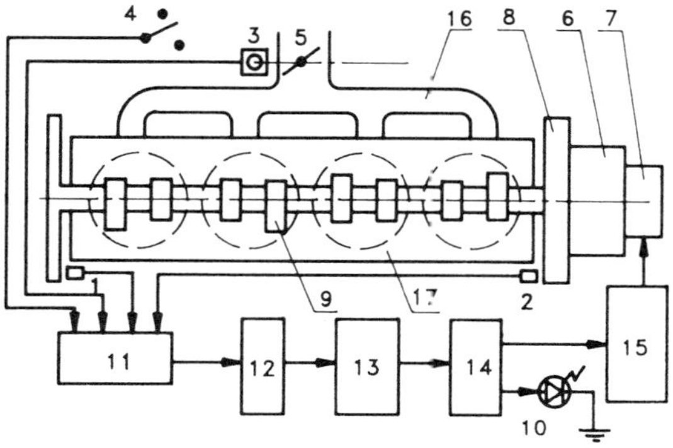

1. ENGINE SPEED SENSOR (1)
2. ENGINE SPEED SENSOR (2)
3. ENGINE LOAD SENSOR
4. STARTING SWITCH
5. THROTTLE
6. HARMONIC DRIVE UNIT
7. STEPPER MOTOR
8. DRIVE PULLEY
9. CAMSHAFT
10. WARNING LAMP
11. DATA GATHERING CIRCUIT
12. I/O CIRCUIT
13. 8098 CPU
14. I/O CIRCUIT
15. MOTOR DRIVING CIRCUIT
16. AIR INTAKE PIPE
17. ENGINE CYLINDER HEAD

Fig. 1 — Entire system layout

Fig. 2 — View of the engine CA488, the entire stepper motor and harmonic drive assembly

THE ACTUATOR

The camshaft phasing mechanism's primary function is to provide the desired angular positioning of the camshaft during dynamic — operating conditions. The heart of the phasing mechanism, similar to the Elrod — Nelson Mechanism[2], is a harmonic drive, a mechanical power — transmission unit with a very high gear — reduction ratio (100:1). The harmonic drive was manufactured by Changchun Optical and Fine Machine Research Institute. Since the CA488 adopts overhead camshaft layout, the camshaft is slender and long (shaft neck Φ24mm, length 468.5mm), and the cam is driven by a cam timing belt through tooth — shaped pulley accomodated in the front of the engine, there is the possibility and necessity to use a harmonic drive unit which has a compact form, high — reduction ratio, and a high degree of accuracy and repeatability.

A stepper motor is used to provide inputs to drive the power — transmission unit. The entire stepper motor and drive assembly rotate synchronously with the camshaft, as a solid body, except when valve — timing adjustments are being made. When signalled by the electronic control unit, the harmonic drive acts as a differential gear box to cause an angular shift of the camshaft relative to the drive pulley.

The harmonic drive consists of three basic components: a rigid circular member (circular spline) with an internal gear, a cup — shaped flexible member (flexspline) with an external gear, and an elliptical wave generator. The circular spline is attached to the camshaft's drive pulley, the flexspline is connected via mechanical elements directly to the camshaft, and the wave generator rides within the flexspline and is driven by a stepper motor. See Fig. 3. When the harmonic drive is assembled, the elliptical wave generator deforms the flexspline along the ellipse's major axis causing the flexspline's external gear to mesh at two places with the circular spline's internal gear. With no input to the wave generator this meshing of the gears at the major axis of the wave generator non—rotatably links the camshaft to the drive pulley so that they rotate in unison.

The harmonic drive's flexspline has two fewer teeth around the circumference than the number of teeth forming the circumference of the circular spline's internal gear, and this differential performs a significant function in achieving controlled—relative rotation between the camshaft and its drive pulley. As the elliptical wave generator rotates, the gear meshing places rotate with it. Because of the smaller number of teeth on the flexspline, it indexes two teeth around the circular spline for each complete revolution of the wave generator relative to the drive pulley. This creeping of the flexspline about the circular spline produces relative rotation between the camshaft and its drive pulley. This results in changing the inlet—valve—close phasing.

Optimizing control signals according to the operating status of the engine as indicated by sensors monitoring various operating parameters and operator—supplied inputs, such as depressing an accelerator pedal, can be provided to the electronic control unit to control the stepper motor. These inputs cause the motor to respond according to predetermined microcomputer programs. The method allows control of camshaft phasing over a ±360 crank degree range. However, a practical adjustment is much smaller. For safety and reliability, the mechanical actuator was designed to allow the phasing adjustment within ±30 crankshaft degree range.

Since the cam phasing is adjusted by the harmonic drive unit which is driven by a stepper motor, the torque output of the harmonic drive will decrease drastically if the stepper motor loses power due to lack of electric current. For safety, the authors have designed an arrow washer. The arrow washer limits the cam phasing adjustment within ±30 crankshaft degree range. The arrow washer is attached to the camchaft by the key groove on the camshaft so that the arrow washer and the camshaft rotate as a solid body. The movement of the arrow washer can be stopped by two screws on the drive pulley. If the stepper motor loses power, the relative motion of the camshaft to its drive pulley will be limited within the range between the two screws—30° angle, corresponding to 60 crank degree range. See Fig. 4.

The advantages of this cam—phasing method are:

(1) the phasing mechanism can optimise IVC phasing for any operating conditions;

(2) the method allows virtually infinite control of camshaft phasing over a ±360 crank degree range;

(3) the actuator has a mechanical self—protection device;

(4) the actuator adds little vibration/noise to the engine;

(5) the actuator layout next to the cooling fan of the engine ensures that the actuator can be cooled by air when the engine runs.

CONTROL UNIT HARDWARE

The electronic control unit(ECU) consists of two speed sensors, a load sensor, a starting switch, a

Table 1. CA488 ENGINE PRINCIPAL SPECIFICATIONS

Item	Specifications
Model	CA488(Former Chrysler 488)
Type	4—cylinder, in—line, 4—stroke, water—cooled, spark ignition, gasoline engine
Bore/Stroke	87.5mm/92.0mm
Total displacement	2210cc
Compression ratio	8.1:1
Rated power	65kW/4500rpm
Maximum torque	157N · m/2800rpm
Minimum B.S.F.C.	292g/kW · h
Fuel type	Octane Number 90
Ignition order	1—3—4—2
Valve mechanism	Overhead camshaft
Valve—timing angle	IVO: BTDC 0° EVO: BBDC 44° IVC: ABDC 56° EVC: ATDC 8° Overlap: 8°

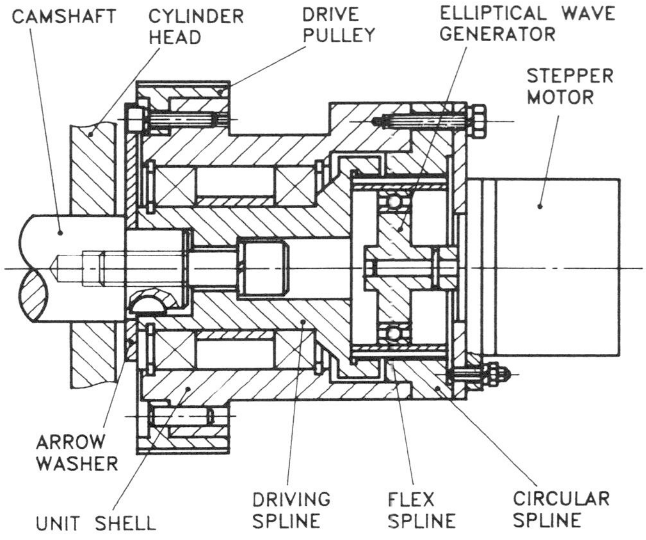

Fig. 3—Structure of the harmcnic drive mechanism

single—chip microcomputer of the Intel 8098, and I/O circuit.

In addition to the CPU, the 16—bit microcomputer which is used contains a 8K—byte ROM, a 232—byte RAM, two 16—bit timers for real—time tasks, four 10—bit A/D convertors, and 10 external interrupt sources. Figure 5 shows the circuit diagram of the control unit. The individual input and output variables are given on the left—hand side. The data map and the program for controlling the valve—timing are stored in an EPROM.

Engine speed is sensed directly at the camshaft by means of an engine speed sensor which uses a magnet attached to the camshaft. By means of this triggering of the magnet passing the sensor, the sensor generates one square wave for each complete revolution of the camshaft. As Fig. 6 shows, the two engine speed sensors (1) and (2) generate continuous signals (in the form of square waves) for Intel 8098 from the camshaft and its drive pulley, respectively. The throttle angle signal is taken from an engine load sensor, a potentiometer connected to the throttle spindle. One of the four A/D channels is permanently allocated to the throttle position signals.

Two important functional outputs for engine operation are the signals for the stepper motor and the signal of failure — detection for warning lamp. The output signals for the stepper motor are generated by the 8098 CPU through HSO. 0~HSO. 3 in the form of square waves according to predetermined control programs, then drive the motor via four power amplifiers. The electric current is conducted to the coils of the motor through carbon brush and conductive copper rings attached to the motor's shell. See Fig. 2. The output signal for the warning lamp is generated by the 8098 CPU via HSO. 5 in the form of square wave.

The engine speed and cam — phasing angle are determined by the signals from the two engine speed sensors (1) and (2). When the engine runs, the two speed sensors will generate a series of signals in the form of square waves alternately and continuously. The signals are conducted into the Intel 8098 CPU through HSI. 0 and HSI. 1. When a square wave enters into the 8098 CPU, its positive jump edge will trigger a time register to read and register the current

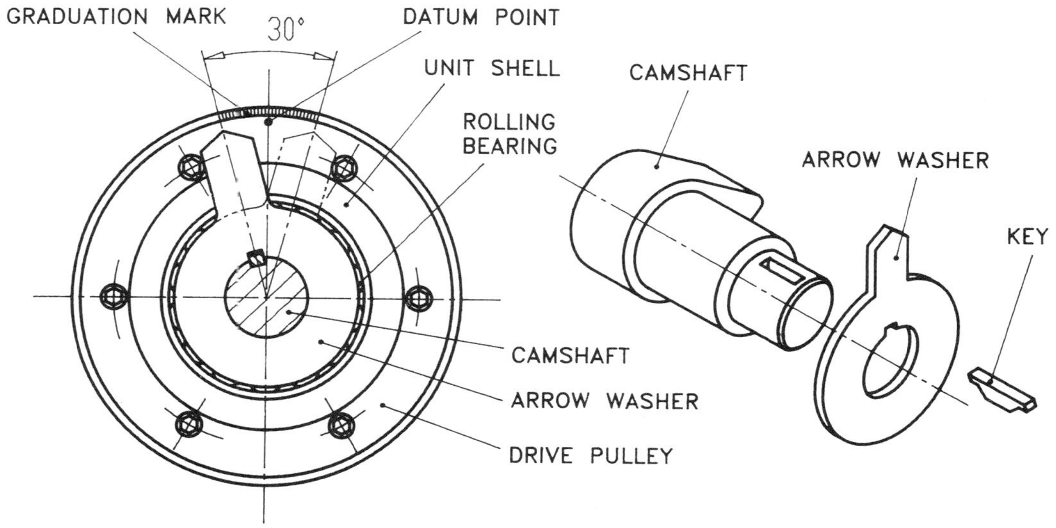

Fig. 4 —Arrow washer and its assembling

time value of the 16-bit timer. As the engine runs continuously, the time register of CPU will record a series of non-dimensional time values: t1, t2, t3, t4, t5, t6, t7, …. See Fig. 6. For these values, t1, t3, t5, t7, …, are triggered by the signals occurring each 720° of crankshaft rotation from camshaft, while t2, t4, t6, t8, …, triggered by signals from drive pulley each 720° of crankshaft rotation. The timer functions as a counter with counting interval of 4 microseconds. The engine speed, n, in revolutions per minute (rpm) can be computed from the time values by the formula below:

$$n=\frac{30\times10^6}{t_{i+2}-t_i} \qquad i=1,3,5,\cdots \tag{1}$$

The current cam-phasing of the engine, α, in crank degree is calculated from the time values by:

$$\alpha=\frac{360\times[(t_{i+1}-t_i)-(t_{i+2}-t_i)/2]}{(t_{i+2}-t_i)/2} \tag{2}$$

$i=1,3,5,\cdots$

The advance or lag of the cam-phasing can be decided by the positive/negative value of α. When $\alpha=0$, the engine runs at the original cam-phasing. By using digital filtering, the effect of fluctuation of engine speed on α can be removed.

ACCURACY OF CAM-PHASING ADJUSTMENT

Since the stepping angle of the stepper motor is 0.9° and the harmonic drive has a high gear-reduction ratio of 100:1, the minimum amount of cam-phasing adjustment is 0.018 crank angle degree. Since the MCVVT-system adopts feed-back links in its control strategies, the accuracy of cam-phasing adjustment can reach 0.01 crank degree angle. For quick system response, the authors have set a clearance band—±0.2 crank degree angle at each desired value for cam-phasing adjustment. The adjustment stops at the very moment when the practical IVC-timing measured by means of the two engine speed sensors enters into the clearance band at the desired value. Therefore, the practical accuracy of cam-phasing adjustment is 0.2 crank degree angle.

SOFTWARE DESIGN

The microcomputer performs four tasks:

(1) to register the measured data received from the sensors,

(2) to calculate the practical camshaft angles and optimum timings on the basis of data maps and characteristic curves,

(3) to generate the output signals for the stepper motor,

(4) to diagnose the defects in the sensors, I/O circuit and actuator.

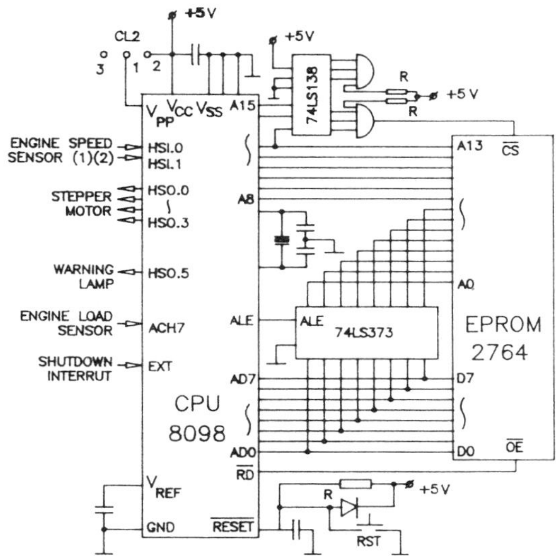

Fig. 5—Schematic of control unit

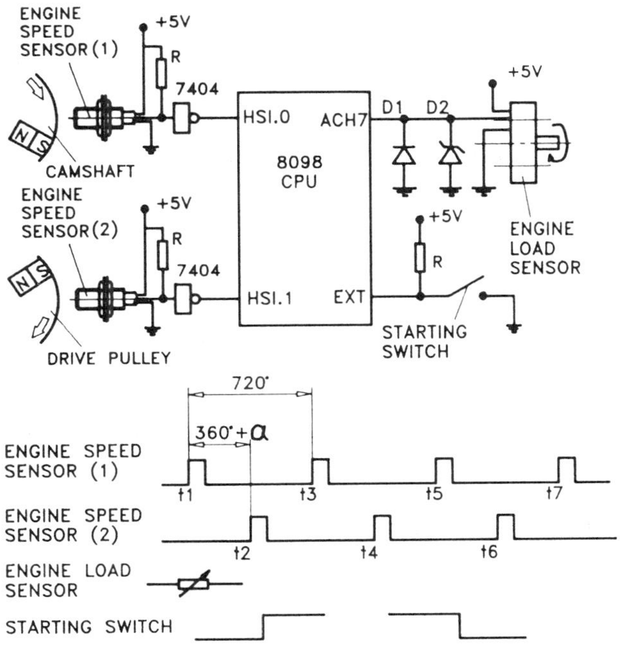

Fig. 6—Sensors and input signals for CPU—Intel 8098

See Fig. 7. A data map for optimum IVC-phasing advance versus engine speed and load was developed through extensive engine testing on a dynamometer under various operating conditions. See Table 2. To respond instantly, the system programs adopt modularized structure, employing five modules and one interrupt service routine:

— Engine speed and cam — phasing measuring module,
— Engine load measuring module,
— Stepper motor control module,
— Cam—phasing adjusting module,
— Self—diagnosis module,
— Engine shut—down interrupt service routine.

The system is under open — loop control of the same microcomputer which may also schedule the engine fuel and spark. According to the system programs, the engine runs under three modes: ACCEL/DECEL mode, STARTING — IDLE mode, and NORMAL OPERATING mode. When the engine runs, the programs continuously store and update the throttle position and engine speed, compare the current and former values, compute the rate of change and thus sense any rapid opening or closing motion of the throttle and any rapid increase or decrease of engine speed. Under steady engine operating conditions, the control programs jump into the NORMAL OPERATING mode, using feed — back links to secure that the camshaft is accurately positioned. A rapid throttle open/close rate or engine speed increase/decrease causes a jump into the ACCEL/DECEL mode, under which the control strategy employs a knowledge base which gives quick system response without cam — phasing angle feed — back. When the engine starting switch is turned on/off or the engine runs at speeds below 900rpm, the cam — phasing angle will be fixed at the largest IVC—phasing advance angle to achieve better engine performance. This part of the control strategy is identified as STARTING—IDLE mode.

Cam position is mapped into a map as a function of engine speed and load. The lookup, cam—phasing computation, and control algorithm add little to the system's response time, which is dependent on the running speed of the stepper motor and the gear — reduction ratio of the harmonic drive, and is affected by the engine speed. According to the predetermined microcomputer programs, the 8098 CPU generates signals in the form of square waves up to 1000Hz for the stepper motor. Driven by these signals, the stepper motor runs at speed of 900° per second. Correspondingly, the cam — phasing angle varies by the amount of 18 crank angle degrees per second. As Table 2 shows, the maximum amount of cam—phasing adjustment at a time is 16 crank angle degrees. Therefore, running under this control strategy, the MCVVT system achieves quick time response ranging from 0. 2 to 0. 9 second. The response time can be further reduced by increasing the frequency of the square waves for the stepper motor, or by decreasing the gear—reduction ratio, or by both.

DIAGNOSIS AND FAIL—SAFE SYSTEM

The MCVVT system contains a self—diagnosis

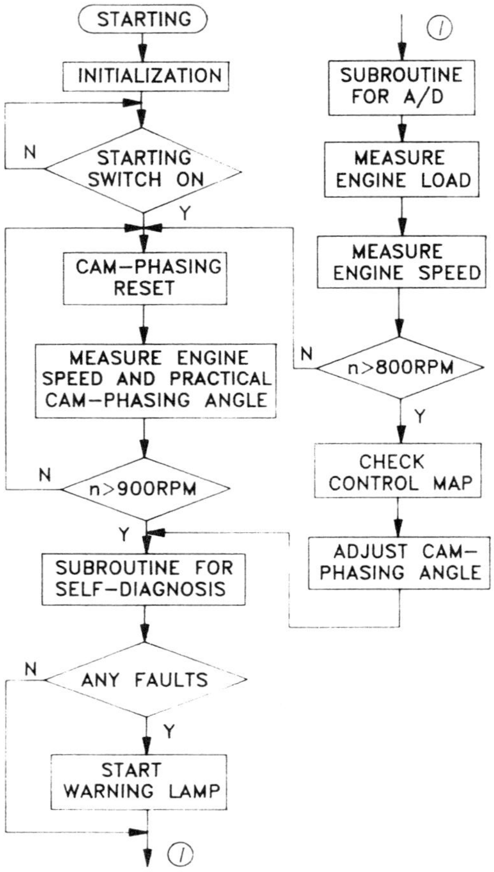

Fig. 7—Flow diagram of main control program

module for sensor testing, output stage testing, and internal diagnosis. Should any faults occur in the sensors, circuit or actuator, the microcomputer is in some cases able to recognise these as impossible operating conditions and take counter — measures to protect the engine. This part of programme runs every 0. 2 to 2 seconds. It can detect defects in the control — unit hardware including engine speed sensors, engine load sensor, stepper motor, I/O, and external circuit in the following ways:

(1) Identification of a steady — state deviation, i. e. the desired value of cam—phasing angle and the actual value measured do not coincide under engine steady operating conditions;
(2) Signal Range Check, i. e. a certain input signal is checked as to whether it falls within a defined reasonable range;
(3) Self — Confidence Check, i. e. the software arithmetic, the program branch and the I/O circuit are monitored as to whether it works reasonably.

Table 2. IVC—PHASING CHARACTERISTICS MAP

IVC-phasing Advance (in crank degree)		Engine Load (Sensor Output, in Volt)					
		≤0.7	≤0.8	≤1.2	≤1.9	≤3.0	≥3.0
Engine Speed	≤1000	16	16	16	16	16	16
(N, in rpm)	≤1400	16	12	12	16	16	16
	≤1800	16	8	8	16	16	16
	≤3000	12	8	8	12	12	12
	≤3400	12	12	8	8	8	12
	≤3800	8	8	4	4	8	8
	≤4600	4	4	0	0	4	4
	4600≤	0	0	0	0	0	0

When a system fault is detected, the microcomputer will automatically turn on the warning lamp on the instrument panel to inform the driver of the necessity of inspection.

ENGINE TEST RESULTS

Engine testing was conducted on a dynamometer to determine the effects of the MCVVT system. The engine CA488 was tested with and without the MCVVT system respectively. Figure 8 gives the full load curves for the CA488 engine with and without the MCVVT system. As can be seen, with the MCVVT system, the actual engine performance was improved throughout the whole range of engine speed, but more significantly at low speed. The torque increased by 6.29% to 149.1 N • m at 1600rpm. The brake specific fuel consumption (BSFC) decreased by 6.68% from 314g/kW • h to 293 g/kW • h at 1200rpm, by 1.8% averagely throughout the whole range of engine speed. The maximum torque increased by 2.5N • m to 161.3N • m. The minimum BSFC dropped by 4g/kW • h to 279g/kW • h. At 4800rpm, the engine's performance did not improve any more. Part—load testing results show that the MCVVT system does not achieve remarkable improvement for the engine running under middle and low loads.

Engine emissions at idle speed were also tested on the dynamometer. Figure 9 shows that as the IVC—phasing advance angle increases, the HC and CO emissions drop slightly.

The fact that the torque and fuel economy can not be improved by changing cam-phasing angle at high engine speeds indicates that the valve-timing of the baseline engine is fixed for high engine speeds. Because of the inertia effects of gases being inducted into the cylinders, the inlet valve closing is held later than normal to achieve greater volumetric efficiency and a corresponding increased power output at high engine speeds. However, during low— and mid—speed operation, holding the inlet valve closing later can result in large pumping losses and cause backflow of the air/fuel mixture into the intake manifold and decrease in cylinder pressure, which would result in reduction in power and higher HC and CO emissions. As the engine steady—state testing for searching for optimum IVC—phasing advance versus engine speed and load shows, at each load/speed point, there is an optimum IVC—phasing angle. The test data is shown in Table 2. A further increase in IVC — phasing advance angle implies a deterioration in engine perfor-

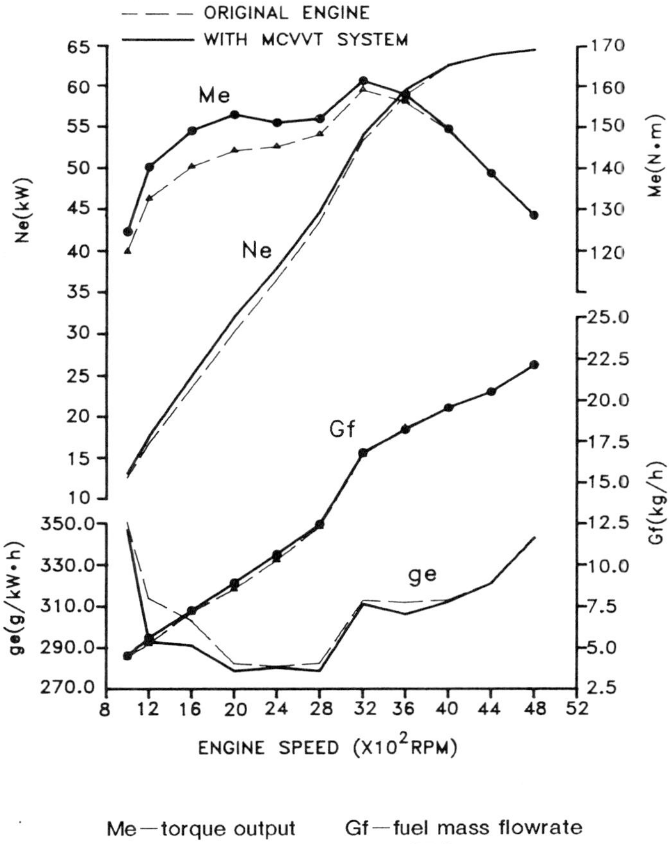

Me—torque output
Ne—power output
Gf—fuel mass flowrate
ge—BSFC

Fig. 8—Comparison of engine performance with and without the MCVVT system

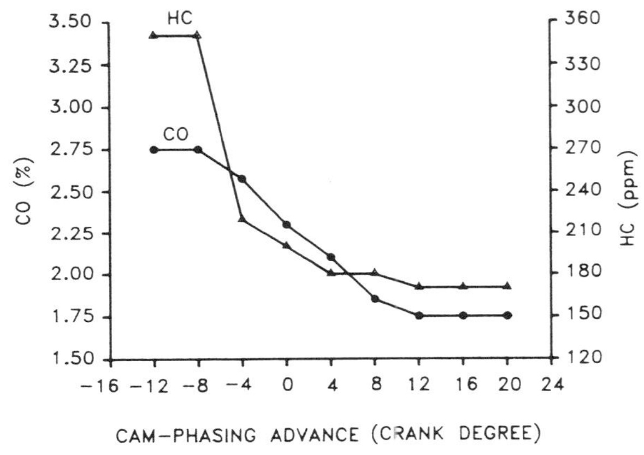

0 —valve-timing of the baseline engine
+△—amount of the valve-timing advance
−△—amount of the valve-timing retard

Fig. 9 —Effect of IVC—phasing on engine idle emission

mance and fuel economy. A simple explanation of the deterioration is as follows: due to the single — cam engine layout, the exhaust valve timing varies together with the inlet valve timing by the same amount. A too early opening of the exhaust valve at low engine speed allows an early escape of the burned gases and reduces the cylinder pressure excessively at the end of the expansion stroke. Therefore, a too early inlet valve closing causes a reduction in the single—cam engine's torque and power. Since the MCVVT—equipped engine has the same throttle opening as the baseline engine at each load/speed point, the amount of air/fuel mixture inducted into the MCVVT — equipped engine and therefore fuel mass flowrate remain unchanged in comparison with the baseline engine. For the MCVVT—equipped engine, the significant increase in low speed torque could result in reduction in BSFC. Therefore, controlling the amount of charge inducted into the engine by varying the inlet valve closing can eliminate the efficiency loss caused by the throttling process of the standard timing and reduce engine emissions at low—speed operation.

The test results indicated that with the proper control strategies, the MCVVT-equipped engine yields improvements in low speed torque and fuel economy as well as reductions in engine HC and CO emissions at idle speed. The control system works well under all engine—operating conditions. In actual road tests made by factory personnel of the First Automobile Works, the light—duty truck CA1046L, which is driven by the MCVVT-equipped CA488 engine, has instant driveability once it starts and faster acceleration.

CONCLUSIONS

The current technology of fixed valve — timing compromises the cam — profile shapes and camshaft phasing best suited for operation at various combinations of load and speed. To overcome this limitation, electronically controlled variable valve — timing is needed in order to optimize engine—operating characteristics at each dynamic operating condition.

In this work, a variable valve — timing control system has been constructed using a microcomputer. The IVC—phasing advance map has been established. The MCVVT system has been running successfully in a light—duty truck CA1046L. In order to confirm its effects on engine performance, acceleration and mode running test were carried out on the test bench and an actual vehicle. It was found that

(1) with the MCVVT system, the engine torques, fuel economy, and idle emissions are improved more significantly at low engine speed;
(2) the response time in vehicle acceleration is reduced;
(3) the IVC-phasing advance is more sensitive to engine speed;
(4) both bench test and road test results show that the fail—safe system works effectively.

The remarkable improvements of power and fuel economy of the CA488 engine at low engine speed indicate that the valve — timing angle of the baseline CA488 engine was fixed for its power output at high speed.

As the work reported in this paper shows, the IVC — phasing approach is applicable to engines with single camshaft arrangements, though the fixed valve — overlap periods limit the benefits to be achieved through its use.

ACKNOWLEDGEMENT

The authors wish to thank the engineers and technicians at Changchun Automotive Research Institute involved in the project for their assistance and help in the engine tests.

REFERENCES

1. C. Gray, " A Review of Variable Engine Valve Timing." SAE Paper No. 880386.
2. Michael T. Nelson and Alvon C. Elrod, "Continuous — Camlobe Phasing: An Advanced Valve — Timing Approach." SAE Paper No. 870612.
3. Sam Sapienza, Ben Shirey, and Nic van Vuuren, " A Electronically Controlled Cam Phasing System." SAE Paper No. 880391.
4. F. Freudenstein, E. R. Maki, and Lung — Wen Tsai, " The Synthesis and Analysis of Variable Valve — Timing Mechanism for Internal Combustion Engines." SAE Paper No. 880387.

960038

Idle Valve Pintle Design by Numerical Flow Simulation

V. Mahieu, B. Leduc, and P. DeHout
ULB

P.-Y. Geels and M. Morenville
Philips

1. ABSTRACT

Several idle control actuators are based on a solenoïd technology but a more efficient solution is to use stepper motors. Ehanced knowledge of the idle air flow parameters allows optimization of the stepper motor design. The objective of this paper is to describe this optimization process, with the help of experimental data and numerical simulation.

2. INTRODUCTION

Philips Automotive Electronics develops throttle bypass valves actuated by stepper motors (Fig.1). The flow trough the valve is characterized by the pintle shape and the bypass design. The air flow control mainly allows the engine idle speed to be stabilized and reduced. It also enables an improved cold start strategy and damping instantaneous air flow variation induced by rapid throttle movement. In order to get a better air flow control, it is useful to work on the pintle design.
In this way, both Philips and the University of Brussels decided to develop a flow bench with high vacuum performances (Fig.2), giving realistic characteristics of idle by-pass: flow function of the opening and/or of the manifold vacuum.

At the same time as the experimental work, numerical flow simulations were carried out on workstations, trying to describe the flow parameters.

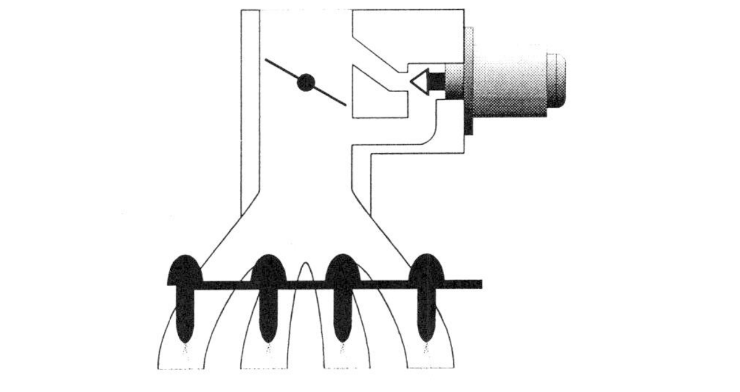

Fig.1: *Bypass implementation*

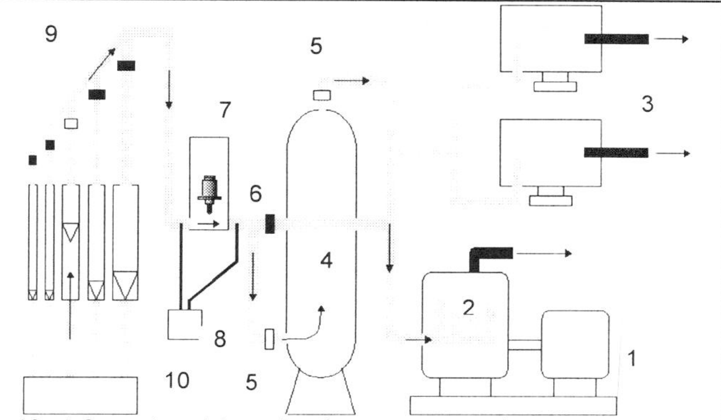

1°: AC motor driven by frequency variator
2°: driven engine (vacuum pump 1)
3°: vacuum pumps (vacuum pumps 2 & 3)
4°: pressure waves reducer
5°: valve (N.O.)
6°: valve (N.C.)
7°: test zone
8°: P,T° recording system
9°: flowmeters battery
10°: air filter

Fig.2: *Flow bench*

This development is helpfull for engineers to design new idle pintle and bypass geometry.

Simulation is of great interest for different reasons: few costs, flexibility, access to global and local characteristics of the flow (pressure distribution, flowrate, velocity vectors, Mach number, and so on).

This study aims to show the feasability of permeability simulation (flow trough the idle valve as a function of the opening at constant pressure drop).

3. FLOW BENCH EXPERIMENTAL DATA

Fig.3 shows the results for a test performed with a standard pintle on a 10mm diameter seat. Air flow is measured from 0 to 250 steps (24 steps = 1 mm). Pressure drop was tuned at 600 hPa to get a sonic air flow.

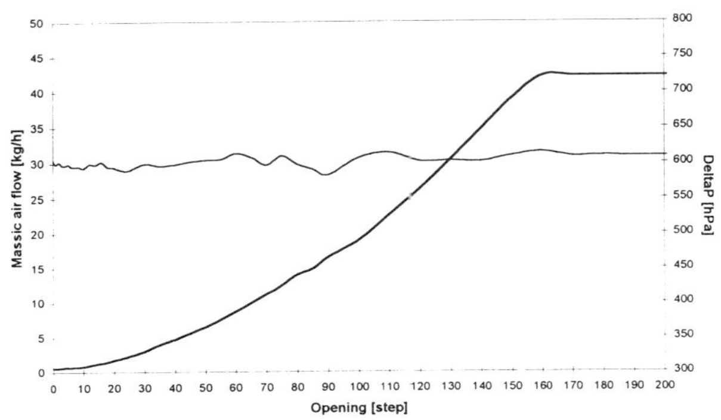

Fig.3: *Air flow experimental data*

As a first step in the numerical simulation process, simplified pintle geometry will be considered (Fig.4).

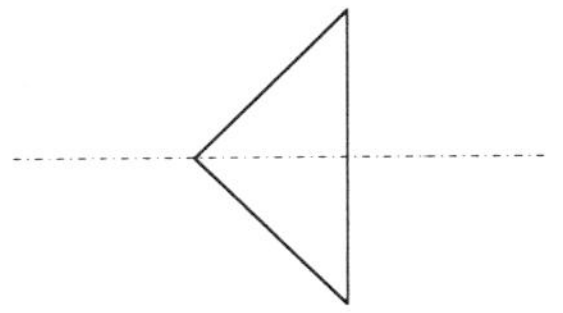
Fig.4:*simplified pintle shape*

Experimental results on flowbench (permeabilities) will be compared to the simulation.

4. SIMULATION

Steps in the simulation development are:

- Wide opening axi-symetric simulations
- Wide opening three dimensionnal simulations
- Different openings axi-symetric simulations

The code used is dedicated to fluid flows, with or without heat transfer, mass transfer or chemical reactions. Flows equations are solved by a finite volume method.
The flow is described by the mass, the energy and the momentum conservation laws. The turbulence model is a (k,ε) type. The finite volume - or control volume - allows use of a non-regular grid, wich can be closer to the real problem geometry.
All the corresponding equations are directly written for each volume defined by the grid. The resolution is equivalent to discretising the integral form of the fluid equations.

We will take an example to illustrate this concept. The mass conservation law, under his differential form, is written:

$$\frac{\partial \rho}{\partial t}+\frac{\partial(\rho u_i)}{\partial x_i}=0$$

This form comes from the mass conservation law under his integral form:

$$\frac{d}{dt}\int_V \rho dV+\int_S \rho(\bar{v}.\bar{n})dS=0$$

where **v** is the velocity vector and **n** the normal at the surface.
If we consider one element of the grid, the equation become:

$$\frac{d}{dt}\int_{Vc}\rho dV+\int_g \rho(\bar{v}.\bar{n})dS+\int_d \rho(\bar{v}.\bar{n})dS+\int_h \rho(\bar{v}.\bar{n})dS+\int_b \rho(\bar{v}.\bar{n})dS=0$$

The other conservation laws follow the same way.

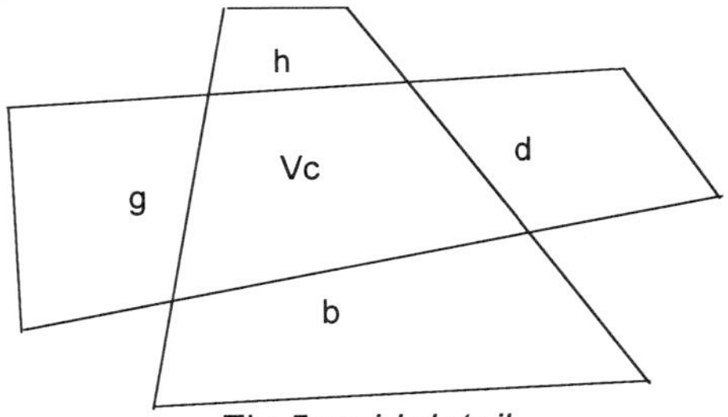

Fig.5: *grid detail*

5. AIR FLOW CHARACTERISTICS

Three steps must be defined: flow type definition, the geometry choice, and boundary condition.

5.1. Flow type definition

As stated formerly, flow trough the valve is transonic. It means that a compressible calculation must be used. Actually incompressible calculation is still needed to get an initial solution for the compressible one.
Is the flow laminar or turbulent? Reynolds number indicates that flow conditions are turbulent.

5.2. Geometry

Air flow is definitely tridimensionnal (Fig.6), but in order to simplify the grid generation and to save calculation time, 2D axi-symetric air flow was considered (Fig.7).

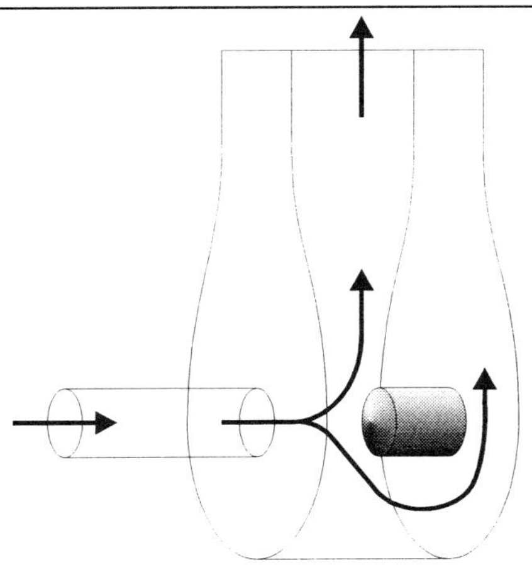
Fig.6: *Real air flow*

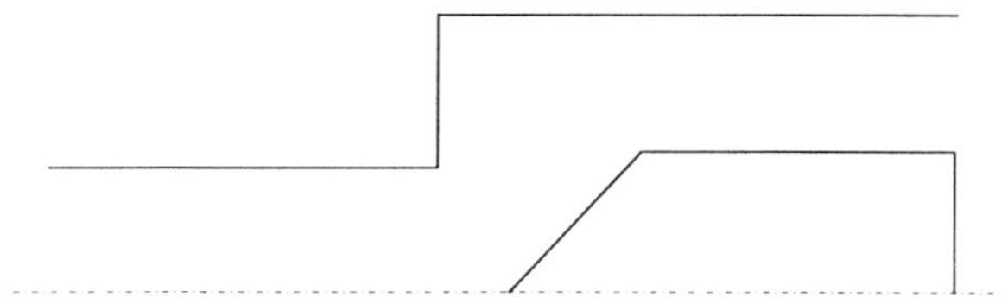

Fig.7: *2D axi-symetric geometry*

5.3. Boundary conditions

Upstream total pressure and downstream static pressure of the flow bench are chosen as boundary conditions. Moreover, wall velocity and turbulence level have to be precised.
For compressible calculation, total temperature upstream and downstream must be known. The wall heat transfer is considered as negligible.

6. WIDE OPENING THREE DIMENSIONAL SIMULATION

For the same geometry, the simulation error due to the axi-symetric approximation must be known. The three dimensional geometry (Fig.6) is generated from the two dimensional geometry, and the inlet and outlet sections are the same than on the real flow.

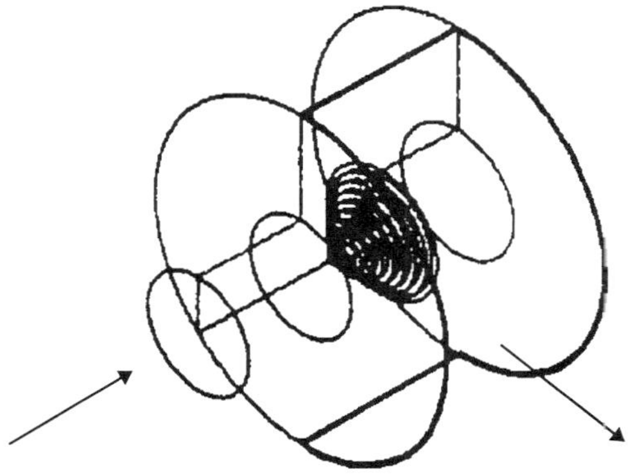

Fig.8: *three dimensional geometry*

Using three dimensional simulation is much more demanding: the number of cells and thus the size of equations system grows consequently. The main consequence is a calculation time multiplication by 7.

Hopefully, it appears that 3D calculation does not give significant deviations (less than 1%). The difference is negligible due to the very wide outlet areas. And this is true for incompressible and compressible calculation.

The 2D axy-symetric hypothesis is then considered as accurate.

7. CALCULATION RESULTS AND EXPERIMENTAL DATA COMPARISON.

Incompressible and compressible simulation were done for several openings (24,48,72,120,144,168,192 and 218 steps). Fig.9 shows the grid for the 72 steps opening.

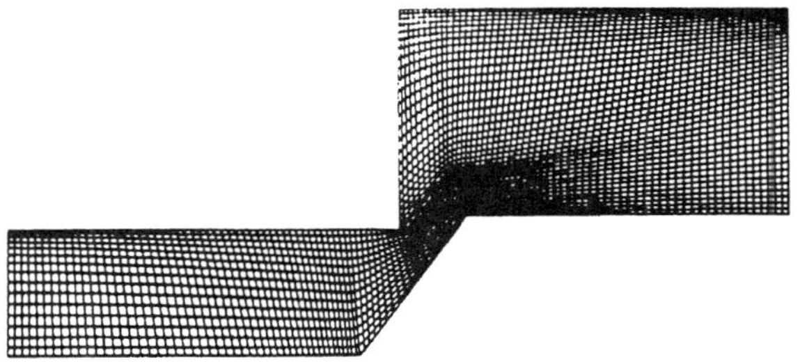

Fig.9: *72 steps opening grid*

Comparison between these two simulations and experience shows that compressible calculation gives best results (Fig.10). Both simulation are above experimental values. This is due to boundary conditions values. Actually, on the flow bench, pressure sensors (used to give the boundary conditions) are located not as close to the seat as expected.

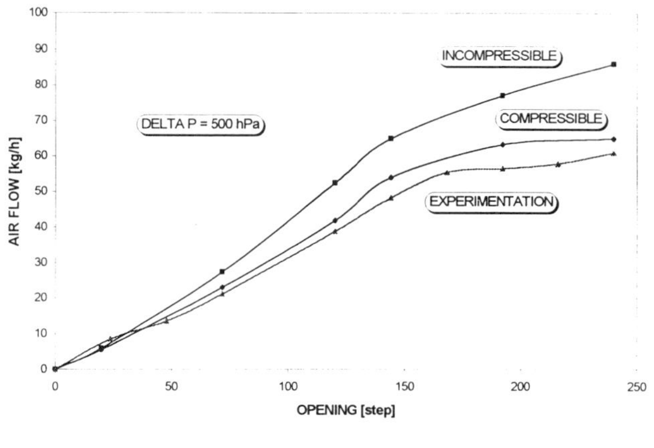

Fig.10: *simulated and experimental air flow*

To the calculation, it means that the inlet pressure is surestimated and outlet pressure is underestimated:

DeltaP used for simulation > real DeltaP
and thus air flow calculated > experimental air flow.

This deviation will increase with the air flowrate and thus with the pressure drop.

8. ILLUSTRATIONS

Up to now we have been focused on global parameters values that can be matched with experimentation, but studying local evolution of flow paramaters could be extremly interesting. This micro analyse was realized on a industrial two slopes pintles which has a slightly cut nose and a flat base.

For one opening (24 steps), different figures will be described.

The Mach number (Fig.11): the flow is transsonic, the Mach number value is close to 1 at the minimum cross section. It justifies the compressible calculation.

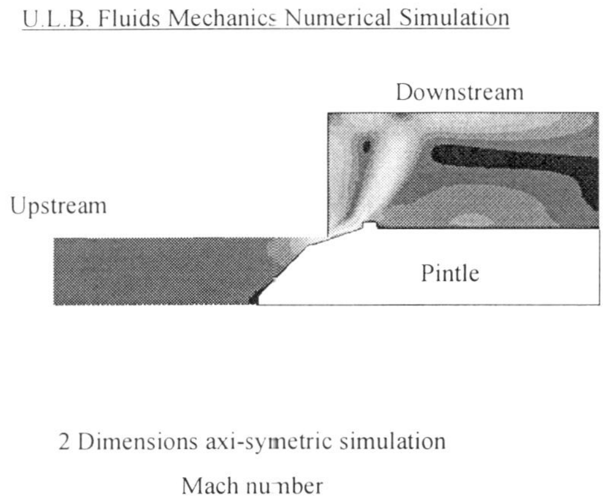

Fig.11: *Mach number*

The stream functions (Fig.12): two recirculation areas can be observed. This is due to boundary layer effect. Flow separation from the wall and the pintle gives strong vortex.

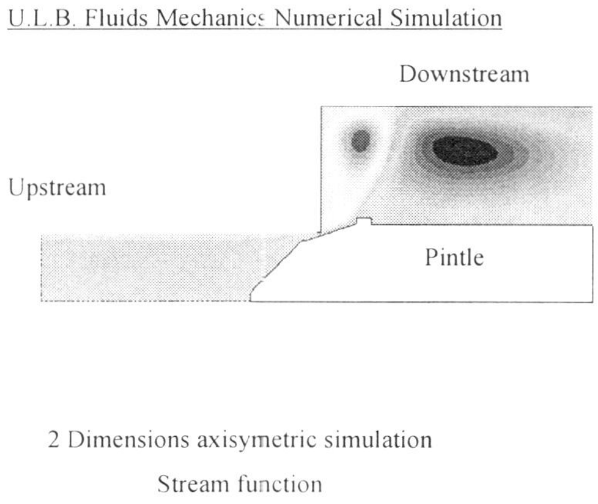

Fig.12: *stream function*

Static pressure (Fig.13): great difference of static pressure is observed as expected. The pressure drop magnitude indicates whether the flow is transsonic or not.

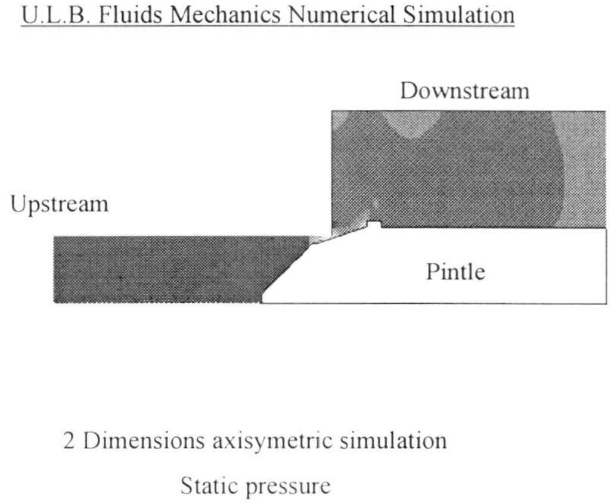

Fig.13: *static pressure*

9: CONCLUSIONS

Using numerical flow simulation seems to be efficient for predicting air flow rate. Compressible calculation is necessary due to transsonic air flow. The precision of calculation and experimental results matches the industrial requirements. Local distribution of static pressure, velocity, Mach number highlight the pintle design influence on the air flow, especially sharp edge influence on boundary layer effect.

Combination of numerical simulation and experimental measurements will lead to an improved pintle design process, saving many hours of research work on the flow bench.

10.REFERENCES

Idle air control recomendations for use.
Philips Automotive Electronics.

Cousteix J.
Aérodynamique: Turbulence et couche limite.
Cepadues Editions, 1989.Toulouse. Col. La Chevêche.

Hirsch C.
Numerical Computation of Internal and External Flows.
John Wiley & Sons, Chichester - New-York, 1988.

Hirsch C.
Discretization Methods for the incompressible Navier-Stokes equations. Von Karmann Institute . Lecture Series, 1990.

ACTUATORS IN BRAKE AND CRUISE CONTROL SYSTEMS

970814

Electromechanical Brake System: Actuator Control Development System

Christof Maron, Thomas Dieckmann, Stefan Hauck, and Hubertus Prinzler
Continental AG

ABSTRACT

Continental Automotive Systems started development of an electromechanical brake-by-wire system (EMB) 2 years ago. A major part of the development deals with the control of the brake actuator. For the development of control algorithms a special test stand was built. It consists of the seat capsule, the actuator and the PC-based electronic control unit. As the electronic unit also performs a real time vehicle and actuator simulation a complete Hardware-in- the-Loop system supports simultaneous engineering within this project.

This paper describes the Hardware-in-the-Loop development system and shows first results obtained in an early state of the development process.

INTRODUCTION

The braking equipment employed in today's automotive vehicles stems from the history of automotive development. The basic principles of brake actuation have remained unchanged for years. They are based on the hydraulic transmission of brake force from the brake pedal to the wheel brakes. In the course of the development of braking systems, additional components such as force assistance units and force proportioning devices have been added along the transmission path.

With demands rising with respect to driving safety, the functional scope of brake systems has also expanded and now includes equipment such as anti-locking devices and traction control. Alongside electronic components like sensors and electronic control modules, intricate hydraulic assemblies like solenoid valve units, hydraulic pumps and in certain cases pressure accumulators have enlarged the scope of brake systems. With further safety upgrading in the pipeline, tomorrow's hydraulic brake systems will be increasingly complicated.

A logical step in the direction of overcoming these disadvantages is the development of EMB, an electro-mechanical brake system. Intelligent electronic control of electronic actuators on the wheel brakes unleashes the potential for a decisive advance in automotive development. By transmitting brake actuation signals electronically, components such as power assistance units, brake master cylinders, hydraulic control units and brake lines can be eliminated.

Functions such as anti-lock and traction control can easily be included in the EMB electro-mechanical brake system, which allows decentralized brake actuation functioning individually for each wheel. Future developments aimed at enhancing active driving safety (e.g. active braking, intelligent cruise control, vehicle dynamics control (VDC),...) can be achieved by simply adapting the software and without the need for any additional hardware. This results in a considerable reduction in the costs entailed in incorporating additional safety devices in future brake systems and makes it possible for smaller passenger car categories to share in this development.

Conventional hydraulic brake systems in vehicles with electronic transmission control require auxiliary electric power to boost braking force. In view of the demand for the introduction of so-called "zero emission vehicles", in particular, the upcoming years can be expected to witness an increase in the number of electric-powered vehicles. By employing EMB, a vehicle's electric power can be used both as auxiliary power and directly as a means of slowing down the vehicle. Thanks to an intelligently controlled brake system, EMB makes it possible to include the electrically powered vehicle motors in braking deceleration so as recuperate energy in overall control loop of the braking unit. This has the effect of improving the overall energy balance of this vehicle and helps the automotive industry move forward on the road to "zero emission vehicles".

BRAKE BY WIRE CONCEPT

With the EMB project, Continental Automotive Systems has taken a fresh approach in the realization of a brake by wire (EMB) system. This approach is marked by the following:

> the use of conventional disc brakes
> decentralized actuation devices at the wheels
> the use of completely "dry" gear concepts.

The fact that hydraulic components are not used here, means there is no mechanical link between the pedal and the brakes, like as it is found in many other BBW systems. This was already taken into appropriate consideration in the system's structural concept (Fig. 1). A structure was created that in itself guarantees a high degree of availability while minimizing the likelihood of operational failure.

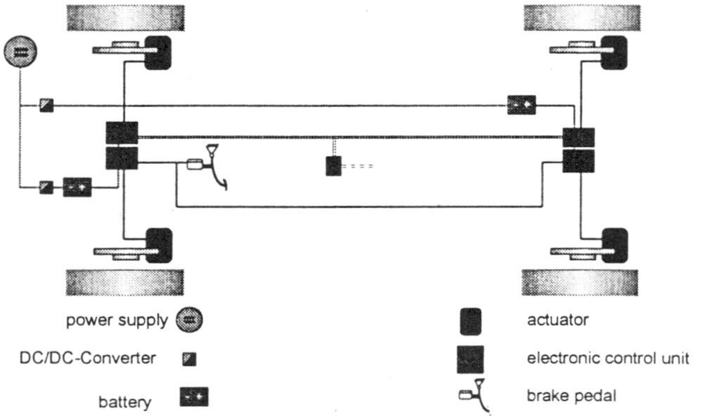

Fig. 1: Structure of the EMB system

A separate electronic module is assigned to each actuator. This module is responsible for the individual brake control on the respective wheel. It communicates with a central module, which sees to general vehicle control tasks such as axial distribution of braking force or VDC. The internal electronic communication functions by means of a bus system.

In a completely electric braking system of this kind, power is supplied by means of two completely independent power circuits, each of which has its own energy accumulator. The functioning of the braking system is thus not crucially affected by a breakdown of the vehicle electric system due to defects in the generator and electric system battery.

Insofar as there is a separate flow of power and information inside the BBW system, a differentiated approach must be taken in considering the system's inherent redundancies. Whereas, for instance, the power supply is dual-circuited, the actuator has four circuits due to the fact that each wheel functions with a separate electronic system equipped with its own intelligence. Another key point in the EMB concept is that the components used allow not only for large-scale but also for cost-efficient production.

BRAKE BY WIRE DEVELOPMENT SYSTEM

<u>MOTIVATION</u> - The development of the electromechanical brake system started from scratch. As already the design of the wheel brake actuator is a very complex task, it was necessary to ensure simultaneous engineering from the beginning of the project in all fields of activity as there are:

> design of actuator mechanics
> design of electronic units
> implementation of control software
> development of sensor technology
> design of an energy management
> design of the pedal unit

One powerful tool for simultaneous engineering is a "hardware in the loop" test stand which provides the engineers with a test bed for all the different units. This brake development system will now be described.

<u>SEAT CAPSULE AND PEDAL UNIT</u> - The brake pedal is a very important component for a brake by wire system. Contrary to what is the case with conventional braking systems, the pedal restoring force must be generated here artificially, as there is no mechanical link between the pedal and the wheel brake. The pedal's force/travel characteristics determine the driver's subjective braking experience. Even more important than the so-called pedal feeling is, however, a reliable link-up of the pedal sensor. The driver's braking intention is transmitted via an electric signal from the pedal to the brake system. This represents, from a safety viewpoint, a bottleneck in the whole system, as even a fourfold redundancy in the wheel brake is of no help in the absence of any pedal signal.

In the vehicle itself, the overall development of the pedal unit has shown itself to be a very involved and arduous undertaking. For this reason, a seat/pedal/steering wheel arrangement corresponding to that of the prototype vehicle was constructed, as shown in Fig. 2.

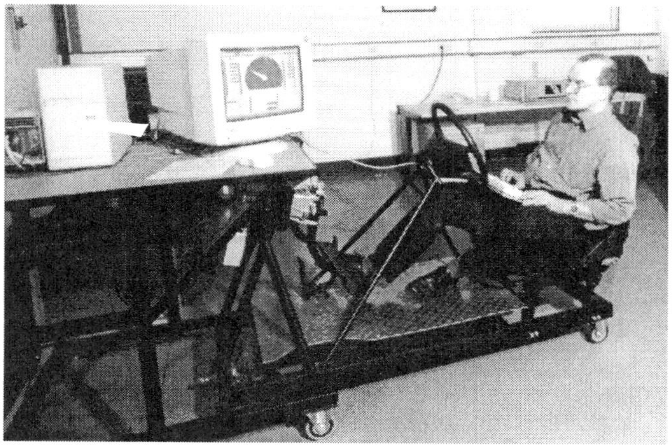

Fig.. 2: Seat/pedal/steering wheel arrangement EMB test bench

This so-called seat capsule allows the use of different pedal feeling simulators, thereby minimizing the work of adapting the individual pedal for actual use in a vehicle.

A passive spring damping system is used for the pedal feeling simulator. It makes minimal demands on space and can even be positioned in front of the scuttle, i.e. in the vehicle interior.

Fig.. 3: Pedal unit

For safety's sake, a triple redundant system must be arranged around the pedal for the measurement of the pedal position. In laboratory tests, only a twofold redundant angular sensor was used on the pedal unit. In the vehicle an additional distance sensor is used, so that the pedal position can be fed into control units threefold. Fig. 3 shows the sensor hook-up and the pedal feeling simulator.

ACTUATOR - The flexibility attainable with the hardware-in-the-loop (HIL) test bench pointed early on to the advisability of attaching importance to a modular and easily modifiable structure. Easily interchangeable components facilitate the possibility of running comparative tests, resulting in the end in the optimization of the actuator system, consisting of the drive, in certain cases a gear, and a rotation translation converter. A modular concept of this kind presents a packaging disadvantage for the actuator, i.e. the flange connections stand in the way of reducing the space requirement. This, however, takes a backseat to the optimization potential.

It was thus possible in a simple manner to run comparative tests on various motors, intermediate transmissions and spindles and to evaluate new individual components in a familiar setting. Good experience was made, for instance, with the interchangeability of the drive motor: For one thing prototypes of the motors planned for later installation in the vehicle could be tested. It was, in addition, also possible to use an "ideal" motor - i.e. one that was known and easy to control. Motors of this kind are, for instance, commercially available as tooling machine drives. With an "ideal" motor of this kind it is possible to determine particularly closely the mechanical parameters of the actuator system. For the following representations, a combination of a drive motor and an intermediate transmission in front of the spindle was selected (Fig. 4).

Fig. 4: Example of an EMB lab test actuator

The drive motor is an conventional commercial AC servomotor. It was used as "ideal" drive for identifying friction, hysteresis and moment of inertia of the following arrangement. The results show the values determined for a high-ratio compact gear connected to a planetary roller spindle. All of the components are available commercially and have not been modified.

ELECTRONIC CONTROL - To control the actuator's AC motor, a commercially available AC servo unit was used. A set torque value given as analog value is specified as input value for this servo unit. This set torque value is generated with a PC-based system exhibiting maximum flexibility and a very comfortable operator/programming interface.

The system consists of a 486 PC with a dSPACE signal processor card and a corresponding I/O link (Fig. 5). MATLAB and SIMULINK are used as programming environment. The functions designed under SIMULINK can be down-loaded onto the signal processor card by just a mouse-click. They run on this card automatically in real time.

The pedal's sensor signals are evaluated via analog inputs. The brake application force is measured with a force measuring ring. The servo unit provides the system with the motor's actual speed. All measured variables for actuator control are thus available.

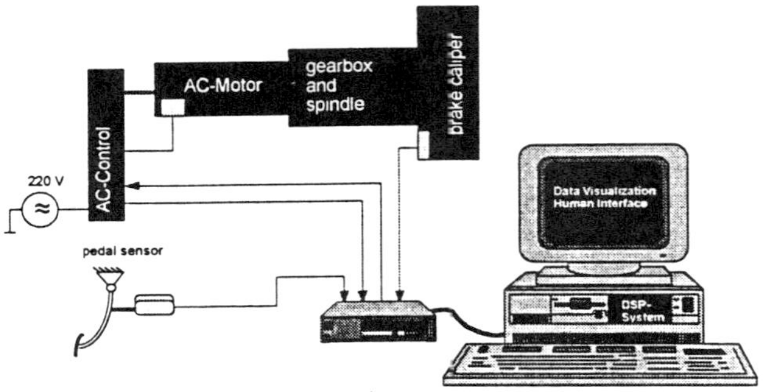

Fig. 5: Diagram of the electronic control

The signal processor card is so powerful that it was likewise no problem to realize in real time the simulation of the longitudinal movement of a vehicle, including a simplified tire model as shown in Fig. 6. A genuine hardware-in-the-loop test bench is thus available for functional development.

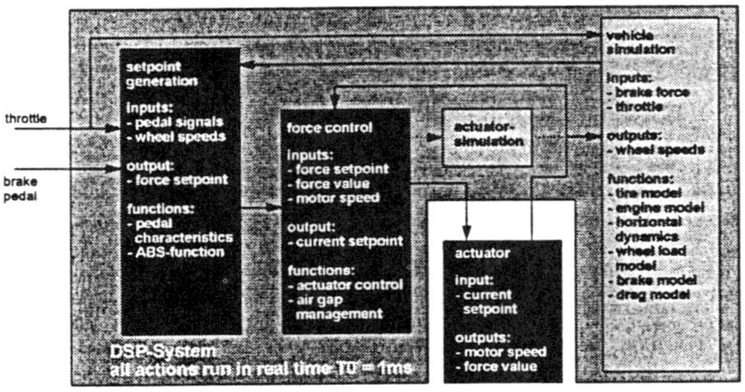

Fig. 6: Tasks of the signal processor system

FUNCTIONAL DEVELOPMENT

ACTUATOR MODEL - For the functional development it is helpful to have a mathematical model of the process to be controlled. For one thing, such a model provides very useful insight into the functioning of the process. What's more, processes and algorithms can first be tested on the model before being used in the real process. The physical modeling provides the block diagram shown in Fig. 7.

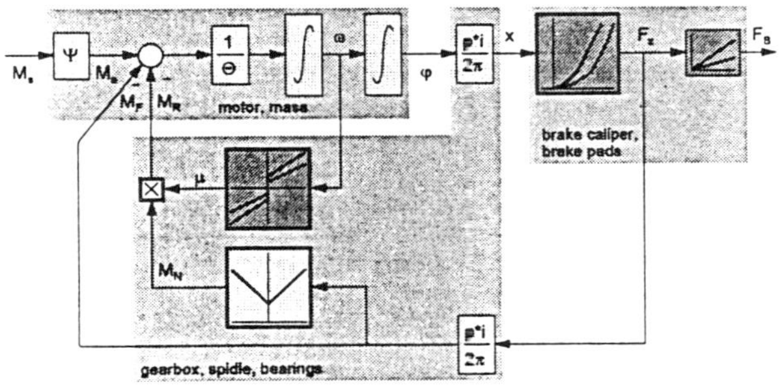

Fig. 7: Block diagram of the EMB actuator model

The model consists of a concentrated mass encumbered with friction, the gear spindle transmission ratio, and, finally, the nonlinear characteristic curve of rigidity of the brake caliper and the brake pads. Attention must be given to the friction due to the brake application force arising from the axial support of the force. With the help of current regulation, the power electronics immediately adjust to the specified torque value. In view of the fact that the electric time constant for setting the torque is small vis-à-vis the mechanical time constant, a simple gain can be assumed here. The later tests provide an indication of the quality of the model.

FORCE CONTROL - Force control is realized with a cascaded force/speed control loop, as shown in Fig. 8.

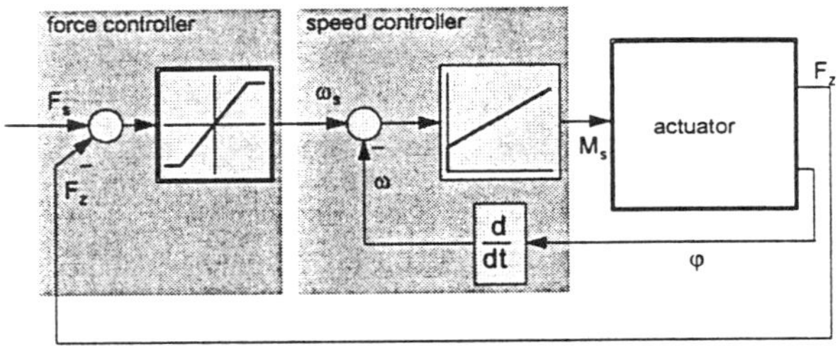

Fig. 8: EMB force control loop

A simple PI-control unit is used as speed controller. Output quantity for the speed control unit is a nominal torque value immediately set by the power electronics. Friction is compensated implicitly with the integral part of the controller. The speed setpoint is given by a simple P-Controller with limitation.

AIR GAP MANAGEMENT - Air gap is the distance of the brake pad from the brake disc. As no direct clearance sensor is planned, clearance is calculated on the basis of the readings. To do this the rotor position is evaluated after the force has fallen below a defined threshold.

So as not to have to alter the structure of the control loop, clearance is given in the form of a negative brake force as shown in Fig. 9.

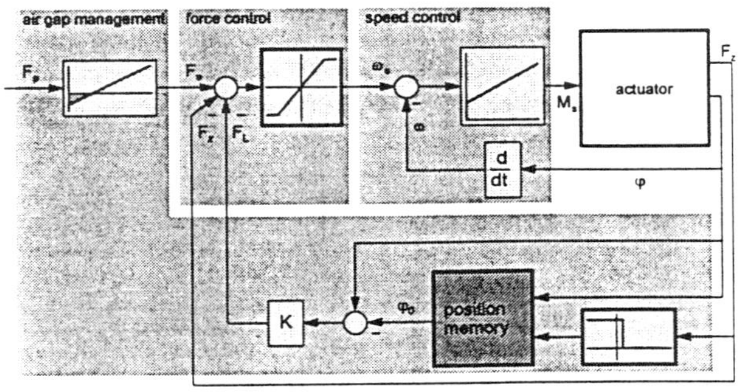

Fig. 9: Force control loop with clearance management

As the air gap is being controlled, the remaining friction during normal driving can be minimized. In order to minimize the time for overcoming the air gap at braking start, the size of the air gap can be adjusted if the driver changes from gas to brake pedal.

ABS FUNCTION - A simple ABS function was implemented for the HIL test bench. The simplified block diagram of this function is shown in Fig. 10. It is a simple slip and acceleration control which puts the nominal torque value for the actuator drive at zero if certain limit values for the slip and acceleration of the wheel are exceeded (see Burckhard, 1993). The test results provide proof that good results can be attained with such a simple ABS algorithm. In the current work more complex traction control algorithms are used in place of this simple ABS function.

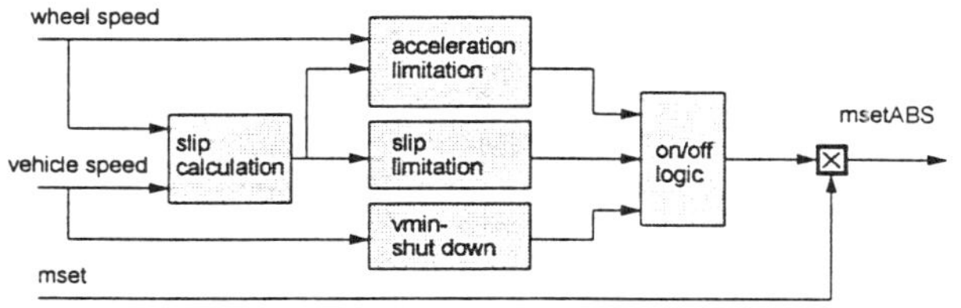

Fig. 10: Simplified EMB ABS structure

TEST RESULTS

FRICTION, HYSTERESIS, INERTIA - With a method described by Held and Maron (1988) the moment of inertia of the motor itself and its internal friction was next determined by evaluating a triangular rotational speed control. Fig. 11 shows the corresponding data. After the gear spindle unit and the brake caliper are flange-mounted, both the moment of inertia and friction increase, as would be expected. Static friction, in particular, seems to increase greatly. This can be attributed to the limited efficiency of the workings of the actuator. On the basis of these readings it was at this point possible to adjust an optimal speed controller.

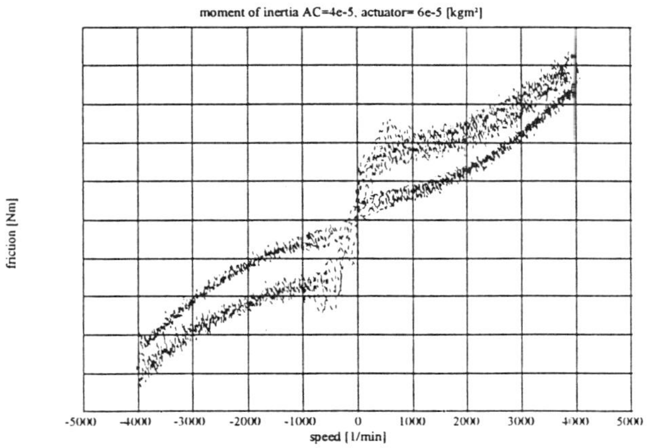

Fig. 11: Moment of inertia and friction curve of an EMB actuator

FORCE CONTROL - After the speed control loop was satisfactorily working, it was possible to close the force control loop. As the force control unit only requires the setting of an amplification factor, it was possible to have it working satisfactorily in just a short time. The result of the force control is shown in Fig.12 for an arbitrary set force value curve. One can see how little variance there is between desired values and actual values. Further more, the force can follow a higher frequent setpoint signal which is a hint for the suitableness for an anti lock system.

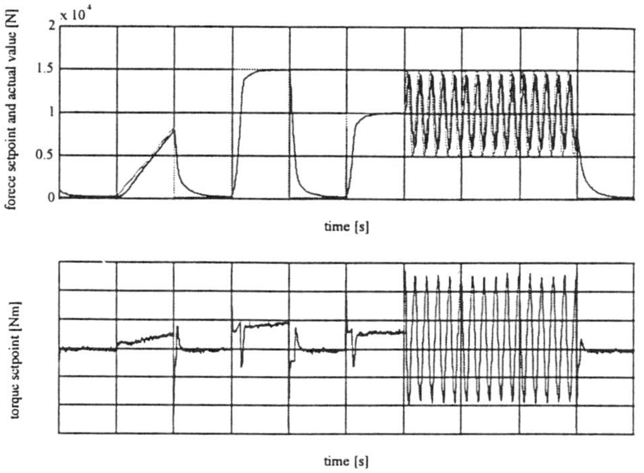

Fig. 12: Force control results

Note should be taken of the fact that (in contrast to what holds for a hydraulic system) the motor is actively reversed when the brakes are opened. What's more, the dynamic behavior is roughly the same when the brakes are applied and when they are released.

A further measurement in the case of more or less static excitation with a very gradual sinusoidal signal, as shown in Fig. 13, provides, by showing clamping-force as a function of the speed control output, the friction loss hysteresis of the whole actuator. This hysteresis is plotted in Fig. 14.

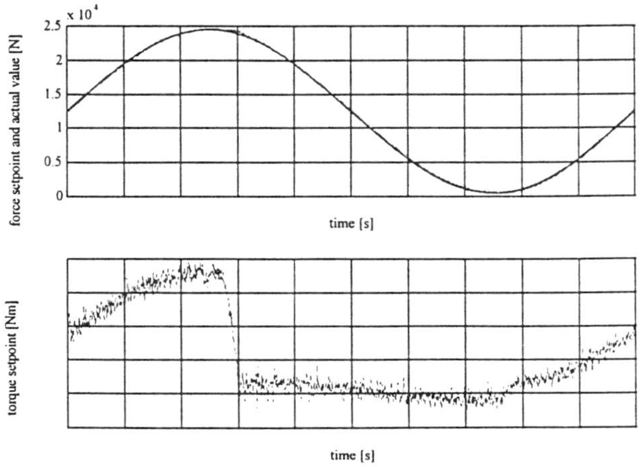

Fig. 13: Measuring signals in the case of more or less static excitation

Obviously the used gear-spindle combination and the concomitant efficiency bring the actuator very close to being self-locking. Insofar the efficiency must be increased for an application in a real vehicle.

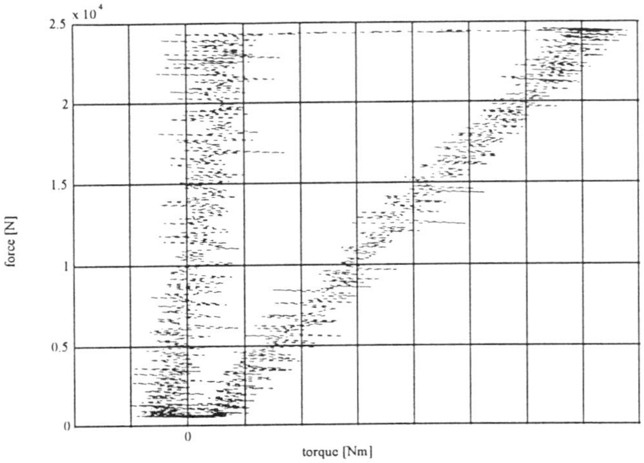

Fig. 14: Loss hysteresis of the actuator

ABS BRAKE ACTUATION - The ABS function can be tested in a meaningful way only in connection with the vehicle itself. Fig. 15 shows a braking test in which a vehicle moving at a speed of 150 km/h comes to a complete stop on a dry road surface. Only the front axle of the vehicle is braked. One actuator exists physically and works on a non rotating brake disc, another unit of the HIL-test bench. The second actuator is simulated in real time with the signal processor system, so is the vehicle dynamics.

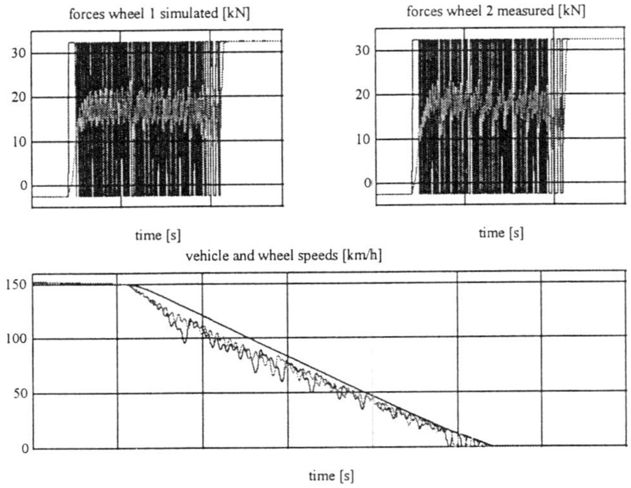

Fig. 15: Simulated ABS braking actuation on a dry road

The ABS operating principle can be clearly recognized. The algorithm adjusts the nominal torque on the motor each time to 100% torque or 0% torque. The mechanism cannot cope with this rapid switching back and forth and adjusts to a median braking force that is just enough to prevent the brakes from locking.

From this experiment one can conclude that the simulation of the actuator works quite good, as the 2 wheels behave almost the same, both qualitatively and quantitatively, even though one actuator exists in reality and the other is only simulated.

For the sake of clarity, Fig. 16 shows an ABS brake actuation for just one wheel with an up-down change in the road pavement coefficient of friction from μ=1,0 to μ=0,1. After the wheel locks briefly, the ABS algorithm performs on the low-μ road with a much more gradual control frequency.

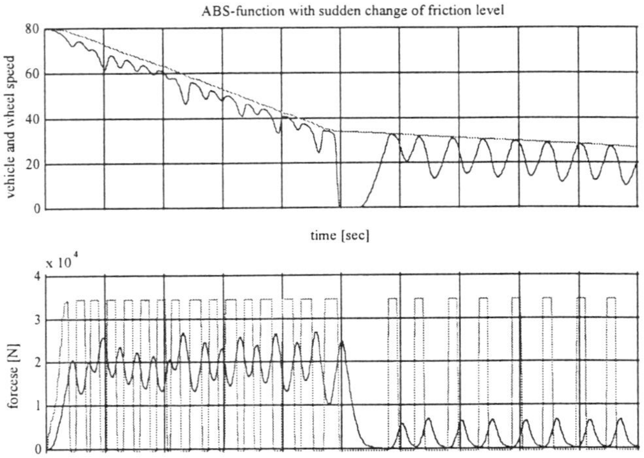

Fig. 16: Simulation of an ABS braking actuation with a sudden μ-change from 1.0 to 0.1

CONCLUSION

The EMB development system described shows how modern engineering tools can be used to achieve rapid advances in development both in the areas of hardware and of software. The fact that the control algorithms can be developed very early on is not the sole advantage. As a consequence of simultaneous engineering, the knowledge so gained flows into the design and development of the system's mechanical components.

The test results presented here show that with this technical approach new solutions in automotive brake technology are possible. The realization of a complete electro-mechanically brake system is an immense technical challenge. It makes major demands on automotive control engineering. As brakes have an impact on safety a high development expense is required to guarantee the operational safety.

REFERENCES

Held, V. and C. Maron (1988): Estimation of Friction Characteristics, Inertial and Coupling Coefficients in Robotic Joints based on Current and Speed Measurements. IFAC Symposium on Robot Control (SYROCO), Karlsruhe .

Burckhardt, M. (1993): Radschlupf Regelsysteme. Vogel Verlag, Würzburg.

981109

Brake-by-wire without Mechanical Backup by Using a TTP-Communication Network

B. Hedenetz and R. Belschner
Daimler-Benz AG

ABSTRACT

By-wire systems have been established for several years in the area of aircraft constructions. There is the visible trend to realize by-wire applications without mechanical or hydraulic backup systems in vehicles. The required electronic systems must evidently be available and safe. This paper addresses a new automotive architecture approach using the time-triggered fault-tolerant TTP protocol that has been designed for class C safety related control applications, like brake-by-wire or steer-by-wire, due to the SAE classification [1]. As an example we present this approach within a brake-by-wire research car (case study) without mechanical backup. The intention of this architecture is to tolerate one arbitrary fault - excepting faults of actuators - without any effects of the brake performance. For this purpose we use redundancy in communication (TTP) and electric components like sensors, actuators and power supply.

1. INTRODUCTION

In the past few years there is the tendency in vehicle construction to increase the safety of vehicles by introducing intelligent assistance systems (e.g., Anti Blocking System ABS, Brake-Assistant (BA), Electronic Stability Program (ESP), etc.) that help the driver to cope with critical driving situations. Typically for these functions is the active control of the driving dynamic by distributed assistant systems, which therefore need a communication network. Faults in the electronic components, which control these functions are safety critical. However the assistance functions deliver only an add-on service in accordance with a fail-safe strategy for the electronic components. If there is any doubt about the correct behavior of the assistance system, it will be shut down. For by-wire systems without a mechanical backup a new dimension of safety requirements for automotive electronics is reached. After a fault the system has to be fail-operational until a safe state (e.g., vehicle stands) will be reached.

To reach the fail-operational requirement an additional effort for redundancy in the control components, sensors, software, power supply and the communication system has to be done. Nevertheless, there are still good reasons to introduce by-wire functions for brake systems.

<u>Assumed advantages:</u>

- Assistance functions (ABS, BA, ESP,...) could be realized only by software and sensors without additional mechanical or hydraulic components.
- Electrical interfaces instead of hydraulic interfaces, allow easier adapting of assistance systems.
- Reduction of packaging problems.
- No brake fluid, no bleeding, simple maintenance, ecological.
- Nearly rest torque free.
- No mechanical links between the brake components and the engine compartment, improving passive safety.
- No perceptible noise emission when braking.
- Reduced costs for assembly during line production.

2. REQUIREMENTS FOR SAFETY CRITICAL BY-WIRE ARCHITECTURES

Distributed by-wire systems consist of electric components, sensors, control software, power supply, actuators, and a communication system. For correct function any part of the system must work in an intended manner. It is not sufficient to improve only one part of the system. The correct function depends on the whole assembly of the subsystems.

2.1 CLASSIFICATION OF BRAKE-BY-WIRE SYSTEMS

There are two ways and philosophies to realize brake-by-wire. In Figure 1 the different systems are characterized. On one hand, the system is based on the traditional hydraulic brake system [13]. The by-wire function is realized through hydraulic pumps and additional electric controlled valves, the system is called Electric Hydraulic Brake (EHB) [13]. In an EHB system a hydraulic backup can be realized. After one fault is detected the complete electric hydraulic system will be shut down and with the help of some valves a direct hydraulic brake circuit will be closed. The hydraulic backup system can only deliver an emergency function with reduced brake force.

On the other hand the brake-by-wire system is based on electric mechanical actuators, called Electric Mechanical Brake (EMB) [18,21]. In an EMB system the brake force and brake control is realized by electric components. In an EMB system, no mechanical backup can be realized. The system must be fault-operational after one fault.

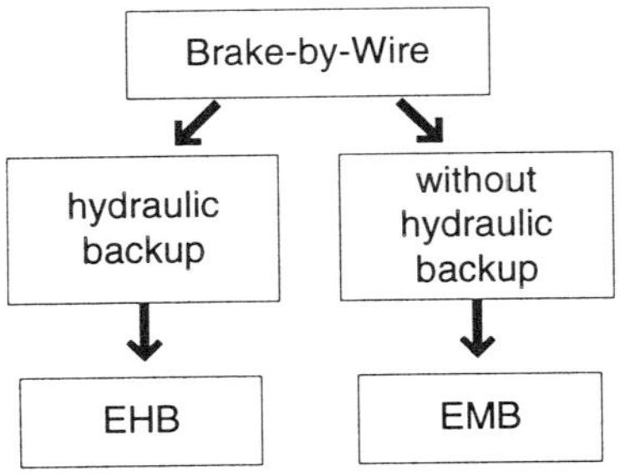

Figure 1: Classification of Brake-by-wire Systems

2.2 REQUIREMENTS FOR BRAKE-BY-WIRE SYSTEMS WITHOUT MECHANICAL BACKUP

There are several requirements for brake-by-wire systems, which are listed below (the terms Safety, Reliability, Maintainability and Availability are used in accordance with the definition of Laprie [14]).

Common requirements for automotive by-wire systems:

- Safety; after one arbitrary fault the system has to be available in a satisfying manner e.g., the brakes have to work with an adequate brake force.
- Reliability; the reliability of a by-wire system has to be at least as high as a comparable mechanical system.
- Availability; the availability has to be at least as high as present systems.
- Maintainability; the maintainability intervals have to be at least as long as present systems.
- Lifetime; equal to present systems.
- The Society of Automotive Engineers (SAE) classified in-vehicle communication systems into three categories, class A-C [1]. For safety related communication systems, class C is required.
- Costs; have to be equal to those of conventional systems.
- Compartment; has to be small enough for an easy integration of the components.
- Legal aspects; have to be fulfilled e.g., EU-Guideline 71/320/EWG.

Special requirements for brake-by-wire systems:

- Actuators; the brake actuator has to be free of braking torque in case of power loss.
- Sensors; triple redundancy is needed for the pedal sensors to get one valid pedal measurement by the loss of one sensor.
- Power supply; two independent power supplies are needed.
- Communication; the communication system between the components has to be fail-operational after one fault.
- Software; the software has to be certified.

To reach these requirements new technologies have to be developed for actuators, communication and electronic components. Redundancy is needed in sensors, electric components, communication and power supply.

3. THE TIME-TRIGGERED PROTOCOL TTP

By-wire functions are distributed in the vehicle and need networks that guarantee a reliable and predictable communication. This type of communication belongs mainly to class C of the SAE classification [1].

3.1 DEMANDS FOR CLASS C COMMUNICATION SYSTEMS

When introducing such new by-wire systems without conventional backup for safety related functions into vehicles, the following new requirements for the communication system arise:

- The communication over the network has to be strongly deterministic.
- The communication system has to support composability [7], e.g., subsystems can be developed independently and integration is done by simply plugging the subsystems together.
- The communication system has to support the connection of the replicas.
- The communication system has to support the distribution of the replicas to avoid common mode failures.
- Communication over the network has to be guarded by an independent device to avoid *babbling idiots*, e.g., one node permanently monopolizing the bus.
- Distributed control loops, e.g., braking, have to be synchronized and therefore require a common global time base in all participating nodes.
- A distributed application has always to know the current status of all connected components. The provision of this knowledge is called *membership service*. The communication system is responsible for providing a consistent membership service in each node.

- The communication system should support gateways to combine safety related and non-safety related applications as well as low and high speed control loops.

None of the commonly used in-vehicles communication systems (CAN [17], A-BUS, VAN, J1850-DLC, J1850-HBCC [2]) meet the requirements for safety related by-wire systems, because they were not designed for this case [11]. They are all lacking in being deterministic, in synchronization and fault tolerant characteristics.

These missing properties are the motivation for finding new approaches for in-vehicle communication systems. As a new start we examine the Time Triggered Protocol (TTP) developed by the University of Vienna and Daimler-Benz Research. TTP is especially designed for safety related applications and fulfills the mentioned requirements.

3.2 TTP OVERVIEW

TTP is an integrated time-triggered protocol that provides:

- prompt transmission of messages with high data efficiency,
- a membership service,
- a fault-tolerant clock synchronization service (global time-base),
- mode change support,
- error detection with short latency
- distributed redundancy management.

A comprehensive description of the TTP protocol is given in [6,7,8].

The overall TTP hardware architecture is characterized by both the TTP system architecture and the TTP node architecture as shown in Figure 2. A TTP real-time system consists of a host subsystem, which executes the real-time application and a communication subsystem providing reliable real-time message transmission. The interface between these subsystems is realized by a dual ported RAM (DPRAM) called *Communication Network Interface (CNI)* [9,10]. The assembly of host and TTP-controller is called *Fail Silent Unit (FSU)*. Two FSU´s form a single redundant *Fault Tolerant Unit (FTU)*. The physical layer consists of two independent transmission channels.

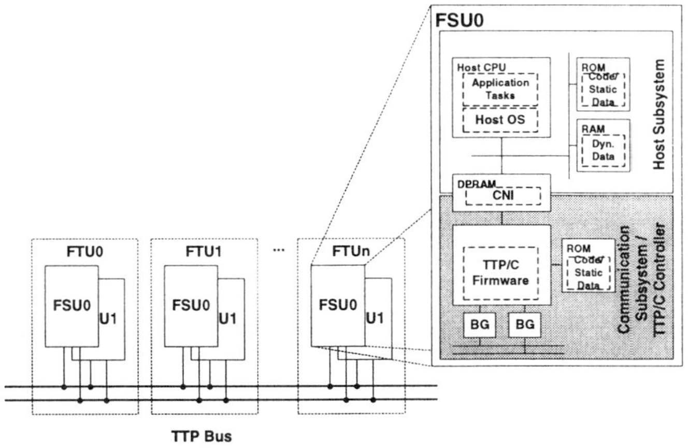

Figure 2: Architecture of a TTP-based Fault-tolerant Real-time System

In an event triggered architecture a host subsystem enforces the communication subsystem to send a message. On the contrary, in a time triggered architecture the communication system decides autonomously when to transmit a message. To do this the controller of the communication subsystem needs its own control data, and a message schedule, that specifies at which instant a message has to be transmitted by the controller. The message schedule is called *Message Description List (MEDL)* and is stored in every TTP controller [16].

A TTP controller hardware (see Figure 3) is supplied with an autonomous device called a *Bus Guardian (BG)* [22], which protects the busses from a timing failure of a controller by allowing access only at one predefined sending slot within a TDMA (time division multiple access) cycle. The bus guardian enables and disables the bus driver to avoid the disruption of concurrent data transmission when a node violates its transmission schedule. In order to fulfill this task the bus guardian obtains knowledge about the message transmission schedule, which enables it to act as an autonomous instance. Additionally, the bus guardian concept solves the babbling idiot problem.

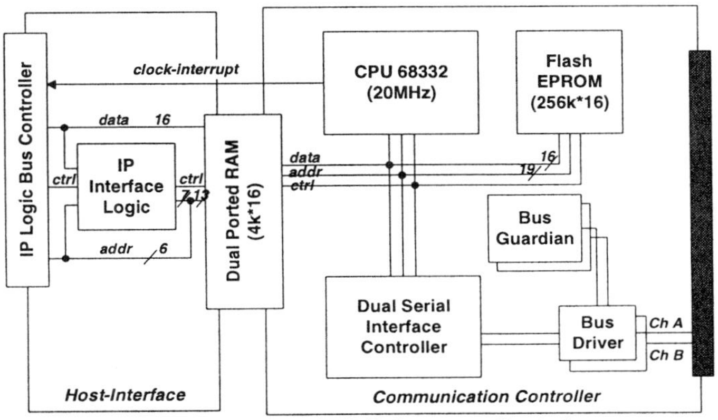

Figure 3: Block Diagram of the TTP/C Prototype Controller

3.3 NODE ARCHITECTURE

The node architecture is designed to be flexible and is based on the fail-silent assumption. The fail-silent assumption means that the output of a node has to be correct or otherwise the node has to be *silent*. A silent node does not interfere with the other system parts.

To reach the fail-silent property a node may consist of one fail-silent unit with fault detection mechanism in software. A node may be realized as a duplex system of two computers, which control each other. Also a node architecture based on a triple-modular-redundancy (TMR) approach is possible.

In applications with strict high fault-tolerant requirements two fail-silent nodes can be connected to form one fault-tolerant pair. Each single node is completely physically separated. If a single fault occurs in one of the nodes, the faulty node will become silent and the fault-free node will continue to deliver the intended services. The global time-base delivered by the communication system simplifies such implementations. Using two fail-silent nodes to tolerate one fault has a number of advantages:

- Efficiency; there is no need for explicit design mechanisms to tolerate faults. Each node has only to focus on the detection of faults. Once a fault is detected the node is shut down.
- Simplicity; by using the fail-silent property as the underlying concept for fault tolerance, a clean architecture with a minimum of cross coupling and special mechanisms like voters is obtained. Furthermore, the concept of a node that is either correct or silent is easy to communicate within the organization and to sub-suppliers and partners.
- Testability; the fail-silent property is a testable property since it is well defined. The testing can be concentrated on verifying the fault containment performance, e.g., test that a node does not violate the fail-silent property.

3.4 CURRENT STATE OF TTP/C COMPONENTS

The Technical University of Vienna and Daimler Benz Research have developed a prototype implementation of a TTP communication controller (see Figure 4), which is now available. These evaluation boards are used to verify the concepts of TTP-based applications. In this paper we present a case study for the investigation of the design of TTP-based real-time systems, e.g., brake-by-wire.

In 1997 the Esprit Project „Time Triggered Architecture (TTA)“ has been started with the intention to develop a prototype TTP controller chip. The first samples are expected to be available in the middle of 1998.

Figure 4: TTP/C Module

3.5 PROJECTABILITY OF TTP-BASED REAL-TIME-SYSTEMS

The projectability of TTP-based real-time systems mainly depends on the powerfulness and the efficiency of the available software development environment. A tool for the definition of the Message Description List (MEDL) [16] is available as well as a primitive C-language-based framework for application development. Tools for the design, integration and test of TTP systems are in the developing.

4. A FAULT-TOLERANT ARCHITECTURE FOR A BRAKE-BY-WIRE VEHICLE

Based on the requirement list in the second chapter our main focus is to develop an architecture that guarantees safety. Our intention is that after one arbitrary fault - with the exception of the faults of actuators - the full brake performance is still available. For this purpose we used redundancy in sensors, power supply, electric components, software, and communication. We do not use redundancy for the actuators. Only the loss of two actuators (e.g., loss of both right or both left actuators) may cause a safety critical situation. Table 1 shows an overview of the fault tolerant characteristics of our brake-by-wire case study.

„O“ means fully operational
„o“ means fully operational for a predefined time interval
„R“ means operational, but functionally reduced
„r“ means operational, but functionally reduced for a predefined time interval
„S“ means „Safe State“, e.g., vehicle stands.

Fail sources	Functional Degradation	Prototypical Realization
Actuators	FS	1 ECU
Sensors brake pedal	FO/FrS	3 sensors
Power supply	FO/FrS	2 batteries
Fault Tolerant Unit (FTU)	FoS	2 Fail Silent Units (FSU)
Communication	FO/FrS	2 busses

Table 1: Fault Tolerant Characteristics, Brake-by-wire Case Study

From our point of view the TTP protocol itself is less relevant. Our work concentrates on the development of safety related applications like brake-by-wire, based on a time triggered network. Therefore we have laid the main focus in the development of the *Brake-By-Wire Manager* and the *Power Manager* depicted in Figure 5, while the other subsystems are developed by other departments or outhouse partners.

4.1 OVERVIEW

Our fault tolerant architecture consists of a set of two redundant ECU's (electronic control unit) for both the *Brake-by-Wire-Manager (BBW-Manager, BBWM)* and the *Power-Manager (PM)*, and 4 single ECU's, one for each brake (see Figure 5). The ECU's are connected by two replicated busses. In this case study the brake ECU's are not designed redundant, in order to reduce costs, since the failure of a single brake is not considered to be as severe as the failure of the BBW-Manager, or the Power-Manager.

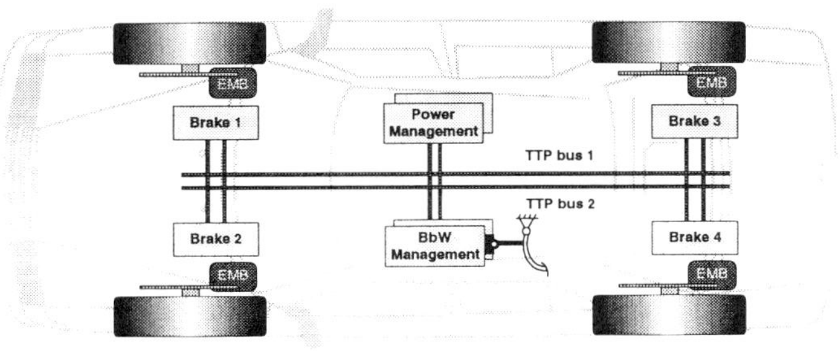

Figure 5: An Example for an EMB Architecture

The functionality of the BBW-Manager is to read the sensor values of the brake pedal, the revolution counters of the wheels, the yaw-sensor, the acceleration sensors, and to calculate from these signals the brake force set points for the four brake actuators. The BBW-Manager should also manage higher assistance functions like ABS, traction and driving dynamic control. The Power-Manager controls the charge of the batteries, proceeds active power management and monitors the status of the power supply. If the generator or one of the redundant power circuits fails or the charge of the batteries is going low the Power-Manager generates a warning signal. Each of the four brake actuators have one brake electronic to control the electric motor of the brake actuator. The brake electronics get the brake force set points from the BBW-Manager.

The whole communication between the BBW-Manager, the Power-Manager, and the brake electronics is based on the fault tolerant TTP bus.

4.2 SENSORS

In our research car we use three sensors, in order to get a verified pedal signal. We use the specification of the sensors, the history of the values and knowledge about maximal physical changes of values to detect sensor errors [15]. Both BBW-Manager get all three sensor signals, to fulfill the fail-operational assumption.

4.3 ACTUATORS

In our case study we use electromechanical actuators. If there is a failure in a brake actuator the brake has to pass into a fail safe state. This state is defined as *open brake*, with no brake torque. A closed actuator can cause a safety critical situation. We get the actuators from our outhouse partners.

4.4 POWER SUPPLY

The main reason for break-down of today vehicles are trouble with the power supply (transmission belt, battery, generator, starter, electrical consumers), [19,24]. The causes are really simple e.g., forget to switch off the lights. Usually the loss of power does not cause a safety critical situation. In the case of a by-wire vehicle a loss of power means the loss of the by-wire function e.g., loss of brake or steering function. The power supply for a by-wire system has to be very reliable. There are several requirements for power supply architectures for by-wire systems:

- Reliability; has to be increased in general. Two separated power supplies are needed to guarantee the by-wire function after one fault in a power circuit.
- Increase of the power capacity. The power requirements of by-wire systems could exceed the power capacity of today systems.
- Higher voltages; in order to reduce the current and to increase the efficiency of electric mechanical actuators.
- Power management; automatic shut down of not safety important consumers e.g., seatheater, in case of a lack of power.
- Charge control of the batteries.

Other new electric systems e.g., active body control, electric hydraulic steering, electrical air-condition, etc. also need some of these requirements. You find a good overview of requirements for future power supply systems in [20].

4.5 TTP COMMUNICATION

In contrast to event triggered systems a TTP system is built up by defining first the static message schedule. Figure 6 depicts the result of this phase, namely the communication matrix of the above example on a brake-by-wire network. The lower part of Figure 6 shows the static synchronous TDMA schedule and its constraint that each subsystem has to send exactly once in a TDMA cycle. The messages marked with 'I' are called I-Frames, used for reintegration of lost members and do not transmit information for the application layer. Hence, they are not shown in the upper part of the matrix, where the receivers of the sent messages are located.

The FSU's of the BBWM and the PM send their messages in vice versa time slots. In this way short burst errors can be recovered. A FSU slot has a length of about 1.2 msec. New brake force set points could be sent every 10 msec, which is needed for an ABS control loop. The brake control ECU's send their status and the current brake force. These messages are not so time critical as the transmission of the brake force set points. The brake ECU's send their messages only once in a cluster cycle, each 16 FSU slots. In the leaving FSU slots the brake ECU's send I-Frames for the network management.

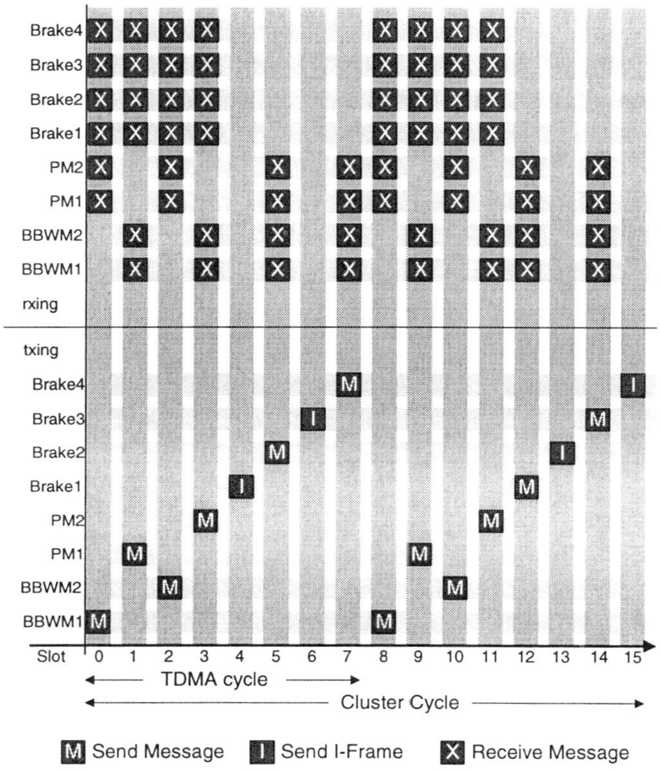

Figure 6: Communication Matrix of the Brake-by-wire Case Study

4.6 HARDWARE ARCHITECTURE

In our research study we use commercial off-the-shelf computers. We do not develop special hardware. We try to reach the fail-silent assumption for a single FSU by using software fault detection methods and low cost hardware mechanisms (such as watchdogs).

4.7 SOFTWARE ARCHITECTURE

In order to reduce the production costs of by-wire systems a minimum number of hardware should be used. In [5] it is shown that a high degree of fault detection can be reached with software fault detection mechanisms. We follow this approach by using software mechanisms to reach the fail-silent assumption.

4.7.1 Overview

The overall aim of our software architecture is to serve as a base for safety related applications. To reach this aim our software architecture is divided into three subsystems:

- Communication Subsystem; this part is responsible for the communication between distributed components.
- Fault Tolerant Subsystem; this part contains safety critical and fault tolerant mechanisms. The safety related application is handled by this subsystem.
- Safety Related Application; this part includes the safety related tasks, which build the application.

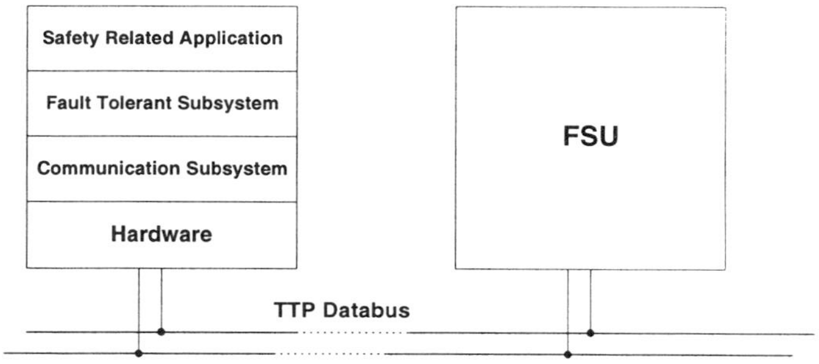

Figure 7: Software Levels of the TTP System

4.7.2 Communication Subsystem

The host subsystem must read and write the new messages synchronous to the TDMA cycles of the TTP communication system. Also the host has to update a TTP watchdog every TDMA round to show the TTP communication system, that he is still alive. The communication subsystem guarantees that the operating system is synchronous to the TTP communication system and handles additionally the services, provided by the TTP (membership, user interrupts, time interrupts, error interrupts).

4.7.3 Fault Tolerant (FT) Subsystem

The concept of TTP mainly depends on fulfilling the fail-silent assumption. The benefit of the fail-silent assumption is that faults are isolated in the single components and must not be handled on the system level. In accordance with our fault model we want to tolerate one arbitrary fault, without effects on the application performance. After a fault is detected the faulty FSU is rebooted and tries to reconfigure.

The safety related application is running exclusively under the control of the FT subsystem. The functionalities of the FT subsystem can be divided into three main groups:

- The *Prime-Level Scheduler* (PLS); is a special scheduler, which handles the tasks that form the safety related application. The FT-Tasks are nonpreemptive and independent of each other in the sense that they do not share common resources on an ECU. They are only activated by the progression of time and are repeated periodically. Especially for digital control loops these properties are quite suitable for the development of applications.
- Data Consistency Checks; this functionality ensures that the read and write access to certain data structures are monitored in order to protect the data structures against undesired changes.
- Replicated Execution; allows to detect transient and permanent faults. This mechanism is used to protect safety related tasks against faults.

4.7.4 Software States of the FT Subsystem

Figure 8 show the different software states of the FT subsystem. After *Power Up* the FSU is in the *Initializing* state. In this state the devices are initialized and the host subsystem tries to synchronize itself to the TTP communication system. After the correct initialization and synchronization the FSU switches into the *Application* state. The host and communication subsystem are now synchronous and the Prime-Level Scheduler (PLS) starts to execute the application tasks. If a fault is detected the FSU switches into the *Exception Handling* state. The exception task sets all output channels into a safe state. After this the FSU reboots, in accordance with the fail-silent assumption. If the FSU is initialized in a correct way after a reboot, the node tries again to reintegrate in the TTP communication system and tries to switch into the *Application* state. If the system can not be initialized correctly, a permanent fault is diagnosed. In this case a the fault conditions are stored permanently and the FSU switches into the *Shut Down* state. In some cases, faults can occur permanently after the system has initialized correctly. To discover these faults additionally a fault counter is incremented after every detected fault and is saved permanently [25]. If the fault counter has exceeded a limit in a defined time interval a permanent fault is also diagnosed and the FSU switches into the *Shut Down* state.

A fail of the TTP communication is handled in the same manner. The FSU switches into the *Exception Handling,* the FSU is rebooted and tries after successful initializing of the devices to synchronize to the TTP communication system. If the FSU can synchronize to the TTP communication system the FSU switches into the *Application* state.

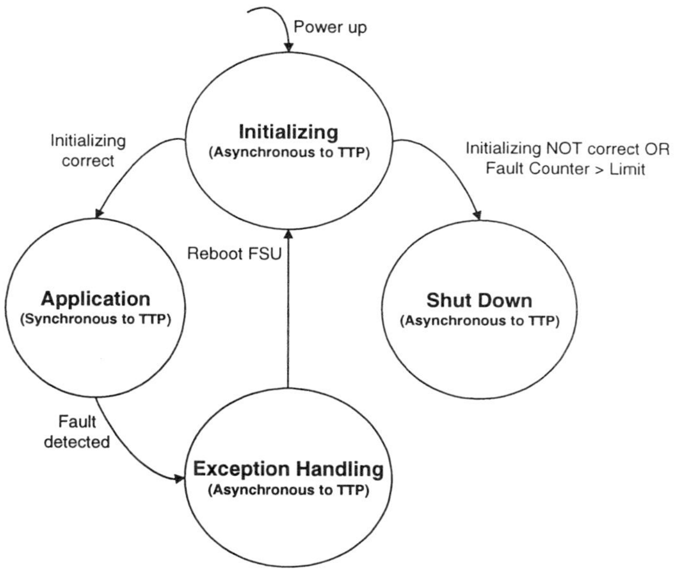

Figure 8: Software States of the FT Subsystem

4.7.5 The Prime-Level Scheduler (PLS)

The PLS is responsible that the time constraints of the FT-Tasks are met [12] and observes the deadline violations of the safety related tasks. The PLS is a time driven scheduler, which is integrated in a special interrupt service routine (ISR). This ISR is linked exclusively to the highest interrupt level of the processor, which ensures that it will be never delayed due to any other CPU activity.

To monitor the time constraints of the FT-Tasks the PLS needs several parameters of the FT-Tasks. These parameters are:

- The period (starttime).
- The offset, which denotes a delay in relation to the beginning of the period.
- The deadline in relation to the earliest possible starttime of the FT-Task.
- The name of the FT-Task.

These parameters will not change over the operational phase of the system. Due to this fact a static scheduling, which is generated off-line is used. For FT systems this has the advantage that a deviation of the system from the predefined behaviour can be detected in real-time. Figure 9 illustrates the functionality of the PLS.

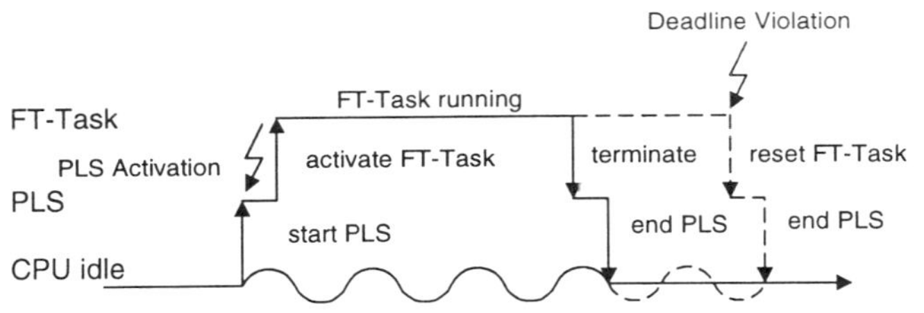

Figure 9: PLS Functionality (the dotted Lines illustrate the faulty Case)

By using the PLS in combination with the TTP communication system the interrupts, which activate the PLS are driven through the TTP system. This feature makes it easy to realize Replica Determinism [4], the property of replicated nodes to perform the same state changes at about the same time. Replica Determinism is needed to realise *hot* redundancy with only two FSU's.

4.7.6 Data Consistency Checks

For safety related software systems we use different levels of data protection (see Figure 10):

1. All safety related variables (variables which are used within the safety critical computation path) are stored with an additional proof variable, depending on memory resources. This proof variable can be a second instance of the safety related variable itself or a generated CRC (cyclic redundancy check) [23] of the variable, which shall be protected (variable level).

2. All TTP application messages are protected by an additional CRC transmitted within the last 2 bytes of each message because that information is mainly exposed to dangerous disturbances during the transmission phase on the bus. On the one hand, the protection of the remaining 14 application data bytes of a message is increased. On the other hand, this method asserts data integrity on the path from the transmitting host memory to the receiving host memory (message level).

3. The TTP messages, which are sent via the TTP bus are protected by the protocol CRC. The TTP message frame (N-Frame) consists of a frame header having a length of 4 bits and an application data field whose length can be from 2 bytes up to 16 bytes. The frame header and the application data field is protected by a 2 byte CRC (protocol level).

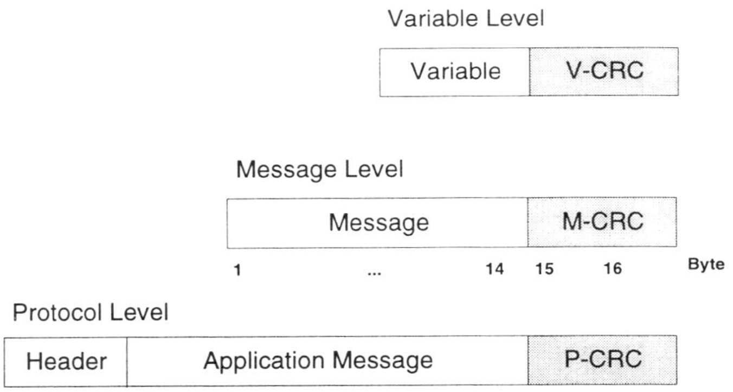

Figure 10: Different Levels for Data Protection

4.7.7 Replicated Computing

Replicated computing is a fault discovery mechanism, which allows the detection of transient or permanent faults as well as the correction of discovered faults. The ladder needs 4 computations of a task while pure detection manages on only 3 computations. All tasks within the safety critical path are executed replicated. The method is illustrated in Figure 11 (fault detection, no correction). In a first step an output data vector d_out_1 is computed for the current input data vector d_in_1, followed by the computation of the vector d_out_2 for an apriori known reference input data vector d_ref_in. The reference output data of d_ref_in is d_ref_out, which is separately stored. Finally d_out_3 is computed using the d_in_1 vector.

Based on output vectors d_out_1, d_out_2, d_out_3 and d_ref_out, a voting process is executed. Between the following cases is distinguished:

1. d_out_1 = d_out_3 AND d_out_2 = d_out_ref: no fault occurred: result is d_out_1

2. d_out_1 NOT = d_out_3 AND d_out_2 = d_out_ref: transient fault during phase 1 or 3

3. d_out_1 = d_out_3 AND d_out_2 NOT = d_out_ref: permanent fault during phase 1 to 3

4. d_out_1 NOT = d_out_3 AND d_out_2 NOT = d_out_ref: multiple fault

Depending on these faults and the application requirements for safety and availability, actions like task shut down or system reboots are initiated depending on the current application state.

Fault correction would require a 4th computation of the task with a modified voting afterwards. Case 2 in the above list would result a voted output vector.

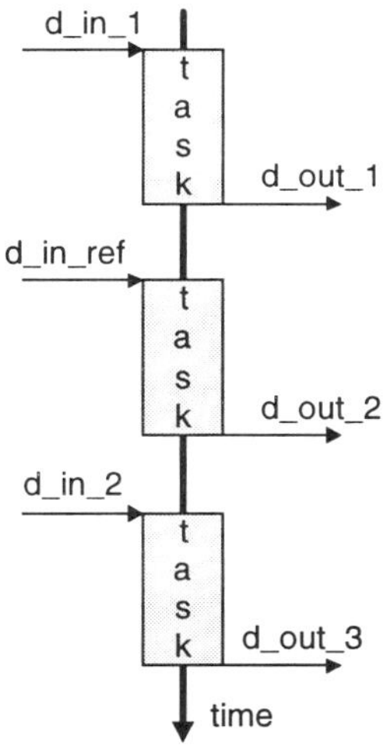

Figure 11: Replicated Computations to protect the Correctness of Computations

A critical part is the voter task, which decides the correctness of replicated execution. If the voter has a failure e.g., decides that correct data are incorrect and vice versa while the remaining tasks work correctly the faulty voter decides that correct data are faulty. In this case a fault is detected and the FSU is rebooted. To cause a critical situation an additional second fault (e.g., the brake force control calculates wrong data) must have arisen at the same time. This attempt is only correct if no common mode fault disturbs voter and application tasks in the same manner.

4.7.8 Safety Related Application

As one example of a component with fail-silent property we describe in this chapter the realization of the BBW-Manager. The other components (Power-Manager and the four brake ECU's) are realized in a similar manner. The BBW-Manager has the duty to calculate the four brake force set points. The brake force set points are safety related and has to be protected from transient and permanent faults. Therefore we use the different mechanism described in the chapters above.

Figure 12 shows the schedule, which is periodically executed through the PLS in the *Application* state on the BBW-Manager:

1. Pedal Signal Measurement; of the pedal signals from the three pedal sensors.
2. Pedal Signal Plausibility Checks; from the three pedal signals one valid value is calculated. This task is executed three times to detect faults.
3. Voter; a voter task votes from the result of the three plausibility check tasks.
4. Brake Force Control; the brake forces for the four actuators are calculated. This task is also executed three times.
5. Voter; a voter task votes from the result of the three brake force control tasks.
6. TTP-Communication; the brake forces are send to the brakes via the TTP communication network.
7. Diagnose; a diagnose task is executed.
8. Diagnose Output; the diagnose values are transmitted to an external diagnosis device.

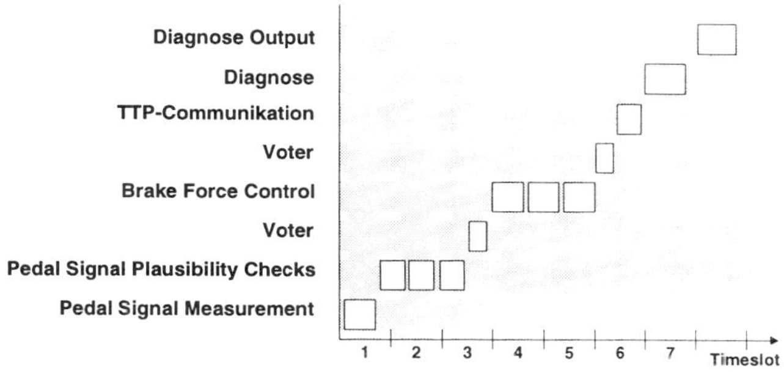

Figure 12: Local Task Schedule of the BBW-Manager

The Pedal Signal Measurement and the Brake Force Control are safety related tasks, which are executed three times to discover faults. Both tasks are safety related, because they manipulate data variables. The other tasks (Pedal Signal Measurement, TTP-Communication, Diagnose and Diagnose Output) only handle messages, which are protected through CRC's. A critical part is also the Voter task.

5 CONCLUSION

In this paper we show a way, in which a brake-by-wire application without mechanical or hydraulic backup can be realized by using the TTP communication system. In detail we describe the realization of the Brake-By-Wire Manager. We did not describe the realization of the other components (sensors, actuators, power supply, brake ECU's, Power Manager, and overall safety strategies) in detail, because that would exhaust the volume of this paper.

The concept of TTP depends on the fail-silent property of the communication nodes. We describe several software mechanisms (Prime-Level Scheduler, Data Consistency Checks and Replicated Computing), which help to fulfill the fail-silent property.

TTP is also in discussion in the Brite-EuRam Project „Safety Related Fault Tolerant Systems in Vehicles" (acronym: „X-By-Wire") [3] to be proposed as a European or International Standard. The X-By-Wire consortium consists of the following partners: Daimler-Benz Research, Fiat Research Center, Ford Europe, Volvo, Bosch, Magneti Marelli, Mecel, University of Chalmers and the University Vienna of Technology.

In parallel, in 1997 the Esprit Project „Time Triggered Architecture (TTA)" has been started with the intention to develop a prototype TTP controller chip. Other aims of the TTA project are the development of tools for the design of TTP systems and the formal verification of parts of the TTP protocol e.g., the clock synchronization algorithm. Partners of the TTA project are: Daimler-Benz Research, British Aerospace Airbus, Alcatel Austria, AMS, Genias Software GmbH, University Vienna of Technology, University of Ulm and the University of York.

REFERENCES

1. SAE, „Class C Application Requirement Considerations", SAE Recommended Practice J2056/1, SAE, June 1993.
2. „Survey of Known Protocols", SAE Information Report J2056/2, SAE, April 1993.
3. E. Dilger, L.A. Johansson, H. Kopetz, M. Krug, P. Lidén, G. McCall, P. Mortara, B. Müller, U. Panizza, S. Poledna, A.V. Schedl, J. Söderberg, M. Strömber, T. Thurner: „Towards an Architecture for Safety Related Fault Tolerant Systems in Vehicles", Proceedings of the ESREL´97 International Conference, 1997.
4. S. Poledna, „The Problem of Replica Determinism", Fault-Tolerant Real-Time Systems, Kluwer Academic Publishers, 1996.
5. J. Karlsson, P. Folkesson, J. Arlat, Y. Crouzet, G. Leber, „Integration and Comparison of Three Physical Fault Injection Techniques", Predictably Dependable Computing Systems, Springer Verlag 309-329, 1995.
6. H. Kopetz, et. al., „A Prototype Implementation of a TTP/C Controller", Proceedings SAE Congress 1997, Detroit, MI, USA, Febr. 1997. Society of Automotive Engineers, SAE Press. SAE Paper No. 970296.
7. H. Kopetz, „Real-Time Systems - Design Principles for Distributed Real-Time Systems", Kluwers Academic Publishers, 1997.
8. H. Kopetz, G. Grünsteidl, „TTP - A Protocol for Fault-Tolerant Real-Time Systems". IEEE Computer, pages 14-23, January 1994.
9. A. Krüger, „Interface design for Time-Triggered Real-Time System Architectures", Ph.D. thesis, Institut für Technische Informatik, Vienna University of Technology, 1997.
10. A. Krüger, H. Kopetz, „A Network Controller Interface for a Time-Triggered Protocol", SAE Symposium on Future Transportation Electronics: Multiplexing and In-Vehicle Networking, SAE, 1995.
11. Markus Krug, Anton V. Schedl, „New Demands for Invehicle Networks", Proceedings of the 23rd EUROMICRO Conference, 601-606, 1997.
12. Markus Krug, „FondOS: Making OSEK Based Operating Systems Fault-Tolerant", 11th International Symposium on High Performance Computing Systems HPCS´97, 1997.
13. BOSCH, „Kraftfahrtechnisches Taschenbuch", 22. Aufl. VDI-Verl., 1995.
14. Jean-Claude Laprie, „Dependability - Its Attributes, Impairments and Means", in B. Randell, J.-C. Laprie, H. Kopetz and B. Littlewood, editors, Pridictably Dependable Computing Systems, chapter II, pages 27-40, Springer Verlag, 1995.
15. Keith Marzullo, „Implementing Fault-Tolerant Sensors", 10th Real-Time Systems Symposium, Los Angles, December 1989.
16. R. Nossal, „MEDL/MODL Specification", Technical Report 25/95, Institut für Technische Informatik, Vienna University of Technology, 1996.
17. Control Area Network, „An Invehicle Serial Communication Protocol", SAE Information Report J1583, SAE Handbook, 1990.
18. Jürgen Balz, Karlheinz Bill, Jürgen Böhm, Peter Scheerer, Martin Semsch, „Konzept für eine elektromechanische Fahrzeugbremse", ATZ Automobiltechnische Zeitschrift 96, 1996.
19. Pannenstatistik 1995, „ADAC-Motorwelt", Ausgabe 5/96.
20. R. Schöttle, D. Schramm, J. Schenk, „Future power supply systems for cars", VDI Berichte No. 1287, 1996.
21. Siemens Automobiltechnik ATMK, „Fahrwerk-Systeme, Systeme für sicheres Fahren", 1997.
22. Ch. Temple, „Bus Guardian Principle of Operation TTP/C IP V1 Prototype", Technical Report, Institut für technische Informatik, Vienna University of Technology, Vienna, Austria, 1996.
23. W. Wesley Peterson, E.J. Weldon Jr., „Error-Correcting Codes", MIT Press, Second Edition, 1981.
24. A. Williams, J. Smith, H. Evans, J. Scholfield, J. Elvidge, „Advanced Electrical Architecture Vehicle Development", Automotive Engineering, 1995.
25. A. Bondavalli, S. Chiaradonna, F. Giandomenico, F. Grandoni, "Discriminating Fault Rate and Persistency to Improve Fault Treatment", in 27th Int. Symp. on Fault-Tolerant Computing (FTCS-27), (Seattle, WA), pp 354-362, IEEE Comp. Soc. Press, June 1997.

Development of Hydraulic Booster Unit for Active Brake Control

Yasuo Konishi, Masayoshi Hattori, Masakazu Sugisawa and Michiharu Nishii
AISIN SEIKI Co., Ltd.

ABSTRACT

Active brake controls such as vehicle stability control, VSC, panic brake assist, PBA, etc. have recently been applied to many vehicles. We have developed a new hydraulic brake actuator for active brake control systems. This new hydraulic brake actuator is composed of hydraulic brake booster unit, power supply unit and control valve unit. This paper describes the design of the new hydraulic brake booster unit.

INTRODUCTION

In recent years, chassis control systems for active safety, which improve vehicle stability or assist driver's operation, have been installed on many passenger vehicles. To achieve these active brake functions, an integrated hydraulic actuator including the hydraulic brake booster unit has been proposed and put into practical use [1][2]. We have further developed the second generation of hydraulic brake actuator with small size and light weight.

AIM OF DEVELOPMENT

We have developed the hydraulic brake booster unit focusing on the following objectives :

(1) High brake boost ratio and fast pressure response

-To improve brake pressure response and ensure assist force in the high deceleration region using the accumulator pressure

(2) To ensure flexibility of brake booster characteristics

- Brake characteristics flexibility equivalent to the conventional vacuum booster regarding brake boost ratio and jump-up characteristics

- Development of variable boost mechanism for adaptive brake characteristics

(3) To be easily applied to vehicle control systems

(4) To minimize actuator size

-To create a unit that is adaptable to a wide range of vehicles.

HYDRAULIC UNIT CONFIGURATION

The outline of the hydraulic brake actuator including VSC and PBA functions is shown in Fig. 1. The hydraulic construction of the brake actuator is shown in Fig. 2. The actuator is divided into three sections: hydraulic brake booster, power supply and control valve. Minor modifications can be made for adapting to required functions and designs.

The brake booster unit, master cylinder and regulator are placed on the same axes. The power supply unit is composed of pump, electric motor and high pressure accumulator. This unit supplies the required pressure to the brake booster.

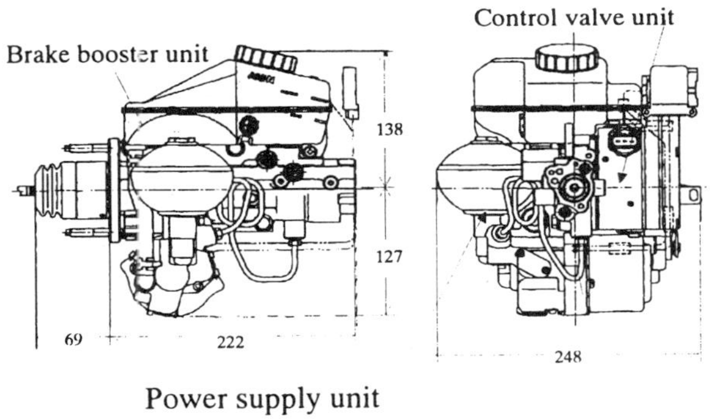

Figure 1. Outline of hydraulic brake actuator

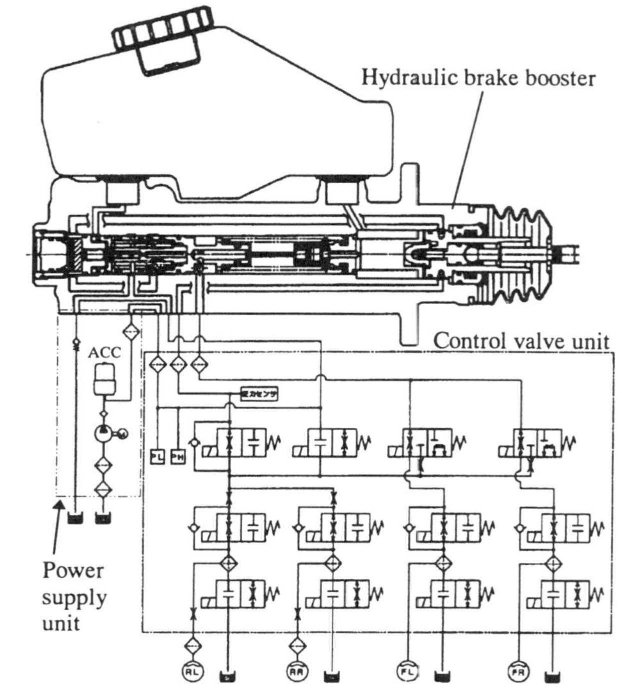

Figure 2. Construction of hydraulic brake actuator

The control valve unit contains switching and control solenoid valves. Master cylinder pressure, power pressure, regulated through the hydraulic brake booster, or accumulator pressure is selected through the switching solenoid valves to apply brake pressure to the wheel cylinders. The control solenoid valves optimally control the brake pressure of each wheel. This configuration can provide active brake pressure to each wheel cylinder.

GENERAL FEATURES OF THE HYDRAULIC BRAKE BOOSTER

The hydraulic brake booster generates a hydraulic assist force according to the brake pedal force applied by the driver. Figs. 3 and 4 show brake booster characteristics and brake pressure response, respectively, compared with the conventional vacuum booster. The hydraulic brake booster has higher brake boost ratio and faster brake pressure response since it utilizes high accumulator pressure.

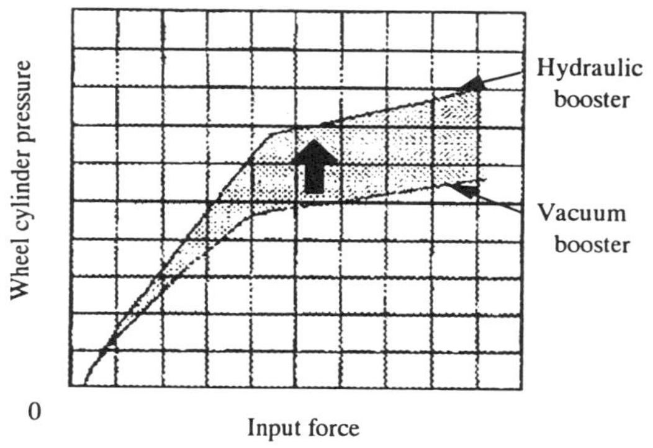

Figure 3. Brake characteristics

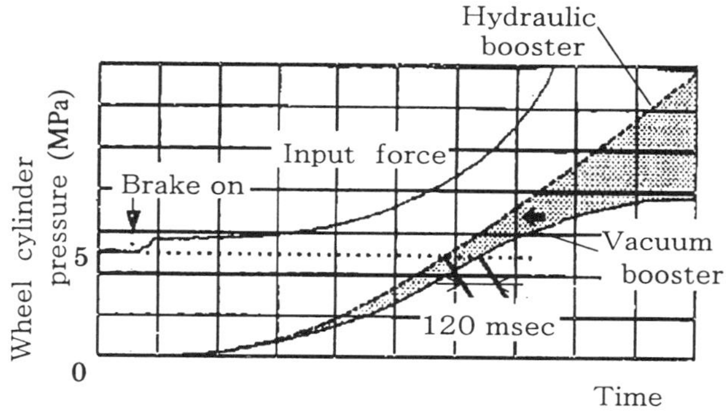

Figure 4. Brake pressure response

DEVELOPED HYDRAULIC BRAKE BOOSTER

Fig. 5 shows a basic configuration of the hydraulic brake booster. To make the brake booster unit simple and compact, the regulator was placed in the same bore as the master cylinder. The power pressure, which is related to the master cylinder pressure through the regulator, is fed back to the booster chamber which is located behind the master cylinder. In addition, the front brake circuits are connected to the master cylinder chamber and the rear brake circuits to the booster chamber. Therefore, two separate hydraulic circuits (front / rear split) with a shortened brake pedal stroke are created.

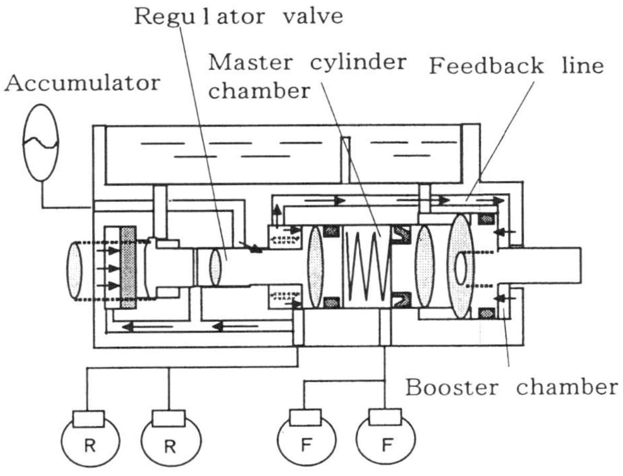

Figure 5. Hydraulic brake booster unit

The regulator is composed of regulator piston, spool valve and reaction providing mechanism. Fig. 6 shows the operation of the regulator. When the driver operates the brake pedal, the spool valve advances with the regulator piston. The accumulator pressure is controlled by the spool valve and introduced into the regulator chamber. The regulated pressure is supplied into the booster chamber via the feedback line to generate assist force.

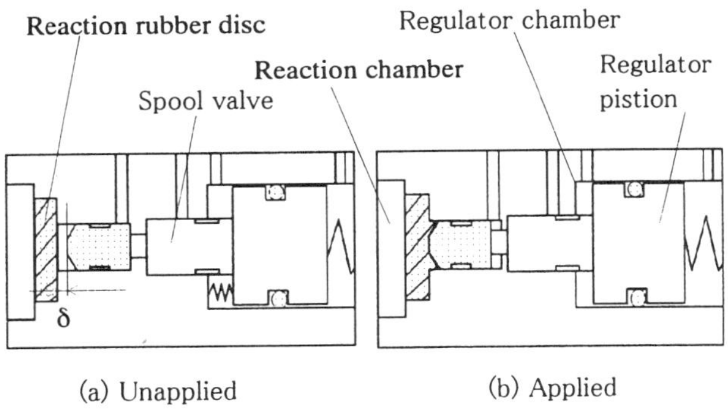

Figure 6. Regulator operation

The reaction providing mechanism, which is located at the end of the regulator spool valve, provides flexibly of the brake booster characteristics. The regulated pressure is also supplied to the reaction chamber. The reaction rubber disc becomes deformed and fills up the space between the disc and plunger. As the space is filled up, the regulated pressure pushes back against the plunger and the boost ratio is changed. This uses similar principle to the current vacuum booster design to provide boost gain and jump-up flexibility.

The above-mentioned design makes the brake actuator simpler and more compact compared with conventional booster/MC units (Fig. 7). This allows the hydraulic brake actuator to be easily packaged in the vehicle.

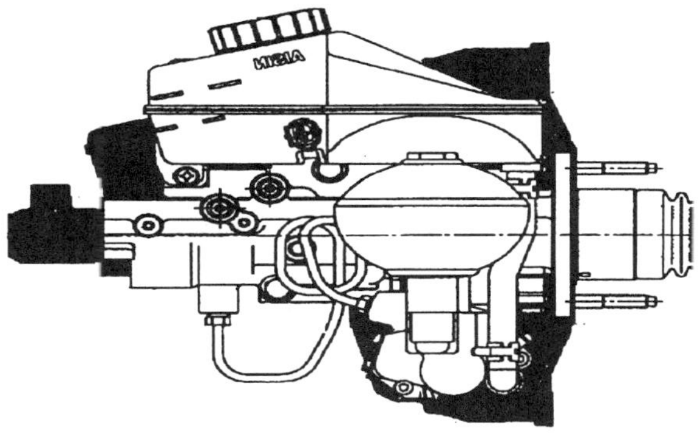

Figure 7. Comparison between hydraulic brake actuator and conventional master cylinder with vacuum booster

BRAKE BOOSTER CHARACTERISTICS

1 . Hydraulic brake booster model

Fig. 8 shows the hydraulic brake booster model for calculating the brake booster characteristics. The symbols in the figure are:

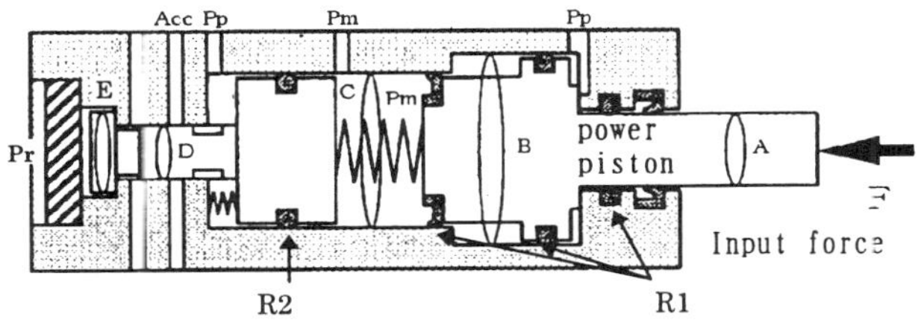

Figure 8. Model of hydraulic brake actuator

A: Piston rod area

B: Piston area

C: Regulator piston area

D: Spool valve area

E: Reaction rubber disk contact area

Pm: Master cylinder pressure

Pp: Power pressure regulated through the regulator

Pr: Reaction chamber pressure

a: Pressure ratio between master cylinder and reaction chambers ($Pr = \alpha * Pp$)

R1: Power piston sliding resistance coefficient

R2: Regulator piston sliding resistance coefficient

The boost ratio, $\Delta Pm/\Delta F$, is approximately obtained with the following equations based on the force balance between the power piston and regulator piston.

$$\frac{\Delta Pm}{\Delta F} = \frac{1}{C - \beta \cdot (B - A - R1)} \qquad \text{(Eq. 1)}$$

where,

$$\beta = \frac{C}{C - D + \alpha \cdot E + R2} \qquad \text{(Eq. 2)}$$

2. When reaction providing mechanism is unapplied (in the case of $\alpha = 0$)

Fig. 9 (a) shows brake characteristics when the reaction providing mechanism is unapplied (pressure ratio, $\alpha = 0$). The boost ratio, ΔPm/ΔF, is constant through the entire hydraulic range. As can be seen from the equation (1), the area ratio of A, B, C, and D determines the boost ratio, ΔPm/ΔF. It would be necessary to change the areas to change the boost ratio, ΔPm/ΔF without the use of the reaction providing mechanism.

3. When the reaction providing mechanism is applied (in the case of $\alpha = 1$)

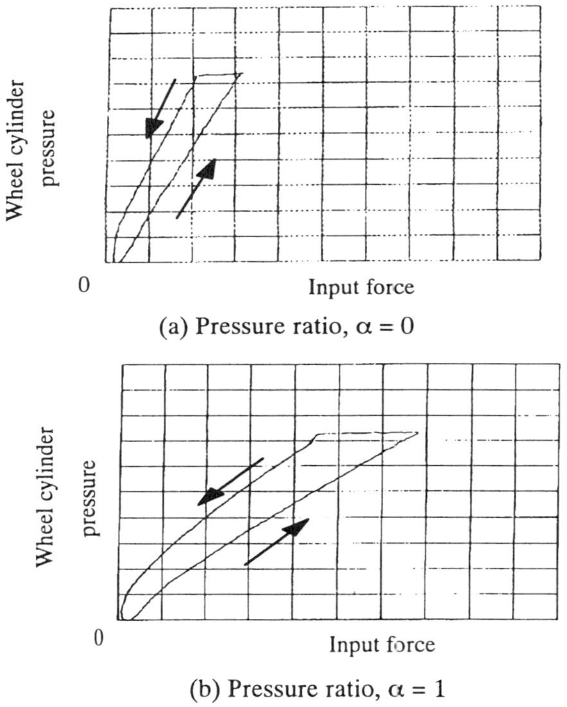

Figure 9. Brake characteristics with and without reaction providing mechanism

Fig. 9 (b) shows brake characteristics when the reaction chamber and master cylinder chamber are constantly coupled (pressure ratio, $\alpha = 1$). In this case, the power pressure is directly applied to the reaction rubber disc. The boost ratio can be changed according to the hydraulic pressure since the reaction rubber disc deforms according to the reaction chamber pressure and the contact area E is gradually changing. Deformation of the reaction rubber disc due to the reaction chamber pressure are shown in Fig. 10.

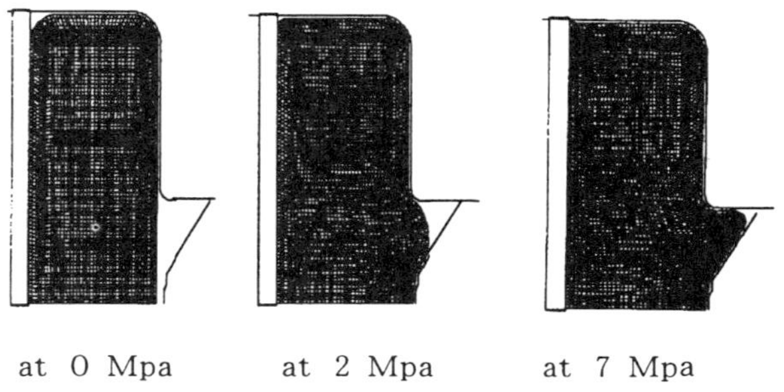

Figure 10. Reaction rubber disc deformation due to pressure (Simulation)

Even if areas A and D are constant, the brake characteristics can be adjustable by changing area E and the clearance between the plunger and reaction rubber disc, δ (Fig. 6). Thus, adaptability of the brake pedal feel can be adjusted so that reaction feel is provided in the medium to high deceleration regions while brake effectiveness is ensured at low deceleration.

APPLICATION TO VARY BOOST MECHANISM

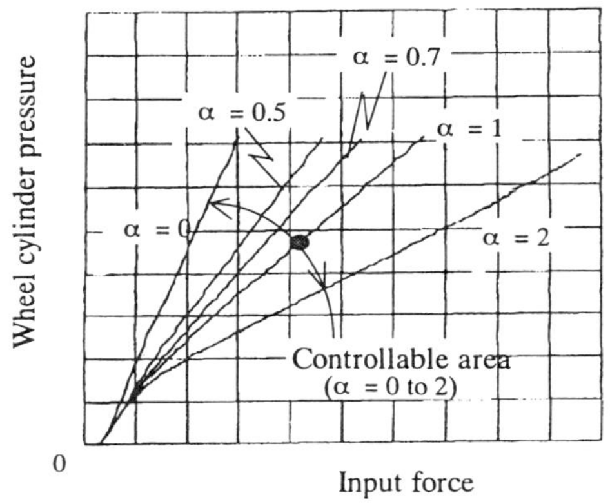

Figure 11. Brake characteristics when pressure ration, a, is controlled

Fig. 11 shows brake characteristics when pressure ratio, α, is changed. Any boost characteristics can be selected by changing pressure ratio, α. Brake characteristics can be controlled during braking by managing the reaction chamber pressure, Pr. Fig. 12 shows variable boost ratio construction. The reaction chamber pressure, Pr, is set by feeding back power pressure, Pp. Therefore, the brake effectiveness can be actively controlled according to the vehicle state such as the loaded weight, road surface friction coefficient, and pad friction coefficient changes to achieve the optimum braking effect.

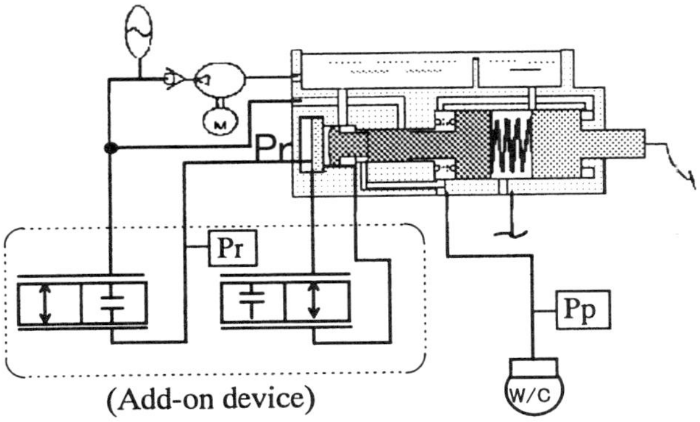

Figure 12. Variable boost ratio construction

REFERENCES

1. Y. Yasui, et. al. : "Improvement of the Vehicle Directional Stability for Transient Steering Maneuvers using Active Brake Control", SAE960485, 1996
2. M. Nishii : "High Pressure Control Technology in Hydraulic Booster Operating Mechanism", Nikkei Mechanical, Nov. 2, 1992

980238

Controlled Brake Component Development Process

Huang-Tsang Chang, Gary C. Fulks and Jeffrey A. Heinrichs
Delphi Chassis Systems

ABSTRACT

A system engineering process has been applied to brake system hydraulic components, allowing rapid prototype development with minimal hardware iterations. Early application and validation of math based models, together with the use of 'flexible' hydraulic configurations, provides for a wide range of functionality. This allows the limits of performance to be established early in the development process. The process has been used to develop a variety of brake system hydraulic components, including proportional valves, pressure relief valves and pumps.

INTRODUCTION

To remain competitive in today's automotive market, one must be able to quickly develop new technology at the lowest possible cost, with maximum confidence. A process developed by the authors to meet this requirement is outlined in Figure 1. From the initial vehicle level specifications, system modeling identifies key subsystem or component design parameters. From these parameters, modeling and simulation further define the set of sub-components that will be capable of meeting the system requirements.

Once a reasonable degree of confidence is reached, initial prototype sub-parts are built, not for complete component test, but predominantly for model confirmation. It is generally faster and more cost effective to build and modify models than physical parts. Additionally, the constraints of high volume, low cost production design preclude the ability to test and measure internal parameters (e.g., force, pressure, etc.). This drives the need for sub-component modeling and prototype testing.

The temptation to rush ahead and build production-like parts and wait for the model to follow is great. This should be resisted. Modeling in advance of prototype parts supports development, while modeling and analysis after prototypes are produced generally supports failure analysis. Once we have a complete set of sub-component parts and corresponding math based models, there will be a greater degree of confidence that the final production components will perform within defined limits.

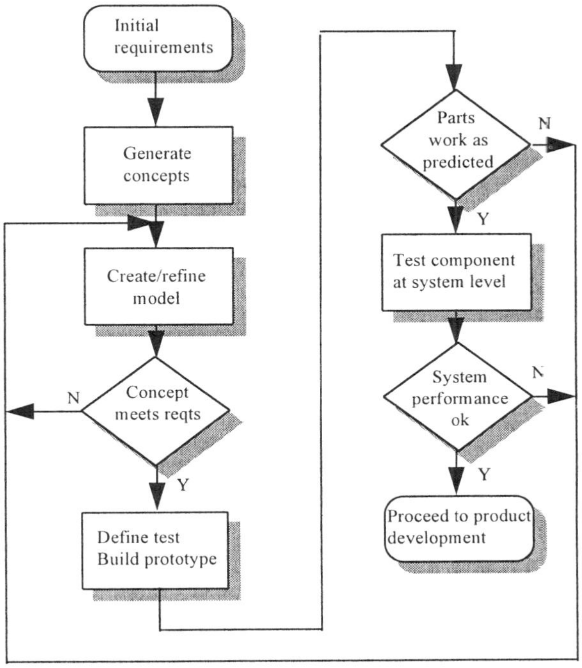

Figure 1. Concept Development Process

SYSTEM SIMULATION

Due to interaction among different physical systems, mathematical modeling and simulation of a typical hydraulic control device (such as an ABS solenoid valve) is complicated and highly nonlinear. Figure 2 shows the block diagram expressing the interaction between armature dynamics, electro-magnetics and fluid dynamics for an ABS solenoid valve.

Consider a proportional solenoid valve during steady state "pressure control mode". A hydraulic flow pulse causes the armature to move. This introduces a disturbance to the magnetic field, causing a change of electric current, and then affecting the dynamics of the armature. Any approach that represents the electric, hydraulic and mechanical systems independently, does not effectively take into account the non-linear interaction between them.

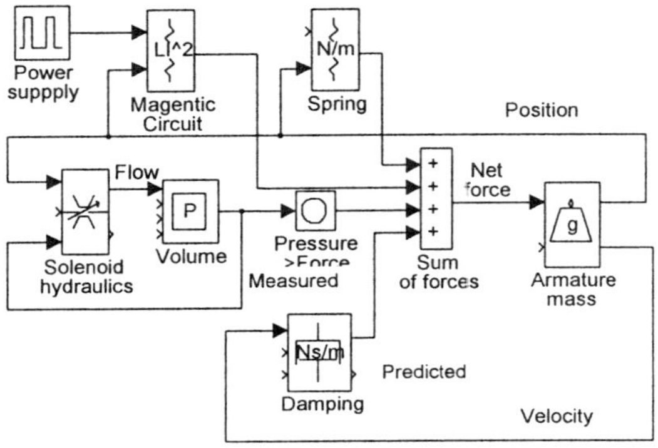

Figure 2. Mixed physics valve model

A significant influence on the simulation technique employed by the authors is the availability of symbolic tools that may be used to represent non-linear differential equations. In the same way that electronic circuits may be represented by a wiring diagram, a mixed physics system may be represented by linking the energy/work flow paths between symbols. Figure 2 shows the network of connections that link the flow and distribution of energy both internal and external to the valve. The mass of the plunger unifies the model as the nodal point for all forces. It interacts with each energy field through its displacement, velocity, and acceleration.

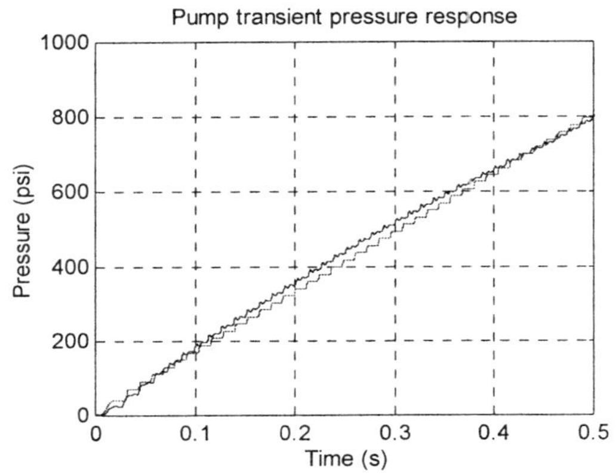

Figure 3. Simulation validation

The characterization of hydraulic flow, magnetic field, mechanical dynamics and damping forces is complex. Theoretical formulation can be cumbersome. Invariably it over-simplifies the system and may not be feasible in many applications. The technique utilized by the authors combines Finite Element Analysis (or test data) for electromagnetic and hydraulic characterization, and mathematical modeling to achieve a simulation tool that thoroughly describes the nonlinear mixed physical system behavior. An example of mixed physics simulation applied to an ABS pump can be seen in Figure 3. The accuracy of such models effectively streamlines the hydraulic component design process.

SYSTEM EQUATIONS

In the electro-magnetically energized, hydraulically driven mechanical system, the electro-magnetic field is described with Equation (1). The dynamics of the mechanical component (such as the armature in a solenoid valve) are described in Equation (2).

$$Li'_t + iL'_x\dot{x} + ri = V \qquad (1)$$

$$M\ddot{x} + b\dot{x} + k(x - l_o) = F_e + F_h \qquad (2)$$

where ***M*** is the armature mass, ***k*** is the return spring constant, ***b*** is the viscous coefficient, ***L*** is the inductance, l_o is the initial plunger position measurement, ***r*** is the coil resistance, ***x*** is the solenoid valve state (armature's position, velocity), ***i*** is the electric current, and ***V*** is the supply voltage. On the right hand side of Equation (2), F_e is a function of the armature state and electric current, it takes into account the magnetic force, and can be computed via Equation (3).

$$Fe = \frac{1}{2} i^2 L'_x = \frac{\mu_o a N^2 i^2}{2(x - l_i)^2} = F(x, i) \qquad (3)$$

where μ_o is the permeability of free space, ***a*** is the flux path cross section area (m^2), l_i is the maximum plunger travel position measurement, and ***N*** is the number of turns in the coil. During the course of the experimental development, the magnetic force transfer function can be accurately integrated with the system simulation model.

The second term on the right hand side of Equation (2), F_h, characterizes the fluid induced flow force through the valve opening. It can be further described by Equation (4).

$$F_h = pA_f + F_{hf}(q, \mu, x, \dot{x}) \qquad (4)$$

The first term on the right hand side of Equation (4) represents the hydraulic force caused by the static pressure differential, and the second term, F_{hf}, accounts for the force induced by the fluid flow. The latter is a function of flow rate, fluid viscosity, and the armature state. For a simple ball type valve as shown in Figure 4, F_{hf} can be approximated using Equations (5-7)

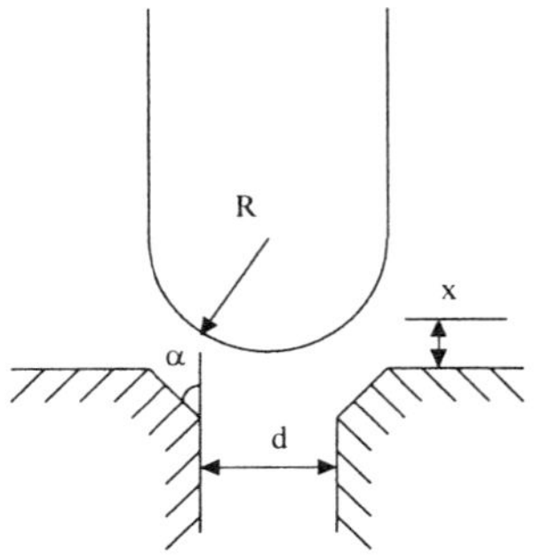

Figure 4 Ball type valve

Effective flow area: $A = \pi x \sin 2\alpha \left(\frac{2R + x\sin\alpha}{2} \right)$ (5)

Flow: $$q = C_d A \sqrt{\left(\frac{1}{2\rho}(P_1 - P_2) \right)} \quad (6)$$

where ρ is fluid density. Force is defined by:

$$F_{hf} = \pi\rho L C_d R \sin 2\alpha \left[\begin{array}{l} \sqrt{\left(\frac{1}{2\rho}(P_1 - P_2) \right)} \frac{dx}{dt} \\ + \frac{x \sin 2\alpha}{\sqrt{2\rho(P_1 - P_2)}} \frac{d(P_1 - P_2)}{dt} \end{array} \right] \quad (7)$$

where ***L*** is equivalent characteristic length at entrance. The characterization of flow discharge coefficient C_d is imprecise even for a valve with conventional seat geometry. For this reason the authors integrate FEA and experimental results, utilizing them as part of the input database for the discrete time solution.

FINITE ELEMENT COMPUTATIONAL FLUID DYNAMICS (FEA/CFD)

The geometry of a practical valve seat and flow channel is complex. As a result, mathematical representation may not be achievable. FEA technology, notable for its accurate representation of complex vector fields, is used in parallel with experimental development. This describes the flow behavior over the valve plunger operating range. In fluid mechanics, the flow behavior can be mathematically described with continuity Equations (8) and momentum Equations (9) (Navier-Stokes).

$$u_{i,i} = 0 \quad (8)$$

$$\begin{array}{l} \rho_o(u_{i,t} + u_j u_{i,j}) - [-P\delta_{ij} + \mu(u_{i,j} + u_{j,i})]_{,j} \\ + \rho_o f_i = 0 \end{array} \quad (9)$$

The Galerkin method can be applied to represent the governing equations in the weak form.

$$-[\int \varphi \psi^T_{,i} dx] u_i = 0 \quad (10)$$

$$\begin{array}{l} [\int \rho_o \psi \psi^T dx] u_{i,t} + \{\int [\rho_o \psi (\psi^T u_j) \psi^T_{,j} dx \\ + \mu \psi_{,j} \psi_{,j} dx]\} u_i + [\int \mu \psi_{,j} \psi^T_{,i} dx] u_j \\ -[\int \psi_{,i} \varphi^T dx] P = \int \rho_o \psi f_i dx + \oint \tau_i \psi ds \end{array} \quad (11)$$

$$u_i(x,t) = \psi^T u$$
$$P(x,t) = \varphi^T P$$

where ψ and φ are the vectors of the interpolation function (weight functions) approximating the pressure and velocity fields, and x denotes the location over the space. Equations (10) and (11) are further discretized and combined to become the system Equations (12) representing the entire flow field.

$$[M]\begin{Bmatrix} \dot{u}_i \\ \dot{P} \end{Bmatrix} + [C]\begin{Bmatrix} u_i \\ P \end{Bmatrix} + [K]\begin{Bmatrix} u_i \\ P \end{Bmatrix} = \begin{Bmatrix} F \\ 0 \end{Bmatrix} \quad (12)$$

Depending on the nature of the physical system, various commercially available CFD programs have been used by Delphi Chassis' engineers to accurately describe the flow phenomena in a fluid control element. FEA in computational fluid mechanics is a reliable process but can be very time consuming. In order to speed up this process, assumptions are often made for the model to effectively represent the valve flow geometry in a simplified manner.

One example of this is the simplification of the flow surface geometry to allow the choice of two dimensional rather than three dimensional analysis. For example, looking at a typical brake system hydraulic component shown in Figure 5, the non-symmetrical aspect of the radial ports would typically call for a full three dimensional analysis. However, the advantage of being able to substantially reduce the modeling turn around time encourages the practicing engineers to make a carefully justified assumption, and use two dimensional modeling during the hydraulic component design process.

The process of making good assumptions for the valve flow characterization also helps the engineers to achieve a more robust solenoid valve design. For example, choosing to model the outlet ports as described above implies a final dimensioning of the ports in keeping with the assumptions made. Furthermore, the speed and cost of today's personal computer makes it possible for the practicing engineer, without formal analysis expertise, to perform reasonably sophisticated FEA work.

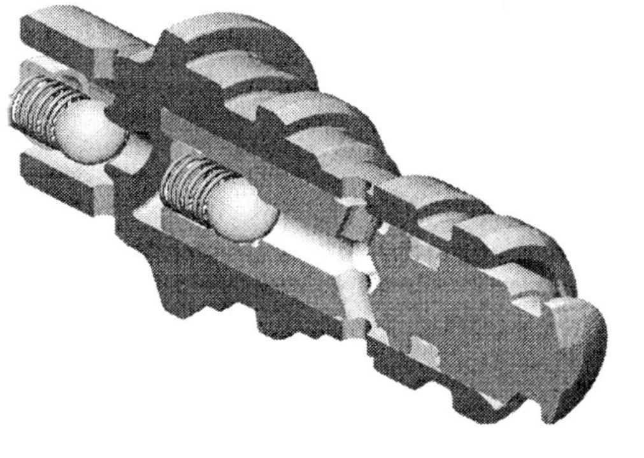

Figure 5. Non-axisymetric geometry

Damping Simulation

Figure 6a shows the flow pattern through a typical control valve obtained from a two dimensional FEA. The level of the valve's hydraulic damping is a function of valve seat geometry, ball velocity and pressure drop across the inlet and outlet chambers (Equation (7)). This is an example where, after careful consideration, two dimensional analysis was applied. Subsequent analysis and test showed that the error introduced through this assumption was small (less than 10%).

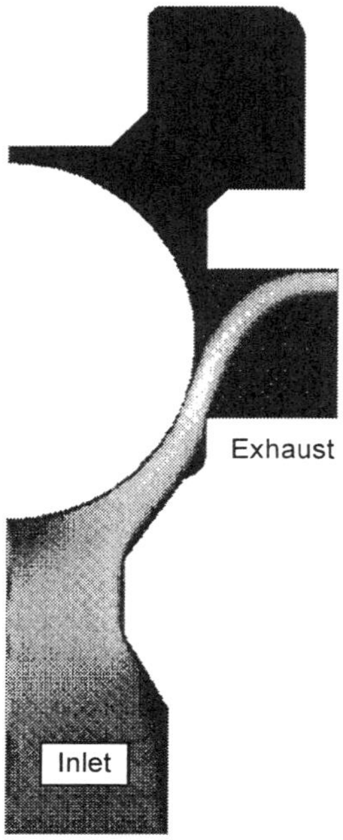

Figure 6a
Normal flow

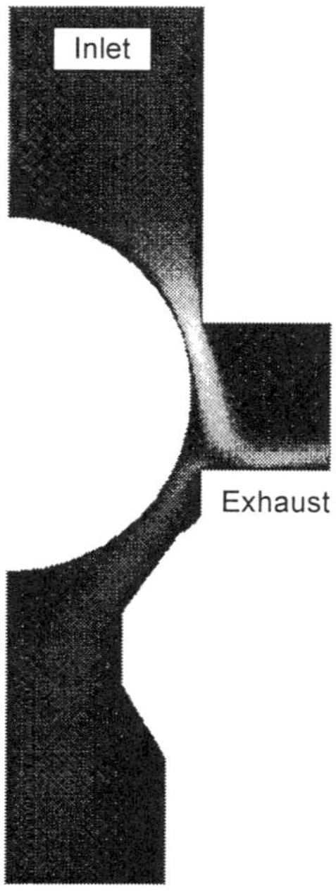

Figure 6b
Damping flow

In order to take into account the ball velocity effect, the authors take an innovative approach to quickly compare relative levels of damping by reversing the fluid flow. Figure 6b shows how the model is "inverted" and the fluid defined as originating from the chamber behind the ball instead of its normal entry port. The fluid flow is defined as a function of the assumed velocity of the ball. In this manner a steady-state fluid model is used to quantify dynamic behavior. A series of steady state models representing various ball positions can be used to provide a spectrum of data relating the flow to pressure and ball position. These relationships are then integrated with the system simulation model described in the previous section, and used to effectively characterize the dynamic damping behavior.

Magnetic Force Simulation

The magnetic field vector potential A can be mathematically described with the diffusion form of the governing equation expressed in Equation (13).

$$\frac{1}{\mu} A_{,ii} = -J_o + j\omega\sigma A \qquad (13)$$
$$\nabla \times A = B$$

Where μ is the magnetic permeability, ω is the angular frequency, σ is the electrical conductivity, J_o is the applied current density, and B is the magnetic flux density. The Galerkin method is applied to further describe the governing Equations (13) in the weak form.

$$\frac{1}{\mu_o} \int (\Omega'_i A'_i + \Omega'_j A'_j) ds + j\omega\sigma \int \Omega A ds = J_o \int \Omega ds \qquad (14)$$

where Ω_i is the vectors of interpolation functions.

$$A = \Omega_i A_i \qquad (15)$$

The Magnetic force can be computed via

$$F_i = -\left(\frac{1}{\mu} A^T S A \right)_{,i} \qquad (16)$$

where

$$S_{ij} = \int \frac{1}{\mu} (\Omega_{i'i} \Omega_{j'i} + \Omega_{i'j} \Omega_{j'j}) ds \qquad (17)$$

Figure 7 is an example of the steady state equi-potential flux plots produced from the FEA. The steady state magnetic transfer function can be derived via the FEA simulation of the solenoid armature over different positions and currents. Since it is difficult to accurately model the geometry of a magnetic field, and compute realistic force with small air gaps, another source of data is required to ensure the accuracy of the electro-magnetic response. Along with the development of the system simulation technology, a magnetic force measurement system was designed to enhance model accuracy.

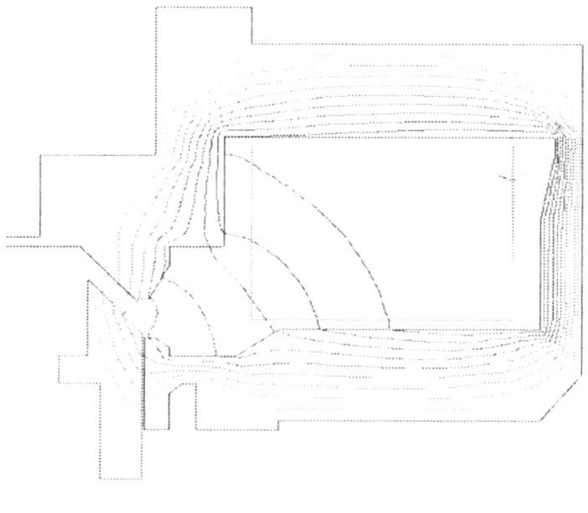

Figure 7. Magnetic FEA

MODEL CONFIRMATION

In adopting the system engineering process, the need to develop a test system concurrent with component design becomes paramount. Not only must one confirm functional requirements but also be able to measure such physical parameters as necessary to confirm model accuracy.

FLEXIBLE PROTOTYPES

Along the path to developing a successful valve are pitfalls that can result in many design iterations and a large number of prototypes. One technique that may be adopted to circumvent these pitfalls is the use of 'take-apart' or flexible prototypes. Figure 8 shows a typical flexible valve. This valve has been designed for ease of assembly. It allows engineers to partially disassemble the valve and quickly modify the design parameters.

With this configuration, it is possible to implement a variety of valve types, either normally open or normally closed. These may be further optimized for either 'on/off' or proportional control. The major force producers (hydraulic, magnetic, etc.) may be rapidly modified through component substitution. Spring pre-load and plunger travel may be adjusted while the valve is being tested.

Early availability of this type of valve allows many design iterations which gives confidence when writing production valve specifications. While the flexible valve can be expensive, typically it is not prohibitive for proof of concept vehicle level systems. In terms of confirming system requirements and performance variation flexible valves are invaluable. With a typical modification, build, and test cycle of 15 minutes, many tests per day are possible.

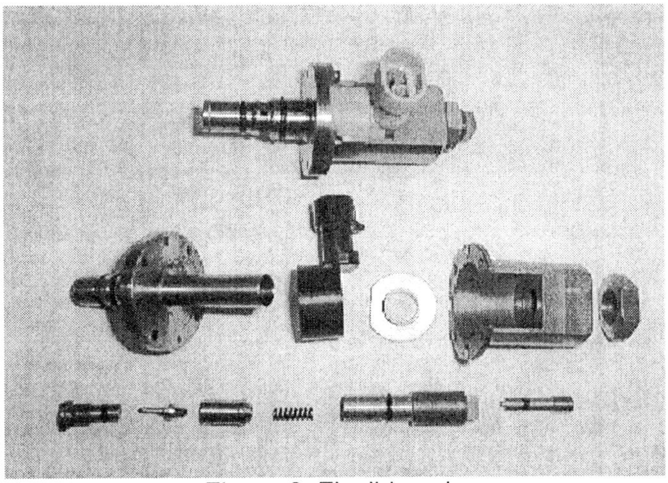

Figure 8. Flexible valve

MAGNETIC FORCE MEASUREMENT

Measuring the magnetic force within a solenoid valve, while easy to envisage, can be difficult in practice. A number of issues need to be addressed. The first and most important is that the mechanical system must be rigid and free from unwanted motion. A second and more subtle problem is that of unwanted Coulomb friction. An instrument was designed to alleviate these potential problems and is shown in Figure 9.

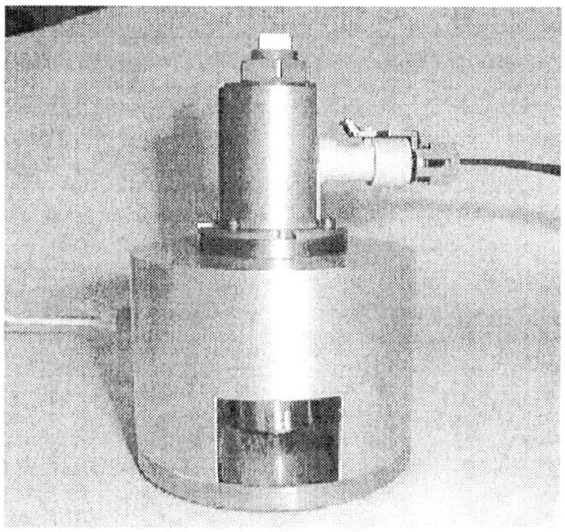

Figure 9. Magnetic force measurement

This fixture was designed to use as many of the flexible valve components as possible. For this test, only the seat and plunger need to be replaced by magnetic force fixture components. Using the components of the actual magnetic circuit allows unexpected flux leakage or mechanical instability to be accurately represented. Figure 10 shows the first step in the model confirmation process. The test data shows a much higher force at small gaps than the finite element model. From this point we would correct any geometric errors and further refine the model and/or test system.

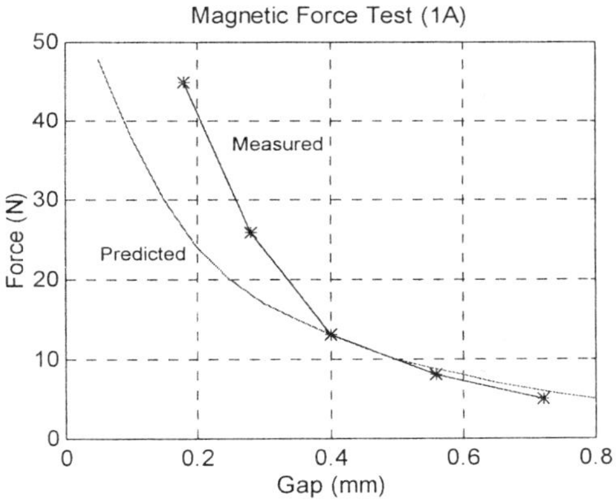

Figure 10. Step 1 model confirmation

A typical valve is characterized with armature travel of approximately 1mm. As the air gap approaches zero, the magnetic force can be so high (100N or more) that magnetic gains in excess of 200N/mm result. For this reason, any backlash or free motion in the fixture or sensor installation can produce significant error. They must be eliminated through the design and setup of the magnetic force measurement system.

HYDRAULIC FORCE MEASUREMENT

Measuring the hydraulic force component can also be difficult. The philosophy of using actual parts at working pressures, while not being invasive, heightens the problem. Figure 11 shows a test fixture designed for this purpose. With this fixture the authors were able to characterize numerous seat designs and verify FEA predictions.

Figure 11. Hydraulic force measurement

The fixture mounts a 'flexible' valve body together with seat and plunger in a specially made manifold block. The manifold block provides inlet and exhaust ports together with an accurately positioned load cell. The load cell is connected by a close fitting stem to the valve plunger assembly. This close fit must achieve a compromise of low friction together with a degree of hydraulic seal. The scavenge chamber acts as a holding volume for leakage flow and is fitted with a vacuum port to maintain reasonable fluid levels and prevent damage to the load cell. The fixture accurately positions the plunger assembly relative to the valve seat. Special care was taken to ensure good alignment and hence low Coulomb friction.

A large capacity pressure controlled pump provides steady state input flow. Repeated test runs at different plunger positions fully map the valve seat hydraulic characteristics. Figure 12. compares simulation and test results for one such valve seat.

DAMPING FORCE MEASUREMENT

Noise has become an overriding concern in today's

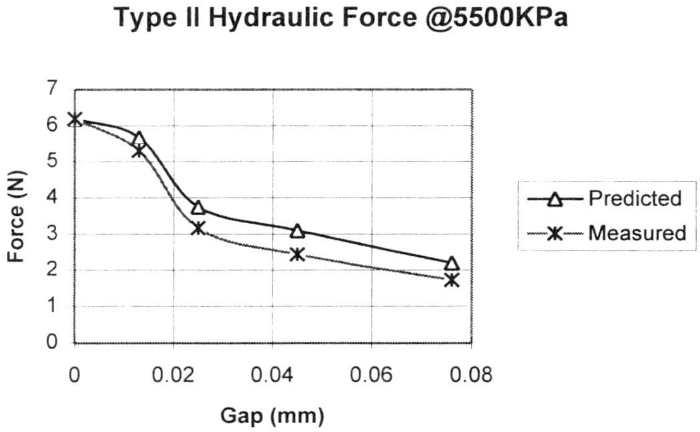

Figure 12. Measured and predicted Hydraulic force

passenger vehicles with ever-reducing acceptable levels. Indeed, 'low noise' has become a marketable feature of many brake systems. Eliminating noise problems later in the product development process is difficult and can be expensive. The system or component designer is well advised to ensure that 'low noise' is built into the design at the onset of product development.

In active hydraulic components, the major noise source is invariably due to the rate of change of flow transmitted through direct (hydraulic) or indirect (airborne or conducted) means. Insofar as flow modulation is achieved through a mechanical component, velocity control of that component is vital.

The 'flexible' valve design allows us to quickly construct a velocity controlled valve. With a suitable test fixture, the damping force can be determined. Figure 13 depicts the test unit developed by the authors. A motor driven eccentric acts on a connecting rod with the actual valve damping component attached. A spring provides the return stroke force. This mechanism provides a sinusoidal displacement from which velocity curves may be derived.

Figure 13. Damping test

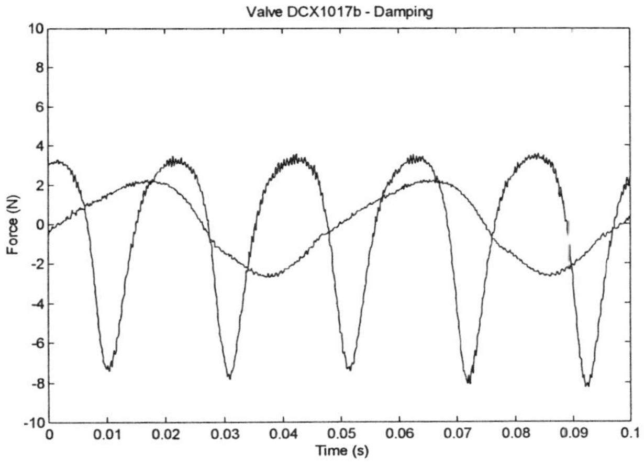

Figure 14. Damping force

One aspect of this type of damping which is often overlooked is the non-symmetrical nature of the damping forces. Figure 14 shows two plots taken from the test stand. At low velocities the force is symmetrical, but at higher velocities the return force is limited. The reason for this is that the damping force results from pressure developed across the damping piston. In the compression mode the pressure is only limited by applied force. In the rebound mode the pressure is limited to that of vacuum. This places limits on damping levels. In practice it is better to limit damping path restriction such that damping forces remain symmetrical at peak operating velocities.

SOLENOID VALVE BENCH TESTING

After the design parameters have been chosen to optimize the hydraulic, magnetic, and spring forces, and an appropriate level of damping force is determined, the valve can be assembled and tested in the lab. This can occur independently of the system level design, and should be well underway before the remainder of the hydraulic system is finalized. A test bench (Figure 15) is

Figure 15. Hydraulic flow bench

constructed to evaluate the valve through its entire operating range. The bench facilitates valve testing as a component (for performance characterization) and as a part of the system within which it will function.

An industrial pump, with output capability an order of magnitude higher than that required by the system, is used as a pressure source. A large accumulator is fitted at the pump output to minimize pressure surges. In addition, a flow meter is placed in series with the valve under test. Pressure transducers are used to measure the source pressure as well as the differential pressure across the valve.

To fully characterize the valve, source pressure is ramped from zero to an appropriate system maximum, while flow and differential pressure across the valve are recorded. (Figure 16) This routine is repeated with current levels increased in step fashion to generate a family of curves for each valve. At maximum plunger displacement, a valve will demonstrate an "effective orifice" characteristic in the delta pressure versus flow curve. The comparison of each flow curve between the closed and open position reveals a great deal about the behavior of the valve. In other words, an 'on/off' valve will generate a very different family of curves from that of a proportional valve. A pressure regulating valve will differ from a flow control device. These curves can be interpreted and component level modifications can be considered based on the sub-component testing described earlier.

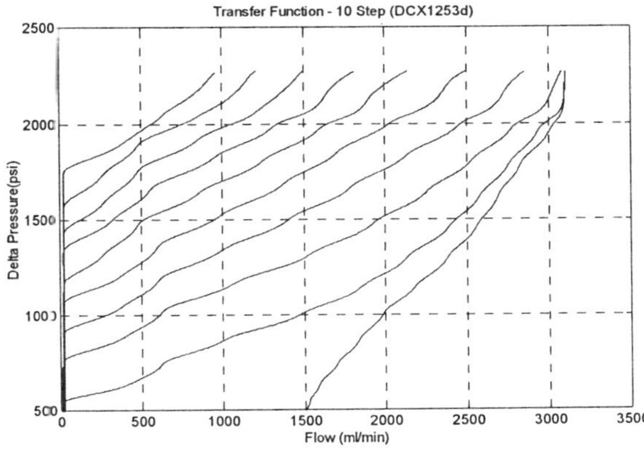

Figure 16. Hydraulic transfer function

Finally, the valve can be connected to a hydraulic circuit that describes a portion of the system in which it will ultimately be used. Interactions with other components can be understood early in the design process, before major investments in time and money have been made.

CONCLUSIONS

Adherence to the system engineering process described within this paper has allowed the development of numerous hydraulic components, which in many cases have no existing knowledge base upon which to build.

The process from initial modeling to final vehicle level test provides a high degree of confidence in the robustness of the design. It ensures a detailed understanding of the functionality of the component and its complex interactions. As we further explore the limits of cost, noise, mass and volume reductions in our brake systems, we invariably move beyond the realm of what is 'known' or 'conventional'. This forces us to look past the need to develop technology and consider the development process itself, assuring its sophistication and reliability are cost effective and fast.

REFERENCES

1. "The Finite Element Methods in Heat Transfer and Fluid Dynamics", J.N. Reddy & D.K. Gartling, CRC Press, 1994.
2. "Electromechanics and Electric Machines", S.A. Nassar & L.E. Unnewehr", John Wiley&Sons, 1979.
3. "Hydraulic Component Design and Selection" Seminar Text, E.C. Fitch & Ing T. Hong, BarDyne Inc., 1997.
4. "Hydraulic Control System" Hebert E. Merritt, MIT press, 1965.
5. "The Modeling and Simulation of a Proportional Solenoid Valve" N.D. Vaughan & J.B. Gamble", ASME Winter Annual Meeting, Dallas, Texas, Nov. 25-30, 1990.
6. "High Speed Solenoid Control Techniques" G.C.Fulks IEEE Paper, Automative Power Electronics, August 1989
7. "Finite Element Analysis of Electrical Machines", Sheppard J. Salon, Kluwer Academic Publishers, 1995.

ACKNOWLEDGEMENT

The authors would like to thank the staff of Rapid Design Services for their support .

Author's Address

Huang-Tsang Chang
Delphi Chassis Systems
Engineering Technical Center
M/C E.520
1435 Cincinnati Street 45408
PO Box 1245
Dayton, Ohio
45401-1245 USA
Phone: [1] 937.455.5749
Fax: [1] 937.455.6798
Email: LNUSDYT1.FZG48X@gmeds.com

Gary C. Fulks
Delphi Chassis Systems
Engineering Technical Center
M/C E.515
1435 Cincinnati Street 45408
PO Box 1245
Dayton, Ohio
45401-1245 USA
Phone: [1] 937.455.6059
Fax: [1] 937.455.6798
Email: LNUSDYT1.ZZF5MP@gmeds.com

Jeffrey A. Heinrichs
Delphi Chassis Systems
Engineering Technical Center
M/C E.520
1435 Cincinnati Street 45408
PO Box 1245
Dayton, Ohio
45401-1245 USA
Phone: [1] 937.455.6307
Fax: [1] 937.455.6798
Email: LNUSDYT1.VZG1LL@gmeds.com

960958

Development of Four Solenoid ABS

Takeshi Naito, Hiroaki Takeuchi, Hiromu Kuromitsu, and Kiyoshi Okamoto
Aisin Seiki Co., Ltd.

ABSTRACT

To develop Low-cost system, we used matrix function under the condition of the vehicle having symmetrical brake conduit and four-wheel control capability. As a result, 3 SOL system is a likely candidate for rear-wheel drive vehicles, and a 4 SOL system for front wheel drive vehicles by using a orifices instead of the Normal-Open solenoid.

In developing a logic system, we used the learning control function with a duty pointer containing four hydraulic wave modes.

The actuator weighs 2.2kg and has a volume of 1.2 liters (W 90 × H 105 × L 133)

I. INTRODUCTION

Due to an increased awareness of safety in recent years, automobile manufacturers around the world have shown greater interest in developing ABS. As ABS grew in popularity, manufacturers competed through cost reduction. With this global trend in mind, we at Aisin strengthened our work in the active safety arena, and began developing a lower cost ABS in order to make a higher level of safety available to a larger number of drivers.

In this report, we are proud to present our latest achievement.

I-1. CONCEPT - One naturally assumes that an ABS must be the most up-to-date to provide the user with a high level of preventive safety. In addition to providing high quality, the system must be made simple and at a low cost.

The development concept for this system must meet the following prerequisites: adaptable to a conventional brake system; possess a dual plunger pump that is adaptable to diagonal brake piping and front-rear split piping; and provide for four-wheel modulation.

I-2. PLANNING - Considering the usefulness of the matrix thinking process in simplifying the hydraulic unit, we investigated various types of hydraulic systems from the two-solenoid to the eight-solenoid type. (Fig. 1) From our investigation, we found that there is no need for a reservoir. We adopted a three-solenoid system for front-engine, rear-drive (FR) vehicles, and a four-solenoid system for front-engine, front-drive (FF) vehicles. However, these systems cannot be used on vehicles with more than 15000N GVW because of their cost-performance constraints.

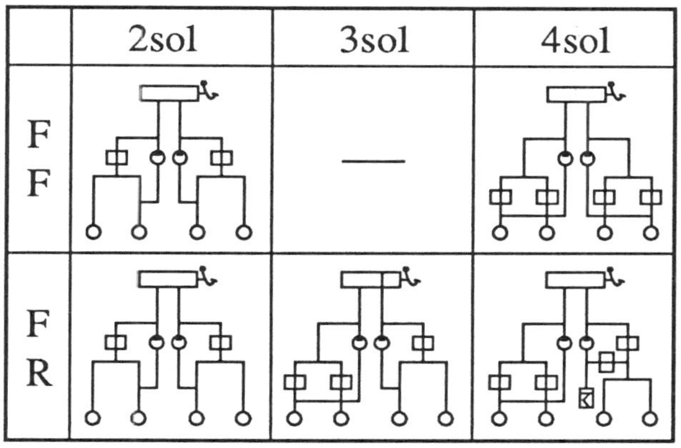

Fig. 1 Solenoid System Matrix

I-3. SYSTEM SIMULATION - We referred to Bosch's thesis for determining the relationship between torque and wheel acceleration, and linearized the build and decay characteristics to symplify the evaluation of the system's modulating capability.

$$J\dot{\omega} + MR\dot{V} = -T_{(t)}$$
$$\dot{\omega} = (\mu RMg - T_{(t)})/J$$
$$T_{(t)} = T_{min} + at$$
$$T_{(t)} = T_{max} - bt$$
$$R\omega i = (1 - \lambda i)\, V_i$$
$$V_i = V_{i-1} + dV/dt|_{i-1}\Delta t$$
$$V_i = R\{-\int T_{(t)}\, dt + J\omega_o + MRV_0\} / \{J\ (1 - \lambda i) + MR\hat{\ }2\}$$

Then, we found that modulation is possible if the decay rate is more than 25 Mpa/s. (Fig. 2,Fig. 3)

```
10 PRINT"abs simyulation"
20 G=9.8
30 J=0.75
40 M=310
50 R=0.28
60 K=2*0.26*0.114*PI/4*(0.054)^2
70 dpdtI=30*10^6
80 dpdtD=22*10^6
90 V0=10
100 W0=V0/R
110 WD=W0
120 VD=V0
130 a=K*dpdtI
140 b=K*dpdtD
150 Z=J+M*R^2
160 MYU=0
170 V=V0
180 um=0.5
190 ur=0.40
200 s1=0.2
210 TIME=0
220 DTIME=0.01
230 t=0
240 dt=0.01
250 O=0
260 MBmin=0
270 FOR I=1 TO 500
280 TIME=TIME+DTIME
290 IF O=2  THEN 520 ELSE 300
300 IF O=1 THEN 350 ELSE 310
310 O=1
320 t=0
330 V=VD
340 MYU=MYUD
350 CI=J*WD+M*R*VD
360 t=t+dt
370 MB=MBmin+a*t
380 X=-MBmin*t-a/2*t^2+CI
390 GOSUB 710
400 WI=RW/R
410 VI=V
420 TI=t
430 MYUI=MYU
440 MBmax=MBmin+a*TI
450 IF RAM>0.2THEN 480 ELSE 460
460 GOSUB 790
470 GOTO 660
480 O=2
490 t=0
500 V=VI
510 MYU=MYUI
520 CD=J*WI+M*R*VI
530 t=t+dt
540 MB=MBmax-b*t
550 X=-MBmax*t+1/2*b*t^2+CD
560 GOSUB 710
570 WD=RW/R
580 VD=V
590 TD=t
600 MYUD=MYU
610 MBmin=MBmax-b*TD
620 IF RAM<0.05 THEN 640 ELSE 630
630 GOTO 650
640 O=0
650 GOSUB 790
660 PRINT" V";ROUND( V,-3);"
670 PRINT"RW";ROUND(RW,-3);"
680 STOP
690 NEXT I
700 STOP
710 RAM=1/J*(Z-R*X/(V-MYU*G*dt))
720 V=R*X/(J*(1-RAM)+M*R^2)
730 RW=(1-RAM)*V
740 GOSUB 790
750 DRW=R/J*(MYU*R*M*G-MB)
760 DV=-MYU*G
770 P=MB/K/1000000
780 RETURN
790 x=RAM
800 IF x>0.2 THEN 830 ELSE 810
810 MYU=-um/s1^2*(s1-x)^2+um
820 IF x<-0.2 GOTO 840
830 MYU=-(ur-um)/(1-s1)^2*(1-x)^2+ur
840 RETURN
```

Fig. 2 Program

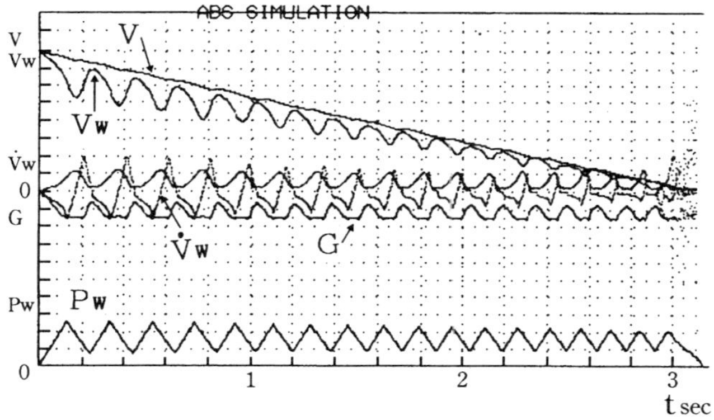

Fig. 3 Control diagram

I-4. HYDRAULIC CHARACTERISTICS - We obtained data on a disc brake volume and established a rough relationship between volume and pressure, as given below.

The hydraulic characteristics of the Fr boosting and Rr reducing phase can be obtained in the following equations.

$$\mathcal{V}_f = A\sqrt{P_F},\ \mathcal{V}_r = B\sqrt{P_R}$$
$$Q_I = Q_F + Q_D$$
$$Q_P = Q_D + Q_R$$
$$Q_I = \phi\sqrt{(P_M - P_F)}$$
$$Q_D = \chi_F\sqrt{(P_F - P)}$$
$$Q_R = \chi_R\sqrt{(P_R - P)}$$

$$dP_F/dt = 2\sqrt{P_F}/A * (\phi\sqrt{(P_M - P_F)} - E)$$
$$dP_R/dt = -2\sqrt{P_R}/B * (Q - E)$$
$$E = \chi_F\,(\chi_F Q - \chi_R * \sqrt{Q^2 - (\chi_F^2 - \chi_R^2)\,(P_f - P_r)})\,/\,(\chi_F^2 - \chi_R^2)$$

The Hydraulic characteristics of the simultaneous Fr and Rr reducing phase can be obtained in the following equations.

$$Q = Q_F + Q_R$$
$$Q_f = \chi_f\sqrt{P_f - P}$$
$$Q_r = \chi_r\sqrt{P_r - P}$$
$$dP_F/dt = -2\,\chi_f\sqrt{P_f}\sqrt{P_f - P}/A$$
$$dP_R/dt = -2\,\chi_r\sqrt{P_r}\sqrt{P_r - P}/B$$

The flow is related to the lead wheel pressure decay and build characteristics. Thus, we can observe both pressure levels chronologically, and each orifice can be adapted to the vehicle wheel's volume requirement. (Fig. 4)

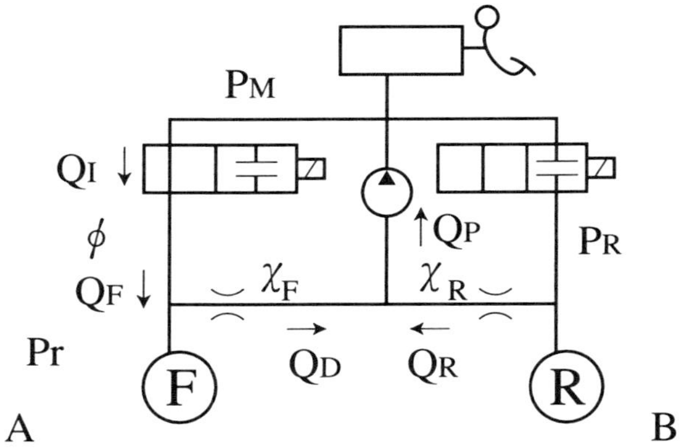

Fig. 4 Hydraulic model

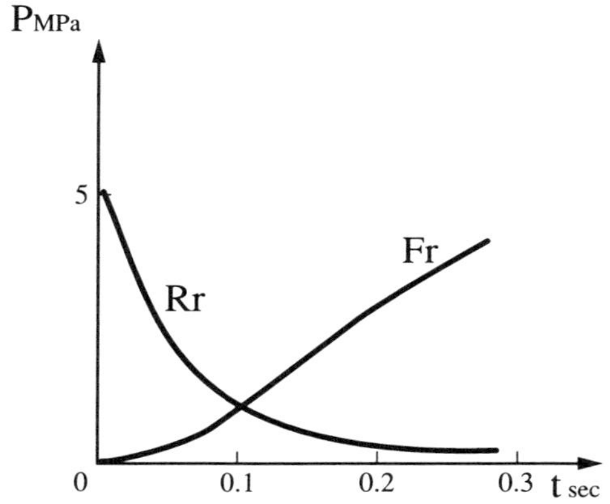

Fig. 5-a Fr boosting and Rr reducing characteristics

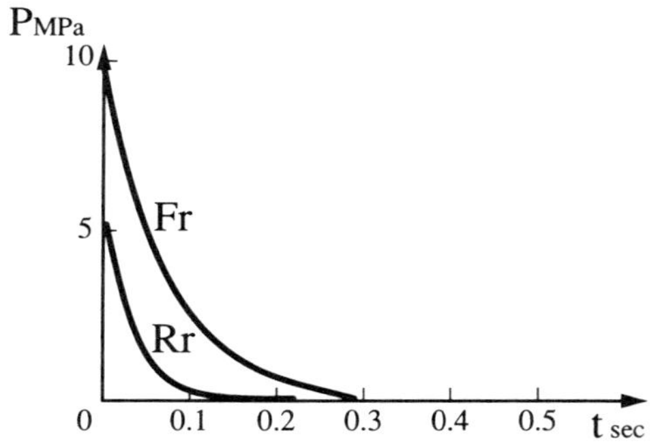

Fig. 5-b Simultaneous Fr and Rr reducing characteristics

II. ALGORITHM

The hydraulic pressure of the ABS must be increased and decreased appropriately in order to maintain an optimal slip rate. Because the present ABS lacks a hydraulic pressure hold mode, we investigated a method for optimally controlling the pressure through the combination of rapid pressure increases and decreases.

CONCEPT OF DUTY CONTROL-The duty control must be variable based on four hydraulic pressure gradients: rapid increase, rapid decrease, hold, and gradual increase (in which the hold and gradual increase are to be created by combining the pressure increase and decrease). The outline of the four-pressure gradient duty control is shown in Fig. 6.

The syetem is in the pressure-increase mode, if the slip rate, λ, is higher than S1 (#1; except during the initial control when the slip rate must be S2 or above, in order to prevent improper operation). When the wheel acceleration rate $\dot{V}w$ is lower than the established acceleration rate G1 (#2), the system will be in the pressure reduction mode.

After that, as the rear wheels start to recover and $\dot{V}w$ becomes higher than the established acceleration rate G2 (#3), or the slip rate becomes lower than S2 (#4), the system will be in the duty-hold mode and will wait for the wheels to recover.

When λ becomes lower than the established slip rate S1 (#5), the pressure will approach the time-rapid-increase output lock pressure that corresponds to the maximum reversion acceleration rate during duty-hold and will continue its duty pressure increase output until a condition is established to start a pressure reduction mode.

By repeating these four modes, the system will be able to provide an optimal slip rate. (Fig. 7)

Through the combination of pressure increases and decreases, the present system provides a duty control based on a four-pressure gradients. Although its basics are the same as those of the ordinary four-pressure gradient control (consisting of rapid increase, rapid decrease, hold, and gradual increase modes), pressure gradients for the hold and gradual increase must be prepared through a combination of pressure increases and decreases. In this case, a problem arises: even though the output is kept at a given rate, the pressure gradient is altered due to brake pedal effort fluctuations, or road-induced control pressure fluctuations. To solve this problem, the system must be able to learn pressure increase/decrease duty ratio values that are appropriate for the changes in the control pressure. This learning method is described below.

II-1. INCREASE IN DUTY PRESSURE - Through an increase in duty pressure, it is possible to establish a pressure gradient at will by changing the ratio of the pressure-increase time and pressure-decrease time within a given cycle. However, an increase in duty pressure has the disadvantage of corrupting the established pressure gradient when exposed to pedal effort fluctuations (Fig. 8a) and control pressure fluctuations (Fig. 8b).

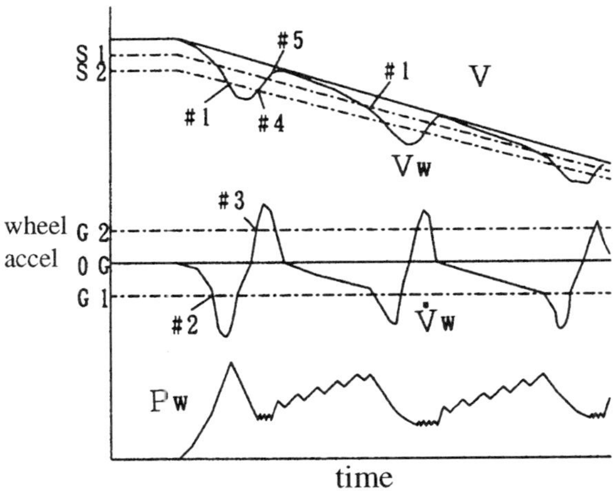

Fig. 6 Concept of duty control

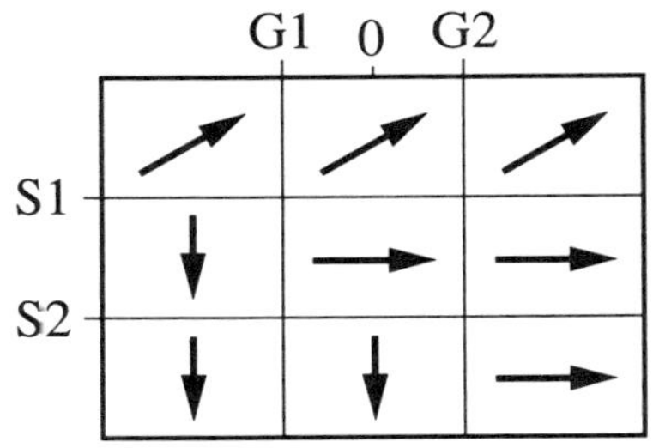

Fig. 7 Pressure gradient control

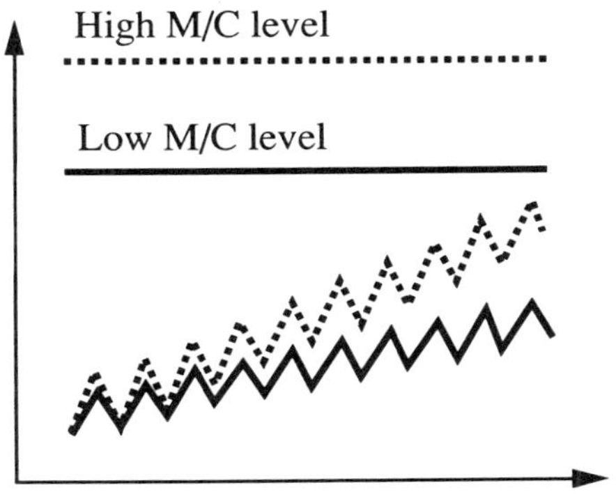

Fig. 8a Variation of W/C pressure by M/C

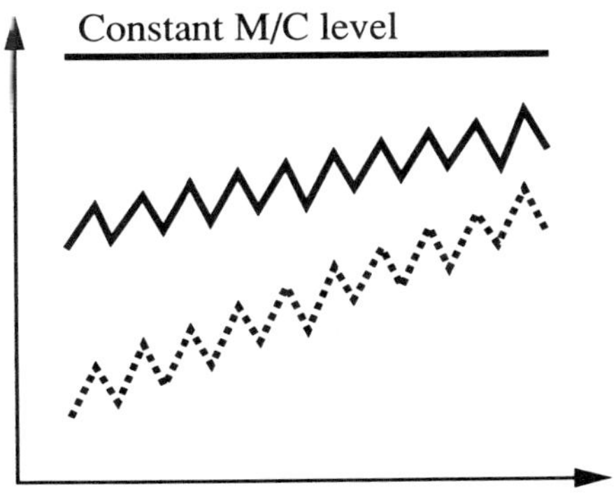

Fig. 8b W/C Saturation phenomena

In other words, even if the system attempts to gradual increase the output at a given duty ratio, once the pedal effort is increased, the pressure gradient becomes sharper. This is because the amount of pressure increase in a cycle becomes greater due to a larger difference that is created between the master cylinder and wheel cylinder pressures.

Conversely, once the pedal effort is decreased, the pressure gradient becomes milder. Similarly, when the control pressure is increased, the pressure gradient becomes milder due to a smaller difference that is created between the master cylinder and wheel cylinder pressures. When the control pressure is decreased, the pressure gradient becomes sharper.

Because an optimal slip rate cannot be ensured unless the pressure gradient remains constant, a learning control function will be needed to ensure a uniform pressure gradient.

II-2. ESTABLISHING DUTY POINTERS - A duty ratio map, such as the one shown in Fig. 9, must be made before implementing duty ratio learning control. The ratio must be set so that the smaller the duty pointer, the shorter the pressure increase time in a cycle will be, and that the larger the duty pointer, the longer the pressure increase time will be. Hence, when the pressure is increased with the duty ratio of a small pointer, the resultant pressure gradient will be milder. Likewise, with the duty ratio of a larger number, the pressure gradient will be sharper. Thus, by using different pointers, the system will be able to dampen the changes in the pressure gradient that are caused by control pressure fluctuations.

II-3. LEARNING DUTY POINTERS - In order for the system to learn the duty pointers, the system must be able to determine the sharpness or mildness of the pressure gradient. A method that uses control cycle times is applied for this purpose. As shown in Fig.10, if the pressure gradient is sharp, the pressure rise time to the lock pressure is shorter (T_1) , thus reducing the time for the duty pressure increase mode. Conversely, if the pressure gradient is mild, the time for the duty pressure increase mode increases (T_2) . This characteristic is applied in implementing learning control. Thus, a smaller pointer is provided if the duty pressure increase time is shorter than the established time, in order to make the subsequent gradient milder; a larger pointer is provided if the duty pressure increase time is longer. Pointers are provided in this manner so that the duty pressure increase time remains constant at all times, thus restraining pressure gradient fluctuations.

III. HYDRAULIC MODULATOR

III-1. CONSTRUCTION - A compact, lightweight, yet highly reliable hydraulic modulator is used. The hydraulic modulator is also designed for ease of installation. In order to reduce its overall length and to ensure the freedom to attach piping in any direction, the solenoid valve is mounted downward (Fig.11).

Fig.11 shows the construction of the modulator. The modulator includes a solenoid valve to control the brake pressure and a motor-driven pump to return the brake fluid to the master cylinder.

In addition, this modulator contains a changing-orifice valve in the solenoid valve.

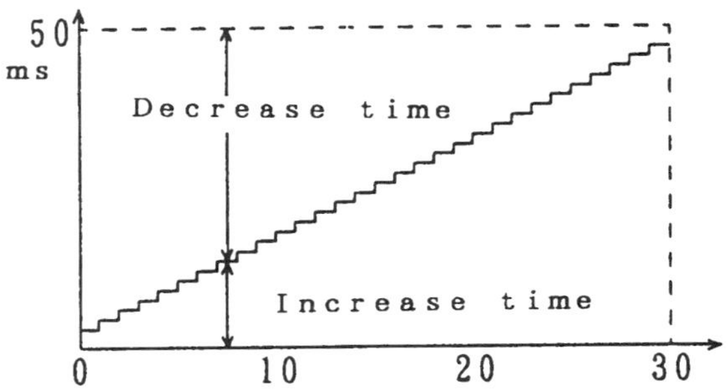

Fig. 9 Map of duty pointer

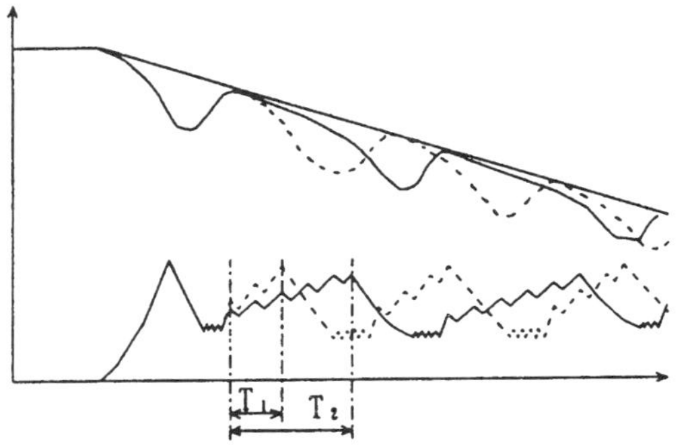

Fig. 10 Modulating chart

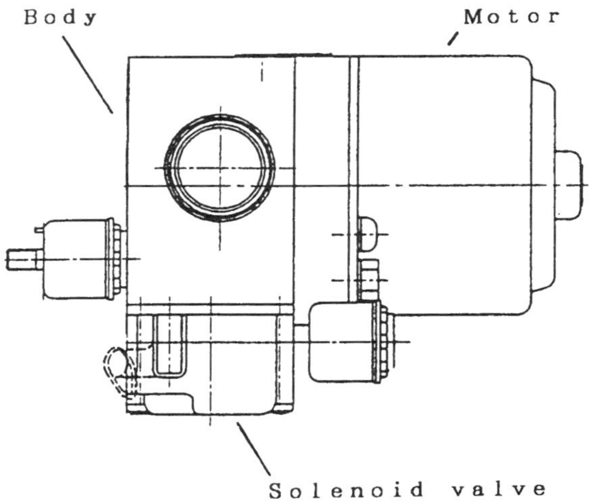

Fig.11 Figure of hydraulic modulator

III-2. CHANGING-ORIFICE VALVE - To reduce the pump discharge volume and fluid pulsation sounds, the changing-orifice valve limits the amount of pressure increase during ABS activation.

This valve was designed so that it can be placed inside of a conventional solenoid valve.

Operation - Fig.12-A shows the changing-orifice valve during normal braking (when the ABS is not activated). The brake fluid from the master cylinder passes through a clearance between sheet and valve "1", which provides a sufficiently wide passage, and is fed to the wheel cylinders without encountering any resistance along the way.

Figs.12-B,C show the changing-orifice valve during ABS activation. Fig.12-B shows the changing-orifice valve when the pressure is decreased. Because the solenoid is ON, the rod in the solenoid valve moves downward with the piston, and valve "1" is closed by the retraction force of the solenoid, thus closing the passage between the master cylinder and the wheel cylinders. Accordingly, the fluid from the wheel cylinders are fed back toward the master cylinder via a pump, thus decreasing the pressure applied to the wheel cylinders.

Fig.12-C shows the changing-orifice valve when the pressure is increased. Because the solenoid is OFF, the rod in the solenoid moves upward, and the fluid from the master cylinder is fed to the "V" chamber. The fluid then passes through the orifice and is fed to the wheel cylinders, thus increasing the pressure applied to the wheel cylinders. (The valve "1" is closed because the master cylinder pressure is greater than the wheel cylinder pressure.)

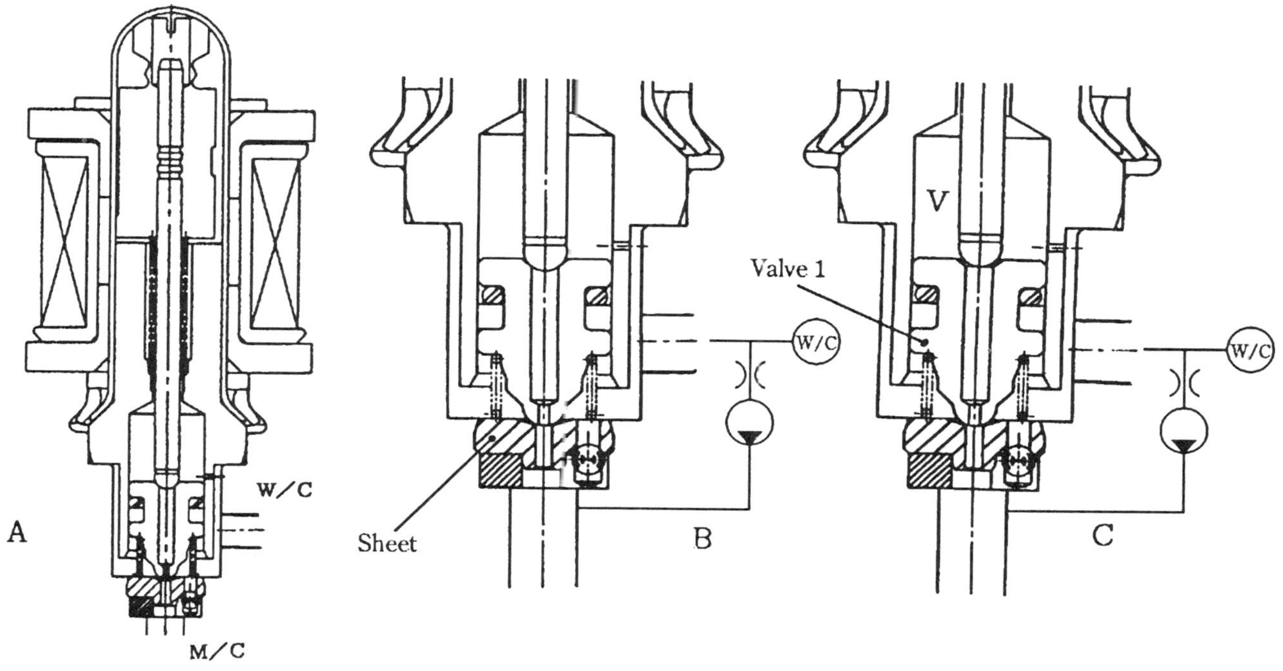

Fig.12 Operation of valve

IV. ELECTRONICS CIRCUIT CONCEPT

The electronic controller consists of a voltage controller, input signal conditioning circuit, solenoid driver circuit, CPU, and a monitoring memory. (Fig.13)

The voltage controller include functions to regulate and monitor voltage and monitor the CPU.

The function of the signal conditioning circuit is to convert the velocities of the wheels and discrete signals, to detect speed sensor faults, and to suppress excess voltage applied by analog input signals.

The function of the solenoid driver circuit is to drive the solenoids, to detect solenoid faults, and to monitor the CPU for any abnormal conditions.

The monitoring memory is an EEPROM that can store diagnostic codes even without power supply.

The diagnostic codes are output through a warning lamp and a serial interface.

IV-1. SIGNAL PROCESSING - Each wheel speed sensor generates a sinusoidal AC voltage according to the wheel velocity.

The noise in the AC voltages are suppressed with low-pass filters and are converted into rectangular-wave signals by an input signal conditioning IC.

Each valve is energized by the solenoid driver in the electronic controller. Any fault on the part of a solenoid or solenoid driver is isolated from the others.

The valve and motor relay is energized by the integrated power driver with over-current and thermal shut-down functions. Any relay fault is monitored by the CPU.

The solenoid predriver and voltage controller monitor the operating status of the CPU through the CPU run-pulse output. These circuits stop valve relay operations and reset the CPU in case any abnormal condition is detected.

Diagnostic codes are stored in the EEPROM and indicated by way of a warning lamp or translated into the serial interface. Furthermore, the serial interface is also useful for translating other types of information into external equivalents.

IV-2. RELIABILITY CONCEPT - In order to achieve a high level of safety and reliability, the open status of speed sensor is detected by the input signal conditioning circuit and short status is detected through CPU digital processing based on the difference of four wheel speeds. The driver circuit detects the open and short status of the solenoids and the leak status of the driver FET.

If a critical fault is detected, the ABS operation is interrupted by switching off the valve relay. When this occurs, the conventional braking function takes over, without any restriction of its function. The warning lamp alerts the driver that the ABS is currently not in operation.

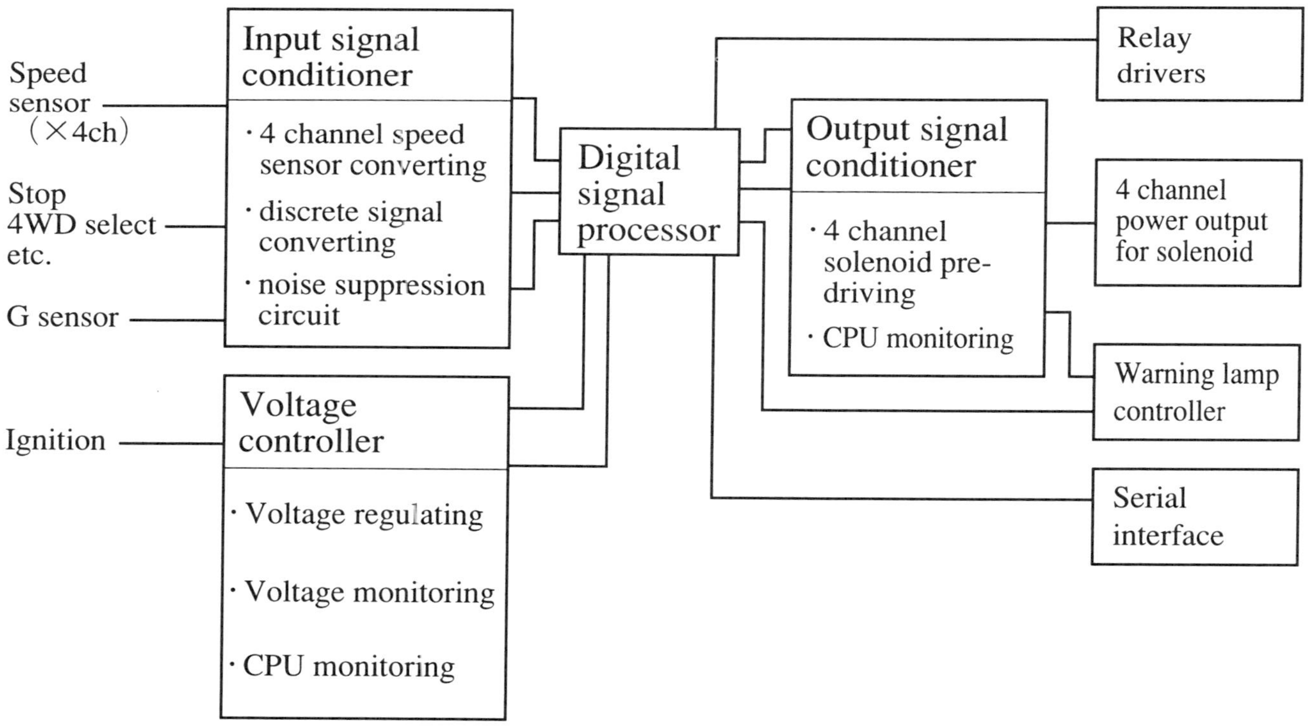

Fig.13 Block diagram

V. EVALUATION

We installed the 4SOL system on a compact-class vehicle and evaluated its on-vehicle performance in our test course.

V-1. ON-VEHICLE MEASUREMENT SYSTEM - We used a measurement system consisting of an evaluation data converter and a personal computer that can perform on-vehicle, real-time software-based evaluation.

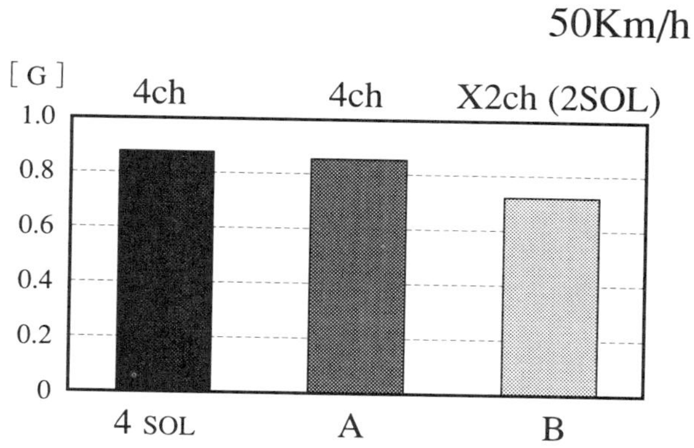

Fig.14 Decelerate on dry road

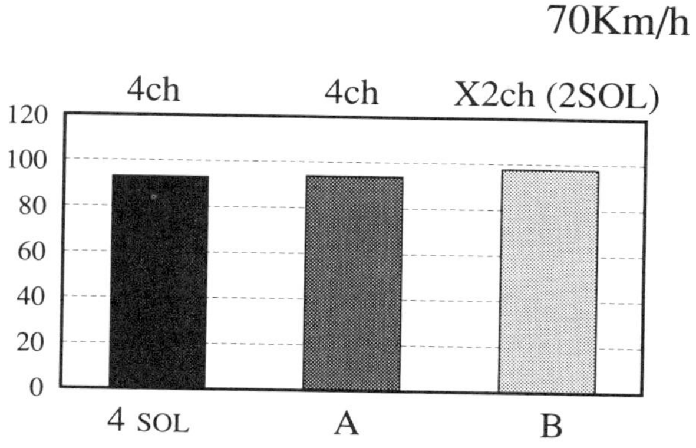

Fig.15 Lock ratio of low-μ road

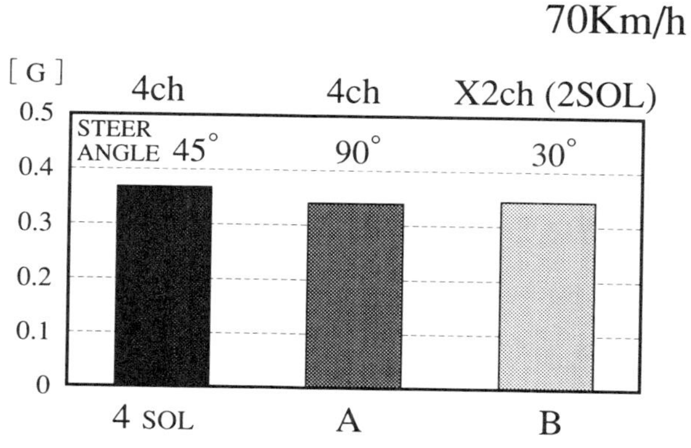

Fig.16 Decelerate on split-μ road

V-2. ON-VEHICLE PERFORMANCE -

1) Dry road: achieves the same level of deceleration as the systems of other manufacturers. (Fig.14)
2) Low μ road: ensures the same level of lock ratio performance as the systems of other manufacturers. (Fig.15)
3) split-μ road: achieves a higher level of deceleration, with minimal corrective steering, than the systems of other manufacturers. As a result, improved stability is ensured. (Fig.16)
4) Cornering and rough road etc.: achieves the same level of performance as the systems of other manufacturers.

V-3. CONTROL WAVEFORM - Learning control takes place in the initial portion of the control pressure waveform. After that, a stable deceleration force is ensured, a fact that is quite evident when the wheel velocity is applied over the waveform. (Fig.17,18)

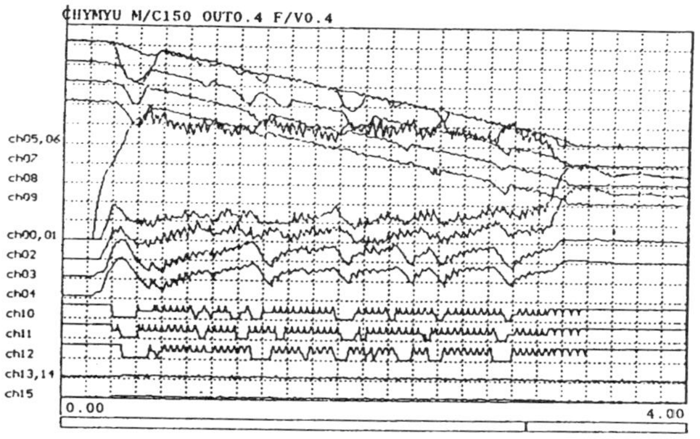

Fig.17 Wet road data

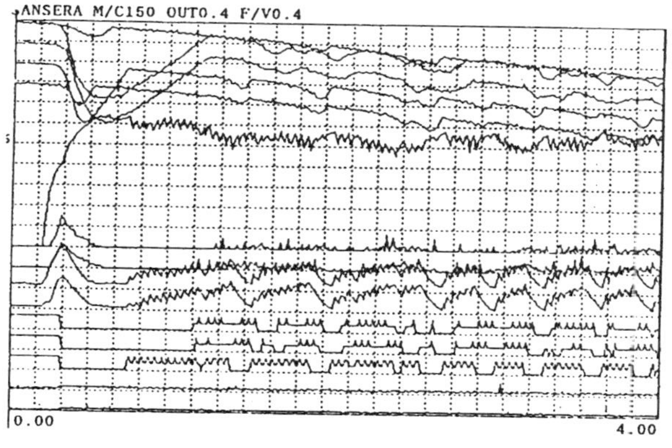

Fig.18 Low μ road data

SUMMARY

In this development, we investigated the issue of active safety for compact size vehicles, in order to be able to offer the car user an ABS at a lower cost. Through active safety, we are confident that this product will contribute to making traffic safer for the 21st century.

REFERENCES

[1] Bosch tecnishe berichte
(English special edition(FRBR.1982) ISSN 0006-789X)
[1] Development of 4 wheel ABS (TOYOTA) 842061(OCT.1984)
(Journal of the society of automotive engeering of JAPAN)

NOMENCLATURE

J .... Moment of inertia of the wheel
ω .... Angular velocity of the wheel
M .... Vehicle mass
R .... Wheel radius
V .... Vehicle speed
T .... Brake torque
a .... Build rate of torque
b .... Decay rate of torque
λ .... Brake slip ratio
v_f, v_r ...... Volume of brake
A .... Volume rate of brake
B .... Volume rate of brake
Q_I .... Flow of N.O. solenoid
Q_D .... Flow of front drain orifice
Q_R .... Flow of rear drain orifice
ϕ .... Rate of N.O. solenoid orifice
χ_F .... Rate of front drain orifice
χ_R .... Rate of rear drain orifice
P_M .... Master cylinder pressure
P_F .... Front brake pressure
P_R .... Rear brake pressure
P .... Pump suction pressure
tTime

950762

Intelligent Braking for Current and Future Vehicles

Donald Schenk, Richard Wells, and Jon Miller
Delco Chassis

INTRODUCTION

The automotive and light truck market is continuing to sharpen its focus on fuel efficiency and safety. Vehicle brake system design is evolving to meet these needs but the demand continues for enhanced level of performance and value. The challenge facing the brake system designer is to engineer systems that appropriately address these needs, plus more.

Future vehicle brake technology must comprehend the increasing requirements for (1) closed loop control of the normal braking event to minimize variability and provide active control of front-to-rear and side-to-side brake balance; (2) active proportional, independent wheel brake control to produce ABS and traction control, handling management, and interface with collision avoidance systems; (3) controllable input ergonomics to provide optimum driver control and comfort and; (4) appropriate interfaces with other vehicle operating systems including powertrain and, in some cases, energy recovery systems., In short, the future demands intelligent brake control.

A number of brake systems have been developed at Delco Chassis to address these needs for current and future vehicles. These systems utilize the proven technology of ABS VI, including truly proportional actuators controlled by sophisticated algorithms. These technologies extend driver performance during normal operating conditions, and optimize vehicle performance when extreme situations occur.

This paper reviews two variants of intelligent brake control. The first consists of a four-wheel electro-hydraulic apply brake system. The second utilizes a front electro-hydraulic apply system with rear electro-mechanical wheel brakes (known as rear electric brakes). Examples will show how the systems fulfill the previously mentioned evolving requirements for brakes.

ELECTRO-HYDRAULIC APPLY

The basic electro-hydraulic brake apply system is an application of full electronic brake control based upon motor/piston technology with hydraulic "push through" for "no power" redundancy. It offers full four-wheel control with features that enhance and augment a vehicle's braking capabilities. Since most vehicles require augmentation to the modulator to provide ABS and traction control function, it was only logical that other brake system functions be combined for mass, cost, performance and packaging advantages. Power assist and traction assist are two examples. The implementation of micro-processor managed electronic modulation of brake pressure also facilitates communication with other vehicle systems including the propulsion system for regeneration and adaptive cruise.

The electro-hydraulic apply system shown in Figure 1 operates in the following manner:

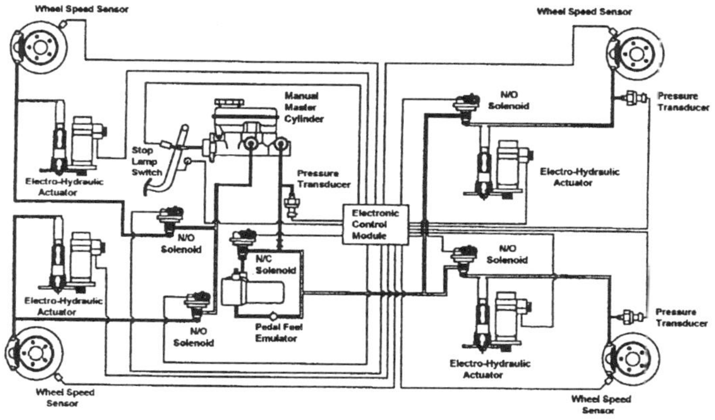

Figure 1 - Four Wheel Electro-Hydraulic Apply

When the brake pedal is depressed, the stop lamp switch closes and a signal is sent to the electronic control module (brake controller) to close the normally-open solenoids. The hydraulic fluid pressure resulting from the brake pedal force is measured by pressure transducers located upstream from the solenoids. The electrical signal from the transducers, proportional to the pedal force, is read by the brake controller and the piston displacement/motors are actuated to produce proportional downstream pressure to the front friction brakes. Pressure is modulated by the signal from the pressure transducers. The rear friction brakes are applied proportionately to give ideal braking balance based on a complex slip control algorithm. An example of brake proportioning with dynamic control is shown in Figure 2.

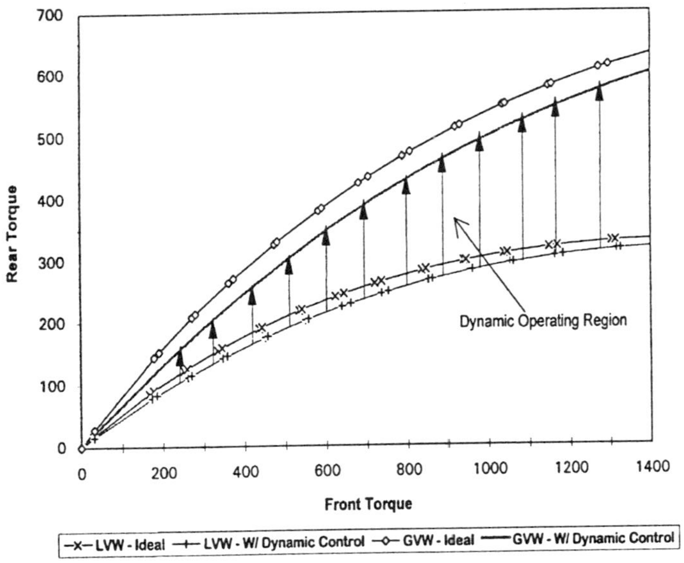

Figure 2 - Brake System Dynamic Proportioning

Pedal feel during normal braking is achieved by displacing the fluid from the master cylinder during a brake application into a pedal feel emulator device. The closed solenoids in the system prevent master cylinder fluid from being displaced to the wheel brakes during normal system operation. The pedal feel emulator provides a predefined force/displacement relationship which gives the driver the feel of a conventional brake system. In addition, the pedal force/deceleration relationship is tunable by the brake algorithm. Thus pedal ergonomics can be tuned for virtually any desired feel.

During normal power brake operation, hydraulic fluid is displaced to the wheel brakes only by the modulator actuation (motor/piston) as commanded by the controller based on driver input. If a fault is detected within the brake system that merits its being shut down, the normally-open solenoids return to the open position. This then allows brake fluid to be displaced directly from the master cylinder to the wheel brakes (no power manual brakes). The total system is designed to meet all "no power" stopping requirements.

The twin downstream pressure transducers read actual wheel brake pressure and feed that information back to the brake controller to ensure that actual pressure is equal to command pressure. Thus the system assures braking stability and balance.

While the brakes are actuated for an extended period of time and the vehicle is stationary, the electric motors are shut off and locked from back-driving by deactivating an electro-mechanical friction device. Thus, generated wheel brake pressure remains constant. This conserves energy while maintaining braking at the wheels. Any change in driver input causes the system to revert to the normal braking mode.

Normal ABS and traction control functions are achieved by overriding the driver input with an appropriate brake control command, based upon wheel speed. These functions and corresponding algorithms are similar to current production ABS/TCS systems.

A second mechanization is the application of the rear electric brake.

REAR ELECTRIC BRAKE

The rear electric brake system option is particularly well suited for those vehicles which focus on low mass and high fuel economy. Typical mass savings range from 3 to 6 kg. by eliminating the park brake cable mechanism. The rear electric brake system can be used in conjunction with a front electro-hydraulic system, or with conventional front hydraulic brake systems. A sample system schematic showing front electro-hydraulic apply and rear electric brakes is shown in Figure 3.

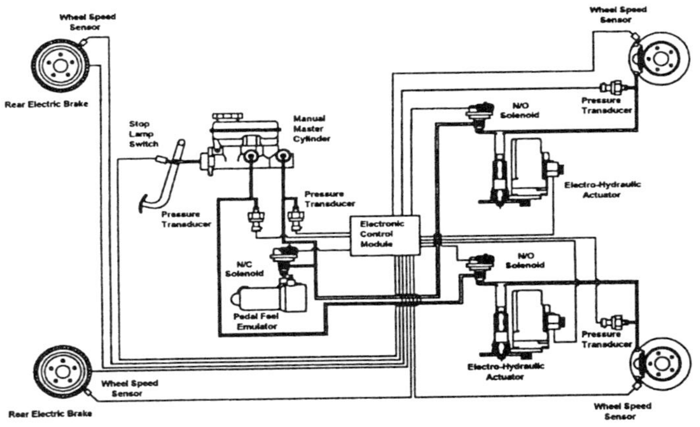

Figure 3 - Electro-Hydraulic with Rear Electric Brake

The rear electric brake is a high gain electro-mechanically actuated design. A high gain wheel brake maximizes the torque capability and minimizes electrical energy requirements. Dynamic stability of the gain is achieved via closed loop control technology. For example, at an operating torque of 400 NM each wheel brake can respond in closed loop control mode at rates up to 4000 NM/second to continuously (>10 ms loop time) adjust dynamic brake output.

Integral to the wheel brake is a permanent magnet DC motor, gear train, and ball screw/nut mechanism. The ball screw converts rotational motion/torque to linear motion/force. This, in turn, actuates a conventional friction surface brake mechanism through a system of levers. A diagram of the wheel brake is in Figure 4.

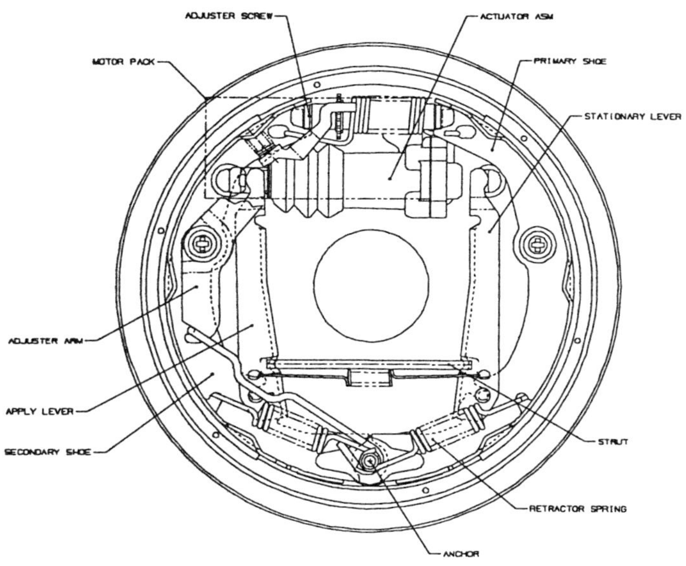

Figure 4 - Rear Electric Brake

Two key requirements drive the basic design concept. The first requirement is precise proportional control facilitated by use of a highly efficient motor, gear train, and lever system. Sliding mechanisms and associated friction losses are minimized by design. Bearings position and support all rotating shafts. A ball screw provides high efficiency and robustness to contamination. The levers utilize pivoting, rather than sliding, contact at the high load connection points. This reduces sensitivity to contamination from debris and corrosion.

The second key requirement is driven by default mode concerns. The wheel brake must mechanically reduce braking to an acceptable level upon power removal; the wheel brake must not remain energized in cases of power interruption during braking. This requires a highly efficient system with a return spring mechanism to "backdrive" the brake without the assistance of electrical power. The "backdrive" capability necessitates a separate park brake latch mechanism.

The rear electric brake system park brake latch mechanism is a bi-stable, spring loaded, clutch device which locks the main motor shaft upon command. The wheel brake mechanically latches at a maximum holding capability until the park brake switch deactivates. The system requires ignition "on" to deactivate the park brake.

NO electrical power is necessary to retain park brake engagement.

MEETING INCREASING REQUIREMENTS

The application of intelligent brake control meets the increasing requirements of future vehicles including (1) minimized variability through closed-loop control; (2) active proportional brake control for ABS, traction control, handling management, and brake activated collision avoidance, (3) tunable ergonomics to provide optimum driver control and comfort and; (4) appropriate interfaces with other vehicle operating systems including powertrain and energy recovery systems.

CLOSED LOOP CONTROL

Closed loop system control greatly reduces the effects of component and operating condition variability. Differential wheel speed information feeds the algorithm for compensation from changes in wheel brake output and vehicle load distribution. Front-to-rear balance due to load variation is maintained at 90 plus percent braking efficiency above 0.2 g deceleration. Test results with and without such dynamic control are depicted in Figure 5.

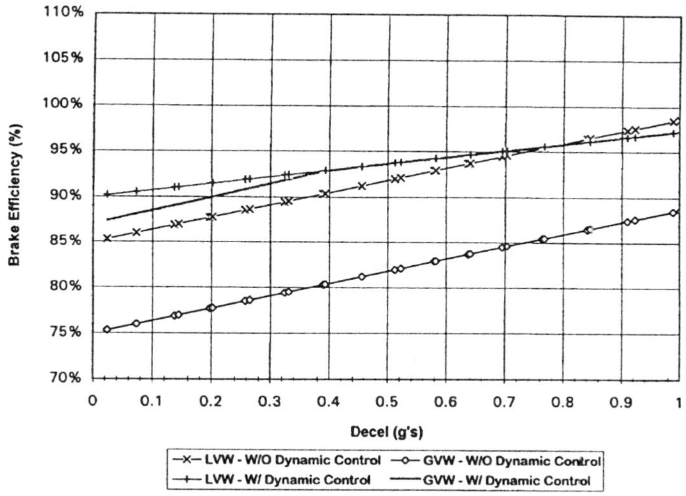

Figure 5 - Brake Efficiency

This same basic control algorithm also compensates and thus reduces side-to-side variability. These variations become transparent as algorithm control alters command torque. Figure 6 depicts two wheel brakes with intentionally differing open loop output being compensated for braking torque by closed loop control. Notice in this example the left rear brake torque reduces while the right brake torque increases to match the total desired axle torque and provide commanded deceleration. The intelligent algorithm memorizes these wheel brake characteristics to compensate during initial activation of subsequent stops.

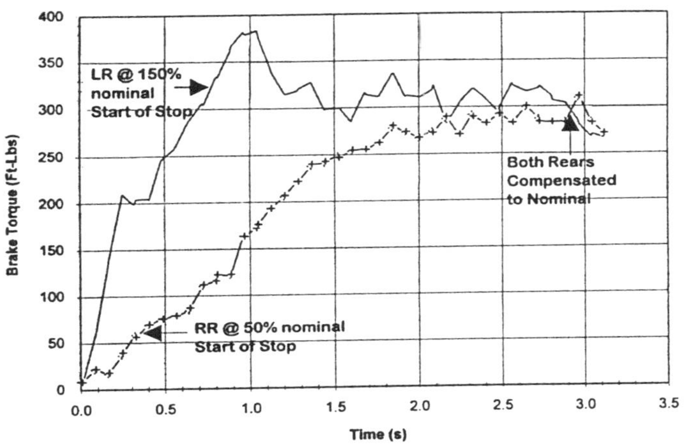

Figure 6 - Side-to-Side Compensation

ACTIVE CONTROL

The system has the capability to assume full four-wheel independent control for ABS, traction control, adaptive cruise, handling management, and automated highway functions. For example, brake augmented handling management optimizes vehicle response by independently activating any or all of the wheel brakes. The resultant brake generated yaw forces improve vehicle maneuverability and handling predictability. Figure 7 depicts differing vehicle yaw control with and without brake system intervention.

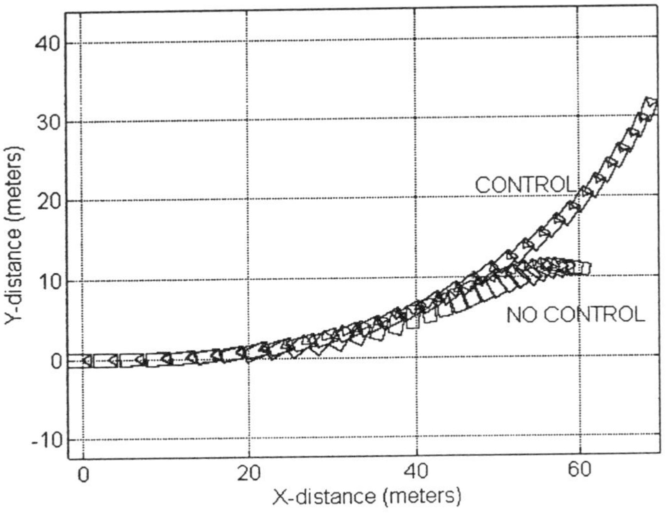

Figure 7 - Yaw Control

CONTROLLABLE ERGONOMICS

Tunable pedal feel provides optimum driver ergonomics. The algorithm has full control of apply system gains. Thus, pedal travel, force, and damping characteristics are easily calibrated. This ensures optimum performance independent of wheel brake parameters. If desired, ergonomics can be mechanized to allow driver adjustment within acceptable limits.

APPROPRIATE INTERFACES

Integration of brake controls with other vehicle systems opens numerous possibilities of greater vehicle performance. An example is the greater electric vehicle operating range obtained by integrating brake with powertrain controls. Full utilization of regenerative braking increases electric vehicle operating range by 25%. However, available regenerative braking varies upon several operating conditions such as state of battery charge and vehicle speed. Integration of brake and propulsion controls allows maximum energy recovery with minimal friction braking. This dynamic interaction between the vehicle's brake and propulsion systems is transparent to the driver. Figure 8 shows an example of a uniformly decelerating vehicle with varying retardation from the propulsion and wheel brakes.

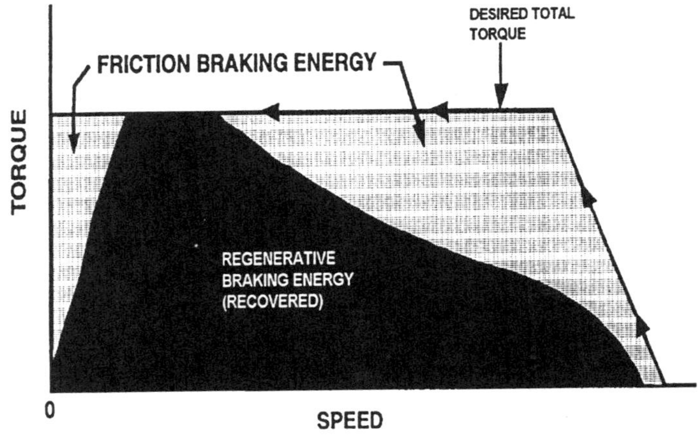

Figure 8 - Blended Regenerative and Friction Braking

SUMMARY

Intelligent brake control systems will provide the high levels of performance, value, low mass, low parasitic losses, and design flexibility required by future vehicles. Some of these features are now on the road in applications such as the GM Impact Preview Fleet, and are daily demonstrating their capability.

As new vehicles are designed, we strongly believe the advantages of these brake systems will set the standard for next levels of vehicle performance.

980606

An Adaptive Cruise Control Using Wheel Torque Management Technique

Satoru Kuragaki, Hiroshi Kuroda, Toshimichi Minowa, Mitsuo Kayano, Tokuji Yoshikawa, Hiroshi Takenaga, Kouzou Nakamura and Kazuaki Takano
Hitachi Lid.

ABSTRACT

This paper describes a longitudinal control method with an Adaptive Cruise Control(ACC) system using a wheel torque management technique. The wheel torque management technique can control vehicular speed by the following procedure without tuning parameters. First, the ACC module calculates a command speed from a desired headway distance and from output data of the radar sensor. Secondly, it calculates a required wheel torque to take the command speed, current speed and running resistance into consideration. Thirdly, the management module controls actuators based on the command wheel torque and characteristics of each vehicle. If the required wheel torque is positive, the management module orders adjustment of the throttle opening position and a change of the gear ratio in the automatic transmission. If the command wheel torque is negative, the management module activates the electronic brake in accordance with the magnitude of the command wheel torque.

This ACC system was installed on a test vehicle. Experiments showed the vehicle follows a preceding vehicle at the speed of the command speed within ± 0.3[m/s] deviation.

INTRODUCTION

Recently, ITS(Intelligent Transport Systems) technology is being developed for manufacturing plants and other users. ACC(Adaptive Cruise Control) system which provides safety and convenience is one of the most important examples of ITS technology. ACC keeps a safe headway distance and provides comfortable driving.[1],[2] A valuable component of ACC is longitudinal control.[3]-[6]

A control technique for the power train has also been developed, based on wheel torque.[7] This technique actuates an electronically controlled throttle valve and automatic transmission cooperatively in order to improve acceleration feeling and get better shifting performance and fuel economy.

This paper describes longitudinal control of a prototype ACC system which adds a speed controller and a headway controller to the system based on the wheel torque management technique. The new method controls has a special feature to control speed without using tuning parameters.

A PROTOTYPE ACC SYSTEM CONFIGURATION

The configuration of a vehicle with ACC is shown in figure 1. The ACC-installed system is designed to have the same configuration manual driving. For manual driving based on a wheel torque management technique, the required wheel torque is determined from the operating stroke of an accelerator pedal. The engine and the transmission are controlled cooperatively by the required wheel torque. For the ACC, the engine and the transmission are cooperatively controlled to correspond with a desired headway distance which is changed by the vehicle speed and the measured headway distance detected by a mm-wave radar sensor. The purpose of ACC is to keep a safe distance from a preceding vehicle by controlling the brake and the power train, which consists of the engine and the transmission. That is to say, expansion from manual driving to ACC using the same configuration can be realized, since the controlling logic of the power train is coordinated for manual driving and the ACC.

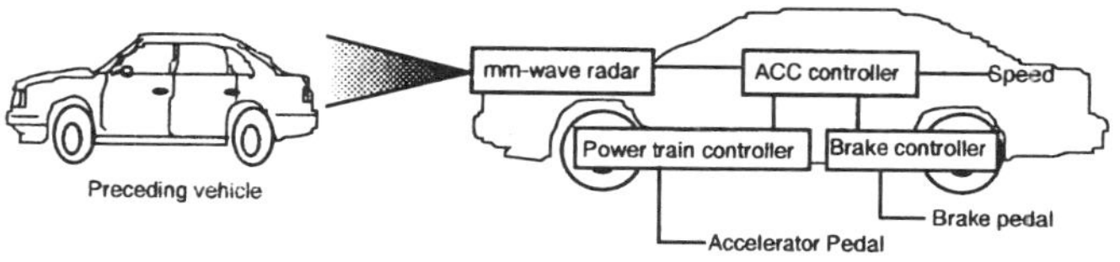

Figure 1. ACC system configuration

BUILDING UP ACC SYSTEM

A block diagram of the proposed ACC system is shown in figure 2. Its function is based on the wheel torque management technique and it has special features to control vehicular speed which are described in the following.

Firstly, the wheel torque management is shown in the lower part of the block diagram. The wheel torque manager selects from an electrically controlled throttle, an automatic transmission or an electrically controlled brake to send commands depending on the required wheel torque. In the following two cases, the required wheel torque is chosen alternatively by the driver or the speed controller. In the manually driven system, the throttle valve opening and transmission ratio are calculated from the required wheel torque T_{rd} which is determined from accelerator pedal angle, vehicular speed and vehicular acceleration (or deceleration). In the other drive system which is treated later, the required wheel torque T_{rc} is calculated from a speed controller. When the selected required wheel torque T_r is larger than a threshold, the throttle valve opening and transmission ratio are calculated and the valve is operated. When the required wheel torque T_r is less than the threshold, the electrically controlled brake is required. These two cases of driving operations are executed by the same software that selects actuators to work. That is to say, it is not necessary to change the software program in the wheel torque manager when the ACC or other systems are added to the manually driven system. By using the wheel torque management technique, the ACC and the manually driven system can be realized with the same architecture.

Secondly, a conventional cruise control function is achieved by adding a speed controller to the wheel torque manager. The speed controller calculates the required wheel torque T_{rc} from the command speed v_c and a speed of the installed vehicle v_o (own speed). If a driver uses the cruise control function in the midst of manual driving, the driver pushes a switch to start the cruise control function. At this time, the command speed v_c would be set to v_o and the speed controller would try to

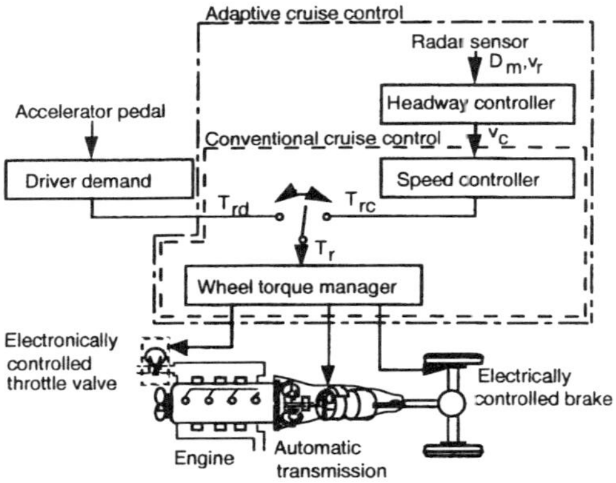

Figure 2. Building up an ACC

regulate speed of the vehicle toward v_c. The speed controller computes the required wheel torque T_{rc} and sends it to the wheel torque manager continuously.

Finally, the ACC function, which consists of a radar sensor and a headway controller, is added to the conventional cruise control function. The headway controller sets itself a desired headway distance D_d which corresponds to the speed of the installed vehicle v_o. The mm-wave radar can measure headway distance D_m and relative speed v_r of the installed and preceding vehicles. The headway controller computes a command speed v_c with a desired headway distance D_d, a measured headway distance D_m, the speed of the installed vehicle and a characteristic of the speed controller. If the ACC function is started, the headway controller calculates the command speed vc and sends the vc to the speed controller. Therefore, the vehicle tries to keep the desired headway distance D_d.

DESIGN OF SPEED CONTROLLER USING WHEEL TORQUE MANAGEMENT TECHNIQUE

SPEED CONTROLLER-- As stated above, the speed controller calculates the required wheel torque T_{rc} from the command speed and the speed of the installed vehicle. The wheel torque manager actualizes wheel torque based on the required wheel torque. Then, the vehicle has the speed that the actualized wheel torque is balanced with the resistance of the driving vehicle. In order to design the speed controller, transfer function of the power train and resistance of the driving vehicle are necessary.

The transfer function of the power train is determined as follows. The required wheel torque which shapes the step function is set for the wheel torque manager, and the wheel torque sensor measures step response of the wheel torque. Then, the transfer function of the power train is described by equation (1) with first order delay and lag time.

$$G_{PW}(s) = \frac{e^{-Ls}}{1+sT_1} \quad \cdots\cdots\cdots\cdots (1)$$

In the test vehicle, L=0.12 and T_1=0.08.

The running resistance is calculated from variance of speed which is measured in the following experiment. The test vehicle is driven at a speed which is the fastest speed possible to make the running model. Recording transition of speed begins at the time to change wheel torque to zero. The transition of acceleration(deceleration) is computed from the recording, and running resistance from to calculated vehicle weight and the transition of acceleration (equation (2)).

$$R(v) = a_1 v^2 + a_2 v + a_3 \quad \cdots\cdots\cdots\cdots (2)$$

In the test vehicle, a_1=-0.3327, a_2=-19.708,

a_3=-222.07.

In order to facilitate use of equation (2), it can be approximated as equation (3) using a Taylor series.

$$R(v) \approx a_4 v + a_5 \qquad \cdots\cdots\cdots\cdots (3)$$

An example of approximating speed is 22[m/s], a_4=-34.35 and a_5=-60.97.

A speed controller can be designed to make a feedback loop with equations (1) and (3). In this case, the speed controller has an I-PD structure as shown in figure 3.[8],[9] Its characteristics of open loop gain and phase are shown in figure 4. Gain margin is 27.2 dB and phase margin is 60 deg.

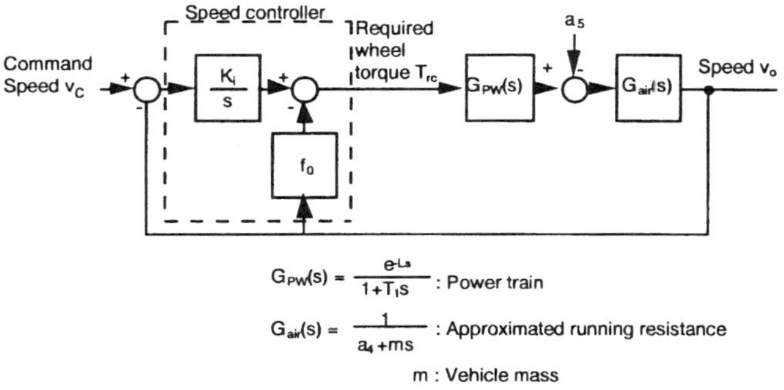

Figure 3. Block diagram of speed controller

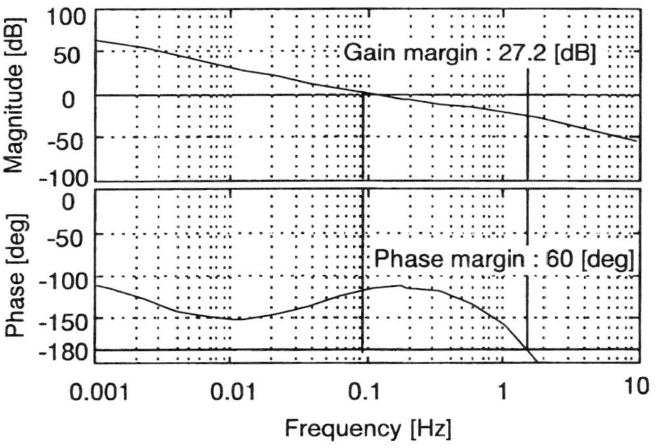

Figure 4. Open loop characteristics of the speed controller

HEADWAY CONTROLLER - The headway controller basically calculates command speed, using equation (4). It has an I-PD structure as shown in figure 5.

$$V_c = K_j \frac{1}{s}(D_d - D_m) - \left(F_1 D_m + F_2 s D_m + F_3 s^2 D_m\right) \ \cdots (4)$$

Relative speed vr is differentiation of measured headway distance D_m. The mm-wave radar can measure relative speed and headway distance simultaneously. Therefore, equation (4) can be changed to equation (5).

$$V_c = K_j \frac{1}{s}(D_d - D_m) - \left(F_1 D_m + F_2 V_r + F_3 s V_r\right) \ \cdots\cdots\cdots (5)$$

Each coefficient K_j,F_1,F_2 and F_3 is determined using computer simulation in consideration of power train, running resistance and the speed controller. Under the condition that the preceding vehicle decreases its speed from 20[m/s] to 11[m/s] with -2.5[m/s^2] deceleration, a set of coefficients that headway distance does not overshoot the desired headway distance is selected. The headway controller's characteristics of open loop gain and phase are shown in figure 6. Gain margin is 30 dB and phase margin is 63.4 deg.

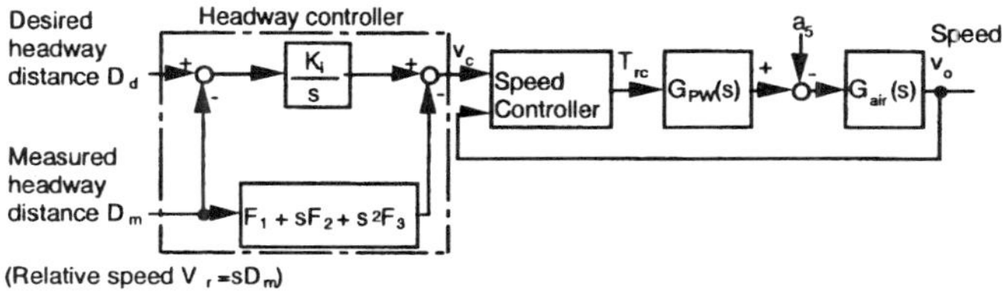

Figure 5. Block diagram of headway controller

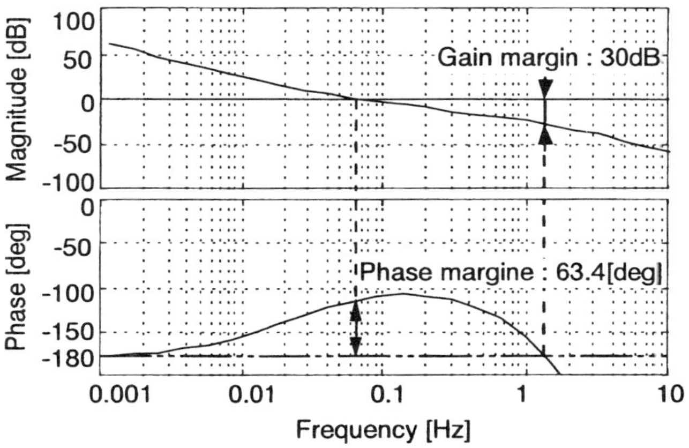

Figure 6. Open loop characteristics of the headway controller

TEST RESULTS

CONVENTIONAL CRUISE CONTROL RESULTS- This ACC system was implemented in a test vehicle. The test results of the conventional cruise control function are shown in figures 7 and 8. For the function, important requirements are stability and transition characteristics of speed. When the command speed jumps to 3[m/s] discontinuously, the target time of accelerating to the command speed is set between 7 to 10 seconds, so that a driver does not feel any time lag or roughness. Under the fixed command speed condition, the target value of speed deviation is set to ±0.3[m/s], so that a driver does not sense any variation of speed.

Target characteristics of the speed when the command speed v_c is changed from 22[m/s] to 25[m/s] are shown in figure 7. The vehicular speed vo follows v_c

by increasing required wheel torque T_{rc} smoothly. In this case, it takes 7.7[s] to increase the vehicular speed v_o to the command speed v_c.

An example of stability characteristics of the speed when the running load changes quickly is shown in figure 8. In this case, the data are gotten on a curved road with a bank when the test vehicle climbs up a slope of 1.5%. The speed controller regulates the vehicular speed v_o within $\pm$0.3[m/s] of the v_c while the vehicle runs on the curved road.

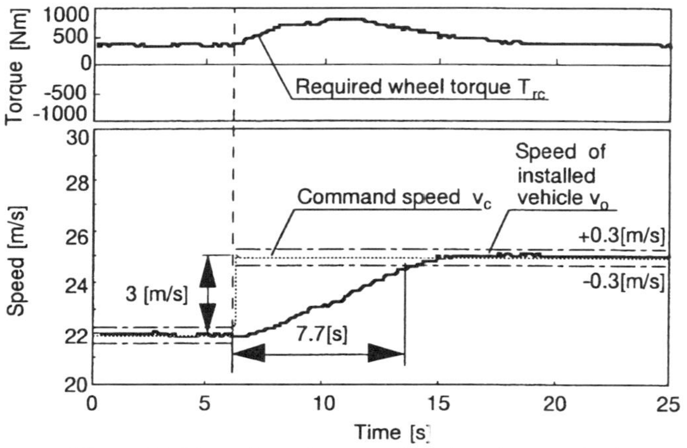

Figure 7. Experimental results of cruise control (Command speed is changed)

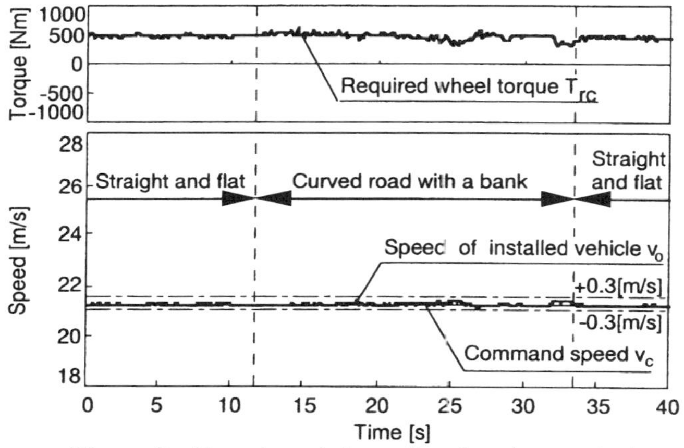

Figure 8. Experimental results of cruise control (Command speed is fixed)

ACC RESULTS– Results of ACC characteristics of required wheel torque, speed and headway are shown in figure 9 for the ACC system. The ACC is designed to follow the speed of the preceding vehicle v_p and to keep a proper headway distance at the same time. At time t =0 the ACC is started, at time t_1 the preceding vehicle increases its speed v_p, at t_2 the preceding vehicle decreases speed v_p by braking and at t_4 v_p is kept at a constant speed. After starting the ACC, the preceding vehicle accelerates and v_p is increasing. The required wheel torque T_{rc} shows its peak value at the beginning of acceleration, and it keeps a constant value during the preceding vehicle's acceleration. During this acceleration, v_o and D_m follow v_p and D_d, respectively. On and after t_2, the preceding vehicle decelerates with about -2.5[m/s^2], the v_p is decreased and T_{rc} is less than zero. The electrically controlled brake operates to keep the measured headway distance the same as desired it. In this situation, the deviation from D_d is -3 meters which is the largest number in this test. On and after t_3, the preceding vehicle accelerates gradually, and deviation from D_d is 3 meters.

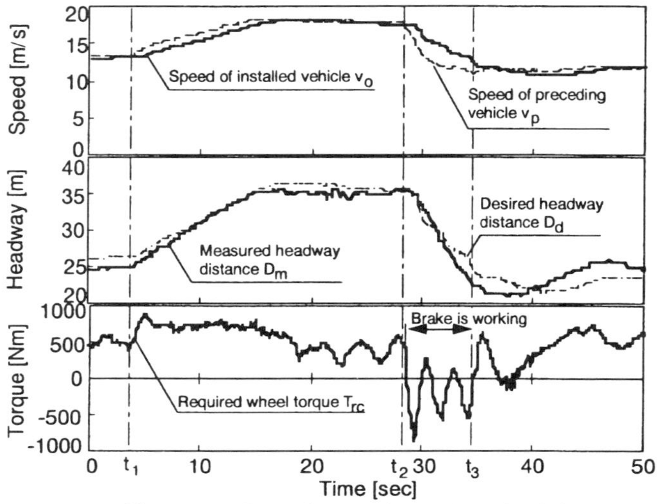

Figure 9. Experimental results of ACC

CONCLUSION

An ACC using a wheel torque management technique has been developed and its performance has been investigated by using a test vehicle. The results of the investigation are summarized as follows.

(1) The ACC system can be built with three controllers. First, the headway controller calculates a command speed from a desired headway distance and output data of the radar sensor which are measured headway distance and relative speed. Secondly, the speed controller calculates a command wheel torque to take the command speed, current speed and running resistance into consideration. Thirdly, the wheel torque manager controls actuators based on the required wheel torque and characteristics of each vehicle.

(2)When using the test vehicle, the conventional cruise control with the wheel torque management technique regulated the vehicular speed v_o within the command speed v_c $\pm$0.3[m/s] deviation.

REFERENCE

[1] Watanabe, T.; Kishimoto, N.; Hayafune, K.; Yamada, K.; Maede, N., "Development of an Intelligent Cruise Control System", Proc. of the 2nd World Congress on Intelligent Transportation Systems, 1995

[2] Hayashi, Y.; Hayafune, K.; Yamada, K., "System Safety Study on Intelligent Cruise Control",Proc. of 4th

World congress on Intelligent Transportation systems, 1997

[3] Fancher, P.; Bareket, Z., "Evaluating Headway Control Using Range Versus Range-Rate Relationships", Vehicle System Dynamics, 1994, pp.575-596

[4] Fujioka, T.; Baba, J., "Control System for platooning -Comparison between Sliding Control and PID Control -"(in Japanese), Proceedings of JSAE, 1995, pp.49-52

[5] Hoess, A.; Hosp, W.; Doerfler, R.; Rauner, H., "Longitudinal Autonomous Vehicle Control Utilizing Access to Electronic Throttle Control, Automatic Transmission and Brakes", SAE961009, 1996

[6] Winner, H.; Witte, S.; Uhler, W.; Lichtenberg, B., "Adaptive Cruise Control System Aspects and Developing Trends", SAE961010, 1996

[7] Minowa, T.; Kimura, H.; Ozaki, N.; Ibamoto, M., "Improvement of Fuel Consumption for a Vehicle with an Automatic Transmission Using Driven Power Control with a Powertrain Model", JSAE Review 17, 1996, pp.375-380

[8] Kitamori, T., "A Method of Control System Design Based upon Partial Knowledge about Controlled Processes" (in Japanese), Transactions of the Society of Instrument and Control Engineers vol.15 No.4 pp.549-555, 1979

[9] Shigemasa, T.;Takagi, Y.;Ichikawa, Y.;Kitamori, T., "A Practical Reference Model for Control System Design" (in Japanese), Transactions of the Society of Instrument and Control Engineers vol.19 No.7 pp.592-594, 1983

ACTUATORS IN STEERING SYSTEMS

981114

Design of Actuator for Active-Rear-Steer System

Akiya Taneda , Toshihiko Yamanaka
AISIN SEIKI CO.,LTD

ABSTRACT

In recent years improved emissions control, energy saving devices and safety features have become the driving forces in automotive manufacturing and technology.

We have developed the Active-Rear-Steer system with an electro-mechanical actuator, which is an improved safety feature. It also features availability to mass production and a compact size. The system is highly responsive, increasing vehicle performance and stability, thus, enhancing the active safety of the vehicle.

INTRODUCTION

We have been developing the four wheel steer system, 4WS, which has used an electro-mechanical actuator for more than 10 years. Our first generation system improved vehicle stability at mid- to high speeds and reduced the turning radius at low speeds. The actuator had a steering control range of +/-5 degrees. The system was a "steer-by-wire" system so assuring reliability was a very important issue.

Therefore the system had 3 Central Processing Units (CPUs), two redundant units and one back up. Also, an extra motor was included in the actuator to return the rack to the neutral position if one motor failed. As a result, the system became extremely expensive and the cost-to-performance trade-off was poor.

We have taken a new look at the system, basing our concept on the ability of a four wheel steer system to maintain high vehicle stability. The reduced turning radius feature at low speeds has been eliminated and the rear steering angle control is limited to within +/-2 degrees. (This system is called the Active-Rear-Steer system, ARS,) Thus it is not necessary to return to the neutral position if the system fails since a vehicle with a rear steering angle of 2 degrees can still be safely driven at reasonable vehicle speeds. Fig.1 shows the ARS system ECU configuration. The system has a control CPU and a monitoring CPU. Returning the rack to the neutral position has also been eliminated, thus simplifying the system and greatly improving the cost-to-performance ratio.

ARS SYSTEM CONFIGURATION

The system includes sensors, an electronic control unit, ECU, and rear steering actuator. The sensors include:

1) Wheel speed sensors : measures the wheel rotational speed and is installed at each wheel.
   These sensors are essential for ABS.
2) Speedometer sensor : measures the vehicle speed and usually is installed on the transmission.
3) Steering sensor : measures the driver's steering wheel operation and is installed on the steering column.
4) Yaw rate sensor : measures the vehicles rotational angular velocity around its vertical axis.

The sensors installed on the actuator follows.

5) Motor pulse sensor : measures the position of the rear steering actuator motor.
   The sensor is attached in the motor.
6) Neutral position sensor : measures the rear steering rack position around the neutral position.

The sensor is installed inside the actuator, and is used for calculating the rack position.

Signals from these sensors are transmitted to the ECU. The ECU consists of the sensor signal interface, motor drive power electronics and interface, a processor which calculates the target rear steering angle and the rear steering actuator position command. In the ECU, vehicle speed, yaw rate, steering wheel angle and rear steering angle are processed using the sensor signals. The target rear steering angle is calculated from the vehicle speed, yaw rate and steering wheel angle using the following equation:

$$\delta_R^* = -K_1(V)\cdot\delta_F + K_2(V)\cdot\gamma$$

where

δ_R^* : Target rear steer angle

δ_F : Steering angle

γ : Yaw rate

$K_1(V)$, $K_2(V)$: Control gain

V : Vehicle velocity

K_1 and K_2 are set to minimize the vehicle body slip angle as a function of vehicle speed. The target rear steering angle is compared with the actual rear steering angle and the deviation between the target and actual values is calculated. The actuator is controlled based upon the deviation.

The ARS actuator does not have any mechanical linkages with the steering wheel. This is the so-called "steer-by-wire" system. The actuator is an electro-mechanical type which is driven by a motor. The rotational motion, which is controlled by the ECU, is transformed to the linear motion of the rear steering rack through a screw and nut mechanism. The rear steering angle is limited within 2 degrees.

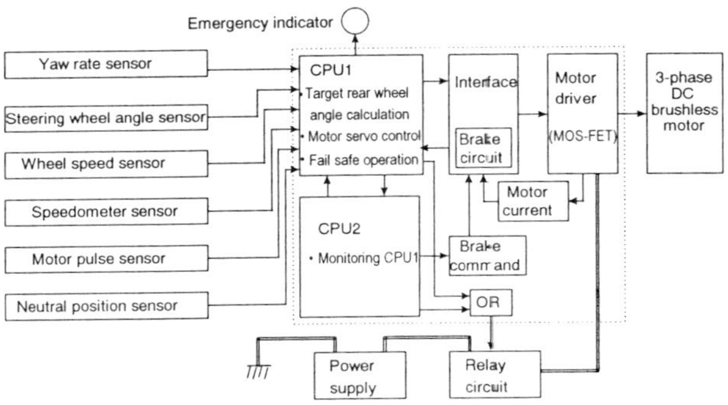

Fig.1 :ARS System ECU Configuration

STRUCTUER OF ARS SYSTEM AND ACTUATOR

SAFETY CONCEPT - The vehicle can safely be driven in case the system, or a part of the system, fails.

Our fundamental safety concepts for the ARS system are:

1) to prevent the sudden rear steer mode where the system steers needlessly without the driver's intention.

2) to fix the rear steering angle when the system fails.

Details of our safety concepts are described below.

A redundant system can eliminate most fatal failures. However, minor failures may increase due to the large number of parts. Through our fundamental design concept to increase reliability, we reduced the number of parts in order to decrease the failure probability of the system.

Therefore, a minimum number of components including sensors, ECU and actuator were used to achieve better system redundancy.

The ARS actuator is designed based upon this safety concept.

REQUIRED SPECIFICATION OF ACTUATOR - The ARS actuator is required to be highly responsive, and the steering angle needs to be very accurate, and to be stable enough to resist wheel disturbances.

The following are the required values for the actuator, using a servo simulation:

Control resolution : Max. 0.055mm(Max. 0.025deg)

Servo frequency response : Max. 3.2Hz

ACTUATOR STRUCTURE - The ARS actuator's simplicity and use of an electric motor provides easy installation. We selected the brushless DC motor in keeping with our design concept: make a system that is fail-safe and simple. Table 1 shows the comparison of motors.

Table 2 shows the comparison of the gear types and the functional ability. We selected the planetary gear set and trapezoidal screw for its concentric layout and its 0% transfer efficiency from the wheels.

The ARS actuator structure is shown in Fig.2. The brushless DC motor and 2 sets of planetary gears are mounted on the same axis. The rotational motion of the motor is converted to the lateral movement of the rack through the screw and nut mechanism. The spline at the end of the rear steering rack prevents rotation of the rack, thus the rotational motion of the motor is only transformed to the lateral motion of the rack. The opposing force from the wheels will not be applied to the

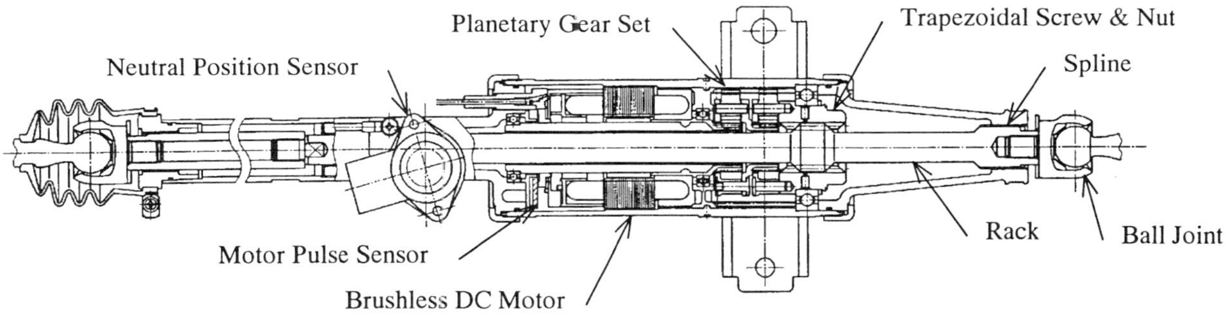

Fig.2 :ARS Actuator Structure

planetary gear sets and motor since the nut is supported by the bearing. And the neutral position sensor is set at the end of the motor.

Table 1 : Comparison of Motor Types

Mechanism	Size	Efficiency	Response	Noise	Fail-safe
DC Brush Motor	Good	Good	Fair	Good	Poor
DC Brushless Motor	Good	Good	Exel.	Good	Exel.

Table 2 : Comparison of Gear Types

Mechanism	Trans. effi. from motor	Trans. effi. from tires	Concentric layout	Noise Vibra.
Rack & Pinion + Hypoid Gear	Fair (40%)	Exel. (0%)	Impossible	Fair
Cam + Worm Gear	Fair (40%)	Exel. (0%)	Impossible	Fair
Planetary Gear Set + Ball Screw	Exel. (90%)	Poor (90%)	Possible	Good
Planetary Gear Set +Trapezoidal Screw	fair (40%)	Exel. (0%)	Possible	Good

Table 3 shows Actuator specification.

Table 3 : Actuator Specification

Parts	Items	Specification
Assembly	Steering transmission ratio	2.2 mm/deg
	Distance between ball joints	805 mm
	Max stroke	± 4.4 mm
	Control resolution	0.0442 mm/step

Fig.3 shows rack stroke speed and rack stroke force.

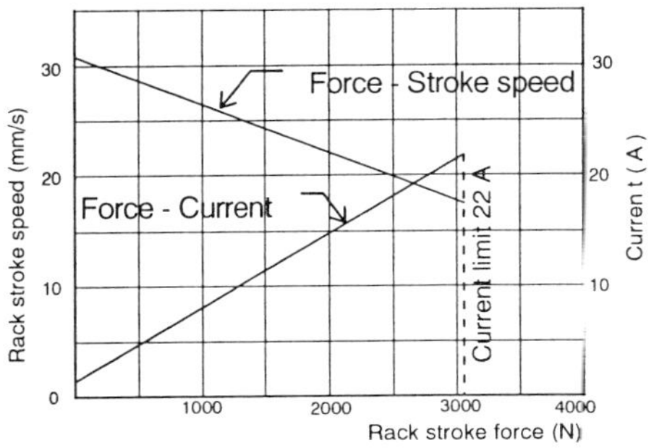

Fig.3 :Actuator Force Performance

BRUSHLESS MOTOR - The brushless motor was an excellent choice not only for its durability but also for its simplification of the fail-safe system. For the motor drive circuit, a redundant system is not necessary since a brushless DC motor is used. Fig.4 shows the motor servo diagram. The current pulse of the motor coil is controlled by a signal that is the difference of the target rear steering angle and the actual rear steering angle.

Fig.5 shows a comparison between a conventional DC motor and a brushless DC motor. The CPU observes the rotation of the rotor by signals from the rotational sensor. The brushless DC motor is driven at small rotational steps. The brushless DC motor can run only when the pulse signals are synchronized. Therefore the motor does not turn in the case of an ECU failure (such as a motor drive circuit) since the synchronized pulses are not transmitted to the motor. The motor drive signal is always monitored by the ECU and anything unusual is easily detected. The failure of the motor coil due to an open- or a short-circuit makes the system stop since the motor driving current of each phase is monitored. Therefore, the sudden rear steer mode (system uncontrollable state) is prevented.

Naturally the failure is easily detected since motor driving pulse and the motor coil current are monitored by the CPU. If an open- or a short-circuit makes the system stop, the failure of the ARS system is indicated to the driver.

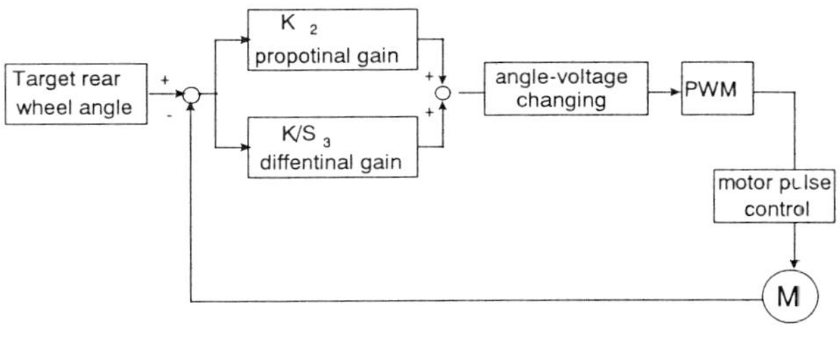

Fig.4 : Motor Servo Diagram

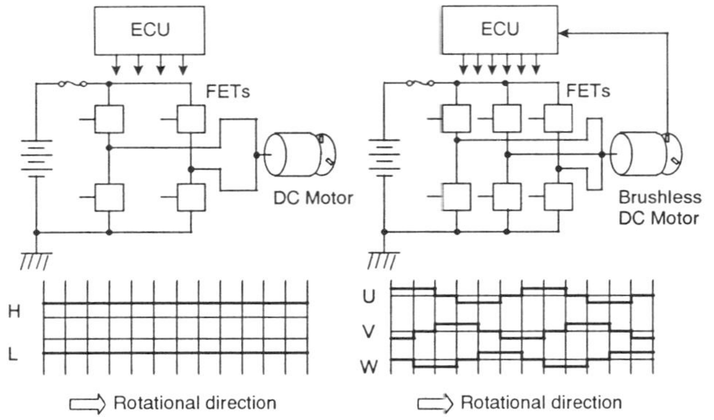

Fig.5 : Comparison of DC Motor and Brashless Motor

PLANETARY GEAR SET - A two stage planetary gear set is located on the same axis of the rack and motor. We made the actuator simple and small by making the planetary gear set of plastic. The gear ratio is 1/3.5 on one stage and 1/12.25 for both stages. The load from the wheels is not applied to this planetary gear set using the trapezoidal screw.

TRAPEZOIDAL SCREW - A trapezoidal screw and nut is used in order to change the motor revolution to an axial stroke and to keep toughness. The trapezoidal nut is mounted by the bearing and coupled to the rack, connecting the output axis of the planetary gear set. The screw and nut are designed to be self-locking in order to hold the rear steering angle fixed in the case of system failure. They also protect the planetary gear set from the force of the wheels.

Fig.6 shows the measured rear steering rack force when the rear wheel hits a side-stone. The screw and nut is designed to withstand a reverse rack pulse force of up to 30 kN. The rear rack is also self-locking against vibration through the suspension members due to the residual force of the motor magnet being amplified by the reduction of the planetary gear set.

FUNCTIONALITY AND DURABILITY

BENCH TEST - Table 4 shows the bench test results of the rack stroke speed and the frequency response of the actuator. The simulated load of the vehicle from the suspension is increased proportionally according to stroke. There is no load at the center position of the rack.

Fig.7 shows the measured rack stroke speed without load. This test sample marks the Max. at 30 mm/s (= 13.6 deg/s) at 6.6 mm of the step target stroke.

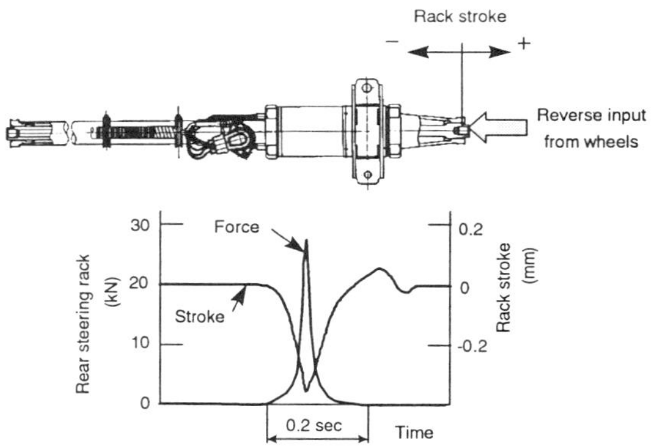

Fig.6 : Road Force Applied to Actuator due to Side Stone

Table 4 : Actuator Functionality

Current	Max. 2.5 A	at no load
Stroke speed	Min. 24 mm/s	at no load
Rack force	Min. 3300 N	at Max. current 22A

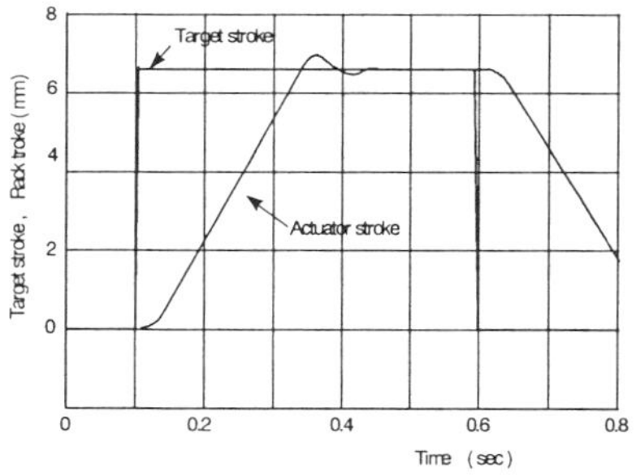

Fig .7 : Rack Stroke Speed

Fig.8 shows the frequency response of the actuator to a sine wave for various target steer angles. Maximum load is 860 N at ± 1.54 mm stroke.

At 4.3 Hz this test sample made the 45 deg point of the phase where the rack stroke is ± 1.1 mm (= ± 0.5 deg).

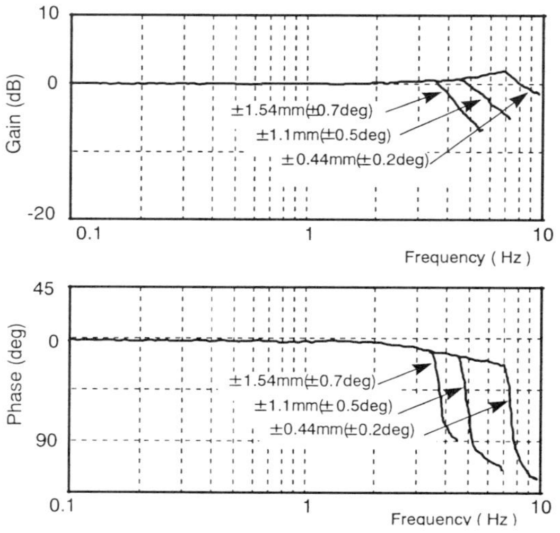

Fig .8 : Frequency Response Characteristic

The above results confirm that the actuator secured the desired response positioning accuracy, and achieved the desired performance of the ARS system.

VEHICLE PERFORMANCE - Fig.9 shows the obstacle avoidance maneuver of vehicles equipped with and without the ARS system. It shows the ARS-equipped vehicles avoided the obstacle at a small yaw rate and at a small vehicle slip angle. This demonstrates that the ARS system has increased the safety limit of the vehicle. Fig.10 shows experimental results of the ARS system for city, highway and mountain driving modes. Table 5 shows the temperature of each part of the system for each driving mode. These results indicated that the circuit breakers which prevent thermal destruction of the motor and FETs were not necessary. The protection circuits were minimized to avoid unnecessary costs.

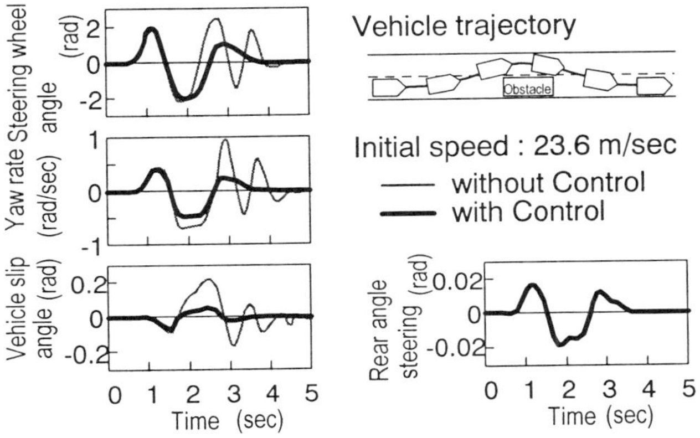

Fig.9 : Obstacle Avoidance Maneuver

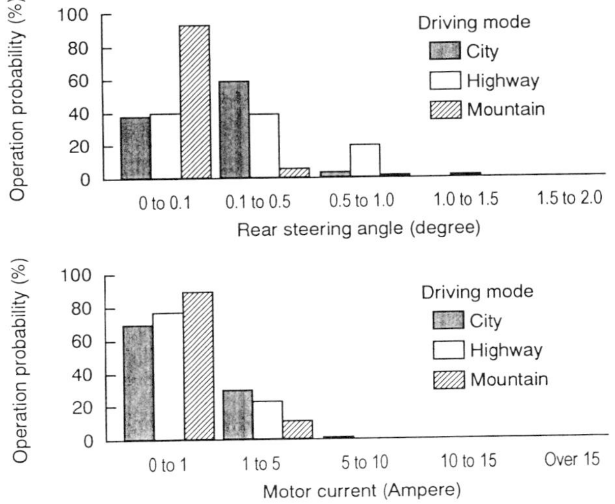

Fig.10 : Operation Probability for Each Driving Mode

Table 5 : Temperatures of Each Driving Mode

Temperature(℃) \ Driving mode	City	Highway	Mountain
Atmosphere	27.0	28.6	26.0
Cabin	28.0	28.2	26.0
Engine compartment	85.8	85.2	70.7
FET	33.0	35.0	32.2
Motor case	37.4	36.3	32.5
Motor coil	27.4	35.8	34.1

CONCLUSION

The test results confirm that the ARS system, which is extremely compact and simple, made a significant contribution to the active safety of the vehicles.

We are convinced that the ARS system should be installed in more vehicles because of its high performance and its easy installation. We firmly believe that the ARS actuator is an item that makes cars safer.

960932

A Study on the Efficient and Non-Interactive Power Transmission Mechanism for the Actuator of a Four Wheel Steering System

Toshiro Ichikawa and Hideaki Nakamura
UNISIA JECS Corp.

Abstract

Non-interactive power transmission mechanism is usually realized by using high ratio worm gearing or hypoid gearing. However, these types of high ratio gears have weak points such as very low efficiency in power transmission when a high reduction ratio is not required. The purpose of this study is intended to realize a non-interactive, or irreversible, power transmission mechanism having a considerably high transmission efficiency with sophisticated arrangement of planetary gearing and wrapped-spring clutch. The experimental result shows that this mechanism can realize a high efficiency more than twice that of comparable worm gearing.

1. Introduction

A 4WS system is one of the systems which are used mainly for position control with a very large dynamic range of load. Thrust forces induced in the tie rod connecting the steering actuator and the knuckle arm of the rear road wheel have been measured under the condition where a small 4WS car is driven on a winding road in mountain area, city road and highway. The thrust force varies in a wide range from 0 to several thousand (N) depending on cornering force and irregularity of the road surface. In order to maintain a precise position of rear steering position, a large power source is required. However, a large power source source is not recommended for energy saving.

In a 4WS system, a large disturbance steering force is applied to the road wheel, when a car runs over an obstacle on the road, it is better to keep the position of rear steering position unchanged. From the viewpoint of safety, it is important to keep the rear steering position unaffected by disturbance force when the driving power source fails. In such a case, it is better from the viewpoints of energy-saving and safety to incorporate an non-interactive power transmission mechanism into the actuator to mechanically isolate the load from affecting the rear steering position.

The non-interactive power transmission mechanism considered in this paper is provided with a special power transmission characteristics. The power inputted from the driving power source to the input shaft is transmitted to the output shaft whereas the torque from the output shaft is not transmitted in the reverse direction. In the latter situation, the output shaft is not allowed to rotate and being kept in its original position.

The above-mentioned function can be realized by applying conventional devices such as worm gearing and hypoid gearing of a high gear ratio which have been used for controlling a large inertia load. However, since such non-interactive power transmission gearing characteristics depend upon the friction of tooth surfaces and lead angles to obtain the irreversibility, they have such a weak point that their power transmitting efficiency is very low.

Non-interactive power transmission characteristics is performed through using a differential gear characteristics of a planetary gearing and control of an output shaft locking with a pair of wrapped-spring clutch. By using a planetary gearing, power transmission efficiency of this system is improved to significant level comparing a worm gearing.

In addition, a driving test is presented by a compact car of 2 litre class equipped with this power transmission mechanism which has a DC motor of 90 W output and a newly developed 4WS control algorithm for low speed.

2. High-efficiency, non-interactive power transmission mechanism

2-1. Configuration of this system

Fig.-1 is a sectional view of this non-interactive power transmission mechanism. Fig.-2 is an exploded view showing the internal construction of this mechanism.

This mechanism consists of a planetary gearing and

a pair of wrapped-spring clutches. The sun gear of the planetary gearing is fixed to the input shaft of this mechanism, connected to the electric motor as shown in this figure.

The internal gear is placed in the housing and supported in a manner rotatable around the center axis of the unit. The planet gears are supported by the shafts provided on the planet carrier via bearings so as to be rotatable around their own axes. These planet gears are engaged with the sun gear and the internal gear. The planet carrier constitutes the output shaft and supported by the housing via bearings so as to be rotatable around the axis of this unit.

Two wrapped-spring clutch hubs (rotatable hubs) formed on the planet carrier are arranged symmetrically in front and rear of the planet gears. Facing these rotatable hubs, two stationary hubs are provided constituting two pairs of rotatable and stationary hubs, with a slight clearance between the ends of paired hubs. The stationary hubs are fixed to the housing with bolts, not shown in the figure.

Each of these two pairs of rotatable and stationary hubs are wrapped with a helical spring, forming a wrapped-spring clutch. The helical springs used in these wrapped-spring clutches are of the same winding direction. The springs are made of a wire having a rectangular cross section for maintaining the contact area with respective rotatable hubs for a long time. The end of each spring at the stationary hub side are hooked so as to engage with the holes provided on the stationary hubs.

Two projections (pins) are provided on the internal gear, one for each spring. The projections (pins) provided on the internal gear and the ends of the spring wire face each other with a slight clearance in the relative position.

Each spring is surrounded by a cylindrical sleeve with a slight clearance. These sleeves are supported by bearings provided in the housing in a manner rotatable around the axis of the unit, or may be fixed to the housing.

In addition, this mechanism has a small amount of silicone oil for lubricating the wrapped-spring clutches.

2-2. Operation of this mechanism

Each wrapped-spring clutch is provided with self-energizing characteristics by the binding effect of the helical spring, functioning as an inherent one-way clutch.[1] The wrapped-spring clutches are so constructed that the rotatable hub (planet carrier) on the motor side and the same on the output shaft side are rotatable only in respective arrowed directions (Fig.-2).

By applying a very small torque to the end of the spring, the self-energizing characteristics of the wrapped-spring clutch is released, and thus, the one-way clutch function become ineffective. The reason for this is that, when the end of the spring wire is pushed in the direction of expanding (increasing) the coil diameter, the hub is made free from wrapping effect of the spring, releasing the contact between the two. Each wrapped-spring clutch can be controlled independently by selectively pushing the ends of the respective springs by the pins provided on the internal gear.

To understand the function of this mechanism, it is necessary to divide the operation into the following modes.

A. No torque is inputted to the input shaft.

B. A torque is inputted to the input shaft.

B-a. Driving state

This state is further divided into two cases, first case is the direction of input torque is opposite to that of load torque and second case where the rotating speed of the output shaft exceeds the input shaft.

The former case includes a state where the

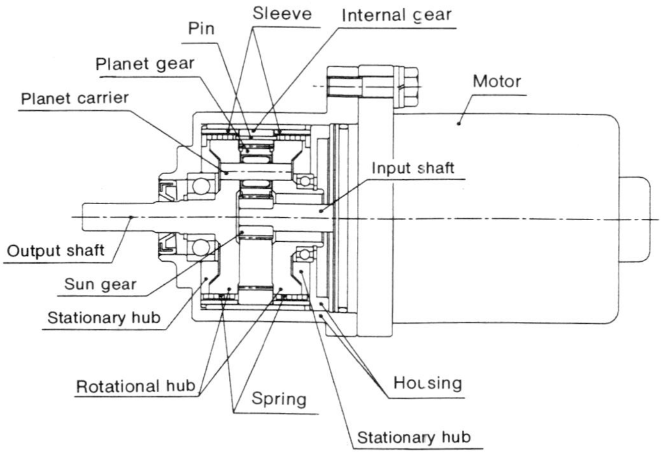

Fig. 1 Configuration of Non-Interactive Power Transmission Mechanism

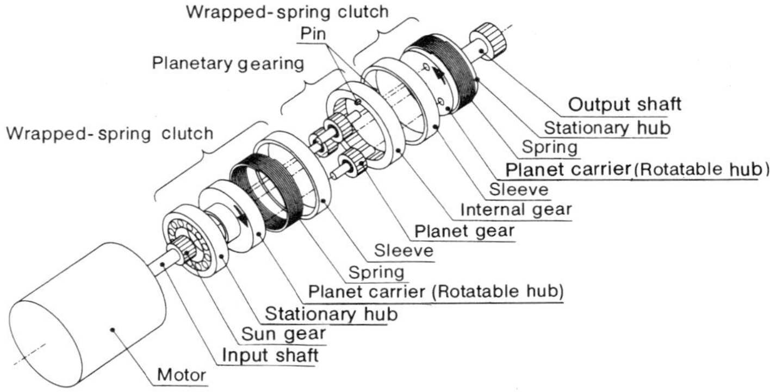

Fig. 2 Configuration of Non-Interactive Power Transmission Mechanism

output shaft cannot be driven because the input torque is smaller than the load torque.

B-b. Braking state

This is a state where the input torque and the load torque are in the same direction and the rotating speed transmitted from the input shaft to the output shaft is smaller than that of load.

2-2-A. No torque to the input shaft

The hubs are tightly wrapped around with two springs by the contracting force of the springs. These two springs are coiled in the same direction. Relative position between the stationary hub and the rotatable hub within one set is the opposite to that of the other set. The rotating direction of the planet carrier when locked by one wrapped-spring clutch (one-way clutch) is the same as that when not locked by the other wrapped-spring clutch. Therefore, the rotation of the output shaft is locked in both directions.

This locking torque is represented by the maximum locking torque of the wrapped-spring clutch which is given by the following equation.[1)]

$$T_b(max) = \frac{2 D_d E I \delta}{D_m^2 D_f} \left(e^{2\pi \mu N} - 1\right)$$

Where,

Tb(max) : Maximum locking torque (Nm)
Dd : Hub diameter (m)
μ : Coefficient of friction
E : Young's modulus (Pa)
I : Geometrical moment of inertia of spring (m^4)
δ : Diametral margin between spring and hub (m)
Dm : Mean coil diameter of spring in driving state (m)
N : Number of effective turns of spring on the hub
Df : Mean coil diameter of spring in free state (m)

When a disturbance torque exceeding this locking torque, the amount of torque over the locking torque is transmitted to the input shaft.

2-2-B. Torque to the input shaft

a. Driving state

At first, the condition where the direction of torque applied to the input shaft is opposite to that of load torque is explained, with the state where no load is applied to the input shaft being defined as the initial state. When a torque is given to the input shaft, the planetary carrier does not rotate because the output shaft is locked by two wrapped-spring clutches. Instead, the planet gears rotate around their own axes and the rotation of the input shaft is transmitted to the internal gear.

The internal gear rotates in the direction opposite to that of the input shaft and the end of spring wire of first wrapped-spring clutch is pushed by the projection (pin) on the internal gear to expand the spring diameter. First wrapped-spring clutch is determined automatically according to the input rotational direction. After expanding the spring diameter, so it becomes loose state between the spring and the hub, that wrapped-spring clutch is released.

The reaction force of the expanded spring to return to its initial diameter increases slightly in accordance with the rotation of the internal gear as shown in the equation below. At the moment when the spring comes into close contact with the inner surface of the sleeve, the reaction force increases suddenly because the expansion of the outer diameter of the spring is restricted there.

The reaction force of the spring applied to the internal gear is represented by the following equation. The inter-wire friction of the spring and the viscous friction between the hub and the spring caused by silicone oil are neglected.

$$Tr = \frac{EI}{Dm}\left(-\frac{\phi}{\pi Na} + \frac{2\delta}{Dm}\right)$$

Where, Tr : Reaction force of spring (Nm)
ϕ : Rotation angle of internal gear
Na : Total number of turns of spring on the hub

When this reaction torque of the spring is almost balanced with load torque on the output shaft via the planetary gearing, the rotation of the internal gear stops and the rotation of the input shaft is transmitted to the output shaft (planet carrier).

At this time, the second rotates together with the output shaft accompanied with slight friction because the direction of rotation is freewheeling direction of this clutch. When a disturbance torque larger than that transmitted from the input shaft to the output shaft is inputted to the output shaft, the one-way clutch functions by virtue of the self-energizing characteristics of this wrapped-spring clutch, prevents the output shaft from being pushed back by the disturbance torque (braking effect).

The freewheeling torque of the wrapped-spring clutch is expressed by the following equation.[1)]

$$To = \frac{2EI\delta}{Dm^2}$$

To :Freewheeling torque of wrapped-spring clutch (Nm)

When the torque inputted to the input shaft and the load torque are of the same direction and the rotating speed of the output shaft transmitted from the input shaft is larger than that of the load, namely;

$$|\omega o| < \frac{|\omega i|}{\eta} \quad \text{(when } \omega o < 0 \text{ and } \omega i < 0, \text{ or } \omega o > 0 \text{ and } \omega i > 0)$$

where, η : Reduction ratio between input/output shafts when the internal gear is fixed
|ωi| : Absolute value of input shaft rotating speed

$|\omega_o|$: Absolute value of output shaft rotating speed,

then, the internal gear tends to rotate in the direction opposite to that of input shaft corresponding to the speed difference between input/output shafts by virtue of the working characteristics of planetary gearing. If a torque large enough to rotate the internal gear and to expand the spring is generated between input/output shafts, the spring is expanded to allow transmission of the power from the input shaft to the output shaft.

b. Braking state

The condition for this state is as shown by the following equation.

$$|\omega_o| > \frac{|\omega_i|}{\eta} \quad \text{(when } \omega_o < 0 \text{ and } \omega_i < 0, \text{ or } \omega_o > 0 \text{ and } \omega_i > 0)$$

The condition where no torque is applied to the input shaft is defined as the initial condition in the following explanation.

When a torque is applied to the input shaft, the planet gears rotate around their own axes and the rotation of the input shaft is transmitted to the internal gear, instead the planetary carrier does not rotate because the output shaft is locked by two wrapped-spring clutches. The internal gear will rotate in the direction opposite to that of the input shaft, pushing the end of one spring wire by the projection (pin) of the internal gear to expand the diameter of the spring. Thus, the clutch which has been locking the rotation of the output shaft is released as in the case of the driving state. When the contact between the spring and the hub is loosened and the locking torque is reduced to a level below the load torque, the output shaft starts to rotate.

Since the rotating speed of output shaft which is transmitted from the input shaft is smaller than that of the load according to the above-mentioned condition, the internal gear rotates in the reverse direction at a speed equivalent to the speed difference between input/output shafts by virtue of the working characteristics of planetary gearing. The spring whose contact being kept loose moves to the engaging direction to increase the friction between the spring and the output shaft and decreases the rotating speed of the output shaft. In this way, the speed difference between the load and the input shaft is adjusted automatically. In other words, the wrapped-spring clutch is functioning properly against corresponding speed difference.

Thus, this mechanism mechanically feeds back the rotating speed of the output shaft to autonomously control the wrapped-spring clutch (brake).

2-3. Results

The typical dimensions and performance of this non-interactive power transmission mechanism are shown in Tables-1 and 2.

The angle of transmission delay shown in Table-2 is defined in the following paragraph.

It becomes to transmit the input power through this mechanism after releasing one-way clutch function after one-way clutch is released by expanding the radius of wrapped-spring clutch according to input shaft rotation. There is a transmission delay of input shaft rotating angle from a point where the input shaft starts its rotation to another point where output shaft receives its rotation.

This angle can be expressed by the following equation.

$$\phi_d\,(max) = (Dr-(Dm+h))*2\pi\, NaZc/DmZa$$

Where, ϕ_d : Angle of delay of torque transmission

h : Height of spring wire (m)

Dr : Inner diameter of sleeve (m)

Za : No. of teeth of sun gear

Zc : No. of teeth of internal gear

The angle of transmission delay shown in Table 2 includes the gear backlash in the planetary gearing.

The transition of torque transmission efficiency obtained through an endurance test conducted under the following conditions is as shown in Fig.-3.

Loading conditions;

Maximum loaded torque : ±0.52Nm (loaded with a spring)

Output shaft rotating angle : ±740°

Operating condition : 0.1Hz

Number of endurance repetition : 6×10^4 times

In this mechanism, an efficiency of approx. 72%, about 2 times better than that of worm gearing is obtained

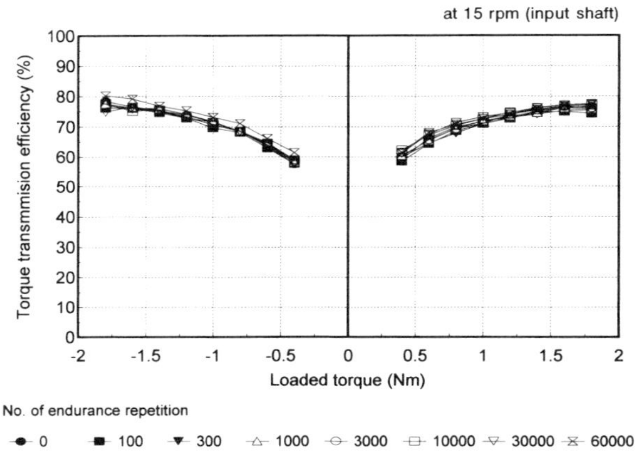

Fig. 3 Result of torque transmission efficiency measurement

Table 1 Dimension of Non-Interactive Power Transmission Mechanism

Hub dia.	Dd= ϕ 45.7mm
Sleeve inner dia.	Dr= ϕ 48.9mm
Mean dia spring coil	Df= ϕ 46.2mm
Wire dia. of spring	b x h=1.5mm x 1.5mm
No. of effective turns	N=4.4 turns
Total No. of turns	Na=6.0 turns
No. of teeth of sun gear	Za=24 (module=0.5)
No. of teeth of planet gear	Zb=30 (three gears)
No. of teeth of internal gear	Zc=84
Reduction ratio	4.5

Table 2 Current performance level of Non-Interactive Power Transmission Mechanism

Max lock torque	Tmax≧100 Nm [Spring deformation torque 45 Nm(Calculated value)]
Play of output shaft	θ max=1.0°
Delay angle of torque transmission	φ max=24° (Input shaft)
Torque transmission efficiency	η=72% (At loaded torque of 1Nm)

constantly. (The torque transmission efficiency of worm gearing and the like is usually at the level of 30% at the best with enough irreversibility.)

2-4. Comparison with conventional devices

This mechanism can provide a high power transmission efficiency since the wrapped-spring clutch employed in this mechanism can control the self-energizing characteristics of its one-way clutch function with a small amount of torque, the loss of input torque due to this operation is quite small. This small loss of input torque described above and the use of high-efficiency planetary gearing (spur gears) for power transmission, this mechanism can provide a high power transmission efficiency.

The comparison between worm gearing and this system described so far can be summarized as shown below.

Worm gearing;
- is low in power transmission efficiency,
- can provide a large reduction ratio with only one step,
- requires skewed input/output shafts configuration (90° in general), and
- cannot provide irreversibility in the range of low reduction ratio.

This mechanism;
- is high in power transmission efficiency (more than two times that of worm gearing),
- cannot provide a high reduction ratio with one step as that of worm gearing,
- can provide irreversibility irrespective of the level of reduction ratio, and
- has an inherent transmission delay angle.

Because of the above, it is considered that this mechanism is effective in a system which requires irreversibility in the range of low reduction ratio where worm gearing and the like cannot provide irreversibility.

3. Application of this mechanism to 4WS

The following is an example of this system applied to an electric 4WS system.

3-1. The purpose of 4WS system

A 4WS system, in which the rear wheels are steered in proportion to the steered angle of the front wheels, gives the driver a disorderly feeling after the start of turning at a low speed ; because the rear of the vehicle goes out of the turning circle. To reduce this feeling of disorderliness, and thus, to improve the driveability, several approaches have been tried.[2)]

The 4WS control system described in this paper is also developed to achieve the similar goal.

3-2. 4WS control by means of traveling distance algorithm

It is considered as possible to reduce this feeling of disorderliness by controlling the outward sway of vehicle rear end, right after the vehicle started turning, to equivalent level of 2WS vehicle.

When a turning locus of a 2WS vehicle is geographically examined, it is found that the maximum outward sway point of vehicle rear end from the turning circle occurs at the point where the rear end reaches the point where the rear wheel located initially as shown in Fig.-4. If the rear wheels are steered so that the vehicle rear end moves along the tangent line drawn from this peak point to the turning circle, it is possible to limit the amount of the sway so that equivalent to 2WS vehicle, and it become possible to reduce the turning radius.

An usually 4WS vehicle employs an algorithm of proportional steering angle. A proportional constant is determined as a parameter depending on each vehicle type and vehicle speed and the target rear steering angle is determined through multiplying the proportional constant by the function corresponding to the distance algorithm.

The rear wheel steering angle target is calculated according to the following formula.

$$\delta r(x) = -\frac{K}{\tau}\,\delta f(x)\,e^{-(x-Lr)/\tau}$$

Where, δf : Front wheel steering angle
δr : Target rear wheel steering angle
Lr : Distance between rear axle and vehicle rear end (m)
K : Ratio between front/rear wheel steering angles
x : Vehicle travel distance (m)
τ : Constant of primary delay (m)

3-3. Construction of the actuator and its function

3-3-A. Configuration of major components

The drive-by-wire system with DC motor is employed because it is impossible to realize such a control if the rear wheel steering mechanism is mechanically linked with the front wheels. It has a steering angle generating unit and a steering angle range setting unit having an individual power source against system failure. When one of the motors fails, this system is so constructed that the remaining motor can fix the rear wheel steering angle in its neutral position.

Configuration of major components are explained below, referring to Fig.-5 and Fig.-6.

The steering rod is connected with steering knuckle arms of the vehicle through tie rods. A stub shaft is fixed to the steering rod at a right angle to the moving direction of the latter. An arm is engaged with this stub shaft through a bush in the manner freely rotatable around the stub shaft.

A fulcrum is provided so as to move only in the direction (Axis-j) perpendicular to the moving direction of the steering rod. This fulcrum, the stub shaft provided on the steering rod, and an eccentric cam are interconnected with an arm which has two grooves, one in which the fulcrum slides and the other in which the cam follower of the eccentric cam slides.

3-3-B. Function of major components

Steering angle generation unit :

The eccentric cam swings to left or to right of the axis-j as indicated by arrows and slides in the groove of the arm. The arm swings around the fulcrum according to the movement of the cam as shown in the Fig.-5, and thus, transmits the movement of the cam to the steering rod and displaces it. The thrust force applied to the steering rod is transmitted to the steering knuckle arms through the tie rods to give a steering angle to the rear wheels (steering angle generation unit). This steering angle generating unit, which requires a high speed response, is provided with the above-mentioned non-interactive power transmission mechanism between the eccentric cam and the motor as shown in Fig.-6.

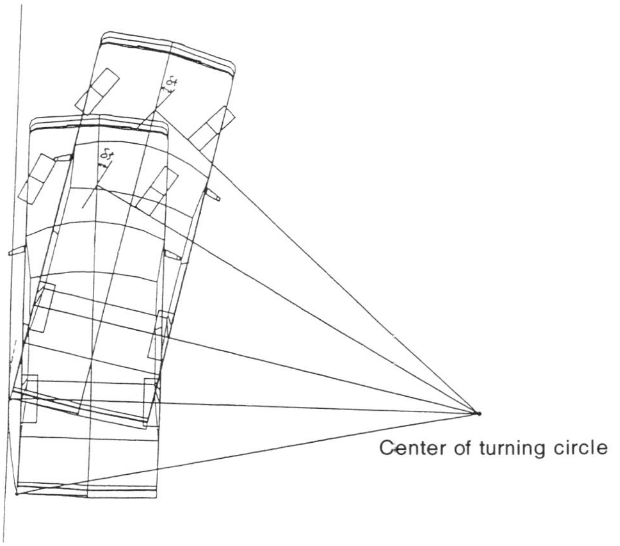

Fig. 4 Turning locus of a 2WS vehicle

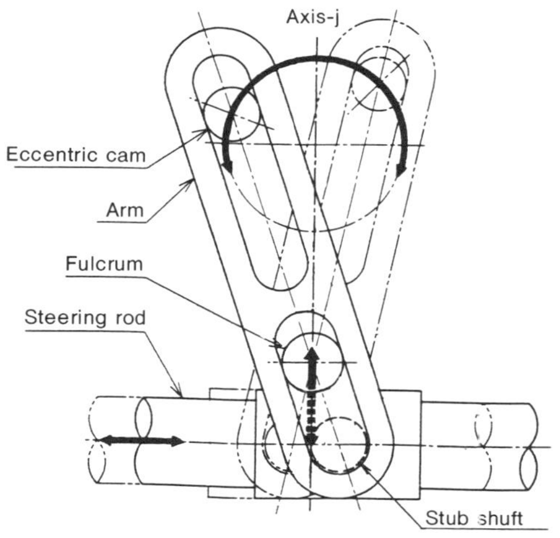

Fig. 5 Principle of steering angle generating mechanism

Steering angle range setting unit :

Otherwise, when the position of the fulcrum is moved in the direction away from the steering rod, the movable range of steering rod become wider. On the contrary, when the former is moved closer to the steering rod, the latter become narrower so as to determine the range of steering angle. A worm gearing is employed for the range of steering angle control because it does not require a high speed response comparing the steering angle generation unit and it is better to employ a large reduction ratio with compactness.

To return to the neutral position :

When the fulcrum is lowered to the position of the stub shaft on the steering rod, the steering rod is fixed in its neutral position wherever the eccentric cam, namely the input force point, positioned because the fulcrum is placed exactly at the position of the output force point. Since the fulcrum (unit for setting the range of steering angle generation) and the eccentric cam (steering angle generation unit) are driven independently by respective DC motors as shown in Fig.-6, the steering rod can be returned to its neutral position independently by these two units.

3-3-C. General specifications

Because of this setting of steering angle range, the overall reduction ratio from the motor to the rear axle and the maximum thrust force applied to the steering rod also

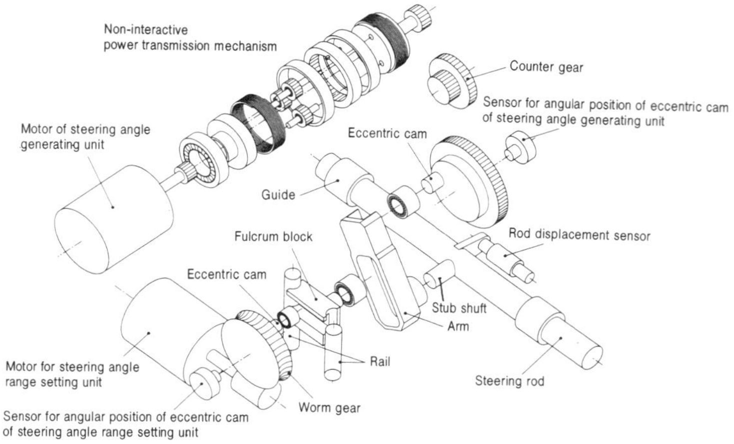

Fig. 6 Schematic drawing of actuator configuration

change according to the vehicle speed. The thrust force of the steering rod generated by the actuator and the range of steering angle are shown in Table-3.

At a low vehicle speed where the steering load is small and the range of steering angle is wide, the reduction ratio of the steering angle generation unit is low and the thrust force of the steering rod is also small. On the contrary, at a high vehicle speed where the range of steering angle is narrow and a large cornering force is applied from the tires, the reduction ratio of the steering angle generation unit becomes larger, and consequently, the maximum thrust force applied to the steering rod also becomes larger. Generally, stationary steering (steering operation while the vehicle being stop) requires a very large thrust force. In this system, however, the rear wheels are not steered at the time of stationary steering operation by virtue of "4WS control by means of traveling distance algorithm".

3-4. System configuration

Fig.-7 is the configuration of this system, showing major component units contained in this system.

In this system, it is necessary to obtain the pulse signals corresponding to the travel distance. For this purpose, the wheel speed sensor by means of magneto-resistance effect device for ABS is employed, it can detect stabilized

Table 3 General spec. of 4WS actuator

Power source	DC motor of 90(w) output with brushes;2 units (One each at steering angle generating unit and steering angle range setting unit)
Size	W260 x H140.5 x D134(mm)
Weight	Approx 8.5(kg)
Thrust force of steering rod(At neutral position)	Approx 7500(N) (Steering angle range ±1.3°) Approx 2400(N) (steering angle range ±7°)
Rigidity of steering rod (At neutral position)	Approx 3500(N/mm) (Steering angle range ±1.3°) Approx 1200(N/mm) (Steering angle range ±7°)

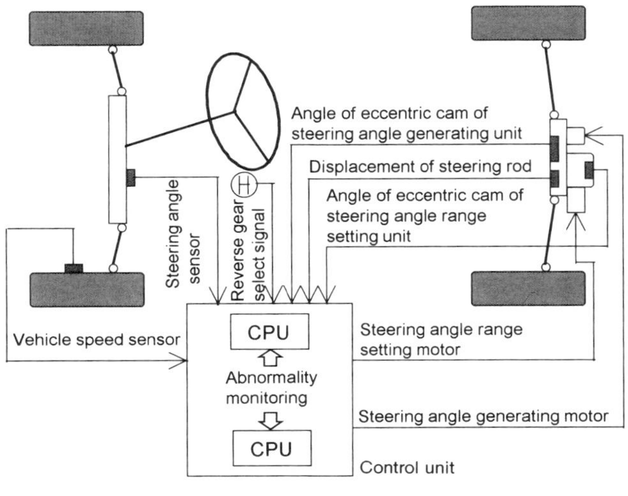

Fig. 7 System configuration

signal until vehicle stops. The signal from the wheel speed sensor and the front wheel steering angle sensor are inputted to the electronic control unit to determine the target values for the steering angle generation unit and the steering range setting unit, and an electric current of corresponding level is supplied to each motor. Three sensors provided in the actuator, namely a sensor for rotating angle of the eccentric cam in the steering angle generation unit, a sensor for rotating angle of the eccentric cam in the steering angle range setting unit and a sensor for displacement of the steering rod, detect and process it in the electronic control unit.

This control is ineffective in case of reverse gear position, while proportional steering angle control is effective even in case of reverse gear position.

3-5. Block diagram of 4WS control system

Fig.-8 shows the block diagram of this 4WS system.

The positioning control surrounded by broken line in this figure is controlled in time basis while steering angle control part is processed by the traveling distance algor-

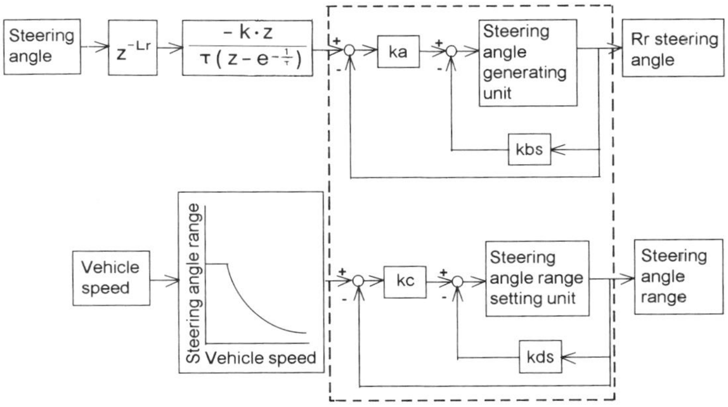

Fig. 8 Control block diagram

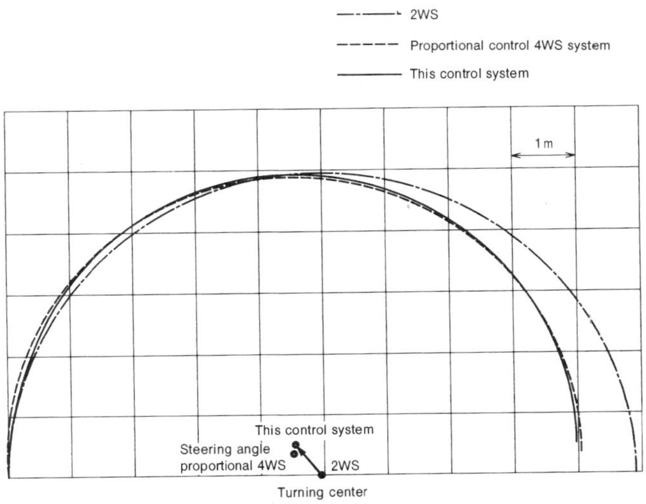

Fig. 9 Locus of rear outer wheel at vehicle turning

ithm with a difference method. In the unit shown at left for calculating the target values, the calculation is executed through an external interrupt routine by using the pulse obtained from the wheel speed sensor. In the steering angle generation unit, the target value is calculated by multiplying the front wheel steering angle by a value obtained through the traveling distance algorithm, whereas in the steering angle range setting unit, the target value is determined based upon the table of steering angle setting corresponding to vehicle speed.

3-6. Result of the test by actual vehicle

Three systems, namely, 2WS, steering angle proportional control 4WS and new 4WS control are compared by using an actual vehicle.

Since the maximum steering angle of rear wheel is 5° in the vehicle equipped with the steering angle proportional control and 7° in the vehicle equipped with new 4WS control, the minimum turning radius of the latter is the smallest among all. When the locus of the outer rear wheel at vehicle turning is examined, it is known that the locus in the case of new 4WS is almost the same as that of 2WS at the initial stage of turning, and then, gradually approaches the locus of the steering angle proportional control 4WS. (Fig.-9)

Assuming the operation of our 4WS at a high vehicle speed, the response of the actuator is checked under the condition where the steering angle range is set at ±1.5°. When the frequency response at rear wheel steering angle

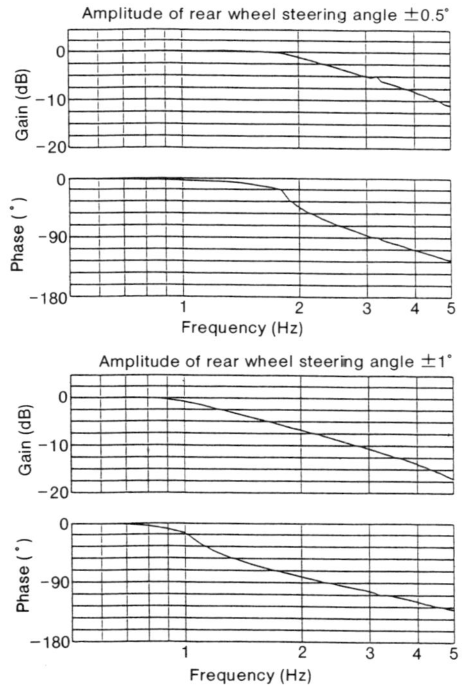

Fig. 10 Actuator response characteristics

amplitude of ±1.0° and ±0.5° was measured, cutoff frequency values of approximately 1 Hz and 1.9 Hz are obtained respectively as shown in Fig.-10, indicating that this system is also applicable to a 4WS system in which the vehicle posture at a high vehicle speed is taken into consideration.

There is no outward sway of vehicle rear wheel in the initial stage of turning when maximum rear steering angle 7° and smooth driveability is obtained. Usually, hydroulic or electric (several hundred W DC motor) power source is required for a 4WS of small passenger car, this system has a 90W DC motor for steering rear wheel.

4. Summary

A system is intended to improve the efficiency of a power transmission mechanism ; which is effective in an application where irreversibility is required in the range where a fairly large reduction ratio is not required.

It is able to propose a 4WS system driven by 90W DC motors by employing an actuator consisting of a high-efficiency non-interactive power transmission unit and a variable steering angle range unit, and "4WS control by means of traveling distance algorithm".

Further task with regard to this new system is to find new applications and to reduce the angle of transmission delay to enlarge their application.

References:

1) R.D.Lowery and A. W. Mehrbtodt, How to do more with Wrapped-Spring Clutches. MACHINE DESIGN, July 22, (1976)

2) Shinsub Jung and Dennis A. Guenther, An Examination of the Maneuverability of an All Wheel Steer Vehicle at Low Speed. SAE Paper 910241 (1991)

940867

Magnetic Power Steering Assist System - MAGNASTEER

Andrzej M. Pawlak
NAO Research and Development Center

David W. Graber and Dennis C. Eckhardt
Saginaw Div.

ABSTRACT

All commonly used speed sensitive power steering assist systems for cars are based on hydraulics. These systems change the pressure or modulate the flow of the conventional hydraulic power steering system. They require some means of actuation in the form of an electromagnetic device such as a solenoid actuator working as a flow or pressure regulator or brushless motor driving a hydraulic pump. Usually they are pulse width modulated controlled by a microprocessor. Therefore, these systems can be quite complicated.

This paper describes MAGNASTEER, a novel magnetic speed sensitive system developed at Saginaw Division in cooperation with the NAO Research and Development Center, which is an add on to the basic hydraulic power steering system. The effort variation provided by MAGNASTEER is the result of an electronically controlled electromagnetic torque, which acts as an addition or subtraction to the torsion bar torsional rate, effectively varying the feel of the hydraulic system. Unlike some other variable effort speed-dependent systems based on hydraulics, the MAGNASTEER system doesn't require additional components such as sensors, hydraulic pump controls, or additional hydraulic plumbing. This simple and highly reliable system doesn't interfere with the existing hydraulics, and is fully integrated into -the system.

INTRODUCTION

Basic hydraulic power steering systems provide a fixed steering assist offset curve, regardless of the vehicle speed. As a result, a compromise has to be made between the high assist steering effort required during parking maneuvers and the "on-center" steering feel for highway driving. Speed variable steering effort can provide light steering efforts during parking maneuvers and more precise efforts at highway speeds.

All conventional speed sensitive power steering assist systems change the pressure or modulate the flow of the basic hydraulic power steering system. Therefore for this purpose the basic hydraulic power steering system is modified to provide controllable assist levels. Common approaches may use electronic controlled hydraulics, a power steering pump with a linear actuator, a hand-wheel speed sensor, or a power gear with a specially modified valve or any combination of the above. Based on vehicle speed and the signal from a hand-wheel sensor if used, the controller varies the steering assist by pulse width modulating the current in the actuator, thereby generating variable levels of steering assist at varying vehicle speed.

MAGNASTEER SYSTEM

The MAGNASTEER system simplifies the speed variable steering assist function and is based on the novel concept shown schematically in Figure 1. Ref. [1,2].

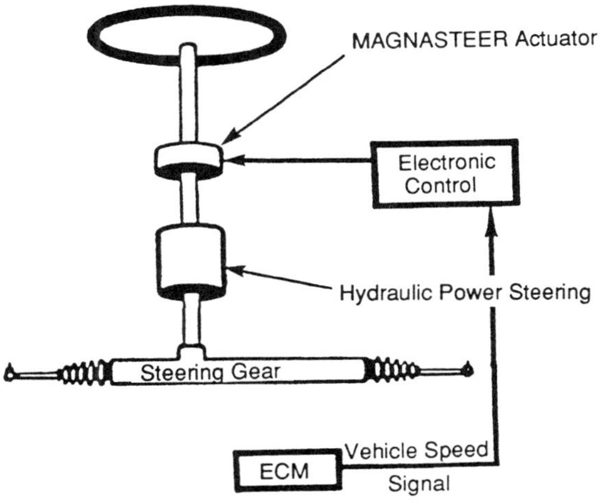

Figure 1. Schematic of the MAGNASTEER System.

The system consists of a homopolar rotary actuator with a multipole ring permanent magnet and a controller. This system provides additional control of the hydraulic power steering valve which is the result of an electronically controlled, vehicle speed and valve position dependent electromagnetic torque developed by the rotary actuator which acts in addition or subtraction to that of the torsion bar. The electromagnetic torque varies with the torsion bar rotation and enhances the torsion bar effective torque for highway speed vehicle operation while reducing it for parking maneuvers, as shown in Figure 2. Both, torque and position angle values are per unit (p.u) which are specified in nomenclature.

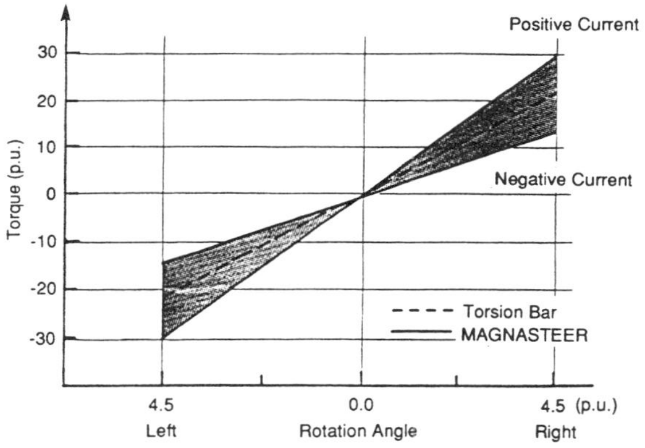

Figure 2. MAGNASTEER Torque versus Rotation Angle.

The simplicity of the MAGNASTEER system relies on the fact that there is no need for additional sensors or pump controls unlike some other systems. Further, this system doesn't interfere with the existing hydraulic system. Moreover, the rotary actuator can be accommodated in the available existing space around the steering gear and, thus, can be used in cars without any major steering system design changes. The MAGNASTEER components and location within a rack and pinion steering gear is shown in Figure 3.

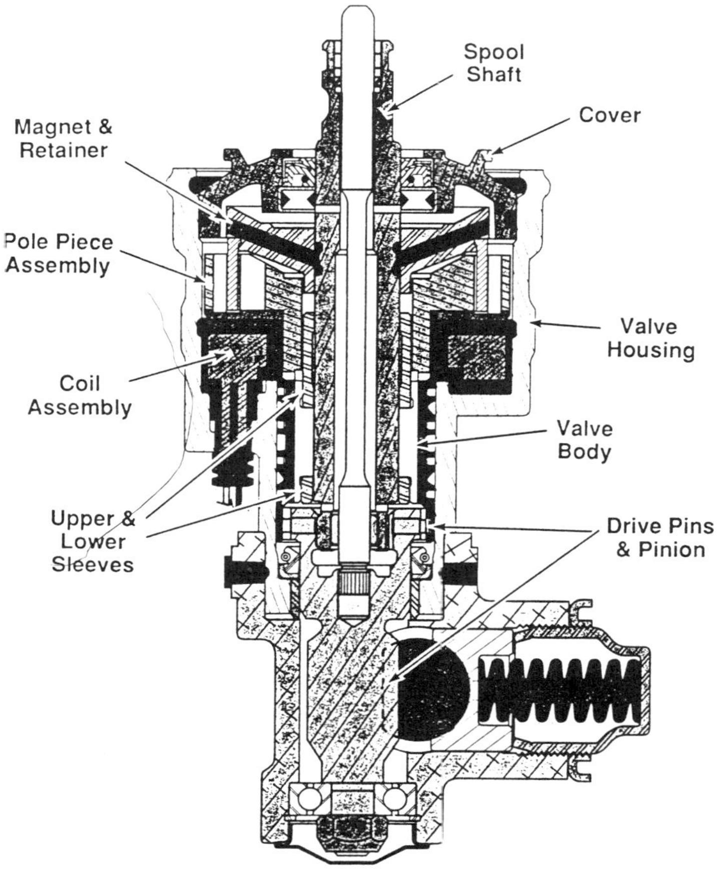

Figure 3. The MAGNASTEER System Components.

The actuator is mechanically attached to the steering gear elements. The stationary external magnetic circuit with an excitation coil is attached to the steering gear valve housing. Both internal rotary actuator elements, a toothed pole magnetic circuit and a permanent magnet ring can rotate with the steering wheel rotation. In addition to the full rotary motion of the actuator center parts, there is a limited angular motion of the permanent magnet ring attached to the spool shaft with respect to the toothed pole magnetic parts attached to the pinion. This limited movement is associated with the hydraulic valve rotation and is normally controlled by the torsion bar stiffness. Generally the valve motion is limited mechanically to several degrees, and therefore the valve and an actuator are optimized for efficient control of motion of up to few degrees in both the left and the right directions of rotation, as shown in Figure 4. The effect of the magnetomotive force (current) is also indicated.

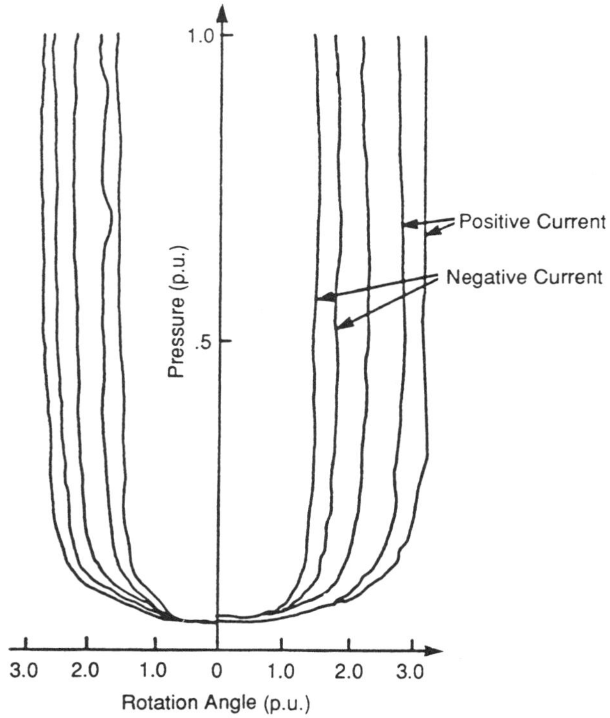

Figure 4. Valve Pressure versus Position

ROTARY ACTUATOR

The homopolar rotary actuator presented in this paper consists of stationary and rotary sections, as shown in Fig. 5. The stationary electromagnet section comprises an external magnetic circuit and an excitation coil attached to the stationary parts. The rotary section consists of two mechanically coupled toothed magnetic structures and a multipole ring permanent magnet inserted between them. The number of poles of the permanent magnet is double the number of salient teeth of the toothed magnetic circuit parts and depends on the required angle of limited angular motion. In addition to the full rotary motion of the actuator rotary section, this configuration allows for a limited angular motion of the permanent magnet ring with respect to the toothed magnetic parts. This motion is mechanically limited for both the left and right directions of rotation. In order to satisfy a desirable symmetrical torque requirement and provide a stable equilibrium at the neutral

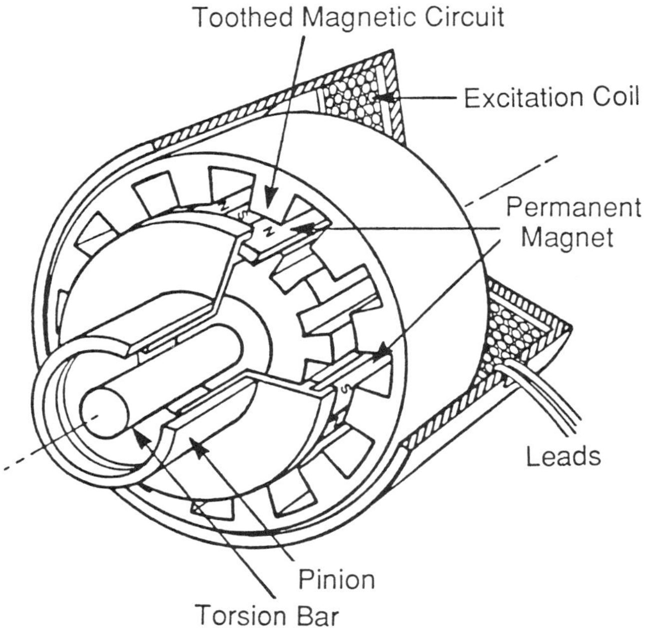

Figure. 5. Homopolar Rotary Actuator Configuration.

permanent magnet position, a unique arrangement of the magnetic circuit was devised. A stable equilibrium for both coil magnetomotive force (MMF) and magnet was established with a half-tooth overlap arrangement and with the magnet located in the middle of this shift, as shown in Figure 6.

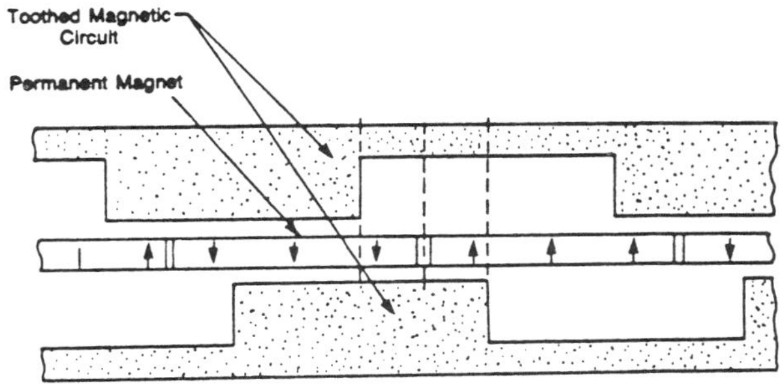

Figure 6. Stable Equilibrium Position Arrangement.

The specific location of the different actuator parts, indicated in Figure 6, is essential to provide a stable equilibrium position for the permanent magnet in its neutral position. The pole teeth on both sides of the magnet are offset by one-half of a tooth width while the permanent magnet pole is centered over the overlapped teeth part in the stable equilibrium position since in this arrangement all forces exerted on the permanent magnet are balanced. If the permanent magnet rotor is moved to the left or to the right of this neutral position a net reluctance force is produced to return it to the stable position. The torque developed by the permanent magnet can be altered by controlling the magnitude and direction of the excitation coil current while the equilibrium position remains unchanged.

The external magnetic circuit provides the same polarity, depending on the direction of the excitation current, to all teeth of the toothed magnetic part regardless of its angular position. This homopolar actuator configuration enables the reversal of torque by simply reversing the direction of the excitation current. The MAGNASTEER rotary actuator develops these attractive and repulsive forces with the minimum number of parts.

Unlike other electromagnetic devices this actuator configuration utilizes both the external and internal permanent magnet surfaces for torque development, thus providing a high torque density structure. Proper balance of both magnetic flux levels (flux densities), one due to the internal permanent magnet and the second due to the stationary external coil, is the most important task during the actuator's design optimization. This can be achieved by establishing a magnet magnetomotive force based on material and magnet thickness, which corresponds to the external coil magnetomotive force based on the nominal current and a number of coil turns. The torque developed by the actuator is a function of the coil MMF and the permanent magnet magnetic field interaction. The total actuator torque consists of two components: the permanent magnet variable reluctance cogging-type torque and the mutual synchronous-type torque due to the interaction of the coil MMF and the magnet. The total torque developed is a superposition of these components:

$$T_t(\theta) = T_m(\theta) + T_r(\theta) \qquad (1)$$

The torque variation with the position is a function of the tooth geometry and the permanent magnet field distribution. For a rectangular toothed structure, sinusoidal magnet magnetization, and nominal coil excitation, the total torque can be expressed as a sum of two sinusoidal components, Ref. [3,4,5]:

$$T_t(\theta) = K_m I_n \Phi \sin\theta - K_r \Phi^2 \sin 2\theta \qquad (2)$$

Reversing the current direction, only the mutual torque component changes its direction and the total torque can be expressed as:

$$T_t(\theta) = -(K_m I_n \Phi \sin\theta + K_r \Phi^2 \sin 2\theta) \qquad (3)$$

The variable reluctance torque component has double the frequency of the mutual torque component and affects both the total torque profile and its maximum value, see Figure 7.

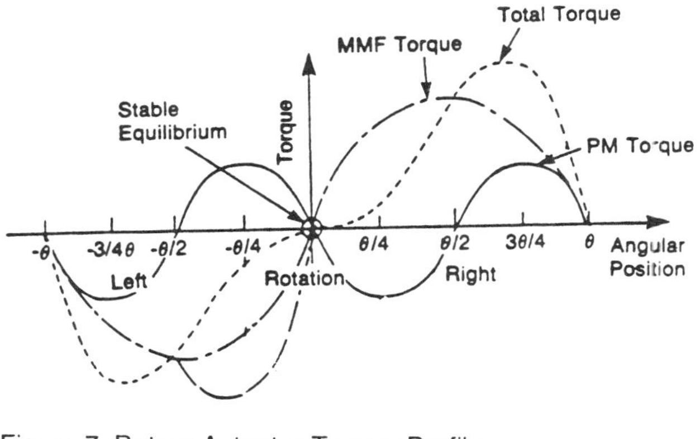

Figure 7. Rotary Actuator Torque Profile.

The variable reluctance torque is determined by the tooth geometry, the air gap length, and the magnet magnetization. High energy density Neodymium based ring magnets with radial magnetization were developed for the purpose of this system. Magnets with similar magnetization are finding applications in rotary actuators, stepper motors, and brushless motors, Ref. [6]. The magnetic flux distribution of these permanent magnets depends upon the magnet material and the applied strength of the magnetizing field. The magnetizing field strength affects both the magnet flux density distribution and its saturation level.

ACTUATOR ANALYSIS

The rotary actuator has a complicated 3-D geometry which was modeled using the 2-D finite element software ANTIC, Ref. [7,8]. Two symmetries were considered, a cylindrical symmetry to establish the air gap flux density of the actuator, and a Cartesian symmetry to calculate the forces developed by the actuator, Ref. [9].

The magnetic force calculations are based on one tooth pitch segment, which extends over a tooth and adjacent slot, using the Maxwell stress tensor over the surface of the permanent magnet two-poles section, defined by:

$$F_t = \iint_S \overline{\overline{T}} * ds \qquad (4)$$

The total force developed by the actuator was calculated as the sum of the forces developed by each tooth pitch. Assuming geometric and magnetic symmetry of the device, the total force developed by the actuator is:

$$F_T = N * F_t \qquad (5)$$

This novel rotary actuator features high performance torque output since the actuator's topology provides what can be called a "torque gearing affect" because torque level is proportional to the number of independent magnetic sections which are equal to the number of teeth or pole pairs. Unlike most of electromagnetic devices, this actuator utilizes both the inside and the outside magnet surfaces, thus featuring a high force density "transverse flux" configuration. The magnetic flux distribution between the magnet and teeth structure was studied. The results presented in Figures 8, and 9. Figure 8 show the magnetic circuit without coil excitation. The magnet is in its equilibrium position, so the resultant force is zero at this position.

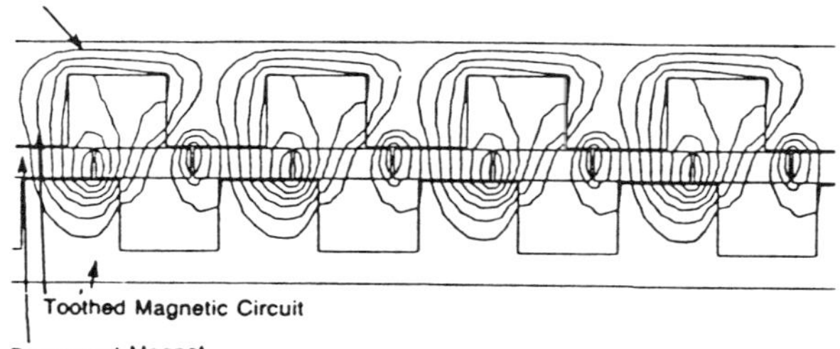

Figure 8. Magnet Flux Lines Distribution.

When the magnet is moved with respect to the toothed structure and the coil is excited, the magnetic flux-line distribution indicates the direction of the developed magnetic forces. Notice the 3-D character of the coil flux distribution as shown in Figure 9.

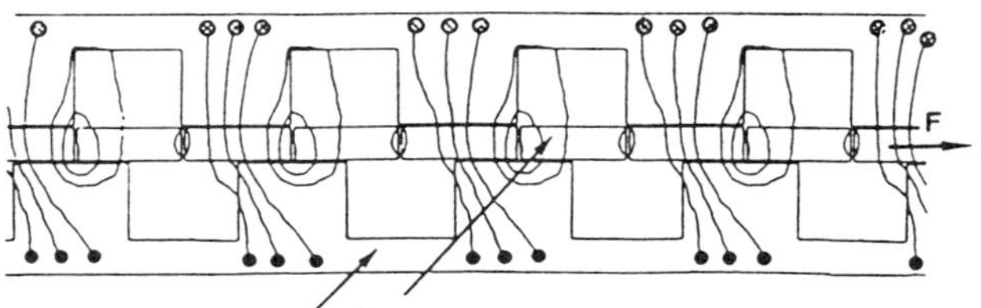

Figure 9. Flux Distribution for Excited Actuator.

Validation of the model, based on both test and analytical results, proved the model to be useful for engineering design purposes. More accurate analysis can be accomplished with the 3-D finite element software ANSOFT, Ref. [10]. Using this mathematical model, parametric studies are being performed to optimize the geometry of the rotary actuator. The goal of the actuator design optimization is to maximize the torque developed for nominal coil excitation current. The optimization is following an iterative process, with each iteration consisting of a three step parametric study for: the excitation coil (wire size, number of turns, resistance), the permanent magnet (material magnetic properties, geometry), and the magnetic circuit (geometry of the toothed structure, magnetic circuit path).

For the excitation coil the nominal voltage and the coil size were predetermined by the electronic controller and envelope size specifications. The coil magnetomotive force was determined from the selected coil parameters, based on the thermal dissipation capacity at a given operating temperature.

Permanent magnet parameters were established based on actuator requirements and constraints. The number of poles was determined by the required angle of rotation, while the actuator geometry was determined by the specified envelope size. This also determined the magnet geometry but not its thickness. The thickness of the permanent magnet is an additional air gap added to the main air gaps for the magnetic flux developed by the excitation coil. Such a large total air gap adversely affects the magnetic flux level in the main air gaps, thus affecting the force developed. The magnet thickness, however, determines the magnet strength for a given magnet material, since the magnet is magnetized. Therefore, there is a trade off between magnet strength and coil field strength based on the magnet thickness, which is to be optimized. Several Neodymium based Magnaquench, Ref. [11], magnet materials were investigated: MQ1, MQ2, and MQ3, or their equivalents, with 7, 15, and 30 MGsOe energy product, respectively, see Figure 10.

The electromagnet magnetic circuit geometry was based on the flux saturation level of the different parts for the selected magnetic material. The toothed structure is similar to variable-reluctance actuator configurations where typically the ratio of tooth height to tooth width has to be optimized. Since tooth width is determined by the required angle of limited rotation as for the number of magnet poles, the tooth height is the only variable of this ratio which was optimized.

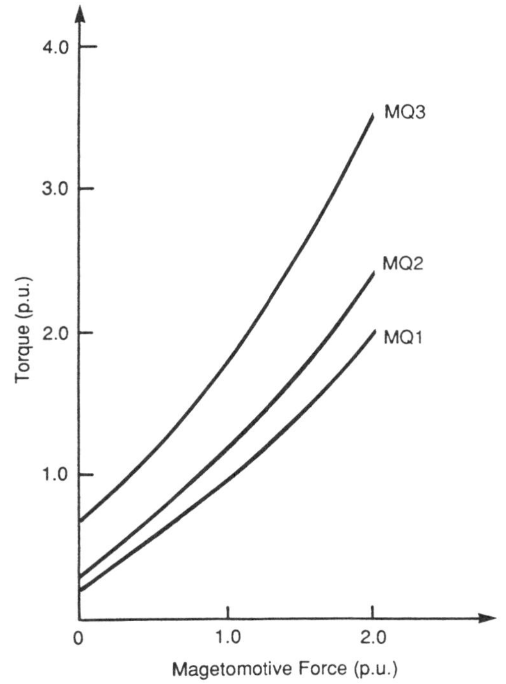

Figure 10. Actuator Torque vs Magnetomotive Force.

Based on this analysis a number of MAGNASTEER actuator components with the optimized geometry and selected magnetic materials were manufactured. Figure 11 shows the components of the MAGNASTEER actuator.

Figure 11. Rotary Actuator Model Components.

The magnetic toothed structure which is physically one piece actually consists of two magnetic toothed-ring sections held by a nonmagnetic section manufactured with advanced powder metal technology. The radially-oriented high-energy ring permanent magnet is manufactured using novel technology which also might serve some other applications. Part of the ring magnet is covered by a foil indicating the multipole magnetization. The coil is also manufactured with use of advanced precision winding technology and is plastic overmolded.

All actuators components are fully integrated within a rack and pinion gear where a number of structural and hydraulic parts carry more than one function. For instance, the housing in addition to its important structural function, is an essential part of the magnetic flux path. The crossection of the assembled MAGNASTEER model is shown in Figure 12, where all components of the system are identified.

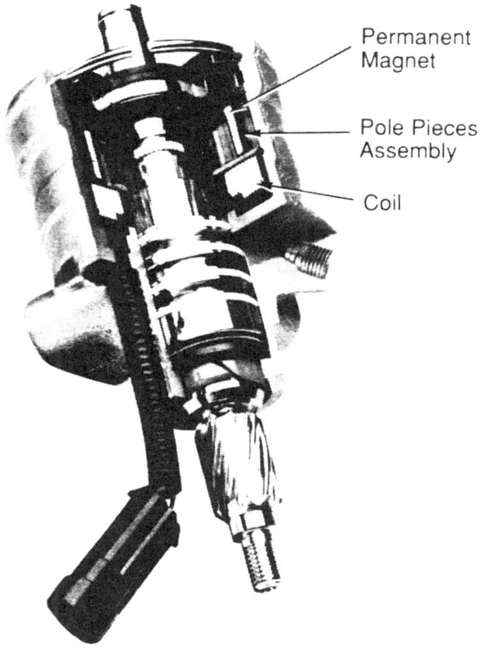

Figure 12. MAGNASTEER Actuator Assembly.

5. TEST RESULTS

The analysis is well supported by test results, see Fig. 13, which were predicted within several percent. Test results also indicated that symmetrical magnetic forces in both directions of rotation were achieved with a stable equilibrium position for both the permanent magnet and the excitation coil MMF.

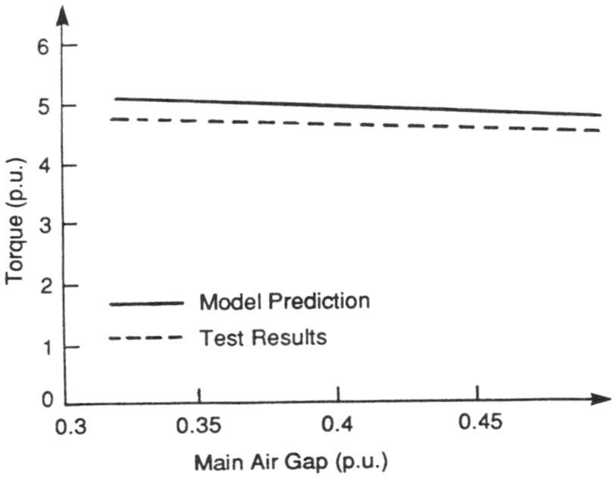

Fig. 13. Torque, Test and Analysis Comparison.

A number of tests were performed on passenger vehicles equipped with the MAGNASTEER system. The MAGNASTEER system was compared with other systems and proved that it meets our objectives for a world-class system as it is shown in Figure 14. Figure 14 identifies the improved ride and handling of the tested vehicles after the MAGNASTEER system was installed and used on these vehicles.

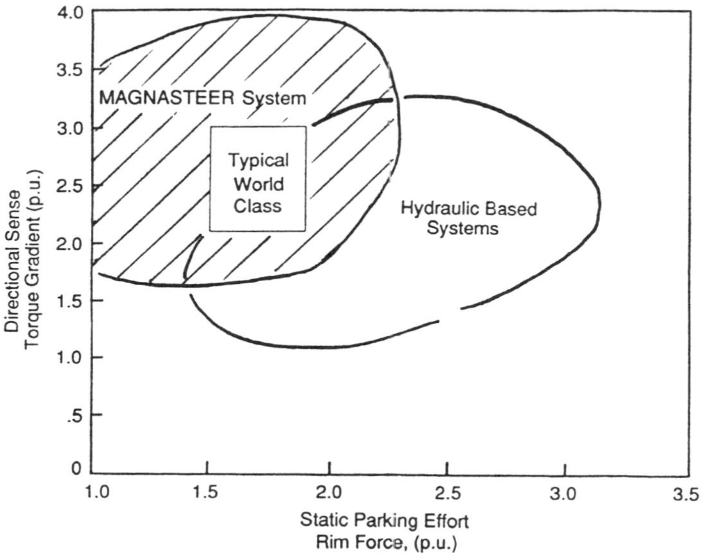

Figure 14. Measured Steering System Performance.

SUMMARY

The magnetic speed dependent variable assist steering system MAGNASTEER was developed for passenger vehicles. This high performance and simple system features a novel homopolar rotary actuator having a multipole ring permanent magnet. This novel rotary actuator features high performance torque output since the actuator's topology provides a "torque gearing affect" with high force density "transverse flux" configuration utilizing both outside and inside magnet surfaces. Using mathematical models, the MAGNASTEER rotary actuator geometry was optimized. In its final design the MAGNASTEER rotary actuator features three main parts fully integrated within a steering gear. Each of these parts; a magnet, a coil, and a tooth structure features leading-edge world-class technology.

An analytical study conducted to design and optimize the actuator geometry and performance is well supported by test results, see Figure 13. The tests results indicate that the MAGNASTEER system achieved the required torque level, variable effort, conrollability, and symmetrical performance which is sufficient to provide a controllable variable effort for most passenger cars.

NOMENCLATURE

$T_t(\theta)$ [Nm] is total torque developed,
$T_m(\theta)$ [Nm] is mutual torque,
$T_r(\theta)$ [Nm] is variable reluctance torque, and
θ [deg] is a position angle,
K_m, K_r are constants for the mutual and the reluctance torque, respectively,
I_n [A] is the nominal current in the coil, and
Φ [Wb] is the flux developed by the magnet
$\underline{F}_T$ [N] is the total magnetic force,
$\overline{\overline{T}}$ is the Maxwell stress tensor in dyadic form,
s [m2] is the surface of integration
F_t [N] is one pole pitch magnetic force, and
N is the number of teeth

REFERENCES

1. J.A.Zuraski, A.M.Pawlak, D.W.Graber, J.M.Babineau, "Electromagnetic Control Apparatus for Varying the Driver Steering Effort of a Hydraulic Power Steering System," US Patent 4,871,040, issued October 3, 1989.

2. D.C. Eckhardt, D.W.Graber, , A.M.Pawlak, B.L. Faist , "Electromagnetic Control Apparatus for Varying the Driver Steering Effort of a Hydraulic Power Steering System," US Patent 5,119,898, issued June 9, 1992.

3. M. R. Harris, at al.,"Static Torque Production in Saturated Doubly-Salient Toothed Structures," Proc. of IEEE, Vol. 122, pp.1121-1127, October 1975.

4. H. D. Chai, "Permeance Based Step Motor Model Revisited," MOTION, pp. 14-28, March/April 1986.

5. H. D. Chai, "Cogging Torque of PM Step Motors," Proc. of 1984 IMCSD Symp., Univ. of Illinois, pp. 163-166, May 1984.

6. Hitachi Metals Ltd., Electronic Materials and Parts," Catalog HE-B46.

7. T. W. Nehl, A. M. Pawlak, and N. Mikhaeil Boules, "ANTIC85 a General Purpose Finite Element Package For Computer Aided Design and Analysis of Electromagnetic Devices," IEEE Transaction of Magnetics, vol 24, no.1, pp. 358-362, Jan 1988.

8. A. M. Pawlak, and T. W. Nehl, "Transient Finite Element Modelling of Solenoid Actuators: the Coupled Power Electronics, Mechanical, and Magnetic Field Problem," IEEE Transaction of Magnetics, vol 24, no.1, pp. 270-274, Jan 1988.

9. A.M. Pawlak," Novel Homopolar Rotary Actuator," Proceedings of ICEM'90 International Conference on Electrical Machines, August 13-15, 1990, pp. 78-83, MIT Boston, USA.

10. Ansoft Corporation, "Maxwell 3D Field Simulator," Users Reference, USA February 1991.

11. B.H.Carsile,"Neodymium Challenges Ferrite Magnets," Machine Design, pp. 24-30, January 9, 1986.

ACTUATORS IN SUSPENSION SYSTEMS

930268

An Actively Damped Passenger Car Suspension System with Low Voltage Electro-Rheological Magnetic Fluid

Andrew Pinkos, Emil Shtarkman, and Thomas Fitzgerald
TRW, Inc.

ABSTRACT

In the realm of "semi-active" suspension systems, this paper describes a truly "active damping" system. Thorough research and development of low voltage Electro-Rheological Magnetic (ERM) fluids, a revived rotary shock absorber suspension philosophy and a newly developed control algorithm have resulted in an active damping system that resolves the selection between ride comfort and handling performance.

The ERM fluid technology, rotary shock absorber configuration and Digital Signal Processor (DSP) technology have advantages of transmitting high forces in a small compact package. The DSP controls the elements that digitize analog sensor signals and through software, result in active damping. We call the actuator the Rotary Active Control Damper (RACD, pronounced "rack-dee"). The variety of on-board sensors makes it possible to adjust vehicle performance through software.

Mathematical analysis, computer simulations, and actual vehicle testing of the developed algorithm in which RACD is the damping actuator have shown the advantage of the developed technology. The system can be operational in either an open or closed-loop form for optimum continuous control of the active damper. Actual vehicle installation of RACD's with extensive testing have shown that this is a practical alternative to today's multi-damping shock absorbers.

ADAPTIVE DAMPING SYSTEM

This innovative suspension system is called the Adaptive Damping System (ADS), and is made up of three elements: (1) Electro-Rheological Magnetic (ERM) fluid, (2) rotary shock absorber, and (3) a control algorithm. The unique design and operating characteristics of the computer controlled Rotary Active Control Damper (RACD), assembled with ERM fluid and associated sensors, allows for continuous adjustment of damping forces applied to the vehicle. The low voltage and low amperage power source required to operate the RACD actuators is supplied by the existing vehicle electric system.

Adaptive control of the damping force further refines both vehicle handling and stability. These improvements may mean a safer and more comfortable passenger ride.

A unique feature of ADS is the demand on the vehicle electrical system. The power consumption, for each RACD, is based on vehicle reaction . Only the quiescent power for the Electronic Control Unit (ECU) is required on a full time basis. Peak power demands, as understood, will occur only when high damping forces are required.

Another feature of the ERM / RACD system is the ability to "self setup" damping forces without transient movement. RACD is a velocity independent actuator.

MAGNETO-RHEOLOGICAL FLUID - The distinct characteristic of rheological fluids is, when exposed to the appropriate energy field, solid particles in the fluid align. When this alignment occurs, the ability of the fluid to flow, or shear, is substantially reduced. Two types of rheological fluids have been researched. Each is based on different applied energy fields. One fluid is responsive to a voltage field while the other is rheological to a magnetic field. Electro-Rheological (ER) fluids, that is, fluids that are responsive to a high voltage, low current electric field, have been extensively investigated. Some application restrictions are they require thousands of volts for operation, and yield low shear stresses. Safety and packaging are clearly design problems. On the other hand, ERM fluid, responsive to a magnetic field, operates on vehicle battery voltage and generates high fluid shear stress. As shown in Figure 1, the value of ERM fluid shear stresses developed are extremely high in comparison to ER fluid.

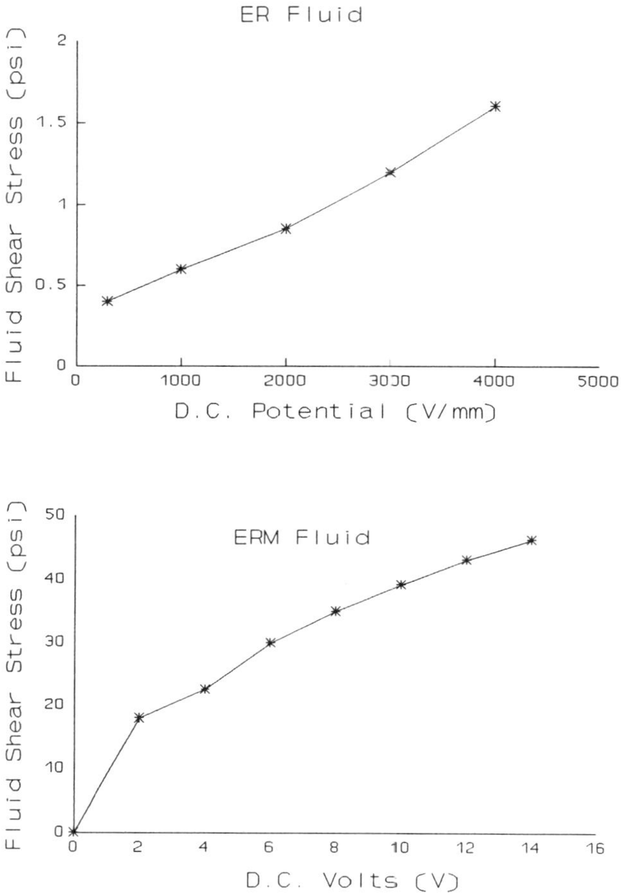

Figure 1 - Shear Stress Comparison Between ERM & ER Fluids

ERM fluid has some intriguing automotive applications, and a partial list includes shock absorbers, clutches, engine mounts, and active bushings.

The ERM fluid is comprised of magnetizable particulate, TRW fibrous carbon and a carrier vehicle. An example of a solid magnetizable particle is carbonyl iron. There are other ferro-metallic particulate and compounds which may also be used. The carrying vehicle can be silicone oil. The fluid may also contain a surfactant to keep the solid particles in suspension.

The application requirements for field responsive fluids include very low shear resistance at zero field, high shear stresses at maximum applied field, very low hysteresis, chemical inertness, temperature stability, and fast response time. ERM fluid meets all these criteria.

ROTARY ACTIVE CONTROL DAMPER - RACD was used in a vehicle for providing damping forces. A typical installation is shown in Figure 2. It is comprised of a housing connected to the vehicle chassis and a shaft having an axial end connected to the lower control arm. With this arrangement, vehicle design and styling can be enhanced. It allows for a reduction in vehicle hood height and an increase in the passenger compartment and/or trunk size.

Figure 3 shows a cross-sectional view of RACD. The housing consists of two halves which define at least one chamber for containing the ERM fluid. The shaft is supported for rotation, about its longitudinal central axis relative to the car body, by a pair of bushings. There are elastomeric bonded seals around the shaft and between the two halves of the housing. Rigidly attached to the shaft is a blade that is disposed in the chamber. The blade is angularly movable in the chamber. Electromagnetic coils are embedded into each half of the housing. The magnets are arranged in an opposing pole configuration through electrical connections to a DC power supply. When the coils are energized, the applied magnetic field causes the fluid to "solidify" in the chamber. The shear resistance of

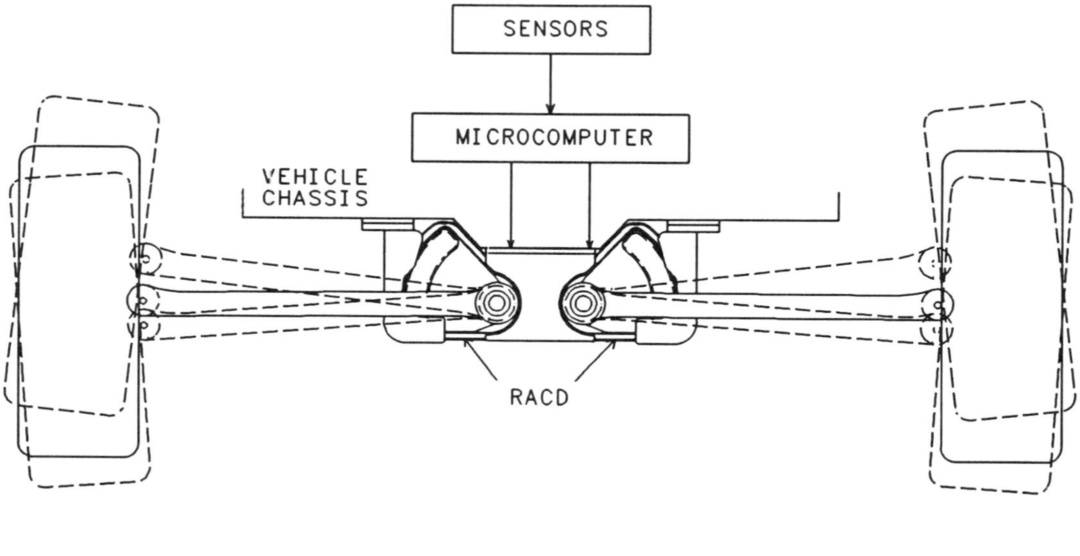

Figure 2 - RACD Installation

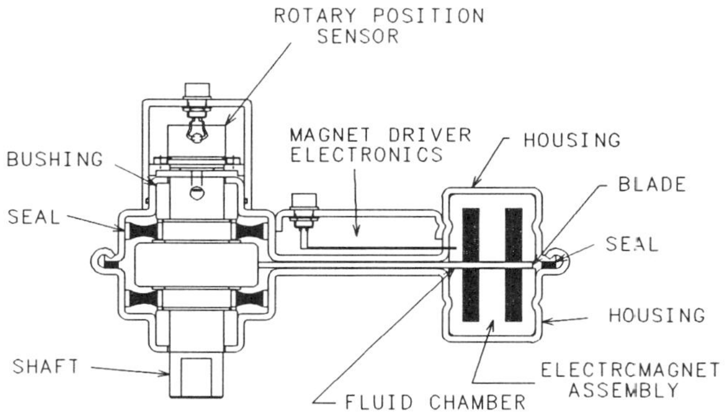

Figure 3 - Cross Section of RACD

the fluid varies as a function of the applied energy field. When the angular movement of the blade in the chamber is resisted by the fluid, a reaction torque is developed.

A rotary position sensor for measuring angular displacements is mounted on one end of the shaft and protected by a sealed cover. Embedded in the housing, next to the electromagnets, are electronics that convert a command signal from the Digital Signal Processor (DSP) into a corresponding energy field produced by the electromagnets.

CONTROL ALGORITHM - The nature of the ERM / RACD system requires a new control algorithm for continuous control of the damping forces. Vehicle dynamics and road disturbance, are defined for multi-degrees of freedom by a set of non-linear differential equations. In their simplest form there are no closed form solutions. Therefore, formulation of the control outputs must be computed in very small time increments. Based on system stability analysis and convergence of a solution, the algorithm is classified as robust. The unique formulation of this control strategy is based on adaptive gain control and vector summation of weighted sensor measurements. Each corner of the vehicle is treated independently, but the total control output is calculated from information on vehicle behavior. Separate calculations are produced for ride comfort, roll, dive, squat, pitch, heave, yaw. The vector summation of these calculations produces an output signal to the corner of the vehicle.

Software - The algorithm is transformed into a set of software instructions that the DSP processes. There are three main sections to the software program: (1) initialization, (2) interrupt processing, and (3) main loop processing. The initialization software, default gain parameters, ECU setup and output command signal, is processed upon a power-on reset. Thereafter, incremental changes to gain parameters are derived from the default set. Interrupt processing is used for fault determination and to control main loop processing time. The main loop is executed every millisecond and must be precisely controlled, a requirement for software filtering. Sections processed within the main loop include input from all sensors, proper signal filtering, all calculations, output of command signals, and serial link interface.

Prioritization - The algorithm calculations are prioritized based on safety related vehicle behavior. Any calculations related to vehicle handling are completed first. Then software decisions within the main loop are processed. These decisions are for implementation of an Adaptive Gain Control (AGC). AGC values are used for full vehicle modal control. Also, gain sensitivity is affected by the decisions.

Sensors - Various sensors accurately measure vehicle behavior. The left hand side of Figure 4 is the sensors used in ADS. In this installation, thirteen sensors are used. In addition to vehicle speed, braking, and

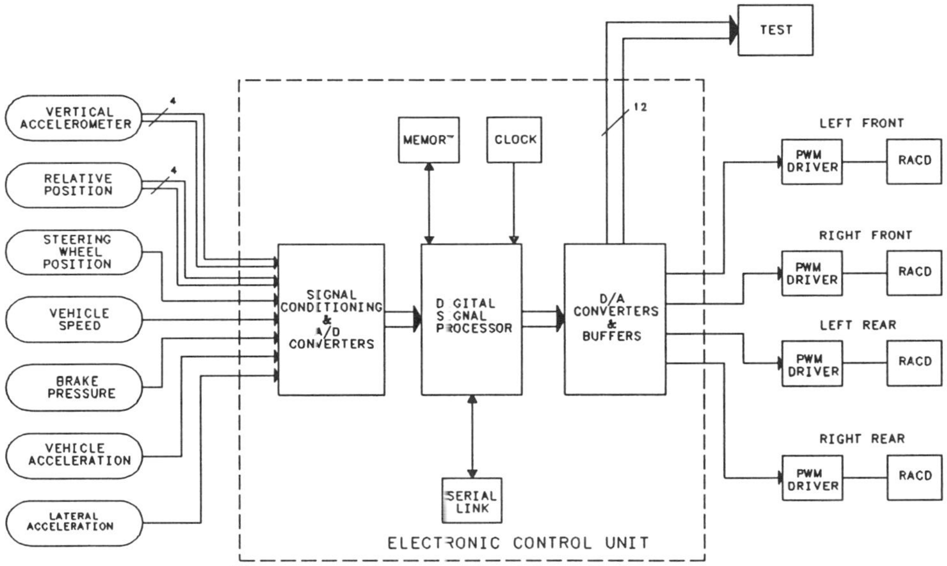

Figure 4 - System Electrical Schematic

acceleration, ten sensors are added. These include vertical accelerometers on the sprung mass, angular position between vehicle body and wheel, lateral acceleration, and absolute steering wheel position.

ELECTRONIC CONTROL UNIT

The Electronic Control Unit (ECU) is the brain of the system (See Figure 4). It contains the sensor interface and signal conditioners, Analog to Digital (A/D) convertors, EPROM and RAM memory, a Digital Signal Processor (DSP), Digital to Analog (D/A) convertors, and output buffers.

The analog sensor signals are hardware anti-aliased filtered before being converted into digital words by using several fast, complete 12-bit, successive approximation A/D convertors with a DSP interface. Each sensor is then software filtered. The filter is implemented by using an Infinite Impulse Response (IIR), or recursive filter. Processing the control algorithm requires several of these sensor signals to be phase lag-lead software filtered.

The heart of the system is a Texas Instruments TMS320C25 DSP, operating at a clock frequency of 40.96 MHz. Also on this board is the serial interface and 16K words of RAM and 8K words of EPROM memory.

The calculated command signal, to the RACD power driver electronics, is converted from digital to analog by using 12-bit convertors. The analog command signals to each corner RACD are buffered.

Each RACD electromagnet are connected to a set of driver electronics. A pair of power MOSFET's configured as high-side switches are used to control the amount of current calculated. The analog command signal from the ECU is converted into an appropriate duty cycle for the switches, more commonly known as Pulse Width Modulation (PWM). The PWM technique controls the MOSFET's turn-on time, thereby controlling current to the electromagnets. The power source for each RACD is a fused connection directly from the vehicle battery.

The ECU has the capability for fault detection. There are two levels of ECU diagnostics, hardware and software. Hardware diagnostics include RACD electrical failure, a watchdog timer, and low power supply voltage levels. Software diagnostics include sensor checks for out-of-range signal levels, memory checks, A/D and D/A operation, and serial interface tests.

The ECU in the prototype vehicle was mounted in the trunk on compliant mounts. Precautions for RFI / EMI interference were taken which included a "sealed" card cage housing the ECU boards.

MATHEMATICAL MODELLING

Computer modeling and simulation were done to analyze the capabilities of ADS. Both quarter-car and half-car vehicle dynamic models were examined using the TRW algorithm. In schematic form, Figure 5 shows the half-car model used in simulation which represents a model for four degrees of freedom system. The prime areas of this study included low and high speed bumps, as well as multiple bumps. Results compared suspension travel, vehicle accelerations, and pitch modes to the classical damper. A FORTRAN program was utilized to produce the response to different road inputs. A subroutine was developed to accurately simulate the instantaneous force developed by RACD in response to wheel and body displacement. Suspension springs include elastic bump-stops.

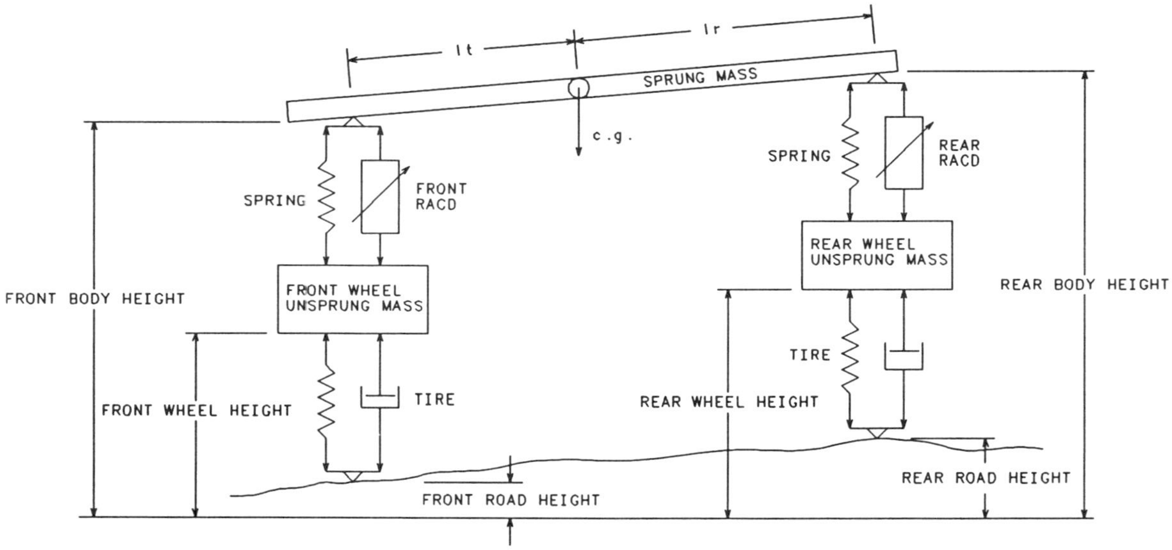

Figure 5 - Half Car Model Schematic

Response to two road disturbances are presented in this paper: (1) a 0.152 m (6") "bump" and (2) a set of three 0.114 m (4.5") railroad tracks. The vehicle was "driven" over the disturbances at different speeds each accentuating certain vehicle behavior.

The single bump disturbance has the shape of a normal or Gaussian curve. It has a superficial resemblance to a speed bump, but more importantly, it gives a Gaussian distribution of energy in the frequency domain. Also, it contains much less energy at high frequencies than step changes in road height. Therefore, any high frequencies must result from non-linear effects in the suspension system.

The other disturbance is a three-humped sinusoidal curve, similar to the road profile for three separate sets of railroad tracks. The combination of track spacing and car speed produces a disturbance with its energy concentrated near the body heave and pitch resonance frequencies.

VEHICLE SIMULATION - The quarter-car results are shown in Figures 6 and 7, while the half-car results are Figures 8 and 9.

Figure 6 shows the quarter-car response to a 0.152 m (6")

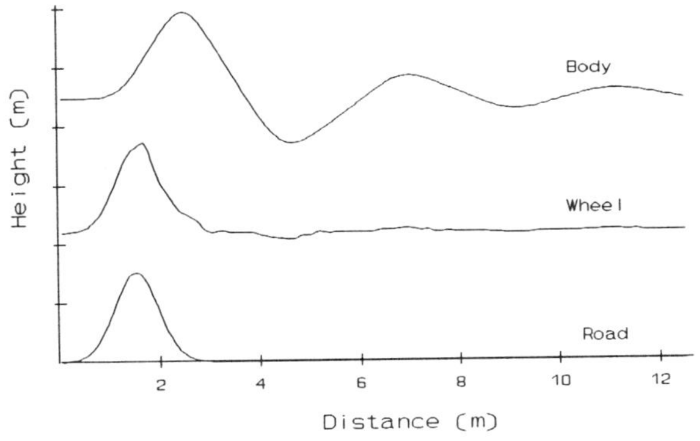

Figure 6 - Quarter Car Response To Gaussian Bump At 16 kph

bump with the vehicle travelling at 16 kph (10 mph). With the algorithm and RACD supplying the damping force, body motion is damped very quickly.

In Figure 7, the road disturbance is the simulated railroad tracks. The car speed is 56 kph (35 mph). The energy of the disturbance is selected to be at the body resonance frequency. As shown, ADS quickly returns the body position to equilibrium.

The results for the half-car model include both the front and rear are superimposed on the same graph.

Examination of Figure 8 for the half-car response to the railroad tracks shows that each wheel responds about the

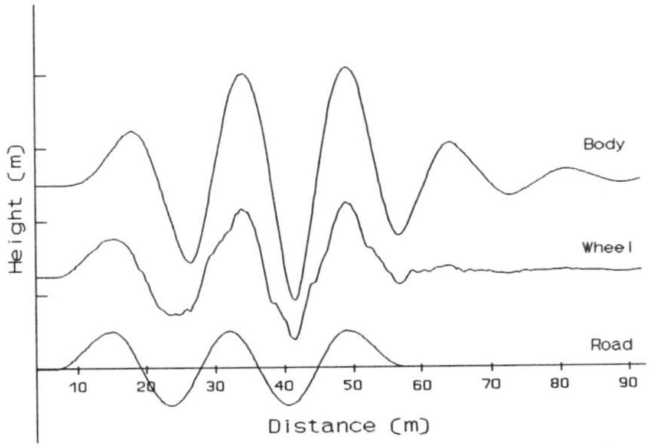

Figure 7 - Quarter Car Response to Railroad Tracks at 16 kph

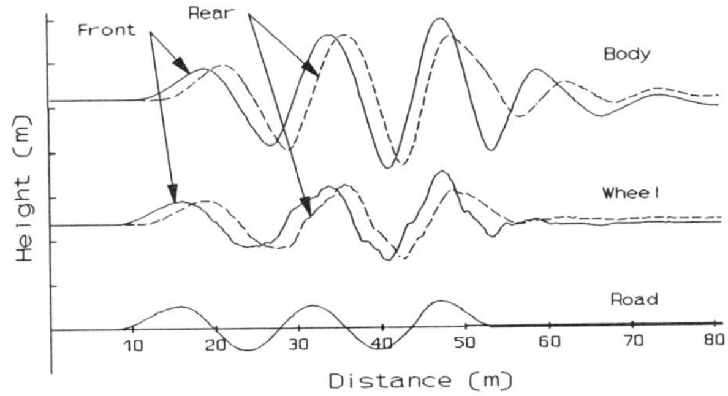

Figure 8 - Half Car Response to Railroad Tracks at 56 kph

same as the quarter-car model predicted. The curves are simply offset by time.

Taking advantage of the velocity independence feature RACD has to offer, different vehicle modes can be controlled. On the half-car simulation, where pitch mode can be studied, an enhancement to the algorithm results in pitch mode control. Figure 9 demonstrates the effect of

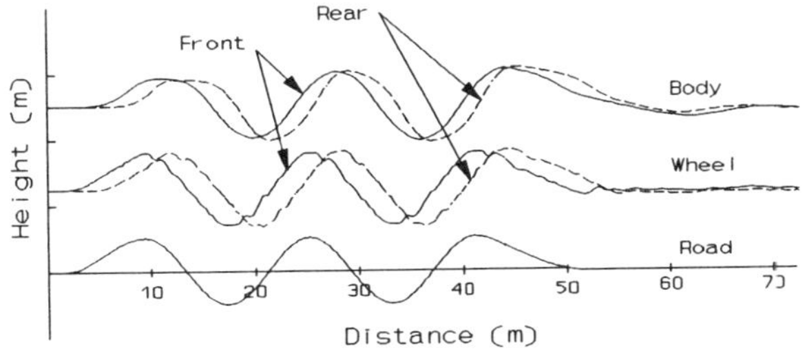

Figure 9 - Half Car Response to Railroad Tracks at 56 kph, with Pitch Control

adding the pitch control feedback loop. It calculates the instantaneous pitch angle and affects both front and rear RACD forces to bring the body axis back to a neutral position.

VEHICLE TESTING

The mathematical model was verified through actual vehicle testing. Vehicle testing was performed on two initially identical vehicles, baseline and ADS equipped. The baseline vehicle is a standard production version. Installation of RACD and electronics were the only modifications on the ADS equipped vehicle. For testing purposes, both vehicles were equipped with identical sets of sensors and electronics and driven over the same road surfaces at the same speed. Various road surfaces were chosen for comparisons of vehicle performance. Some of the measurements recorded include relative wheel travel and vertical, roll, and pitch accelerations.

INSTRUMENTATION - A Humphrey, Inc. gyroscopic instrument package was installed during testing. The system consists of two cageable gyroscopes, one horizontal and one vertical. Tied to the horizontal gyroscope is a pitch and roll platform that contains three mutually orthogonal servo accelerometers. Both platforms, during vehicle testing, remain in the uncaged position allowing the gyroscope to keep the platforms stabilized to the inertial, or earth, reference.

The package was securely mounted to the vehicle body on the fore-aft centerline directly behind the c.g. of the vehicle.

The following is a list of vehicle tests that have been performed relating to ride comfort:

- Body Resonance
- Wheel Hop Resonance
- Sine Wave Road
- Washboard Road
- Cobblestone Road
- Inverted Chuckhole Road
- Resonance Bumps
- Random Input Inverted Chuckhole Road
- "Typical Rural Road"

The following is a list of vehicle tests that have been performed relating to vehicle handling performance:

- Serpentine
- Slalom
- Lane Change
- Evasive Maneuver
- Constant Radius Circle
- J-Turns
- Deceleration
- Acceleration

Ride comfort performance, namely bounce, pitch and roll, relates to body and other resonances. Figure 10 shows the comparison between the baseline and ADS vehicles at low frequency or body resonance. Each vehicle has a slightly different resonant frequency. A drastic reduction in

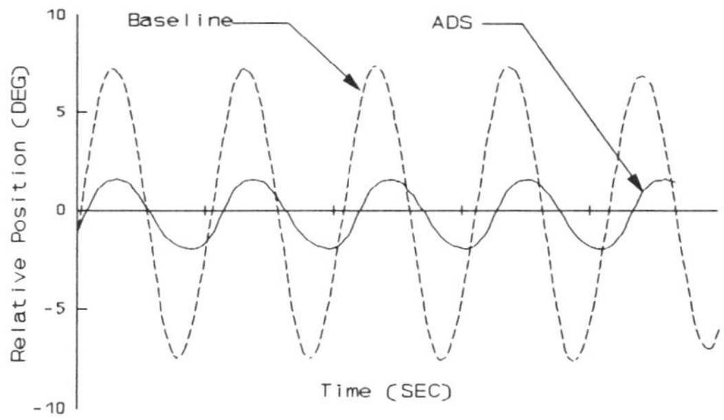

Figure 10 - Body Resonance

suspension travel in the ADS vehicle was observed.

A reduction in suspension travel at high frequency, or wheel hop resonance, is shown in Figure 11. An ability to control both key resonances is an essential component of ride performance.

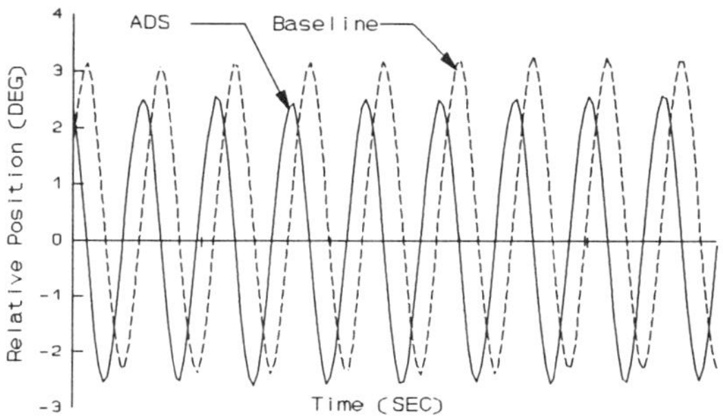

Figure 11 - Wheel Hop Resonance

Actual test track handling maneuvers were recorded and two tests are shown, serpentine and evasive maneuver.

The serpentine course consisted of ten cones spaced 32.8 m (100 ft) apart in a straight line. The indicated vehicle speed, at entry, was 105 kph (65 mph). As shown in Figure 12, there is a drastic reduction in roll motion.

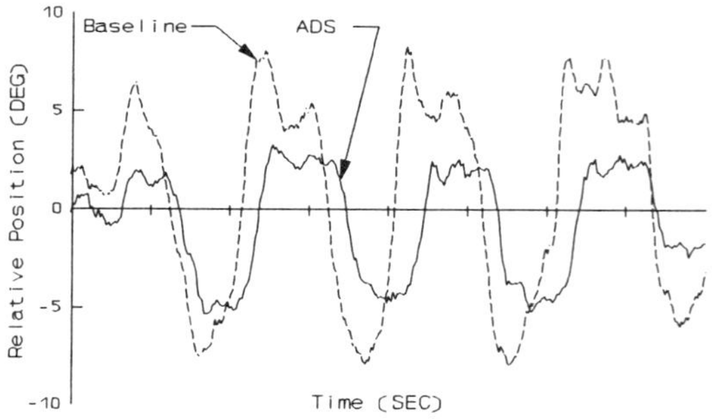

Figure 12 - 105 kph Serpentine

The evasive maneuver course consisted of an entry chute that was 3.9 m (12 ft) wide. Then the vehicle performed a single left lane change through a 24.6 m (75 ft) opening, a straight stretch of 24.6 m (75 ft), and a right lane change, mirror image to entry. Indicated entry speed was 128 kph (80 mph). Figure 13 shows that even during this maneuver the ADS vehicle body shows less roll and is much more responsive.

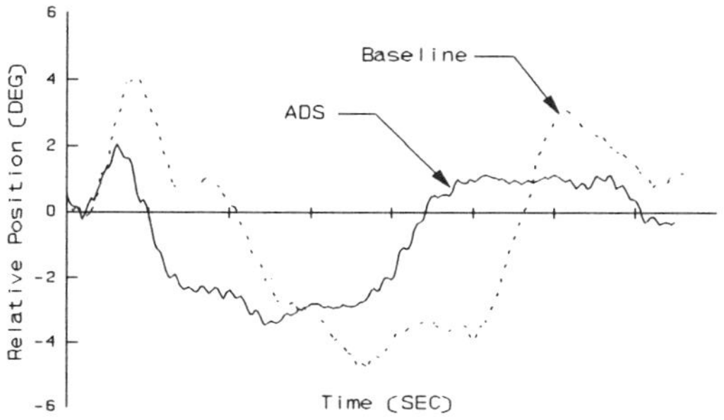

Figure 13 - 128 kph Evasive Maneuver

SUMMARY

The rotary shock absorber configuration coupled with ERM fluid provides a unique solution for use in a passenger car suspension system. This Adaptive Damping System means greater flexibility in the styling of exterior and interior envelopes, improves passenger safety and comfort, and provides easily adjusted vehicle suspension control.

The shear resistance of the ERM fluid varies as a function of the applied magnetic field. Therefore, damping force of the rotary shock absorber varies as a function of the applied magnetic field.

The resistance to relative movement within the RACD creates the damping force. For different ERM fluid compositions the reaction torque value per amp of input energy ranges from 5.65 N-m (50 lbs-in) to over 113 N-m (1000 lbs-in). ERM fluid meets all of the criteria necessary for automotive applications, including low voltage and low power consumption.

RACD is an active damper. The feature of being velocity independent allows early activation of necessary damping force for better handling performance.

The formulation for vector summation of sensor measurements is the key to total vehicle performance. Thorough simulation and actual vehicle testing have shown the advantages of ADS. Based on the comparison of all testing data reviewed, ADS has demonstrated broad range control capabilities proven through software programming.

920275

An Electronically Controlled Shock Absorber Using Electrorheological Fluid

Nicholas K. Petek
RheoActive, Inc.

ABSTRACT

Electrorheological (ER) fluids, sometimes called "smart" fluids, are materials having flow properties that can be modified with electric field. This unique behavior is utilized in the design of a shock absorber with fast-acting electronic control. The damping force of the shock absorber is adjusted by selectively applying voltage to the ER fluid. In this paper, data are provided to illustrate that the damping forces needed for an automobile suspension are achieved using a device with realistic package size and power requirements. The data also demonstrate that the prototype shock absorber performs consistently in a durability test where the conditions are typical for life testing of commercial shock absorbers. The prototype is observed to operate successfully on a vehicle as a retrofit for an original equipment shock absorber with variable damping.

A SUSPENSION SYSTEM is designed to control the ride and handling of a vehicle for the comfort and safety of its passengers. A conventional *passive* suspension, consisting of a passive spring and damper, must compromise its ability to control a vehicle. Generally, low damping is preferred for good isolation of noise, vibration, and harshness; high damping is desired for control of body and wheel motion during cornering, acceleration, and braking. An *adaptive* suspension utilizes a passive spring and an adjustable damper with "slow" response to improve the control of ride and handling. The level of damping at all wheel positions can be selected by the driver, or it can be automatically tuned for variations in vehicle speed, steering angle, throttle position, brake pressure, or accelerometer signals. A *semi-active* suspension is similar, except that the adjustable damper has "fast" response (less than 10 ms) and each wheel position is independently controlled in real time. A *fully-active* suspension replaces the damper with a hydraulic actuator which can achieve the optimum in vehicle control, but at high cost.

A semi-active suspension is shown in the literature to theoretically approach the performance of a fully-active suspension at a fraction of the cost. The preferred control algorithm, known as the "sky-hook" model, consists of a feed-back loop that adjusts damping level based on 1) the absolute vertical velocity of the vehicle body, and 2) the relative velocity between the body and the wheel. A key component of the system is an adjustable damper with sufficiently fast response, and in recent years much development has been occurring in the design of fast-acting valves. One means of achieving variable damping is using an electro-mechanical variable orifice to alter the flow resistance of a conventional hydraulic fluid [1-3]. Another means is using an electrically-charged fixed orifice to change the flow resistance of a field-responsive fluid, such as electrorheological fluid [4-6].

This paper discusses an electronically controlled shock absorber which uses ER fluid for variable damping and which is intended to ultimately be incorporated in a semi-active suspension system. The objective here is to demonstrate the viability of ER technology in a damper application. Several features of ER offer advantages over conventional technology in the design of an adjustable shock absorber: 1) the ER effect is efficient, typically requiring only a few watts of electric power; 2) the device construction is simplified due to the non-mechanical nature of ER technology, potentially decreasing manufacturing cost and increasing reliability; and 3) the ER effect provides the fast response needed for control of a semi-active suspension. The complete suspension system will consist of adjustable shock absorbers filled with ER fluid, power supplies to energize the fluid, sensors to provide feed-back, and a central control module.

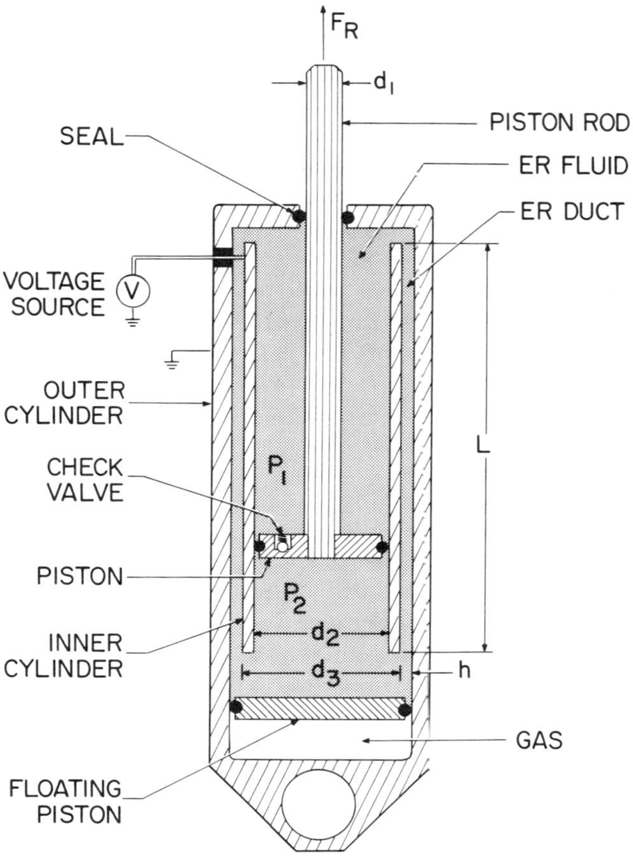

Figure 1: ER fluid-filled shock absorber schematic

DESCRIPTION OF CONCEPT

A concept sketch of the shock absorber is shown in Figure 1. The motion of the piston pushes ER fluid through the stationary, annular duct via openings at either end of the duct. The gas reservoir accommodates the fluid volume displaced by the piston rod and by thermal expansion. When voltage is applied across the duct, an electric field is created perpendicular to the fluid flow. The ER fluid is assumed to instantly change from Newtonian to Bingham-plastic, exhibiting a yield stress which increases its resistance to flow through the duct. This causes an increase in pressure drop across the piston which increases the force required to move the piston. Thus, the zero-voltage damping is due entirely to the viscosity η of the fluid and is proportional to its flow rate Q through the duct. The pressure drop ΔP_η across a duct for the Newtonian case is:

$$\Delta P_\eta = 12\eta \frac{L}{bh^3} Q \quad (1)$$

where L = length of duct
b = width of duct
h = separation of duct walls

When voltage is applied, the yield stress τ_y of the ER fluid theoretically increases the damping force by an amount that is constant regardless of piston speed. This is fundamentally different than the performance of an adjustable viscous damper. The pressure drop ΔP_{ER} across a duct due to the yield stress is:

$$\Delta P_{ER} = 2\frac{L}{h}\tau_y \quad (2)$$

These expressions for pressure drop are simple approximations of the ER fluid behavior. While more detailed modeling of ER fluid flow is possible, the development here is sufficient for an estimation.

To calculate the expected rebound damping force, assumptions are made that there is no flow across the piston in rebound and that the resistance losses of the fluid entering and leaving the duct are negligible. Thus, referring to Figure 1, the total pressure drop across the duct is the sum of the viscous and yield stress effects:

$$P_1 - P_2 = \Delta P_{ER} + \Delta P_\eta \quad (3)$$

In this geometry, the duct width is:

$$b = \pi d_3 \quad (4)$$

The flow rate is proportional to the piston velocity V_p:

$$Q = \frac{\pi}{4}(d_2^2 - d_1^2) V_p \quad (5)$$

By substitution, the pressure P_1 above the piston can be determined:

$$P_1 = 2\frac{L}{h}\tau_y + 3\frac{L}{d_3 h^3}(d_2^2 - d_1^2)^2 V_p \eta + P_2 \quad (6)$$

By assuming that P_2 equals the gas charge pressure and by balancing the forces on the piston rod, the rebound damping force F_R can now be calculated:

$$F_R = \frac{\pi}{2}\frac{L}{h}(d_2^2 - d_1^2)\tau_y + \frac{3\pi}{4}\frac{L}{d_3 h^3}(d_2^2 - d_1^2)^2 V_p \eta - \frac{\pi}{4} d_1^2 P_2 \quad (7)$$

A further assumption regarding the ER fluid is that its current density j is a fundamental fluid property which depends on electric field E, temperature, and shear rate. For a given set of operating conditions and a known duct geometry, the current and power draw of the shock absorber are calculated:

$$i = \pi d_3 L j \quad (8)$$

$$p = \pi d_3 L h \phi \quad (9)$$

where $\phi = jE$ is defined as the power density of the fluid.

In suspension system design, the desire is generally to have higher damping force in rebound (extension) than in jounce (compression). So, in order to decrease the jounce damping force relative to rebound, a check valve is included in the piston. During jounce, this valve allows some fluid flow through an orifice in the piston, thereby reducing the total resistance to flow and reducing pressure drop across the piston. The gas reservoir is pressurized to raise the total pressure inside the shock absorber which prevents the pressure above the piston from dropping below atmospheric. This prevents the fluid from cavitating during jounce and prevents air from being drawn past the rod seal and into the housing. The amount of gas charge must be greater than the maximum pressure drop expected across the piston. An alternative to the floating piston-type accumulator is to contain the gas charge in a reserve tube similar to a conventional twin-tube shock absorber.

PROTOTYPE DESIGN

For the purpose of demonstrating an application of ER technology in an automotive shock absorber, a prototype ER shock was designed as a retrofit for a commercial shock absorber having variable damping. The target vehicle chosen was the 1990 Ford Thunderbird Super Coupe which is equipped with an adaptive suspension system. The goal was to retrofit the original equipment (OE) rear shock absorbers with ER prototypes having the same package size and stroke, thus demonstrating an ER damper that compares favorably with a conventional damper and that offers the potential for semi-active control. A photo of the OE and ER shock absorbers is shown in Figure 2. The total stroke of these units is 165 mm.

Room temperature damping force data for the OE shock absorber are shown in Figure 3. The target specifications for the ER prototype in *soft* mode are identical to the OE. In *firm* mode, the ER effect causes behavior similar to a friction damper, so the behavior of a viscous damper such as the OE unit cannot be duplicated. Thus, the decision was to duplicate the OE damping force at an intermediate speed of 0.4 m/s. This corresponds to the maximum piston rod speed during a sinusoidal oscillation of half the total stroke of the damper, occurring at the approximate natural frequency of the car body in vertical bounce mode (1.5 Hz). With the actual prototype device, the damping force could be easily adjusted to different values by changing the voltage applied to the ER fluid. In addition, a maximum of 1 mA current draw while

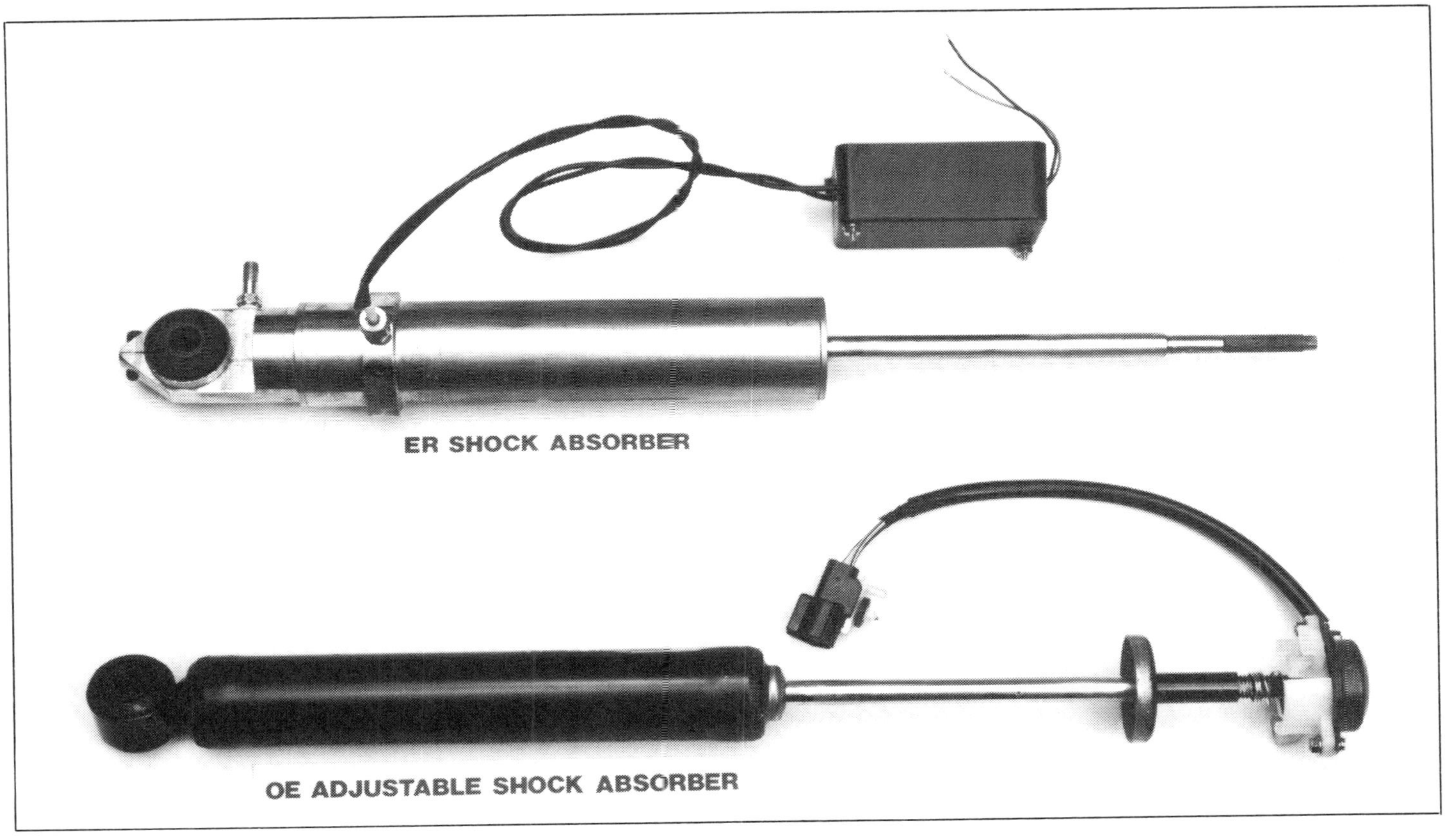

Figure 2: ER prototype and OE Thunderbird electronically controlled shock absorbers

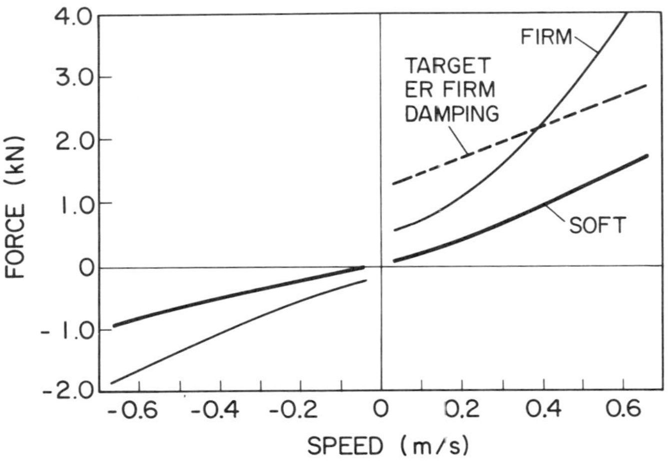

Figure 3: OE Thunderbird shock absorber damping force vs. piston speed

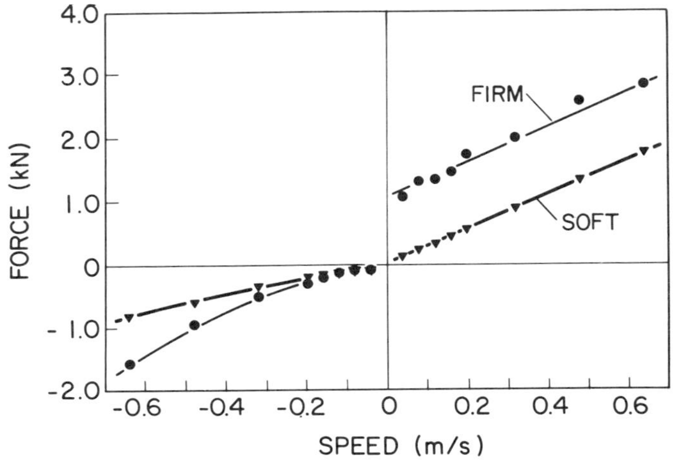

Figure 4: ER prototype shock absorber damping force vs. piston speed

in *firm* mode was targeted as realistic.

The target ER fluid properties for this application were chosen based on known values for experimental fluids:

η	< 50 cP	at 25°C
τ_y	> 4 kPa	at $E < 10$ kV/mm
j	< 100 mA/m^2	at E = max

The temperature range for a commercial application is generally considered to be -40 to 120°C. Performance over this wide range is presently the major challenge for ER fluids.

The dimensions of the prototype shock absorber were chosen to fit the required package size and to remain within realistic limitations of machinability. Thus, using Equation 7 and allowing for some overdesign, the relevant dimensions for the prototype were determined:

d_1 = 12.7 mm
d_2 = 42.5 mm
d_3 = 46.6 mm
L = 63.5 mm
h = 0.5 mm

PROTOTYPE PERFORMANCE

DAMPING CHARACTERISTICS - Figure 4 shows the Force vs. Speed data for the ER shock absorber, indicating that the target room temperature forces for *soft* and *firm* damping as shown in Figure 3 are achieved. To evaluate durability, the prototype was tested using a procedure that is representative of life testing of commercial shock absorbers. The test consisted of a sinusoidal oscillation at ± 50 mm and 1.5 Hz, a duty cycle where the damping is switched between *soft* and *firm* every five oscillations, and a temperature of 70 to 80°C on the housing surface. The target was to maintain consistent performance for two million oscillations. Figure 5 shows Force vs. Time data for the prototype that demonstrate nearly equal performance at the beginning and end of the durability test. The current draw of the prototype in the *firm* mode was less than 1 mA using a 4kV applied voltage, so the power draw was less than 4 W throughout the test.

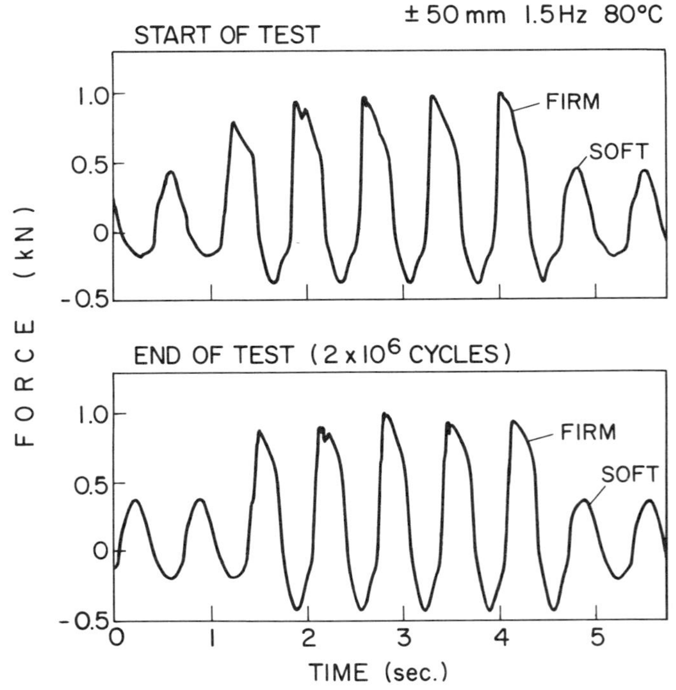

Figure 5: ER shock absorber durability test data, damping force vs. time

ROAD TESTING - A Thunderbird Super Coupe equipped with ER shock absorbers on the rear suspension was road tested using an accelerometer to measure vertical acceleration of the body near the point where the shock absorber is attached. Also, a displacement transducer located next to the shock absorber measured relative displacement between the wheel and body. Figure 6 is the power spectral density measured over a rough road surface. This illustrates the increased accelerations associated with *firm* damping compared to *soft*, occurring at frequencies above the primary resonance. However, Figure 7 illustrates that *firm* damping provides better wheel and body control than *soft* when driving over a bump. These same tests conducted with the OE shock absorbers indicate that the ER prototypes compare favorably.

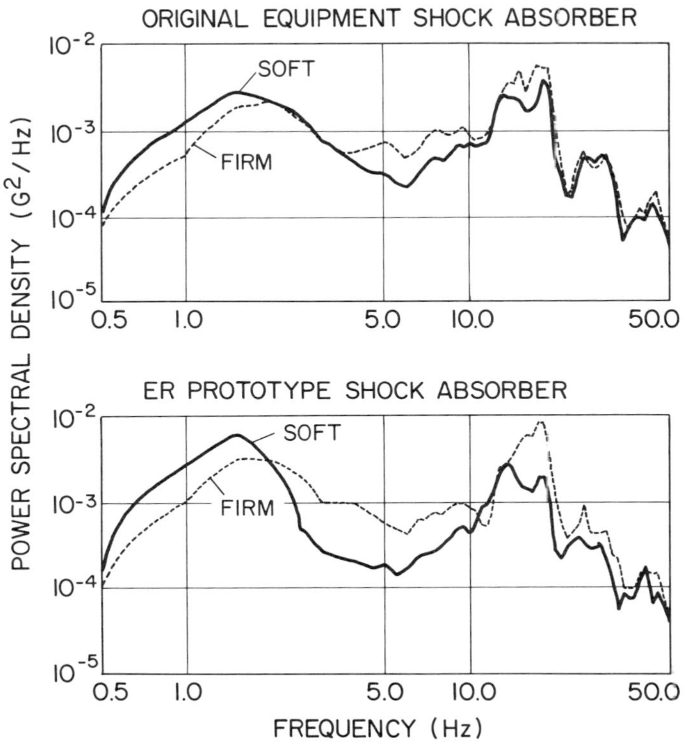

Figure 6: Rough road power spectral density

RESPONSE TIME - As mentioned, fast response in switching between *soft* and *firm* damping will be needed to include the ER shock absorber in a semi-active suspension system. To illustrate the potential of ER technology in this area, a prototype was tested using an ER fluid known to be fast at room temperature. Figure 8 is an example of Force vs. Time data for the shock absorber oscillating sinusoidally at ± 12.5 mm and 1.5 Hz. The voltage is being switched between *off* and *on* at 15 Hz, ten times the frequency at which the shock absorber is oscillating. The time required to switch between *soft* and *firm* damping is about 5 ms. Figure 9 shows Force vs. Displacement data for the same conditions, further illustrating the ability of the shock absorber to quickly switch between *soft* and *firm*. In these response time tests, the bushings at the ends of the shock absorber were omitted so that their compressibility would not be a factor.

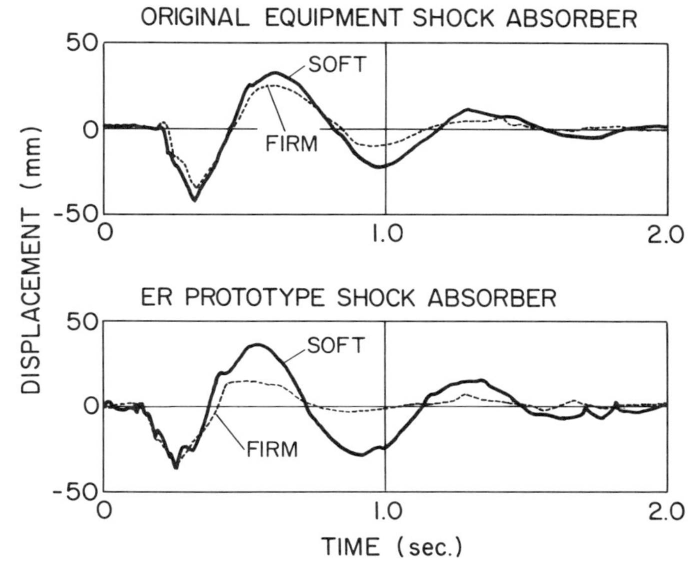

Figure 7: Relative displacement between wheel and body while driving over bump

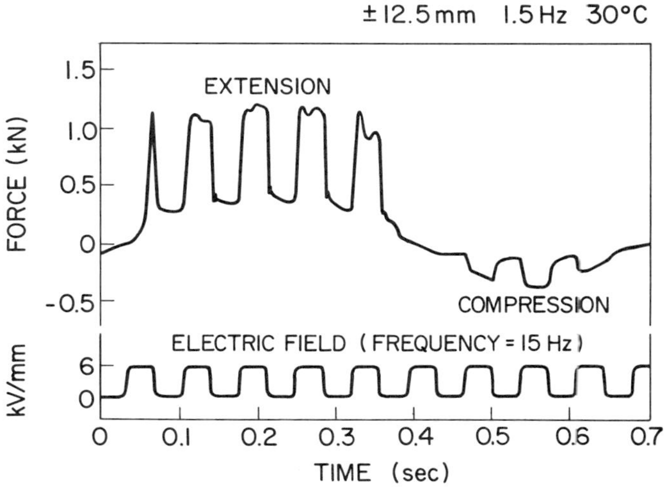

Figure 8: ER shock absorber response to square wave voltage control, damping force vs. time

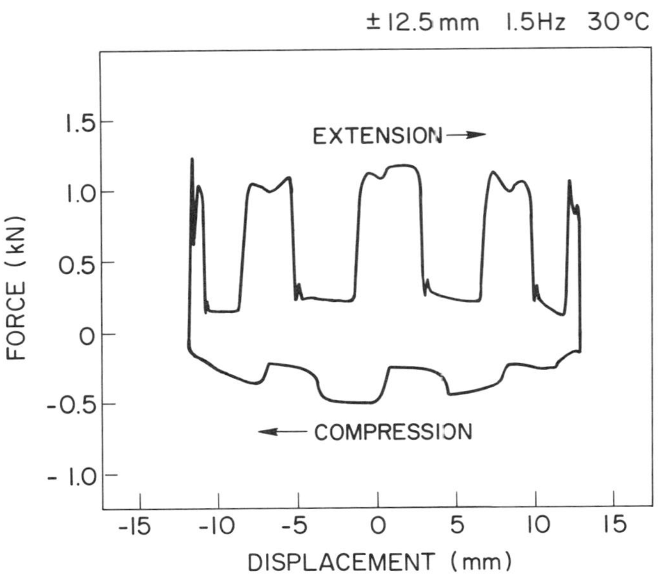

Figure 9: ER shock absorber response to square wave voltage control, damping force vs. displacement

CONCLUSION

A shock absorber containing electrorheological fluid was demonstrated which offers a viable means for fast-acting electronic control of damping force. A prototype was designed to retrofit the adjustable shock absorbers on an automobile without modifying the existing package constraints. The prototype achieved damping forces comparable to the original equipment shock absorbers and performed successfully in a vehicle test. The ER shock absorber also maintained consistent force in a durability test while requiring only a few watts of electric power. The target 5 ms response time in switching between *soft* and *firm* damping will allow this shock absorber to ultimately be incorporated into a semi-active suspension system. Advances in the ER fluid formulation will be required to achieve the desired performance over the entire range of operating conditions of a commercial application.

ACKNOWLEDGEMENT

This work was completed with the support of The Lubrizol Corporation which supplied the experimental ER fluids.

REFERENCES

1) M. Shiozakai, S. Kamiya, M. Kuroyanagi, K. Matsui, and R. Kizu, "High Speed Control of Damping Forces Using Piezoelectric Elements," Society of Automotive Engineers Paper No. 910661 (1991).

2) Y. Guy, M. W. Kerastas, and R. E. Bruckman, "A Solenoid-Actuated Pilot Valve in a Semi-Active Damping System," Society of Automotive Engineers Paper No. 881139 (1988).

3) L.R. Miller and C. M. Nobles, "The Design and Development of a Semi-Active Suspension for a Military Tank," Society of Automotive Engineers Paper No. 881133 (1988).

4) S. Morishita and J. Mitsui, "Controllable Shock Absorber System (An Application of Electro-Rheological Fluid)," Society of Automotive Engineers Paper No. 910744 (1991).

5) R. Stanway, J. Sproston, and X. Wu, "Variable Suspension Damping Using Electro-Rheological Fluids," Institution of Mechanical Engineers Paper No. C382/034 (1989).

6) Z. P. Shulman, B. M. Khusid, E. V. Korobko, and E. P. Khizhinsky, "Damping of Mechanical Systems Oscillations by a Non-Newtonian Fluid with Electric-Field Dependent Parameters," Journal of Non-Newtonian Fluid Mechanics (1987), vol. 25, pp. 329-346.

970385

An Energy Efficient Electromagnetic Active Suspension System

Francis B. Hoogterp
US Army TARDEC

Joseph H. Beno and Damon A. Weeks
The University of Texas at Austin

ABSTRACT

The technology thrust to develop an effective electromagnetic actuator for application in an active suspension system has precipitated a fresh look at the active control schemes in an effort to reduce the required force levels of the actuator. The resulting "near constant force" control algorithm is described and its ability to greatly reduce vehicle sprung mass motion is documented through simulation and single wheel station laboratory test stand results. The vehicle power and energy requirements associated with this unwanted vehicle vertical are analyzed and comparisons between the corresponding passive and active systems are presented. The success of the active system leads naturally to the conclusion that a passive suspension equipped vehicle will become power limited at a much lower speed than will this active system when traversing severe cross-country terrain.

INTRODUCTION

The possibility of utilizing an electromagnetic (EM) actuator in a high-wheel travel active suspension system has recently gained considerable interest in both the commercial and the military sector [1-2]. The bus developers are interested in the potential benefits of active suspension for ride comfort and vehicle roll control as well as for the accompanying height and attitude control. Height and attitude control can be used to greatly simplify potential solutions to the handicap access requirements the bus manufactures must meet. The interest in an EM actuator based system is predicated on the availability of an energy efficient, and appropriately packaged system, that if exists would eliminate the need for a complex hydraulic system.

A passive suspension system is simply a traditional suspension system composed of springs, dampers (i.e. shock absorbers), and various structural linkages [3]. The springs and dampers of a passive system are appropriately tuned for the expected input frequencies, to attenuate the terrain input (as seen by the tires). The purpose of the shock absorbers (or dampers) is to dissipate the energy that is input into the chassis by the ground through the suspension system. The more energy that is allowed to be input to the chassis, the more energy that the shock absorbers must dissipate, and the more energy that the vehicle's power train must provide. A semiactive suspension system is defined here as a passive system which has adaptive damping characteristics. The system has the traditional springs and dampers, but the vehicle motion is constantly monitored, and the damper characteristics are reduced (or set to zero) when the damping forces would increase chassis motion. It has been demonstrated that semiactive suspension systems can moderately increase the ride quality of a vehicle while simultaneously slightly decreasing the energy dissipated by the suspension system [4-5].

An active suspension system represents a distinct departure from traditional suspension system concepts. Springs and dampers are no longer required (though for practical reasons springs are often still included). The suspension forces that are normally provided by the springs and dampers are now provided by some form of actuator (e.g. hydraulic, pneumatic, or EM) based on the output of vehicle motion based control algorithms. The active suspension control research for rough terrain negotiation has primarily focused on more sophisticated approaches to the attenuation of the ground induced inputs [6-7]. The success of active suspension has been quite impressive in the areas of on-road stability and handling and has been demonstrated repeatedly. The cross-country ride problem, however, has not proven as easy to conquer to date.

The military's interest is active suspension systems also begins with significantly improving vehicle ride and stability, but for the military, cross-country ride is the focus of interest. The Army's thrust to remove hydraulics from its ground combat vehicles precipitated its interest in EM actuator based active suspension systems. The technology thrust to develop an electric drive combat vehicle has only served to heighten this interest since adequate electric power would be available on the vehicle and the addition of a hydraulic system for active suspension would only add to the overall vehicle complexity.

The intended application of such a system would be to obtain maximum high-speed cross-country performance desired by military applications. The major barrier to making such an application practical has been the weight and power requirements of an electromagnetic actuator based system. Since the actuator weight for such a system is driven largely by the actuator force requirements and therefor indirectly by its power demands, minimizing the required active suspension power will also greatly reduce the system weight.

An active suspension control algorithm is presented that requires significantly less power (and correspondingly lower forces) than do other algorithms. This "near constant force" strategy attempts to prevent the excitation of the vehicle's sprung mass rather than to attenuate the motion once it has begun. The resounding success of this algorithm is demonstrated through a series of simulation and laboratory test stand results. The laboratory test stand results include test data obtained from a single wheel station, rotary EM actuator, supporting a sprung mass of 4500 kg as well as test data from a reduced scale, 4 wheel representation of a high mobility off-road vehicle. Simulation results corresponding to the test data are also presented.

The power and energy requirements of this active suspension configuration are actually shown to be significantly lower than those for the comparable passive suspension configured vehicle. In addition the acceleration levels of the sprung mass are also dramatically reduced resulting in the operational maximum cross-country speed being significantly increased over, not only the passive system, but over other proposed active suspension configurations as well.

NEAR CONSTANT FORCE ACTIVE SUSPENSION CONTROL ALGORITHM

The underlying philosophy for the "near constant force" (or NCF) active suspension system algorithm is quite simple. Since the purpose of a suspension system is to isolate the chassis (or sprung mass) from the motion of the wheel (or unsprung mass) which is trying to follow the terrain, a perfect suspension system (in terms of isolation) would always generate only the static force to the chassis. With a constant vertical force on the chassis, the chassis would then be subjected to zero vertical acceleration and therefor a perfect ride. In practice, of course, this is simply not practical since the vehicle must go up and down hills and the suspension travel is limited to a few inches at best.

The idea persists, however, that the best active suspension control algorithm would be the one that would provide the least variation in suspension forces while minimizing the prospect of exceeding suspension travel. Thus was born the NCF control algorithm. The perfect solution of this problem would require prior knowledge of the terrain, but the development of the NCF algorithm was developed without the assumption of this prior knowledge and yet has been shown in the laboratory to perform at an extremely high level of efficiency.

A model of the original laboratory test stand for which the NCF control scheme was developed, is shown in figure 1. This model represents (as does the test stand) one of the fourteen roadwheel stations and one-fourteenth of the hull mass of a main battle tank (the M1 Abrams). The road-wheel is mounted on a trailing arm and the hull mass is constrained so that it can move only in the vertical direction. The suspension vertical force to the hull is actually applied as a torque between the road arm and the hull at the road arm pivot. Since this model is restricted to vertical motion, it can be viewed the same as a 1/4 car model would be for a four wheeled vehicle. Figure 2 presents the predicted motion of this model's hull mass for the standard torsion bar suspension system on the M1 tank to the terrain input shown. This terrain is extremely rough as indicated by its terrain roughness index of 8.9 cm rms. The gradual hills were removed from the terrain prior to the simulation to comply with the travel

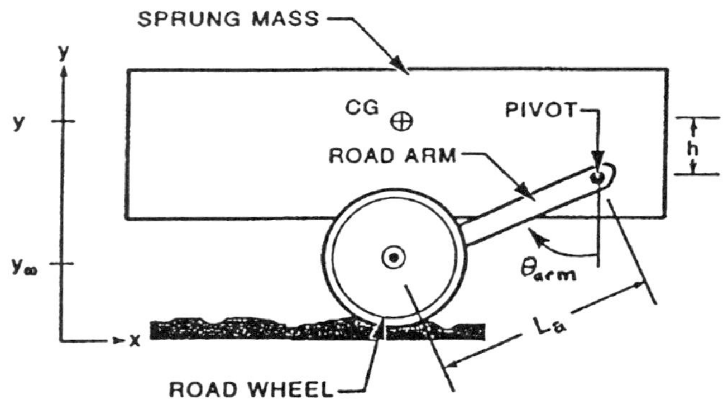

Figure 1 One-dimensional tank suspension model

limits of the laboratory terrain generation hardware. This simulation represented a vehicle velocity of 16 kph and it can be seen from the plots that for this very stiff passive suspension system, the hull tends to follow the terrain.

Any practical active suspension design will include a passive spring to statically support the vehicle and to reduce the overall power requirements of the active system and the EM actuator based system is no different. The demonstration rotary actuator built on a main battle tank scale, incorporated air spring for this purpose. This permitted the spring rate to be kept quite low while still providing the required static force. The first task of the NCF algorithm is to negate the effect of this low rate spring which varies with the arm angle of the trailing arm suspension system. The NCF algorithm was developed based on the premise that it would have available only measurements that could be obtained from the vehicle itself. These could include arm angle and velocity making the cancellation of the spring force a simple matter, and could also include, as needed, chassis (or hull) vertical, pitch, and roll accelerations, velocities, and displacements (or estimations thereof). The resulting actuator

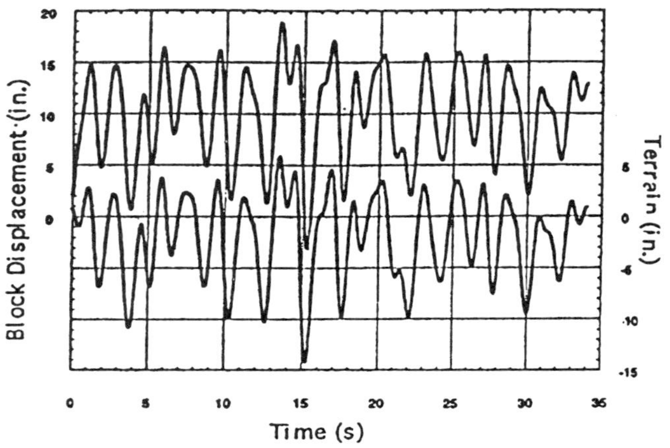

Figure 2 Simulated hull motion at 16 kph for passive M1 torsion bar system

force from the NCF algorithm for a 1/4 car model (i.e. considers chassis vertical motion but not pitch or roll) is given by:

$$F_a = -F_s(\theta_{arm}) - k_1[y - y_{ref}] - k_2 y' - k_3 \theta_{arm} - k_4 \theta'_{arm} \quad (1)$$

where $F_s(\theta_{arm})$ is the force required to offset the air spring's additional force (as indicated, F_s is a function θ_{arm}), y is the vertical displacement of the chassis, θ_{arm} is the angle of the roadarm, y_{ref} is the height of the desired chassis vertical trajectory, and k_1, k_2, k_3, and k_4 are appropriate gains. Of course the displacement of the chassis, y, and of the desired reference trajectory, y_{ref} , are not readily available to the controller and so the gain k_1 is set equal to zero in the NCF algorithm. The term is left in equation (1) to emphasize the fact that it is ideally what is desired to be controlled. However much can be discerned about the nature of the term $[y - y_{ref}]$, from the more easily measured signals y' and θ_{arm} and feedback based on these terms is used instead of on $[y - y_{ref}]$.

Figure 3 shows the hull motion for a 16 kph simulation as did figure 2, but in this case an active suspension system was assumed and the suspension force is determined by the one-dimensional NCF algorithm. Also, as was done for figure 2, the gradual hills had been removed from the terrain. Figure 4 shows the actual results from the laboratory test stand corresponding to the predicted results of figure 3. The comparison is quite comforting and lends credibility to the attempt to greatly decrease hull motion in response to severe cross-country terrains.

The appropriate choice of gains in (1) allows the NCF algorithm to actually respond to hills while effectively filtering out the higher frequency inputs inherent in rough cross-country terrain. Figure 5 presents the simulated hull response to the same terrain used in the previous examples, but where the gradual hills have been left in the terrain. This simulation was performed at a vehicle speed of 64 kph and shows almost a total attenuation of the higher frequency inputs.

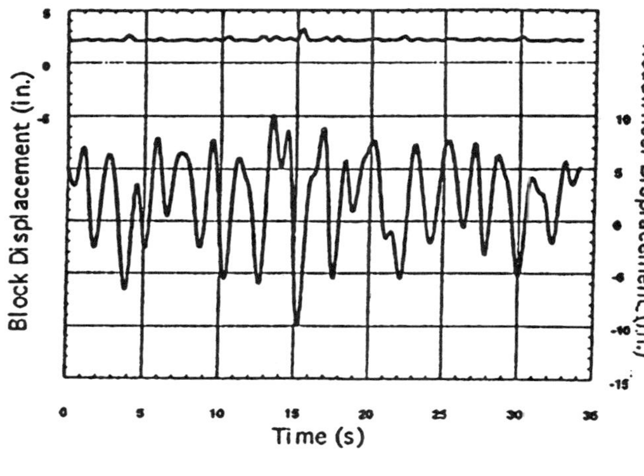

Figure 3 Simulated hull motion at 16 kph for NCF active suspension system

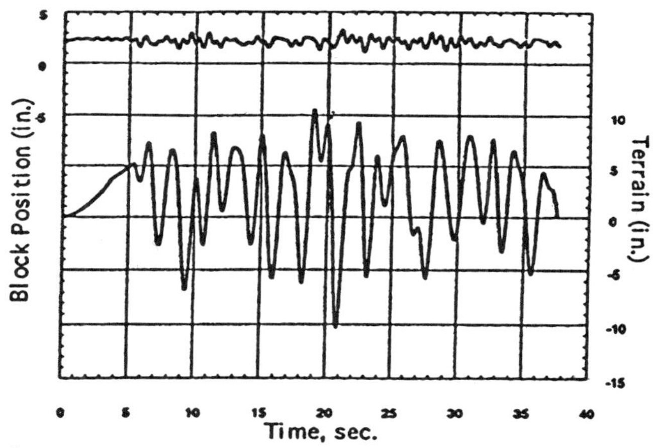

Figure 4 Laboratory test stand hull motion at 16 kph for NCF active suspension system

The version of the NCF algorithm presented in (1) is concerned with minimizing vertical chassis motion only. The extension of the NCF algorithm to a full 3-dimensional vehicle model must also consider the chassis pitch and roll motions in calculations. The NCF actuator force for this full vehicle model can be similarly be written as:

$$F_a = -F_s(\theta_{arm}) - k_1[y - y_{ref}] - k_2 y' - k_3 \theta_{arm} - k_4 \theta'_{arm} - k_5 \theta - k_6 \theta' - k_7 \alpha - k_8 \alpha' \quad (2)$$

where θ and θ' are the chassis pitch angle and pitch rate, α and α' are the roll and roll rate, and the additional k's are appropriate gains as before. This 3-d version of the NCF active algorithm was implemented to control a 30 % scale, 4 wheeled active suspension test rig representing the Army's High Mobility Multi-purpose Wheeled Vehicle (or HMMWV). Preliminary results from this test rig were similar in terms of chassis motion reduction, to those discussed earlier for the single wheel station system.

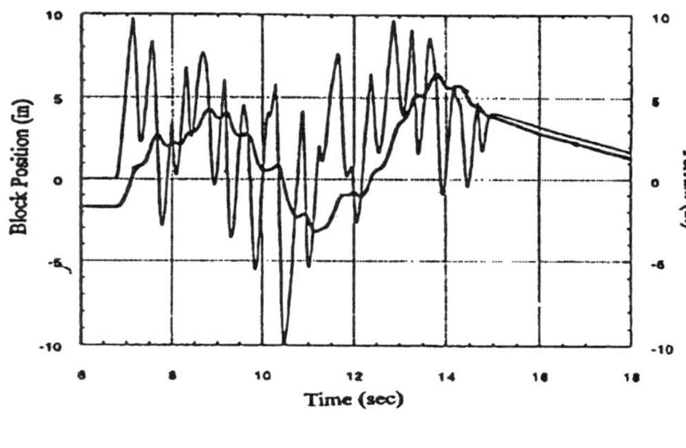

Figure 5 Simulated hull motion at 16 kph for NCF active suspension with hill following employed

POWER COMPARISONS FOR ACTIVE AND PASSIVE SUSPENSION SYSTEMS

It is clear that when the sprung and unsprung masses are moving up and down, that some of the power that was intended to drive the vehicle forward is being expended in the vertical direction. This power is, of course, supplied by the vehicle's engine and is delivered through the inefficiencies of the drive train to provide motion to the vehicle. Power expended in exciting the vehicle vertically is totally wasted, and if it becomes excessive will significantly limit the overall performance of the vehicle. It will be demonstrated that this is in fact the case for the passive suspension system on the M1 when trying to negotiate very rough terrain.

The unsprung mass (roadwheel or tire) is intended to follow the terrain and thus must be continuously accelerated in the vertical direction when traversing rough terrain. The unsprung mass, however, is usually less than 10 % of the sprung mass and thus is a rather minor drain on the system power than is the sprung mass at the same excitation levels. The active suspension system cannot, therefor, effectively reduce the unsprung mass motion of the power required of the engine to produce that motion.

The sprung mass on the other hand has no requirement to follow the high frequency component of the terrain input, and in fact, the suspension systems very purpose is to keep it from doing so. Unfortunately, the passive system, as was demonstrated in figure 2, is not very effective at accomplishing this. In the passive system the power that is input into the unwanted vertical (as well as pitch and roll) motions is ultimately dissipated by the shock absorbers and lost. In order to compare the vertical motion power between the active and passive suspension systems, vertical sprung mass power was computed as:

$$\text{Power} = F\,v = F\,a\,v \tag{3}$$

where F is force, v is velocity, m is mass, and a is acceleration. The resulting comparisons that are shown below are for a single wheel station as represented in figure 1, where the sprung mass represents 1/14th of an M1's sprung weight (the M1 has 14 road-wheels).

Figures 6 and 7 depict the power required to accelerate the sprung masses for the passive and NCF active suspension systems at 4.47 m/s and 17.8 m/s. The simulated results are over the same severe cross-country terrain that was used in the previous examples. The passive system employed the standard M1 torsion bar and rotary damper whereas the active system was modelled with the electromechanical suspension active suspension system described in [1] and utilizing the NCF control algorithms described above. At the lower speed, figure 6 shows that the rms and peak powers associated with both systems is quite small. At 17.8 m/s, however, the difference in power between the two systems is dramatic as demonstrated in figure 7. The passive system requires 540 kW peak and 71 kW rms power just to accelerate the sprung mass of a single wheel station along its vertical trajectory. Projecting this rms power over the 14 road-wheels of the M1 would result in a total sprung mass acceleration rms power of 994 kW, or almost 90 % of the M1 engines rated gross power. It is realized that not all the M1 wheel stations are identical (only 6 of the 14 have dampers) and vertical motion of the sprung mass is somewhat abated over the length of the vehicle, but this power figure does not include contributions from the hull pitch and roll motions either.

It should be noted that the passive M1 could not attain nearly this speed (i.e. 17.8 m/s) across this terrain even if the vehicle and crew could stand the accelerations generated by such a run. The vehicle is in fact power limited long before this speed even though this is approximately the level surface top speed of the M1.

Figure 8 portrays the peak and rms power levels for this same pair of suspension systems as a function of vehicle speed. The terrain is the same cross-country terrain that was used previously and the plots were generated from simulation results using the same single wheel station model. The item of major interest here is the extreme divergence in power requirements at higher vehicle speeds between the passive and the NCF algorithm based active suspension system. Thus high speed travel on such a

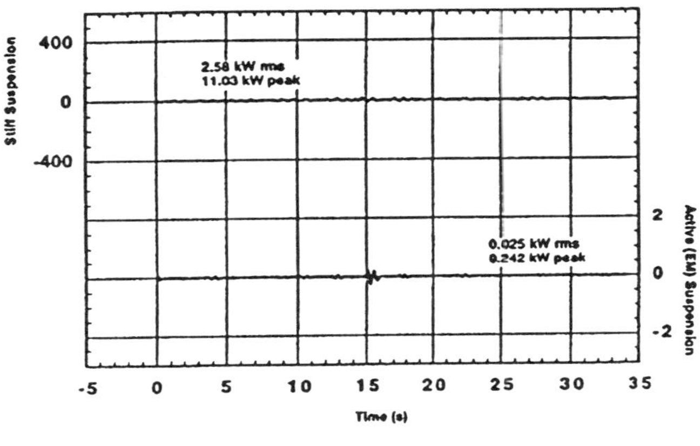

Figure 6 Sprung mass vertical power requirements for active and passive systems at 16 kph

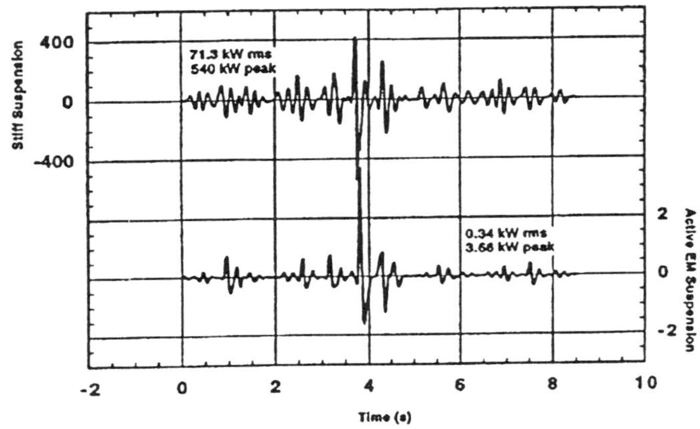

Figure 7 Sprung mass vertical power requirements for active and passive systems at 64 kph

terrain simply cannot be accomplished by this passive suspension system (or probably any other practical passive system). It will require instead, not only an active suspension system, but an active system that significantly reduces unwanted vehicle hull motions.

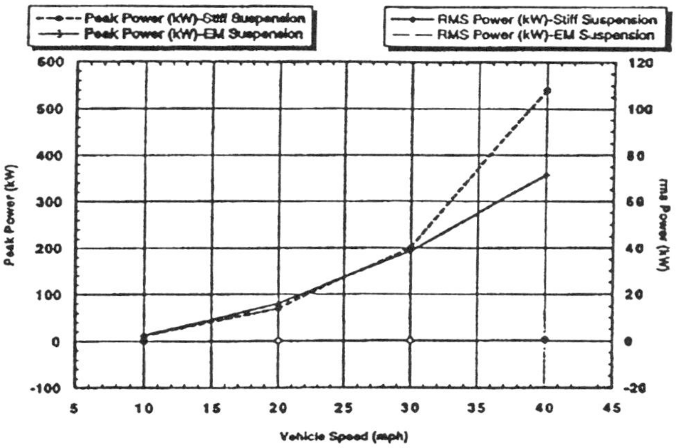

Figure 8 Power consumption between active and passive systems as a function of vehicle speed

ENERGY USAGE

When considering energy consumption of the passive and active suspension systems, the vehicle drive train and active suspension power delivery system efficiencies must be considered. At the 17.8 m/s speed, the passive system required 994 kW rms of power to provide the sprung mass vertical acceleration of the total vehicle. Since the power to accomplish this motion must be provided through the vehicle's drive train, the associated additional power losses can also be attributed to the excessive vehicle motion. If it is assumed that the vehicle automatic transmission and differential each operate at 90 % efficiency, these losses are approximately 233 kW.

These power train losses would represent the total power lost if the energy imparted to the sprung mass were completely recoverable, however this is not the case. Separate simulations show that an individual M1 shock absorber can dissipate as much as 95 to 100 kW of power when the vehicle is driven at 17.8 m/s over this same terrain. It should be noted that there are six shock absorbers on the M1 (on the front two and the rear wheel stations on each side) but the other wheel stations also have some inherent damping associated with them also.

The rms power required for the active system to accelerate the sprung mass was shown to be only 5 kW. This power can be applied by one of two paths to the sprung mass. As in the passive case, the power can be delivered through the vehicle drive train with the partially misdirected goal of vehicle forward motion in mind. Alternatively, the power to accelerate the sprung mass can be delivered directly through the active suspension system actuators. If in this latter case, one assumes that the active suspension power conversion device (for either mechanical to hydraulic or mechanical to electrical) and the suspension force actuator are both 90 % efficient, then the overall efficiency of the power delivery process is 81 % regardless of whether it is delivered through the vehicle drive train or through the active suspension system. Thus the total power required from the engine to provide the sprung mass motion for the NCF active system is only 6 kW even if none of the sprung mass motion energy is recoverable.

The power losses attributable to the unsprung mass vertical motions have not been considered here but they are not avoidable since the wheels must be made to follow the ground in order to be able to provide the necessary hull (or chassis) support. Ground contact is also required, of course, to permit adequate vehicle stability and handling. In either case, the wheel motions of both the active and the passive systems should be quite comparable making the power lost do to unwanted sprung mass motion the major power differential between the two systems.

CONCLUSIONS

The attempt to negotiate severe cross-country terrain at high speeds with conventional suspension equipped vehicles has been shown, to not only impart excessive motion to the vehicle hull or chassis, but to also require such high levels of available engine power that the higher speeds are simply not possible. The "near constant force" (or NCF) active control algorithm has been defined, and when coupled with an EM suspension system has been shown to reduce the sprung mass motion power by a factor of about 200. The NCF approach to active suspension control requires much lower actuator force levels than do previous active control schemes, thus making the prospects of a competitively sized EM actuator quite plausible.

REFERENCES

1. J.H. Beno, F.B. Hoogterp, D.A. Bresie, S.K. Ingram, D.A. Weeks, and W.F. Weldon, "Electromechanical Suspension for Combat Vehicles," SAE Techninal Paper No. 950775, March 1995.

2. F.B. Hoogterp, J.H. Beno, and M.R. Whitmore, "The State of Electromechanical Active Suspension Technology," Proc. of First International Conf. on All Electric Combat Vehicle (AECV), Haifa, Israel, May 1995.

3. D.A. Crolla, D.N.L. Horton, R.H. Pitcher, and J.A. Lines, "Active suspension control for an off-road vehicle," IME Vol. 201, No. D1, pp. 1-10, 1987.

4. L.R. Miller and C.M. Nobles, "The Design and Development of a Semi-active Suspension for a Military Tank," SAE Technical Paper No. 881133, 1988.

5. N.L. Temple and F.B. Hoogterp, "Semiactive Suspension: A Mobility Enhancement for Combat Vehicles," Proc. of ISTVS Conf. on Off-Road Vehicles, London, England, June 1992.

6. D. Hrovat, "Applications of Optimal Control to Advanced Automotive Suspension Design," Trans. of ASME Journal of Dynamic Systems, Measurement, and Control, vol. 115, no. 2B, pp. 328-342, June 1993.

7. P.W. Claar II and J.M. Vogel, "A Review of Active Suspension Control for On and Off-Highway Vehicles," SAE Technical Paper No. 892482, 1989.

950588

A New Class of Smart Automotive Active Suspensions Using Piezoceramic Actuation

Sridhar R. Thirupathi and Nagi G. Naganathan
The University of Toledo

ABSTRACT

Active suspension for automobiles involves an active control of wheel motions. This work investigates the possibility of realizing controlled wheel motion for an active suspension using a class of *smart* materials, namely piezoceramics. The proposed concept deviates significantly from conventional practice of using heavier, hydraulic actuators. Rather than accomplishing the wheel motion using the direct stroke of an actuator, the motion is to be realized by actively modifying the effective stiffness of the suspension system using piezoceramic sensing and actuation. Piezoceramic actuators have been typically waived as totally unacceptable for such large stroke, load carrying applications since these actuators have always been limited to micro-motion ranges. In this research, a strategic combination of elastic and piezoceramic layers is shown to have the potential to meet this challenge. Preliminary analysis and experimentation have been conducted to demonstrate the feasibility of the above concept. A candidate design is analytically investigated using composite piezoceramic suspension elements. It is shown that it piezoceramic actuation has the potential to control suspension system deflections through sufficient motion ranges necessary for this application. A prototype actuator demonstrating larger stroke actuation has also been designed, fabricated, and tested.

INTRODUCTION

Active suspension for automobiles is an emerging technology intended to provide an excellent ride in all driving conditions. An active suspension system continuously monitors the bounce, pitch, roll, and yaw motions of the vehicle and generates necessary wheel motion compensations so that disturbances due to terrain irregularities and inertia forces are minimized. It also corrects errors in steering (under steer/over steer) and tire contact forces resulting from undesirable changes in weight distributions during severe manuevers. Today's active suspension technology uses hydraulic actuators at each one of the wheels to actively adjust the position of the wheel with respect to the car chassis. A typical system consists of a hydraulic power unit, a hydraulic control unit, and a set of wheel actuators. Additionally, these systems are heavy and demand a significant fraction of the engine power output (McElroy, 1986).

OVERVIEW OF THIS RESEARCH

This work proposes a new concept in active suspension technology where we use the sensing and actuation capabilities of a class of smart materials, namely piezoelectric materials. When the piezoelectric materials are strategically bonded to elastic members and are electrically excited appropriately, they exert manipulative forces on the elastic members. The shape of the elastic members will then change; hence, the deflection of the suspension system. As a result, we have a suspension system that behaves as a system of variable effective stiffness. Note that the stiffness of the suspension element (spring element) is not changed. It is the composite stiffness of the entire system that changes since we derive varied deflections for a given external load.

In the following sections, we present some background on piezoelectric technology and conventional active suspension technology. In addition, we will discuss how the piezoelectric behavior might be used to produce a new class of smart automotive active suspension systems.

BACKGROUND

Piezoelectric Technology

Piezoelectrics are pressure-sensitive materials that generate electric signals upon application of mechanical pressure. They also exhibit the converse property whereby they undergo mechanical strains when under an applied electric field. Their response time to any external loading is of the order of 10 milliseconds or less. As an actuator, they can offer controllable displacements in the micrometer range while exerting significant surface forces of the order of 30 MPa (Takahashi, 1985). Also, they produce less heat and noise during operation and thus, operate with higher efficiency. Piezoelectric materials are both naturally occurring and man-made. Anisotropic materials such as Quartz, Tourmaline, and Rochelle salt are some of the naturally occurring piezoelectric materials. Synthetic piezoelectric materials are termed piezoceramics. Lead Zirconate Titanate (PZT), Lead Magnesium Niobate (PMN), and Poly-vinylidene Fluoride (PVDF) are some of the commercially available piezoceramics.

Piezoceramics have become increasingly popular in many engineering applications. These include: piezoceramic stacks, bimorphs, precision positioning tables, ultrasonic sensors, motors, and cleaners. Of recent interest has been the use of piezoceramic actuation for active control of structures. Since they can sense mechanical strain conditions and respond quickly to command excitations exerting relatively large forces, their use is being studied extensively in active vibration control of flexible structures.

Commercial Automotive Active Suspension

Recent trends in the development of automotive active suspension have been due to a considerable amount of research and development efforts both within and outside automotive companies. At the present time, active suspensions are available for selected, high performance automobile models.

Dominy and Bulman (1985) suggested the feasibility of using a semi-active suspension for Formula-I race cars and demonstrated by analysis that it was possible to maintain a constant ride height and attitude during cornering, braking, and acceleration while minimizing the chassis response to road bumps.

In the General Motors-Lotus System, conventional springs and shocks are replaced by hydraulic actuators (Johnson, 1989). The suspension travel is sensed and fed to an on-board computer, which controls the actuator's hydraulic pressure that maintains operator selected ride characteristics. Active suspension can be set for any selected spring rate, damping, or roll mode. It can generate several ride modes simultaneously. General Motors developed a Corvette with an active suspension and a significant improvement in ride and handling quality was recorded (McElroy, 1986).

Nissan's Infinity Q45 luxury performance sedans have been fitted with active suspension systems (Keebler, 1990). Nissan's system retains conventional coil springs and incorporates a power source that constantly generates actuating power and controls wheel movements continuously. While the GM Lotus system is capable of responding to ride disturbances up to 30 Hz, the bandwidth of the Infinity system is restricted to 5 Hz.

Aura systems, Los Angeles, claims that it has developed a suspension design that provides the same handling and ride benefits as the Lotus active suspension at one-tenth the cost and with only 7% of the power consumption (Keebler, 1990). Aura's system has an electromagnetic shock absorber with a soft-ride spring. The electromagnetic system regulates body motion and the low-cost spring handles wheel motion. This is unlike the Lotus system which controls both body and wheel motions. According to Aura, it is the control of the wheel motion that requires higher power levels and sophisticated electronics and hydraulics.

As seen from the above survey, a typical active suspension system involves multiple sensors and actuators, the functions of which are effectively coordinated and regulated using appropriate control algorithms. The focus of our research is the use of piezoelectricity as the source of actuation. In the following section use of piezoelectricity for active control will be discussed.

Applications of Piezoelectricity in Active Control

Since piezoceramics offer large actuation forces over small motion ranges, their initial applications in the area of active control have been in vibration control of structures. Piezoceramic actuators have been popular in active structural control applications for more than two decades now. Selected references that investigate the role of these actuators are cited in this section.

Das Gupta (1979) applied plate theory in calculating equivalent forces and moments that a piezoelectric layer would generate when excited electrically. He solved for deflections and contact forces due to these equivalent mechanical loads using finite element software SAP IV. Crawley and Luis (1987) studied the actuation capabilities of piezoelectric layers and segmented actuators bonded to or embedded in elastic members. They obtained a pair of fourth order equations on piezoelectric and substructure strains and by using the strain boundary conditions derived closed form solutions for the strain distributions within the piezoelectric actuator and the substructure.

Lee, et al. (1989) developed a dynamic model for smart flexible structures with piezoelectric control layers. They employed a variational theorem in their finite element formulation for analyzing the dynamics of such flexible structures for use in mechanical linkages and robotic systems. They have also demonstrated through examples with one-dimensional beams that vibration amplitude could be actively controlled using their methodology.

Dimitriadis, et al. (1991) demonstrated potential applications of surface bonded patch type piezoelectric actuators in active vibration control of flexible structures. They studied the effect of piezoelectric patches of different configurations in exciting various two dimensional modes in thin elastic plates. They concluded that it was possible to selectively excite any particular mode in a thin plate and the same behavior could be extended for active vibration suppression.

Newnham, et al. (1991) have developed a large deflection actuator (Moonie) that utilizes the lateral deformations of a circular piezoelectric plate to cause large axial motion. In their device, they attach a pair of thin plates on either side of a piezoelectric plate and cause them to buckle. This produces an amplified motion along the axis of the actuator. However, the 'buckled' state of the thin elastic member will not allow any significant axial force carrying capacity.

In the above studies, piezoceramic actuators were limited to providing either small controllable displacements with large actuation forces or slightly larger strokes with smaller resistive forces to overcome. Hence, they cannot directly replace the hydraulic actuators that are presently used in automotive active suspensions. Their utilization has so far been limited to the role of controlling damping (Shiozaki, 1991).

Use of Piezoceramics in Automotive Suspension

Shiozaki, et al. (1991) proposed the use of a piezoelectric sensor-actuator system to effect a semi-active suspension by varying the damping force of a shock absorber. In their semi-active system they used a piezoelectric sensor to generate signals proportional to road conditions. The signals were fed to an electronic control unit that generated appropriate high voltage signals for a piezoelectric actuator. The actuator was used to control the oil flow channel in the shock absorber to produce the desired damping effect. Naganathan and Thirupathi (1993) have proposed seven different configurations for active deflection control of suspension mechanisms using piezoceramics. They used lateral and shear deformations of piezoceramic segments bonded to or embedded in suspension components to derive the needed deflection control for active suspension.

While Shiozaki's work utilized the piezoelectric material to regulate the damping in the system, it appears one would be able to realize a more effective suspension by using the piezoelectric material to actively control the stiffness of the suspension system. Thus, by employing the piezoelectric material in a manner where it is most effective, a fully-active system could be realized. The active control of stiffness would then enable the vehicle to be controlled in any desired ride mode. This control when realized from piezoelectric actuation, is expected to overcome the excessive noise and weight handicaps of the presently available hydraulic systems. The focus of the applied research component of this work is to prove the feasibility of this concept and develop a candidate design for active suspension based on this concept.

SCOPE OF THIS WORK

Following activities are undertaken to meet the objectives of this work:

1) Demonstration of the concept of controlled effective stiffness with experiments.
2) Development and analysis of a candidate design based on this concept.
3) Development of a prototype, demonstrating macro-motion potential using piezoceramics together with its analysis and testing.

ILLUSTRATION OF THE CONCEPT

If a piezoceramic specimen is subjected to varied electric fields in the presence of a mechanical load, the deflections of the specimen will be observed to vary. Thus, in the presence of mechanical loads, the system tends to behave as a system of varied equivalent stiffness. If one were to be able to prescribe the required electrical excitations as a function of the mechanical load and the desired equivalent stiffness, the above piezoceramic based system can serve as a system of variable stiffness. An automotive active suspension requires a capability to actively vary suspension deflections for the same applied load. This can be accomplished using a smart material system featuring piezoceramics. This concept was investigated both analytically and experimentally on a piezoceramic bimorph. A commercial piezoceramic bimorph manufactured by Morgan Matroc, Inc. (Bedford, Ohio) was modeled using our finite element method and also was investigated experimentally. The bimorph has three layers having two piezoceramic layers of 242 μm thickness on either side of a brass layer of 127 μm thickness. A schematic of the bimorph is shown in Figure 1.

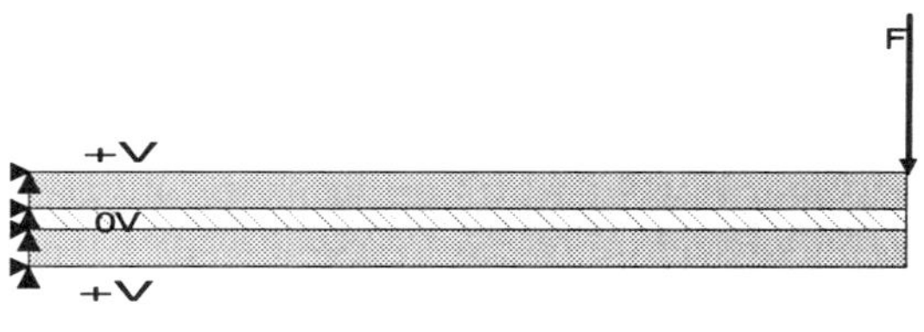

Figure 1: Schematic of Bimorph

Analytical Study Illustrating Concept

The eight-noded shell element developed by the authors of this paper (Thirupathi and Naganathan, 1993a) was used to model the bimorph to demonstrate the concept. Stiffness and permittivity constants of the material were chosen from Mason (1964). The piezoelectric strain constants for the material were chosen from the manufacturer's catalog (Morgan Matroc, 1992). The structure was modeled using 3 shell elements and 18 nodes. Various electrical excitations and forces were applied to study the static response. Voltages and forces consistent with the values applied during experiments were used. End deflections were computed for comparison with the experimental observations described below.

Experimental Study of Bimorph

Figure 2 shows a schematic of the experimental set-up. The set-up was designed giving due considerations for the required power, mechanical loading, and gaging requirements. The experimental set-up consisted of a high voltage power source (Model 609 C-6 from Trek, Inc.) with an output voltage range of -3000 to +3000 volts and a variable gain of 15 to 300. Fiber-optic displacement sensors were used for non-contact measurement of deflections. A data acquisition board (Metrobyte DAS-20) was used to generate reference voltages for the Trek amplifier. The fiber-optic probe signal is converted to digital data using the A/D capabilities of the DAS-20 board and the corresponding microdisplacements of the specimen are recorded on the computer. Consistent with the electric field rating of the bimorphs used, excitations were applied in steps to the specimen in the range of -28V to +28V. Piezoceramic materials have both electrical and mechanical load limits beyond which the material exhibits failure due to misalignment of internal poling directions. This is called *depoling*. In order not to exceed the failure and depoling strain limits of the ceramic, a mechanical load of 0.1N was applied and the corresponding deflections of the bimorph system were measured and recorded in microns.

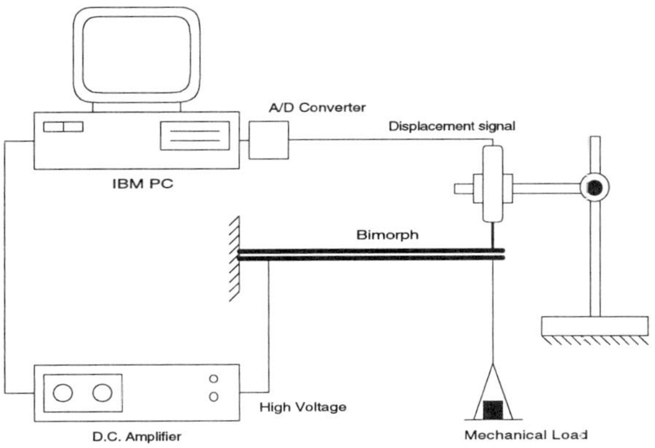

Figure 3: Schematic of Experimental Set-up

Results of Bimorph Investigation

Free end deflections of the cantilever under applied loads were used to calculate the equivalent stiffness of the bimorph-system in the lateral direction. Analytical and experimental results were plotted indicating the variation of equivalent lateral stiffness with respect to electrical excitations for different loads (Figure 3). The results are seen to be in good agreement.

These results indicate that it is possible to obtain different free end deflections for the same mechanical load by varying electrical excitations - a characteristic that can be used to realize variable stiffness properties through controlled excitations for active suspension. It is to be noted here that an automotive suspension requires much higher displacement and load capacity than those displayed in the above example. An objective of the further work described next is to show analytically that the above concept can be extended to meet the requirements of this application.

DEVELOPMENT OF A CANDIDATE DESIGN

An Automotive Active Suspension Using Piezoceramic Smart Actuation

Present designs for automotive active suspension carry significant weight handicaps in order to control wheel motions at each of the wheels using hydraulic means. The thrust of our work is on an efficient realization of the wheel motion with minimal weight increases for the vehicle. The new concept for realizing the required wheel motion compensation is as follows. The required wheel motion is to be accomplished, not by directly lifting or lowering the wheel, but through a variation of the effective stiffness of the suspension system by using the sensing and actuation capabilities of

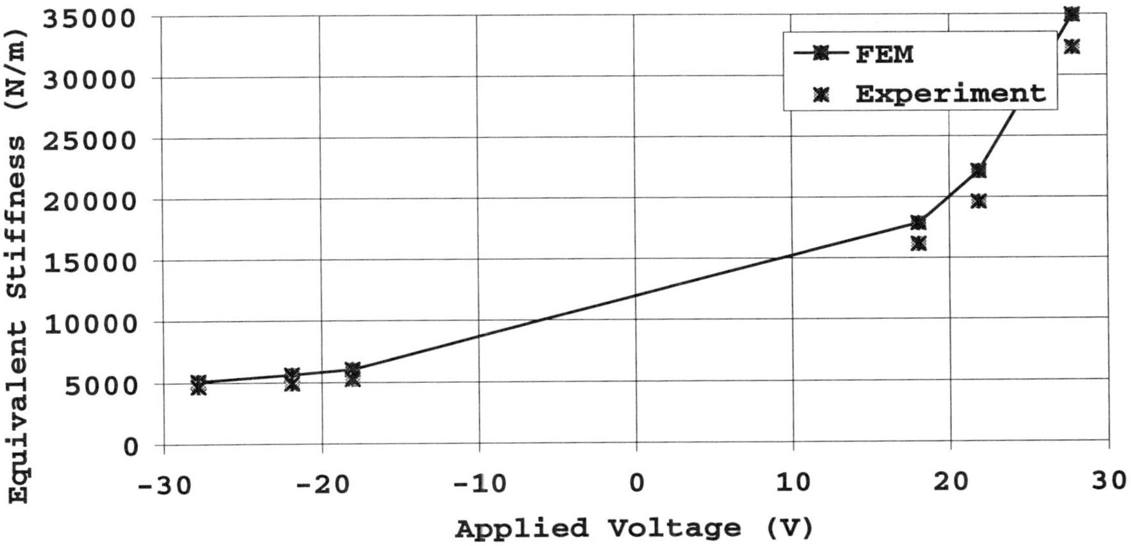

Figure 2: Bimorph: Variation of Equivalent Stiffness

piezoceramic actuators. This variation is to be accomplished by using elastic members of certain geometry along with piezoceramic control layers. *The piezoelectric layers will be energized to exert manipulative forces on the elastic layers. This will result in a controlled motion of the elastic layers resulting in variable deflections for the piezo-elastic composite system. Since these deflections occur for the same applied load and no physical rigid body motion is effected by direct actuation, the entire system (and not the elastic layer) tends to display a characteristic of variable stiffness.*

Figure 4 illustrates one such configuration. As a compliant suspension member, it has an array of composite spring leaves. Each composite leaf consists of piezoceramic control layers bonded on either side of a conventional steel leaf. Using the sensing capabilities of the bonded piezoceramic layers, instantaneous deflections of the members will be sensed. Appropriate electrical excitations will then be computed using an on-board computer and applied to the piezoceramic layers. This will result in the desired stiffness value at that instant of time. By suitably controlling this variable stiffness property of such members, a light-weight, high performance active suspension system may be realized. Some of the advantages of using piezoceramics for active suspension are reduced weight, higher bandwidth due to rapid response of piezoceramics, and lower power consumption. The combination of lower weight and higher efficiency should normally result in lesser power extracted from the engine's output as compared to today's conventional active suspensions.

Automotive suspension require a travel in the range of a few inches. Piezoelectric specimens normally offer displacements of the order of a few microns. Hence, designers have merely stacked them with hundreds of layers to obtain motion in sub-millimeter ranges (Takahashi, 1985). Such motion ranges are still inadequate for this application. One possible technique to realize a larger motion output is to admit a nominal compliance to an elastic support structure and actively modify its compliance using the actuation capabilities of piezoceramics. Actuation modes of piezoceramics in such designs will be largely in flexure. A pilot study and a preliminary analysis demonstrating the feasibility of this concept for automotive active suspension has been presented previously by the authors (Thirupathi and Naganathan, 1992). Results indicated that it would be possible to realize suspension travel in the range of inches using a stack of elastic support elements coated with piezoceramics. In this work, an attempt is made to design one such configuration for automotive active suspension (see Figure 5).

Figure 5 shows an arrangement using a parallel spring system consisting of a stack of composite belleville springs with active piezoceramic

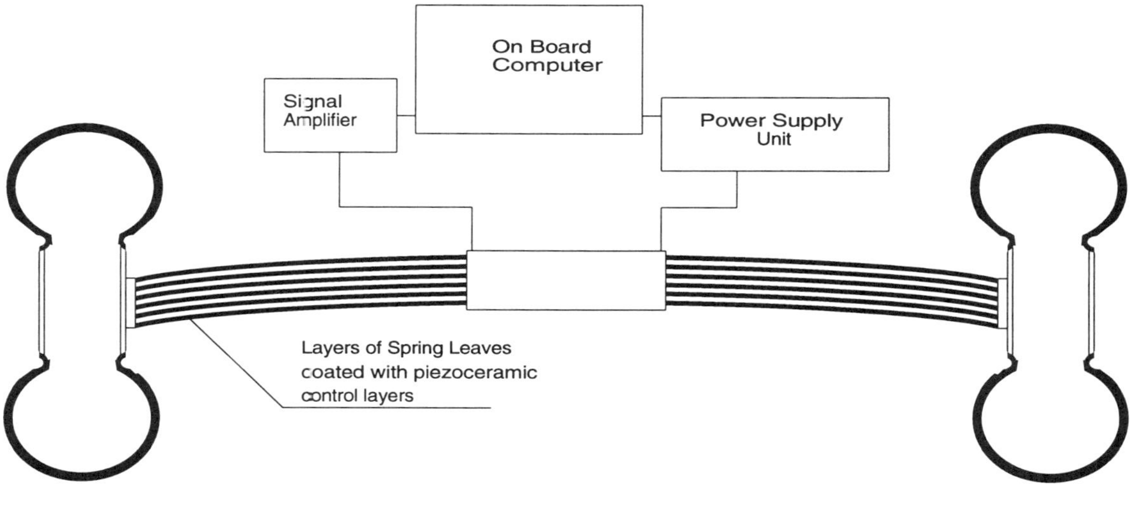

Figure 4: A Smart Structure Concept for Active Suspension

segments. Elements of this system are to be electrically excited so as to cause flexural deflection of the belleville elements (Figure 6). Individual elements of any particular design on this configuration shall carry the load and the stack will have the required number of elements to deliver the stroke requirements. The system as a whole will have signals from G-sensors processed through an on-board CPU to calculate the deflection requirements of this active member. Computer-controlled excitations to this stack of belleville geometries will result in dynamic changes in effective deflections of this smart structure; hence, its effective dynamic stiffness. Thus, the motion compensations required for active suspension will be realized. Finite element analysis of one particular design based on this configuration is presented in the following section.

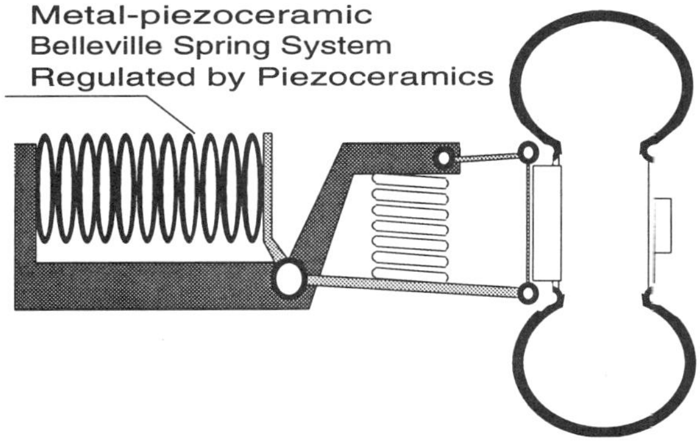

Figure 5: A Belleville Spring System for Active Suspension

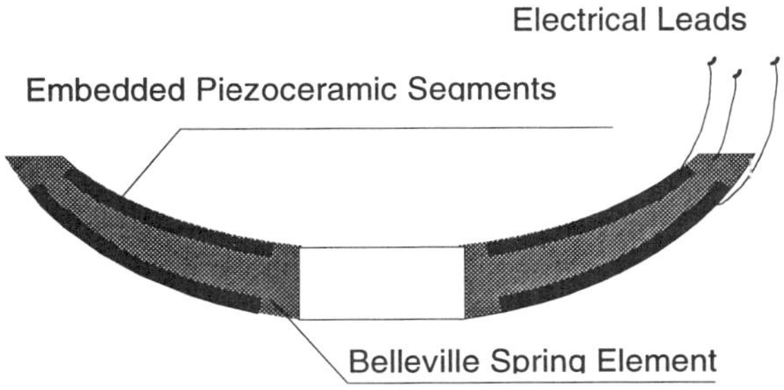

Figure 6: A Typical Belleville Composite Element

FINITE ELEMENT ANALYSIS OF THE DESIGN

Each composite element of the belleville system (see Figure 6) is to be electrically excited so as to cause flexural deflection of the element. In an earlier publication by the same authors (Thirupathi and Naganathan, 1993a), development of a composite shell element to analyze such geometries was described. One quarter of a belleville element is analyzed here using the shell elements. Table I shows the chosen specifications for the belleville element. These dimensions are typical for an automotive suspension. The shown thickness of 0.0043m includes a metallic (steel) layer and two piezoceramic layers of equal thicknesses. The finite element mesh adopted is shown in Figure 7. In addition to x,y,z coordinates for this model, the node point data will include direction cosines of a normal vector at each of the nodes. Exploiting symmetry, we are able to analyze the element by modeling only a quarter section of the geometry. Symmetry plane boundary conditions were imposed for nodes on the two symmetry planes to represent the entire structure. As discussed in the element development section, the response of the structure due to piezoelectricity is studied by modeling the system as a system subjected to piezoelectric initial strains.

Table I: Belleville Element Dimensions

Material	PZT5H
Inner radius	0.0508m
Outer radius	0.1016m
Spherical radius	1.016m
Composite Thickness	0.0043m

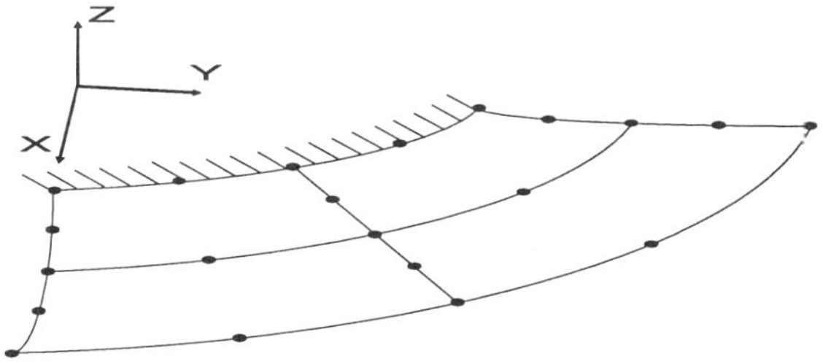

Figure 7: Finite Element Mesh of the Element

A typical light weight automobile weighs about 10,000N (app. 2250 lb.). Hence a mechanical load of 2500N was applied to this belleville suspension element. Various electric field excitations in the range of -400V/mm to +400V/mm were also applied simultaneously to study the variable deflections/stiffness behavior of this suspension component. The above voltage value is based on a nominal limiting electric field of 400 V/mm for this material. As seen from Table II, it would be possible to actively regulate deflections from 2.481 mm to 3.352 mm, which is a range of 0.871 mm. By a series stack of a number of such elements, active deflection control over the needed range will be possible for this application. For an actuation range of 20 mm, we can find that we will need 23 such belleville elements. Mass of an assembly of 23 belleville elements is found to be 19.8 Kg. This would fit in a cylindrical space of 102 mm diameter by 164 mm axial length. Likewise, a ±20 mm stroke range will need 328 mm for the axial length and a 102 mm diameter. On-going developments in piezoceramic material technology will help us in realizing more compact and lighter geometries.

It should be noted here that the configuration, geometry, curvature, piezo-to-metal volume fraction in the belleville elements, etc. are not yet optimized for volume, weight, and performance in this work.

Table II: Results on Belleville Element Deflections

Applied Field (V/m)	Deflection (mm)
-400	2.481
-300	2.591
-200	2.700
-100	2.808
0	2.917
100	3.027
200	3.134
300	3.243
400	3.352

DEVELOPMENT OF A PROTOTYPE AND RESULTS

While the belleville composite element seen in the last section is promising in terms of its motion and load carrying capabilities, it is not possible yet to realize such curved ceramic geometries in their physical form [Near, 1994]. Hence, for the purpose of the prototype development, the belleville geometry is to be approximated using flat bimorph strips arranged strategically. One such configuration is shown in Figure 8. A concept prototype of the suspension structure described in the last section has been fabricated and tested. The prototype consists of eight layers of such flat bimorph strips arranged as shown in Figure 8. When in operation, each of the piezoceramic bimorphs on the device will function as a simply supported beam with a load vector at the middle. When excited electrically, the flat bimorph strips will undergo positive and negative curvatures alternatively. Thus the flexural deflections of the individual elements will add up and the arrangement will function similar to the stack of belleville elements which it approximates.

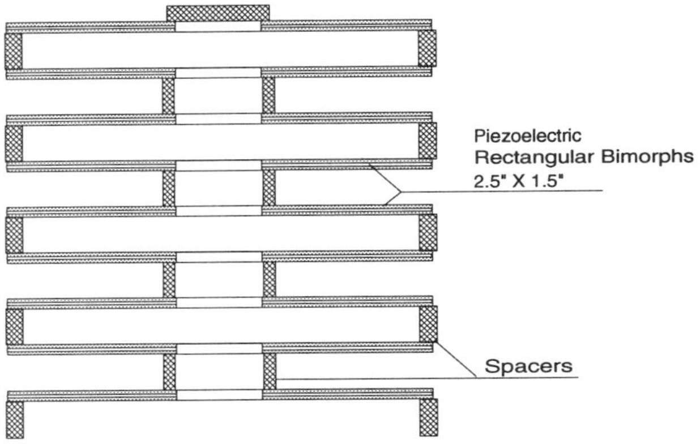

Figure 8: Prototype Schematic

The following sections outline the fabrication of the prototype, its finite element analysis, and experimental results.

Fabrication of The Prototype

A rectangular plexiglass shell was constructed to house the arrangement of rectangular bimorphs. Since the ends of each one of the bimorph element will undergo rotations when functioning as simply supported beams 3.175 mm diameter aluminum rollers were used as end support spacers. Another set of 3.175 mm thick flat spacers were used for the supports at the middle. This is illustrated in Figure 9.

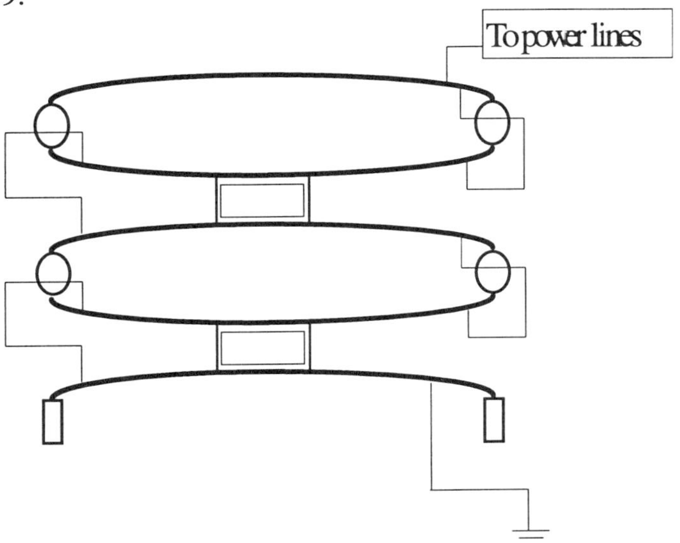

Figure 9: Wiring Diagram and End Supports Arrangement

Sufficient insulation was used between the aluminum and the ceramic to avoid short-circuiting by the spacers. Piezoceramic bimorphs supplied by Morgan Matroc, Inc. were used. Their specification is given on Table III.

Table III: Specifications of Bimorph Elements for Prototype

Material	PZT5H
Length	2.5”
Width	1.5”
Thickness	0.020”

Finite Element Analysis of the Prototype

Since the device uses a series stack of bimorphs, analysis was performed on a single layer of piezoelectric bimorph for static behavior evaluation. The structure was modeled as a simply supported beam with a concentrated load at the center. The finite element mesh used is shown in Figure 7, excepting for the use of non-zero rotational degrees of freedom at both ends to represent the simply supported condition here. An electrical exciting field of 400 V/mm (material limit) was applied and the response was studied with and without mechanical loads. A mechanical load of 1N. was used for the purpose of demonstration. This is to make the loading consistent with the experiments. The predictions of the shell element are recorded in Table IV. The prototype was subjected to the same loading conditions experimentally as described in the next section. A comparison of results is shown on Table IV.

Experimental Studies on the Prototype

The test set-up established for the studies described under Bimorph illustrations was used for displacement measurements of the device. Measurements were made both with and without the mechanical loads for the same 400 V/mm electric field. Each of the element being 0.020" thick, (0.5mm) the applied voltage for the 8 series layers of piezoelectrics was 1600 V. As outlined earlier, the prototype is wired so that one layer will undergo positive curvature while the adjacent one will display a negative curvature. This cascades the displacement output to yield macro-motion. A configuration file was set-up under the LabTech Notebook data acquisition software to generate excitation signals which are amplified by the Trek power amplifier. This configuration file was set-up to produce sinusoidal signals of upto 5Hz frequency. These signals are amplified by the Trek power supply to excite the device sinusoidally with in maximum limits of + and - 400 V/mm. Results and a comparison with the FEM predictions are shown on the Table IV. Results are indicated here only for the maximum values. Case I corresponds to electrical excitation without loads and Case II is with a mechanical load of 1N, for demonstration purposes. We can see that the actuator is capable of delivering the stroke under the presence of resistive loads as well. This mechanical resistive loads will be limited by the failure strain limit on the ceramic. This was ensured while demonstrating the performance of the belleville design.

Table IV: Results on Prototype Analysis and Experiments

Test Case	Max. Deflection: FEM	Max. Deflection: Experiment
I	2.485	2.594
II	2.479	2.586

CONCLUSIONS

A new concept in automotive active suspension technology was proposed and analytically illustrated and presented in this work. This uses smart structures featuring piezoceramics bonded or embedded. The concept was illustrated via analytical and experimental procedures. A mechanism to accomplish macro-motion ranges using smart materials was developed. A candidate design configuration using composite belleville elements was analyzed. This offers us one potential means of delivering required motion ranges for an automotive active suspension system. A demonstrative prototype device capable of macro-motion outputs using piezoceramic technology was also built and tested. Results of our past and present efforts suggest a promising future for this technology.

ACKNOWLEDGMENTS

The authors acknowledge support of this research from NSF (Grants: MSS-9215656), The University of Toledo URAF Program, Ohio Board of Regents Research Challenge Grants Program, and the Department of Mechanical Engineering at The University of Toledo.

REFERENCES

Crawley, E.F. and De Luis, J., Use of Piezoelectric Actuators as Elements of Intelligent Structures, *AIAA J.,* October 1987, Vol. 25, no. 10, pp. 1373-1385.

Das Gupta, S., On Bending of Piezoceramic Multilayer Plates, *IEEE Trans Components Hybrids Manufacturing Technology,* December 1979, Vol. CHMT-2, pp. 537-540.

Dimitriadis, E. K., Fuller, C. R., and Rogers, C. A., Piezoelectric actuators for distributed vibration excitation of thin plates, *J. of Vibration and Acoustics*, January 1991, Vol.113, pp. 100-107.

Dominy, J. and Bulman, D.N., An active suspension for a formula one Grand prix racing car, *J. of Dynamic systems, Measurement and control*, March 1985, Vol.107, pp. 73-78.

Johnson, R., Lotus sees GM leading with active suspension, *Automotive News,* September 11, 1989, pp. 44.

Keebler, J., Active suspension refines Q45's road manners, *Automotive News*, April 16, 1990, pp. 20.

Lee, C.Y., Choi, S.B., Thompson, B.S., and Gandhi, M.V., A Variational Formulation for the Finite Element Analysis and Control of Linkage and Robotic Systems Featuring Smart Materials Incorporating Piezoelectric Materials, *Proc. of the 1st National Applied Mechanisms and Robotics Conference*, Vol. 1, 89AMR-2C-1, November 1989.

Mason, W.P., Physical Acoustics: Principles and Methods: Vol. 1 Part A, New York: Academic Press, 1964.

McElroy, J, GM's first Lotus blossom active suspension Corvette, *Automotive Industries*, September 1986, Vol.166, pp. 108-110.

Morgan Matroc, Guide to Modern Piezoelectric Ceramics, Product Catalog, 1992.

Naganathan, G. and Thirupathi, S.R., Light-Weight Automotive Suspension Using Piezoceramics, Patent application in progress, 1993.

Near, Craig, Morgan Matroc, Inc., Private Communications, 1994.

Newnham, R., et al., "Moonie Actuator," *Proc. of the 1991 Adaptive Materials Conference*, Alexandria, VA, November 1991.

Shiozaki, M., Kamiya, S., Kuroyanagi, M., Matsul, K. and Kizu, R., High speed control of damping force using Piezoelectric Elements, Vehicle Dynamics and Electronic controlled Suspensions, SAE International, SP-861, February 1991, pp. 149-154.

Takahashi, S., Development of New Piezoceramic Actuators, *New Materials & New Processes*, 1985, Vol.3, pp. 63-67.

Thirupathi, S.R. and Naganathan, G., Use of Piezoceramic Actuation for Automotive Active Suspension Mechanisms - A Feasibility Study, *Proc. of the 1992 ASME Design Technical Conferences - 22nd Biennial Mechanisms Conference*, Scottsdale, Arizona, September 13-16, 1992, DE-Vol. 45, pp. 233-241.

Thirupathi, S.R. and Naganathan, G., A Composite Shell Finite Element for the Analysis of Smart Structures, *Proc. of the SPIE's 1993 Conference on Smart Structures and Materials,* Albuquerque, New Mexico, February 1-4, 1993.(1993a)

Thirupathi, S.R. and Naganathan, G, Finite Element Analysis of Smart Structures and Designs for Automotive Active Suspension, *Proc. of the SPIE's 1993 Conference on Smart Structures and Materials*, Albuquerque, New Mexico, February 1-4, 1993. (1993b) □

ACTUATORS IN VIBRATION AND NOISE CONTROL

961810

Application of Rare Earth Material Based Actuators for Reduction of Vehicle Vibration

David M. Dozor and Jerome E. Kiley
SatCon Tech Corp.

ABSTRACT

The control of noise and vibration will improve the achievable return from agricultural, construction, and industrial equipment by improving equipment reliability and operator comfort. To achieve a reduction in the transmitted noise and vibration on these vehicles, actuation technologies that require low power and possess high reliability must be employed.

This paper will discuss the transmission of noise and vibration from source to driver. A discussion of actuator location and sizing will also be presented. In addition to discussion of active vehicle suspension, the initial results from ongoing development of a rare earth magnetostrictive device will be presented.

INTRODUCTION

Vehicle vibrations may arise from a number of sources. These include terrain induced vibration and equipment induced vibration. Sustained exposure to these types of vibrations may result in physiological damage and reduced operator performance. The effects of vehicle vibration of the operator may include slowed reaction time, loss of foot pressure and errors in compensatory tracking [1]. A considerable quantity of literature is available concerning the ride quality of wheeled vehicles [2,3,4].

Vibration levels measured at the driver are often sufficient to determine ride performance. The "absorbed power" [5] is a scalar quantity proposed to measure the average rate of energy dissipated through the damped elastic properties of the human anatomy. For multi-axis vibrations, the summation of the absorbed power due to each axis equals the total average power absorbed. For off-road vehicles, an average absorbed power of 6-10W is considered acceptable [6].

In the past, methods to reduce vibration have typically been passive in nature. The use of energy absorbing materials has been employed to reduce transmitted vibrations as well as reducing the instantaneous force at couplings. Examples of these couplings include motor mounts, shock absorbers, etc. Often, these vibration control elements are placed in a series path between the vibration source (the engine or terrain) and the remainder of the structure.

Passive methods are also employed to isolate the effects of transmitted noise and vibration. In general, to reduce the effects of airborne vibration, sheets of vibration absorbing material are bonded to membrane-like structures, such as the inside of vehicle cabins. These methods do not arrest the vibration, but increase the acoustic impedance in a parallel path of the mechanical circuit.

As with any passive method, performance limitations exist. Employing active methods can greatly decrease the transmitted noise and vibration by allowing the amplitude, phase, and frequency of the vibration to be more accurately controlled.

The following paper will address vibration isolation from ground excitation and attenuation of vehicle component vibration using two types of vibration control actuators: Permanent Magnet Active Struts (PMAS) and Magnetostrictive Reaction Mass Actuators (RMA). PMAS are utilized to isolate the vehicle from terrain induced vibrations while RMAs may be used to cancel vibrations transmitted equipment induced vibrations. Both are based on rare earth material technologies. The advent of these materials have allowed low mass, low power consumption devices to be designed.

ISOLATION FROM GROUND EXCITATION

TERRAIN MODELING - Terrain profiles exciting vibration are often expressed in terms of their power (PSD) and cross (CSD) spectral densities, under the assumption of undeformable terrain. Rakheja, Afonso, and Sankar summarize these excitations [3]. Spectral densities that closely approximate experimental terrain roughness are described by the following expression [7].

$$s_P(\Omega) = s_P(\Omega_o)\left[\frac{\Omega}{\Omega_o}\right]^{-w}$$

where $s_p(\Omega)$ is the spatial displacement PSD, $s_p(\Omega_o)$ is the roughness coefficient, Ω is the spatial frequency, Ω_o is a constant, and w is the terrain waviness. A comparison of various displacement spectra, that may be described (piecewise) by the above expression, is shown below in Figure 1.

The temporal spectral density $s_p(f)$ is related to the spatial spectral density by the vehicle speed V as follows.

$$s_P(f) = \frac{1}{V} s_P(\Omega)$$

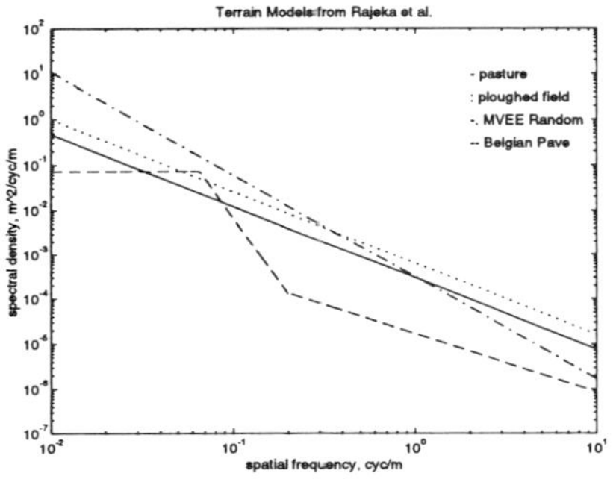

Figure 1: Comparison of Displacement Spectra

Under the assumption that all wheels follow the same path, the cross spectral densities are a function of the time delay τ between wheels i and j.

$$\tau_{ij} = \frac{L_{ij}}{V}$$

where L_{ij} is the length between wheels i and j. The CSD is given by [8].

$$s_{ij}(f) = s_{jj}(f)e^{\frac{-j2\pi f L_{ij}}{V}}$$

VEHICLE MODELING - Designing semi-active struts for demanding applications requires optimal use of the strut materials for strength, heat transfer, and force/damping production. To adequately accomplish this design, it is useful to have a vehicle terrain model to understand the requirements of the suspension/strut system. A lumped parameter vehicle model, which considers the 6 rigid body degrees of freedom of the vehicle, as well as those (4 or 6) of the wheels, will be used for this purpose. The model is to be used to assess the power/force requirements of the vehicle during expected maneuvers.

For brevity a quarter car model is shown below in Figure 2, followed by a transfer function matrix describing the model.

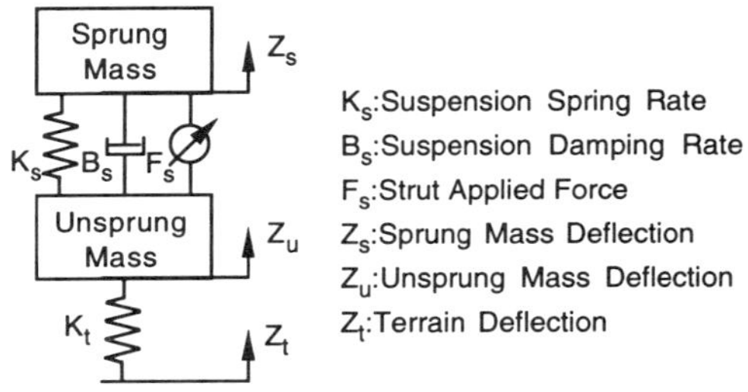

Figure 2: Quarter Car Active Suspension Model

$$\begin{bmatrix} Z_s(s) \\ Z_u(s) \end{bmatrix} = \begin{bmatrix} \dfrac{m_u s^2 + K_t}{DEN(s)} & \dfrac{K_t(B_s s + K_s)}{DEN(s)} \\ \dfrac{-m_s s^2}{DEN(s)} & \dfrac{K_t(m_s s^2 + B_s s + K_s)}{DEN(s)} \end{bmatrix} \begin{bmatrix} F(s) \\ Z_t(s) \end{bmatrix}$$

$$DEN(s) = m_s m_u s^4 + B_s(m_s + m_u)s^3 + [K_t m_s + K_s(m_s + m_u)]s^2 + B_s K_t s + K_s K_t$$

The model shown is a good starting point for investigation of the semi-active strut requirements. Based on a terrain excitation model, feedback control laws may be studied to determine the actuator force required to deliver the desired isolation of the sprung mass. A time response simulation will allow a force trajectory to be determined for proper design of the actuator in terms of force, steady state temperature, etc. However, the simple quarter car model does not allow cornering enhancements, through semi-active control, to be studied.

A whole car model may be developed, allowing not only the longitudinal dynamics, but also the lateral dynamics to be studied under active control. This will allow the actuator requirements to be better understood, resulting in a more optimal design.

ATTENUATION OF COMPONENT VIBRATION

Presently, reaction mass actuators (RMAs) are being explored for use as attenuators of noise resulting from helicopter gearbox vibration. By strategically placing RMAs at different locations inside the helicopter cabin, a substantial reduction in the noise transmitted to the passengers can be achieved. This section describes a magnetostrictively-based RMA design for this purpose.

Reaction mass (or inertial force) actuators are commonly used to produce forces at frequencies above the DC level. RMAs produce forces that are substantially greater than their weight, thus minimizing the effect of mass loading on the structure under control.

While voice coils and piezoelectric elements have been used for similar applications, the magnetostrictive material Terfenol-D™[1] exhibits properties that are more favorable for this application. Specifically, Terfenol-D™ has energy and power densities rivaling piezoelectric materials and has no known fatigue limits. These properties allow the active magnetostrictive element to be driven at resonance without failure. Use in this manner minimizes the power required by the actuator and results in an extremely high specific power.

ACTUATOR DESIGN - Like most magnetic actuators, magnetostrictive actuator design is multi-disciplinary, involving electrical, magnetic, mechanical, and thermal subsystems. A schematic diagram of the prototype actuator is shown in figure 3. The magnetically actuated terfenol rod is used to drive a mass. The acceleration of this mass creates a reaction force at its base.

A specially shaped terfenol rod is placed within an aluminum bobbin wound with 875 turns of 24AWG copper wire. The terfenol rod drives a mass fabricated from a magnetically permeable tungsten alloy, which also serves as a magnetic return path for magnetic flux. A preload washer fabricated from Viton™ is used to load the tungsten reaction mass against the terfenol rod creating a compressive preload.

Since the terfenol rod expands for positive and negative applied fields, a magnetic bias is utilized to achieve bidirectional force. The terfenol is magnetically biased by creating a bias current in the actuator winding.

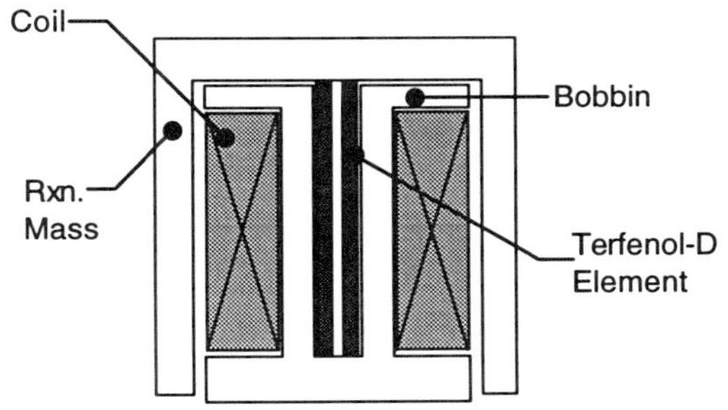

Figure 3: Schematic Diagram of the Terfenol-D Based RMA

ACTUATOR MODELING - The actuator transfer function may be expressed in the following form.

$$\frac{X(s)}{\Phi(s)} = \frac{1}{d_{33}m\left[s^2 + \left(\frac{b}{m}\right)s + \left(\frac{k}{m}\right)\right]}$$

where m is the reaction mass, b is the equivalent viscous damping, k is the equivalent spring rate of the Terfenol-D rod, d_{33} is the ratio of Terfenol strain to magnetic field, and Φ is the magnetic flux.

The transfer function given above shows that the system behaves as a pure 2nd order oscillator. In order to achieve this behavior, a properly designed flux loop is required. The pure 2nd order behavior implies that closing a flux loop will act to make the RMA act as a force actuator. Figure 4 below compares recently acquired experimental data with that of the model presented above.

---

[1] Terfenol-D™ is an alloy of terbium (Tb), iron (Fe), and dysprosium (Dy) ($Tb_{0.3}Dy_{0.7}Fe_{1.9}$) capable of strains of approximately 1.6×10^{-3}.

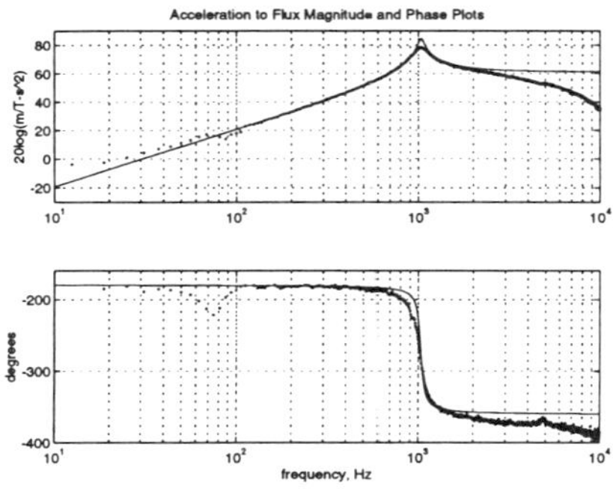

Figure 4: Experimental vs. Theoretical Transfer Functions

SUMMARY

This paper presents actuator technologies that may be applied to control of vibration in off-road vehicles. Rare earth based permanent magnet active struts can be designed to isolate a vehicle from terrain induced vibrations. Terfenol-D based reaction mass actuators may be employed to reduce vibration due to machinery induced vibrations.

Use of active control loops based on the models presented for the terrain and actuators will reduce noise and vibration, thus reducing detrimental physiological effects.

REFERENCES

[1] Hornick R.J., "Effects of Vibrations on Operators," Agricultural Engineering, Dec. 1961, p674.

[2] Rula A.A., Nutall C.J. Jr., "An Analysis of Ground Mobility Models," WES Technical Report No. 71-4, US Army Engineer Waterways Experiment Station, Vicksburg MS, July 1971.

[3] Rakheja S., Afonso, M.F.R., Sankar, S., "Dynamic Analysis of Tracked Vehicles with Trailing Arm Suspension and Assessment of Ride Vibrations," Int'l Journal of Vehicle Design, vol. 13, #1, 1992.

[4] Glotzback, R.W., Wentz, R.A., Mehta, N.C., "Ride Comfort in Medium and Heavy Duty Trucks," Passenger Vibration in Transportation Vehicles, ASME, AMD vol. 24, 1977.

[5] Lee, R., Pradko, F., "Analytical Analysis of Human Vibration," SAE Transactions #680091, 1968, pp. 346-370.

[6] Stikeleather, L.F., "Review of Ride Vibration Standards and Tolerance Criteria," SAE Transactions #760413, 1976, pp. 1460-1467.

[7] Robson, J.D., "Road Surface Description and Vehicle Response," Int'l Journal of Vehicle Design, vol. 1, #1, 1979.

[8] Dokainish, M.A., Elmadany, M.M., "Dynamic Response of Tractor-Semitrailer Vehicle to Random Input," 5th Proceedings VSD-2nd IUTAM Symposium of Dynamics of Vehicles on Roads and Tracks, Technical University Vienna, Austria. Sept 1977.

970851

Active Vibration Control of a Thin Plate Using Piezoelectric Actuators/Sensors - Multi-Input Multi-Output Control

Chul Ki Song
Kia Motors Co.

Jang Moo Lee, Jin Kwon Hwang, and Jong-Ho Choi
Seoul National Univ.

ABSTRACT

This paper presents the vibration control of a thin plate by multiple piezoelectric actuators and sensors which are co-located. To use piezoelectric materials efficiently, the Linear Quadratic Gaussian (LQG) controller is designed on the basis of a Multi-Input Multi-Output system so that each sensor monitors all the modes to be controlled and each actuator excites all the modes to be controlled. The robust LQG controller is then applied for attenuating vibration of a plate that is clamped along its edges and disturbed by a band-limited pseudo-random noise. This controller uses two sensor signal filters in order to control the first three natural modes of the plates. The results of the experiment show that the vibration of controlled modes are attenuated by 10 to 20 dB without destabilizing the other modes.

INTRODUCTION

Since a thin plate is an infinite dimensional system, control methods for lumped-parameter systems can not be directly applied for vibration control of the plate. In the process of control, spill-over effects can result in some uncontrolled modes becoming unstable. Because of spill-over effects, the positions of actuators and sensors are very important, since uncontrolled modes should not be monitored and should be actuated as little as possible. The co-location of actuators and sensors is also important since the system becomes minimum phase with co-location. The minimum phase system is much easier to control. In the cases of noncollocated actuators and sensors, the design of a controller is very difficult due to the uncertainty and the variation of modal parameters [1, 2]. The basic control methods for distributed systems are presented well by Meirovitch [3].

Vibration control methods for a plate with co-located sensors and actuators were studied in Hong [4], Tzou and Fu [5] and Falangas et al. [6]. Due to uncontrolled modes, it is still difficult to control the vibration even though the actuators and the sensors are co-located at positions with desirable characteristics. A closed-loop system must be designed to be stable even for the uncontrolled modes. Falangas et al. [6] used shaping filters in H-inf. control for an all-clamped plate and Sepe [7] determined the weight matrix for states in a LQG controller by using the right and the left eigenvectors of a flexible structure for robustness.

In this study of vibration control of a thin plate, actuators and sensors for control are piezoelectric materials, which are co-located at the plate. Disturbance is applied by a piezoelectric actuator excited by a band-limited pseudo-random noise. The design of controllers are on the basis of a Multi-Input Multi-Output (MIMO) system with several modes to be controlled [8]. In the MIMO control method, each actuator drives all the modes to be controlled and each sensor senses all the modes. The advantage of this control method is an efficient use of piezoelectric materials in comparison to the other methods using one actuator and one sensor for each mode. Before developing a robust LQG controller, a velocity feedback controller and a classical LQG controller were designed and experiments were performed. Both controllers were not successful in controlling the vibration due to problems such as the spill-over effect and high frequency noise in sensor signals. Descriptions of the experiments will be presented later in detail. To resolve the problems of these control methods, a robust LQG controller is designed, using two sensor signal filters. One of the filters is a lowpass filter. This filter can attenuate high frequency noise and uncontrolled modal signals in sensor signals sufficiently enough not to affect the stability of uncontrolled modes due to the spill-over effect. The other filter is a highpass filter, which can make the gain of the robust LQG controller higher. The robust LQG controller is applied for vibration control of an fully-clamped plate. The experimental results show that the vibration of modes to be controlled are attenuated by 10 to 20 dB and some uncontrolled modes near the controlled modes are also attenuated by the spill-over effect. The stability of the closed loop system is investigated by adding uncontrolled modes, one by one, to the model. The stability is closely related to the gain of the robust LQG controller and the uncertainty of uncontrolled modes. To make the closed loop system stable, the gain of the LQG controller has to be reduced when adding more uncontrolled modes to the model. The unstable mode of the closed loop system occurs at frequencies lower than the first mode of the plate. A highpass filter, which has a small gain for frequencies lower than the first mode, is introduced to eliminate this unstable mode and the gain of the LQG controller can be increased significantly while maintaining stability. The vibration control of the plate is simulated under the same conditions as the experiment with the first thirteen modes of the plate. The results of the simulation show good agreement to those of the experiment.

DESIGN OF A ROBUST LQG CONTROLLER

A model of a thin plate with multiple co-located piezoelectric actuators and sensors can be identified in the state space form. Without loss of generality, it is assumed that the number of modes to be controlled is $n \geq 2$, and the pairs of co-located actuators and sensors are $m \geq 2$ in the design of the MIMO LQG controller.

$$\dot{w}(t) = A\, w(t) + B\, u(t) \tag{1}$$

$$z(t) = C\, w(t) \tag{2}$$

Here, $w(t)$ is the $2n \times 1$ state vector of the displacements and velocities of each mode in increasing order of natural frequencies. Input vector $u(t)$ for actuators and output vector $z(t)$ from sensors are the $m \times 1$ vectors, where each i-th element corresponds to the co-located actuators and sensors located at the same position. By the characteristics of the modal motions in a thin plate, the system matrices A, B and C have the following properties. The $2n \times 2n$ matrix A is diagonalized with 2×2 block matrix expressed with the natural frequency ω_i (rad/sec) and the damping ratio ζ_i for the i-th mode, i.e.,

$$\begin{bmatrix} 0 & 1 \\ -\omega_i^2 & -2\zeta_i\omega_i \end{bmatrix}, \qquad i = 1,2,\cdots,n$$

All elements in odd rows of the $2n \times m$ matrix B are zeros, and all elements in even columns of the $m \times 2n$ matrix C are zeros.

In order to remove effects of the uncontrolled modes higher than n from the sensor signals, a lowpass filter is introduced. Its bandwidth contains only the natural frequencies of the modes to be controlled. The lowpass filter can be designed to sufficiently attenuate all the uncontrolled modes. However, the filtering affects the phase and the magnitude of the modes to be controlled. Therefore an augmented system which includes the dynamics of the filter is used in the design of the controller. It is desirable that the augmented system is minimum phase. The Butterworth lowpass filter has this property. The l-th order Butterworth lowpass filters are expressed in the following form:

$$\dot{v}_i(t) = F\, v_i(t) + G\, z_i(t) \tag{3}$$

$$r_i(t) = H\, v_i(t), \quad i = 1,2,\cdots,m \tag{4}$$

where $v_i(t)$ is the $l \times 1$ state vector of the filter for the i-th sensor, $z_i(t)$ is the signal from the i-th sensor, $r_i(t)$ is the scalar output of the filter for the i-th sensor and F, G and H are the system matrices of the filters with proper dimensions. Using equations from (1) to (4), the augmented system is

$$\dot{w}_a(t) = A_a\, w_a(t) + B_a\, u_a(t) \tag{5}$$

$$z_a(t) = C_a\, w_a(t) \tag{6}$$

where $w_a(t) = \left[w^T(t) \;\; v_1^T(t) \;\; v_2^T(t) \;\; \cdots \;\; v_m^T(t)\right]^T$, $z_a(t) = \left[r_1(t) \;\; r_2(t) \;\; \cdots \;\; r_m(t)\right]^T$, $u_a(t) = u(t)$, and the dimensions of the augmented system matrices A_a, B_a, C_a are $(2n + m\ l) \times (2n + m\ l)$, $(2n + m\ l) \times m$, $m \times (2n + m\ l)$ respectively. The augmented system matrices are

$$A_a = \begin{bmatrix} A & O & \cdots & O & \cdots & O \\ GC_1 & F & \cdots & O & \cdots & O \\ \vdots & \vdots & \ddots & \vdots & \ddots & \vdots \\ GC_i & O & \cdots & F & \cdots & O \\ \vdots & \vdots & \ddots & \vdots & \ddots & \vdots \\ GC_m & O & \cdots & O & \cdots & F \end{bmatrix}$$

$$B_a = \left[B^T \;\; O\right]^T, \; C_a = H$$

where C_i is the i-th row of C matrix and O denotes a zero matrix with a proper dimension. Including the process noise and the measurement noise, the physical equations of motion for the augmented system can be expressed as

$$\dot{w}_a(t) = A_a\, w_a(t) + B_a\, u_a(t) + \xi(t) \tag{7}$$

$$z_a(t) = C_a\, w_a(t) + \theta(t) \tag{8}$$

where $\xi(t)$ is a $(2n + m\ l) \times 1$ process noise vector and $\theta(t)$ is a $m \times 1$ measurement noise vector. $\xi(t)$ generates the vibration in the plate which needs to be attenuated. The modes to be controlled are selected as those that mostly contributes to the vibration of the plate by $\xi(t)$. The co-located actuators and sensors must be bonded at the positions where the modes to be controlled can be excited and monitored easily. To design a LQG controller for the augmented system, the covariance matrices of $\xi(t)$ and $\theta(t)$ must be found. Since the actual value of $\xi(t)$ is not measurable, the covariance matrix Ξ of $\xi(t)$ is set as follows.

$$\Xi = p\, B_a\, B_a^T \tag{9}$$

Here, the noise to the filter is assumed to be zero. There are many different ways of setting Ξ, and equation (9) is one of them. The power of the noise is adjusted by the design parameter p, and this adjustment is only to make the controller robust. By making p smaller, the controller becomes more robust while the performance of the system becomes worse. Another covariance matrix Θ of $\theta(t)$ can be computed from the resolution of a measurement device, i.e., from the voltage resolution of an A/D converter which is used to sample sensor signals. It is assumed that $\theta(t)$ is a uniform distributed probability density function. The covariance matrix Θ can be computed as

$$\Theta = \frac{V^2}{12 \times 2^{2b}} I \tag{10}$$

where V is the total voltage range of the A/D converter, b is the number of bits for the A/D converter and I is the identity matrix.

The design procedure of LQG controller can be found in Doyle and Stein [9], Stein and Athans [10], and Maciejowski [11]. By the separation principle, the design of a LQG controller is divided into two sub-problems. One is the Linear Quadratic Regulator (LQR) design and the other is the Linear Quadratic Estimator (LQE) design. In the sense of minimizing a given cost functional in a given noise environment, the LQR is the optimal state feedback controller and the LQE is the optimal state estimator. Here, for the design of the LQG controller, the cost functional is

$$\lim_{T\to\infty} E\left[\frac{1}{T}\int_0^T \left\{w_a^T(t)\, Q\, w_a(t) + u_a^T(t)\, u_a(t)\right\} dt\right] \tag{11}$$

where Q is the $(2n + m\ l)\times(2n + m\ l)$ matrix of weights for states. Q can be set in a manner that the state of the mode to be attenuated more is given more weights than the others and the states for modal displacements are weighted more than the states for modal velocities. When p in equation (9) is given, the solution for minimizing equation (11) can be obtained from two Algebraic Riccati Equations (ARE). One is the Control ARE where the LQR gain matrix can be obtained. The other is the Filter ARE where the Kalman filter gain matrix can be obtained. The resulting form of the LQG controller is

$$\dot{\hat{x}}_a(t) = A_a\, \hat{x}_a(t) + B_a\, u_a(t) + K_f [y_a(t) - C_a\, \hat{x}_a(t)] \tag{12}$$

$$u_a(t) = K_c\, \hat{x}_a(t) \tag{13}$$

where "^" denotes an estimated signal, K_f is the Kalman filter gain matrix and K_c is the LQR gain matrix. The suitable value for the design parameter p is found experimentally. In the beginning of the controller design, a small value is used for and the experiment is performed with the controller. If the control system is stable and the performance is not satisfactory, p is increased and the process is repeated until satisfactory results are obtained. While a larger value of p gives better performance in attenuating the vibration for a model consisting of only the controlled modes, it may not guarantee the stability in the experiment due to the uncontrolled modes. This is because the gain of the LQG controller is proportional to $\sqrt{p}$. Therefore, p should be chosen in the vicinity of the largest value which gives the robust stability.

EXPERIMENTAL SET-UP AND A MODEL OF THE PLATE WITH PIEZOELECTRIC SENSORS AND ACTUATORS

The dimension of the fully-clamped plate is 0.360 m $\times$ 0.300 m $\times$ 0.0008 m. The piezoelectric materials used are G1195 of Piezo System Co. Two pairs of actuators and sensors are co-located at the positions, where the first three modes are equally actuated and monitored. These three modes are the mode to be controlled. For this experiment, the disturbance is generated with another disturbance actuator by using a DSP program. The signal for the disturbance actuator is amplified 34 times with a high-level voltage amplifier to drive the disturbance actuator. The position of the disturbance actuator is shown in Figure 1.

The LQG controller is digitally implemented with TMS320C30 DSP by using its assembly language. Two control signals for the actuators and the signal for the disturbance actuator are multiplexed through one D/A converter, which is connected with a 1-by-3 analog demultiplexer. Each control signal for the actuators is amplified 11 times through its high-level voltage amplifier to drive the piezoelectric actuator directly. Each sensor signal is amplified 3 times through its low-level voltage amplifier before it is sampled by its A/D converter. One D/A converter and two A/D converters have 12-bit resolution and the maximum absolute voltage for these converters is 5 V.

Unfortunately, data sampled with the A/D converters have a small DC bias. This is due to voltage deviations for the A/D converters, and the ground voltage difference between electrical circuit boards and the plate where piezoelectric actuators and sensors are bonded. The DC bias drives the LQG controller, which has an integrator, to the saturation limit of the D/A converters. Therefore, the DC bias must be removed after the A/D converters by using digital highpass filters. The cut-off frequency of the highpass filters must be small enough compared to the natural frequency of the first mode so that their dynamics can be ignored. The digital highpass filters are programmed with the DSP assembly language and they are designed as second order Butterworth highpass filters with a 10 Hz cut-off frequency. The sampling rate for this experiment is 4.868 kHz and the continuous-time LQG controller is transformed into the discrete-time LQG controller using Tustin's rule.

The system matrices with the first six modes for the configuration plate is desirable to control many modes of the plate and to have the order of the lowpass filters high for robust control. But these make the dimension of the LQG controller very large and the sampling interval is no longer fast enough. The first three modes with natural frequencies, 68.68 Hz, 125.16 Hz, 155.01 Hz are the modes to be controlled in the experiment. The model for the design of the LQG controller are

$$A = diagonal(A_1,\ A_2,\ A_3),$$

$$A_1 = \begin{bmatrix} 0 & 1 \\ -1.8619 \times 10^5 & -1.8768 \end{bmatrix},$$

$$A_2 = \begin{bmatrix} 0 & 1 \\ -6.1840 \times 10^5 & -2.3140 \end{bmatrix},$$

$$A_3 = \begin{bmatrix} 0 & 1 \\ -9.4863 \times 10^5 & -3.0408 \end{bmatrix},$$

$$B = \begin{bmatrix} 0 & -288.66 & 0 & 553.01 & 0 & -2250.6 \\ 0 & -280.93 & 0 & -1239.1 & 0 & 791.16 \end{bmatrix}^T,$$

$$C = \begin{bmatrix} 0.72281 & 0 & -0.29284 & 0 & 1 & 0 \\ 1 & 0 & 1 & 0 & -0.35961 & 0 \end{bmatrix}$$

THE RESULTS OF EXPERIMENTS AND ANALYSIS

VELOCITY FEEDBACK CONTROL AND CLASSICAL LQG CONTROL - Before the experiments for the robust LQG control, a velocity feedback control and a classical LQG control without the lowpass filters were conducted experimentally. The velocity feedback control method does not require a model for the plate with actuators and sensors. A velocity signal for a certain mode can be obtained by differentiating the sensor signal which senses displacement of that mode. This control method increases the modal damping ratio effectively. This controller can be easily implemented with only a simple analog circuit, which is composed of differentiators and high-level voltage amplifiers for the co-located actuators and sensors. Each differentiator for the sensors is connected electrically with a high-level voltage amplifier for the co-located actuator. The gain of each high-level voltage actuator was designed to be adjustable for the experiments. The transfer function of the differentiator was designed to be $s/(\omega_d s+1)$ in the Laplace transform and ω_d (rad/sec) could be also adjustable. The velocity feedback control method was applied for vibration control of the fully-clamped plate with two pairs of co-located sensors and actuators. But this control method was not successful in all experiments of different combinations of ω_d's and gains of the high-level voltage amplifiers. The vibration of the plate could be hardly attenuated while maintaining the stabilities of the control system. The critical problem in these experiments was the high frequency noise in the sensor signals. Differentiating the noise made some high frequency modes unstable. Besides this problem, there are several problems in using piezoelectric actuators and sensors. Piezoelectric materials have a limited-bandwidth of 3 to 5 kHz. This can introduce phase distortion of signals for actuators and sensors. The phase distortion can drive high frequency modes unstable. Another problem is induced from not collocating sensors with actuators exactly. It is not desirable that a sensor is exactly co-located with its actuator because it senses the signal driven for its co-located actuator rather than the modal displacements in the plate. Hence, the piezoelectric sensor had better be co-located slightly apart from the boundary of its corresponding actuators. This co-location can cause phase difference for high frequency modes. This phase difference can also drive high frequency modes unstable.

Next, a classical LQG control, where the lowpass filter for robust control were not used, was applied to control the vibration of the plate. The modes to be controlled were the first three modes. Like the velocity feedback control method, all experimental results were not good due to uncontrolled modes as well as the problems in the velocity feedback control method. The unstable mode of the vibration in the experiments occurred around 530 Hz which corresponds with the 13th natural frequency of the plate.

THE EXPERIMENTAL RESULTS OF THE ROBUST LQG CONTROL - For experiments, the plate is vibrated by the disturbance actuator, driven by a band-limited pseudo-random noise of 180 Hz bandwidth. For simulations, the vibration data are collected from the two sensor signals, denoted as $z_1^d(t)$ and $z_2^d(t)$. The analog lowpass filters used for robust control are the second order Butterworth filters of 160 Hz bandwidth [12, 13]. The bandwidth contains the natural frequencies of all modes to be controlled. The system matrices of the filter are

$$F = \begin{bmatrix} -1.412 \times 10^3 & -1.000 \times 10^6 \\ 1 & 0 \end{bmatrix}$$

$$G = \begin{bmatrix} 1 & 0 \end{bmatrix}^T, \ H = \begin{bmatrix} 1 & 1.588 \times 10^6 \end{bmatrix}$$

The weight matrix Q of the states are chosen as equation (14) so that the lower frequency modes are weighted more. Also, the state of the modal displacement is weighted more than the state of the modal velocity.

$$Q = diagonal(1600,\ 160,\ 400,\ 40,\ 100,\ 10,\ 10,\ 10,\ 10,\ 10) \tag{14}$$

A few experiments were tried in designing the robust LQG controllers with different values of p, so that the stability of the control system can be maintained. But the experimental results were not good in attenuating the vibration of the plate.

Since the plate has a small damping ratio for each mode, it is desirable that the LQG controller has a high gain to suppress the vibration. The plate has infinite modes and only the first three modes are controlled. Increasing the gain of the LQG controller makes the control system unstable. This is due to unmodelled dynamics. The unstable mode usually occurs at the low frequencies below the first mode of the plate. This indicates that the unmodelled dynamics in the low frequency range affects the stability of the control system. To eliminate this unstable mode at low frequency, additional highpass filters are used besides the digital highpass filters for the DC bias of the A/D converters. The highpass filters are designed to have a cut-off frequency much lower than the natural frequency of the first mode. By doing this, the dynamics of the highpass filters can be also ignored in the design process and the LQG controller can be designed considering only the lowpass filter.

The transfer function of these analog highpass filters are $0.01s$ / $(0.01s + 1)$. These filters are inserted in front of the high-level voltage amplifiers for the actuators. Also, these filters are inserted in front of the two A/D converters. Including the analog highpass filters, the block diagram of the experimental system is shown in Figure 2. Figure 3 shows the closed loop system. The design parameter p is set to $\Theta = 4.97 \times 10^{-7}$ initially. After experimenting several different values of p, it is found that 3.5×10^{-4} is the largest value of to maintain the stability of the control system. Figure 4 shows the frequency responses of the disturbance $z_1^d(t)$ and $z_2^d(t)$. Figure 5 presents the experimental results for the frequency responses when $p = 3.5 \times 10^{-4}$. Comparing the frequency responses in Figure 5 with those in Figure 4, the vibration of controlled modes are attenuated about 10 to 20 dB. Moreover, a few uncontrolled modes near the controlled modes are also attenuated. This phenomenon results from spilling the control signals over the uncontrolled modes. Figure 6 shows the attenuation of the vibration in the time domain when $p = 3.5 \times 10^{-4}$.

CONCLUSION

There are many physical problems in the vibration control of a thin plate using piezoelectric materials. They are the unmodelled dynamics of uncontrolled modes, the limited bandwidth of piezoelectric materials, inexact co-location of the sensors with their corresponding actuators, and the sensor noise. All these problems can result in instability of the control system. To overcome these problems, Butterworth lowpass filters were used for the robust LQG control of vibration in the plate. Also, the highpass filters were used so that the controller could have a high gain. The LQG controller was designed on the basis of the MIMO model, so that piezoelectric materials could be used effectively. Through the experiment of the all-clamped plate where disturbance was caused by the band-limited pseudo-random noise, the vibration of the controlled modes was attenuated by 10 to 20 dB. Moreover, in the experiment the vibration of uncontrolled modes near the controlled mode were also attenuated due to the spill-over effect. The lowpass filter reduced the effect of the uncertainty due to uncontrolled modes in the high frequency range. The highpass filter was effective in decreasing the effect of the uncertainty in the low frequency range. This made the higher gain of the LQG controller possible, which in turn attenuated the vibration significantly.

REFERENCES

1. Cannon Jr,. R. H. and Rosenthal, D. E., 'Experiments in Control of Flexible Structures with noncollocated Sensors and Actuators,' Journal of Guidance, Vol. 7, No. 5, pp. 546-553, 1984

2. Song, C. K., Han, S. W., Lee, J. M., Lee, S. W., Hwang, J. K., 'Distributed Modal Sensitivities of Piezoelectric Materials for Active Vibration Controls,' Proceedings of 1995 KSME, International Technical Conference, Part 1 Automotive Technology, San Francisco, CA, Aug. 31 ~ Sep. 4, pp. 263 - 276, 1995

3. Meirovitch, L., 'Dynamics and Control of Structures,' Wiley-Interscience, 1990

4. Hong, S. Y., 'Active Vibration Control of Adaptive Flexible Structures Using Piezoelectric Smart Sensors and Actuator,' Ph. D. Thesis, The Pennsylvania State University, 1992

5. Tzou, H. S. and Fu, H. Q., 'A Study of Segmentation of Distributed Piezoelectric Sensor and Actuator Part II: Parametric Study and Vibration Controls,' Journal of Sound and Vibration, Vol.172, No.2, pp. 261-275, 1994

6. Falangas, E. T., Dcoorak, J. A. and Koshigoe, S., 'Controlling Plate Vibration Using Piezoelectric Actuators,' IEEE Control System, Vol.14, No.4, pp. 34-41, 1994

7. Sepe, R. B., 'Robust LQG Position Control of a Flexible Structure Using Participation Factors,' Proceeding of The 32nd conference on Decisions and Control, San Antonio, Texas, Vol.1, pp. 1396-1401, 1993

8. Hwang, J. K., Choi, J-H., Song, C. K., Lee, J. M., 'Identification of a Thin Plate with Bonded Piezoelectric Actuators/Sensors,' '96 IFAC 13th World Conference, San Francisco, CA, June 30 ~ July 4, 3a-18 5, pp. 95 ~ 100, 1996

9. Doyle, J. C. and Stein, G., 'Mutivariable Feedback Design : Concepts for a Classical/Modern Synthesis,' IEEE Trans. Automat. contr., Vol. AC-26, No. 1, pp. 4-16, 1981

10. Stein, G. and Athans, M., 'The LQG/LTR Procedure for Mutivariable Feedback Control Design,' IEEE Trans. Automat. Contr., Vol. AC-32, No. 4, pp. 105-114, 1987

11. Maciejowski, J. M., 'Multivariable Feedback Design,' Addison-Wesley, 1989

12. Johnson, D. E., 'Introduction Filter Theory,' Prentice-Hall Inc., 1976

13. Ghausi, M. S. and Laker, K. R., 'Modern Filter Design Active RC and Switched Capacitor', Prentice-Hall Inc., 1981

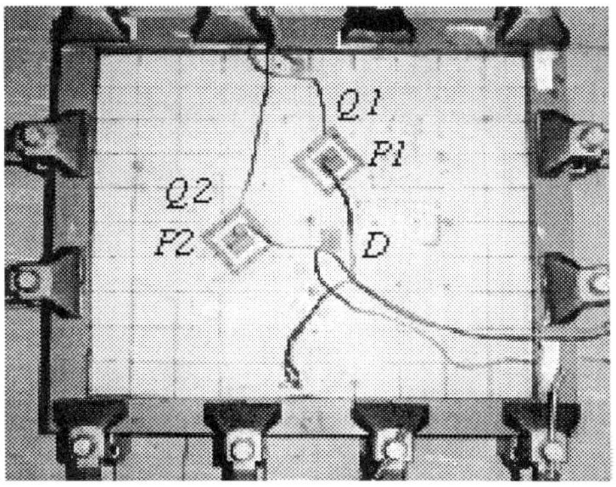

Figure 1. All-clamped plate with two co-located actuators/sensors and one disturbance actuator (P : actuator, Q : sensor, D : disturbance actuator)

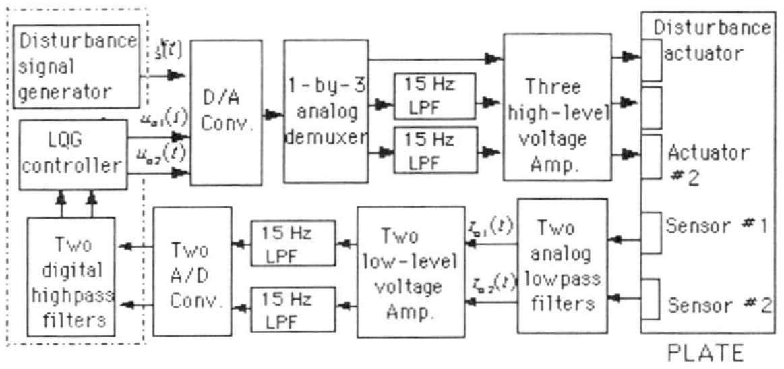

Figure 2. Block diagram of a robust LQG controller

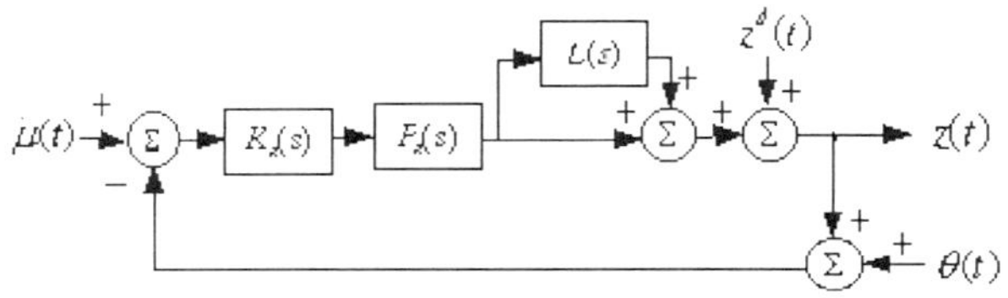

Figure 3. Block diagram of the closed loop system

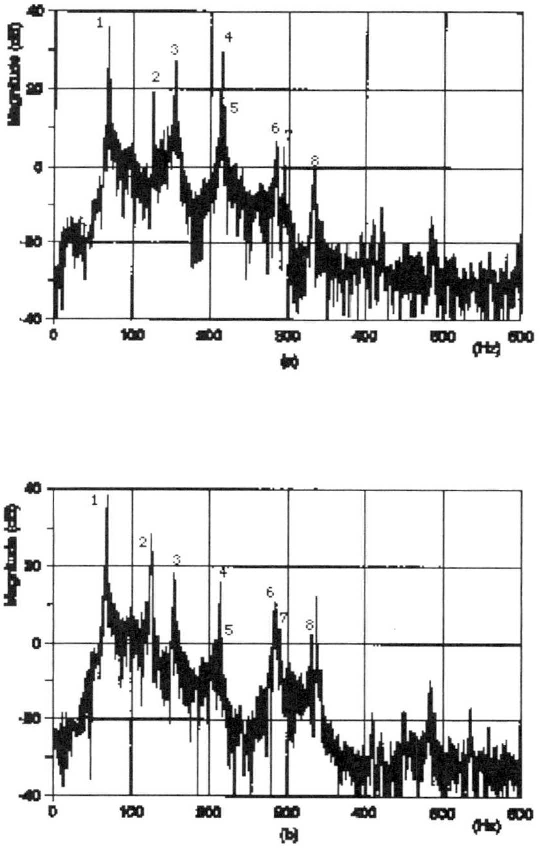

Figure 4. Frequency responses - uncontrolled
(a) $z_1^d(t)$ (b) $z_2^d(t)$
(The numbers denote the modal numbers)

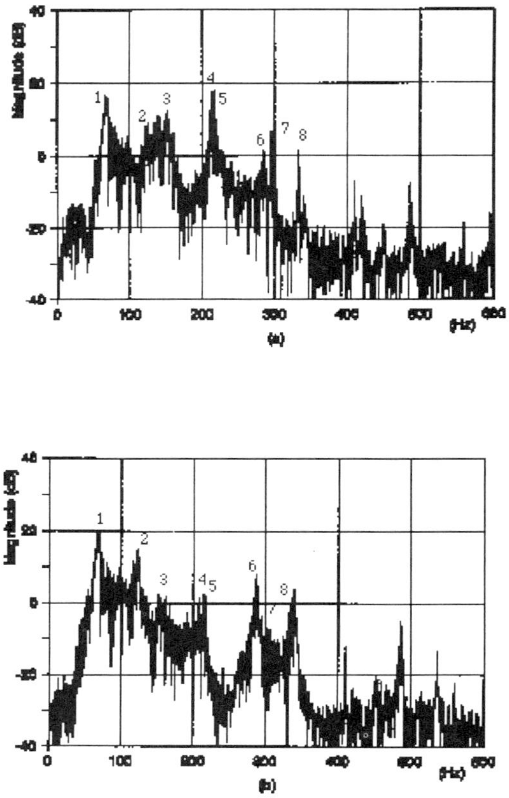

Figure 5. Frequency responses - controlled
(a) $z_1(t)$ (b) $z_2(t)$

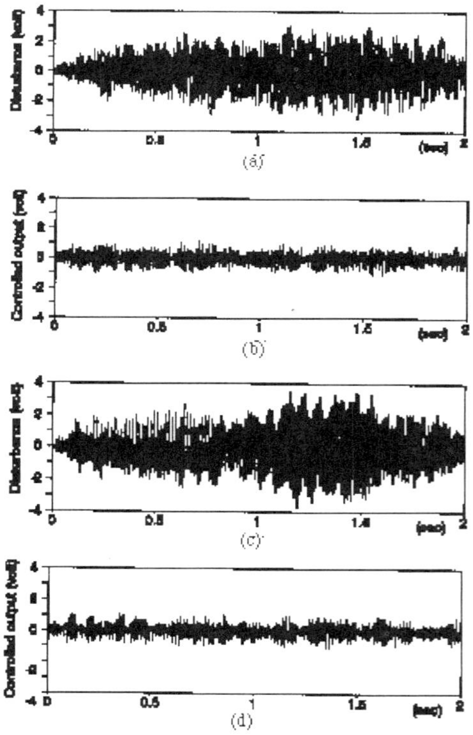

Figure 6. Time responses - uncontrolled and controlled
(a) $z_1^d(t)$ (b) $z_1(t)$ (c) $z_2^d(t)$ (d) $z_2(t)$

932432

Active Engine Mounts for Vehicles

Douglas A. Swanson
Lord Corp.

ABSTRACT

This paper discusses how active mounts can reduce vibration and noise caused by engine vibration in vehicles. In recent years, vibration and noise problems have been exacerbated by the trend towards lighter, more fuel efficient vehicles. The connection of more powerful engines with fewer cylinders to lighter vehicle frames has resulted in vibration problems that passive elastomeric and hydraulic mounts cannot adequately address. Active mounts, which consist of passive mounts, force generating actuators, sensors, and electronic controllers, can achieve significant performance improvements over passive mounts. The superior vibration isolation capabilities of active mounts may also allow for the elimination of the engine's balance shaft, reducing engine weight, height, and cost, and helping to achieve further fuel efficiencies. In this paper, several different active mount configurations, actuation technologies, and sensor and controller choices are discussed. Also, a mount effectiveness analysis demonstrates the performance improvements of active mounts versus passive elastomeric and hydraulic mounts.

INTRODUCTION

The primary benefit of engine mounts is to reduce the vibration that is transmitted by the engine to the vehicle chassis. Reciprocating engines generate unwanted low frequency vibration in the form of inertia forces and torques, as well as firing pulses. This vibration can also cause a "booming" sound inside the vehicle interior. In the past, passive elastomeric and hydraulic engine mounts have successfully reduced engine vibration in vehicles [1-7]. However, the demand in recent years for lighter, more fuel efficient vehicles has resulted in increased vibration problems that passive mounts cannot adequately solve. Other passive techniques such as stiffeners and damping materials carry unacceptable weight penalties. One solution to this problem is active mounts, which use the addition of energy to provide significant vibration reductions over traditional elastomeric and hydraulic mounts [8-13].

Figure 1 shows an engine that is supported by one front mount and two rear mounts. The engine mounts must satisfy two competing functions. At low frequencies (below 20 Hz), the mounts should be stiff and possess a large amount of damping to hold the engine to the chassis at a small relative displacement. At higher frequencies (20 to 200 Hz), where engine vibration occurs, the mounts should be soft to isolate the chassis from engine vibration. Passive elastomeric and hydraulic mounts offer a compromise between these two competing functions. Active mounts, however, can eliminate this tradeoff by being very stiff at low frequencies, yet very soft at high frequencies. Active mounts incorporate a force generating actuator into a passive mount, which can be elastomeric or hydraulic. The passive mount limits the engine motion at low frequencies, and provides a fail-safe feature in the event of an actuator failure. At high frequencies, a closed loop controller uses sensor information to command the force generating actuator to reduce vehicle vibration.

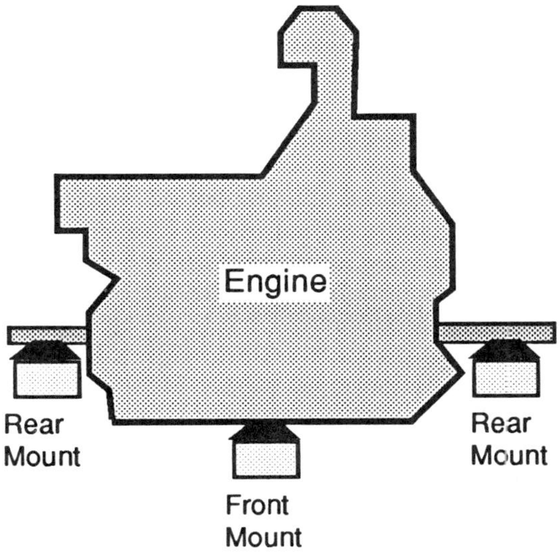

Figure 1. Front view of an engine supported by three mounts

In this paper, the basic mount design tradeoffs are expressed in terms of the dynamic stiffnesses of the engine, vehicle structure, and mounts using an effectiveness model. Mechanical models of elastomeric, hydraulic, and active mounts are then shown. From these models, dynamic stiffnesses for each mount are shown which illustrate the significant performance improvements that active mounts can achieve versus passive mounts. Also, different active mount configurations, actuation technologies, and controller choices are discussed.

ISOLATION SYSTEM MODEL

In this section, a mount effectiveness analysis is used to derive the performance of a mount in terms of the dynamic characteristics of the engine and vehicle structure. The mount effectiveness method is explained in detail in Sykes [14] and Swanson and Miller [15] for single mount isolation systems, and Swanson et al. [16] and Miller et al. [11] for multiple mount isolation systems. Mount effectiveness is a frequency domain analysis method that relates the performance of an isolation system in terms of the dynamic stiffness (or some other frequency response function) of the engine, mount, and structure. The effectiveness model of a general isolation system is shown in Figure 2. The isolation system consists of an engine connected to a structure through an engine mount. For simplicity, only a single mount is shown.

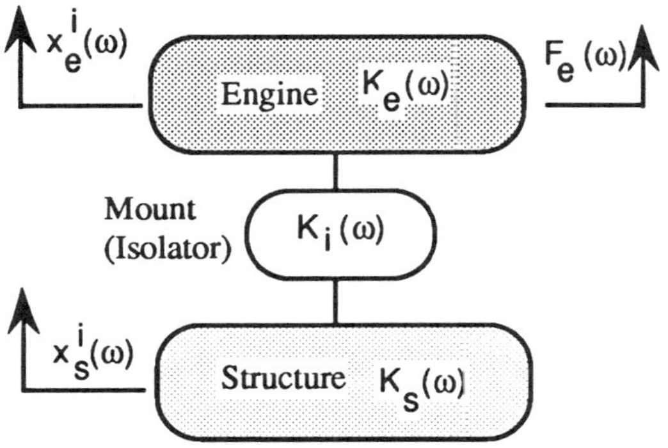

Figure 2. Isolation system model.

In Figure 2, the engine motion is represented by $x_e^i(\omega)$, the structure motion is represented by $x_s^i(\omega)$, the vibration disturbance force generated by the engine is represented by $F_e(\omega)$, where ω is the vibration frequency. The dynamic stiffness of the engine at the mount location is represented by $K_e(\omega)$, the dynamic stiffness of the mount, or isolator, is represented by $K_i(\omega)$, and the dynamic stiffness of the vehicle structure at the mount location is represented by $K_s(\omega)$. Each dynamic stiffness is a complex function of frequency (called a frequency response function, or FRF). FRF's are useful for representing the dynamic response of an object in the frequency domain. If an object is linear and elastic, a sinusoidal input will produce a sinusoidal output at exactly the same frequency, but with a different magnitude and phase. The FRF consists of the magnitude change and phase shift of the output to the input. FRF's can be determined through analytical models, finite element models, or experimental measurements. Dynamic stiffness is defined as the ratio of an output force to an input displacement. Other frequency response functions can be defined for other choices of input and output variables.

As discussed in the introduction, at low frequencies (below 20 Hz), the mounts must limit engine motion. Low frequency force sources include the engine weight, engine torque, vehicle maneuvering (acceleration and deceleration, braking, cornering, etc.), engine starting and stopping, and shocks caused by uneven road surfaces. Swanson and Miller [15] show that the engine motion for the isolation system shown in Figure 2 is equal to:

$$x_e^i(\omega) = \left(\frac{F_e(\omega)\frac{1}{K_e(\omega)}\left(\frac{1}{K_s(\omega)} + \frac{1}{K_i(\omega)}\right)}{\frac{1}{K_e(\omega)} + \frac{1}{K_s(\omega)} + \frac{1}{K_i(\omega)}} \right) \tag{1}$$

At low frequencies, the isolation system model can be simplified further by assuming that the engine is a mass ($K_e(\omega) = -m\omega^2$), the structure is rigid ($K_s(\omega) = \infty$), and the mount consists of a spring k and viscous damper c connected in parallel ($K_i(\omega) = k + cj\omega$). These assumptions are commonly used in modeling engine mounts, and are primarily valid for low frequencies only. Substituting these relations into EQ (1) results in the following expression for the engine motion:

$$x_e^i(\omega) = \left(\frac{F_e(\omega)}{-m\omega^2 + k + cj\omega} \right) \tag{2}$$

The static engine motion (ω=0) is then equal to:

$$x_e^i(0) = \frac{F_e(0)}{k} = \frac{W}{k} \tag{3}$$

where W is the engine weight. EQ (3) shows that the engine motion is smallest when the engine mount is stiffest (k is large).

The simplified spring-mass-damper model of the isolation system can also be used to examine the engine's motion at resonance. From EQ (2) we see that at resonance ($\omega = \sqrt{k/m}$) the engine motion is equal to:

$$x_e^i(\sqrt{k/m}) = \frac{F_e(\omega)}{cj\sqrt{k/m}} \tag{4}$$

To damp this resonance (called the engine "bounce"), the viscous damping coefficient c for the mount must be large. When this occurs, the mount will adequately damp out transient effects caused by sudden acceleration and deceleration, braking, cornering, riding on uneven roads, etc. For many off highway, automotive, and truck/bus applications, the mount stiffness k is designed so that the engine bounce resonance occurs at approximately 6 to 12 Hz.

The mounts objective changes from limiting engine motion to reducing the vibration transmitted from the engine to the vehicle structure at higher frequencies. In reciprocating engines, vibration is caused by mass imbalances in the rotating and reciprocating parts of the engine (piston, connecting rod, crankshaft), as well as firing pulses resulting from combustion forces. These components generate vibration that is primarily correlated to the engine speed and its harmonics (called orders). The engine speed is the first order; the first harmonic of the engine speed is the second order; the second harmonic of the engine speed is the third order, etc. Fractional orders can also occur due to gearing. Generally, the contribution from the different orders depends on the number of cylinders in the engine and whether the engine is a two-stroke or four-stroke engine. For example, a four cylinder, four stroke internal combustion engine with speeds ranging from 600 to 6000 rpm (10 to 100 Hz) generates vibration primarily at the second order (20 to 200 Hz). A six cylinder, four stroke diesel engine with speeds ranging from 600 to 2100 rpm (10 to 35 Hz) generates vibration primarily at the third engine order (30 to 105 Hz).

In general, at the lower engine speeds (near idle), engine vibration results in an annoying shaking of the vehicle. At higher frequencies (between 50 and 200 Hz), a booming sound is created inside the vehicle compartment when engine vibration coincides with an acoustic resonance of the passenger compartment.

The vehicle structure vibration can be expressed in terms of either the structural displacement $x_s^i(\omega)$, or the transmitted mount force, $f^i(\omega)$. Swanson and Miller [15] show the structure displacement as:

$$x_s^i(\omega) = \left(\frac{F_e(\omega)\left(\frac{1}{K_s(\omega)K_e(\omega)}\right)}{\frac{1}{K_e(\omega)} + \frac{1}{K_s(\omega)} + \frac{1}{K_i(\omega)}} \right) \quad (5)$$

Swanson and Miller [15] also show the transmitted force as:

$$f^i(\omega) = \left(\frac{F_e(\omega)\left(\frac{1}{K_e(\omega)}\right)}{\frac{1}{K_e(\omega)} + \frac{1}{K_s(\omega)} + \frac{1}{K_i(\omega)}} \right) \quad (6)$$

EQ's (5) and (6) represent the structural vibration in an isolated system. They do not, however, represent the isolation performance of the mount. A good measure of mount performance is called mount effectiveness, which is a nondimensional comparison between the structure vibration in a hard mounted system to the structure vibration in an isolated system. From EQ's (5) and (6), we can solve for the structure displacement $x_s^h(\omega)$ and the transmitted mount force $f^h(\omega)$ when the engine is hard mounted to the structure ($K_i = \infty$).

$$x_s^h(\omega) = \left(\frac{F_e(\omega)\left(\frac{1}{K_s(\omega)K_e(\omega)}\right)}{\frac{1}{K_e(\omega)} + \frac{1}{K_s(\omega)}} \right) \quad (7)$$

$$f^h(\omega) = \left(\frac{F_e(\omega)\left(\frac{1}{K_e(\omega)}\right)}{\frac{1}{K_e(\omega)} + \frac{1}{K_s(\omega)}} \right) \quad (8)$$

The effectiveness of a mounting system in reducing vibration is defined as:

$$E(\omega) = \frac{f^h(\omega)}{f^i(\omega)} = \frac{x_s^h(\omega)}{x_s^i(\omega)} = 1 + \frac{K_{eq}(\omega)}{K_i(\omega)} \quad (9)$$

where $k_{eq}(\omega)$ is the series dynamic stiffness of the engine and structure, defined as:

$$K_{eq}(\omega) = \left(\frac{1}{\frac{1}{K_e(\omega)} + \frac{1}{K_s(\omega)}} \right) \quad (10)$$

To achieve a high degree of vibration isolation, the effectiveness $E(\omega)$ must be much greater than one (because then the structure vibration in the isolated case will be much smaller than the structure vibration in the hard mounted case). EQ (9) shows that a mount is "effective" when it is much softer than the series dynamic stiffness of the engine and structure.

$$|K_i(\omega)| << |K_{eq}(\omega)| \quad (11)$$

Therefore, for the isolator to reduce vibration, its stiffness at the vibratory disturbance frequency must be much lower than either the engine's or the structure's (chassis) dynamic stiffness. This criteria is getting more and more difficult to achieve with passive mounts because of the current trend towards lighter vehicle frames.

ELASTOMERIC MOUNTS

Elastomeric mounts have isolated vehicle structures from engine vibration since the 1930's. A physical model of an elastomeric mount is shown in Figure 3. The elastomer is bonded to the metal in this conical shaped mount.

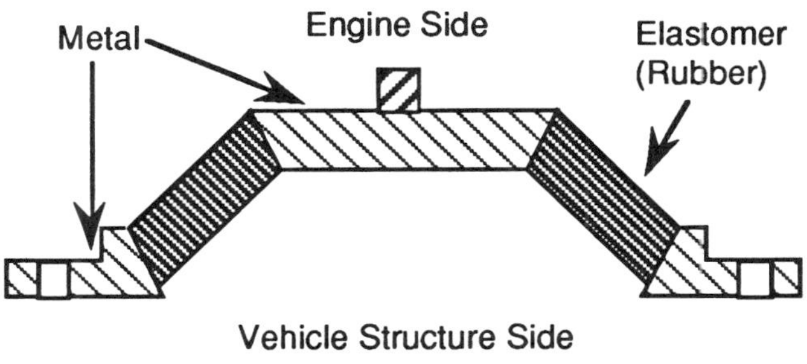

Figure 3. Physical model of an elastomeric mount.

The elastomeric mount can be represented by the familiar Voigt model, which consists of a spring K_r and a viscous damper B_r connected in parallel. This representation is shown in Figure 4.

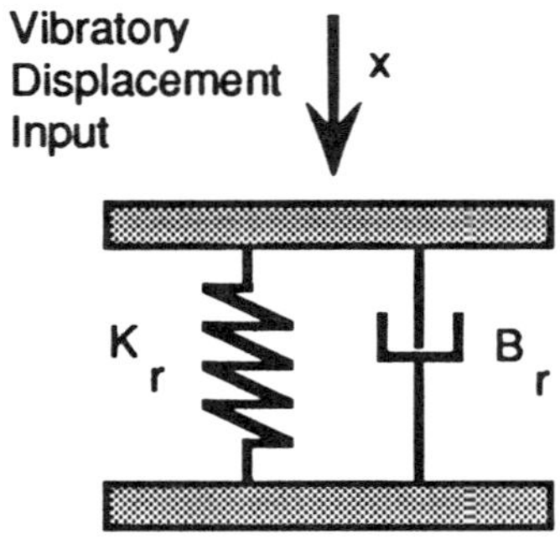

Figure 4. Mechanical model of an elastomeric mount.

The dynamic stiffness of a typical elastomeric automotive engine mount is shown in Figure 5. Notice that the damping causes the elastomeric mount's stiffness at high frequencies to be greater than it's stiffness at low frequencies. In many situations, this characteristic makes it difficult to design a mounting system that satisfies all of the design requirements. For example, providing additional stiffness and damping at low frequency to better control the engine motion will significantly increase the mount's dynamic stiffness at high frequency, decreasing isolation performance. Conversely, decreasing the mount's stiffness at high frequency to improve it's vibration isolation capabilities greatly increases engine motion. Softer mounts also exhibit more creep and set in the elastomer and have a shorter service life due to fatigue. Elastomeric mounts are described in more detail in Schmitt and Leingang [1], Bucksbee [2], and Marshall and Breindel [3].

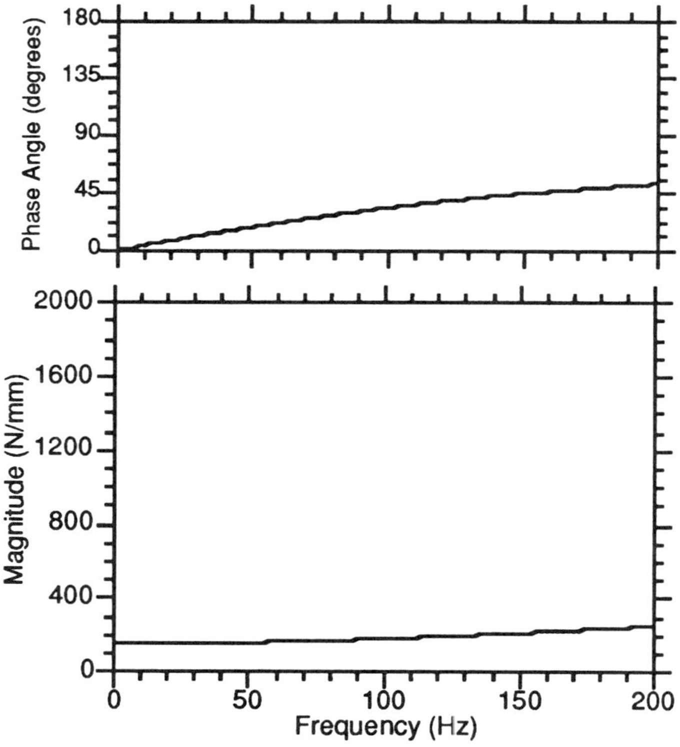

Figure 5. Dynamic stiffness of an elastomeric mount

HYDRAULIC (FLUID) MOUNTS

Hydraulic, or fluid, mounts have gained popularity in the last ten years because they allow a mount's damping and stiffness to be tuned independently. Figure 6 shows a physical model of a hydraulic mount. In addition to an elastomeric mount, the hydraulic mount contains two fluid chambers separated by a spiral shaped inertia track. A very soft rubber bellow contains the fluid in the bottom fluid chamber.

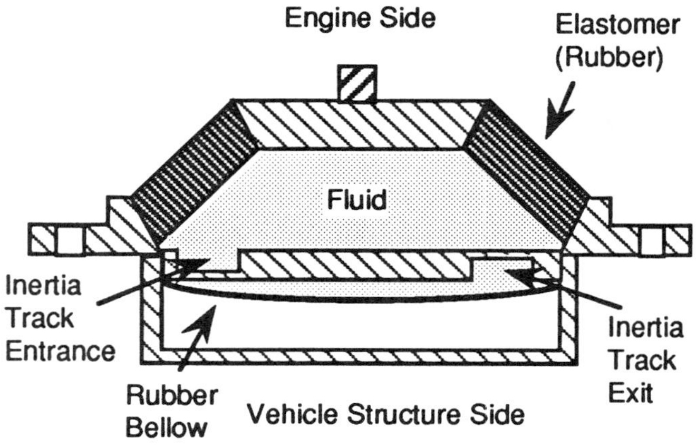

Figure 6. Physical model of a hydraulic mount

Figure 7 shows a mechanical model of the hydraulic mount with no decoupler. Again, the elastomeric top section is modeled using a spring K_r connected in parallel with a viscous damper B_r. The volumetric (bulge) stiffness of the top elastomeric section is modeled using the spring K_t. The volumetric stiffness of the rubber bellow is modeled using the spring K_b. m represents the fluid mass in the inertia track. A_p represents the piston area of the top elastomeric section. A_t represents the cross sectional area of the inertia track. B_0 represents fluid damping in the inertia track. A mechanical lever models the fluid leverage between the inertia track area and the piston area of the mount.

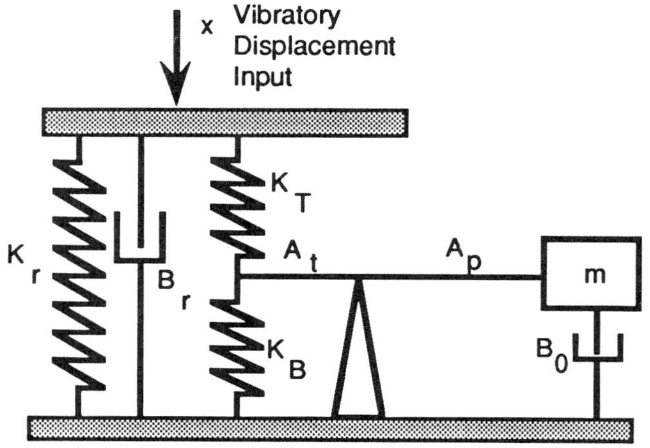

Figure 7. Mechanical model of a hydraulic mount

Figure 8 shows the dynamic stiffness of the hydraulic mount with no decoupler. The hydraulic mount operates as follows. At extremely low frequencies, fluid passes freely from the top chamber to the bottom chamber through the spiral shaped inertia track. Because the bulge stiffness of the rubber bellow is very low, the static stiffness of the hydraulic mount is approximately equal to the static stiffness of the top elastomeric section K_r. At the frequency f_n (called the notch frequency), the fluid mass resonates on the elastomeric sections of the mount. The resonating fluid mass causes the dynamic stiffness to be reduced at the notch frequency. The fluid mass is leveraged by the area ratio A_p/A_t. Typically, because A_p/A_t is greater than one, the leverage effectively increases the fluid mass. As the frequency increases further, the dynamic stiffness increases to a maximum (called the resonance frequency, f_r). Above the resonance frequency, f_r, the inertia track essentially closes off. When this occurs, the stiffness of the mount is approximately equal to the sum of the elastomeric stiffness K_r and the volumetric stiffness K_t.

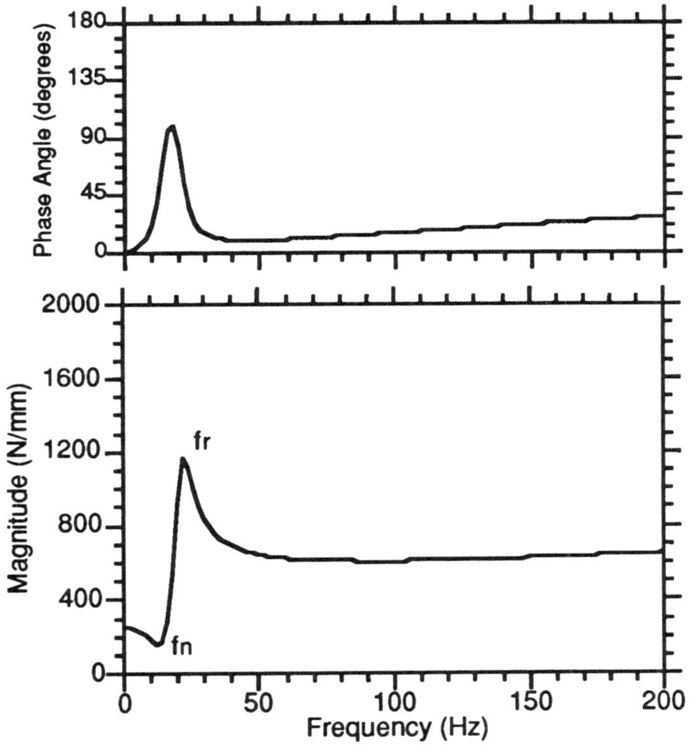

Figure 8. Dynamic stiffness of a hydraulic mount

This type of hydraulic mount can be used in two ways. First, the mount can be tuned to provide exceptional isolation at a single frequency (the notch frequency). The mechanical model of Figure 7 shows the hydraulic mount to be similar to an elastomeric mount with an embedded vibration absorber. Mounts used in this manner are primarily tuned to idle shake frequencies of an engine. This type of mount is also commonly used in aerospace applications where the engine's speed is relatively constant. Second, the mount can be tuned to provide additional damping at a single frequency. Notice from the phase diagram of Figure 8 that the phase angle of the mount's dynamic stiffness approaches 90 degrees at a frequency between the notch frequency f_n and the resonance frequency f_r. At this particular frequency, the mount acts like a viscous damper, or shock absorber. In vehicle applications, irregularities in the road surface such as potholes create large forces in the vertical direction. For this reason, hydraulic mounts are typically oriented vertically to damp out the vertical resonance of the engine bouncing on the mounts, which typically occurs between 6 and 12 Hz.

There is one problem, however, with using the hydraulic mount to provide extra damping at low frequencies. Notice from Figure 8 that the dynamic stiffness of the hydraulic mount is greater than the dynamic stiffness of a comparable elastomeric mount. This indicates that, while the hydraulic mount greatly increased damping at low frequency, it also degraded isolation performance at high frequency. This problem has been solved by incorporating a decoupler into the hydraulic mount. A decoupler is a lost motion device that allows a hydraulic mount to behave like a hydraulic mount at high displacement amplitudes (to provide damping to the engine bounce), yet behave like an elastomeric mount at low displacements (to provide good vibration isolation). A hydraulic mount with a decoupler is shown in Figure 9.

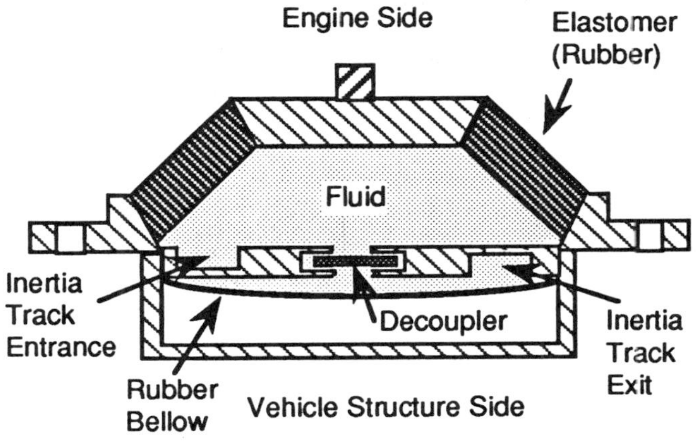

Figure 9. Physical model of a hydraulic mount (with decoupler)

A mechanical model of a hydraulic mount with a decoupler is shown in Figure 10. Except for the additions of the decoupler (mass, lever arm, and lost motion device), the mechanical model shown in Figure 10 is identical to the mechanical model shown in Figure 7.

The hydraulic mount with decoupler operates as follows. At low vibration amplitudes (say 0.1 mm), fluid travels around the decoupler instead of through the inertia track, causing the mount to behave like an elastomeric mount. At high amplitudes (say 1.0 mm), the decoupler is forced against the snubbers, which limit it's deflection. When this occurs, fluid travels through the inertia track, creating additional damping to the engine "bounce". The hydraulic mount with decoupler is currently used in many vehicles (automobiles, trucks/busses, off/highway vehicles). The dynamic stiffness of a hydraulic mount with a decoupler is shown in Figure 11. For more information on hydraulic mounts, refer to Taylor [4], Flower [5], and Marjoram [6]. Singh, et al. [7] show a detailed analysis of hydraulic mounts with decouplers.

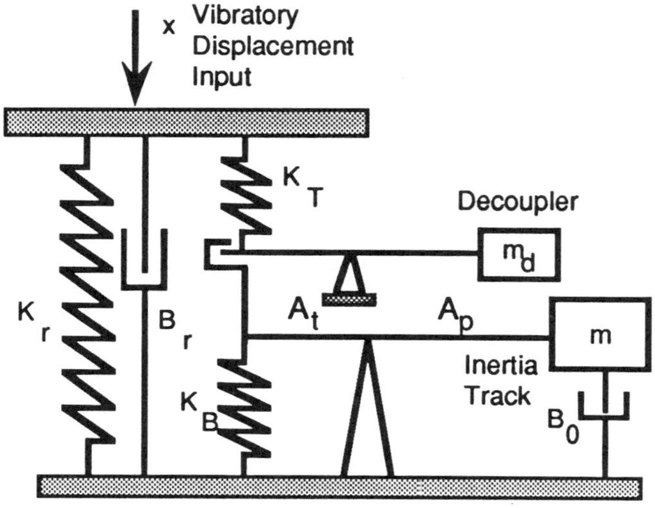

Figure 10. Mechanical model of a hydraulic mount (with decoupler)

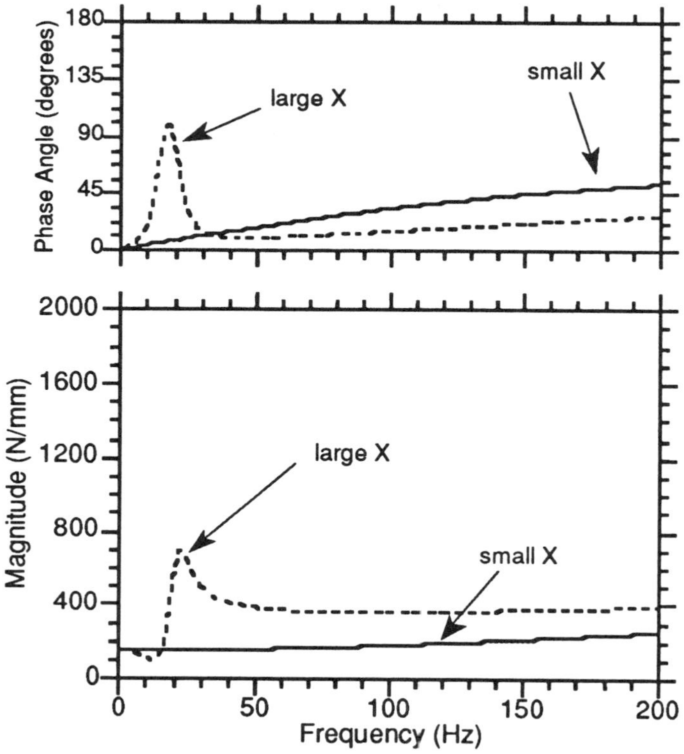

Figure 11. Dynamic stiffness of a hydraulic mount (with decoupler)

ACTIVE MOUNTS

Active mounts consist of a passive mount (elastomeric or hydraulic), a force generating actuator, a structural vibration sensor, and an electronic controller. In this paper, two different active mount configurations will be discussed: the active elastomeric mount, and the active hydraulic mount.

A mechanical model of the active elastomeric mount is shown in Figure 12. The active elastomeric mount consists of a force generating actuator placed in parallel with the passive elastomeric mount. K_r and B_r represent the stiffness and damping in the elastomer. K_a represents the stiffness of the actuator, and F_a represents the force generation of the actuator.

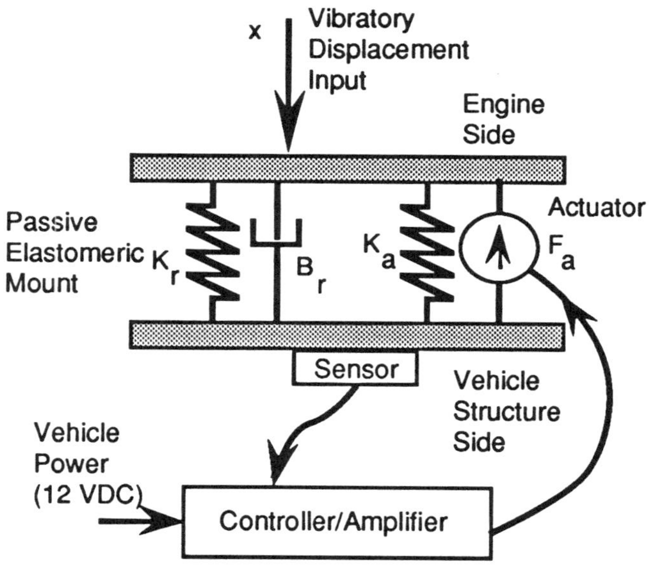

Figure 12. Mechanical model of an active elastomeric mount

Typically, a low stiffness/high stroke electromagnetic actuator is used in this configuration. The actuator provides dynamic forces that, when properly phased by the controller, reduce vibration in the structure. Because the actuator stiffness K_a is significantly lower than the elastomeric mount stiffness K_r, the majority of the static loads are supported by the elastomer. This is an important feature because it ensures that the active mount will support the engine in the event of an actuator failure.

There are several important requirements that must be satisfied when choosing an actuator. These requirements include force, stroke, bandwidth, power limitations, size, weight, service life, temperature, and cost. For example, consider an automotive application where a 4 cylinder, 4 stroke internal combustion engine is running at a constant speed of 1800 rpm. This engine generates vibration primarily at its second order, or 60 Hz. If the engine displaces 0.1 mm, and the mount's stiffness at 60 Hz is 200 N/mm, then the required actuator force is 20 N. Also, because the actuator is placed in parallel with the mount, the required actuator stroke equals the engine displacement, 0.1 mm. The mechanical power generated by this engine vibration can be calculated from EQ 12:

$$P(\omega) = F(\omega)\ v(\omega) = k(\omega)\ x(\omega)^2\ (j\omega) \qquad (12)$$

where $P(\omega)$ is the mechanical power, $F(\omega)$ is the transmitted mount force, $v(\omega)$ is the engine velocity, $k(\omega)$

is the mount stiffness, $x(\omega)$ is engine displacement, $j = \sqrt{-1}$, and ω is the frequency in radians per second. At 60 Hz, the actuator must generate a mere 0.75 watts of mechanical power. Of course, due to inefficiencies in the electrical to mechanical power conversion in the actuator, the amplifier must supply the actuator with more than 0.75 watts. The power necessary to operate the actuators is small, however, if one considers the power savings (in the kilowatt region) that will occur if the engine's balance shafts are removed.

To keep the actuator force/stroke/power requirements small, it is important to design the actuator to provide forces only at the frequencies where vibration occurs. For example, the automotive active mount mentioned above has to provide 20 N of force to reduce vibration at 60 Hz. However, if the actuator were sized to support the engine's weight, it would have to provide approximately 1000 N. This would significantly increase the actuator's size, cost, and power requirements.

Electromagnetic actuators satisfy all of the above requirements. There are two basic types of electromagnetic actuators: voice coils and solenoids. Voice coil actuators have been used for over 100 years in loudspeakers. Voice coils are compact, have a high frequency bandwidth (up to a few kilohertz), and generate a force output that is both linearly proportional to input current and independent of stroke. Voice coils have long strokes (2 mm or more), and can generate the required forces in a reasonable size. Solenoids offer higher force outputs in smaller packages than voice coils. However, solenoids undesirably generate forces that are nonlinearly related to both input current and stroke. Solenoids also have high inductance's and generate significant eddy currents, which reduces their frequency bandwidth. For a more complete discussion of electromagnetic actuators, see Buck, et al. [18].

Other possible actuator technologies for active mounts include servohydraulic [10], electrostrictive and piezoelectric material [19], and magnetostrictive material actuators [20]. Servohydraulic actuators can produce extremely high forces. However, servohydraulics have a limited bandwidth, require high maintenance, require a hydraulic power supply, and are expensive. Electrostrictive and piezoelectric actuators produce a mechanical strain in response to an applied electric field. Magnetostrictive actuators produce a mechanical strain in response to an applied magnetic field. Because electrostrictive, piezoelectric, and magnetostrictive actuators are very stiff, they generate high forces through small strokes. They are also extremely compact, and have very high bandwidths (well above a kilohertz). However, they can be expensive, and produce displacement (stroke) outputs that are too small for many engine mount applications.

Once the actuator is incorporated into the active mount, its forces must be properly controlled to reduce vibration. Hodgson [10], Elliot, et al. [21], Widrow and Stearns [22], Sievers and vonFlotow [23], and Erikkson [24] describe different controllers that are applicable to active engine mounts. Essentially, the controller can be either a feedback or feedforward type. Feedback controllers use structure vibration sensor information to modify the dynamic stiffness of the active mount [10]. The structure vibration sensor can either be a force sensor or an accelerometer.

Feedforward controllers, on the other hand, require an additional sensor that is correlated to the engine vibration (like a tachometer signal). Feedforward algorithms, such as the Filtered-X Least Mean Square (LMS) algorithm and the Filtered-U Recursive Least Mean Square algorithm (RLMS), have gained popularity in recent years because of their adaptive nature. These adaptive algorithms have become practical with the advent of inexpensive, high powered Digital Signal Processors (DSP's). Adaptive feedforward algorithms can follow changes in engine speed, and can tune themselves to different engine mount installations.

The dynamic stiffness of the active elastomeric mount is shown in Figure 13. The active mount's stiffness is equivalent to the stiffness of the passive elastomeric mount at every frequency except where engine vibration occurs (60 Hz in this example). At the disturbance frequency, a tonal controller commands the mount to be very soft (reducing the accelerometer or force sensor signal). The active mount, therefore, can overcome the limitations of the passive elastomeric mounts by being very stiff at low frequencies, yet very soft at high frequencies. Because the active mount provides superior isolation to a passive mount, its static stiffness can be increased, thus offering both improved engine motion control along with improved vibration isolation.

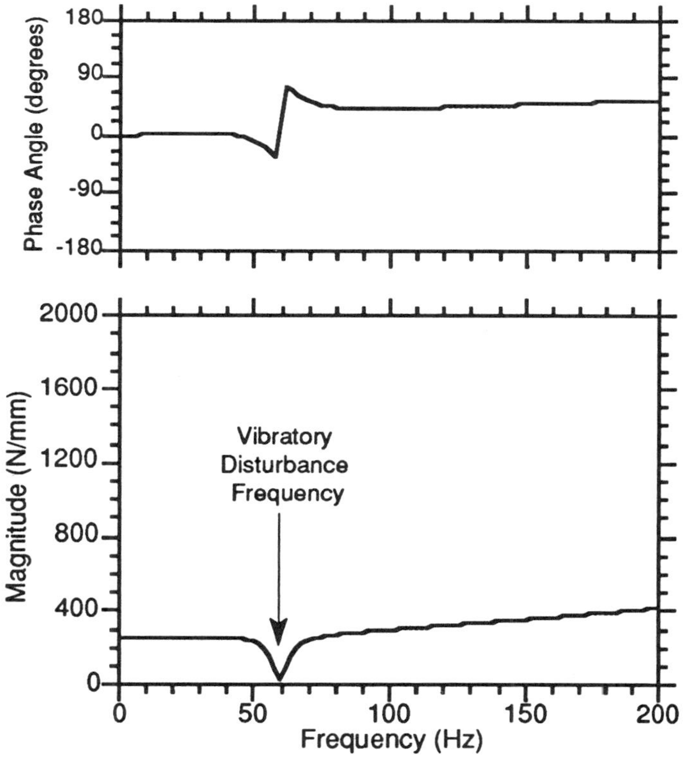

Figure 13. Dynamic stiffness of an active mount (actuator in parallel with passive mount)

An active hydraulic mount can also be created by incorporating an actuator into a passive hydraulic mount. A mechanical model of an active hydraulic mount is shown in Figure 14. In this configuration, the actuator pushes on the fluid mass to create forces across the mount. Because the piston area of the top elastomeric section is typically larger than the piston area of the actuator, the actuator force is magnified. This reduces the force/stroke/power requirements of the actuator. Hodgson [10], and Miller, et al. [11] describe an aerospace version of the active hydraulic mount.

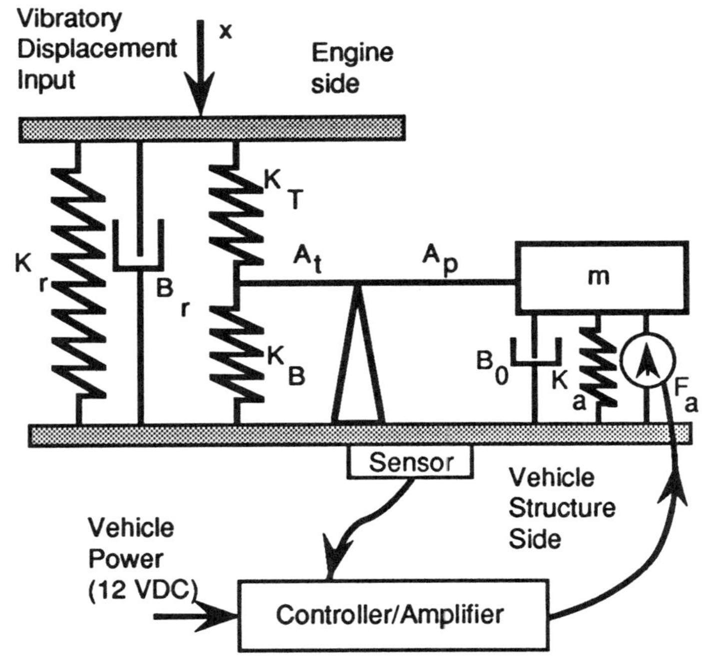

Figure 14. Mechanical model of an active hydraulic mount

Similar to the passive hydraulic mount, the active hydraulic mount can be used in two different ways (only one of which will be shown here). First, the active hydraulic mount can be used to increase the passive isolation at the notch frequency. Because the hydraulic mount acts like it contains a tuned absorber, it's dynamic stiffness is already small at the notch frequency. However, because of damping caused by fluid flow, the dynamic stiffness at the notch frequency, while low, is not equal to zero. The active mount can then use the controller to drive the dynamic stiffness to zero at the notch frequency, achieving perfect isolation. This type of active mount has been proposed for aerospace applications (see Miller, et al. [11] for details). In industrial vehicle applications such as automotive, off-highway, truck and bus, etc., the active hydraulic mount can be used in a different manner. In these applications, the hydraulic mount is tuned to achieve excellent damping at the engine bounce frequency. The active mount then drives the dynamic stiffness to zero at higher frequencies where vibration occurs. This type of mount can be used where adequate damping cannot be achieved with an active elastomeric mount. The dynamic stiffness for this type of mount is shown in Figure 15.

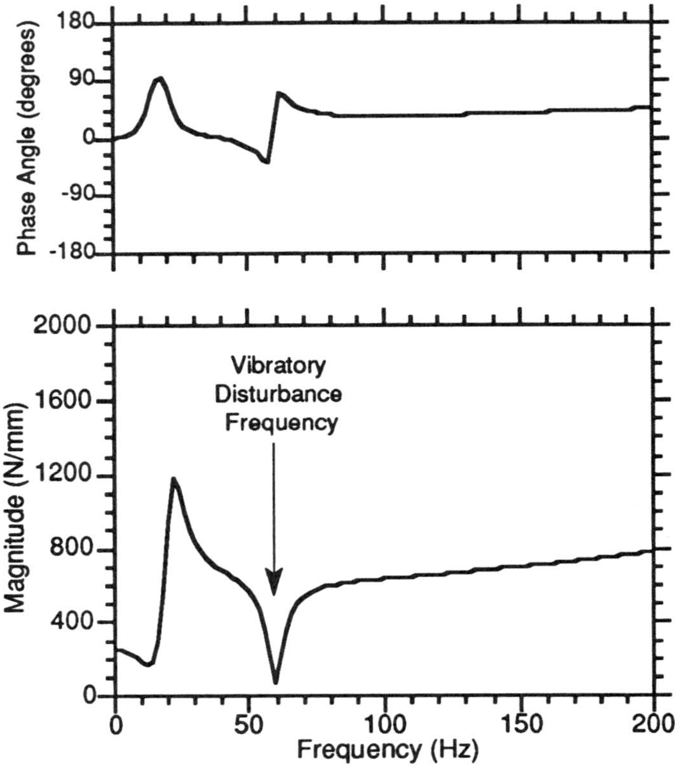

Figure 15. Dynamic stiffness of an active hydraulic mount

BENEFITS OF ACTIVE MOUNTS

A fundamental benefit of active mounts is reduced vibration in the vehicle chassis. By providing superior isolation, active mounts can allow larger engine vibration levels. This reduces the balance shaft requirements, or possibly eliminates the need for the balance shaft (decreasing engine weight). Of course, the balance shaft can only be removed if the engine and the components attached to the engine (alternator, etc.) can withstand higher vibration levels. Superior isolation may also allow the vehicle chassis to be lighter. Donovan and Auslander [25] show how using mounts can result in lighter attachment structures. Removing the balance shaft and decreasing chassis weight reduces vehicle cost. These cost savings are maximized when the active mounts are designed into the vehicle early in the design process.

CONCLUSIONS

Active mounts represent the next generation in engine mounts (compared to hydraulic and elastomeric mounts). The addition of energy allows the active mount to be stiff at low frequencies, yet soft at high frequencies. The improved vibration reductions achieved by active mounts may allow for the elimination of engine balance shafts, decreasing engine size, weight, and cost, and further improving fuel economy.

REFERENCES

[1] Schmitt, R.V. and Leingang, C.J., "Design of Elastomeric Vibration Isolation Mounting Systems for Internal Combustion Engines," *SAE Earthmoving Industry Conference*, Peoria, Il., April 26-28, 1976, SAE Paper 760431.

[2] Bucksbee, J.H., "Use of Bonded Elastomers for Energy and Motion Control in Construction," *Rubber World*, vol. 196, no. 1, April 1987, pp. 38-45.

[3] Marshall, G. and Breindel, W., "Boosting Power Train Life With Engineered Elastomers," *Conference on Diesel and Gas Turbine Progress*, October, 1968.

[4] Taylor, H.J., "The New Generation of Engine Mounts," *SAE Milwaukee Section Lecture Series*, Milwaukee, Wi., April 3, 1986, SAE Paper 86052.

[5] Flower, W.C., "Understanding Hydraulic Mounts for Improved Vehicle Noise, Vibration, and Ride Qualities," *SAE Surface Vehicle Noise and Vibration Conference*, Traverse City, Mi., May 15-17, 1985, SAE Paper 850975.

[6] Marjoram, R.H., "Pressurized Hydraulic Mounts for Improved Isolation of Vehicle Cabs," *SAE Truck and Bus Meeting and Exposition*, Chicago, Il, December 9-12, 1985.

[7] Singh, R., Kim, G., and Ravindra, P.V., "Linear Analysis of Automotive Hydro-Mechanical Mount With Emphasis on Decoupler Characteristics," *Journal of Sound and Vibration*, vol. 158, no. 2, pp. 219-243, 1992.

[8] Quinn, D.C., "Automotive Active Engine Mount Systems," *AVL Conference on Engine and Environment*, 1992.

[9] Miller, L.R. and Ahmadian, M., "Active Mounts - A Discussion of Future Technological Trends," *Inter-Noise 92 Conference,* Toronto, Canada, July 20-22, 1992, pp. 421-426.

[10] Hodgson, D.A., "Frequency-Shaped Control of Active Isolators," *AHS/RAE International Technical Specialists Meeting,on Rotorcraft Acoustics and Rotor Fluid Dynamics*, Valley Forge, PA, October 15-17, 1991.

[11] Miller, L.R., Ahmadian, M., Nobles, C.M., and Swanson, D.A., "Modeling and Performance of an Experimental Active Vibration Isolator," *ASME Winter Annual Meeting*, Anaheim, CA, November 8-13, 1992.

[12] Scribner, K.B., Sievers, L.A., and von Flotow, A.H., "Active Narrow Band Vibration Isolation of Machinery Noise From Resonant Substructures," *ASME Winter Annual Meeting,* Technomic, Dallas, TX, December 1990, pp. 101-111.

[13] Ushijima, T. and Kumakawa, S., "Active Engine Mount with Piezo-Actuator for Vibration Control," *SAE International Congress And Exposition*, Detroit, MI, March 1-5, 1993, SAE Paper 930201.

[14] Sykes, A.O., The Evaluation of Mounts Isolating Nonrigid Machines from Nonrigid Foundations," David Taylor Model Basin, Annapolis, MD, Report 1094, October, 1957.

[15] Swanson, D.A. and Miller, L.R., "Design and Effectiveness Evaluation of an Active Vibration Isolation System for a Commercial Aircraft," *AIAA 2nd Aerospace Design Conference*, Irvine, CA, February 15-18, 1993, AIAA Paper 93-1145.

[16] Swanson, D.A., Miller, L.R., and Norris, M.A., "Multi-Dimensional Mount Effectiveness for Vibration Isolation," *AIAA/ASME/ASCE/AHS/ASC 33rd Structures, Structural Dynamics, and Materials (SDM) Conference*, Dallas, TX, April 13-15, 1992, AIAA Paper 92-2381.

[17] Sakata, T., "Analysis of the Transmission of Engine Vibration to the Body by the Mechanical Impedance Method," *SAE Automotive Engineering Congress*, Detroit , MI, Feb. 25 - Mar. 1, 1974, SAE Paper 740163.

[18] Buck, M., van Namen, F.T., and Weisman, R., "Electrodynamic Linear Motors for Active Vibration and Noise Control," *Proceedings of the 2nd Conference on Recent Advances in Active Control of Sound and Vibration*," Blacksburg, VA, April 28-30, 1993.

[19] Damjanovic, D. and Newnham, R.E., "Electrostrictive and Piezoelectric Materials for Actuator Applications," *Journal of Intelligent Materials, Systems, and Structures*, vol. 3, April, 1992.

[20] Moffett, M.B., Clark, A.E., Wun-Fogle, M., Linberg, J., Teter, J.P., and McLaughlin, E.A., "Characterization of Terfenol-D for Magnetostrictive Transducers," *Journal of the Acoustical Society of America*, Vol. 89, No. 3, March 1991.

[21] Elliot, S.J., Stothers, I.M., and Nelson, P.A., "A Multiple Error LMS Algorithm and Its Application to the Active Control of Sound and Vibration," *IEEE Transactions on Acoustics, Speech, and Signal Processing,* Vol. ASSP-35, No. 10, October, 1987.

[22] Widrow, B. and Stearns, S.D., *Adaptive Signal Processing*, Prentice Hall, Englewood Cliffs, NJ , 1985.

[23] Sievers, L.A. and von Flotow, A.H., "Comparison and Extensions of Control Methods for Narrowband Disturbance Rejection," *ASME Winter Annual Meeting*, Technomic, Dallas, TX., December 1990, pp. 11-22.

[24] Eriksson, L.J., "Development of the Filtered-U Algorithm for Active Noise Control," *Journal of the Acoustical Society of America*, 85 (2), 1989, pp. 797-802.

[25] Donovan, J.B. and Auslander, E.L., "The Use of Vibration Isolators to Reduce Aerospace Subsystem Weight and Cost," *AIAA/AHS/ASEE Aerospace Design Conference*, Irvine, CA., February 16-19, 1993, AIAA Paper 93-1146.

930201

Active Engine Mount with Piezo-Actuator for Vibration Control

Takao Ushijima and Syoichi Kumakawa
Bridgestone Corp.

Abstract

Electronics technology has made remarkable progress in recent years, and in the field of automobile noise and vibration control, it has attracted the interest of engineers due to its potential for use in the active control of vibration isolators.

This report describes an investigation on reducing automobile vibration caused by engine explosion using a piezo actuator(PA) engine mount. Piezo actuator has a very high speed response but the displacement is generally very small. Therefore PA mount requires an amplitude enlargement mechanism. Here, an active mount with an enlargement mechanism incorporated was developed. For the control system, the adaptive control algorithm proposed by S. J. Elliott et al. was employed as a base, and a control rule against limitation in the power of the actuator was used in combination with it. A significant attenuation of the vibration of the dominant harmonic order in idling was achieved. The results are reported in the following pages.

Introduction

Ride comfort and quietness in automobile design have become particularly important technical themes in view of environmental problems and affluent lifestyles. This has made the role of anti-vibration devices more important, and required higher performance from these devices. Engine vibration contributes greatly to overall noise and vibration, and so intensive research is now underway, particularly with respect to high performance engine mounts.

Conventional isolators such as solid rubber mounts and fluid filled mounts are unable to comply with increased demands, so research is looking at isolators which have a control mechanism to externally adjust the features of an automobile depending on the condition of its use. Promising results have been reported with control mounts using electro-rheological fluid(ERF) and electromagnetic actuators. [1],[2]

The piezo actuator(PA), which has considerable potential in the area of active control, will be discussed in this report. PAs are capable of displacement at extremely high speeds according to an applied voltage and are used in the field of OA equipments, with many improvements already undertaken. Thus they are expected to be suitable for active control mounts also.

For the control mount to work effectively, it has to generate a vibration of the same amplitude as the engine vibration and thus cancel the vibration transmitted to the chassis, as shown in the Appendix. However, the PA itself produces very small displacement when voltage loaded, and so some displacement enlargement mechanism is required. Here, the amplitude enlargement was achieved by providing a substantial difference in the effective area between the upper and lower inner-surface of the fluid filled mount.

For vibration control, a simplified adaptive control algorithm was used. This algorithm is designed so that a signal such as, for example, acceleration on the chassis side is treated as an error signal; the control parameter of the signal to be transmitted to the actuator is then calculated from the magnitude of this error signal so as to reduce it to zero. In the present application, the algorithm was used for minimizing the vibration of the second order or dominant harmonic order of the vibration of a four-cylinder engine.

First, the constructions and features of the PA unit and PA control mount, including the theory of the control algorithm and experiments, will be introduced. Next the results will be discussed, showing how the above theory, when put into practice using a control mount installed on an actual automobile, produces good vibration control.

Concept of Active Engine Mount

The fluid filled mount was initially developed as a countermeasure to engine shake, achieving high attenuation in the low frequency range of 7~15Hz, and in recent years, has been widely used. Mounts with decouplers built in to reduce booming noise have also been developed and are often used. However, these fluid filled mounts are not capable of simultaneously resolving all noise and vibration problems, and there still remain various issues concerning tradeoffs such as torque shock at engine ON-OFF and degradation of idle vibration from liquid loading.

Active control can be used instead, but it is very difficult from a cost performance perspective to simultaneously resolve all the problems. It is therefore necessary to use selected features from each type of mounts, for example, some problems solved by conventional mounts and others by active control. The focus of the design was on reducing the particularly important idle vibration, and so the basic design concept required for the active PA engine mount will be described.

The vehicle used for the experiment was a four-cylinder engine passenger car, with the dominant harmonic order of the vibration being of the second order. The high-idle speed at engine start was 1200RPM, and the idle speed at stationary state 700RPM, so that the engine idle speed was within the range of 700~1200RPM. In this frequency region, the inertia of the engine mass is the dominant force balancing engine vibration. The displacement amplitude of the second order is scarcely affected by the stiffness of the engine mount. As seen in Fig. 1, the displacement amplitude is within the range of ±0.05~±0.1mm in this region. The PA is required to generate a displacement of this order. However, the PA employed as dimensionally suitable for practical application is capable of generating a displacement no larger than ±35μm. A mount

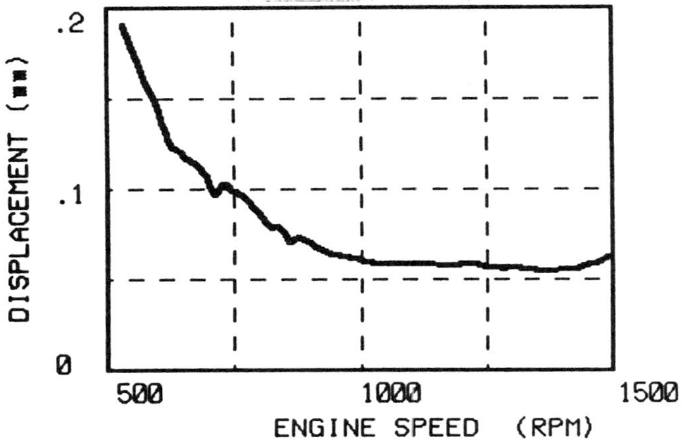

Fig. 1 Displacement amplitude of engine side vibration at idling.

using a liquid amplitude enlargement mechanism, shown in Fig. 2, was therefore developed.

Piezo Actuator(PA)

The PA used in the active control mount in this report is shown in detail in Fig. 3. PAs are a laminate of several alternating layers of ceramics and electrodes, which generate a displacement according to the magnitude of the applied load voltage. The greater the number of layers, the greater the displacement. However, for practical applications the size is restricted. The PA used in this report has a height of 98mm, and its characteristics are as shown below. The fundamental characteristics of PAs are described in detail in Inoi et al. [3], and so only those characteristics relevant to active mounts are discussed here. Some of the data presented here was taken from the NEC's PA Catalogue.

Fig. 4 shows a load-displacement diagram. The curve in Fig. 5 shows the PA response to the applied 1KHz sinusoidal wave voltage. The response speed given is compatible with the speed required of a control mount. One of the notable features of PAs is in this fast response, in the order of tens of microseconds.

Fig. 6 shows displacement as a function of load voltage. Displacement increases in proportion to voltage from 0 to 70μm maximum, obtaining a displacement of ±35μm.

Fig. 7 shows the temperature characteristic of PAs.

For applications such as engine mounts, which are located in high temperature regions, temperature characteristics pose important problems. Yet, the quick response mentioned above makes PAs a very attractive and advantageous proposition in respect of electrically controlled system.

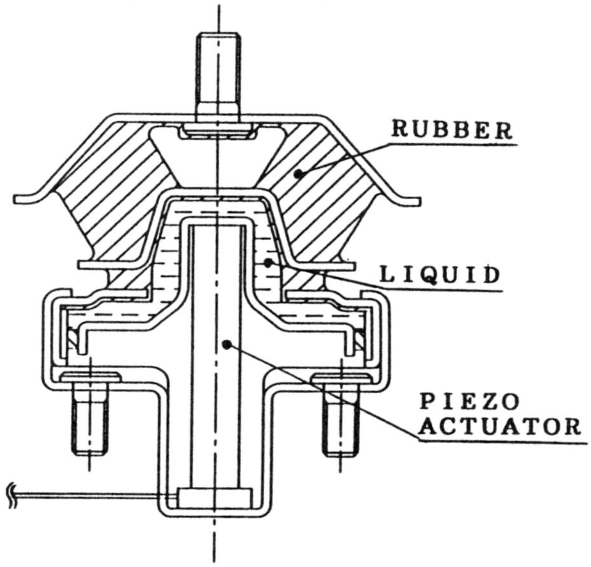

Fig. 2 Schematic diagram of active mount with Piezo Actuator.

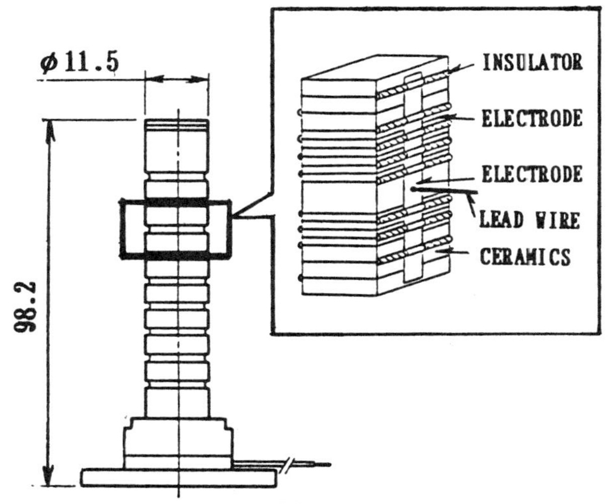

Fig. 3 Detail of Piezo Actuator.

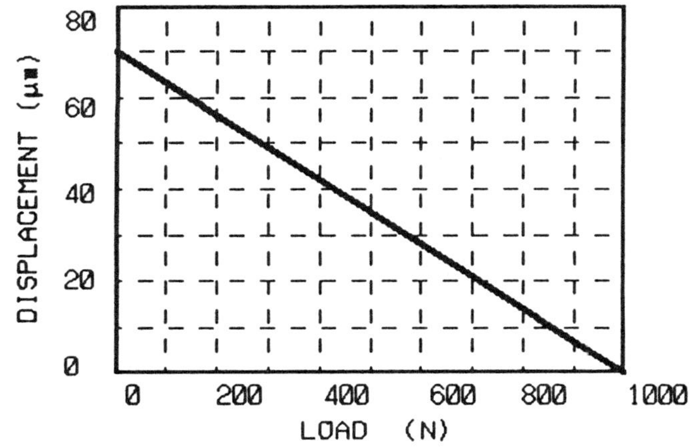

Fig. 4 Load-Displacement curve at applied voltage 150V. DC.

Active Control Engine Mount

Active mounts consist of two parts, each acting independently, as shown in Fig. 2. In the diagram, the upper part is a conventional solid or fluid filled engine mount. Fluid filled engine mounts are designed so as to effectively function against engine shake and other similar phenomenon. The lower is designed for active cancellation of vibrations in the idle vibratory region. This active part must be designed hard enough so that the features of the upper conventional mount are not impaired, and in the prototype used this time, it was designed to a hardness of the about 30 times that of the upper part. The analytical model of the whole control mount system is given in the Appendix. Here, discussion will be restricted to the lower active part, and its amplitude enlargement mechanism will be explained in detail with reference to Fig. 8. PAs generate a displacement according to the applied voltage and waveform. If the displacement is represented by X_{P1}, it can be readily seen that the volumetric change $\triangle V$ is

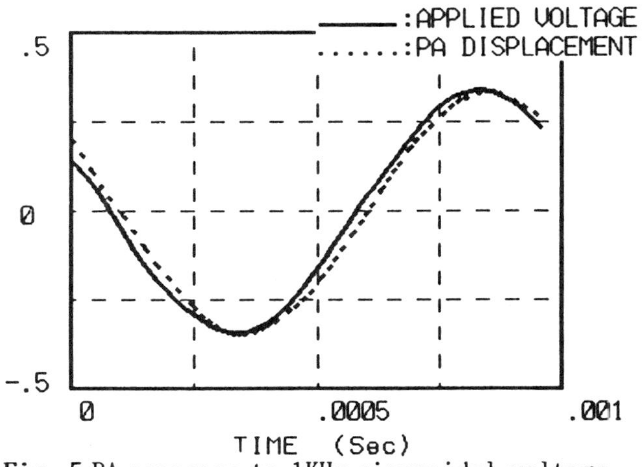

Fig. 5 PA response to 1KHz sinusoidal voltage.

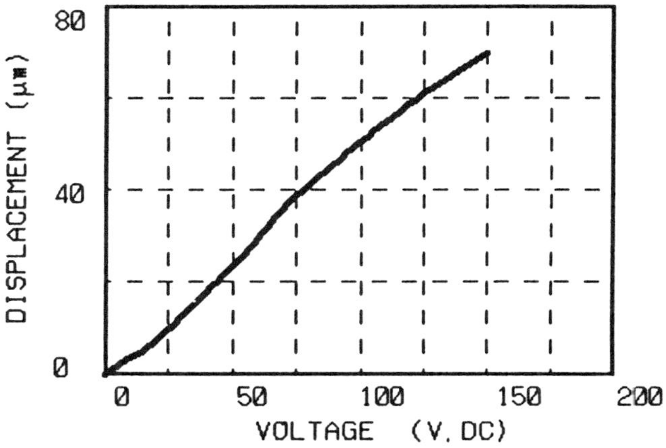

Fig. 6 Voltage-displacement curve of PA.

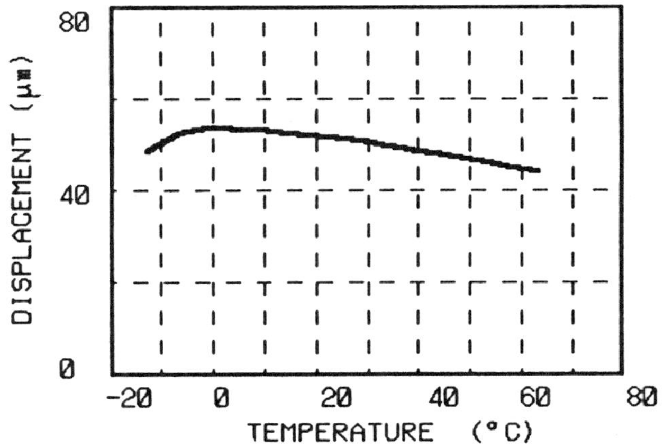

Fig. 7 Temperature dependency of PA displacement.

expressed by A_1X_{P1}. Here, A_1 is the effective cross-sectional area of the piston P1. Assuming that the bulk modulus of the liquid chamber formed by the rubber, piston P1 and piston P2 is equal to infinity, then from the incompressibility of the liquid, the piston P2 is displaced by a value of X_{P2}, satisfying the formula $\triangle V=A_2X_{P2}$. Here, A_2 is the effective cross-sectional area of the piston P2. Consequently, the displacement X_{P2} of the piston P2 is equal to $X_{P1}A_1/A_2$, and so it can be expected that the ratio of displacement enlargement will be given by the proportion of the effective areas of the pistons to each other. With the trial product used here, the proportion is taken at 4, so ideally 4 times the amplitude enlargement is feasible. But the bulk modulus of the liquid chamber is finite, and part of the volume change is absorbed by deformation of the rubber, and thus the actual ratio of amplitude enlargement is about 2, as shown in Fig. 9.

The PA itself is adapted for an amplitude of $\pm 35 \mu m$, and this enlargement ratio is 2, so the total amplitude of the active mount system becomes $\pm 70 \mu m$. This value nearly satisfies the required value described in the preceding section, and it can be said that the design was achieved as hoped.

Adaptive Control Algorithm

This section discusses the method of preparing a control signal for the PA, including constrained adaptive control due to the limit in the output of PA. Several research works have reported on adaptive cancelers, where it is known that a frequency of periodic interference is present in the signal. [4]~[7] This report specially considers active cancellation of engine vibration using the previously mentioned control mounts.

The exciting force F_a from the engine is expressed by use of the angular frequency ω with respect to the explosion period of one cylinder, by formula (1).

$$F_a=\sum_{n=1}^{\infty} \alpha_n SIN(n\omega t+\delta_n) \quad (1)$$

Therefore, the force F_b transmitted to the chassis across the engine mount is expressed as

$$F_b=\sum_{n=1}^{\infty} \beta_n SIN(n\omega t+\eta_n) \quad (2)$$

Roughly speaking, it may be said that by creating a force of out-of-phase to F_b in formula (2), or $-F_b$, by the active mount, the transmitted force can be cancelled.

The desired sinusoidal wave may be produced by adequately adjusting the amplitudes a_n^s and a_n^c of the sine and cosine waves as shown in formula (3) below.

$$-\beta_n SIN(n\omega t+\eta_n)=a_n^S SIN(n\omega t)+a_n^C COS(n\omega t) \quad (3)$$

Based on the foregoing, the adaptive control algorithm will be specifically described with reference to a flow chart, Fig. 10. The force

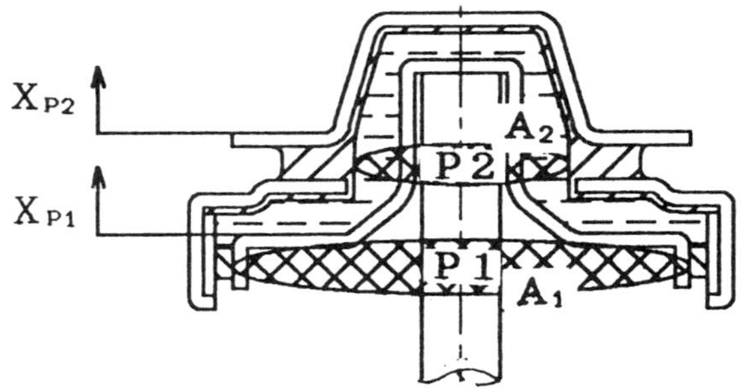

Fig. 8 Detail of amplitude enlargement mechanism of PA.

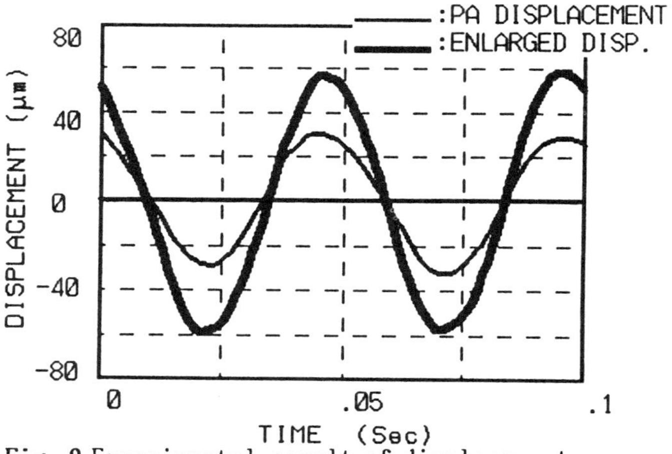

Fig. 9 Experimental result of displacement enlargement.

created by the actuator and the control voltage input to the amplifier, while not exactly the same, are substantially not different from each other, and the same symbols will be used in the explanation. Sine and cosine waves generated in unit amplitude from the ignition pulse are taken as reference signals. The error signal e(t) is the output of a sensor installed at the location where vibration is to be reduced to zero. Here, experiments were carried out with the acceleration on the chassis side at the point of installation of the engine mount. As algorithms for adjusting the amplitudes of the reference signals, integral algorithms of simplified formulas (4) of those proposed by Narendra et al. were applied. [8]

$$a_n^S=-G_n\int e(t)\, SIN(n\omega t)\, dt$$

$$a_n^C=-G_n\int e(t)\, COS(n\omega t)\, dt \quad (4)$$

where G_n is the corrected gain at the n-th order.

Control is via the signal obtained from formula (5) by multiplying the sequentially corrected amplitudes by the reference signals.

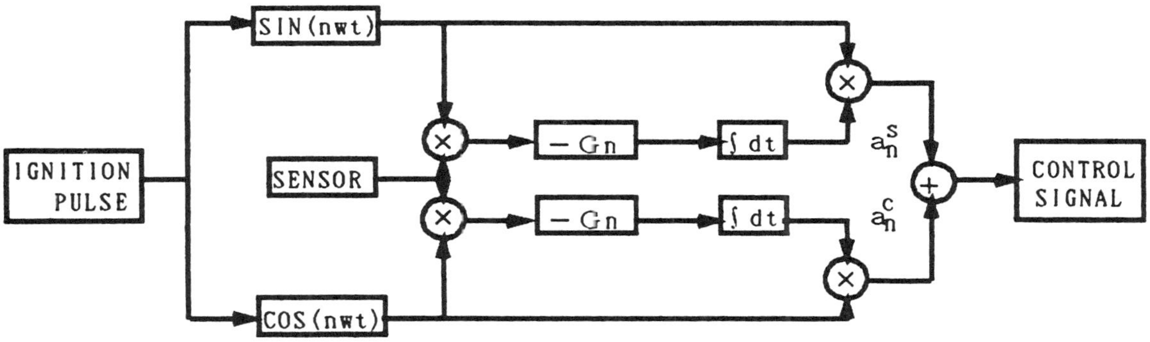

Fig. 10 Flow diagram of control algorithm.

$$a_n(t) = a_n^s SIN(n\omega t) + a_n^c COS(n\omega t) \qquad (5)$$

As the secondary vibration of the four-cylinder engine is taken up as an objective, n in the foregoing formula having the explosion pulse as a reference is given as 4. As is common knowledge, PA output is limited, and so if correction is made according to formula (4), the coefficients a_n^s and a_n^c will diverge. For constrained adaptive control, the following simple method was used. Where the amplitude of the sinusoidal control input value $a_n(t)$, or $\sqrt{((a_n^s)^2+(a_n^c)^2)}$ would exceed the output limit M of PA, the formula was modified as shown in (6) so that the coefficients a_n^s and a_n^c would not diverge.

$$a_n^s = a_n^s M/A, \; a_n^c = a_n^c M/A \quad if \quad A>M \qquad (6)$$

where $A = \sqrt{((a_n^s)^2+(a_n^c)^2)}$

The result of testing the active control mount for control of the idle vibration in vehicles according to the constrained adaptive control algorithm are given below.

Experiment System and Results

First, an overall outline of the experiment, shown in Fig. 11, will be described. The active mount is used as a front engine mount.

In synchronism with the ignition pulse of the engine, reference signals of frequencies having unit amplitudes in the forms of sine and cosine waves are produced in the second order of revolution of the engine and are input to the computer. To execute the algorithm described in the preceding section, the signal of an accelerometer provided on the chassis side of the active mount is taken as an error signal and input to the computer together with the foregoing reference signals. Operations are performed so as to reduce the error signal to minimum, and the resulting operation voltage signal is transmitted to the amplifier for control of the automobile vibration. For evaluation of the control effect, the following two experiments were carried out.

The first was to check the response of the active mount. The check was made through on-off of the control switch while the engine was revolving at a certain speed and from the changing condition of the error signal. Figs. 12 and 13 show differences in chassis vibration with and without vibration control at an engine speed of 750RPM. The lower graphs of Figs. 12 and 13 show respectively the time series of the control signal itself when it was transmitted to the PA amplifier, and the upper graphs show changes in error signal. AS seen from Fig. 13, the response speed from when the control is started until the attenuation effect is provided is very high at 0.1 second or less, and Fig. 14 shows the power spectrums at such time, the upper spectrum showing the data without control and the lower one the data with control. It can also be seen that for the vibration of the second order (25Hz), an effect of 10dB or more is achieved.

The other experiment was to examine follow-up and vibration reducing effects of control when the engine speed changed. A sweep test was conducted, sweeping the engine speed from 700RPM to 1200RPM, with the normal idle speed assumed to be 700RPM and the idle speed at engine start to be 1200RPM. The results of tracking analysis of the second order are show in Fig. 15. The thin curve represents without control, and the thick curve with control. A considerable damping effect of 10～20 dB over the whole range was observed.

From these results, it can be seen that the present control system with active PA mount and in combination with a simplified control algorithm is a very effective system.

Conclusion

In view of the increasing demand for comfortableness in passenger vehicles improving noise and vibration as well as the ride comfort is one of the most important areas of development for today's engineers.

It is expected that the active control will in the near future come into widespread use in addition to conventional passive control. PAs have some difficulties, such as very small amplitude and large temperature dependency but, at the same time, they have significant advantages, including high speed response and large output of force, and active PA control mounts are considered a very attractive option among noise and vibration control devices. In this report, PA application has been limited to idling vibration. But, the high speed response of PAs gives them considerable potential in higher frequency regions of booming noise and road noise as well. It is therefore anticipated that further improvement of the piezo actuator itself and development of control systems utilizing its features will be undertaken in the future.

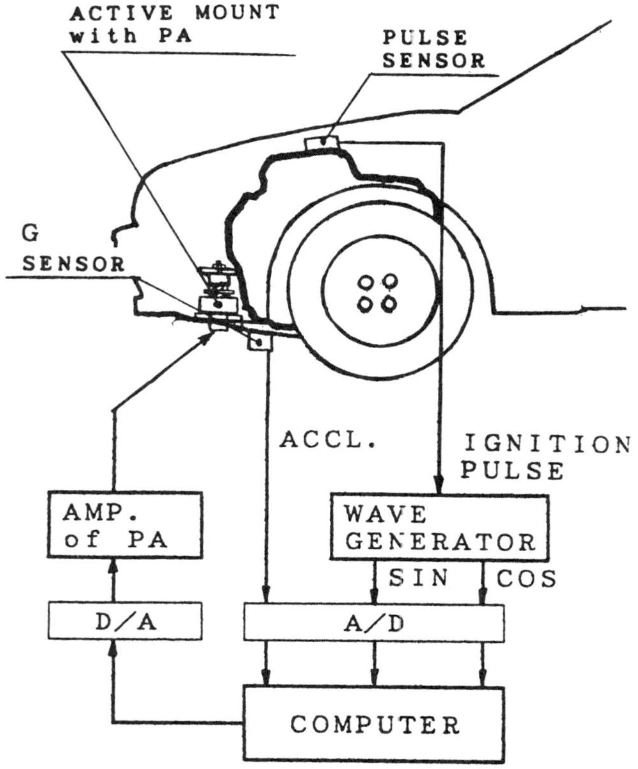

Fig. 11 Apparatus of the active control test using the vehicle.

Appendix:
Analytical Model of PA Mount.

The frequency region of idle vibration is the region where the exciting force of the engine and the force of inertia are equilibrium, and amplitude calculation is unaffected by the stiffness of the conventional engine mount. Therefore, the problem is reduced to the one of forced displacement.

The PA mount model of Fig. 16 can be expressed by a simplified equivalent kinematic model of Fig. 17. For the symbols used in these drawings, refer to the explanation given later.

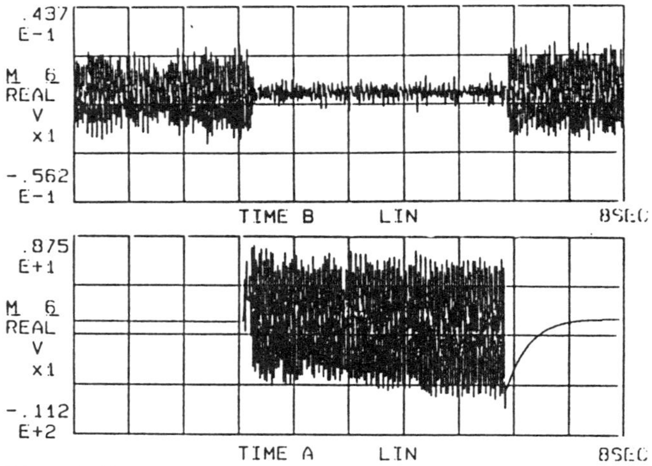

Fig. 12 Time series of chassis vibration with and without control at engine speed 750RPM.

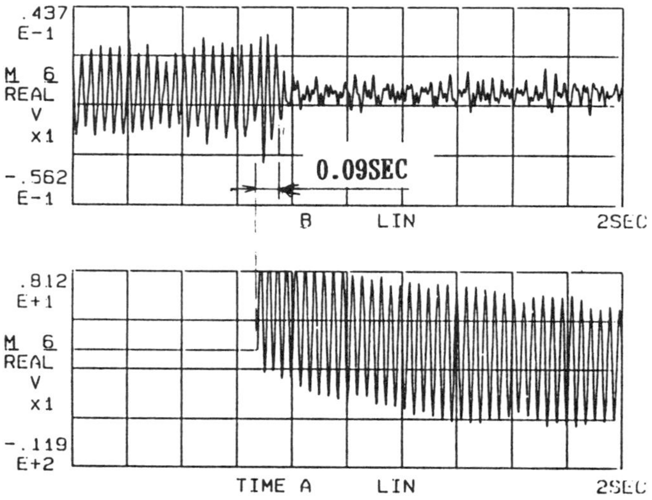

Fig. 13 Time axis elongation from 8sec to 2sec of the same test as Fig. 12.

If the amplitude of engine vibration is X_1, the object is to reduce to zero the force of transmission F to the chassis. Using the symbols in the drawing, the formula expressing the displacement as a function of the force is:

$$K_1(X_2-X_1) + K_2X_2 + K_V\triangle V = 0 \qquad (A1)$$

For the force transmitted to the chassis, the following relation holds:

$$-f_0 - K_2X_2 - K_3X_3 = F \qquad (A2)$$

Balancing of the force transmitted to the PA is expressed as

$$f_0 + K_3X_3 - K_V\triangle V = 0 \qquad (A3)$$

Substituting formula (A3) in formula (A2), the following formula is obtained:

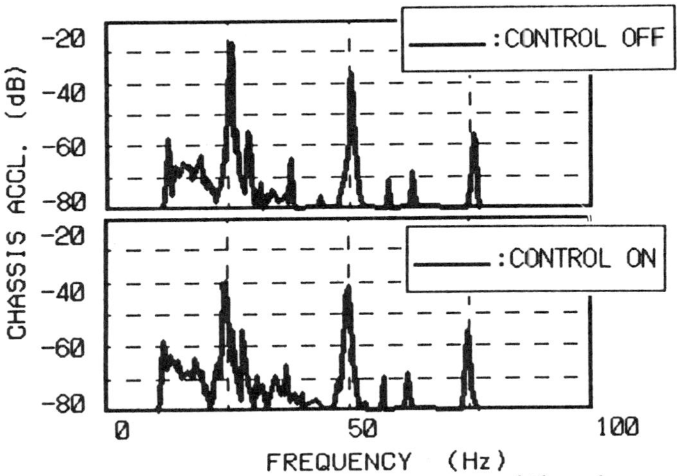

Fig. 14 Comparison of power spectrums with and without control at engine speed 750RPM.

$$F = -(K_2X_2 + K_V\Delta V) \quad \text{(A4)}$$

In order for F=0 to be valid, we must have:

$$K_2X_2 + K_V\Delta V = 0 \quad \text{(A5)}$$

From this formula (A5) and formula (A1), the following is obtained:

$$K_1(X_2 - X_1) = 0\text{, or } X_2 = X_1 \quad \text{(A6)}$$

Then, if PA is controlled so that X_2 is of the same displacement with the forced displacement X_1, it is possible to interrupt the idle vibration.
Here,

K_1 : Stiffness of conventional mount;
K_2 : Stiffness of main rubber in active part;
K_3 : Stiffness of sub-rubber in active part;
K_V : Equivalent bulk modulus of liquid chamber;
X_1, X_2, X_3: Displacements at illustrated points;
ΔV : Volume change of liquid chamber; and
f_0 : Control force of PA.

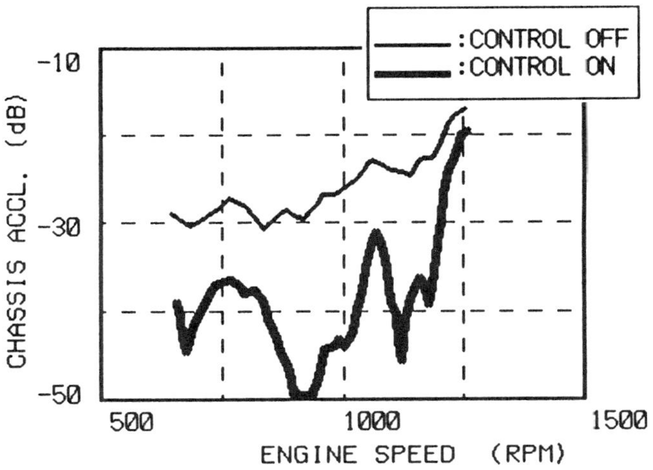

Fig. 15 Vibration reducing effects of active engine mount with Piezo Actuator.

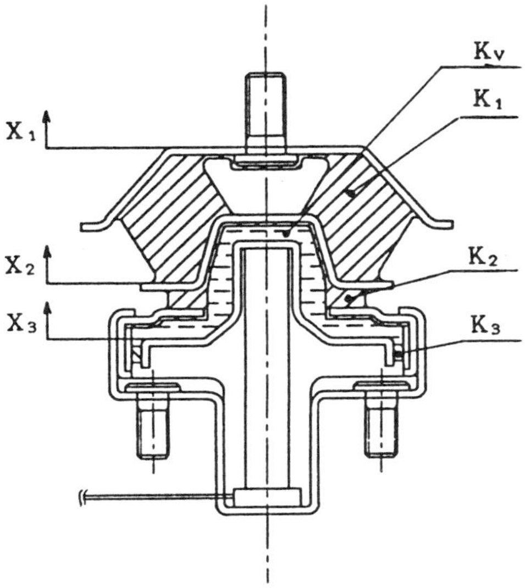

Fig. 16 Analytical model of active mount with Piezo Actuator.

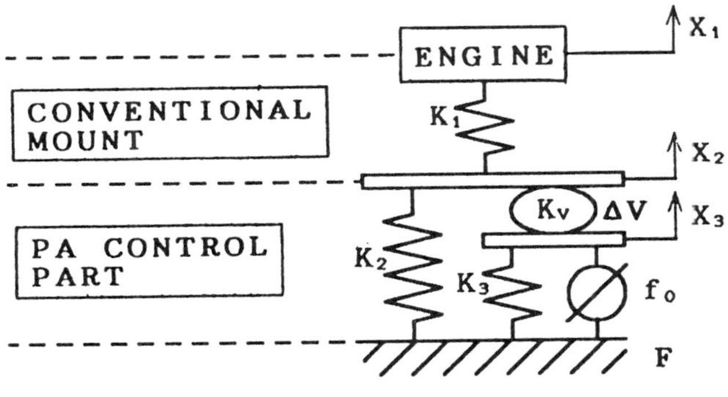

Fig. 17 Simplified kinematical model.

References

1. T. Ushijima, K. Takano & T. Noguchi, 'Rheological Characteristics of ER Fluids and Their Application to Anti-Vibration Devices with Control Mechanism for Automobiles,' SAE Paper 881787.
2. A. M. McDonald, S. J. Elliott & M. A. Stokes, 'Active Noise and Vibration Control within the Automobile,' International Symposium on Active Control of Sound and Vibration, April 9-11, 1991, Pages 147-157.
3. T. Inoi et al., 'Medium Size Metal Case Type Piezoelectric Actuator,' NEC Technical Report, Vol. 43, No. 10, 1990, Pages 130-135.

4. B. Widrow et al., 'Adaptive Noise Cancelling; Principles and Applications,' Proc. IEEE, Vol. 63, 1975, Pages 1692-1716.
5. J.R. Glover, 'Adaptive Noise Cancelling applied to Sinusoidal Interference,' IEEE Trans. Acoust., Speech, Signal Processing, Vol. ASSP-25, 1977, Pages 484-491.
6. S.J. Elliott and P. Darlington, 'Adaptive Cancellation of Periodic, Synchronously Sampled Signals,', Speech, Signal Processing, Vol. ASSP-33, 1985, Pages 715-717.
7. S.J. Elliott, I.M. Stothers and P.A. Nelson, 'A Multiple Error LMS Algorithm and Its Application to the Active Control of Sound and Vibration,' IEEE Trans. Acoust., Speech, Signal Processing, Vol. ASSP-33, 1987, Pages 1423-1434.
8. Y. Lin and K.S. Narendra, 'A New Model for Adaptive System,' IEEE Trans., AC-25-3, 1980, Pages 585-587.

922290

An Electronically Controlled Engine Mount Using Electro-Rheological Fluid

Shin Morishita
Yokohama National Univ.

Jun'ichi Mitsui
Tonen Corp. Research Center

ABSTRACT

This paper describes a newly developed prototype engine mount which is electronically controlled for variable spring and damping force. The controllability was accomplished by applying Electro-Rheological (ER) fluid whose apparent viscosity can be varied by applied electric field strength. Main attractive characteristics of ER fluid are wide variation of its apparent viscosity and quick response time, and one of the promising applications is a controllable damper. In the present paper, a prototype engine mount was constructed and its typical performance was investigated, where all the experiments were conducted under constant vibration amplitude to examine the amplitude dependence of the ER characteristics on ER devices. On the basis of the experimental results, numerical simulations of vibration control with ER engine mount and an adaptive neural-net control system were conducted. It is concluded that the stiffness as well as damping capacity of the present engine mount can be varied by controlling the applied electric field strength. Furthermore, an idealized model was proposed for the present type of engine mount and numerical simulations with the idealized model show the possibility of engine noise reduction with the present control system.

INTRODUCTION

For the reduction of engine noise into the compartment of an automobile, the base isolation of engine vibration is an effective technique. Various kinds of engine mounts such as rubber type, hydraulic type, or hydroelastic type have been developed for this purpose[1-4]. This type of controllable engine mount is expected to be effective not only for the isolation of the engine vibration to the supporting frame, but also the vibration reduction of the automobile caused by the road noise. So the damping as well as the stiffness properties should be varied in accordance with the rotational speed of the engine and, if possible, the condition of the surface of the road. Several authors have reported on controllable engine mounts which were application studies of ER fluid[5]-[7], and concluded that the stiffness and damping characteristics of the ER engine mount can be controlled by the applied electric control signal.

ER fluid is known as a class of highly functional material whose apparent viscosity can be varied in response to the applied electric field[8]-[11]. The attractive characteristics of ER fluid are wide variation of apparent viscosity and quick response time. The viscosity can be varied several tens times compared with the value obtained without any electric field when subjected to some electric field[9]. The response time is said to be as short as several milliseconds.

The authors have been trying to apply ER fluid to mechanical components such as shock absorber system, squeeze film damper bearing, and damping-controllable dynamic damper. The controllable shock absorber system was proved to be effective for improvement of drivability of automobiles[12]. High speed flexible rotor could pass through its critical speeds with the controllable squeeze film damper quite smoothly by getting most appropriate supporting damping for each vibration mode of the rotor [13]. The damping-controllable dynamic damper could reduce the vibration response of a structure simultaneously for several modes with a single dynamic damper [14]. Each application has a damping-controllable component using ER fluid.

In the present paper, a prototype engine mount using ER fluid was constructed and its performance was investigated. The experiment showed that the stiffness as well as the damping properties was controllable. Detailed investigation of the effect of vibration amplitude on the damping characteristics was made. On the basis of the characteristics obtained in the experiment, the present engine mount was modeled as a so-called three-element system. Furthermore, a numerical simulation of single-degree-of-freedom system, which is a simplified model of the present vibration control system with the ER engine mount and an adaptive neural-net controller, is described in the present paper.

EXPERIMENT

CONTROLLABLE ENGINE MOUNT - Figure 1 shows the construction of the present prototype engine mount, which is composed of a top plate, a coil spring, a rubber cover, electrodes, and a diaphragm. The height of the present mount, from the bottom to the top plate, is about 120 mm, and the maximum diameter is 100 mm. The stiffness of the coil spring is 2x104 N/m, and that of the rubber cover is 2x103 N/m in vertical direction.

Inside of the isolator is filled with ER fluid. The base viscosity of the present ER fluid is 0.065 Pa·s at 40 °C, and fine semiconducting particles whose average diameter is 1 mm are dispersed in a base oil.

The top plate is to support the engine on it, and set up on a coil spring, which works as the main stiffness of the mount. Around the coil spring, a rubber cover is arranged for the following two purposes: one is to transmit the volume change of the upper part of the mount to the lower part through the electrodes, and the other is to prevent ER fluid from spilling out of the mount. If the volume change of the upper part of the isolator is not transmitted to the lower part, ER fluid would not pass through the electrodes, then no damping effect would be produced. This volume change is guaranteed by the flexibility of the diaphragm set up at the bottom of the mount.

The detailed photo of the electrodes is shown in Fig.2. The electrodes are made of polyimide resin plated with thin layer of copper, and placed parallel to each other with spacers made of polytetrafluor-ethylene (Teflon). Each electrode is 40 mm in length and 10 mm in width, and the total thickness of one electrode is 0.47 mm including the thickness of polyimide resin of 0.4 mm and the layer of copper of 35 micrometer on both sides. Two kinds of spacers are prepared, 0.4 and 0.8 mm, to investigate the spacing effects on the ER engine mount properties.

All the components excluding electrodes, a rubber cover, coil spring, diaphragm, O-ring, and screw bolts for the present mount are made of polyetherimide resin, a class of engineering plastics, having sufficient stiffness and electric insulation.

EXPERIMENTAL APPARATUS - The experimental apparatus is shown in Fig.3. A mass of 5 kg is put on the engine mount, and the mount is set up on an electromagnetic exciter. The mount is excited vertically by sinusoidal waves produced by a signal generator. The frequency of the exciting signal is varied from 5 Hz to 35 Hz.

To the authors' best knowledge, there have been few investigations of the amplitude effect on ER characteristics, so the present experiments are performed under constant relative amplitude, 0.5 and 0.1 mm, between the mount and the weight. The relative displacement is measured by an eddy-current displacement sensor whose frequency range is from DC to 20 kHz, and the sensitivity is 5 V when displaced 2.0 mm from a target.

The responses to the excitation are measured at

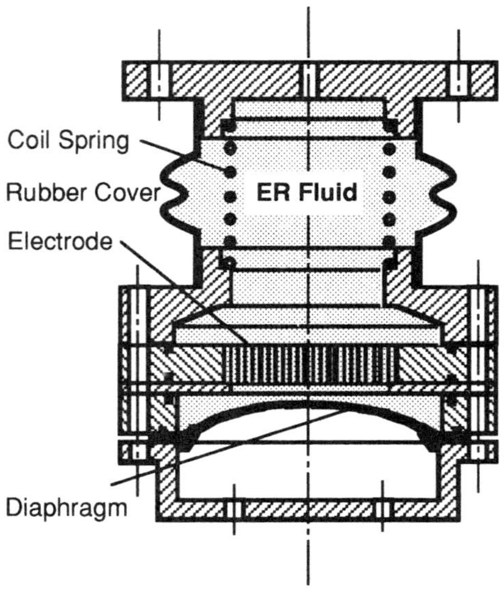

Fig.1 Schematic view of ER engine mount

Fig.2 Electrodes

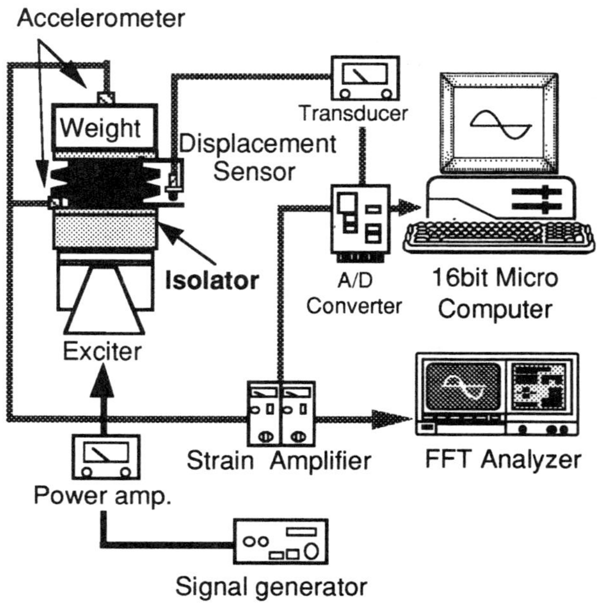

Fig.3 Experimental apparatus

the bottom and the top of the mount by the accelerometers, and the magnification factor is estimated with them.

Because the ER effect is a function of temperature[8,9], temperature control is required to investigate the performance of the present engine mount. The temperature is controlled at 40 °C by blowing hot air heated by a dryer. For the measurement of the temperature of the ER fluid in the mount, a C-C thermocouple is set just behind the electrodes.

The experimental conditions, such as the weight of the mass and the environmental temperature, are not practical for the actual engine mounts, however, these conditions are selected for easy operation because the purpose of this experiment is to get typical characteristics of the ER engine mount.

All the signals from the accelerometers and the displacement sensor are amplified up to +10 V and introduced into a 16 bit microcomputer through a 12 bit A/D converter in the usual manner. The signals put into the microcomputer are processed almost in a real time.

EXPERIMENTAL RESULTS

Experimental results are shown from Fig.4 to Fig.7. Each figure presents the response curve at different electrode-spacing and amplitude.

TYPICAL CHARACTERISTICS OF ER ENGINE MOUNT - Figure 4 shows the response curve measured in the present experiment at the relative amplitude of 0.1 mm when the electrode-spacing was 0.8 mm. Without any electric field, the magnification factor got to the maximum value at the resonance frequency, and decreased as the frequency was increased, which shows that the response was almost the same as tha of 1 degree-of-freedom system. As the electric field strength was increased up to 625 V/mm, the peak value of the response became lower and the shape c the peak became round. This was caused by the en ergy dissipation resulting from the friction of the flow through the electrodes, and the system damping wa increased. At the same time, the peak frequency shifted to higher value from 15Hz to 16.5Hz.

At the field strength of 625 and 875 V/mm, ther appeared two different peaks, which indicated that th system could be modeled by two-degree-of-freedom system. As the field strength was increased further, th steep peak grew gradually, and the system damping was estimated to become smaller. The peak frequenc was about 18 Hz.

The reason why the system damping became sma when the applied electric field strength was greater than 875 V/mm can be explained as follows: As the strength of applied electric field was increased, the apparent viscosity of the ER fluid through the electrodes becomes higher and the resistance to flow wa increased. Because of this resistance, the total volum of the flow through the electrodes was limited to a cer tain degree, and the system damping caused by the EF effect was increased. However, beyond a certain leve of the electric field strength applied, the ER fluid wa stuffed between the electrodes like chains or bridges c particles dispersed in the fluid and there might be littl flow through the electrodes, which caused less systen damping.

In this state, the volume change in the upper part c the mount caused by the change in length of the co

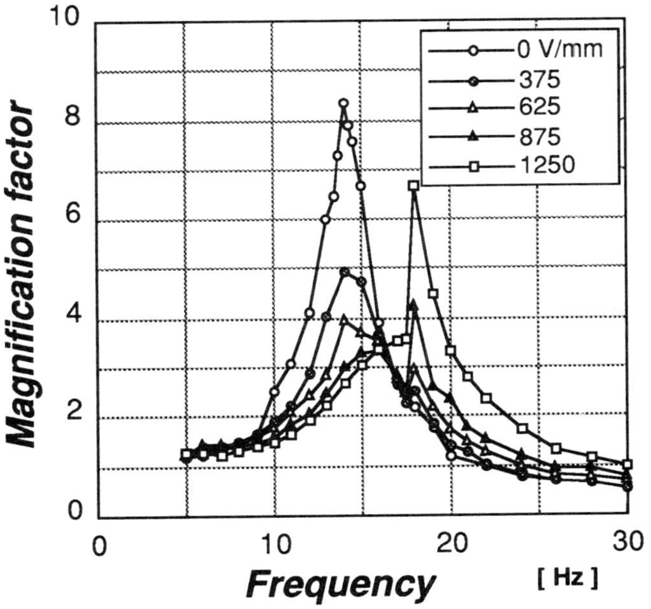

Fig.4 Response curve
(Spacing=0.8 mm, amp.=0.1 mm)

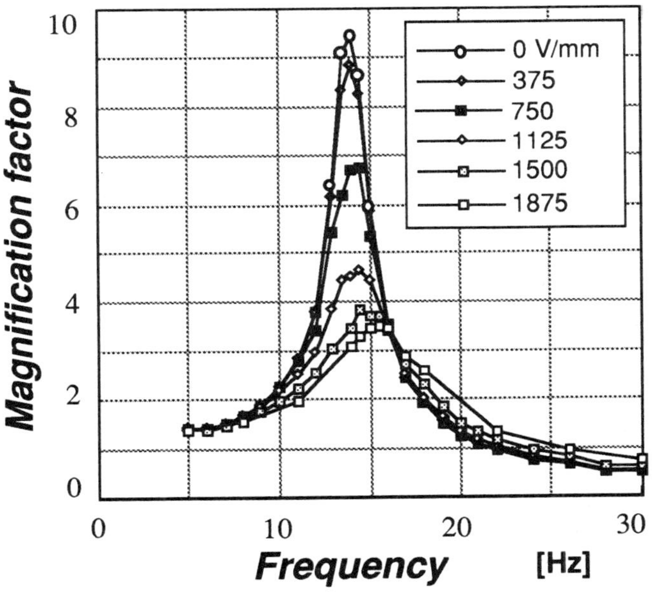

Fig.5 Response curve
(Spacing=0.8 mm, amp.=0.5 mm)

spring was absorbed by deformation of the rubber cover, and the mount oscillated in the upper part of the mount. Though the vertical stiffness of the rubber cover was very small in the first stage, the pressure change in the rubber cover caused a membrane tension in the rubber, and the effective total stiffness of the cover became large.

AMPLITUDE EFFECT - Figure 5 shows the results when the vibration amplitude was 0.5 mm, at the same electrode-spacing as in Fig.4, 0.8 mm. The peak value of the response curve at zero field was a little larger in Fig.5 than in Fig.4, and as the field strength was increased, the peak value became small, which means that system damping became large. However, the applied electric field contribution to the system damping was much smaller in Fig.5 than in Fig.4 at the same level of the electric field strength, and that, in Fig.5, there did not appear another steep peak like in Fig.4 even when the applied electric field strength was more than 1500 V/mm.

These phenomena may be caused by the typical mechanism of the ER effect. Though the mechanism of the ER effect has not yet been described clearly in detail, the ER effect is believed to be produced by the tension force of the chain structures of particles between the electrodes. If this description is true, these chains can be easily broken by strong flow through the electrodes. This is because the system damping was smaller at large vibration amplitude without any electric field, and because the system damping did not become small at the vibration amplitude of 0.5 mm. even when the applied electric field was strong. The ER effect was not strong enough to prevent the fluid from passing through the electrodes.

ELECTRODES SPACING EFFECTS - Figures 6 and 7 show the results when the spacing of electrodes wa 0.4 mm at the relative amplitude of 0.1 and 0.5 mm, re spectively. As shown in Fig.6, the response curve wa almost the same as that of single-degree-of-freedom system at zero electric field. As the field strength wa increased, the peak value of the response curve became small up to the field strength of 750 V/mm. An as the field strength was increased further, the peak value grew gradually and shifted to higher frequency about 20 Hz. In contrast to the results shown in Fig.4 the responses were the same as a single-degree-of-freedom system.

Comparing the results in Fig.6 with those in Fig.4 the peak value at zero field is much smaller in Fig.6 than in Fig.4. This is because the friction loss of th flow through the electrodes is much larger at the spac ing of 0.4 mm than 0.8 mm. From the results shown i both figures, the peak values of the response curve were minimum at almost the same electric field strength, from 625 to 875 V/mm in Fig.4 and 750 V/mn in Fig.6, which means that the ER effect was independ ent of the spacing of electrodes at the vibration ampli tude of 0.1 mm.

On the contrary, the results shown in Figs.5 and are a little different. The peak value of the respons curve was minimum at the field strength of 1500 V/mn in Fig.7, however, as shown in Fig.5, the peak value continued to decrease even when the electric field strength was 1875 V/mm. It is impossible to describ this phenomenon clearly from just these experimenta results, and it can be estimated that it is closely relate with the typical mechanism of the ER effect.

From the results shown in Figs.6 and 7, the sam discussions are possible with those in the previous section. When the vibration amplitude is large, the E effect contribute much less to the system damping o the isolator.

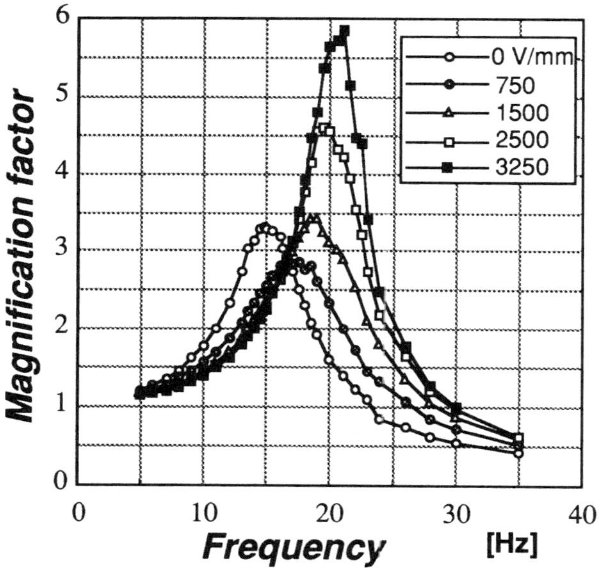

Fig.6 Response curve
(Spacing=0.4 mm, amp.=0.1 mm)

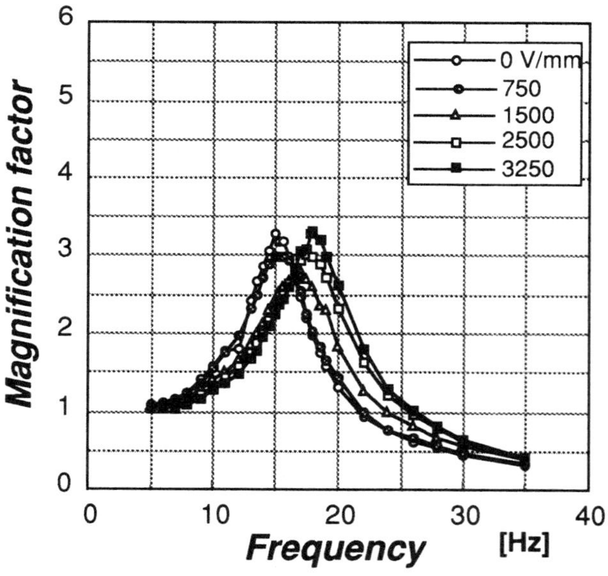

Fig.7 Response curve
(Spacing=0.4 mm, amp.=0.5 mm)

EQUIVALENT MODEL - There have been proposed various kinds of equivalent models of engine mounts[1-4], and hydraulic engine mounts can be modeled by a two-degree-of-freedom system taking account of the inertia force of the liquid. When the spacing of the electrodes is 0.8 mm, the equivalent model can be constructed according to the previous studies, however, the results shown in Figs.6 and 7 are quite different from the previous authors' results. This is caused by the flexible rubber cover around the coil spring.

The experimental results shown in Figs.6 and 7 are also quite different from those obtained in the experiment of ER engine mount by Petek et al.[5], Ushijima et al.[7], or Seto et al.[15], so the authors made an equivalent model of the present isolator as follows. According to the experimental results, the response was quite similar to that of so-called three-element system, a modified single-degree-of-freedom system. The model of three-element system is shown in Fig.8. The system is composed of a mass, a main spring A, a damping component, and a sub-spring B which is connected with the damping component in series. The response curve of this system is shown in Fig.9. The damping component represents the damping characteristic produced by the ER effect. The main spring A of the model system corresponds to the coil spring of the isolator, and sub-spring B to the stiffness of the rubber cover.

The sub-spring B is placed with the damping component in series in spite of the fact that the rubber cover is arranged in a line with the coil spring and that the rubber cover is not placed in a series with the damper. However, this inconsistency can be explained as follows: As the strength of applied electric field is increased, the apparent viscosity of the ER fluid through the electrodes becomes higher and the resistance to flow is increased. Because of this resistance, the total volume of the flow through the electrodes is limited to a certain degree, and the system damping caused by the ER effect is increased. However, beyond a certain level of the electric field strength applied, the ER fluid is stuffed between the electrodes. In this state, there may be little flow through the electrodes, and the volume change in the upper part of the isolator caused by the change in length of the coil spring is absorbed by deformation of the rubber cover, and the isolator tends to oscillate in the upper part of the isolator. Though the vertical stiffness of the rubber cover is very small in the first stage, the pressure change in the rubber cover causes a membrane tension in the rubber, then the effective total stiffness of the cover becomes large. Without any electric field applied, the eigenfrequency of the isolator is calculated from the mass and stiffness of the coil spring and a rubber cover, and is estimated around 14 Hz. The vertical stiffness of the rubber cover is $2x10^3$ N/m, which is 10 percent stiffness of the coil spring, and leads to only 5 percents increase in the eigenfrequency determined by the mass and the stiffness of the coil spring.

This is because the eigenfrequency increased as the electric field strength was increased. From the results shown later that the eigenfrequency was about 20 Hz with an electric field, the effective total stiffness is estimated twice as large as that of the coil spring.

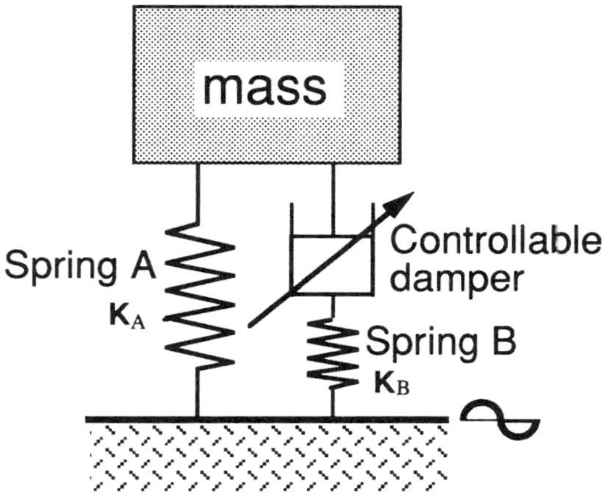

Fig.8 Equivalent model

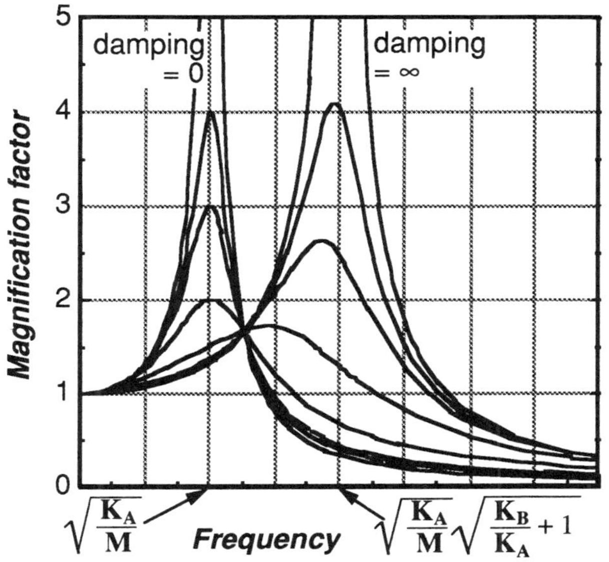

Fig.9 Response of the equivalent model

VIBRATION CONTROL SIMULATION

ADAPTIVE NEURAL-NET VIBRATION CONTROL SYSTEM - For the evaluation of the ER engine mount, simple simulations were conducted with a single-degree-of-freedom system controlled by an adaptive neural-net control system.

Figure 10 shows a schematic structure of the present neural-net adaptive control system. The present control system was composed of a structural object, a forward model network, a controller network and an

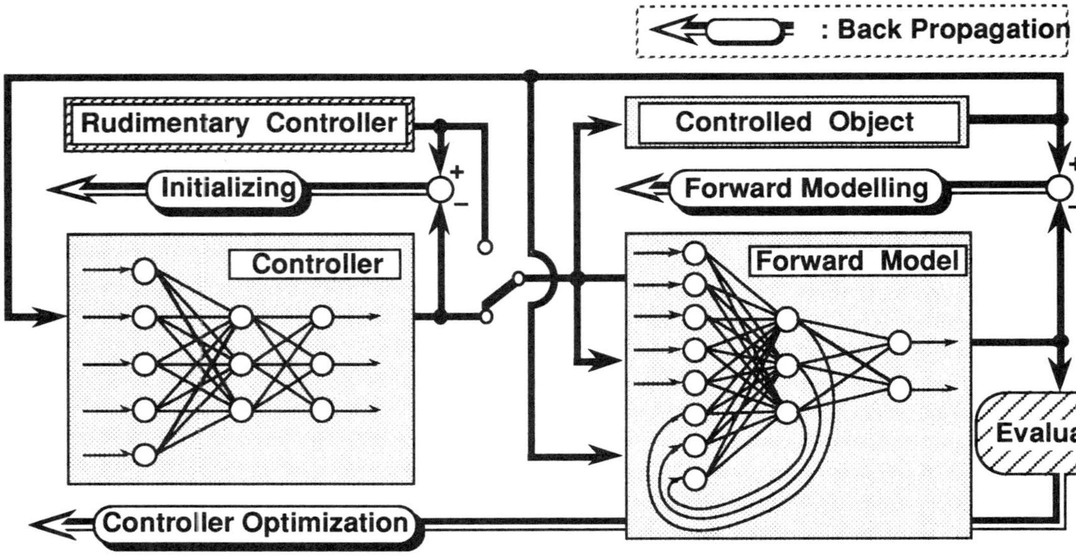

Fig.10 Adaptive neural-net control system

adaptation mechanism. This controller was an application of the neural network controller for underwater vehicles developed by Fujii and Ura [16, 17].

The forward model network represented dynamics of a structure including the ER engine mount, and was composed of three layers of neurons; input, hidden, and output layers. The input signals to the network were accelerations of the engine at the previous time step and corresponding control signal to the ER engine mount. Because the Connectionist Model, which represents the connection between the neurons in different layers, can only express a static system, recurrent connections from the hidden layer to the input layer were important to deal with the time historical order of signals in a dynamic system as shown in Fig.10. Comparing the output signals with the actual ones and making the criterion function minimized, the forward model network could be modified by conventional back propagation algorithm.

The controller network was also composed of three layered neurons whose inputs were accelerations of the engine and the control signal calculated by the controller network. When the minimized output from the forward model was greater than the threshold value,Tc, the controller could be optimized by regarding the controller and the forward model as a single six-layered neural network; this procedure including forward model modification was defined as the adaptation mechanism.

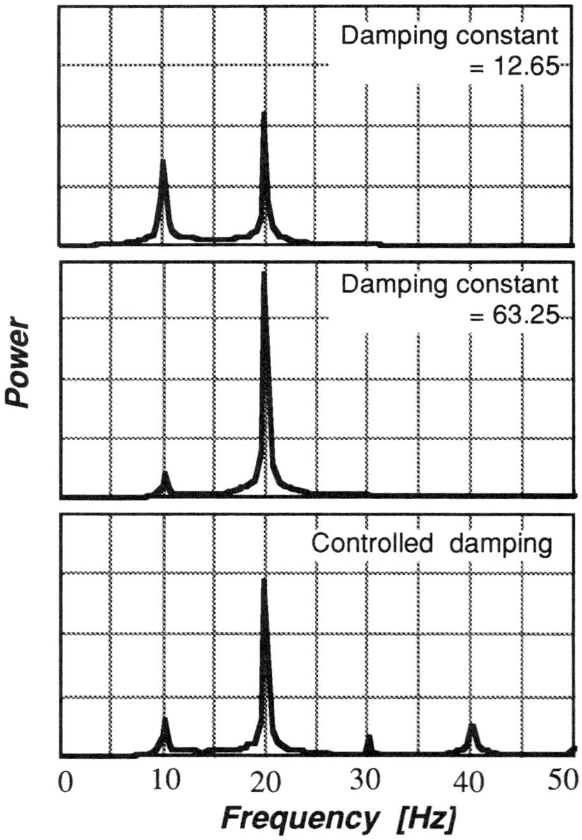

Fig.11 Simulation results

NUMERICAL SIMULATIONS - A numerical simulation was performed with a single-degree-of-freedom system when a forced oscillation was given to the base of the system in two different frequencies simultaneously. The controllable mount was set up to support the mass of the system. The purpose of this simulation is to show the possibility to decrease the response in several frequencies with the aid of a single controllable mount. Two frequencies included in the base oscillation were 10 and 20 Hz.

A neural network generally needs a certain preknowledge before starting the learning process. This approach was found to be effective not only for quick learning but also for avoiding the influence of initial condition of the network on the learning process. So it was necessary to make a rudimentary controller by means of a certain control rule. In the present case, the following rule was chosen:

◊ if $\mathrm{sign}(x_b) = \mathrm{sign}(x_m)$ and $|x_b| < |x_m|$
then control signal, V , should be smaller.
◊ in other case V should be larger.

One of the simulation results is shown in Fig.11 in frequency domain. It was shown that the amplitude in each frequency could be decreased by controlling the damping and the stiffness properties with the control system. With the present control system, even when the engine of an automobile is oscillated in different frequencies such as combination of the engine vibration and the road noise, the neural network can learn the condition by itself and make the appropriate signal for the controllable engine mount.

CONCLUSIONS

In the present paper, an application study of ER fluid to an engine mount is investigated, in which a prototype mount was constructed and its performance was demonstrated experimentally. An equivalent model of the present mount was also discussed, and a numerical simulation of the present engine mount with an adaptive neural-net control system was performed. The conclusions obtained are as follows:

(1) The stiffness as well as the damping characteristics of the present engine mount can be controlled by the applied electric field strength. It was shown that the equivalent model was a so-called three-element system which was composed of a mass, a main spring, a damping component in a line with the spring, and a sub-spring connected to the damping component in a series.

(2) It was shown experimentally that the vibration amplitude affects the characteristics of a controllable damper using ER fluid. This may be caused by breaking the chain structure of the particles dispersed in the fluid by the flow passing through the electrodes.

(3) The present adaptive neural-net control system is effective for the controller of ER engine mount.

The performance of the present type engine mount can be improved by choosing stiffness of the rubber cover, viscosity of ER fluid without electric field, and spacing of the electrodes more properly.

REFERENCES

[1] Bernuchon,M.,1984,"A New Generation of Engine Mounts", SAE paper #840259

[2] Flower,W.C.,1985,"Understanding Hydraulic Mounts for Improved Vehicle Noise, Vibration and Ride Qualities", SAE paper #850975

[3] Clark,M.,1985,"Hydraulic Engine Mount Isolation" SAE paper #851650

[4] Tokushige,M., Funahashi,Y., and Katayama,M., 1990, "New Technology of Antivibration Rubber Products for Automobiles", Japanese Automotive Engineering, Vol.44, No.12, pp.24-30 (in Japanese)

[5] Duclos,T.G.,1987,"An Externally Tunable Hydraulic Mount Which Uses Electro-rheological Fluid", SAE paper,#870963

[6] Petek,N.K., Goudie,R.J., and Boyle,F.P., 1988,"Actively Controlled Damping in Electro-rheological Fluid Filled Engine Mounts",SAE paper #881785

[7] Ushijima,T., Takano,K., and Noguchi,T.,1988, "Rheological Characteristics of ER Fluids and Their Application to Anti-vibration Devices with Control Mechanism for Automobiles",SAE paper #881787

[8] Winslow,W.M.,1949,"Induced Fibration of Suspensions", Journal of Applied Physics, Vol.20 , pp.1137-1140

[9] Klass D.L. and Martinek T.W.,1967,"Electroviscous fluids. I. Rheological Properties",Journal of Applied Physics, Vol.38, No.1, pp.67-74

[10] Block,H. and Kelly,J.P.,1988,"Electro-rheology", Journal of Physics D: Applied Physics, Vol. 21, No.1? ,pp.1661-1677

[11] Jordan T.C. and Shaw, M.T.,1989, "Electrorheology", IEEE Transactions on Electrical Insulation, Vol.24, No.5,pp.849-878

[12] Morishita,S. and Mitsui,J.,1991,"Controllable Shock Absorber System (An Application of Electro-Rheological Fluid)",SAE paper#910744

[13] Morishita,S. and Mitsui,J.,1992, "Controllable Squeeze Film Damper as an Application of Electro-Rheological Fluid",Transaction of ASME, Journal of Vibration and Acoustics, Vol.114, No.3, pp.354-357

[14] Morishita,S., Mitsui,J., and Kuroda,Y.,1991, "Controllable Dynamic Damper as an Application of Electro Rheological Fluid",Proceedings of PVP Conference, ASME, PVP-211,pp.1-6

[15] Seto,K.,Nagamatsu,A.,Ishihama,M.,Doi,M., and Sawatari,K.,1991,"Optimum Design Method for Hydraulic Engine Mount",Transaction of JSME,C,Vol.57, No.534,pp.453-459

[16] Fujii,T. and Ura,T., 1990, "Development of motion control system for AUV using neural nets", Proc. IEEE Symp. on AUV Tech., Washinton DC, pp.81-86

[17] Fujii,T., Ura,T., and Kuroda,Y., 1990," Development of self-organizing neural-net-controller system and its application to underwater vehicles", J Soc. Naval Arch. Japan, Vol.168, pp.275-281

951324

Active Control of Simulated Structure Borne Road Noise Using Force Actuators

Wouter Dehandschutter, Renaat Van Cauter, and Paul Sas
K. U. Leuven

ABSTRACT

A new approach to the reduction of structure borne road noise inside car cabins is developed and verified experimentally. The experiments have been performed on a mid-size station wagon, in laboratory conditions with one wheel excitation. In the frequency range from 100 to 150 Hz, more than 10 dB noise reduction has been achieved using control forces applied to the car body. For maximum noise reduction the control configuration had to be tuned carefully to satisfy the causality constraints for the reduction of broadband disturbances.

INTRODUCTION

Road noise consists of two components, namely air borne and structure borne road noise. Air borne road noise originates at the tire-road interaction, whereas structure borne road noise inside automotive vehicles is generated by road induced excitation forces. This road excitation is transferred to the car body through various transfer paths, and thus generates structure borne noise inside the car cabin (physically, these transfer paths are the connections of the suspension to the car body, i.e. the so-called bushings). All interior road noise below 600 Hz is mainly structure borne. The aim of the work presented here is to reduce the structure borne noise level in the car cabin by modifying the vibration behaviour of the car body using active control techniques.

Active control techniques have already successfully been applied to reduce engine noise which is, however, a pure harmonic disturbance with a fundamental frequency, determined by the engine rpm. The main difference with road noise lies in the randomlike nature of the phenomenon. The road disturbance at the wheels has a *broadband* frequency content, while engine noise is periodic with a fundamental frequency and some harmonics.

When dealing with structure borne road noise the wheel suspension and the bushings play a crucial role. The final design of a car suspension, however, is determined by a compromise between road handling and the acoustic comfort of the passengers. Mainly for road handling purposes, all suspension systems (passive, semi-active, or active) have a limited response bandwidth (0 - 25Hz). Above this bandwidth, all road inputs are transmitted to the car body, which is pernicious for the interior noise.

To improve the acoustic comfort, vibration transmission in the range of 50 to 300Hz should be further reduced. Current solutions for the reduction of structure borne noise inside a car interior rely mainly on the use of rubber bushings ("silent-blocks") at the connection of the suspension elements with the car body. The need for high rigidity at lower frequencies, as required for road handling purposes, in turn limits the isolation capabilities of these bushings. In other words, passive solutions are often not effective.

Due to these limitations there is a clear interest among car manufacturers to search for more efficient techniques, such as active control techniques. Active control techniques differ from passive solutions in that an actuator is used to impose a desired sound or vibration field (e.g. a reduced sound pressure or vibration field). The actuator can be used either to excite the cavity acoustically (with loudspeakers) or to excite the structure with vibration inputs (using shakers).

In the first case, a controller drives a loudspeaker to generate a sound field inside the car cabin that reduces the original sound field. This solution is known as Active Noise Control (ANC). ANC intends to reduce the noise level by generating an acoustic signal with controlled amplitude and phase such that it cancels out the original noise signal. ANC has been successfully applied to the reduction of engine noise in cars and aircraft [6]. Some research on road noise has already been done in this area [9, 10], yielding 7 dB A weighted noise reduction in the range from 100 to 200 Hz.

Active Structural Acoustic Control (ASAC) intends to reduce the noise level by modifying the vibration behaviour of the radiating structure. This can be accomplished using vibration actuators which are connected to the radiating structure. One possibility is to locate the actuators at the most sensitive positions of the vibration transmission, i.e. on the main transfer paths. This approach is fundamentally different from ANC.

This paper presents a feasibility study of Active Structural Acoustic Control of structure borne road noise. The aim is to quantify and validate the potential control performance by means of simulations and experiments in laboratory conditions.

THE EXPERIMENTAL SET-UP

The control objective of ASAC is to reduce the interior noise level by applying control forces to the car body. This is accomplished is by connecting control actuators (shakers) to the car body to modify the vibration behaviour of the car body such that the noise level inside the car cabin is reduced. The control actuators are located at the main paths of vibration transmission (vibration transmission from the suspension to the car body). For the demonstrator car the main transfer paths for the vibration energy flow from the four wheels to the car body have been determined experimentally beforehand by transfer path analysis [11], [12].

To simulate a random road input to the car body, one wheel of the demonstrator car is excited by an electrodynamic shaker (Figure 1). This set-up is a simplification of the real-life situation in the sense that only one wheel is excited in only one direction, while in reality, all four wheels are excited theoretically in 6 directions (3 forces and 3 moments).

Figure 1 presents a schematic overview of the test set-up. The controller processes a reference signal, calculates the control signal and sends it to the control actuator. An error microphone inside the car cabin is used to measure the resulting sound level, which indicates to what extent the controller settings should be further optimised. The elements of the set-up are detailed hereafter :

CONTROL ALGORITHM - An adaptive feedforward controller is used to determine the appropriate actuator signals (a block diagram is presented in Figure 2). The control signals are determined by feeding forward the reference signals through the control filter. The control filter in turn is adaptively optimised using a filtered-X LMS (Least Mean Square) update scheme. This algorithm reduces the least mean square error at the error microphones using information of both the error and reference signals ; and a model of the transfer function of each of the control actuators to the error sensors (so-called secondary paths). These secondary paths are modeled beforehand ("Identification part", figure 2) using either Finite Impulse Response (FIR) or Infinite Impulse Response (IIR) filters. A full description of the algorithm and its robustness properties is given in [6].

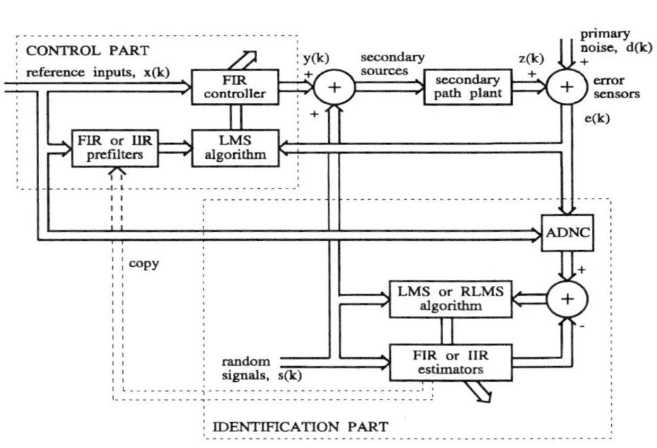

Figure 1. Block diagram of the Filtered-X LMS feedforward control algorithm.

REFERENCE SIGNAL - The choice of an appropriate reference input is of fundamental importance to the final controller performance. In the first place, a high coherence between the reference detection sensor and the error sensor is required. This is necessary because the information in the reference signal must be representative for the noise at the error sensor.

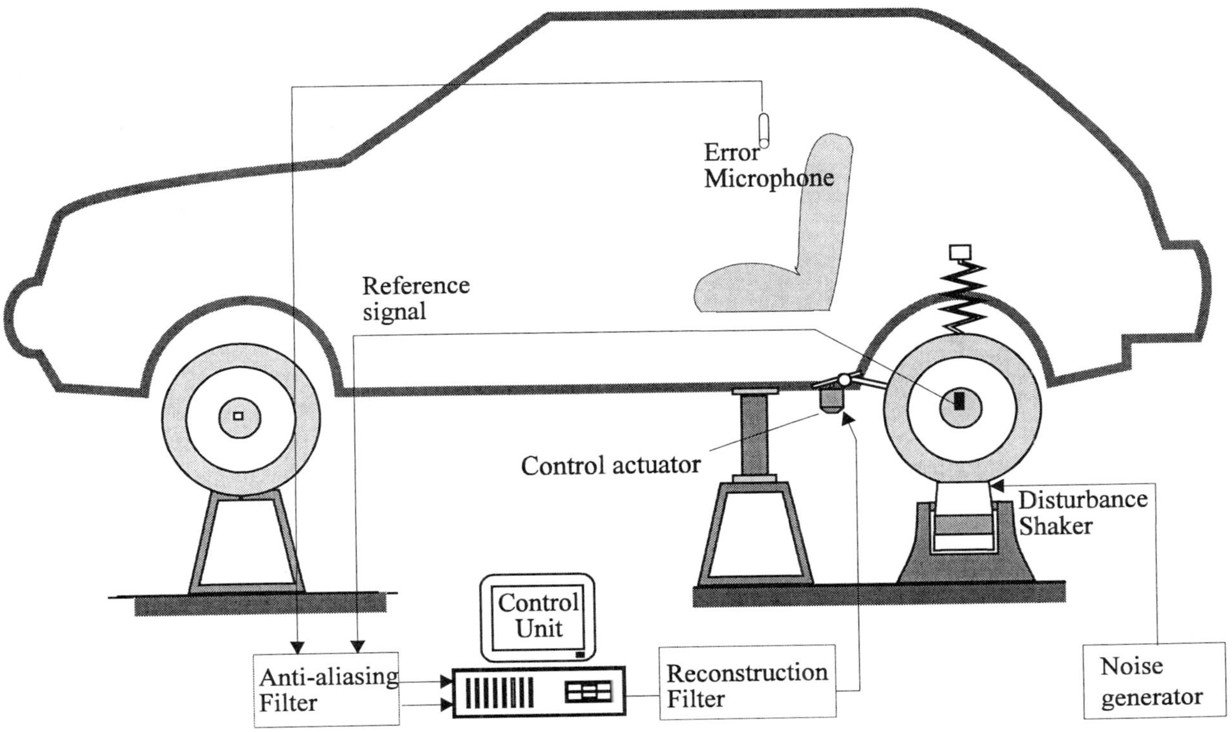

Figure 2. Schematic overview of the control set-up

In this set-up, an accelerometer is placed at the wheel hub. The position of the reference detection sensor is also determined by a second constraint namely the causality constraint. This will be discussed in the next section.

HARDWARE FOR DIGITAL SIGNAL PROCESSING - Since the control of road noise deals with a random input, the controller should have a sufficient bandwidth. This implies a high computational speed. A state of the art DSP-processor is therefore used for the implementation of the control algorithm.

Control and identification programs have been implemented in C-code and run on a DSP Board (Loughborough Sound Images) which is hosted in a PC. The DSP board is constituted of a mother board with one Texas Instruments TMS320C40 Processor Module installed on the board. The C40 is a 32 bit floating point digital signal processor with a clock frequency of 40 Mhz, and a maximum calculation speed of 40 MFLOPS. Additionally, a Daughter Module with 16 bit AD and DA converters is installed on the motherboard.

CONTROL ACTUATORS (SECONDARY SOURCES) - Ideally, the control actuators should be located in the main tranfer paths and supplement, for example, the original -passive- bushings that connect the suspension to the car body. However, this would imply that the actuator would have to withstand superimposed forces of low frequency (e.g. static loads, such as the car weight), as well as control high frequency vibrations (50 to 300 Hz), and at the same time preserve the original road handling characteristics (which are largely influenced by the bushings). It is obvious that it is not an easy task to design such an actuator. An alternative approach is to mount the actuator in parallel with the bushing, but only connected to the car body, and working against an inertia mass (a so-called absolute shaker). The final design of the control actuator is further determined by three design criteria : (i) size, (ii) maximum output force, and (iii) bandwidth. A prototype version of such an actuator has been used for the experiments described here.

LOCATION OF THE ACTUATOR - The main transmission paths of the demonstrator car have been identified beforehand. In the active control experiments only one rear wheel is excited, and therefore, only the main transfer paths of the rear suspension have to be taken into account, namely the connection of the rear axle with the car body (two transfer paths in total).

ERROR SENSOR - The objective of the filtered-X LMS adaptive scheme is to update the control filter such that the mean square sound pressure inside the car cabin is reduced. For this purpose, the sound pressure needs to be fed back to the controller. In the experiments presented here, two error microphones are positioned above the passenger's seats.

PRIMARY DISTURBANCE - A white noise signal with frequency content from 0-320Hz is used as input to the primary shaker. The amplitude of this signal is chosen such that the displacement amplitude of the wheel center is comparable to real driving conditions.

CONTROLLER SPECIFICATIONS - Although the controller operates fully adaptive, there are some parameters left that can be tuned to achieve optimal performance. More specifically, the lengths of the control filter and the secondary path prefilters should be determined, as well as the sample frequency and the adaptivity rate of the algorithm.

FILTERING - All analog signals have to be filtered to avoid various signal processing problems. An anti-aliasing filter is used for all input signals (reference signal and error microphone) to avoid the aliasing effect due to the chosen sample frequency *Fs* (cutoff-frequency should be smaller than *Fs*/2). A reconstruction filter has to be used for all the outputs (control signals, sent to the actuators) to avoid high frequency quantisation noise. These lowpass filters have the disadvantage to introduce considerable time delays due to their phase lags. The phase lag that is introduced in the frequency band of interest (in this case from 50 to 300 Hz), can however be manipulated by tuning the filter cut-off frequency and the filter order. The effect of using different combinations of these filters is examined in the tests.

CONSTRAINTS FOR THE REFERENCE SENSOR POSITION

To derive the control signals, the feedforward controller processes a reference signal, which is representative for the primary noise at the error microphone (the original road noise).

There are two critical requirements for the reference signal, namely the causality constraint and the need for a high coherence between the reference signal and the error sensor signal. These constraints determine the performance of the control algorithm.

CAUSALITY CONSTRAINT - In the case of road noise, a random disturbance signal is picked up by the reference signal sensor and thus it is impossible to predict the time history of the disturbance signal. Obviously, the control action should reach the error microphone *at the same time* as the primary disturbance. In other words, the sum of the *electrical time delay* (the time delay between the measurement of the reference signal and the control action) and the *secondary time delay* (the time required for the control action to "reach" the error microphone) should not be larger than the time required for the primary disturbance to travel from the reference signal location to the error microphone (*physical time delay*) (Figure 3). Theoretically, in cases where this time constraint is not satisfied, the optimal controller would have to be non-causal. However, an non-casual control filter can not be implemented in practice and hence, any causal control filter that is used under these circumstances will converge to a suboptimal solution.

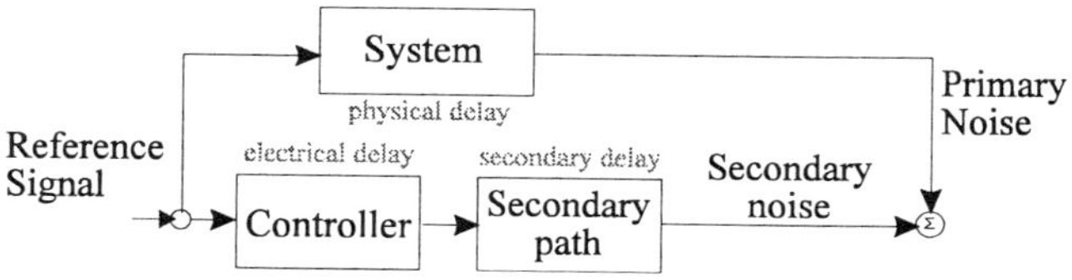

Figure 3. Time delay block diagram

For these reasons it is important to check the timing of the control loop. The reference signal passes through the following cycle between reference sensor and actuator :

reference accelerometer → anti-aliasing filters → ADC → calculation →DAC → reconstruction filter → actuator

The electrical time delay from reference to actuator can be estimated taking into account the phase lag of the different signal processing elements (such as the filters, the DSP etc.). The time delay for the secondary path is measured from actuator location to error microphone and can be derived from the secondary path model. The physical time delay of the primary disturbance has been measured on the demonstrator car [1].

It was observed that the time delays in both loops are of the same order of magnitude. The use of reconstruction and/or anti-aliasing filters influences this time delay considerably due to their phase lag. Therefore, different filter configurations (different filter cut-off frequency, different filter order, etc.) have been tested during the experiments.

COHERENCE ANALYSIS - A high coherence between the error microphone and reference signal is a second requirement for achieving optimal control results. It is well known [2], [10] that the maximum achievable reduction is determined by

$$S_{EE}(\omega)_{\min} = (1-\gamma_{XD}{}^2) * S_{DD}(\omega)$$

with : - S_{DD} the autopower spectrum of the disturbance signal at the error microphone

- S_{EE} the autopower spectrum of the resulting sound pressure at the error microphone with control.
- γ_{XD} the coherence between reference signal and error signal.

For a SISO-case, we can use the single coherence between one reference signal and one error microphone. Results are satisfying as can be seen in Figure 4. This plot shows the coherence between the accelerometer at the wheel axis and the microphones in the car when exciting one wheel with white noise.

CONCLUSION - Two constraints determine the reference position. Since the coherence is more than acceptable in the desired frequency range for the reference sensor position at the wheel center, the time constraint is chosen as large as possible. This means that the reference accelerometer position is chosen at the wheel center.

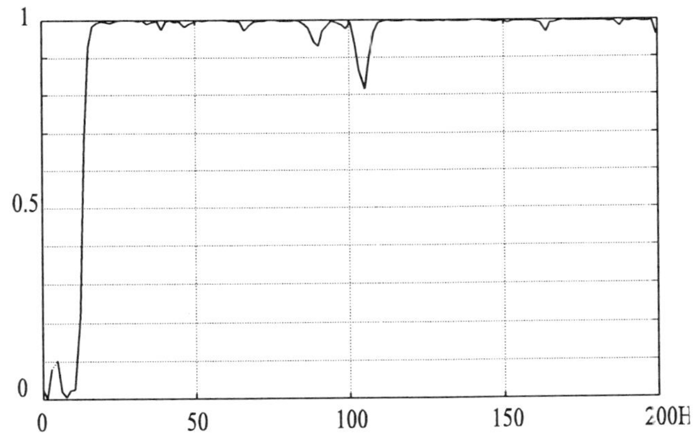

Figure 4. Coherence between reference accelerometer and error microphone

TEST RESULTS

It is the aim of the experiments to validate the control approach (ASAC). For the experiments, the controller is tuned for optimal performance. The parameters are the secondary path and the control filter lengths, the convergence rate and the sample frequency. In theory, large filter lengths would be necessary to perfectly cancel a broadband signal. For causality reasons, a very high sample frequency would be preferable too. Obviously, these requirements can not be satisfied altogether and therefore some optimal setting has to be found. The determining factor in the optimisation process is the computing power supplied on the DSP.

The control performance in broadband conditions was tested using a pure random excitation (white noise from 0 to 320Hz) to simulate the road excitation. The control result is presented in Figure 5. One control actuator has been used, reducing the sound power in one error microphone inside the car cabin. An average noise reduction of 7.13 dB has been achieved in the complete band from 50 to 200 Hz. Between 110 and 160 Hz, an average reduction of more than 10 dB is observed. The limited reduction in the 90 and 110 Hz frequency band, is primarily due to controllability problems. The secondary path transfer function has a large dip around 100 Hz and hence the controller is not able to excite this frequencies appropriately. Also the coherence (fig. 4), which is a measure for maximum achievable reduction, shows a dip around this frequency.

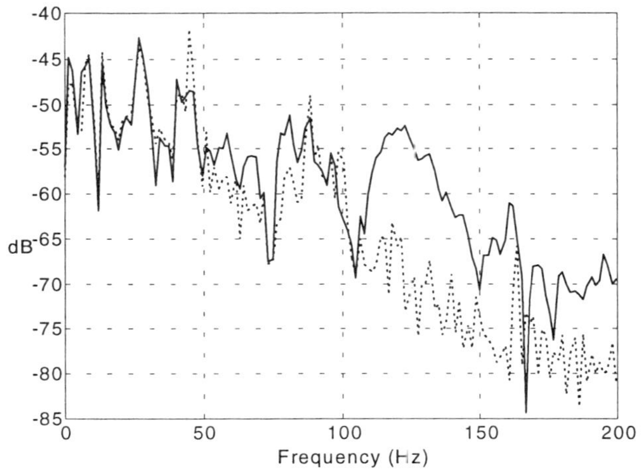

Figure 5. Primary and reduced noise spectrum in one error microphone (solid : no control, dotted : with control). One control actuator was used. Overall reduction from 50 to 200 Hz : 7.13 dB.

Not surprisingly, the overall reduction increases as the control filter length is increased from 2 coefficients (which enables the controller to control no more than one frequency at a time) to 64 coefficients. Obviously, the noise reduction will not increase infinitely as a function of filter length. Typically, for the circumstances encountered here, an optimal number of control filter coefficients is about 64.

When the reconstruction filter for the control signal is not used (decreasing the electrical time delay), overall performance increases slightly, indicating that the physical time delay and the sum of the electrical and secondary time delay are very closely matched. However, in this configuration, additional quantisation noise is injected in the higher frequency range, which is an undesirable side-effect. Alternatively, in order to further reduce the electrical time delay, leaving out the anti-aliasing has resulted in slightly better control performances, however, at the expense of the robustness properties of the controller. In noisy environments, the performance for such configuration would rapidly decrease due to aliasing effects (which could eventually introduce instabilities).

The control filter length and the sample frequency also determine how accurate the controller can model an appropriate broadband control filter and hence they determine the controller bandwidth. This property is tested using a pure random signal up to 640Hz as primary disturbance. The result of this experiment is shown in Figure 6. The bandwidth of reduction covers the whole frequency band 0 - 500Hz. An overall reduction of 3.2dB is achieved in the frequency band from 0 to 500 Hz.

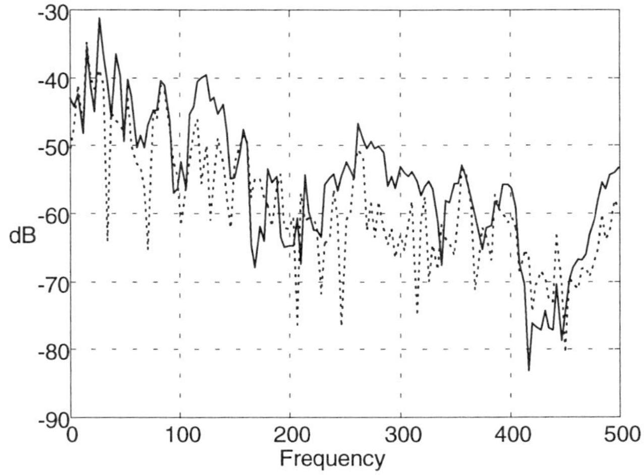

Figure 6. Verification of control bandwidth. Control in one error microphone with one control actuator. Solid : without control, dotted : with control. Overall reduction is 3.2 dB.

In the previous experiments, the reference accelerometer was located at the wheel hub. Obviously, the coherence of reference signal with the error signal can be improved if the reference sensor location is changed [14]. Figure 7 presents the coherence for three different reference sensor positions (position (a), (b), and (c)). The overall values indicate that the coherence increases for position (b) and (c) (although this effect is very limited because the measurements have been performed in laboratory conditions). When changing the reference sensor location, however, the causality constraints are not always satisfied and consequently, the control performance degrades considerably. Figure 8 presents the reduced noise levels inside the car cabin when one error microphone and one control actuator are used in the same configuration as before, but with different reference sensor positions. Although the coherence is slightly better, the overall reduction decreases for positions (b) and (c), indicating that the control configuration does not meet the time constraints any more.

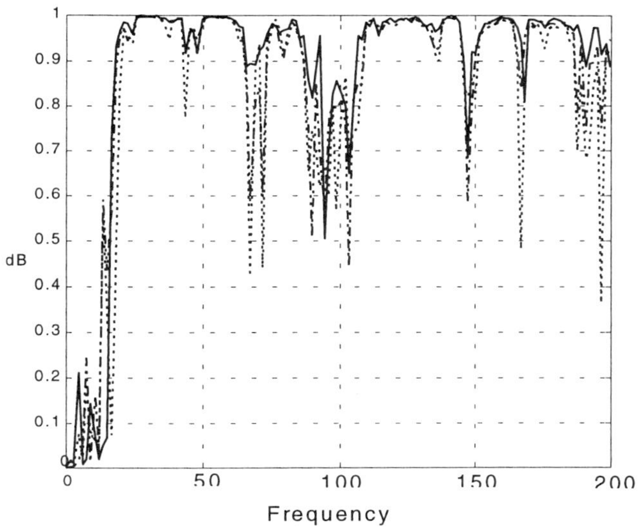

Figure 7. Coherence of the reference signal with the error sensor signal for three different reference sensor positions. Overall values : position (a) : 0.91, position (b) : 0.92, position (c) : 0.94.

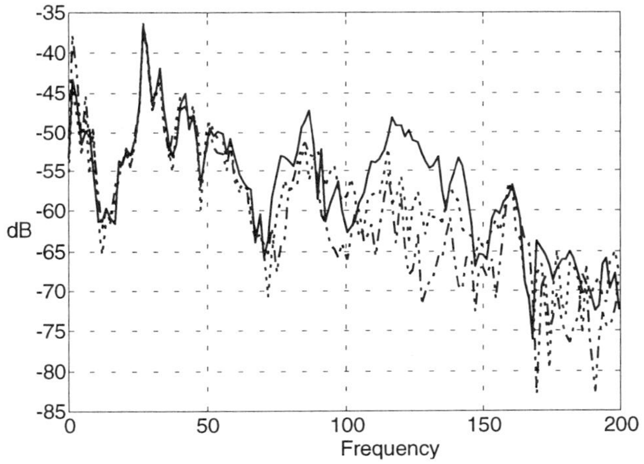

Figure 8. Controlled noise level in the error microphone using one control actuator and three different reference sensor positions. Overall reduction for reference sensor in : position (a) : 7.13 dB, position (b) : 5.44 dB, position (c) : 2.6 dB.

In the next experiments, the controller complexity was increased. First of all, the number of error sensors was set to 2 (one microphone at each rear passengers' seat), such that a larger "zone of quiet" could be created. As a consequence, the maximum achievable noise reduction in each of these error sensors was considerably lower than in the case of one error microphone. Figure 9 presents the result of this experiment.

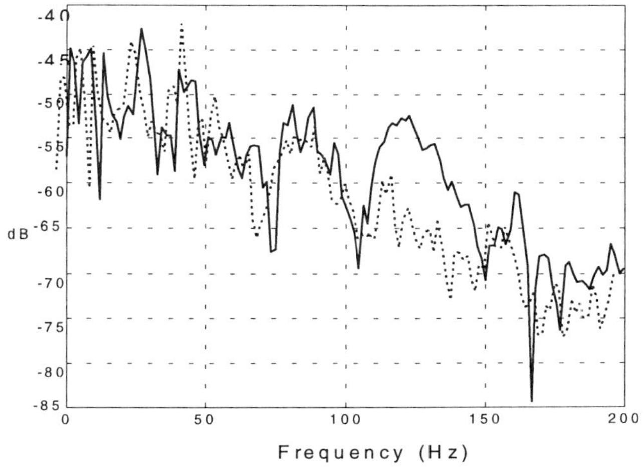

Figure 9. Noise reduction in one error microphone when two error microphones and one control actuator are used. Solid : without control. Dotted : with control. Overall reduction from 50 to 200 Hz is 2.55 dB.

In the frequency range from 50 to 200 Hz, no more than 2.55 dB noise reduction could be achieved with one control actuator.

Hereafter, an additional control actuator was connected to the car body. Figure 10 presents the results of the experiment in which the controller uses both control actuators to reduce the noise level in one error microphone. The overall noise reduction in the frequency band from 50 to 200 Hz is now increased to 9.1 dB (in the frequency band from 100 to 160 Hz, the overall reduction was more than 10 dB). In the second microphone (which was not used as an error microphone for this experiment), the noise level increases slightly.

Finally, two error microphones were used together with two control actuators. The results are presented in figure 11. The overall reduction in the frequency band from 50 to 200 Hz is 7 dB. The reduction in both microphones is of the same order of magnitude. Again, from 100 to 160 Hz, the reduction is more than 10 dB (in both microphones).

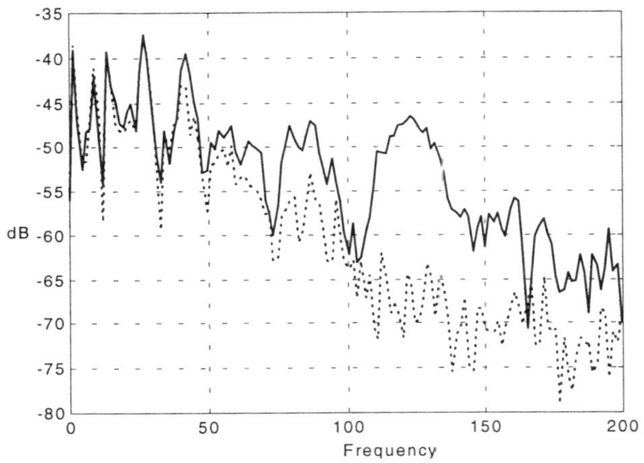

Figure 10. Noise reduction in one error microphone when one error microphone and two control actuators are used. Solid : without control, dotted : with control. Overall reduction from 50 to 200 Hz is 9.1 dB.

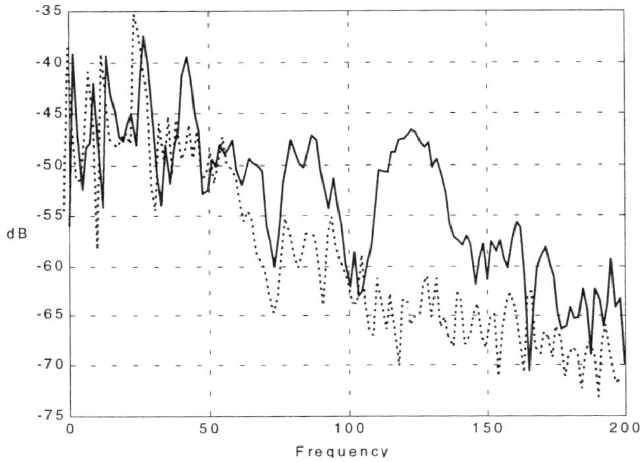

Figure 11. Noise reduction in one error microphones when two error microphones and two control actuators are used. Solid : without control, dotted : with control. Overall reduction from 50 to 200 Hz is 7 dB in one microphone and 6.8 dB in the other.

CONCLUSIONS

An experimental set-up has been developed to investigate the reduction of road noise using active structural acoustic control approach. The controller uses vibration actuators to change the vibration behaviour of the car body such that structure borne road noise generation inside this car body is reduced. The primary disturbance is an electrodynamic shaker that excites the left rear wheel. The control algorithm used for this purpose is a filtered-X LMS adaptive feedforward scheme. In a first series of experiments, a single road input single control output using one error microphone and reference signal has been used, and 7.13 dB noise reduction has been achieved.

The effect of using different loactions for the reference accelerometer has been investigated. The location of the reference sensor influences the total noise reduction because it determines : (i) the coherence of the reference signal with the error signal and (ii) whether or not the causality constraint can be met.

Next, the control configuration was extended from a SISO-system to a MIMO-system with two control actuators and two error microphones. The overall noise reduction inside the car cabin in a frequency band from 50 to 200 Hz was 7 dB. In a limited frequency band from 100 to 160 Hz, more than 10 dB noise reduction could be achieved in both error microphones.

ACKNOWLEDGEMENT

The work reported herein was related to the EC Brite/Euram Research Project ANRAVA (under contract BRE2-CT92-0327). The project is supported by the Directorate-General for Science; Research and Development of the CEC. This text presents parts of the research results of the Belgian programme on Interuniversity Poles of attraction initiated by the Belgian State, Prime Minister's Office, Science Policy Programming. the scientific responsability is assumed by its authors.

REFERENCES

[1] F. Vermersch, P. Veys, "Active Control of road noise in cars: aspects of implementation", Thesis 94ep03, Dept. of Mech. Eng., KULeuven, Belgium, 1994

[2] D. Otte, H. Van der Auweraer, J. Leuridan, "The use of Multivariate Correlation Techniques for Vibro-Acoustic Experimental Data Analysis", Proceedings of the 17th ISMA conference, KULeuven, pp. 117-140, 1992.

[3] "Automotive Active Engine Mount Systems", AutoTechnische Zeitung, Sondernheft Motor und Umwelt, pp. 62-64, 1992.

[4] C. Fuller, C. Hansen, S. Snyder, "Active Control of Sound Radiation from a Vibrating Rectangular Panel by Sound Sources and Vibration Inputs :

An Experimental Comparison", *Journal of Sound and Vibration*, 145 (2), pp. 195-215, 1991.

[5] C.C. Boucher, S.J. Elliott, P.A. Nelson, "Fast Algorithms for Multichannel Feedforward control", Proc. Internoise 93, Leuven, Belgium, Aug. 1993

[6] C. Bao, P. Sas, H. Van Brussel, "Adaptive Active Control of Noise in 3-D Reverberant Enclosures", *Journal of Sound and Vibration*, 161 (3), pp. 501-514.

[7] R. Clark, W. Saunders, "An Intelligent Systems Approach to Active Structural Acoustic Control", SAE paper 931282, 1993.

[8] Clark, R.L., Fuller, C.R., "Active structural acoustic control with adaptive structures including wavenumber considerations," Proceedings of the 1st International Conference on Recent Advances in Active control of Sound and VibrationBlacksburg, Virginia, pp. 93-119, 1991.

[9] W. Brent Ferren, Active Control of Simulated Road Noise, School of Mechanical Engrg, Purdue University, 1991

[10] T.J. Sutton, S.J. Elliott, I. Moore, Use of non-linear controllers in the active attenuation of road noise inside cars, Proc. Recent Advances in Active Control of Sound and Vibration, VA, April 1991, p682-690

[11] K. Wyckaert, Vander Auweraer H., Hendrickx, W., "Correlation ao acoustical modal analysis with operating data for road noise problems", Proceedings of 3rd Int. Congress on air- and structure-borne noise and vibration, Montreal, Canada, June 13-15, 1994, Montreal, Canada.

[12] Wyckaert, K., Hendrickx, W., "Transmission path analysis in view of active cancellation of road induced noise in automotive vehicles", Proceedings of 3rd Int. Congress on air- and structure-borne noise and vibration, Montreal, Canada, June 13-15, 1994, Montreal, Canada.

[13] Burdisso, R.A., Vipperman, J.S., Fuller, C.R., "Causality constraints of feedforward-controlled systems with broadband inputs", *J. Acoust. Soc.Am.* **94** (1), pp. 234-242, July 1993.

[14] Heatwhole, C.M., Dian, X., Bernhard, R.J., "Determination of the number of input transducers required for active control of road noise inside automobiles", Proc. of Noise-Con 93, pp. 207-212, Williamsburg, Virginia, May 2-5, 1993.

ACTUATORS IN RESTRAINT SYSTEMS

971049

Different Approaches for Air Bag Triggering Units Using a Firing Bus

Michael Bischoff, Günter Fendt, and Derrick Zechmair
Temic Telefunken Microelectronic GmbH

ABSTRACT

Safety systems in the automotive area become more and more important. New sensors inside the passenger compartment evaluate additional information about the occupants. Forward-looking sensors identify the crash severity before the crash really happens.

The variety and also the number of safety units per vehicle continues to rise. Some of these units get smarter than the current units. To control the actuators, a firing bus system is designed. The common part philosophy allows an easy adaptation of this system to different platforms. An image of the safety system is stored in the non volatile memory of the Central Control Unit (C^2U). The External Remote Firing Circuits (ERIC) are connected at the desired locations along the bus line(s).

INTRODUCTION

Safety systems become standard in today's new cars and trucks. Driver and passenger airbags prevent injuries, safe lives and are necessary to fulfill the legal requirements in several countries around the world. In the near future the number of actuators will increase very fast. There are two different types of actuators:

- New safety devices
- Current safety devices with smarter functionality

New electronic systems are required to control the high number and different types of actuators. There are several possibilities well known from other applications in the automotive industry or other high volume products. The main requirements are:

- Pricing, it should be possible to optimize the costs for each platform and option.
- Flexibility, because of the possible optimality of the safety functions.
- Fast exchange of firing commands.
- Smart algorithms, the activation of the safety devices should be controlled by all available information about the occupants and the crash scenario.
- Smart protocol, the length of the protocol should be independent of the number of actuators. It should be possible to transmit also slow sensor information like, seat occupied, seat belt status and occupant weight.

Flexibility means, flexible towards the number and location of the actuators, towards enabling and disabling and programming of the safety system after the assembly inside the vehicles. These requirements and the pricing lead in the proposal of the *Firing Bus System.* The components of this system are the common Central Control Unit *(C^2U)* and identical External Remote Firing Circuits *(ERIC)*. This means that the *ERIC*'s are similar parts from the production line of the electronic supplier to the assembly inside the vehicles. They will be programmed with own addresses via the *C^2U*. With this method the *ERIC*'s are 'easy to use', also for the services in the workshops.

The following paper describes different bus topology's and bus interfaces and also a new philosophy for the triggering of the safety components.

BUS TOPOLOGY

There are two possible topologies. One is a tree structure, shown in figure 1. The safety devices are divided in up to five bus lines. The rules for the system design are:

1. The *ERIC*'s should be ordered so that the last one in the line (n, seen from the C^2U) is the first which is triggered in a crash situation. The *ERIC* before the last one (n-1) is the next that is triggered, and so on. It is also possible that two or more neighboring *ERIC*'s are triggered at the same time. This method allows a delay in the triggering of the previous *ERIC*('s), even if the triggered *ERIC*('s) is/are destroyed after the ignition of the safety devices.
2. The busses are assigned to different crash scenarios. This means that one or more lines are

assigned to the frontal, the right and/or left side and the rearward direction and also the over-roll situation. With this assignment the reliability, that the remaining bus lines after the triggering, are still in function for a second impact from another direction.

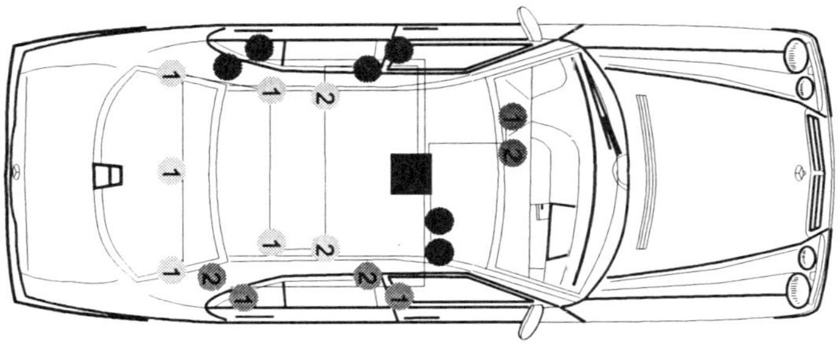

Figure 1: Tree Structure

The second topology is shown in Figure 2. It is a ring structure with a central ring line and for each *ERIC* a spur line. This topology is easier for the system design. An isolation in the spur line prevents the bus communication, if a short circuit at an *ERIC* occurs. If the central ring line will be damaged during the crash the whole system is stopped.

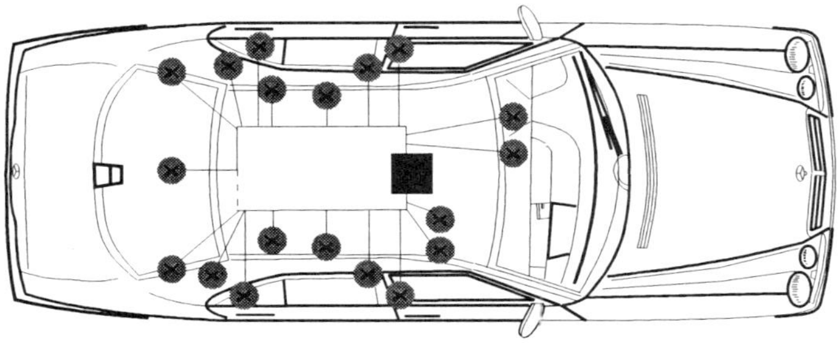

Figure 2: Ring Structure

The electrical connection between the *C²U* and the *ERIC´*s is realized with a two wire harness in both concepts.

The bus interfaces for both topologies are described in the next chapter.

BUS INTERFACE

The electrical interface between the bus lines and the *ERIC´*s can be realized in two ways. One possibility is the *Daisy Chain* interface (Figure 3). The *ERIC* has an input and an output connection. An internal switch joins the input and the output after the initialization of each *ERIC*. In a system with this bus interface the addressing of the external units runs sequentially. This feature allows the programming of the *ERIC´*s after the assembly in the vehicle.

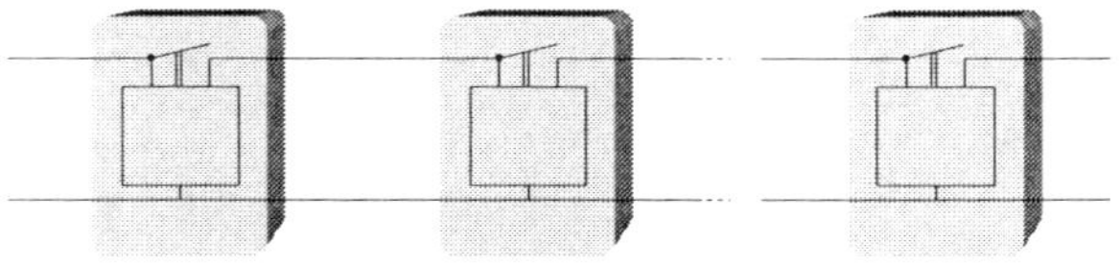

Figure 3: Daisy Chain Interface

If the branches of the tree have no loop back to the C²U and an *ERIC* fails the following parts are also out of function. The other branches are still in function. This can prevent with a loop back to the *C²U*, which allows the transmission also in the other direction.

If the *ERIC´*s are not included in the safety devices they can be identical parts from the production of the device to the assembly line at the car manufacturer and the workshop in the field. There is also no risk to exchange the *ERIC´*s during the assembly of the safety devices or the vehicle. The order in each bus line is mechanically fixed by the wiring harness. The structure of the system is stored in the non volatile memory of the *C²U* at the end of the assembly line.

In a daisy chain system in is easy to find the failure location in the wiring harness. It is possible to qualify the location between two *ERIC*'s. In case of a wiring harness failure the repair cost are reduced to a minimum.

The second way to connect the *ERIC* to the bus line is a parallel interface (Figure 4). Here they are equal partners in the system. This principle requires the addressing of the *ERIC,* before the assembly in the vehicle, at the assembly line of the car manufacturer and also in the workshops.

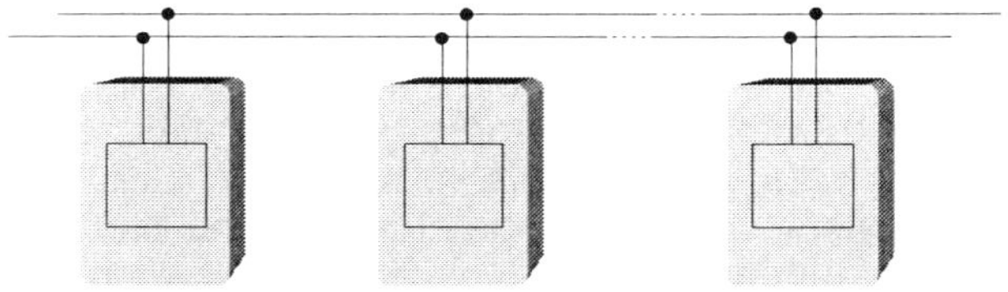

Figure 4: Parallel Interfacing

Because of the small size of the *ERIC´*s it is not possible to have as many mechanical keys at the connector as different safety devices in the vehicle. This results in a certain risk to exchange the programmed parts during the assembly. If the ERIC is part of the Safety device it can be programmed after the assembly of the safety device. This procedure prevent an unwanted exchange, in case of a parallel interfacing.

A proposal for a Daisy Chain connector system is shown in Figure 5. It is divided in two parts. The Branch Terminal is connected with an Insulation Displacement Connection (IDC) to the wiring harness. The power and communication line are interrupted during the assembly of the Branch Terminal. The interface between the Branch Terminal and the

Electronic Connector has 3 pins. The Electronic Connector contains all the electronic components of the *ERIC*. A 2 Pin interface joins the Electronic Connector with the safety device. Depending on the assembly process of these two parts, it is possible to omit the shorting clip inside the safety device.

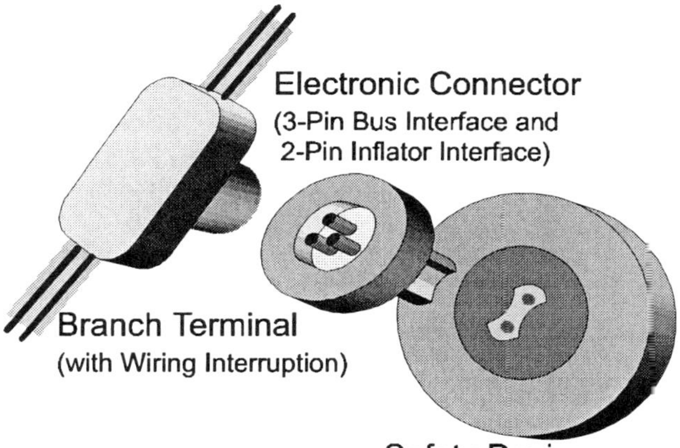

Figure 5: Daisy Chain connector system

In this case the Electronic Connector should be a first time locking device. The input pins of the *ERIC* should be resistant again all allowed electrical signals that can lead to an unwanted deployment. This includes also EMI and ESD requirements.

This connector system allows a successive implementation of the electronic parts inside the safety devices. In this case the Branch Terminal is connected directly to the safety device. This is one of the main goals in the near future, to reduce the costs of the whole system.

After the weighting of the pros and cons the decision for a tree structure with a *Daisy Chain* interface was made. One of the main argument is the assembly of the bus system in the car and the 'easy to use' service in the field. If all *ERIC*'s are included in the safety devices a ring structure with a parallel interfacing seems feasible.

BUS PROTOCOL

The data transmission runs with two different frequencies to reduce the EMI emission if a low speed is sufficient. A low frequency for the initialization, the updating of the firing information and the normal diagnosis. The high frequency is necessary for the fast triggering information. The change in the frequency can be realized in different ways. One possibility is realized with a special Sync-sequence at the beginning of each protocol. With this sequence the synchronization of the *C²U* and the *ERIC* will be done in each protocol. This procedure reduces the effort for the synchronization. Another possibility is a „return to Zero" coding of the information.

The bus protocol contains 4 different information's: address of the *ERIC*, content of the data information, the data information for the *ERIC*'s and a check sum.

There are tree different types of addresses. Each *ERIC* has a default address. Via this address each *ERIC* gets its own address at the assembly line of the car manufacturer or each time after ignition on. The addressing can be done after the assembly of the whole safety system in the vehicle. In the protocol with the default address the ERIC gets its safety device related address. Broadcast addresses will be used for the triggering protocols. With this address it is possible to suppress the triggering of the front airbags in case of an OOP situation. The width of the address information limits the number of safety devices.

The data information is divided in four groups.

- Programming of the *ERIC*. If the address is the default address the data information contains the safety device related address. If the address is the safety device related address, it contains the new triggering information.
- Diagnostic Mode. The data sent to the *ERIC* is the actual triggering information that is stored in the *C²U*. The *ERIC* compares the received information with the stored one. A failure will be detected, if they are not equal. The answer from the *ERIC* to the *C²U* includes the failure register of the *ERIC*. The failure differentiation depends on the requirements of the car manufacturer.
- Triggering mode. This protocol will be sent only from the *C²U* to the *ERIC*. The data information contains the crash scenario, which is currently detected by the crash sensors. It is possible that more than one protocol will be sent, to realize a delay between the triggering of the safety devices. Further protocols and the time delay depend on the crash severity. The information of this protocols is described in the paragraph FIRING PHILOSOPHY
- Sensor mode. With this protocol it is possible to implement slow sensor signals, like seat occupied and belt status, in the cyclic diagnostic routine. It is very efficient, if for example the belt pretensioner and the seat belt status sensor are part of one safety device.

A checksum is added to increase the reliability of the bus communication. If the *ERIC* detects a checksum error he does not answer. The *C²U* repeats the message until a predefined number of times is reached or the *ERIC* answers.

TRIGGERING PHILOSOPHY

For a bus system that should be open for future applications, a new triggering philosophy was developed. The new philosophy leads to a protocol that does not depend on the number of *ERIC*'s.

The triggering information contains the impact direction and the severity of each direction. Depending on the occupant parameters (present, weight, belted,..) the *ERIC´*s are set, to assigned triggering information's. If a crash occurs the *C²U* sent only the measured crash scenario via the bus. Each *ERIC* compares the received information with the stored triggering information and decides then to trigger or not the attached safety device.

After power on of the *ERIC* the triggering information says NO-FIRE in every case. During the power on test of the system, the information will be set to default according to the philosophy of the customer. For example it is possible to deactivate the passenger airbag depending on the child seat priority.

The time for the new configuration of the *ERIC´*s is less than five milliseconds after the *C²U* received a new sensor status.

If the triggering information is set to NO-FIRE, it is not possible to trigger the safety device because of a protocol failure. This is an advantage especially in the case of the passenger airbag and a rear facing child seat mounted on the passenger seat.

LOGISTIC CONCEPT

The above described connector system allows the following logistic concept. The Branch Terminals will be delivered from the connector supplier to the wiring harness manufacturer. Here it will be connected to the harness at the required locations. After the assembly of the harness the functionality can be check in a simple way by a resistance measurement. The assembled wiring harness will be delivered then from the wiring harness supplier to the car manufacturer.

The separate parts of the Electronic Connector will be delivered from the connector supplier to the *ERIC* manufacturer. The electronic components will be assembled together with the connector parts. The complex test of the assembled parts is planned at the electronic supplier plant. This procedure guarantees the same test standards, like in today's Sensing and Diagnostic Module production. The tested *ERIC* will be delivered then to the safety module supplier.

At the plant of the safety module supplier the *ERIC* can be connected during different steps of the assembly process. The optimum is the replacement for the shorting clip, because of the possible cost reduction. The safety module will be delivered then to the car manufacturer.

Regarding the assembly line of the car manufacturer, in comparison to today's systems, is no change in the assembly order and number of parts.

CONCLUSION

This paper shows a comparison of different bus topologies and interfaces. A decision for a tree structure with a *Daisy Chain* interface was made because of the advantages at the assembly line and in the workshops.

With the proposed connector system a successive integration of the external remote firing circuits inside the safety devices can be done. If all *ERIC* 's are integrated in the safety devices, it is possible to change the topology and the interfacing. The bus protocol runs in both topologies and with both interfaces.

The bus protocol and the triggering philosophy demonstrate one possible application of a bus system to control a complex passenger safety system. It is easy to realize different options and optimize the costs per vehicle.

The bus protocol and the new triggering philosophy is open for future applications. The maximum number of safety devices is nearly independent from the length of the protocol. The triggering information is independent from the number of safety devices.

A combination of a slow signal sensor and the firing bus is intended in the protocol.

The described firing bus system fulfills the today's requirements. The modular design opens several possibilities for future application and has the potential to became a world wide standard.

DEFINITIONS, ACRONYMS, ABBREVIATIONS

C²U	Central Control Unit
ERIC	External Remote Firing Circuit
IDC	Insulation Displacement Connection
ESD	Electrostatic Discharge
EMI	Electromagnetic Interference

971053

Bus System for Wiring Actuators of Restraint Systems

Joachim Bauer, Gerhard Mehler, and Werner Nitschke
Robert Bosch GmbH

ABSTRACT

The continuing increase in the performance of restraint systems has led to a drastic increase in the number of actuator devices. The individual wiring of the igniters becomes more and more problematic through the accompanied large number of plug connections and cables. Along with demands for weight and volume reduction, there are requirements for EMI and short circuit protection to eliminate erroneous deployment and misuse.

As a solution, a new multi-protocol dual wire bus system is described that has the capability to supply energy and address multiple peripheral output stages to simultaneously fire any combination of actuators.

INTRODUCTION

The success of air bags with motor vehicles is illustrated by the sudden increase in the number of firing loops for pyrotechnically triggered restraint actuators. Apart from the classical air bags in the front and side areas, belt tensioners are standard equipment in many motor vehicles. The increasing attention given to passenger safety, related to accidents, has resulted first, in an increasing number of new pyrotechnical occupant protection devices and second, in triggering characteristics better adapted to impact situations.

The systems that will prevail in the marketplace will be those with comprehensive interior sensing systems and intelligent occupant protection measures ("smart restraint systems"). For the electronic control unit this means a rapidly growing number of required firing loops. Fig. 1 shows schematically a circuit that could represent a future standard for restraint systems.

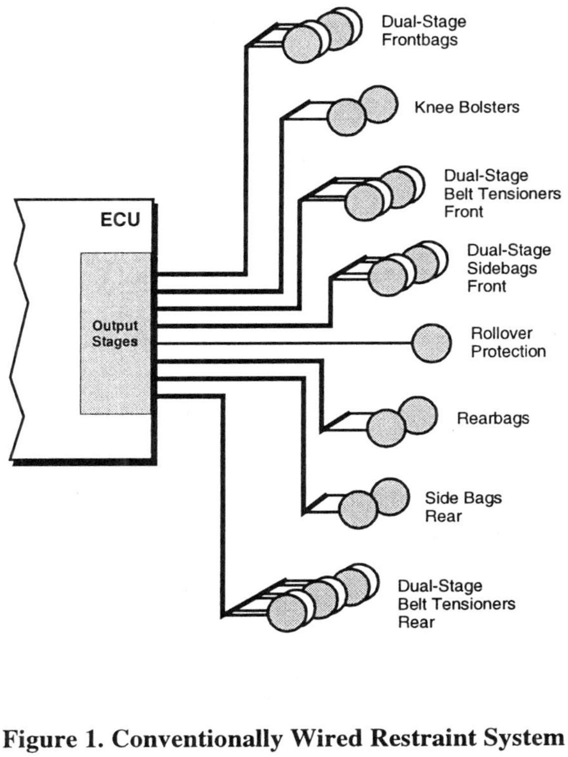

Figure 1. Conventionally Wired Restraint System

Table 1 illustrates the required number of firing loops.

Table 1. Firing loops

Restraint Device	Firing Loops
2-Stage Frontbags (Driver, Codriver)	4
1-Stage Knee Bolsters (Front)	2
2-Stage Belt Tensioners (Front, Rear)	10
1-Stage Rearbags (Passenger 1, 2)	2
2-Stage Sidebags (Front)	4
1-Stage Sidebags (Rear)	2
Rollover Protection	1
total	**25**

A conventionally wired restraint system with the shown fittings requires 25 dual-wire loops to the actuators. Apart from the growing weight and volume problem of the system, further problems concerning handling, logistics and flexibility become apparent.

A senseful approach would therefore appear to be the installation of a bus system, whereby all actuators can be wired in a simple manner. In this way, universal bus-compatible actuators can be connected by means of a simple conduit system with an equally universal central device. The outfitting variations of the restraint systems would then only be activated according to the programmed number of bus participants and triggered from the central control unit according to an appropriate firing scheme.

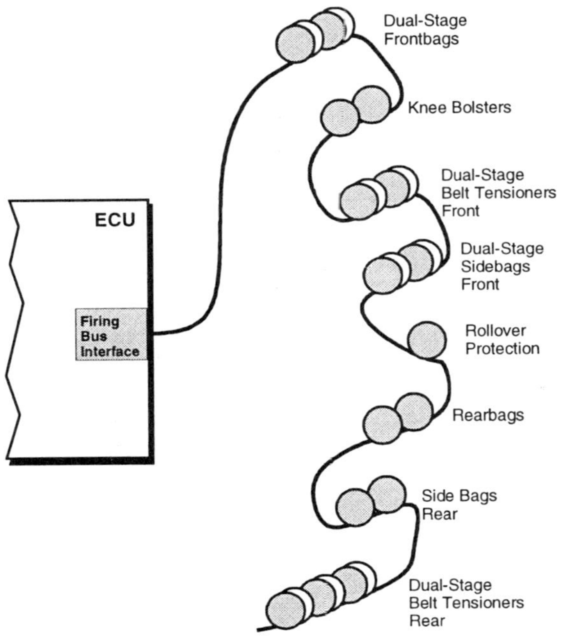

Figure 2. Firing Bus

MOTOR VEHICLE AND BUS SYSTEM

CAN - The wiring of electronic motor vehicle components can be advantageously carried out by means of CAN. This bus system has proven its reliability in the field and is finding increasing acceptance and application in the motor vehicle sector. In addition, a large number of hard- and software tools in development are available. However, it is a fact that a large number of bus participants, while making only minimal demands on the communication system, have simultaneously extremely acute demands on a single firing signal for selective activation of actuators and therefore stand on the debit side of the application of CAN for actuator wiring of restraint systems. The functions that CAN offers, such as Multi-Master Operation, destruction-proof bus arbitration, and command signal controlled addressing, are not actually necessary for the fulfillment of the main demand of the restraint system, that is, the transmittal of a firing combination on all actuators. On the other hand, CAN offers indispensible capabilities, for example, for securing the data framework for the transmittal of the safety-relevant firing information.

A TAILORED BUS SYSTEM - The question of cost is the determining factor in the choice of a suitable bus system supplying all restraint actuators. The demands on the versatility of the bus system are minimal compared with other types of electronic network systems. First of all, the **general requirements** of restraint systems for provisioning actuators are as follows:

- Minimization of the number of transmitting wires.
- The transmission of the firing signal should be as rapid as possible.
- Selective firing, i.e. any required combination of simultaneously triggered actuators.
- High degree of reliability of the firing signal.
- All connected units should be capable of being monitored.
- The EMI must be taken into consideration.
- Protection against a total breakdown because of defects in individual components or circuitry.
- Easy handling for installation into motor vehicles.
- The possibility of standardization must be guaranteed.
- The cost of realization must be as low as possible.

The demand for a low cost system requires the reduction of functions to a minimum with an accompanying exact specification of these remaining functions. The main data of the system were therefore established as follows:

- Possible number of units: 32
- Allowable firing transmission time: 200 μs
- Number of bus lines: 2

To choose the means of transmission, a thorough investigation of the messages to be transferred is necessary. Three different classes of messages can be identified:

- Diagnostic requests
- Diagnostic responses
- Firing

The characteristics of these three classes differ from one another quite considerably. These are:

- The content of information
- The frequency of signals
- The required transmission time
- The safety relevance

Table 2 gives an overview of the facts of the matter:

Table 2. Classes of Messages and Characteristics

Class / Characteristic	*Diagnostic Request*	*Diagnostic Response*	*Firing Command*
Information Content	address	condition	firing combination
Frequency	cyclic	according to requests	sporadic
Transmission Time	not critical	not critical	critical
Safety Relevance	low	low	high

Characteristic for the actuator circuitry is the complete sufficiency of a Single Master/Multiple Slave System. All messages are either issued directly by the Master or a Slave Message is requested by it.

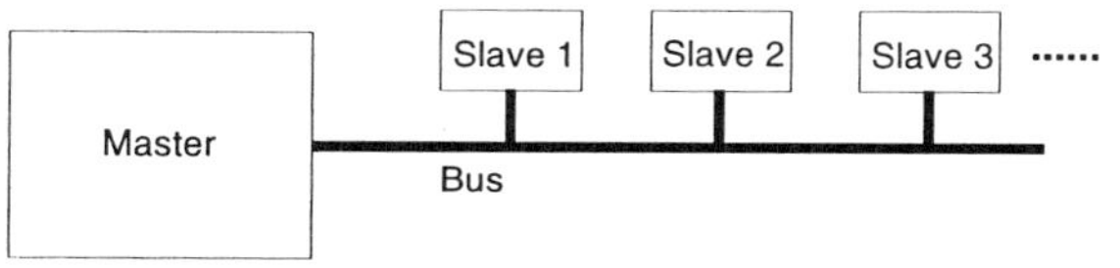

Figure 3. Master/Slave Bus System

The diagnostics is carried out by means of the sequential requesting of all connected units and the analysis of the responses received. The cycle time of this procedure can be kept relatively long because of the small degree of probability of a failure occuring. The implementation of self-diagnostic devices in the bus slaves requires, as response only, a signal concerning its condition in the form of a "sign of life". In operation, a greater content of information is not required. Because of the non-critical cycle time of the diagnostic procedure, the safety relevance of the diagnostic request and response is minimal. An occasional error in a response can be compensated for by result-averaging all diagnostic responses ("noise elimination").

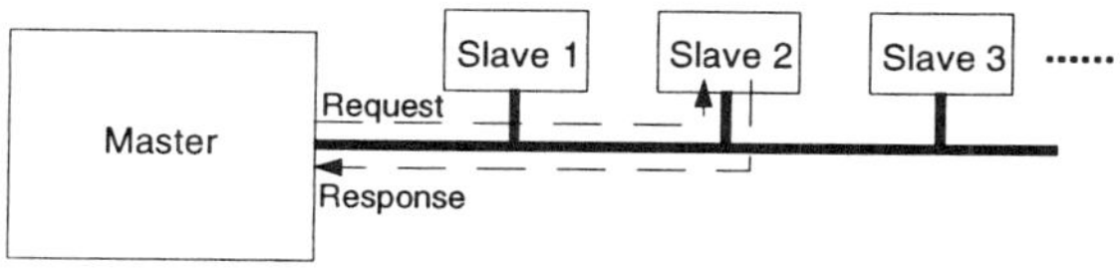

Figure 4. Diagnostic of Slave

The firing commands, however, exhibit a direct contrast. These must reach all units simultaneously, have the highest priority, have the most critical transmission time and possess a high safety relevance. In addition, the transmittal of firing during the lifetime of a motor vehicle would be an extremely rare occurrence.

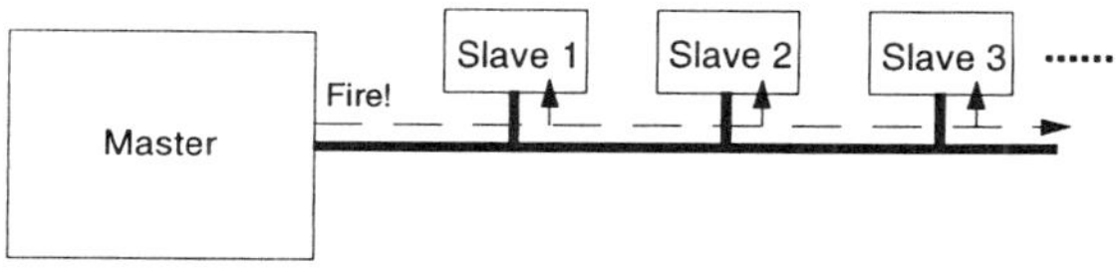

Figure 5. Firing Command

The large differences in the characteristics of the command signals actually require two separate bus systems that satisfy each of the particular requirements. Normally, with respect to the speed and capacity of the transmission of information, the capability of the signalling devices is specificed according to the most critical requirement of any command signal. However, in the case of the firing bus system, the firing command is merely a triggering signal, that is, a signal with minimal frequency. Therefore, a high-performance bus would be dimensioned almost exclusively according to the requirements of simple diagnostic signals with all the consequences resulting from the EMI.

The further possibility is to provide a plurality of channels for the various messages. Dimensioning can then be made to conform with the individual requirements of the different classes of messages.

THE FIRING BUS SYSTEM

The concept of the firing bus system can be illustrated by means of the following modified layer model according to OSI:

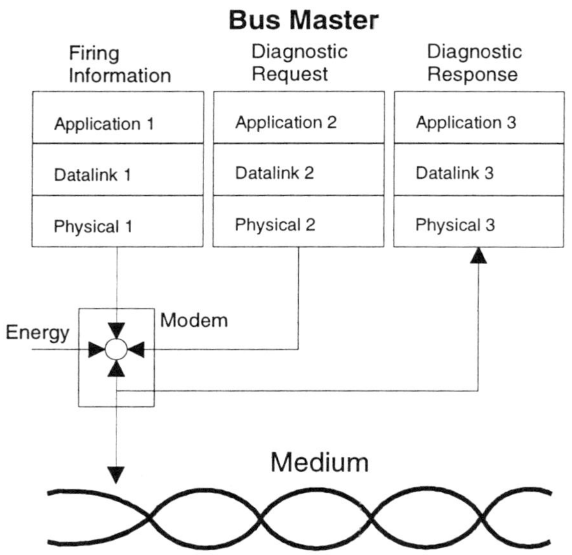

Figure 6. Layer Model of the Bus Master

Accordingly, with the model for the bus Slaves, the directions of the signal and energy flow are reversed.

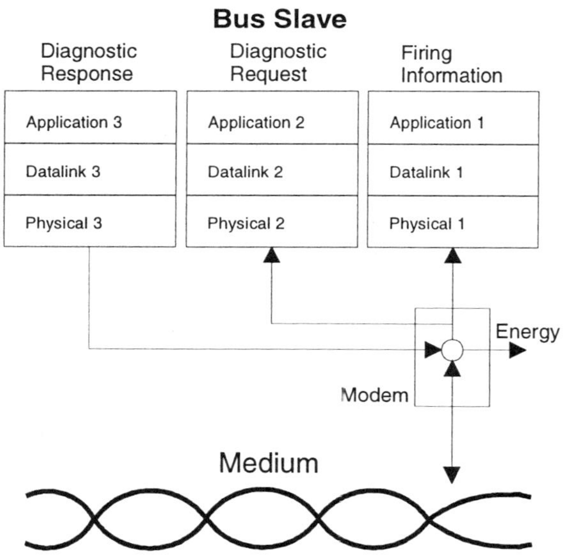

Figure 7. Layer Model of a Bus Slave

Each of the three named classes of messages is provided with its own channel and has at its disposal its own physical signal representation, its own coding system and consequently its own network. The flow direction for each network is marked in. A modem circuit takes care of the integration, that is separation of the physical layers and the supply of energy to the bus slaves. A twisted dual wire conduit serves as transmission medium.

Physical Layers - A physical representation of the signals for three networks is undertaken that conforms with each of the individual requirements. Because of the high cost factor involved with carrier modulation and demodulation, the possibility of the realization of the channels with selective frequency was abandoned. Instead, an amplitude separation was employed. With this, the electrical parameters of current and voltage are utilized for transmitting information. Fig. 8 shows the amplitude coding of the channels assigned to the classes of messages.

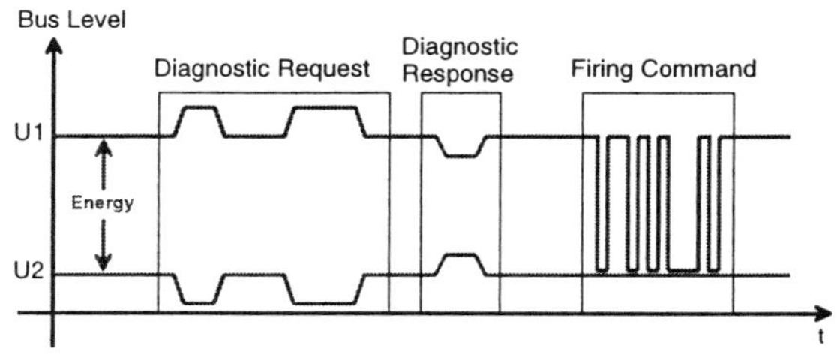

Figure 8. Physical Signal Representation

For the transmittal of diagnostic messages (requests and responses), small amplitudes can be employed because of the minimal demands on the transmission security. This also conforms with the EMI security stipulations, because the diagnostic command signals are carried out cyclically and their signals are continually present on the medium. The modulation of the smaller voltage swings is suitable for application only with the bus master, as it otherwise involves a certain expense for the modulation circuitry (analogue operation). This form of signalling, therefore, finds a favourable employment for **diagnostic requests**. On the other hand, **diagnostic responses** can be easily handled with current modulation. For this, the bus slaves increase their power intake for a short time interval by means of switching a load shunt in parallel. If the Master of the system possesses a low enough internal impedance, a correspondingly small potential difference occurs on the bus lines. The **firing signal** shows the highest degree of modulation. Because not all channels can operate independently with the chosen amplitude separation (on firing, the detection thresholds of diagnostic requests are also bypassed), with the introduction of signals with larger amplitudes, a destruction or interruption of a concurrently transmitted signal with smaller amplitude takes place. This, however, does not affect the firing bus application. On the contrary, a running diagnostic is interrupted by a high priority firing signal.

Because of the priority characteristic of the amplitude mode, the desired independent and, therefore, free channel is placed at the disposal of the firing information. In this way, delays with a single channel system, due to decision making on the availability of the bus, are avoided.

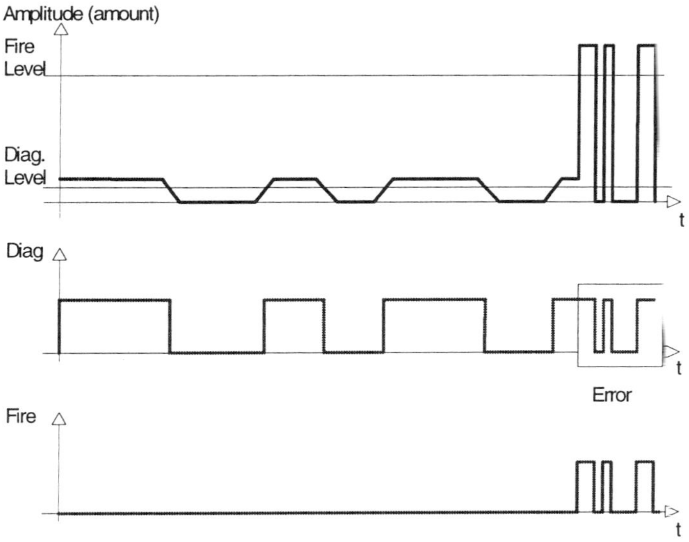

Figure 9. Signal Separation by Slave Receiver

Apart from the amplitude coordination, the chronological representation of the single information bits must be made available. A PWM coding is placed at the disposal of the diagnostic request and firing signals. This offers the advantage of the simultaneous transmission of synchronous information by means of a constant carrier and the simple decoding. By means of these measures the necessity of a precise chronological base or PLL circuit for each slave is avoided.

The transmission rates for the different classes of command signals are also established according to their particular requirements. The transmittal of the firing commands is dependent on a given maximum allowable transmission time and the necessary extent of the framework of the message. An extremely high data rate must therefore be chosen. For carrying out the diagnostic the determining factors are the cycle time of the diagnostic of the entire network and the EMI limiting value. However, an increasingly smaller data rate compared with firing, results in more expense for the bus slaves, as a larger dynamic receiving capacity must be provided.

To increase the resistance to outgoing electromagnetic radiation, the diagnostic requests are carried out with symmetrical operation of the voltage modulation. The diagnostic response (current modulation) is naturally symmetrical because of the floating dual wire system.

Datalink Layers - A protocol can be defined for each signal network. The provisions for the inclusion of data securing bits are thereby again dependent on the transmissive requirements.

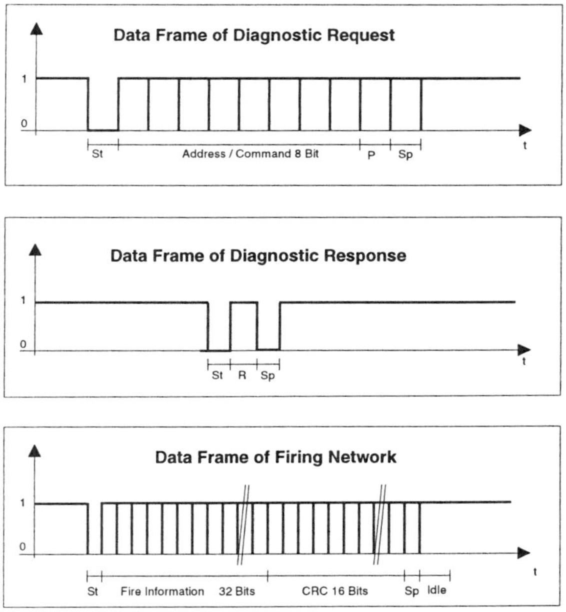

Figure 10. Protocol Layers

The signal separation of the slaves feeds two separate receiving circuits with varying data streams. In order to limit the hardware expense of the bit slaves, the diagnostic requests are packed in a simple standard format (e.g. V.24). Security takes place by means of a parity check, the synchronization with start and stop bits.

The diagnostic response in the form of a "sign-of-life" is provided by merely one start and stop bit and a response bit. This can be realized in the slave with the simplest of provisions. A correct answering "intelligence" is registered by the master through the right chronological sequence of the response format.

A more extensive redundancy protection must be provided for the firing signal, because of the higher security relevance; the 16 bit CRC as fitted in CAN is also applied here for protocol security.

Application Layers - The information extracted from the data framework has the following meaning:

The **8-bit diagnostic request** signal contains one single address and control information for the indicated bus slave. 5 address bits are generated for 32 slaves, the remaining 3 bits serve, for example, for programming the non volatile address memory, the demand for retrieving serial numbers, etc.

The **1-bit diagnostic response** contains only a "sign-of-life" of a slave without explicit address information. Erroneous functions, as for example, a permanent response or a plurality of identical addresses disposed of in the network, can be recognized by scanning the entire address capacity.

The **32-bit firing information** represents the required trigger combination at the moment of transmittal to the restraint actuators. The connected units (slaves) are here addressed according to the setting of a coordinated bit position and after a final check by the CRC, cause the firing of the inflators.

A receipt confirmation after reception is not generated by any of the named messages. With diagnostic requests, there is either no response or there is a response from a wrongly addressed unit. However, by means of averaging the diagnostic responses, a momentary disturbance is repressed.

In the case of an erroneous transmittal of the firing commands, another strategy is employed. Considering the error marking of CAN with which a defective command is returned by means of an error frame, with the Single Master / Multiple Slave System, the same command signal is repeated. As in the event of firing, a plurality of transmittals of the same command results in no negative effect (a slave can only be triggered once), the information is basically repeated for a certain period of time.

Topology - Many possibilities are available for the connection of the bus slaves and the configuration of the conduit systems:

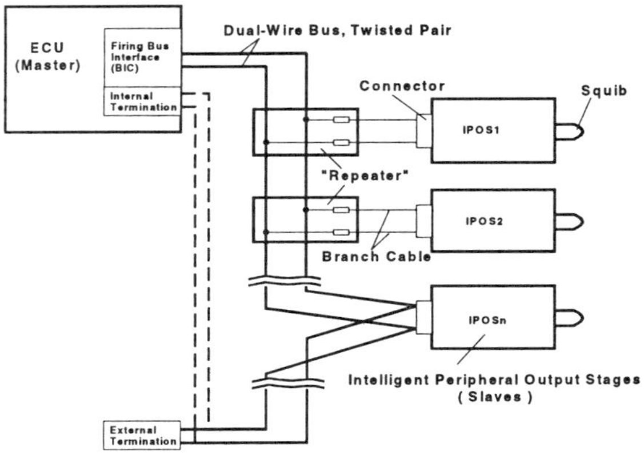

Figure 11. Topological Plan

- branch circuit with decoupling "passive repeater",
- direct (linear) connection,
- bus terminal as external element,
- built-in bus-terminal with the possibility of circuit feeding from both sides.

For the connection of the bus slaves (intelligent peripheral output stages, IPOS), two possibilities are available. With the linear topology, the bus circuits traverse the elements resulting in a branch circuit of zero length. Alternatively, an IPOS can be connected by a branch circuit that is optionally decoupled with passive repeaters. For the especially endangered motor vehicle sector, as for example, doors, the decoupling serves as short circuit protection in the event of a crash. They consist of simple resistors that are dimensioned in such a way that the supply current experiences no serious drop in voltage, the signal current incurs no damping effects and in the event of short circuits the current can be provided by the bus master.

A bus terminal in the form of an RC combination is absolutely necessary with the given bus circuit lengths and transmission rates. Building it into the bus master requires the return lead of the bus circuit. The advantage lies in the possibility of testing the actual functioning of the terminating elements and the feeding of both sides of the bus system in the event of an interruption. The termination can, however, also be realized externally if a return lead out of the steering wheel is unsuitable.

INTELLIGENT PERIPHERAL OUTPUT STAGE (IPOS)

In order to carry out the function of the triggering of the air bags, the bus slaves contain, apart from the communication means, further components. All together, one receives an IPOS with the following functional elements:

- local energy reserve to actuate an igniter,
- energy management,
- firing stage,
- firing circuit diagnostic device,
- timing oscillator for internal operation,
- operation control,
- permanent memory slave addresses,
- a signal separating stage for energy decoupling,
- a diagnostic and firing information transceiver.

Fig. 12 shows a block diagram of an IPOS:

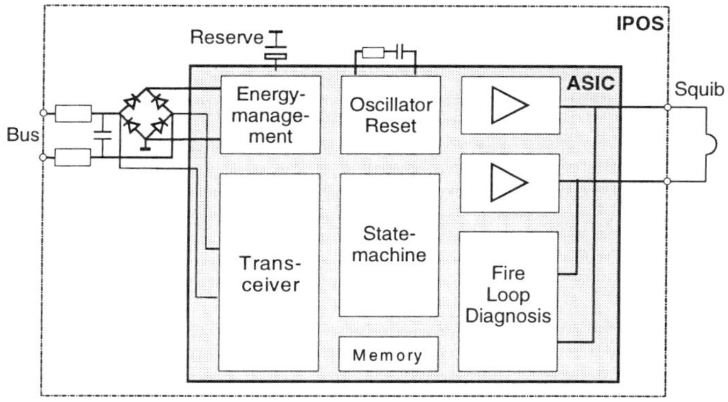

Figure 12. IPOS Block Diagram

The energy management delivers a stable supply voltage for the logic circuits, charges and tests the firing energy reserve and supplies reference potential.

The transceiver contains a signal separating stage, that separates the modulated signals from the supply, a double comparator for demodulating the separated amplitude signals, a load switch for generating the diagnostic response as well as the complete receiving devices with error detection.

The oscillator contains a timing generator for the transceiver and the operation control as well as a power-up reset circuit.

The statemachine contains all logic functions for the control of the internal operations, the diagnostic and the firing operation after addressing.

The permanent memory contains the address explicitly reserved for the slave unit.

The output stages drive the necessary current from the energy reserve for firing through the igniter.

The firing loop diagnosis motivated by the operation control carries out the state of the output stage and the igniter.

The installation of a bridge rectifier renders the IPOS on the bus independent of polarity. Additional resistors protect and decouple the bus circuits from defects in the ASICs, e.g. after a crash and a resultant firing.

TECHNICAL DATA OF THE FIRING BUS SYSTEM

The described firing bus system has the following physical key data:

- Dual wire bus system (twisted pair),
- Signal-on-energy modulation,
- Supply voltage: 24V,
- Supply current IPOS: < 5mA,
- Data quota for firing: 250kBit/sec, PWM,
- Data quota for diagnosis: 15kBit/sec, PWM,
- Amplitude firing command: 24V,
- Amplitude diagnostic request: 2V,
- Amplitude diagnostic response: 5mA,
- Max. number of participants: 32,
- Integration of IPOS into the igniter connector.

CONCLUSION

The introduction of a bus system in the circuitry of actuators solves the problematic points relating to a well fitted restraint system and results in many advantages. The handling of bulky, branched cabling, weight and volume are problems which can only be solved in this way. Further logistical advantages result from the standardization of components. Technical problems, such as faulty triggering by ESD and EMI or the misuse of the inflators, are solved by serial coding. The described bus system fullfills the listed general requirements on a distributed restraint system:

- Minimization of the number of transmitting wires.
  Only two wires for energy and signal .
- Rapid transmission of firing signals.
  Transmission time = 200µs.
- Simultaneous triggered actuators.
  Firing protocol includes bit-position addressing mode.
- High reliability of the firing signal.
  16-bit CRC and high signal/noise ratio.
- Monitoring of all connected units.

Slave response on master request through direct addressing.

- EMI taken into consideration.
  Cyclic messages with low data rates and amplitudes.
- Protection against defects in circuitry and wiring.
  Decoupling with passive elements.
- Easy handling for installation into motor vehicles.
  Uniform slave circuits integrated into squib connector,
  no polarity, end-of-line programming.
- The cost of realization must be as low as possible.
  Minimized functionality, suited for this specific application.

The introduction of the bus system in this sector is a fundamental prerequisite for achieving favourable cost effectiveness in the entire processing chain involving restraint systems.

REFERENCES

[1] Gerdsen, Peter; Kröger, Peter: "Kommunikationssysteme 1"; Springer-Verlag 1994, Berlin.

[2] Kaiser, K.H.; Schramm, D.: "Security Mechanisms of the CAN bus and their inpact on data integrity". International Symposium ATA-EL 95: System Development and Co-Design in Automotive Electronics, Belgirate, May 1995.

981103

Contactless Air-Bag Firing and Signal Transmission on the Steering Wheel with an Inductive Contact Unit

Erich Zabler, Anton Dukart and Thomas Herrmann
Robert Bosch GmbH

ABSTRACT

The electrical connection between chassis and steering wheel which is used for igniting and monitoring the driver's airbag and also for transmitting horn and other signals today takes the form of a galvanic contact — in other words, an electrically conductive connection. In virtually every case multiple elastic contacts resembling clock springs are used but very occasionally slip rings are still used. Contactless electrical connection would present a wide range of benefits here and would not cost any more. In the case of an AC airbag firing system the usual capacitative coupling would be replaced by the new inductive system.

The use of a magnetic compound material means virtually lossless transmission up to higher transmission frequencies and a comparatively simple manufacturing process. The very first prototype is a system which carries out all essential functions on the rotating part of the steering wheel without the use of electronic elements and which is still predominantly analogue in character. A brief treatment is given of the concept of integrated digital transmission systems on which we are already working.

INTRODUCTION: REPLACEMENT OF THE PREVIOUS CLOCK SPRING, DEVELOPMENT OBJECTIVES FOR A NEW SOLUTION

When a driver airbag is fitted, present-day contact units which enable not only the airbag ignition current and the monitoring current but also the horn signal to be transferred from the fixed chassis (steering column) to the rotating steering wheel (steering spindle) in all cases use so-called clock springs (Fig. 1). In the case of the horn, these are also wiper contacts and represent an electrical connection. This connection is however subject to wear and to fatigue and is one of the leading causes of failure. It is not silent in operation nor easy to install (in the production plants or in the workshop).

Fig. 1: Steering wheel contact unit with an electrical coupling

Clock springs typically cost between DM 10 and 20 each to produce, depending on the individual design.

Particularly in view of impending functional expansion — such as additional driver controls on the steering wheel as well as a steering wheel heating unit — automotive manufacturers are looking for new solutions which would at the same time avoid the drawbacks of the old system but without involving higher costs. In discussions of this topic, a contactless inductive (in other words, working on the transformer principle) signal transfer has turned out to be highly efficient and very promising. This approach is currently being examined by ourselves in collaboration with a steering wheel manufacturer and work is going on to develop it into a new product. The basic idea of a transformer-type signal transfer is not new and thus cannot be patented . However when the new requirements are taken into consideration it does have a much better chance of being introduced than was previously the case. When the steering wheel is removed it will no longer be necessary to undo a multi-core plug connection but simply to separate the two contactless and only magnetically coupled halves of a rotary transformer. This means that the idea of a steering wheel which the driver himself can replace enters the realm of the possible and — as far as reliability is concerned — of the acceptable. An inductive transformer system of this type will be particularly simple and inexpensive when the airbag is not fired using direct voltage, as is usually the case, but using alternating voltage. An alternating voltage ignition system of this type is feasible at only a little more cost as far as the central control unit is concerned and will at the same time mean that there is no longer a requirement for the so-called 'Safing' sensor. This is an acceleration switch which prevents the airbag from being fired when the vehicle ignition is turned on or as the result of direct voltage short circuits when the vehicle is stationary.

ROTARY TRANSFORMER FOR AIRBAG AND HORN WITHOUT ELECTRONIC CIRCUITRY IN THE STEERING WHEEL

Our first and 'analogue' approach was to cover only the range of functions which are already available: in other words, transfer of the ignition signal, monitoring the resistance of the igniter and transfer of the horn signal. Initially the use of electronic circuitry in the steering wheel should and can be entirely dispensed with for the time being and as has previously also been the case. This means that although higher demands are placed on the transformer, due to the absence of local electronic circuitry the system will remain simple and very reliable. Further variants and capacity levels will be described below.

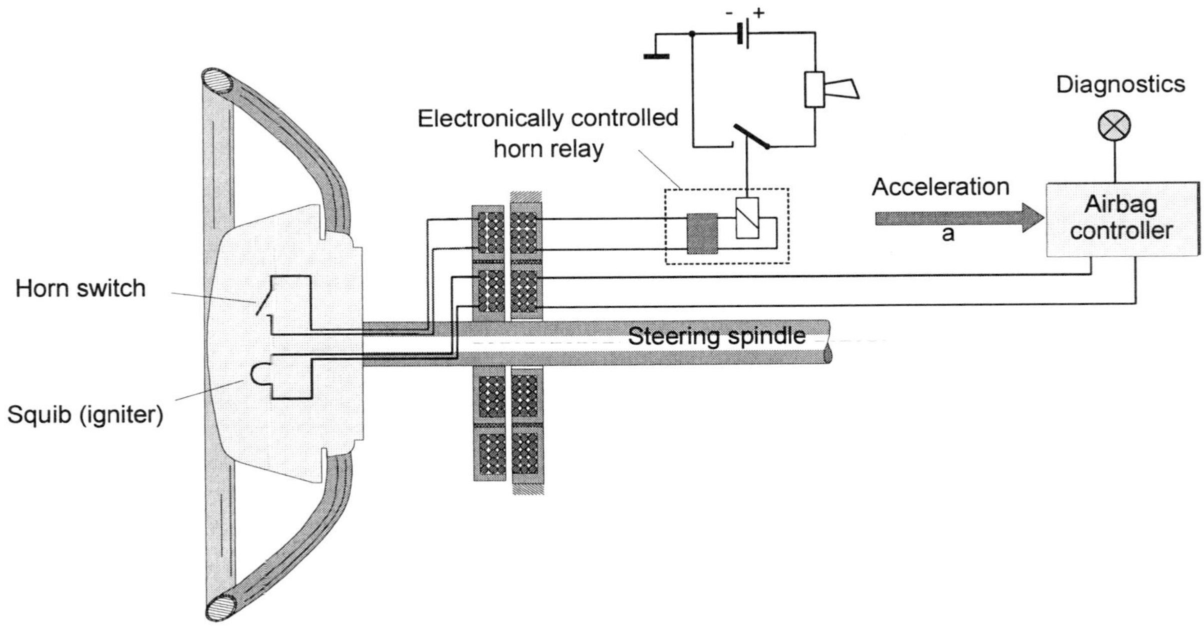

Fig. 2: Contactless signal transfer by transformer on the steering wheel

To realise this system, two separate and concentrically arranged transformers have initially been planned for the airbag and horn functions (Fig. 2). This should ensure the highest degree of reliability for the airbag function. Of course, from the purely technical point of view, the two functions could also be realised, should this be required and permissible, by using a single transformer. The airbag protection system places technical demands according to the following table:

Variable	*Symbol*	*Value*
AMBIENT		
Operating temperature	T	- 40 to 85 °C
AIRBAG IGNITION		
Ignition energy required	E_Z	≤ 15 mJ
Permissible ignition delay	t_Z	≤ 2 ms
Igniter resistance (nominal)	R_Z	2 Ω
No-fire current	I_{nf}	0.2 A
Minimum firing current	I_{Zmin}	1 A
All-fire current	I_{af}	1.75 A
IGNITER DIAGNOSTICS		
Frequency		1 - 5 Hz
Permissible monitoring tolerances	R_{Zu} - R_{Zo}	1.4 - 6 Ω
Permissible monitoring resolution (including temperature influence)	ΔR_Z	± 0.35 Ω
INSTALLATION TOLERANCES: no contact between primary and secondary cores		
Axial	Δz	± 1.5 mm
Radial (air gap)	Δr	± 1 mm
WEIGHT, VOLUME INSTALLED		
Volume	≤ volume of clock spring	
Mass	≤ mass of clock spring (130 g)	
GENERAL REQUIREMENTS		
Any short circuit of the primary supply lines (for example, +12 V and ground) must not result in airbag ignition.		
During installation and removal of the steering wheel, the transformer halves should be separated in such a way that no plugged connection needs to be separated (magnetic interface).		
When the steering wheel has been removed the activated controller must detect as a fault the absence of the secondary half of the transformer (in the same way as if this transformer half had been forgotten during installation).		
When the steering wheel (which contains the ignitable airbag) has been removed it must be ensured that no inadvertent ignition occurs (for example, in response to the stray fields of other devices).		

Table: Demands on airbag protection systems

THE AIRBAG TRANSFORMER

The airbag transformer has been designed such that its magnetic coupling changes as little as possible within the prescribed installation tolerances. This is achieved by using ring cores with a u-shaped cross-section to accommodate the primary and secondary windings, with one core being placed inside the other (Fig. 3). The conical exterior profile of the inserted core and the conical internal profile of the outer core results in additional stabilisation of the coupling factor in the event of the two cores shifting laterally with respect to each other (Fig. 4). With this core shape the magnetic flux changes in the radial direction from one core to the other. The large flux-attenuating air gap arising from the stipulated installation tolerance is compensated for by an increased leg length (Fig. 5).

Fig. 3: Steering wheel contact unit with an inductive coupling

SIMULATION AND CALCULATION OF THE MAGNETIC CIRCUIT - The magnetic response — including that of the secondary electrical load (the igniter) — was simulated and optimised with the aid of a very powerful magnetic field calculation program (MAFIA). This means that new, production-friendlier variants, for example, can be analysed theoretically in a very short time indeed, thus saving valuable time making up and testing hardware variants. Figure 6 shows an example of the profile and intensity of magnetic induction (B-field) in the cross-section of the cores. In addition to arrow diagrams, other, very lucid methods of presenting the material are also possible: for example, coloured plots or so-called intensity plots.

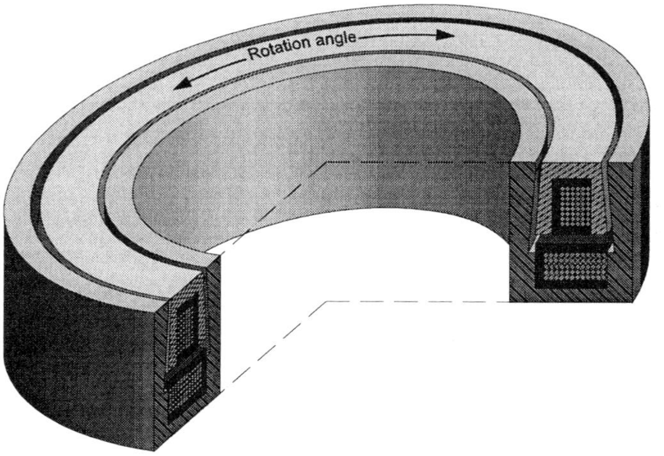

Fig. 4: Rotary transformer for contactless airbag triggering and diagnostics

Figure 7 shows the calculated profile of the primary input inductance L_S during axial shifting of the cores. It is clear that input inductance changes only slightly in response to axial offset of the optimised core shape, and this is confirmed by actual measurements. This is an important requirement if you wish to monitor the input inductance of the igniter resistor (squib).

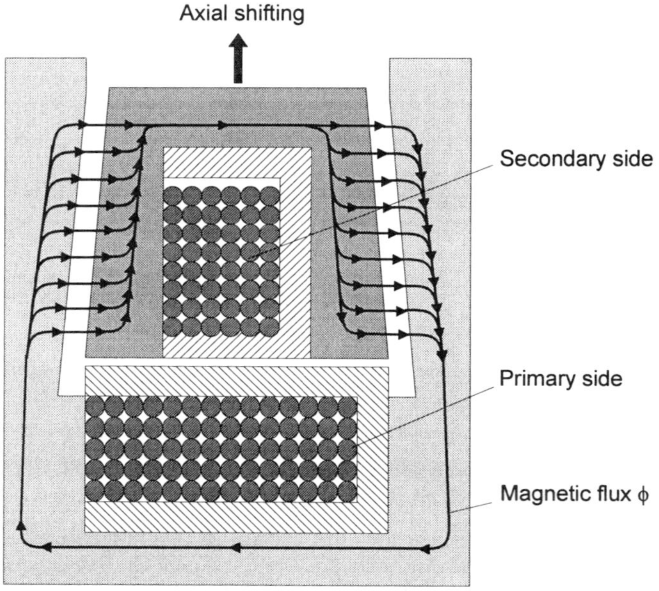

Fig. 5: Airbag transformer

TRANSFORMER MATERIAL - If the transformer cores were to be made of solid iron it would certainly be possible to trigger the airbag but it would not be possible to monitor the secondary igniter resistance. The losses caused by eddy currents in the core

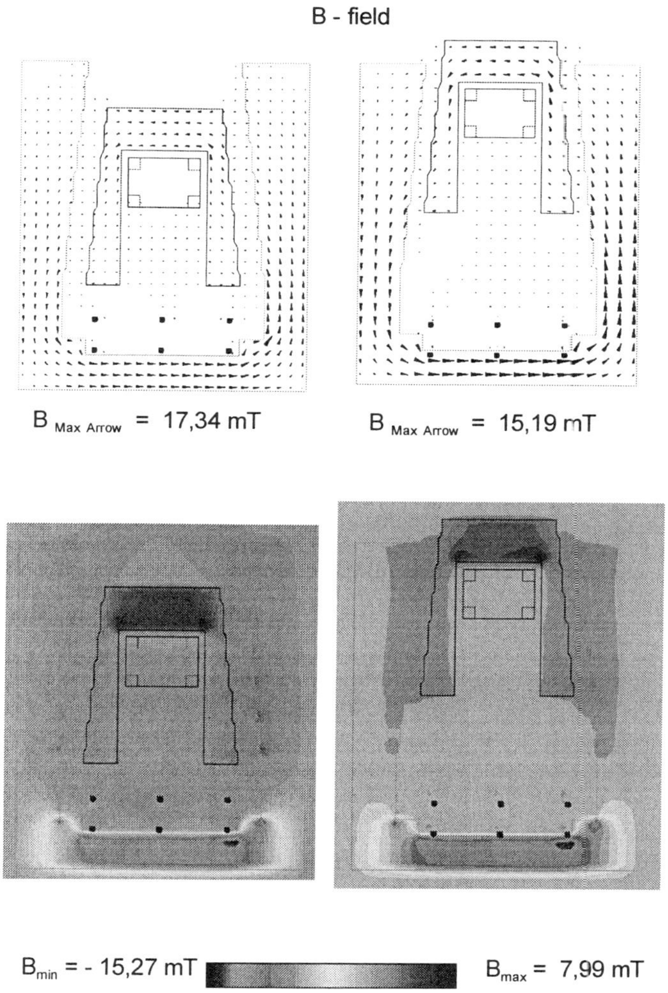

Fig. 6: Calculated magnetic field strength of an airbag transformer in two positions according to mounting tolerances

which are closely linked to temperature and air gap would falsify any measurement made from the primary side to such an extent that any useful diagnosis would be ruled out.

There is a new moulded material whereby the iron particles are covered in plastic and thus insulated from one another. This material is considerably cheaper than the ferrite material usually used for such applications and is virtually loss-free up to high frequencies (100 kHz to 1 MHz). This means that it allows unimpeded observation of the igniter resistance within the required tolerance right through the transformer. Although this new material comes off well when compared with other moulded materials as regards its relative permeability (a value of 160 is typical), it still remains far below the values for normal ferrites. Compensating for this, however, is the fact that the specified permeability figure is to all intents and purposes independent of temperature: in other words, it does not produce any fault effects from fluctuations caused by temperature changes.

Thanks to the low intrinsic losses of this material, a great deal of electric power, considering the size of the cores, can be transformed for specific purposes should this be required — for example, for steering wheel heating. Some very interesting processes have even been developed whereby electric power can be transferred simultaneously with information at a high baud rate.

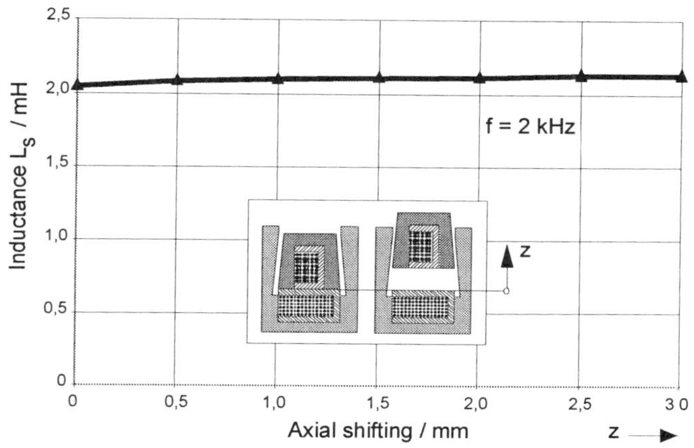

Fig. 7: Airbag transformer, primary inductance $L_S(z)$

MONITORING THE IGNITER RESISTANCE - In the ideal transformer the secondary load resistance (igniter) can also be very simply measured at the primary terminals (Fig. 8) where it may be found, as is well known [1], multiplied by the square of the turn ratio ($u = w_2 / w_1$). With a real transformer, however, this does not apply up to any level of load resistance. If the load resistance is higher than the inductive resistance of the main inductance (ωL_h) the primary resistance will only change a little or will even fall to a lower level in the event of an absence of iron losses. If a relatively high operating frequency is selected (5 to 100 kHz) the primary resistance will change within a wide range in almost perfect proportion to the secondary load resistance (the igniter).

If the lack of unambiguousness at high load resistances (> 100 Ω) makes evaluation of the input resistance R_S problematic, it may be a good idea to use the

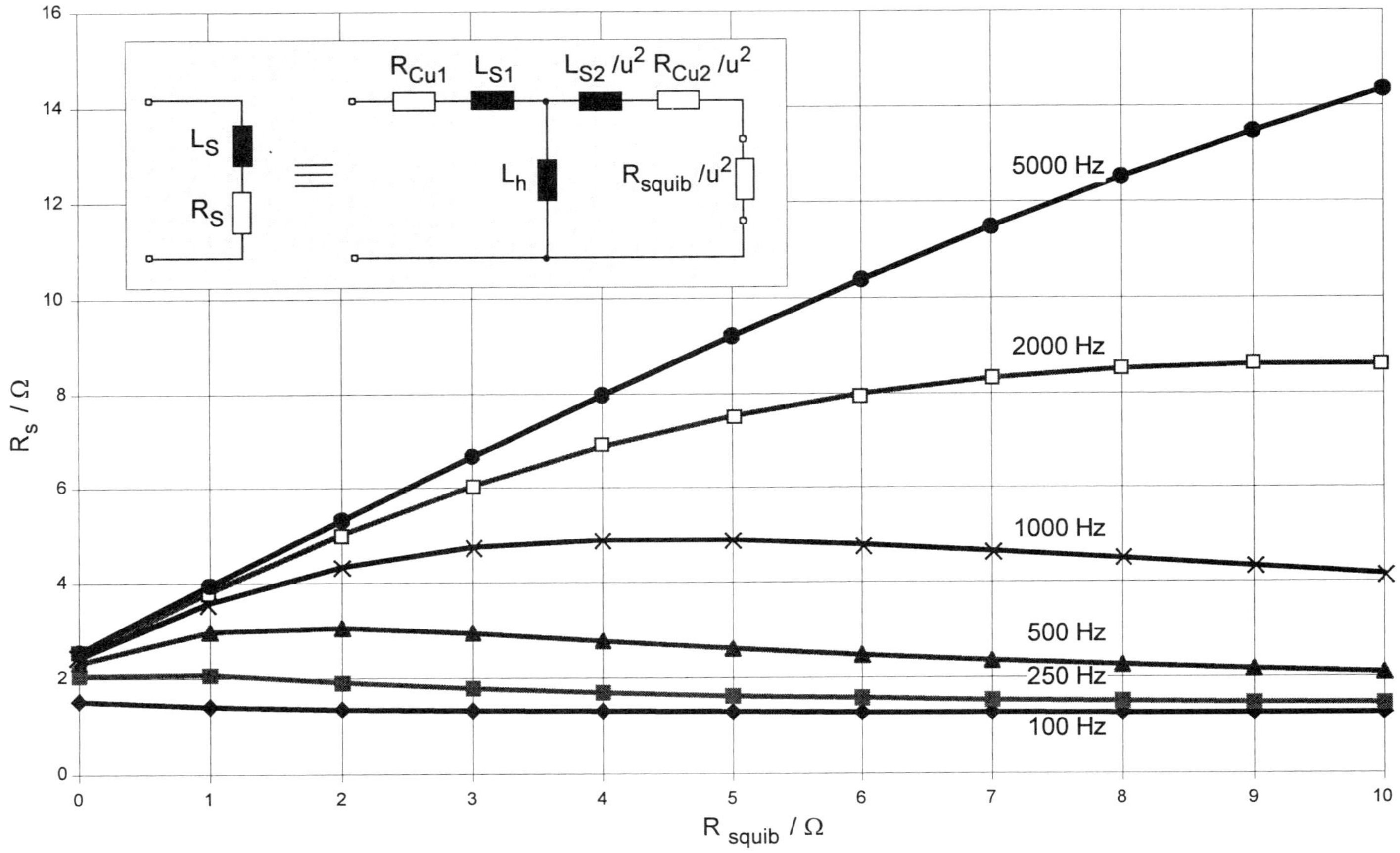

Fig. 8: Measurement on airbag rotary transformer (primary 60 turns; secondary 40 turns; 20°C), series resistance R_S (R_{squib})

inductance L_S to monitor igniter resistance since this inductance can be measured on the primary side. Although it typically changes only in a 1:3 ratio, even at high load resistances it still maintains an unambiguous result (Fig. 9). Here, instead of the secondary stray inductance the far higher main inductance is connected in series with the primary stray inductance by increasing the load resistance. In order to be able to exploit this effect to advantage even at low load resistances (1 - 10 Ω) a fairly low operating frequency (< 1 kHz) should be selected.

MEASUREMENT RESULTS - Figure 8 shows the profile (measured at various temperatures) if the primary-side effective resistance R_S as a function of load resistance as well as the profile measured at room temperature for various axial distances. It is evident that the influencing variables shown do not exert an unacceptable influence on the determination of the igniter resistance. Figure 9 shows the measured profile for inductance values L_S on the primary side. It is clear that thanks to the material's lack of response to temperature changes and its freedom from losses, both temperature and axial shifting only exert a minor influence. These values are not significantly impaired by manufacturing tolerances either.

Due to the closed structure of the iron circuit there is only slight crosstalk between the two transformers which thus effectively excludes mutual interference. When the airbag transformer was tested for electromagnetic compatibility neither the triggering nor the diagnostic functions suffered malfunctions.

A test was conducted whereby the operating frequencies of the airbag were increased up to 1 MHz: no signal losses worth mentioning were detected.

In order to ascertain what power (for example, for steering wheel heating or similar) can still be transferred under continuous operation with the designed transformer made of low-loss moulded material, the power tapped on the secondary side was continually increased by raising the primary feed voltage until the

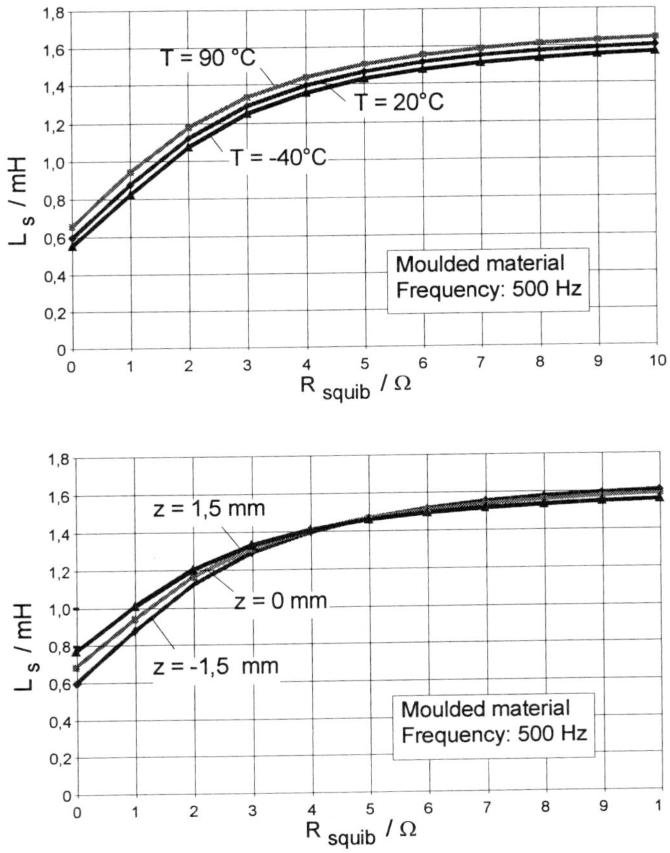

Fig. 9: Series inductance L_S to load resistance R_{squib} at various temperatures and axial distances

transformer reached an overtemperature of around 40 °C. This was only achieved at an output power of approximately 70 W, a figure which may be considered quite high when the volume of the transformer is taken into consideration (steering wheel heating approx. 110 W).

At around 230 g, the overall weight of the prototype (which includes the housing and terminals) is pretty close to the weight of a clock spring (130 g). It appears that further weight savings are possible by further optimisation of the cores.

THE HORN TRANSFORMER

Since the horn transformer, unlike the airbag transformer, does not need to transfer any analogue values but only the binary information 'Horn on' and 'Horn off' it can be made substantially simpler. The core is made of solid iron and in cross-section it has the L-shape which makes for a considerable saving in space (Fig. 10, bottom). A small electronic device is provided on the primary side which in conjunction with the transformer inductance and an additional capacitor can generate a 20-kHz AC voltage (Fig. 2). At 'Horn on' the secondary side is shorted. This de-attenuates the tuned circuit thus created to the point that the oscillator vibrates and the rectified AC voltage signal picks up the horn relay. At 'Horn off' the horn contact is opened. The iron losses which have now increased slow down the response of the tuned circuit to the point that the oscillator cannot vibrate and the horn is not actuated. The small electronic circuitry required for this purpose can easily be accommodated in the horn relay. If an AC voltage relay is used it might even be possible to dispense with the electronic circuitry entirely.

If the steering wheel has been removed or if the transformer half on the steering wheel side has been forgotten, the damping of the tuned circuit will also be decreased due to a substantial reduction in iron losses when the circuit is open. This is similar to when the horn is actuated and in this case too the horn will sound: this is a useful side effect. Figure 10 (top) shows that it is possible to detect to an adequate degree

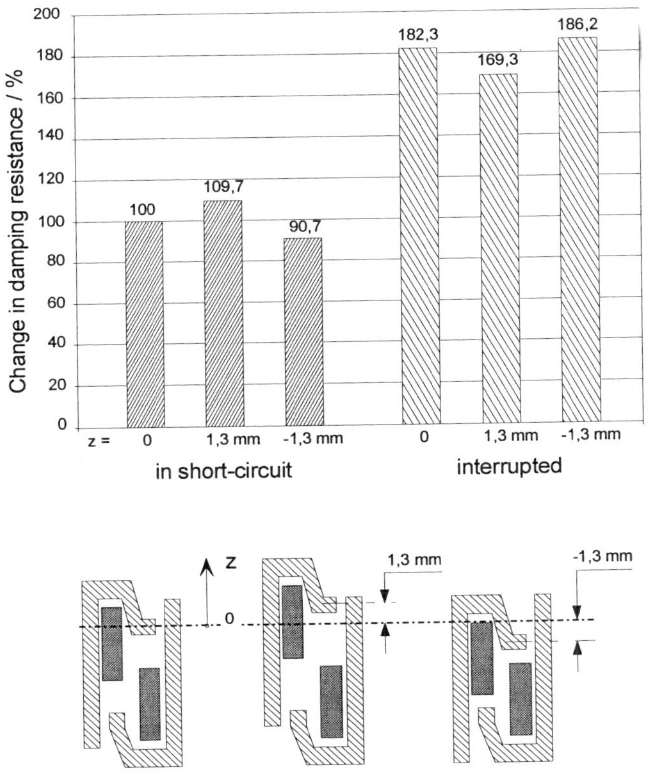

Fig. 10: Horn transformer, influence of axial travel z

of reliability the on and off states of the horn for the whole range of axial play (and also throughout the entire temperature range).

FUNCTION SAMPLES

For initial testing by customers both transformers were installed in a standard housing such as is used for clock springs and as can be fitted in a particular model of vehicle. Figure 11 shows the sample in its open state. It will allow removal of the steering wheel together with the airbag without it being necessary to disconnect any plugged connections (Fig. 12) since the contact unit is only separated magnetically (see under 'General requirements' above).

Fig. 11: Steering wheel contact unit with inductive coupling

For installation in a customer vehicle an airbag controller for alternating current ignition of type AB 6.2 AC was adapted to the transformer coupling with regard to igniter diagnostics but with triggering of the airbag suppressed for reasons of safety. One customer subjected the complete system to an EMC test, which it passed straightaway. (With the clock springs previously used this was not exactly a matter of course, due to the open and asymmetric structure of the springs.) Even previous EMC testing of a more simple nature showed that the transformer does not represent an additional source of problems with regard to EMC interference but is itself very well shielded.

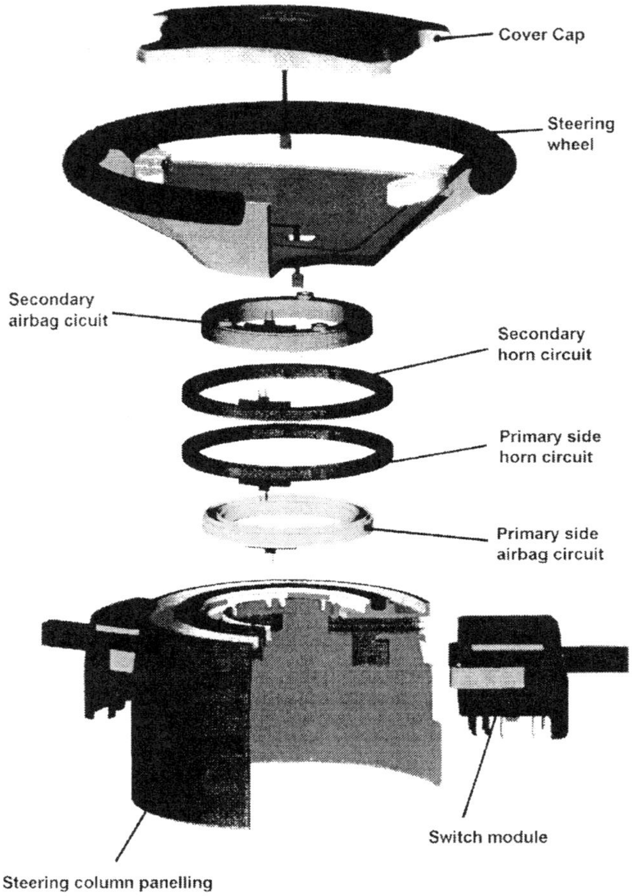

Fig. 12: Assembly drawing of inductive signal and power transformers in the steering column

VARIOUS SOLUTION AND CAPACITY LEVELS

The horn transformer presented here can also be used without any problem for transforming additional switching functions (multifunctional steering wheel). However this will necessitate the installation on the steering wheel of a small electronic multiplexer circuit which will be powered from the same transformer (Fig. 13). The information signals are then superimposed in a suitable manner on the power transfer. The signals are demultiplexed by an electronic circuit fixed to the steering column and fed into a CAN bus, for example. It is even possible for the system to be expanded on the basis of the same basic model to include a complete so-called steering wheel module which apart from performing switching functions also contactlessly transmits signals from, for example, a torque sensor (electric power steering). Once the steering wheel electronic circuitry is in place it can be sensibly extended to take over local monitoring in the steering

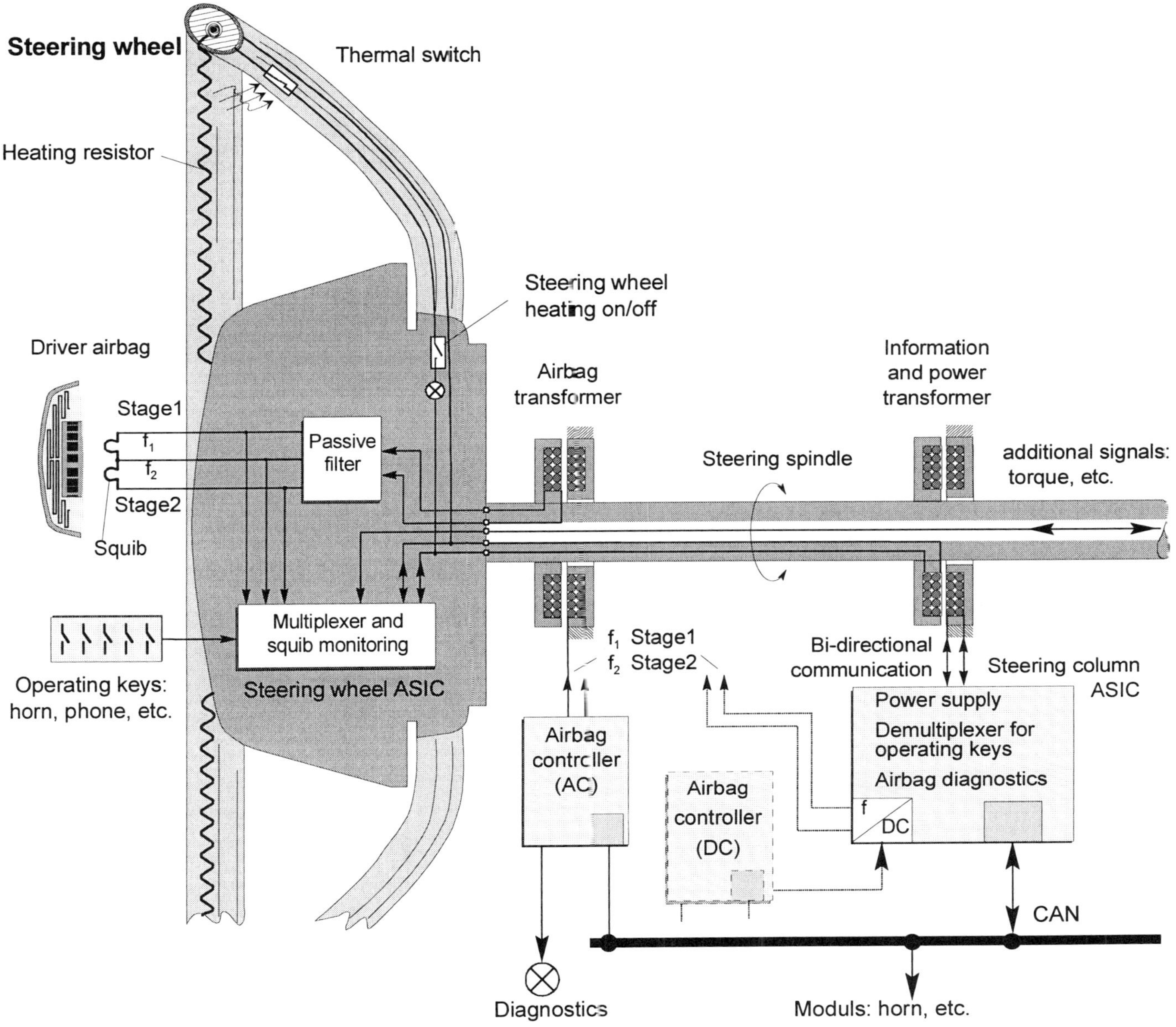

Fig. 13: Inductive contactless signal and power transfer with direct airbag firing

wheel of the igniter(s) as well. This means that now only the results of monitoring need to be transferred to the chassis in the form of a binary good / bad signal. If the power transmitted via this transformer is also enough to feed a steering wheel heater, the transformer core will need to be designed and optimised magnetically and the transformer windings designed and optimised thermally.

Since in a system of this kind the airbag transformer is only used for transmitting the ignition energy it can be considerably simplified and be given a more production-friendly design. Accordingly, the conical shape for compensating for geometric tolerances and which is somewhat difficult to manufacture can be dispensed with. If, as will be the case in the future, the driver airbag is to be ignited in two stages, then the ignition signals can, for example, be frequency-coded [2] and then separated again on the steering wheel by means of a passive frequency separating filter so as to route them to the corresponding igniters.

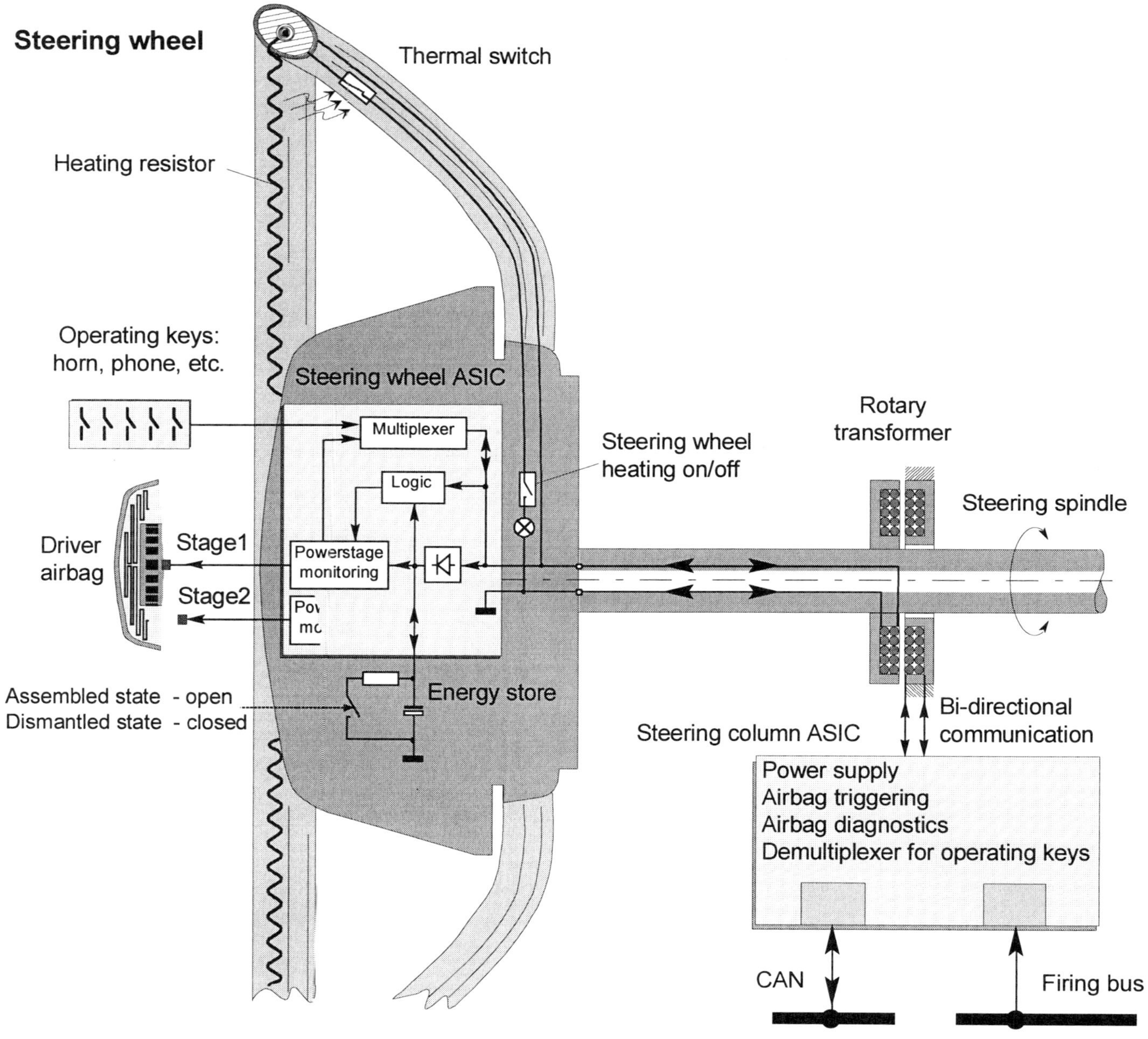

Fig. 14: Inductive power and signal transfer on the steering wheel with digitally coded airbag firing

A transformer which is more production-friendly could also be designed if all information, including the triggering command for the gas generator, were to be transferred in digital (encoded) form (Fig. 14). Here a single transformer would almost certainly be adequate for all functions. However there is now a need for a highly reliable local electronic circuit for airbag triggering as well, and also a remote energy storage facility and remote output stage which is capable of re-igniting the airbag on a direct current basis. The decoding logic and the output stage would then be located so closely to the igniter that false triggering as the result of incorrect contacting would be ruled out. Although the demands on the transformer in this case are considerably lower, more serious system and reliability problems have to be solved in the design of a digital controller, the so-called firing bus. Until this system has actually become reality, the system we have been describing — one with more analogue, contactless airbag contacting — would be able to render very satisfactory service for some time yet.

CONCLUSION

The clock springs normally used today can be advantageously replaced by contactless inductive signal and energy transformers in the steering wheel at approximately the same cost. The benefits deriving from this are essentially:

- An increase in reliability
- Easier to install or remove
  (separation without disconnecting plugs)
- Silent operation

Following more analogue transfer systems with a low proportion of electronic circuitry and without a firing energy reservoir in the steering wheel, the trend will be towards purely digital transfer systems which will ignite the airbag from an energy reservoir installed in the steering wheel and which will probably manage with a single, more simple transformer. The material required for such a development — low losses up to high frequencies and at the same time production-friendly — is already available.

The contactless signal and energy transfer described here could be used in a similarly advantageous manner in other places in the vehicle as well, places where signals and energy must be transferred between the chassis and moving parts. This could, for example, apply to tyre pressure monitoring, to the control of complex functions in the vehicle doors ('inductive hinges') [3] and with moveable, sometimes even removeable, car seats [4]. The advantages of inductive signal and energy transfer are exploited not only in vehicles but also in the articulated joints of robots, for example[5].

ACKNOWLEDGMENTS

The authors would like to thank K. Dobler for contributing valuable design ideas and M. Rosenberger for carefully arranging the layout of text and figures of this paper.

REFERENCES

[1] Fischer, J.: „Elektrodynamik", Springer-Verlag Berlin, Heidelberg, New York 1976

[2] Anthofer, A. et. al.: „System zum Auslösen eines Rückhaltemittels in einem Fahrzeug", Patent (Siemens) DE 19536573C1, 1996

[3] Uhl, G., Schulze, G.: „Induktion statt Kabel", Automobil-Elektronik, April 1993

[4] Zabler, E. et. al.: „Anordnung zum kontaktlosen übertragen von Signalen zwischen gegeneinander linear bewegbaren Fahrzeugteilen", Patent (Bosch) OS DE 19545220, 1997

[5] Esser, A.: „Berührungslose, kombinierte Energie- und Informationsübertragung für bewegte Systeme", Dissertation, Technische Hochschule Aachen, Germany, 1992

MULTIPURPOSE ACTUATORS

970852

Compact Magnetic Solenoid Valves Using a Composite Magnetic Material

Yoshihiro Tanimura, Keizo Takeuchi, Toshiaki Terada, Shinya Sugiura and Yoshitada Katayama
DENSO CORPORATION

Tsutomu Inui
Hitachi Metals, Ltd.

ABSTRACT

The recent progress of electronic control systems in vehicles is remarkable as evidenced by the development of electronic fuel injection systems,(EFI), automatic transmission control systems, and anti-lock brake systems,(ABS). The number of actuators for the systems has been increasing.

Consequently, a need has been identified for a reduction in volume and number of the system actuators for control purposes. A composite magnetic material has been developed with the aim of miniaturizing magnetic solenoid valves for actuator applications. A composite magnetic material is such that both ferromagnetic and paramagnetic sections coexist within a single material, and can contribute to optimization of the magnetic circuit of a solenoid valve.

This paper describes the development of a composite magnetic material, and its resultant characteristics. The chemical composition of the material, essentially 18% Chromium - 8% Nickel stainless steel, required to improve both ferromagnetic and paramagnetic properties and the process of producing composite magnetic properties are reported in detail.

This composite magnetic material exhibited superior properties, having a magnetic flux density B40 (Induction at H of 4000A/m) of 0.8 Tesla in the ferromagnetic section, and relative permeability μr of 1.01 in the paramagnetic section. The application of this material to a magnetic solenoid valve for ABS could reduce by approximately 20% volume.

INTRODUCTION

Magnetic solenoid valves shown in Fig.1 are usually constructed from coil, moving core, stator core, and valve components. The function of the valve is to control the flow of a liquid or gas by moving a core using a magnetic force. In order to seal the liquid or gas, a metal component called a sleeve is used. This component is generally paramagnetic to maximize magnetic flux transfer from the moving core to the stator core.

If this component is constructed from a ferromagnetic section and a paramagnetic section as shown in Fig.2, magnetic efficiency can be improved, and the volume of the valve can be reduced. This may be accomplished by joining ferromagnetic and paramagnetic sections by welding; however, this method is costly, and presents difficulties in terms of sealing.

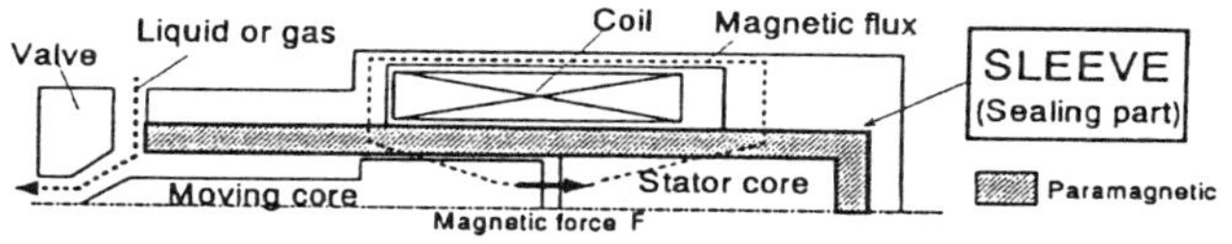

Figure 1. Typical magnetic solenoid valve

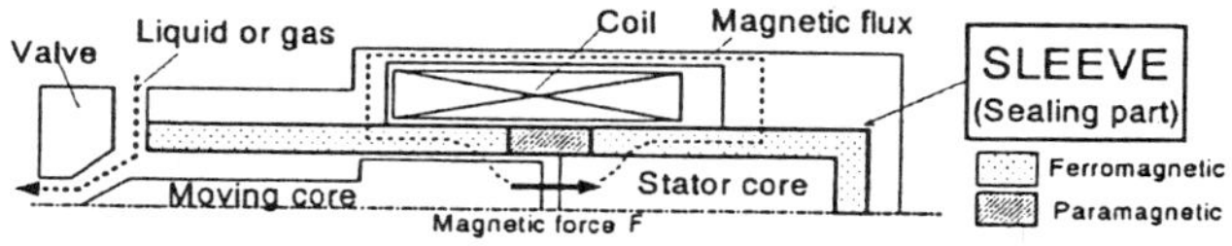

Figure 2. Efficient magnetic solenoid valve

A material has been investigated which may be either ferromagnetic or paramagnetic through the use of simple processing, and which may have both ferromagnetic and paramagnetic sections in coexistence. This material has been designated a composite magnetic material.

CONCEPT OF THE COMPOSITE MAGNETIC MATERIAL

It has been well known that austenitic stainless steels (e.g. AISI type 304) are composed of austenite after solid solution treatment, and that these materials exhibit paramagnetic properties. Furthermore, the austenite phase may be transformed to martensite through cold working, so called deformation-induced martensite, and the resultant martensite phase is ferromagnetic. Therefore, when a part of this martensite is made transform to austenite by heat treatment, a composite magnetic material is obtained as shown in Fig.3.

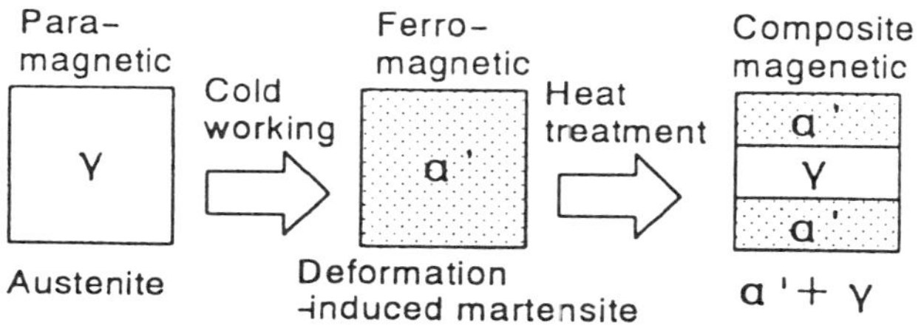

Figure 3. Schematic production process of composite magnetic material

Unfortunately, the amount of deformation-induced martensite formed upon cold working is small in typical austenitic stainless steels, such as AISI type 304, and these steels have insufficient ferromagnetic properties, a low magnetic flux density and a low maximum permeability. The development of a composite magnetic material was based upon the optimization of 18Cr 8Ni austenitic stainless steel with improved magnetic properties.

EFFECT OF CHEMICAL COMPOSITION ON MAGNETIC PROPERTIES

The effect of chemical composition on magnetic properties was investigated using various 18-8 stainless steels as detailed in Table 1.

Magnetic flux density, B40 (Induction at H=4000 A/m), for test samples after 70% cold reduction is shown in Fig.4 as a function of (Ni). Relative permeability after solid solution treatment is shown as a function of (Ni) in Fig.5. (Ni) is Hirayama's Ni equivalent [1,2] and defined as follows:

$$(Ni) = [Ni] + 0.65[Cr] + 0.98[Mo] + 1.05[Mn] + 0.35[Si] + 12.6[C] \quad (1)$$

where [Ni], [Cr], etc. are elemental concentrations in wt% of the steel of interest. It has been known that (Ni) has a close relation to the volume fraction of deformation-induced martensite or athermal martensite.

The magnetic flux density following cold rolling increases with the reduction of (Ni) as seen in Fig.4. Additionally, the relative permeability following annealing decreases with increasing (Ni) as seen in Fig.5. Using these two factors, a chemical composition which optimizes magnetic properties may be arrived at. Magnetic flux density indicates that (Ni) should be held below 21.5 wt%. Relative permeability data indicates that optimum performance is obtained with (Ni) above 20.5 wt%.

Md AND Ms TEMPERATURES

The application of the composite magnetic material to automotive magnetic solenoid valve

Table 1. Chemical compositions of steels examined (wt%)

No.	C	Si	Mn	Ni	Cr	N	(Ni)
1	0.012	0.30	0.52	8.32	17.45	0.0018	20.5
2	0.023	0.30	0.50	8.39	17.58	0.0025	20.7
3	0.050	0.29	0.51	8.36	17.70	0.0033	21.1
4	0.030	0.10	0.50	8.34	17.84	0.0028	20.9
5	0.029	0.46	0.48	8.41	17.70	0.0042	21.0
6	0.031	0.86	0.51	8.33	17.91	0.0034	21.2
7	0.030	0.32	0.12	8.76	17.68	0.0011	20.9
8	0.031	0.31	0.79	8.38	17.72	0.0031	21.2
9	0.031	0.30	0.51	8.01	17.68	0.0028	20.5
10	0.030	0.29	0.51	8.17	17.90	0.0044	20.8
11	0.029	0.30	0.50	8.67	17.99	0.0036	21.4
12	0.030	0.28	0.50	8.37	17.23	0.0032	20.6
13	0.031	0.29	0.50	8.40	17.40	0.0022	20.7
14	0.031	0.29	0.52	8.37	18.32	0.0037	21.3
15	0.032	0.29	0.51	8.32	17.98	0.0120	21.1
16	0.032	0.29	0.50	8.35	17.98	0.0190	21.1

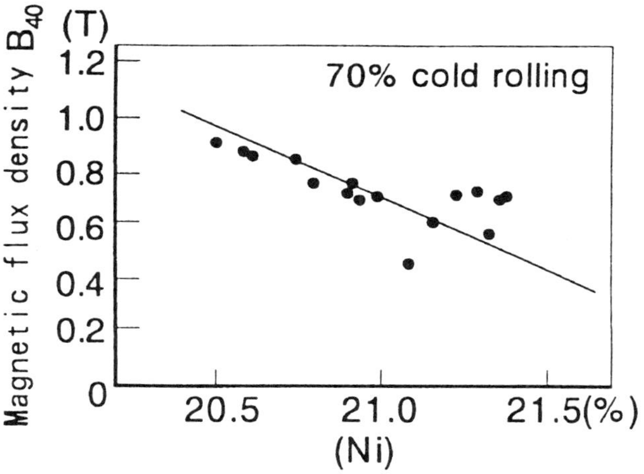

Figure 4. Variation of magnetic flux density as a function of Hirayama's Ni equivalent (Ni)

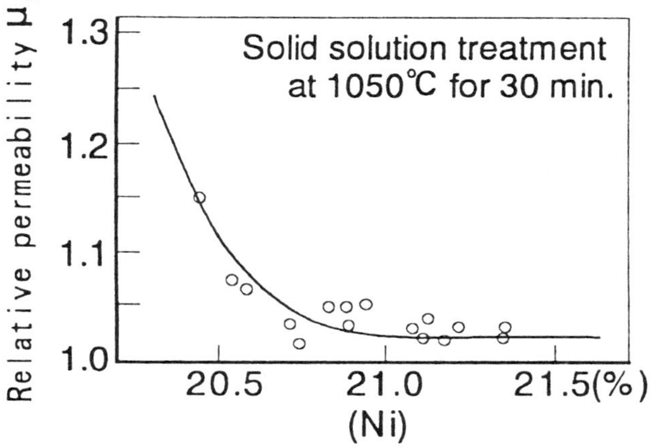

Figure 5. Variation of relative permeability μ r as a function of Hirayama's equivalent (Ni)

actuator sleeves further requires an investigation of the Md and Ms temperatures of the alloy. This is necessitated by the relationship of the Md temperature to ferromagnetism, and the Ms temperature to paramagnetism. Chemical compositions of the steels which were examined are given in Table 2.

Md TEMPERATURE - The Md temperature is defined as the temperature above which austenite will not transform to deformation-induced martensite upon mechanical working[3]. The current application dictates that the material temperature during cold deformation should be kept below the Md temperature so as to enhance the ferromagnetic properties of the material, thus promoting magnetic flux density and maximum permeability.

The martensite content variation as a function of temperature at 30% true strain is illustrated in Fig.6. Deformation-induced martensite fraction decreases rapidly with increasing temperature. The Md temperature can be estimated from Fig.6 by observing the temperature at which the volume fraction of martensite approached zero.

The variation of Md temperature with (Ni) is shown in Fig.7. As can be seen, the Md temperature is approximately 80℃, and is a very weak function to (Ni). This data indicates that working temperature is a critical property for ferromagnetic property development, and is a critical parameter for process control.

Ms TEMPERATURE - Paramagnetic austenite transforms to athermal martensite below the Ms temperature, and the magnetic property becomes ferromagnetic. The Ms temperature of the composite magnetic material should be lower than the expected operating range of the component to inhibit this transformation. If not, the relative permeability of the paramagnetic section would increase, and the advantages gained with respect to magnetic flux transfer between the moving core and the stator core would be lost.

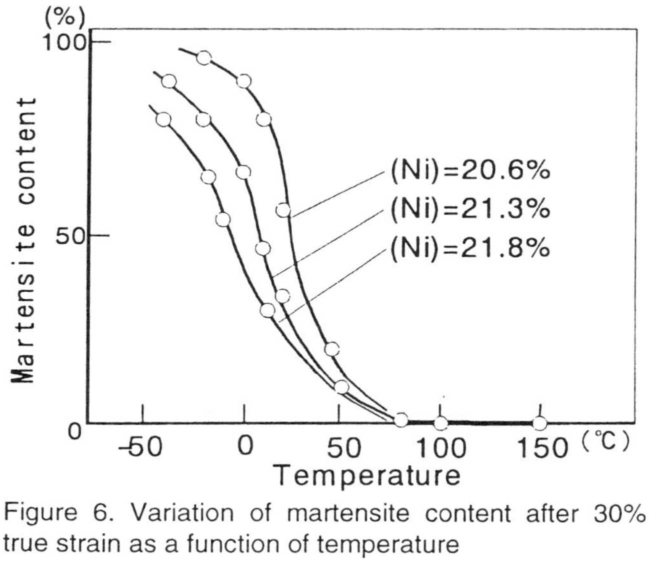

Figure 6. Variation of martensite content after 30% true strain as a function of temperature

Table 2. Chemical compositions of steels examined (wt%)

C	Si	Mn	Ni	Cr	N	(Ni)
0.028	0.17	0.43	8.20	17.75	0.0017	20.6
0.038	0.23	0.52	8.45	18.11	0.0015	21.3
0.038	0.28	0.55	8.75	18.19	0.0017	21.8

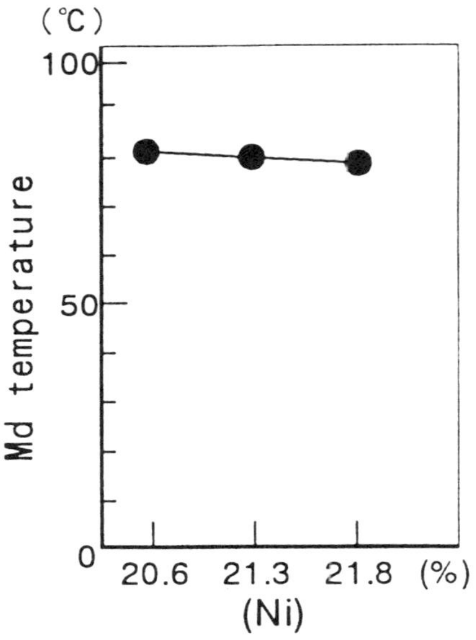

Figure 7. Variation of Md temperature as a function of Hirayama's Ni equivalent (Ni)

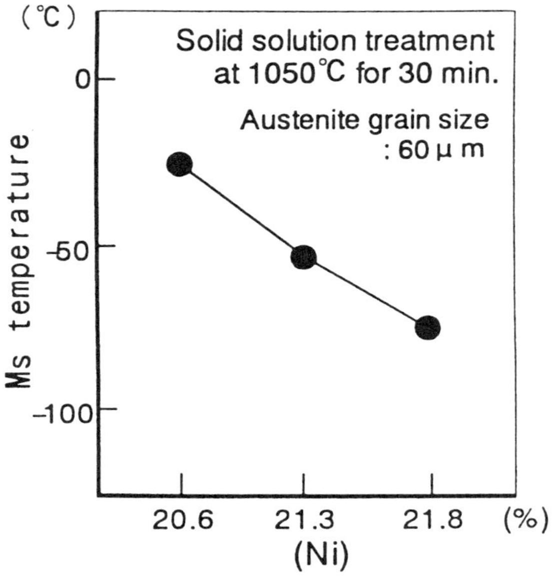

Figure 8. Variation of Ms temperature as a function of Hirayama's equivalent (Ni)

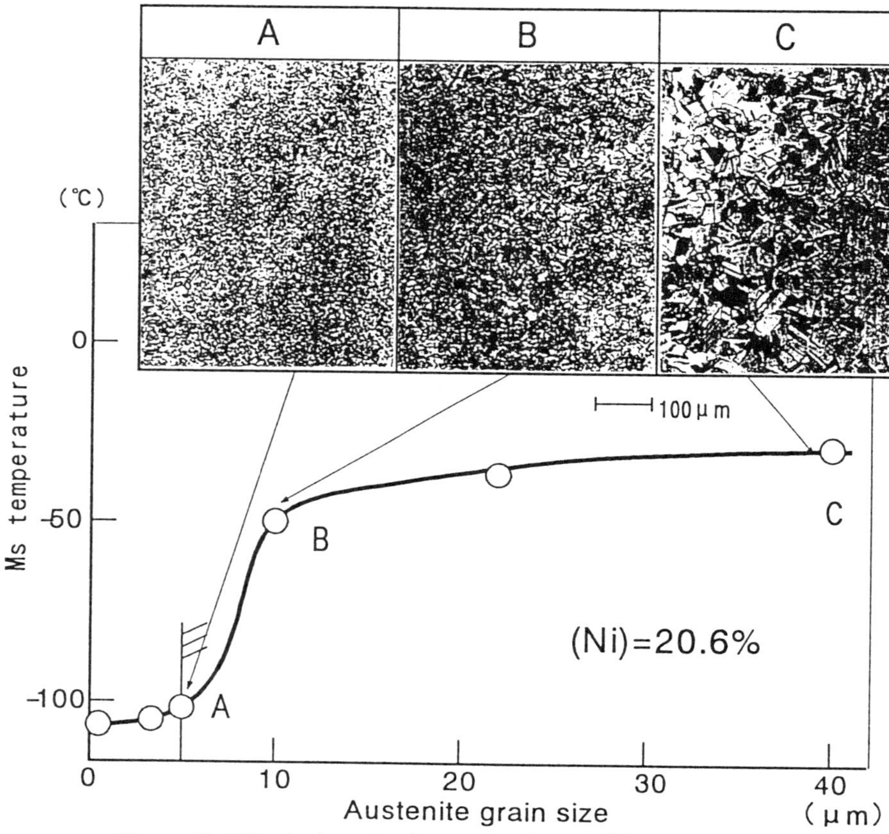

Figure 9. Effect of austenite grain size on Ms temperature

The variation of the Ms temperature with (Ni) for the materials in Table 2 is illustrated in Fig.8. The Ms temperature was measured with the differential scanning calorimeter. Ms temperature is found to decrease as the (Ni) is increased. At a (Ni) of 20.6%, the Ms temperature is found to be -28℃. As the actuator operating temperature may be below -28℃, the Ms temperature must be decreased further than is possible from strictly chemistry control.

There have been numerous studies of the Ms temperature in various steels throughout the literature[4-6], mainly conducted in Fe-Ni and Fe-Ni-C systems. Several studies have been conducted using Fe-Ni-Cr alloys, such as AISI type 304. Previous studies of Fe-Ni and Fe-Ni-C systems had concluded that the austenitic grain size had an effect on the Ms temperature of these steels [7], but that this grain size effect was small in comparison when Fe-Ni-Cr system was considered.

Based on these results, the effect of grain size on the Ms temperature of Fe-Ni-Cr steels was investigated further. The steel considered for this portion of the evaluation had a (Ni) of 20.6%. Austenitic grain size was varied through control of the austenitizing temperature of the alloy.

Fig.9 illustrates the Ms temperature as a function of austenite grain size. The Ms temperature is found to rapidly decrease with decreasing austenitic grain size when the austenite grain size is below about 10 μm. This effect is enhanced by controlling the austenitic grain size below 5 μm, for maximum Ms reduction.

APPLICATION TO ABS MAGNETIC SOLENOID VALVES

The composite magnetic material discussed above has been suggested for application in an ABS magnetic solenoid valve. A schematic of this type of valve is given in Fig.10. The composite magnetic materials described in Table 2 were tested for performance in the sleeve component of the valve. A production process for the manufacture of the composite magnetic sleeve is illustrated in Fig.11.

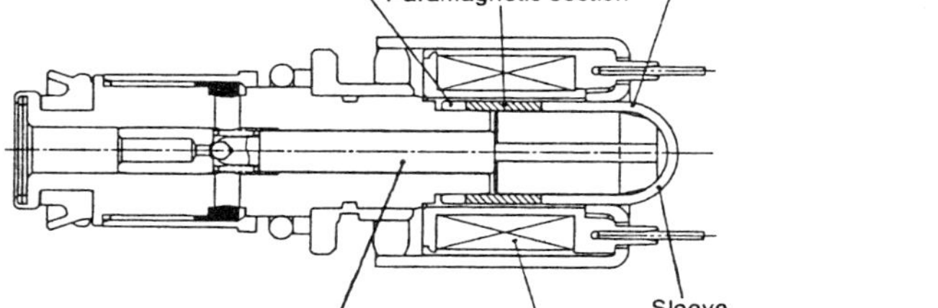

Figure 10. Schematic view of ABS magnetic solenoid valve

The sleeve was formed of the composite magnetic material, having an initial structure consisting of paramagnetic austenite. The working temperature was maintained below the 80 ℃ Md temperature established earlier. Mechanical working under these conditions induces a certain portion of the austenite to transform to ferromagnetic martensite as detailed previously. Subsequently, a section of the sleeve was induction heated into the austenite range to form the paramagnetic region of the sleeve. Induction heating parameters such as frequency, voltage, current, and time were adjusted to yield an austenitic grain size of less than 5 μ m, and thus obtain maximum Ms reduction.

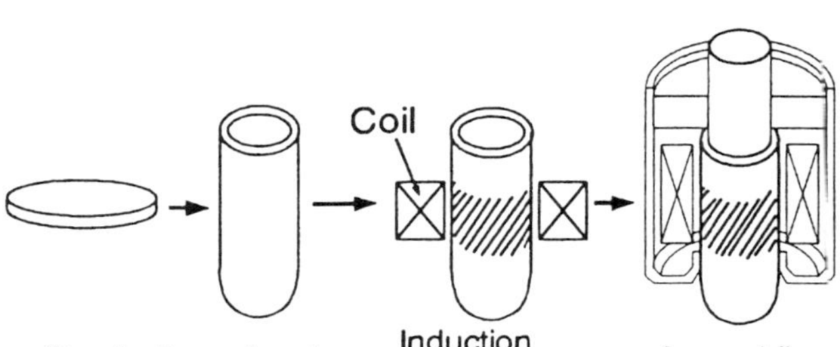

Figure 11. Schematic production process of sleeve of ABS magnetic solenoid valve

Table 3. Magnetic properties and Ms temperature of composite magnetic sleeves

(Ni)	B40(T)	μr	Ms(℃)
20.6	0.38	1.007	-110
21.3	0.60	1.012	-105
21.8	0.80	1.020	-90
Desired value for sleeve of ABS valve	≧0.3	≦1.2	<-40

·B40 of ferromagnetic section Is measured after press forming

·μr and Ms of paramagnetic section are measured after induction heat treatment

Table 4. Comparison of ABS magnetic solenoid valve performance

	Sleeve material	
	Composite magnetic material	AISI type 304
Outer diameter of ABS magnetic solenoid valve (mm)	φ19	φ24
Coil ampere turns (A·turns)	320	452
Magnetic force (N)	30	30

·Composite magnetic material has (Ni) of 21.3%

The resultant magnetic properties of the both ferromagnetic and paramagnetic sections of the composite magnetic material are given in Table 3. The resultant Ms temperature of the paramagnetic section is also given. The sleeve was found to have a magnetic flux density, B40, of 0.38 - 0.80 T, a relative permeability, μ r, of 1.007 - 1.020, and an Ms temperature of -90 to -105℃. These properties satisfied required values.

A comparison of the ABS magnetic solenoid valve with a composite magnetic sleeve and with a sleeve of AISI type 304 is given in Table 4. Application of the composite magnetic material to the sleeve allows a 20% reduction in outer diameter as compared to the AISI type 304 sleeve. ABS valves using the composite magnetic material have been successfully used in field service since 1996.

CONCLUSION

A composite magnetic material having both ferromagnetic and paramagnetic sections has been developed from an austenitic stainless steel.

Results of this development can be summarized as follows:

(1) The coexistence of ferromagnetic and paramagnetic regions within the material may be accomplished by holding the Hirayama's Ni equivalent (Ni) within the range of 20.5 to 21.5%.
(2) Ms temperature of the alloy was substantially reduced through the control of the austenitic grain size.
(3) Application of the composite magnetic material to the sleeve of the ABS magnetic solenoid valve allowed reduction of the valve outer diameter of approximately 20%.
(4) Typical properties of the sleeve component were found to meet product expectations as follows: magnetic flux density, B40, 0.38 - 0.80 T, relative permeability, μ r, of 1.007 - 1.020, and Ms temperature of -90 to -105℃.

REFERENCES

[1] T. Hirayama and M. Odagiri, Journal of the Japan institute of metals, Vol. 34, No. 5, p. 507, 1970.
[2] T. Hirayama and M. Odagiri, Journal of the Japan institute of metals, Vol. 34, No. 5, p. 511, 1970.
[3] T. Hirayama and M. Odagiri, Journal of the Japan institute of metals, Vol. 34, No. 8, p. 826, 1970.
[4] V. Raghavan and A. R. Entwisle, Physical Properties Special Report No. 9, London, 1965.
[5] S. R. Pati and M. Cohen, Acta Met., Vol. 14 , P. 1001, 1966.
[6] A. R. Entwisle and J. Feeney, Institute of Metals Monogragh, No. 33, p. 156, 1968.
[7] M. Umemoto and W. S. Dwen, Metallurgical Transaction, Vol. 5, September, p. 2041, 1974.

940838

A New Multipurpose Servo Valve for Fast Applications

Christian Menard
Thomson-CSF

Abstract

Actuator technology is changing fast and most manufacturers are endeavoring to cut down on costs of studies and produce products in series.

It can be said that actuators govern the action faculties of "muscles" in applications. Depending on the area, actuators are considered to be slow, semi-fast, fast or otherwise control influx : flow, action, pressure, power, strenght.

There is a large family of actuators but it is indeed incomplete for many years because many applications require fast actuators capable of using relatively high flows and pressure levels at acceptable costs. For instance, flows ranging from 0,16 to 0,5 dm^3/s, pressures of 100 to 300 x 10^5 Pa on pressure loads of 20 Hz to 30 Hz bandwidth.

Today, the main clients are users of aircraft and automobile brakes and other industrial areas.

This article is aimed at provided professionals with information about fast acting actuator technology in the form a new multipurpose servo valve for fast applications.

1) Introduction

Today, every engineer knows that the performance of a system depends upon the control devices integrated into control and test equipment.

As far as devices using hydraulic controls are concerned, studies have demonstrated that high performance actuating arrangements are necessary, capable of meeting the demands of "fast real time" applications. For instance, it is known that a servo-valve having a pressure bandwidth included between 20 and 30 Hz would resolve many problems of a braked wheel in contact with a paved and set dry pavement.

However, if the performance of the actuating device is of particular importance, we must look at the conception of a modular design concept so as to scope with the following general objectives :

- *Scope* : reduction of production costs by modular design of actuators, upgradeability.
- *System definition* : splitting the hydraulic or electronic actuators into individual interchangeable functional modules with standardized components and interfaces.
- *Development status* : modular design fonctionnal system in production.
- *Requirements* : reliable functional transmission, interchangeability, compatible with monitoring and control equipment.
- *Assessment criteria* : installation conditions, costs, weight.

2) Description

2.1) Fonction

The function of a three-way proportional actuator consists in establishing correspondence between an electrical input quantity, current or voltage, and two fluid passage sections coming from two opposite directions having a common point related to the load.

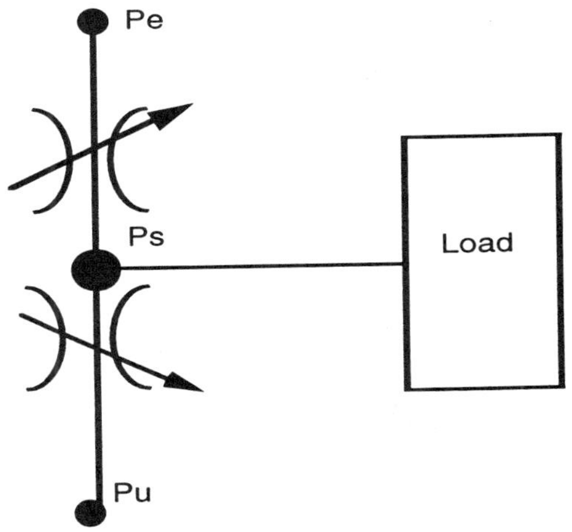

Fig. 1 Functional 3 way

In this way it is possible to modulate the mass incoming or outgoing flow at the load orifice as long as the pressure value is included between the source pressure and the exhaust pressure. The ideal component is characterized by static and linear correspondence. The output magnitude is a flow or a pressure when the device incorporates a servo system intended to control the flow and a servo system designed to control the pressure.

The nature of the incoming magnitude requires including an electric actuator whose position can be tested. For hydraulics, in spite of the existence of fast pressurizing dynamics and high value jet action, it is possible to produce a valve that only has a simple fluid stage in order to improve the overall bandwidth. The hydraulic stage is a slide stage whose positioning dynamic depends partly on the mechanical part and partly on the dynamic that establishes current in the electrical actuator. To decrease the current establishing time a voltage/current power amplifier is used. The actuator, comprising a chamber, means using a three-channel module according to figure 2 for which compliance with the adjustments of the liner and slide do not involve any delicate adjustments.

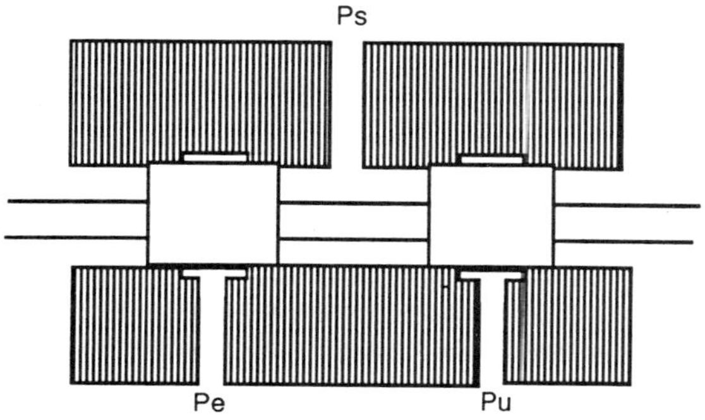

Fig. 2 Actuator 3 way

2.2) Modular design

The design of the servo valve includes :

• a 0 to 45° rotary electromagnetic motor combined with a cursorless potentiometer repeating the position of the electromagnetic motor shaft.

• a mechanical unit through which a hydraulic slide slides.

• a hydraulic unit determining the flow characteristics of the servo valve.

• a mechanical coupling transforms the rotary movement of the electromagnetic motor into a linear movement for the hydraulic slide. This coupling includes a spring designed to bring the electromagnetic motor back to the zero position.

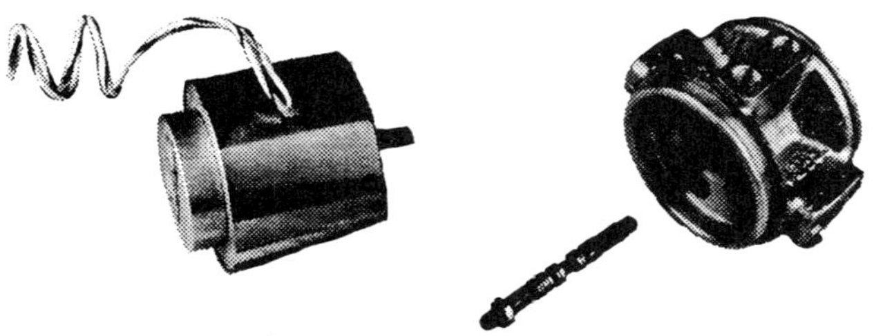

Fig. 3 Basic components

2.3) Main elements

The following description of the main elements indicates the design of the servo valve modularity.

• **Electromagnetic motor**

The electromagnetic motor is a rotor and stator assembly. It is built with a disk-shaped magnet and an axial gap while the stator comprises a SiFE isotropic magnetic molded part supporting four coils.

The rotor is very simple : an iron core is press fitted on the shaft. The gap between the magnet and stator is 0,4 to 0,5 mm.

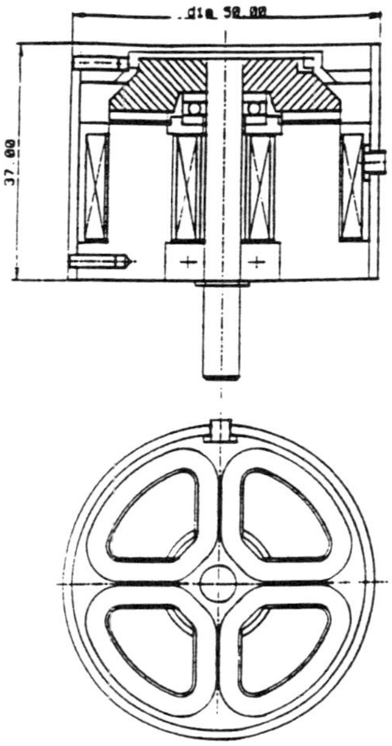

Fig. 4 Electromagnetic motor

The torque is a linear, function of the current up to very high peak current and the relationship torque versus current for different angular positions.

The angle of variation of the motor stroke is included between 0 and 45° so that it is possible to choose the area that is interesting for operation.

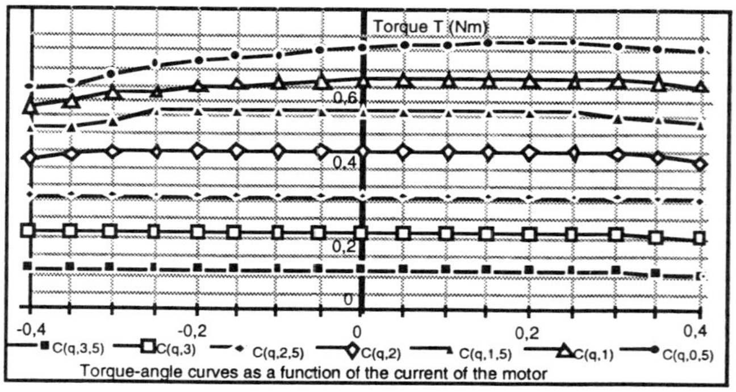

Fig.5 Torque vs angle

Key points are

- good ratio torque/size and torque/mass.
- constant torque versus position for a given current.
- very linear torque versus current in the rated operating range.
- low detent torque.
- high ratio peak torque/rated torque for instance to break the strength and possibility for the same purpose to superimpose a vibration (allows a lower peak torque value).
- large operating range of temperature.
- possibility to integrate the actuator in a die casted valve body (1 bearing only). The only expensive part is still the oriented ring magnet, but its cost is expected to decrease significantly as production grows.
- power supply 12 V.
- two wires
- temperature range - 40°C + 150°C.

• Rotary position sensor

The sensor is a cursorless concept

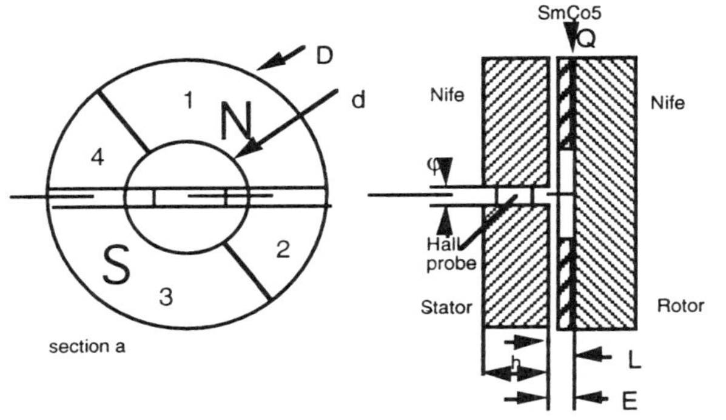

Fig. 6 Rotary sensor

with disc or ring magnet and linear hall device featuring high linearity and resolution at low cost. Position sensor has a capability of operating in a very large range of temperature typically - 40°C to 120°C, peak temperature 150°C.

- large strokes, up to 170° in sizes > 40 mm.
- high linearity, including microlinearity, up to 99,8 %.
- high resolution, down to 0.01 % of the full scall.
- low cost, comparable to that of good multibrush potentiometers.
- same output as potentiometers.
- good design flexibility for disc or ring shaped magnets.
- power supply 5 V.
- output signal 2,5 V to 3,5 V.
- temperature range - 30°C + 120°C (peak 150°C).
- three wires.

• Movement transformation mechanics

The axis of the motor whose angular variation is included between 0° and 45° is associated with a connecting rod, ball end and clevis system transforming the rotary movement into a linear movement designed to move the hydraulic slide.

The advantage of this arrangement is the fact that the slide, depending on the arm length, is adaptable to the geometry of the hydraulic slide, customarily ± 4 mm. The device also features a return spring whose functionality is to bring the electromagnetic motor back to zero.

• Flow characteristic

for information, a flow characteristic is given depending on the stroke of the slide.

Flow 0,16 dm^3/s Flow under $100x10^5$ Pa

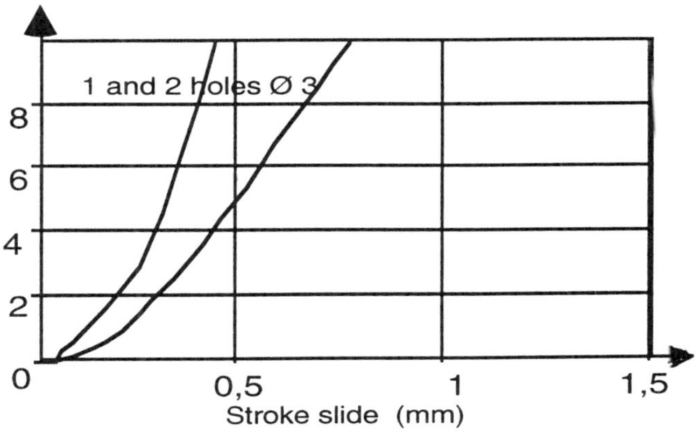

Fig. 7 Flow vs stroke

• Position/flow servo control : flow servo valve

In accordance with the graph, for the movement of the slide, a flow is generated in the circuit.

The electronic control circuit built into the servo valve carries out a slaving function. The block diagram includes a symmetrical current control interface in the electromagnetic motor and threre is an electric motor position comparison loop operating as a function of the required set-point.

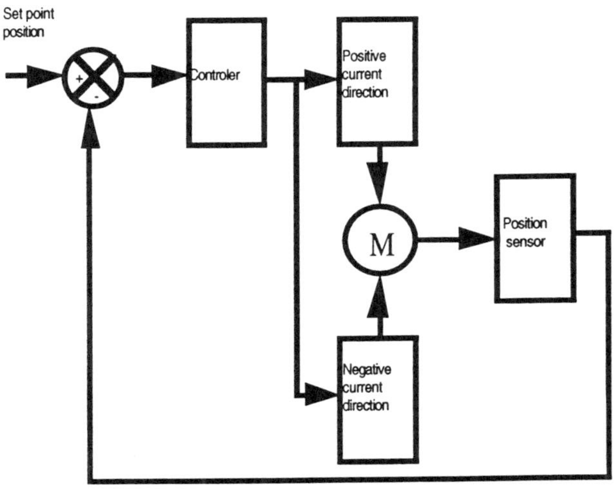

Fig. 8 Slave position control

- **Pressure slave control : servo valve pressure**

Once the flow control has been obtained as a function of the position of the slide, it is obvious that for a given use, when the volume is known, or even variable, it is possible to slave the load pressure. This is carried out by providing the servo valve with a pressure sensor incorporated into the mechanicals. In general the sensor is housed in a zone directly connecting to the load. The graph below gives a characteristic of brake capacity.

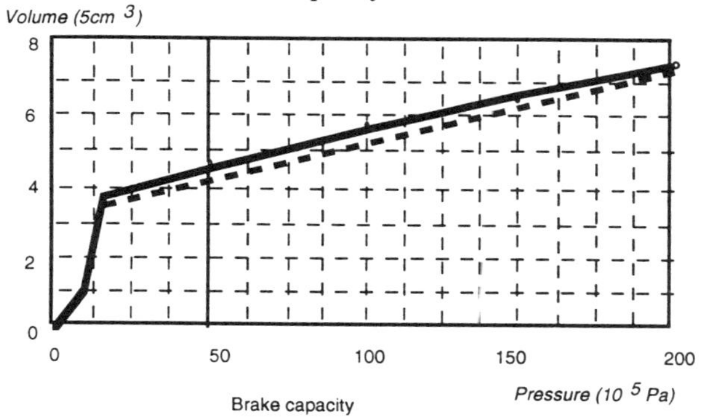

Fig. 9 Brake capacity

3) Digital design servo devices

The digital control of servo devices is provided by a MOTOROLA 68 HC16 microprocessor. The software, written in assembler language, uses a proportional action and derived action.

- **Position slave control.**

The computation elements are :
Proportional action :

$$P(z) = Kp*E(z)$$

Derived action :

$$D(z) = (z-1)/Tz*X(z)$$

where :
Kp = Proportional gain
Kd = Derived gain
T = Sampling period (300 μs)
z = Discret variable
We use weighting on two points which modifies the derived action transfer equation in such a way as to limit the influence due to the derivation of the noise and interference generating :

$$D(z) = (z-1)/Tz*(1/2 + 1/2z)*X(z))$$

In the time-related field we obtain :
Proportional action :

$$P(k) = Kp*E(k)$$

Derived action :

$$D(k) = Kd/2T*(X(k)-X(k-2))$$

The accumulated actions produce :

$$Y(k) = Kp*E(k) +Kd/2T* (X(k) - X(k-2))$$

In order to carry out calculations in the program, we will simplify by :

$$Y(k)= p*E(k) + b*(X(k) - X(k - 2))$$

With :

$$p = Kp \text{ and } b = +Kd/2T$$

- **Pressure slave control**

Pressure slave control incorporates position slave control with a "loop". The pressure "loop" uses the derived action on the error without integral correction so as not to slow down the system.
Proportional action :

$$P(z) = Kp*E(z)$$

Derived action :

$$D(z) = (z-1)/Tz*E(z)$$

where :
Kp = Proportional gain
Kd = Derived gain
T = Sampling period (300 μs)
z = Discret variable
In the time-related field we obtain :
Proportional action :

$$P(k) = Kp*E(k)$$

Derived action :

$$D(k) = Kd/T*(E(k)-E(k-1))$$

The accumulated action produce :

$$Y(k) = Kp*E(k)+Kd/T*(E(k)-E(k-1))$$

To perform calculations in this program, we will simplify matters as :

$$Y(k) = p*E(k) + b*(E(k)-E(k-1))$$

with :

$$p = Kp \text{ and } b = Kd/T$$

- **Data format**

In the multiplication program, whether on 16 bits or with results on 32 bits, the coefficients p, b, the error and the position are on 16 bits and the values read are such that the set-point and position are converted on 8 bits. By means of 8 shifts to the right, we obtain values on 16 bits with a sign extension. The output value is reduced to 8 bits in such as way as to assign the PWMA command to the register.

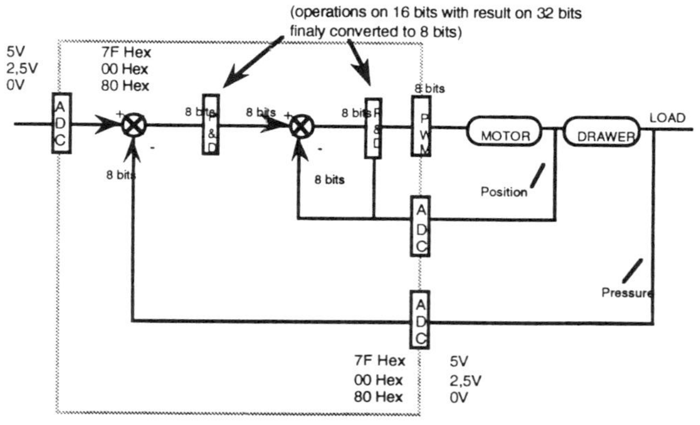

Adapter format

Fig. 10 adapter format

The motor receives a command by a Pulse Width Modulation (PWM) together with management of the motor displacement direction ; the calculated error is loaded into a register which generates a PWM signal. The sign of this error determines the value representing the direction of the current.

These two signals then enter a power interface controlling the motor.

4) ACTUATOR MODEL

Actually, there are several methods of describing physical systems : some mathematics champions prefer Lagrangian's equations, some brilliant engineers would rather choose a self-made theory (understandable only to themselves). We have chosen a simple way to model systems with a graphic method : the Bond Graphs.

- **The Bond Graphs**

Let us consider an example of three engineers : an electronics expert, a hydraulics expert and a mechanics expert who want to model an hydro-electric system. Each speaking his own language makes for difficult communication between the experts. Therefore, their work is slow and the results product mistakes or omissions.

The use of Bond Graphs breaks down these language barriers because of its unified notation for most physical systems. It is based upon power and information flow in systems and on the real effect of each part, expressed in the form of energy (production, conversion, storage, dissipation).

- **The evaluated device**

The device is a hydro-electric actuator consisting of an electromechanical motor coupled to a hydraulic 3-way valve. The two parts were analyzed separately and, thanks to the Bond Graphs, the two models can be brought together without any further worries.

In this example, we are looking at the electromechanical system.

After some basic analysis, we can see that the electromechanical system can be represented in the following figure :

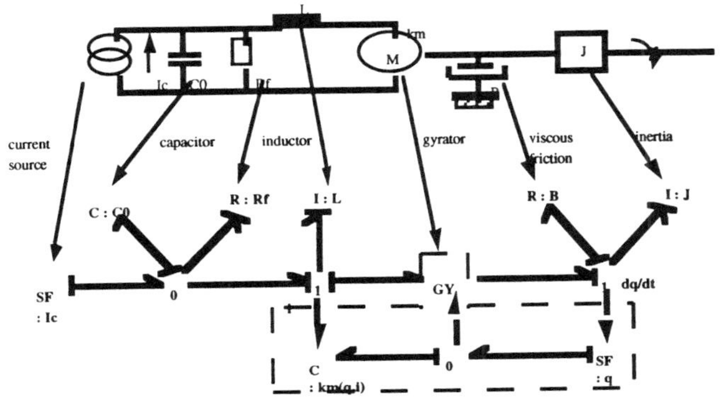

Fig. 11 Bond Graph description of the motor

The system consists of a controlled current source Ic, a smoothing capacitor Co, a leakage resistor Rf, a serial inductor L, a gyrator M (defined by its duration ratio km), by viscous friction B and by shaft inertia J.

- **Model description - Analysis of device**

We can now draw the Bond Graph of this system. Before doing so, we need to list the basic Bond Graph elements we need :

- energy production element (in this case a flow source):current source Ic
- an energy transformation element (a gyrator) : gyrator M
- a potential energy storage element (capacitor) : smoothing capacitor C0
- kinetic energy storage elements (inertias) : inductor L and inertia J
- dissipation elements (resistors) : leakage resistor Rf and the viscous friction device B.

We used a software called Prouesse, working on a Sun station, offering many possibilities of simulations, systems analysis and computer-aided design.

First we draw the graph with a graph editor. The software then translates the graph automatically into an evolved language (C type language). At this stage, we can quickly change some of the parameters in the program if the need arises. To run the program, we simply need to click on an execution icon. After execution, the results can be plotted in a display window or can be printed.

- **Simulation**

This part represents some of the results of simulation.

We have looked at the position of the servo control. The following figures represent a frequency response and a time-related response.

The input (first graph) represents a command (in radians).

The output (second graph) represents the motor shaft position (in radians).

The last graph represents the motor current (in amperes)

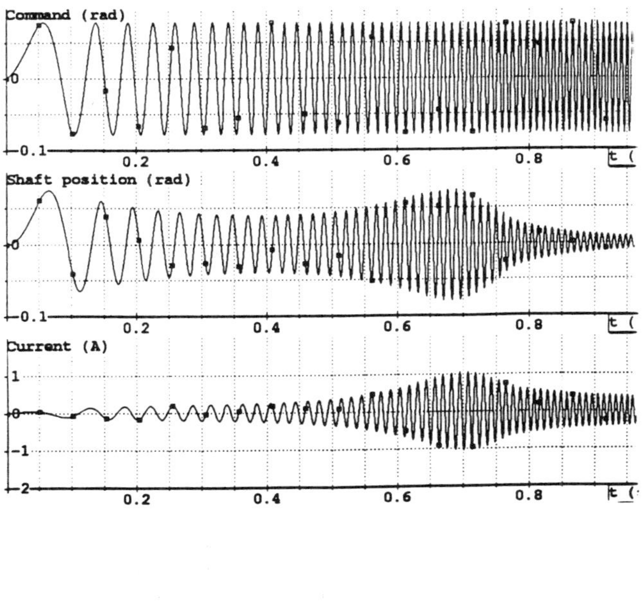

Fig. 12 Frequency response

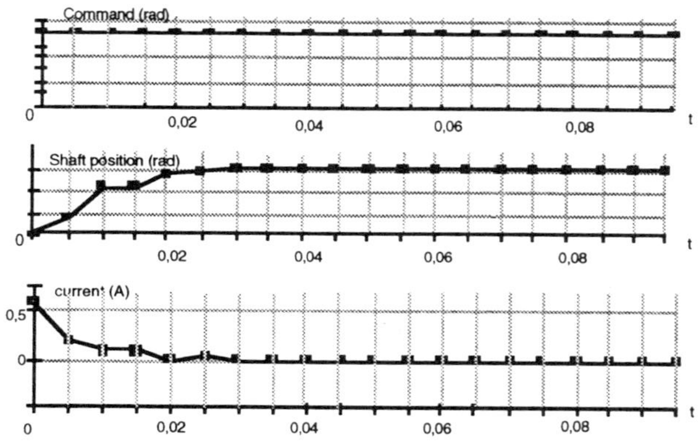

Fig. 13 Time-related response

We model also hydraulic systems, the Bond Graph in this case is.

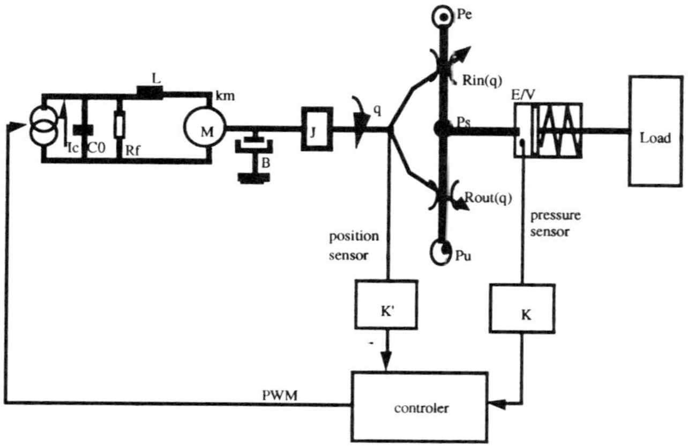

Fig. 14 Hydro-electric system scheme

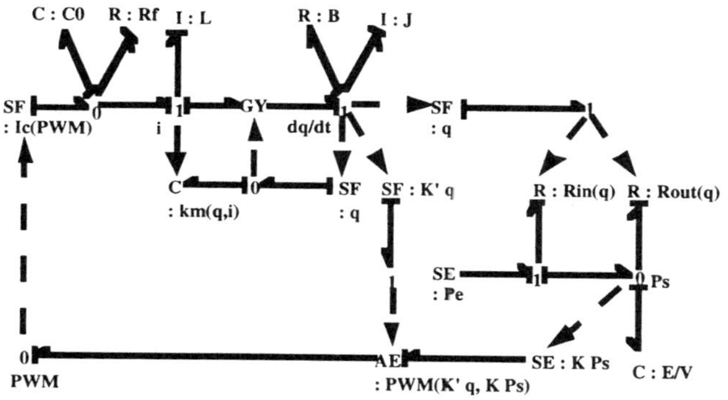

Fig. 15 Bond Graph

5) Performances

- **Position/flow slave control**

The position slaving bandwidth, including the operation of the electromagnetic motor of the potentiometer and the transformation of the movement, are given in the graph amplitude.

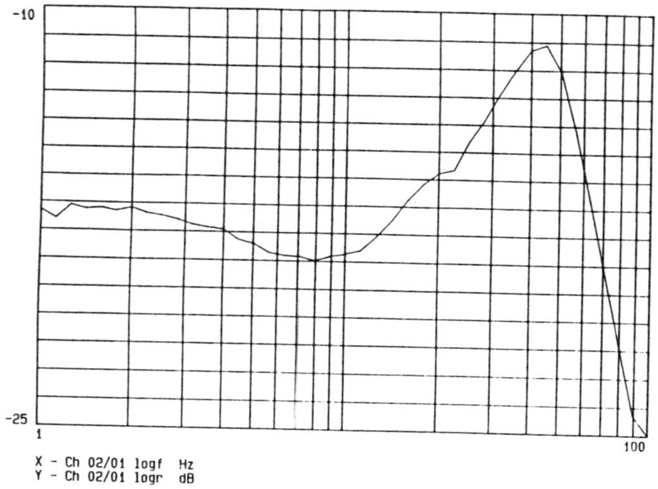

Fig. 16 Bode diagram

- **Pressure slave control**

The pressure slaving bandwidth on an automobile brake for a car having a cubic capacity .2cm3 implementing the entire hydraulic installation fed by a pump delivry 4 cm3/rpm pump with an average speed of 1500 rpm and a maximum speed of 3600 rpm is given in the graph amplitude and phase.

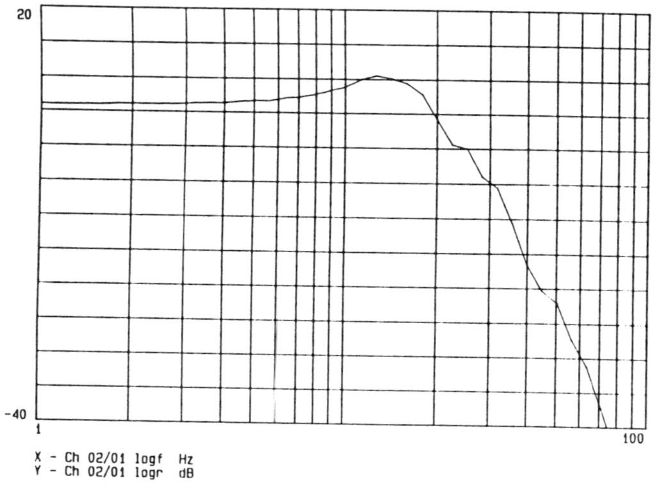

Fig. 17 Bode diagram

6) Conclusion

With a unified approach, we model the electrical, mechanical and hydraulic systems and link them together easily. We can also model thermal and magnetic effects, inducing non-linearties and display all kinds of mechanical kinematic behaviors with equal ease.

With Prouesse software and the Bond Graph theory, we can study a complete system from low power sensors to high power actuators, capable of writing functional specifications.

Taking into account the characteristics of the elements included in the production of a flow overpressure servo valve, it is easy to understand that there is a mutual relation between the simulation and implementation of a software tailored to the requirements of servo valve control and test system requirements.

Furthermore, by the very simple design of the servo valve, there is no longer any need to choose the work-pieces or to carry out individual adjustments.

Finally, the performance obtained opens the way to many applications requiring a fast, simple and low cost actuator that can be integrated into numerical control and test systems.

REFERENCES

(1) Sandra Buettner. Apitech, 12801 "Electrohydraulic suspension offers a unique challenge to the valve designer". Automative design engineering 93 page 266.

(2) Colloque Européen "aimants modernes et nouvelles machines à aimant". Grenoble 13-15 juin 1990.

(3) Conferences capteurs 89 : "Magnetic Field Sensors in Cars".
A. Peterson, Philips Components, pp.233-241.

(4) Solid-State Electronics, Pergamon Press. Vol. 9, pp.471-848. "Applying the Hall Effect to Angular Transducers", R.S. Davidson and R.D. Gourlay. Hughes Aircraft Company.

(5) Bundesrepublik Deutschland, Patent // 29 23 644. Unionspriorität 1978. Positionsfuhler, Diesel Kiki Co. Ltd Tokio.

(6) Honeywell : Solid state sensors. Catalogue E20, Issue 2.

(7) M Hassenforder, D Lienhardt, G Gissinger "PROUESSE : a software tool for simulation and analysis of physical systems, specially designed for automatic applications". IMACS'91-DUBLIN vol3, pp 1296-1299.

(8) M Hassenforder, D Lienhardt, G Gissinger. "Sige a software tool which translates a system of equations into a source C program". IMACS-MCTS'91-LILLE vol 2, pp 140-146.

(9) M Hassenforder, D Lienhardt, G Gissinger "Graphical editor for modeling with bondgraphs in PROUESSE". ICBGM'93-SAN DIEGO 17-20 January 1993.
0000

(10) D Lienhardt. "Exploitation de la représentation d'état linéaire. Modélisation en simulation des systèmes non linéaires décrits par le langage des graphes à liens". Thesis of Université de Haute Alsace, Mulhouse 1989.

(11) C. Oudet. Inductance et constante de temps électrique dans les moteurs à aimant permanent, principalement pas à pas. Compte-rendu des 2èmes journées d'études sur les moteurs pas à pas - Avril 82, Lausanne, p.147-161.

(12) G. Lacroux. Les actionneurs électriques pour la robotique et les asservissements : Chapitre 8 Edition Technique et Documentation Lavoisier - 1985.

961800

High Force Electric Linear Motors

James H. Goldie, Kevin J. Leary, and Michael J. Gerver
SatCon Technology Corp.

ABSTRACT

SatCon is developing electric machines to produce more torque or force within a given volume. Some are novel multi-air gap configurations which employ the same operating principles as conventional machines. Others employ smart materials. As linear motors or actuators, these machines offer an alternative to hydraulics.

Direct drive electric linear motors fail to match the force capability of hydraulics. As a result, rotary motors have been combined with gear reducers and ball or lead screws to create a high force linear actuator. This approach, although effective in many situations, is often too sluggish to employ for applications requiring rapid response and maneuvering. Simply speaking, the inertias of the motor and gear reducer consume an unacceptably large fraction of the force capability of these actuators.

In response to these limitations, SatCon has conceived of a permanent magnet-based linear electric actuator which employs multiple air gaps. The most basic configuration resembles a "sandwich", where permanent magnet (PM) shuttle plates are placed between multiple wound stator plates. Only the end stator plates require return iron. This approach is effective in achieving high torque within a given envelope, because additional force-producing air gaps are provided without the penalty of added return iron and housings.

A second approach employs a magnetostrictive material in an "inch-worm" configuration. The "inch-worm" approach has been implemented with piezoelectrics for low force, high precision positioning. For higher force applications, there has been limited development of magnetostrictive inch-worm devices. Unlike the PM actuator, it holds its position when power is removed. SatCon has developed a novel configuration of this approach, which could serve as an alternative to hydraulic actuators.

INTRODUCTION

Researchers are seeking to develop electrically-powered actuators which can provide greater forces and torques within a given volume. In particular, if electric actuators can achieve force densities (i.e., force/volume) which are comparable with hydraulic actuators, then there are a wide range of potential applications of electric actuation: high force positioning systems, robotics, automated manufacturing, flight control surface actuators, etc. Furthermore, a future market is developing because of the increasing interest in electric and hybrid electric vehicles. In addition to providing traction for these vehicles, electric actuators and motors will be the natural choice for all auxiliary functions such as power steering, air conditioning compressor drives, brakes, and even automatic transmissions.

SatCon is developing a wide range of electric motors and actuators to produce higher forces and torques within a given volume. This paper summarizes work at SatCon in two types of linear electric actuators: multi-air gap permanent magnet (PM) linear actuators and magnetostrictive "inch-worm" actuators. SatCon believes that these actuators offer the means for achieving the above-described goals. This paper includes a description of how these actuators are configured, how they work, analytical predictions of their performance, and experimental results.

PM LINEAR ACTUATORS

This class of actuator employs the recent dramatic increases in the energy product available from permanent magnets. Figure 1 shows schematically the most basic configuration for this type of actuator. It resembles the voice coil actuator, with two exceptions:

(1) The magnet moves rather than the coils.
(2) The coils are installed in slots in the armature iron rather than being placed in the air gap.

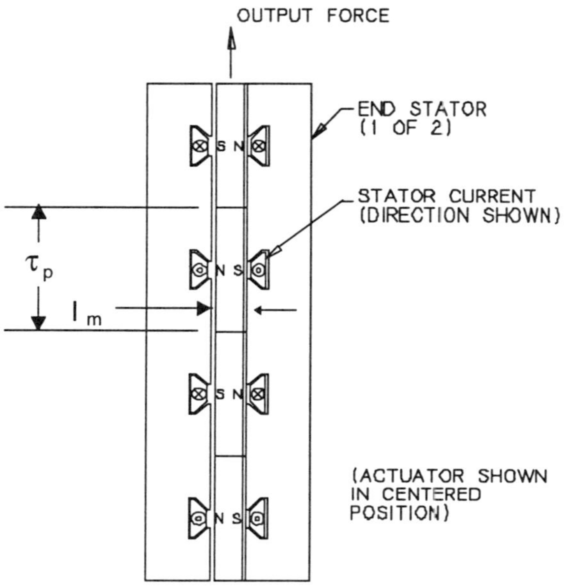

Figure 1. Basic PM linear actuator configuration.

The traditional voice coil is designed to achieve extremely high bandwidths (often as high as many kilohertz), so there is a big incentive to minimize the moving inertia. As a result, the moving element of a voice coil actuator consists entirely of the freestanding coil and no iron. For most actuator applications, however, the inertia of the moving element is not so critical. As a result, we have elected to place the coils inside slots in a silicon steel armature structure. This greatly increases the flux emerging from the magnets, since the magnets need no longer drive flux across the substantial thickness of the coil. In addition, we generally prefer to place the coils on the stator to avoid power transfer to the moving element.

First-order analysis of the PM linear actuator configuration shown in Figure 1 shows that the maximum electromagnetic shear stress (τ_{max}) which it can develop is given approximately by

$$\tau_{max} = H_{max} B_R \left(\frac{l_m}{\tau_p} \right) \qquad (1)$$

where l_m is the thickness of the magnet, τ_p is the pole pitch, B_R is the residual induction of the magnets, and H_{max} is the maximum field intensity imposed by the windings on the magnets. B_R, the residual induction, ranges from 1 to 1.2 T for Neodymium Boron Iron magnets, with the exact value depending on the grade chosen and the operating temperature. Acceptable values for H_{max} decrease as the anticipated operating temperature of the magnets increases. For example, Figure 2* shows that H_{max} = 500 kA/m (6300 Oe) will not demagnetize the magnets for operation of Crumax 3125* at 150°C. At 185°C, however, this value for H_{max} is only marginally acceptable.

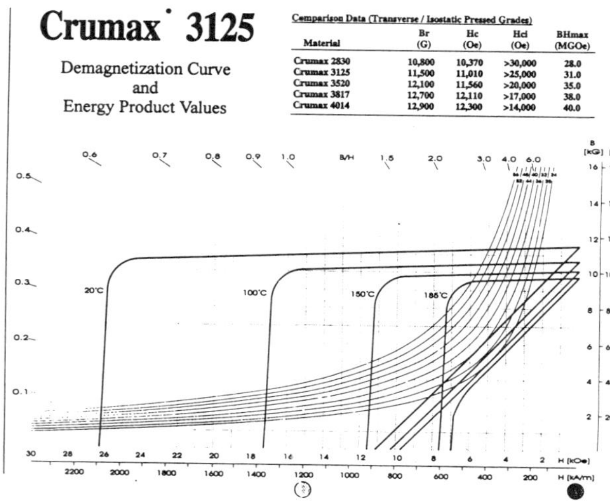

Figure 2. Crumax 3125 (available from Crucible Magnetics) or equivalent has high intrinsic coercivity at elevated temperatures.

As the expression indicates, the ratio of magnet length to pole pitch is also an important parameter with respect to the maximum shear capability. The expression begins to lose validity for l_m/τ_p greater than 0.25, since adjacent magnets begin to short each other out. Therefore, the maximum shear possible is 20 psi or so for a PM linear actuator, with the precise value depending on B_R and H_{max} and, hence, on the permanent magnet material chosen and the anticipated operating temperature.

In its simplest implementation, the actuator possesses a single phase. Force is simply proportional to the current in the armature: This is attractive in that control of the actuator is identical to control of a conventional DC motor. Like the DC motor it will most likely be used within a position feedback control loop. It will dissipate heat whenever it is producing force, even while it is stationary.

EXPERIMENTAL RESULTS - SatCon developed and tested for the Navy the rotary actuator shown in Figure 3. It employs the same general configuration as the linear actuator. It is designed to directly drive control surfaces on Navy vessels, particularly the 21ATV vehicle. The actuator achieves a peak torque capability of 300 in-lbs. and has a travel of ±30° within a 4.5 x 4.5 x 2-inch volume. The peak torque is achieved with an electromagnetic shear of 25 psi. This

---

* Data sheet provided by Crucible Magnetics; Crumax 3125 (available from Crucible Magnetics) or equivalent has high intrinsic coercivity at elevated temperature.

* Crumax 3125 is a Neodymium-Iron-Boron magnet material available from Crucible Magnetics. Similar materials are available from other sources.

confirms the high electromagnetic shear stresses achievable with this approach.

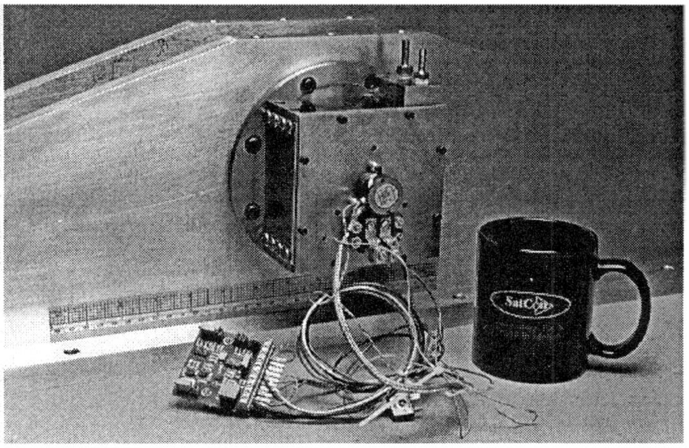

Figure 3. SatCon's rotary PM actuator achieved electromagnetic shear of 25 psi.

STROKE CAPABILITY - As Figure 1 reveals, with only a single phase the stroke of the actuator is limited to roughly the length of the magnet poles (τ_p). Consequently, for increased strokes the pole pitch must be increased. The backiron thickness (T_{bi}) must also be proportionately increased, in order to prevent saturation in the armature backiron:

$$T_{bi} = \frac{1}{2}\left(\frac{B_{gap}}{B_{sat}}\right)\tau_p \qquad (2)$$

where B_{gap} = field in air gap and B_{sat} = saturation field in backiron. At some value of stroke it is worth reducing the pole pitch and using multiple phases. In other words, to use a conventional linear PM motor. In this arrangement the phase currents are properly controlled, based on shuttle position, much like control of brushless DC motors. In addition to the increased electronics complexity, there is some penalty on performance, due to the fact that multiple phases implies that at any instant of time there are windings which are not fully utilized.

MULTIPLE AIR GAP CONFIGURATION - The force per unit volume of the device may be further enhanced by the use of multiple stators and shuttles, as shown in Figure 4. The benefit of this arrangement is that the stator force capability increases proportionately to the number of air gaps without any increase in the stator backiron. As a result, the 6-gap configuration of Figure 5 has three times the force capability of the 2-gap configuration of Figure 1 but only requires roughly two times the volume.

Tables 1 and 2 give requirements, sizes and losses for a multi-air gap actuator for two very different applications. The first is an actuator which could be used in an automobile transmission now under development by an automobile manufacturer. The second is for actuation of and F-18 aileron. The results presented in Tables 1 and 2 reveal that as the size of the actuator is reduced, the losses increase for a given force. This stems from the fact that the only avenue open for size reduction is in the volume available for windings, assuming that the magnetics is designed efficiently. This, of course, increases the current density in the winding and, hence, the ohmic losses. In addition, we looked at the possibility of reducing the stroke by a factor of 2 for the automotive transmission. This resulted in a reduction in the required backiron thickness, leaving more volume for the windings and reducing the ohmic losses. This is consistent with the fact that above a certain stroke, it becomes worthwhile to add the complexity of multiple phases.

The force densities (force per unit volume), specific forces (force per unit mass), and motor constants (force per root watt) realized by these designs is much higher than those achieved by more conventional electromechanical actuators. Table 3 compares these performance parameters for the mult-air gap PM actuator designs given in Table 1 and a voice coil actuator available from a major supplier. The result is that the multi-air gap PM actuator exceeds the voice coil by a factor between 5 and 30 in force density and by a factor of between 2 and 6 in specific force. This suggests that the multi-air gap PM actuator can be practically applied to the applications now reserved for hydraulics.

MAGNETOSTRICTIVE INCH-WORM ACTUATORS

SatCon is also developing a class of electromechanical actuators for those applications which require force densities in excess of those achievable with PM linear actuators. These actuators employ the principle of magnetostriction in which materials expand in the presence of a magnetic field. SatCon has developed magnetostrictive actuators for a number of actuation applications, including vibration suppression in structures, vibration cancellation, and movement of aerodynamic surfaces deployed on helicopter blades. The stroke of these actuators is limited by the magnetostrictive strain of these materials. For example, the most promising material now commercially available, Terfenol-D*, can achieve strains in excess of 1000 ppm. Consequently, a rod length of two inches can only achieve .002 inches of motion. These materials, however, can readily work at compressive stresses as high as 3000 psi. Thus, a cross-section of Terfenol-D of only one square inch can deliver 3000 lbs. of force to a load.

The "inch-worm" approach is a way to combine the inherent high force capability of magnetostriction with an increased stroke capability. This is accomplished by inducing a slab of Terfenol-D to move in a peristaltic manner between two plates, as shown in Figure 6.

* Available from Etrema Products, Ames, Iowa.

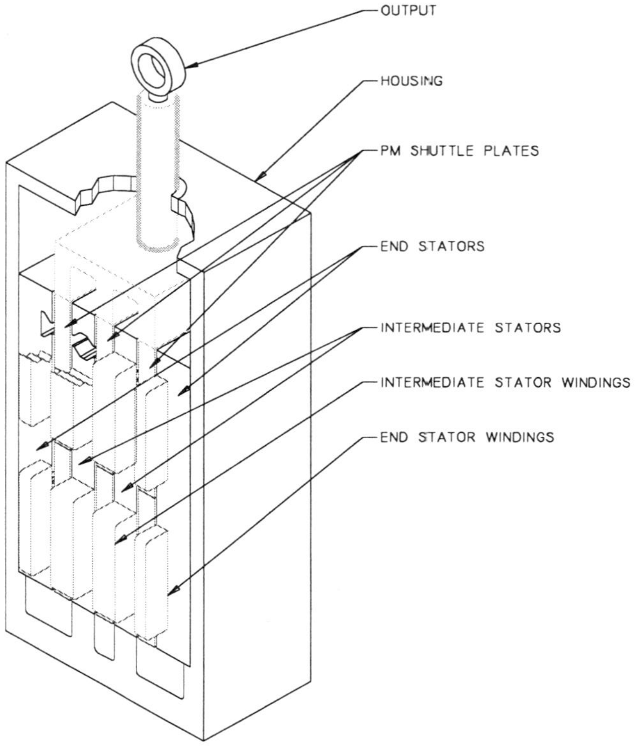

Figure 4. A multi-air gap PM linear actuator provides high force density.

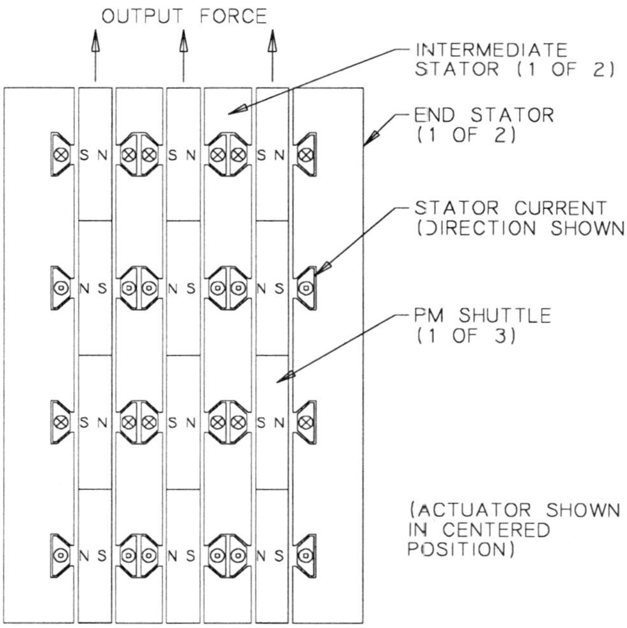

Figure 5. Multiple-air gaps combined with high energy product magnets allow high forces in a compact package.

Table 1a. Requirements for automotive transmission.

Stroke	40 mm
Maximum speed	500 mm/s
Force	250 N

Table 1b. A sample of possible designs for the transmission.

F_{peak} & F_{avg} (N)	Stroke (mm)	L x D x W (mm)	P_{peak} and P_{avg} (W)	Weight (lb.)
250 50	40	100 x 30 x 50	520 21	1.8
250 80	40	100 x 40 x 50	230 24	2.4
250 80	40	100 x 50 x 50	140 15	3.1
250	40	100 x 60 x 60	51	4.4
250	40	200 x 50 x 50	47	6.1
250 80 50	20	100 x 30 x 30	310 32 12	1.1
250 80	20	100 x 40 x 40	84 8.6	2.0
250	20	100 x 40 x 45	66	2.2
250	20	100 x 45 x 45	56	2.5

Table 2a. EMA requirements for the F-18 aileron[*]

Maximum output force	13,107 lbs. (compression) 12, 092 lbs. (tension)
Maximum continuous force	5,000 lbs.
Stroke	4.5 inches (+/- 2.25 inches)
Maximum rate	6.7 inches
Frequency response	as shown in Figures 4 & 5 of the EPAD SOW (roughly 8 Hz. for large amplitudes)

Table 2b. The losses are reduced by increasing actuator size.

Cross-section (inches)	Losses at 5000 lbs. (kW)
6 x 10	12
8 x 12	3.2
10 x 14	1.6

Table 3. Comparison between SatCon's linear multi-air gap PM actuators and commercially-available voice coil actuators

	PM linear actuator	Commercially-available voice coil actuator
Peak force (N)	250	310
Force density[**] (MN/m^3)	0.5-2.8	.1
Specific force[**] (N/kg)	90-500	18
Motor constant ($N/\sqrt{W}$)	11-36	15

[*] Derived from "Electrically Powered Actuation Design Validation Flight Test Program, Electromechanical Actuator Statement of Work"

[**] A range of values is given for force density and specific force, since they differ for the various designs given in Table 1b. As the force density and specific force increase, the motor constant drops. The force density and specific force given for the PM linear actuator do not take into account the volume required for a housing. This additional volume will reduce both these values some.

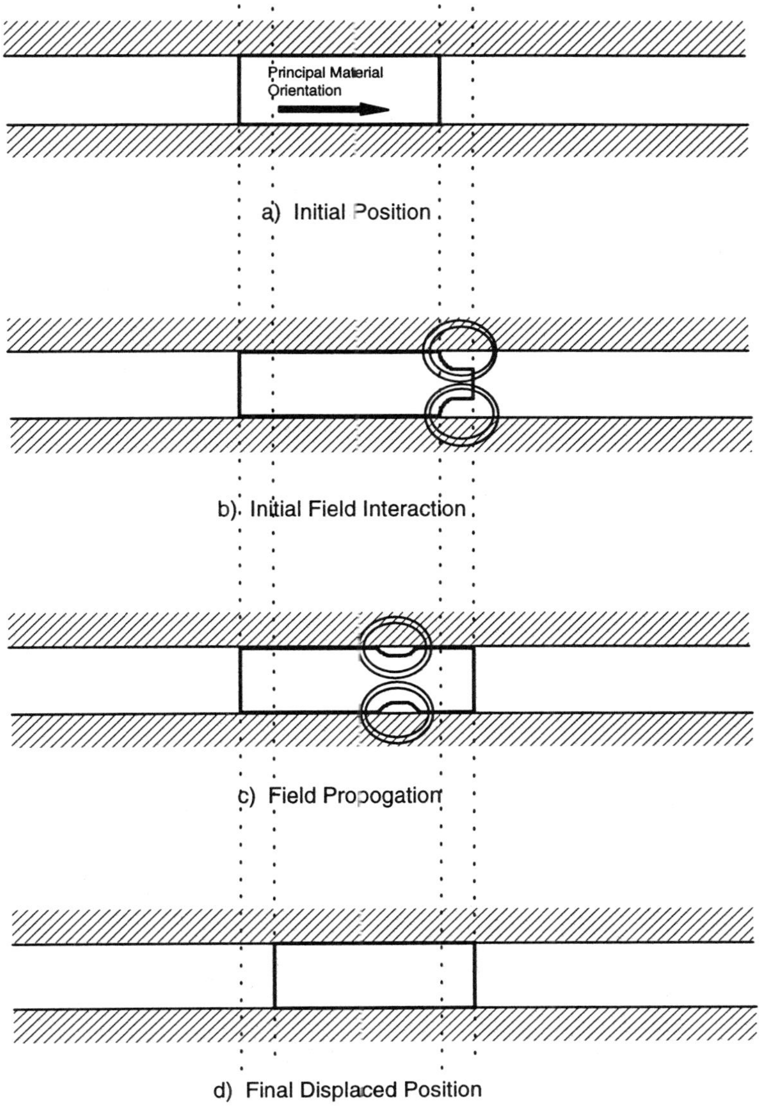

Figure 6. Principle of operation of a linear magnetostrictive actuator.

Prior realizations of this actuator have placed a cylindrical rod of Terfenol-D placed within a tight fitting bore*,**. A set of sequentially-excited cylindrical coils wrapped around the whole assembly provided the necessary field excitation. SatCon instead employs a rectangular slab of Terfenol-D between a pair of flexure-loaded plates and a standard 3-phase winding. The benefit of this arrangement is that the compressive load on the Terfenol-D slab is maintained, in spite of wear at the contacting surfaces. The plate above the Terfenol-D slab is a slotted silicon steel stator which possesses a conventional 3-phase winding in its slots. In order to provide a smooth durable contacting surface these slots are closed with a nonmagnetic material and the surface is ground flat.

* Kuang-Yuh Huang, Lothar Kiesewetter, "Design Guidelines for Terfenol Actuators," Technical University of Berlin, Oct. 13, 1988.

** Lothar Kiesewetter, "The Application of Terfenol in Linear Motors," Technical University of Berlin, approx. 1988.

FIELD PATTERN - Figure 7 shows the field produced by the 3-phase winding at one instant of time, as predicted by finite element analysis. Like conventional motors the 3-phase winding produces a sinusoidally distributed magnetic field which propagates along the direction of motion. In the case of the magnetostrictive "inch-worm" the field lines emerge from the pole of the stator, enter the Terfenol-D, turn to run along the Terfenol-D slab and then turn again to return to the stator at the next pole. At those locations where the field is entering and exiting the Terfenol-D, there is a significant component of the field in the radial direction. This causes the Terfenol-D to grow and snug up against the surrounding contact surfaces. Where the field lines are predominantly oriented in the direction of motion, the Terfenol-D rod tends to both thin and elongate the slab. Since the 3-phase winding causes this pattern to sweep longitudinally down the rod, the desired peristaltic action shown in Figure 6 occurs.

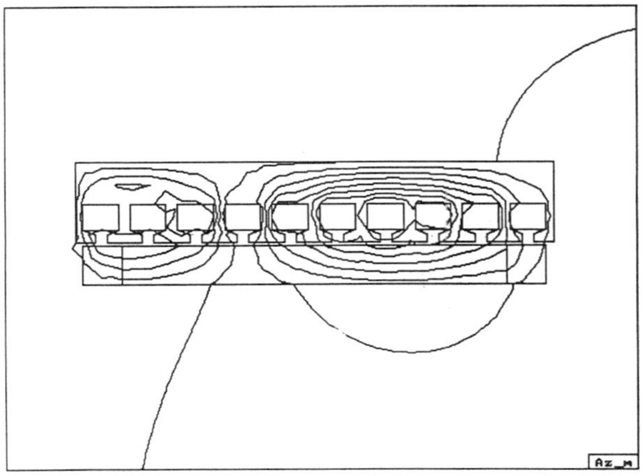

Figure 7. Magnetic field produced by 3-phase winding at one instant of time.

EXPERIMENTAL RESULTS - Figure 8 has two photographs of the proof-of-concept magnetostrictive "inch-worm" actuator currently being tested at SatCon. It resides within a 30 mm x 30 mm x 60 mm volume. Since the prototype was intended as a proof-of-concept, all the components were unlaminated. Consequently, it exhibits high eddy current losses which would not be present in a laminated version of the actuator. Since these losses become prohibitive with speed in the unlaminated prototype, we tested to a top speed of 2.6 mm/s. This speed was achieved at an excitation frequency of 400 Hz. Higher speeds require higher frequencies. In addition, the maximum applied force was 156 N. It was limited by the fixturing rather than the actuator. We expect that the actuator could deliver considerably higher forces. Table 4 shows the performance measured for the actuator and compares it with the expected performance for a laminated magnetostrictive actuator, the PM linear actuator designs given in Table 1b, and the commercially-available voice coil.

SUMMARY AND CONCLUSIONS - SatCon is developing high force electrically-powered actuators which can serve as an alternative to hydraulics and other actuator technologies. These actuators fall into two classifications: permanent magnet (PM) linear actuators and magnetostrictive "inch-worm" actuators. We have successfully tested the PM actuator in a rotary version and the magnetostrictive "inch-worm" with a proof-of-concept prototype.

Both the multi-air gap permanent magnet linear actuators and the magnetostrictive "inch-worm" actuators offer high force densities and specific forces relative to commercially-available electromechanical actuators. As a result, they may offer an electrically-powered substitute to hydraulic actuators in some situations. The PM actuator requires power when stationary, whereas the magnetostrictive "inch-worm" does not. The PM actuator offers higher speeds than the "inch-worm" but does not offer quite as high force density and specific force as the "inch-worm".

(a) Photo showing the major components of the Phase I proof-of-concept actuator.

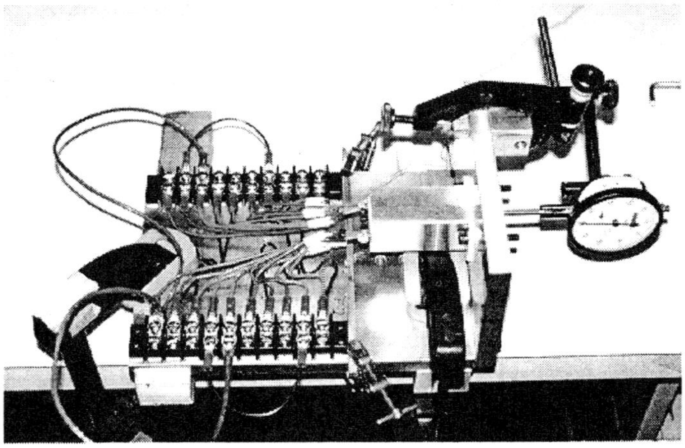

(b) The Phase I proof-of-concept actuator mounted to its test stand.

Figure 8. Photographs of the proof-of-concept magnetostrictive inch worm.

Table 4. Measured performance of the proof-of-concept magnetostrictive "inch-worm" actuator compared with predicted performance of an optimized design and other EM actuators.

	Magnetostrictive actuator (prototype)	Magnetostrictive actuator (optimized design)	PM actuator	Commercially-available voice coil
Max. force (N)	156	220	250	310
Force density (MN/m^3)	2.9	5.9	0.5-2.8	.1
Specific force (N/kg)	520	1070	90-500	18
Motor constant ($N/\sqrt{W}$)	6	12.7	11-36	15
Max. speed (mm/s)	2.6	50	N/A	N/A

MECHATRONICS

970926

Multi-Domain Modeling and Simulation of Mechatronic Systems

Aravind S. Bharadwaj
GM Powertrain Group

ABSTRACT

Automotive applications require complex systems with strong interaction between electrical and mechanical modes of operation. To achieve fast to market products and reduce development costs, math modeling is essential to analyze these complex mechatronic systems and accurately characterize their behavior. One of the tools widely employed by GM to address this need is Saber®. While Saber® is well developed in electrical and mechanical domains, it is difficult to accurately model the magnetic behavior. To alleviate this problem, the magnetic circuit is characterized by the finite element analysis program, Maxwell®, and then integrated with Saber®. This paper discusses the development of models using this process with a transmission application.

1. INTRODUCTION

The complexity of the systems in a vehicle is increasing rapidly due to sophisticated customer demands, and stringent government regulations on emissions and fuel economy. The conventional way of building and testing the prototype for performance is inadequate to optimize the design of these systems. In addition, tough competition is forcing automotive manufacturers to reduce prototype development costs and introduce their products fast to market. These factors have contributed significantly to the increasing use of math modeling techniques for the design and development of mechatronic systems for automotive applications. At General Motors Powertrain Group, Saber® developed by Analogy, Inc., is extensively used to analyze the components, sub-system and system behavior. It should be noted that GM does not endorse any specific software package such as Saber® or Maxwell® as an exclusive tool set even though they are the tools presently being used. The math based analysis enables the product engineer to better understand the behavior and characteristics of the powertrain systems - transmission and engine.

A typical automatic transmission consists of a number of electromechanical components, such as solenoids and sensors. To better understand the dynamic behavior of the system, these components have to be modeled accurately. In the case of the solenoids, the input is typically electronic, analog or digital, and the output is used to control the fluid pressure, etc. The electrical energy is first converted into an electromagnetic force which results in the displacement of the armature. This, in turn, causes a change in the hydraulic pressure, which is the control variable. Saber® is an ideal candidate for the simulation and analysis of such a component, since there are templates available in the software library in all these domains[1]. Similarly, for a sensor, the typical input is in the form of mechanical energy which is converted into the magnetic domain resulting in an electrical output signal.

The analysis of the components in these domains are important for predicting system performance. This paper is limited to the discussion of the modeling of the magnetic behavior exhibited in these components and how it is integrated into a system model. An accurate representation of the electromagnetic behavior in a device is very essential to achieve a robust design of the controller. Also, this results in a better understanding of the force balance required for a stable operation of the mechatronic system.

Traditionally, an electromechanical component such as a solenoid is modeled as a pure resistive-inductive circuit. This fails to comprehend the electromechanical nature of the device since the inductance varies with respect to the motion of the armature. The most common approach in Saber® is to model the magnetic circuit of the solenoid by using the el_magnet template that is available in the mixed discipline mechanical template library[2]. Even though this template is able to provide a good solution to the problem, it fails to take into account the magnetic properties of the material.

A second method to characterize the magnetic behavior in the mechatronic system is to construct the equivalent magnetic circuit using the core template available in the magnetic template library. Since this template models the circuit in the electrical and magnetic domains, the resulting magnetic output needs to be transformed into an equivalent force. Once this force is derived, the model is integrated into the mechanical representation of the device to characterize the appropriate displacement caused by this force. The solution resulting from this process is more accurate since the magnetic properties of the material along with the parameterized geometry of the component are accounted for in the model. The major drawback of this method is the difficulty in accurately representing the magnetic path for complex geometric structures and availability of magnetic properties of materials, thus increasing the modeling effort. In addition, the non linear magnetic behavior is not included in such a model and this leads to inaccurate characterization of the component.

A refinement to the previous approach is to construct the model using the corenl template available in nonlinear magnetic template library. This process predicts the nonlinear behavior based on the Jiles-Atherton magnetic hysteresis model. The component model based on this template characterizes the magnetic behavior with a high resolution of accuracy, provided the parameterized geometry of the magnetic path is defined accurately. A very major drawback of this method is that the model is highly theoretical and the constants and parameters needed for use in the corenl template are rarely available in the industry. Moreover, a comprehensive understanding of the magnetic field distribution is required to make critical assumptions for the development of a high fidelity model. These factors limit the effectiveness of this approach since the development of an electromechanical device in mechatronic systems involve several different organizations.

From the discussion above, the need for the development of a practical user-friendly process for modeling the dynamic behavior of the magnetics accurately in a system is highlighted. Also, this process needs to be less time consuming and should be able to protect the proprietary information specific to each supplier. Very accurate characterizations of electromagnetic devices are normally achieved using finite element analysis (FEA) techniques. In the FEA method, geometry of the device and the material characteristics are the inputs. The solution is obtained by breaking the complex geometry to finite number of small pieces and finding the best possible solution using a rigorous, computation intensive math process. The FEA method is a very time consuming process and until recently has been limited to only static solutions, i.e., the capability to solve only one particular state of the device. Another problem with this approach is that it is limited to the solving of only the magnetic solution. There was no feature available to integrate these accurate solutions into a mechatronic system.

An attempt has been made to overcome these obstacles with the Maxwell® EMSS package developed by Ansoft Corp[3]. This paper discusses the application of this method for the development of a model for an automatic transmission subsystem. Even though this process has been used in the model development for a VR sensor, the discussion is limited to that of a typical solenoid system.

The next section describes such a system followed by a detailed discussion of the steps involved in the development and validation of a model. A generic example for a pressure control solenoid model developed using this process along with results is presented in Section 4. Due to the proprietary nature of the data, the model presented in this section characterizes the behavior of a typical component and does not reflect the behavior of any hardware in GM transmissions. The experimental validations of models for actual components in GM transmissions have been performed and not presented here due to confidentiality agreements with the suppliers. The paper concludes with a brief description of the concerns and lessons learned by adopting this process for the development of a mechatronic system model.

2. SYSTEM DESCRIPTION

A typical mechatronic system containing an electromechanical actuator such as a pressure control system in shown in Figure 1. The input to the pressure control system is a voltage signal that is used to control the torque signal pressure in the transmission. This system is a feedback control system with the hydraulic and electronic signals in the feedback circuit. A pressure control system typically contains the following subsystems:

1. **The Driver Unit**: In a powertrain, this is normally the driver circuitry in the Powertrain Control Module (PCM) that drives the solenoid. These electrical signals are derived by a combination of both analog and digital circuitry. A typical driver signal is a voltage pulse train of fixed frequency with different duty cycles that correspond to the request from the PCM.

2. **The Solenoid**: The solenoid consists of three distinct sub-systems, namely, the electromagnetic, mechanical, and the hydraulic section. As a result of the input from the driver circuitry, an electromagnetic energy

is induced in the solenoid which results in the creation of an electromagnetic force. This force, in turn, is responsible for the movement of the armature which is typically resisted by friction, inertial mass and spring forces. An interaction of the mechanical and electromagnetic force is linked with a change in the hydraulic pressure in the solenoid chamber. The balancing of these forces using appropriate control technology results in the output pressure to be controlled to the desired level.

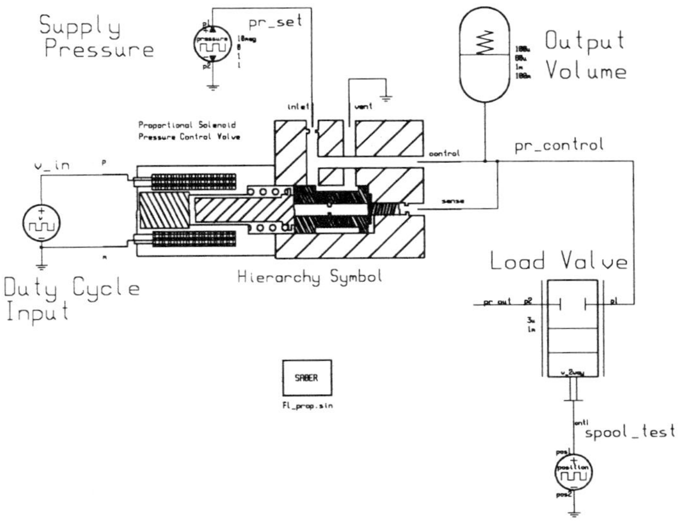

Figure 1: A Typical Configuration of a Pressure Control Solenoid System

3. **Hydraulic Load**: The output from the solenoid, the controlled pressure signal, is utilized to drive the load which is typically a valve that is used to further control the line pressure to improve the shift feel in the transmission. The status of the pressure and current signals alongwith the request from the driver of the vehicle are employed to close the loop for the control system.

It can be noted from the simple description of the pressure control system in the previous paragraphs, the process to develop a high fidelity model describing all these interactions is quite involved. It is important to note that each component in this subsystem needs to be robust for analysis before they can be integrated into the subsystem. Since different suppliers are responsible for the development of these models, it is critical to understand the assumptions and limitations of each component model. This requires the system developer to layout a clear model validation process that is practically feasible. It will be ideal if this validation process can be closely linked with the hardware validation process that is used for these components. The various stages of development of the model of a generic solenoid system and the validation procedure follows in the next section.

3. MODEL VALIDATION PROCESS

The models of different units of the subsystem need to be validated in several stages before they are integrated together. The first step in validating the driver circuit is comparing the time response of the circuit with a simple LR circuit. This experiment can be easily setup in the driver circuit supplier's lab. This step will ensure that the model has accurately captured the time delays, tolerances, temperature effects, etc., in the circuit.

The development of the solenoid consists of two stages. The first step is the development of the magnetic model in Maxwell® EMSS. This model can be experimentally validated with basic mechanical parts in the host domain before being exported into Saber®. The normal procedure for experimental validation is to check the time response of the system for a square input voltage pulse. This pulse causes the armature to move in a fluid free condition. The forces examined in this test are the electromagnetic, frictional, inertial and spring forces. An accurate correlation in this mode will ensure that the time varying nature of the inductance is characterized in the model. This model is then exported and can be tested separately with the driver circuit to resolve for any model integration issues.

The next step in the model validation process is to include the hydraulic and mechanical components using native Saber® templates. This solenoid model has to be tested for performance without the driver circuitry integrated. The source of this model can be an ideal current source to simulate the performance for the entire current range. This will result in a pressure current relationship that should correspond closely to the pressure current curves measured with the hardware. It should be noted that his verification will correlate the mechanical hysteresis of the component. A validation of the model to the experimental setup will be useful to resolve issues related to component specifications. The load for this validation process can either be a dead headed chamber or an accumulator load that is used in a hardware setup.

The next step in the validation process is to integrate the Saber®/ Maxwell® model of the solenoid with that of the driver circuit model. This process requires intensive verification for different analysis setup. The user should be aware of the numerical instability during the time domain simulation of the integrated subsystem. The robustness of this subsytem model should be confirmed before proceeding to integrate the load. Usually a realistic load of this system consists of several valves and hydraulic elements. These devices can either be created using the MAST language or can be implemented using the templates available in the hydraulics library.

4. EXAMPLE

The schematic diagram of part of a typical pressure control solenoid system is illustrated in Figure 2. This system configuration does not represent any current production or future (not in production yet) architecture in a GM transmission. While a typical input to this system is a driver circuit, the model is simulated using an ideal pulse width modulated (PWM) source. The input pressure to the device is simulated using an ideal pressure source feeding into the solenoid chamber. In practical systems, this variable is usually affected by the line pressure that is being controlled. Since the load is not completely represented in this model, the feedback nature of the supply pressure is ignored.

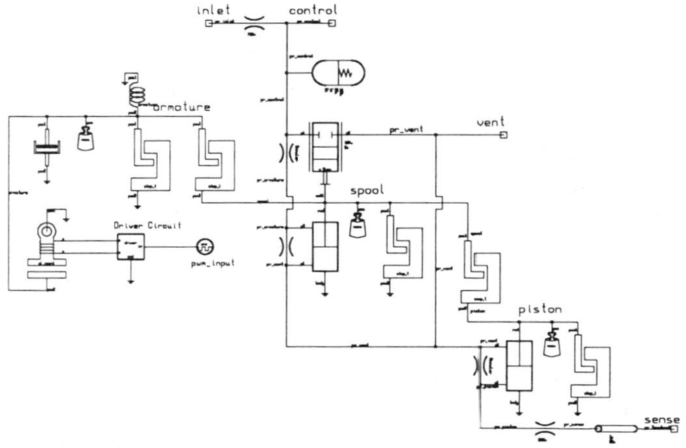

Figure 2: Schematic Diagram of a typical Pressure Control System

As mentioned in the earlier section, the first step in the model validation process is to develop the magnetic model and validate it experimentally in the Maxwell® EMSS package with basic mechanical parts. This magnetic model is exported into Saber® and model integration issues are resolved with mechanical parts as shown in Figure 3. The step response of the fluid free component can be validated with this electromechanical model.

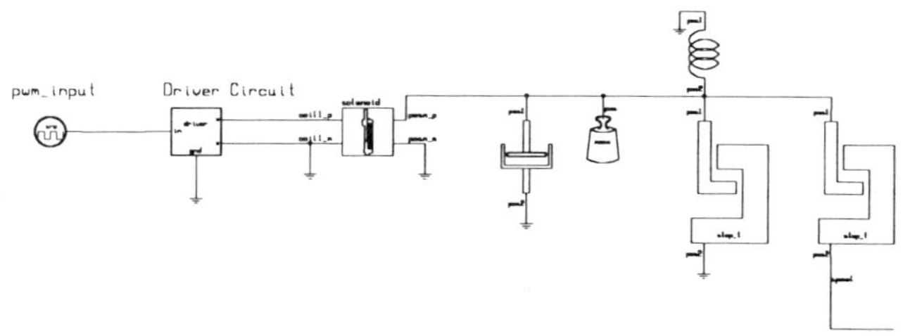

Figure 3: Schematic Diagram of Saber/Maxwell Test Circuit

The step response characteristics of this circuit using the el_magnet template and the Saber®/ Maxwell® model is illustrated in Figure 4. The experimental results are not included due to the proprietary nature of the information. As can be seen from the figure, there is a wide deviation in the response between the el_magnet and the more accurate finite element analysis model. The difference between these two cases will be reflected in the electromagnetic force for different positions of the armature. This will significantly affect the balance of the electromagnetic, mechanical, and hydraulic forces of the system.

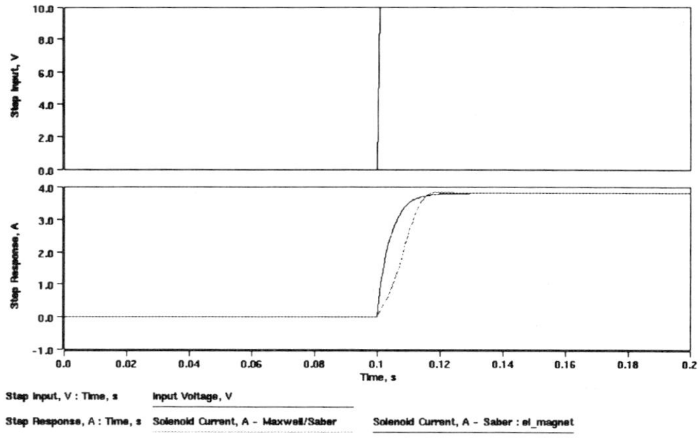

Figure 4: Comparison of Step Response Characteristics of a typical solenoid with Maxwell/Saber and el_magnet Modeling

The validated electromechanical model is then analyzed with the hydraulic components of the intermediate system. To validate the hydraulic circuit with an experimental setup, this model is verified with an ideal current source rather than the driver circuit. This circuit will result in the characterization of the pressure current relationship as shown in Figure 5. This relationship is typically used to specify the behavior of the solenoid to include the variations due to mechanical hysteresis. This validation circuit can also be utilized for tolerance band study to perform variation analysis of the component.

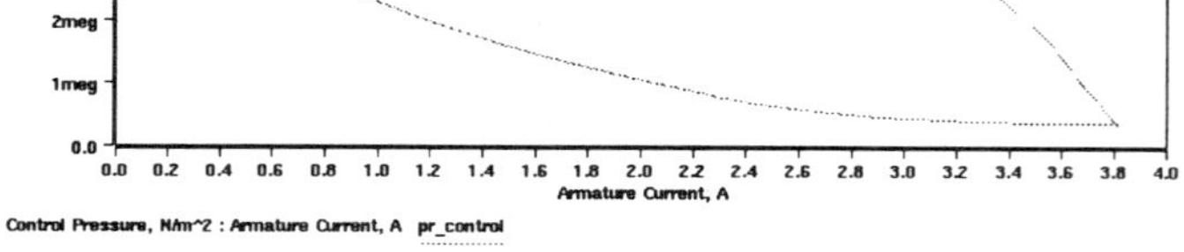

Figure 5: Typical Pressure Current Characteristics for Model Validation

Once the model is validated at this level, the next step will be to integrate the realistic load that includes all the hydraulic interactions. This will include the feedback of the hydraulic and electronic signals in the system to characterize the closed loop nature of the system. A validated model of the system will serve as a tool to refine and optimize the current system and as a benchmark model for design and development of future systems.

The process for modeling and validating a speed sensor system containing a variable reluctance sensor is formulated similar to a solenoid control system. In this case, the magnetic model of the target wheel and the sensor are developed using Maxwell® EMSS package. The results from this model are validated with a simple experimental setup and the model is exported into Saber®. This model is integrated with the PCM model of the electrical load on the sensor. This will involve the accurate modeling of the digital and analog behavior of the processing circuits. Finally, the input of the magnetic model can be integrated with the mechanical model of the transmission, engine crank, or cam shafts.

5. CONCLUSION

A process for the development and validation of models for complex mechatronic systems with accurate characterization of the behavior of the magnetic circuit is described in this paper. An application of this process in an automatic transmission subsystem using Saber® and Maxwell® EMSS is presented and discussed. The pressure control solenoid system in this paper is modeled to closely resemble an existing system in a transmission, but does not reflect the performance of hardware in any GM automatic transmission systems. This process can easily be applied to analyze and validate models for other mechatronic systems such as speed sensor systems, motion control systems, etc.

It can be concluded that the increased accuracy and computation time savings involved in such an integration of models is critical to achieve optimal designs of mechatronic systems. Also, the high fidelity model from this process can assist in improving current designs by evaluating tolerance effects for variation analysis, impact of material characteristics and component geometry on performance, etc. This is very essential in the automotive industry's commitment to reduce prototype and warranty costs while introducing products fast to market. Some important concerns and lessons learned in simulating this system are as follows:

Simulation of mechatronic systems involves integration of models developed in different domains. Time constants in each domain vary in order from microseconds to seconds. This presents a classic dilemma in choosing the optimal time step for integration that is small enough to avoid numerical instability and large enough to save computation time and disk space. Even though Saber® has an adaptive time step feature, it was found inadequate in the solenoid system that included the detailed driver circuit. The error messages from the failure of simulation indicated a failure of a device in one of the subsystem rather than indicating a failure in the numerical algorithm due to sharp discontinuities in system behavior. Hence, choosing the time step requires a good understanding of the system behavior and knowledge of the sharp time discontinuities.

The concern for security of proprietary information in the supplier developed model complicates the process of simulating such a system. Often, details of the model development at the device level are encrypted. These models should either be de-encrypted or steps should be taken to ensure the stability of this model before it is integrated into a subsystem and system level model.

The availability of detailed information and data from a the supplier is a major bottleneck in the model development process. With the involvement of more suppliers in modeling effort, this obstacle will be overcome with the passage of time.

During the experimental validation process, it is critical to know the assumptions and limitations of the hardware setup as well as the model. Errors in establishing clear guidelines and standards for the validation procedure will result in long delays in the development of a high fidelity model.

As the size of the system increases, so does the complexity of the modeling process. This also means that more device models provided by different suppliers are being integrated into a single modeling unit. It is very important to understand the assumptions and limitations of these models in each domain to ensure the integrity of the system model.

Finally, the increasing complexity and level of detail in the model require more computation power to make it practical in terms of time and resources. The rapid growth in computer technology is expected to fulfill this need in the near future.

REFERENCES

1. Analogy, Inc., "Saber® Optional Template Library Reference Manual," 1996.
2. Mike Donnelly and Dale Witt, "Electrohydraulic Component System Design - Effective Hardware Modeling, Simulation and Verification Methods," Proc. International Fluid Power Exposition, April 1996.
3. Ansoft Inc., "Maxwell® ElectroMechanical System Simulator Manual," 1996.

981486

Electrohydraulic Valve with Cylinder Piston Velocity Feedback

Max P. Gassman
Iowa State Univ.

Mark A. Schoessler
Iowa State Univ.

ABSTRACT

The use of mechanical, electrical, electronic, computer, and control engineering concepts combined in a particular machine or process is occurring at an increasingly rapid pace. The integration of these concepts is referred to as *mechatronics*. Engineers and technicians now must have knowledge and skills in all of the areas mentioned. Mobile machines have always incorporated knowledge from a variety of areas as needed to incorporate desired features and work functions. However, the time has come to make a more formal recognition of this integration of ideas.

INTRODUCTION

Design of a machine is very much an iterative process. Use of mathematical simulation allows the engineer to examine the influence of various machine parameters in order to arrive at the desired machine configuration. Use of mathematical simulation, in the design process, generally reduces the amount of time and expense required to build and test prototypes of a machine. The topic covered examines a fluid power system that incorporates an electrohydraulic valve. Conclusions reached in the work indicate the usefulness of the simulation approach, as well as the definition of suitable parameters for the desired machine configuration. The work is presented in a tutorial fashion to facilitate understanding of the method used.

The primary purpose of this topic is to illustrate the merging of equations from different fields, in order to illustrate the process of combining knowledge from different engineering fields in a single machine function. Information on mechatronics may be found in a variety of new textbooks and other publications [1,6,8]. A description and mathematical simulation of a fluid power system with an electronic feedback loop to limit cylinder piston velocity follows. The method used is suitable for instruction at a university as well as actual applications in industry. Good descriptions of electrohydraulic fluid power systems may be found in publications[2,5].

MATHEMATICAL MODEL

Fig. 1, 2, and 3, which follow, provide a graphical definition of the system that was modeled.

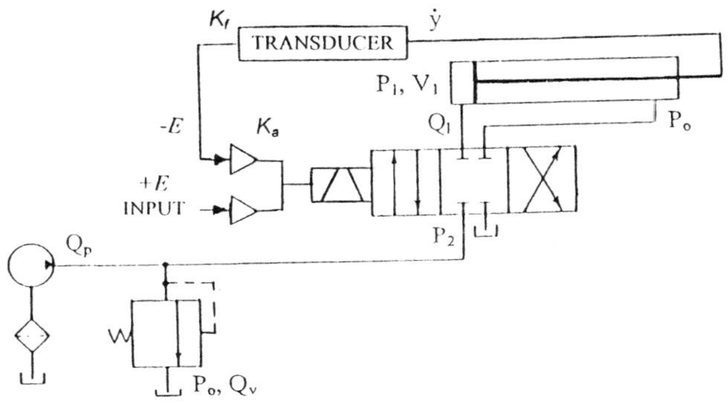

Figure 1: System defined with fluid power graphical symbols

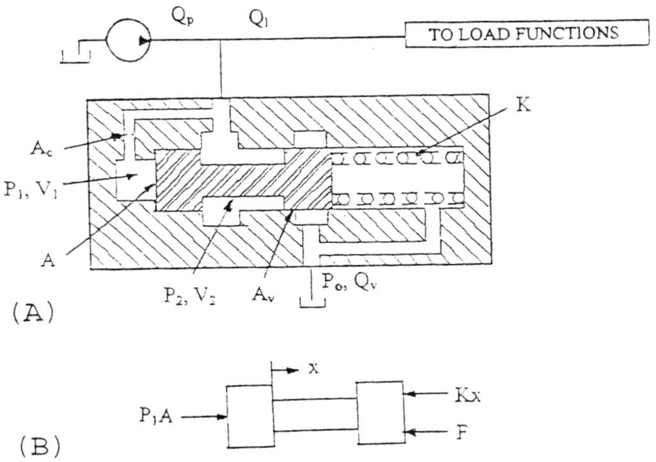

Figure 2: (A) Schematic of pressure regulating valve (B) forces on valve

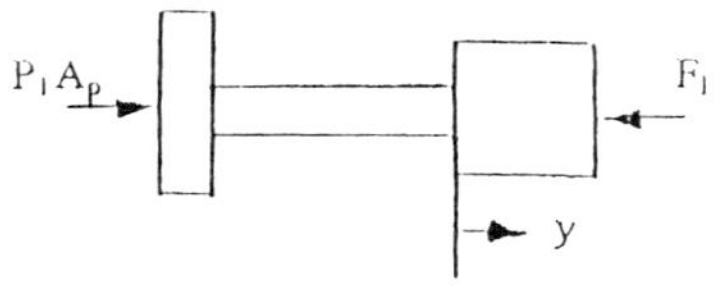

Figure 3: Forces on cylinder piston

To provide a mathematical model of the system, the equations must first be written for the pressure-regulating valve as schematically described in Fig. 2. These equations have been previously presented in SAE Paper 851954[7]. Parameters used in the equations are described in Table 1. Valve spool motion may be described with Newton's second law of motion as

$$\ddot{x} = (P_1 A - F - Kx) / m_v \qquad (1)$$

Pressures, P_1 and P_2, may be determined with use of the continuity equations as

$$\dot{P}_1 = \frac{\beta}{V_1}(Q_c - A\dot{x}) \qquad (2)$$

And

$$\dot{P}_2 = \frac{\beta}{V_2}(Q_p - Q_l - Q_v - Q_c) \qquad (3)$$

Values for flow, Q_c and Q_v, may be determined with the orifice equation as

$$Q_c = C_d A_c \sqrt{\frac{2}{\rho}} \sqrt{P_2 - P_1} \qquad (4)$$

Table 1. Description of parameters

Symbol	Description	Units
A	Valve end area	m^2
A_c	Orifice area	m^2
A_v	Relief valve area	m^2
A_l	Load valve area	m^2
A_p	Piston area	m^2
ρ	Oil density	kg/m^3
β	Oil bulk modulus	N/m^2
C_d	Flow coefficient	
F	Spring preload	N
K	Spring rate	N/m
m_v	Valve mass	kg
P_0	Return pressure	N/m^2
P_1	Control pressure	N/m^2
P_2	Regulated pressure	N/m^2
P_l	Load pressure	N/m^2
Q_p	Pump flow	m^3/s
Q_c	Control flow	m^3/s
Q_v	Valve return flow	m^3/s
Q_l	Load flow	m^3/s
V_1	Control volume	m^3
V_2	Valve volume	m^3
V_l	Cylinder volume	m^3
F_l	Cylinder load	N
m_p	Piston mass	kg
x	Valve spool motion	m/s
y	Piston motion	m/s
E	Control voltage	V

And

$$Q_v = C_d A_v \sqrt{\frac{2}{\rho}} \sqrt{P_2 - P_0} \tag{5}$$

Flow area, A_v, for the regulating valve is

$$A_v = 5.97 \times 10^{-2} x - 3.03 \times 10^{-5} \tag{6}$$

The valve spool has an overlap of 0.508 mm.

Pressure, P_l, in the load cylinder may be determined with the equation

$$\dot{P}_l = \frac{\beta}{V_l}(Q_l - A_p \dot{y}) \tag{7}$$

Cylinder volume, V_l, consists of an initial plus a variable volume

$$V_l = 1.64 \times 10^{-5} + A_p y \tag{8}$$

Flow, Q_l, through the load valve is

$$Q_l = C_d A_l \sqrt{\frac{2}{\rho}} \sqrt{P_2 - P_l} \tag{9}$$

Cylinder piston motion may be described with Newton's second law of motion

$$\ddot{y} = (A_p P_l - F_l) / m_p \tag{10}$$

RESULTS AND DISCUSSION

The mathematical model was derived in a university course on fluid power design and analysis. The specific purpose of the model was to demonstrate how velocity feedback, from the cylinder piston, may be used to limit the piston velocity. The model makes use of electronic feedback. A similar university educational project, which involved a hydromechanical system, was published as SAE Paper 971579[3].

Since design is an iterative process, a variety of parameter values were applied to the model. Results shown on the plots in Fig. 4 make use of the specific values shown in Table 2.

Table 2. Values applied to the solution

Symbol	Values	Units
A	$2.58\text{X}10^{-4}$	m^2
A_c	$4.56\text{X}10^{-7}$	m^2
A_p	$4.56\text{X}10^{-3}$	m^2
ρ	847.5	kg/m^3
β	$1.222\text{X}10^{9}$	N/m^2
C_d	0.6	
F	5226	N
K	210141	N/m
m_v	0.114	kg
P_0	0	N/m^2
Q_p	$3.27\text{X}10^{-3}$	m^3/s
V_1	$4.92\text{X}10^{-5}$	m^3
V_2	$1.64\text{X}10^{-3}$	m^3
F_l	86736	N
m_p	17.5	kg

Velocity of the piston was to be limited to 0.254 m/s. Therefore, the feedback loop was tailored to meet that objective. The load valve flow area, A_l, was given an initial input value of $3.23\text{X}10^{-5}$ m^2, with use of input voltage, E. The flow area was then linearly increased to 6.46×10^{-5} m^2, between time, t, of 0.1 and 0.3 seconds and then returned to the initial value.

The feedback loop, as shown in Fig. 1, contained an amplifier, K_a, with a value of 1000. The transducer, K_f, was given a value of 0 at cylinder piston velocities of 0.254 m/s or less. If the velocity tended to exceed 0.254 m/s the value of -6.414×10^{-6} $m^2/(m/s)$ was applied. These values kept the velocity of the cylinder piston very close to the desired value of 0.254 m/s.

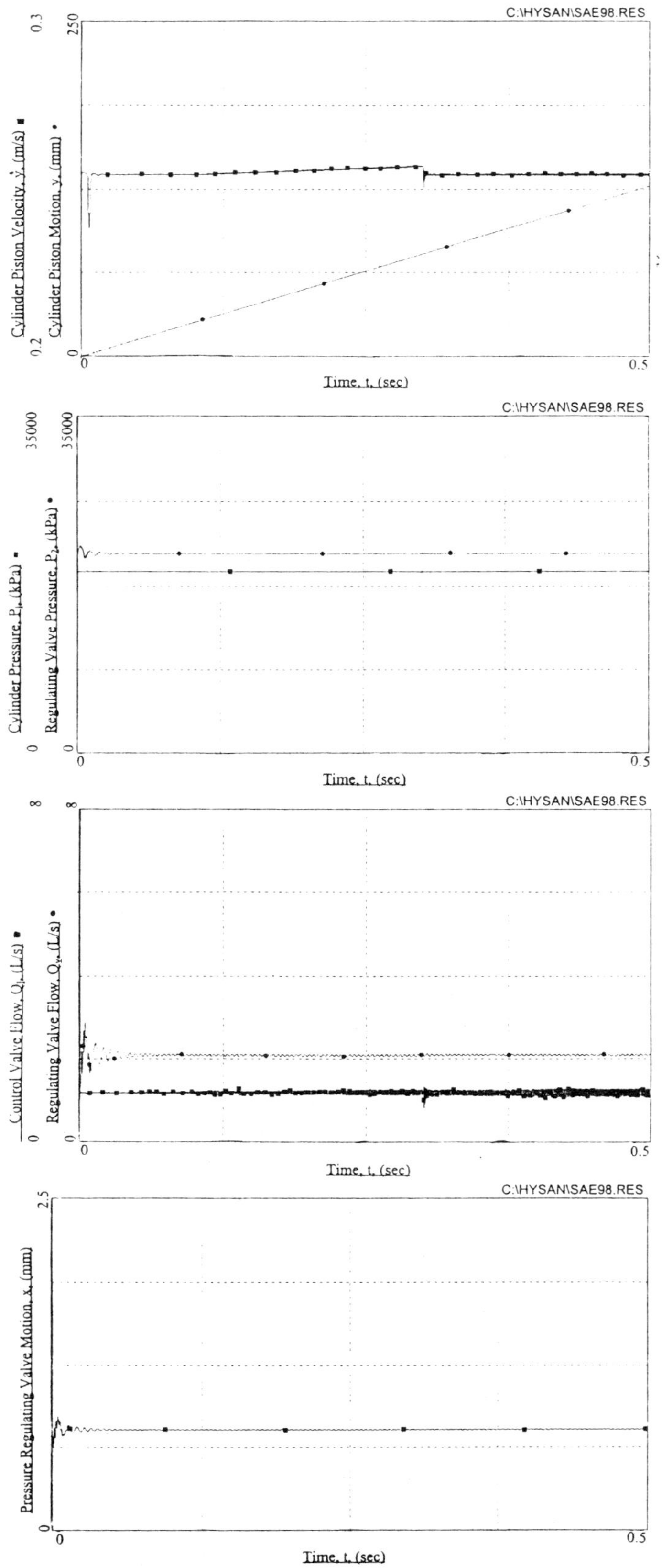

Figure 4: Plotted simulation results

Plotted information in Fig. 4 show cylinder piston motion and velocity, cylinder and regulating valve pressures, control valve and regulating valve flows, and regulating valve motion. Since the cylinder control valve flow area is increased in the time interval of 0.1 to 0.3 seconds, the plot for cylinder piston velocity increases slightly during that time. However, it is the function of the feedback loop to vary the flow area, A_l, to hold the cylinder piston velocity near the desired value of 0.254 m/s.

The mathematical simulation, as shown, allows iterative variation in the amplifier constant and the feedback transducer. Also, the model allows for easy investigation of the effect of other parameters in the system

The magnitude of other plotted variables, shown in Fig. 4, is all within the expected range. Solution of the model was carried out with the use of computer software as noted in the *ACKNOWLEDGEMENT* section. The software used is of the graphical type[4].

CONCLUSIONS

The type of mathematical model presented is very useful in sizing parameters to provide the desired operating characteristics for a fluid power system. Many parameter variations could be applied to the model to achieve desired results.

ACKNOWLEDGMENTS

The project described was used in the fall 1997 course in ME414 Hydraulics Systems and Controls at Iowa State University. Students in the course supplied many interesting and useful insights to the solution presented.

The authors are also grateful to David L. Persson of the *HYDRASOFT* Corporation. Mr. Persson provided considerable help in the use of the graphical software used to solve the simulation model[4].

REFERENCES

[1] Bolton, W.: *Mechatronics*, Addison Wesley Longman Limited, Harlow, England, 1996.

[2] *Electrohydraulic controls for proportional flow valves*, Off-Highway Engineering, Jan/Feb, pp. 43-45, 1997.

[3] Gassman, M. P.: *Mathematical model of a fluid flow control valve*, SAE Paper 971579, Peoria, IL, pp. 187-245, 1997.

[4] *"Hysan software for hydraulic system analysis,"* Hydrasoft Corporation, Kankakee, IL, 1996

[5] Johnson, J. E.: *Hydraulics for engineering technology*, Prentice Hall, Englewood Cliffs, NJ, pp. 187-245,1996.

[6] *Mechatronics,* Machine Design, September, pp. 45-46, 1997.

[7] Morris, H. C.: *Hydraulic system dynamic simulation on a personal computer*, SAE Paper 851954, Long Beach, CA, 1985.

[8] Shetty, D. & Kolk, R. A. *Mechatronics system design*, PWS Publishing Company, Boston, 1997.

981485

Cylinder Piston Velocity Feedback Applied to a Hydraulic Pump

Jeremy D. Rubens
Iowa State Univ.

ABSTRACT

Systems today are designed to be more flexible than ever before. This increasing demand has made electronically controlled systems more popular and brought into being a formal field called *Mechatronics*. The science of mechatronics uses mechanical, electrical, electronic, and computer science concepts in a single machine or system[1,6,8]. The key of mechatronics is that all of these areas are combined in the design stage. By using the ideas of mechatronics, more flexible and efficient machines can be produced. Electronically controlled systems are currently replacing the previous mechanically controlled systems[2,3,5].

INTRODUCTION

Mechatronics emphasizes a quicker, more efficient design. This puts more emphasis on modeling and simulation. Simulation software is one tool that can ease the design of these systems. It allows the engineer to observe the effects of changing specific parameters in order to determine the combination that gives the desired system response. This process also minimizes the amount of time and expenses used for testing and prototyping. This paper introduces a hydraulic system that uses a varying flow hydraulic pump.

The main purpose of this example is to show how mechatronics incorporates information from different fields to design a machine function. A description and mathematical simulation of a fluid power system with an electric feedback loop attached to the pump to limit the piston velocity is presented.

MATHEMATICAL MODEL

Fig. 1, 2, and 3 give a graphical definition of the system that was modeled.

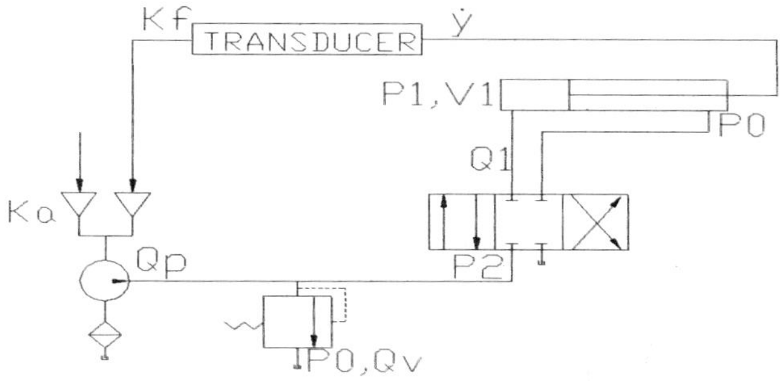

Figure 1: System definition using fluid power symbols

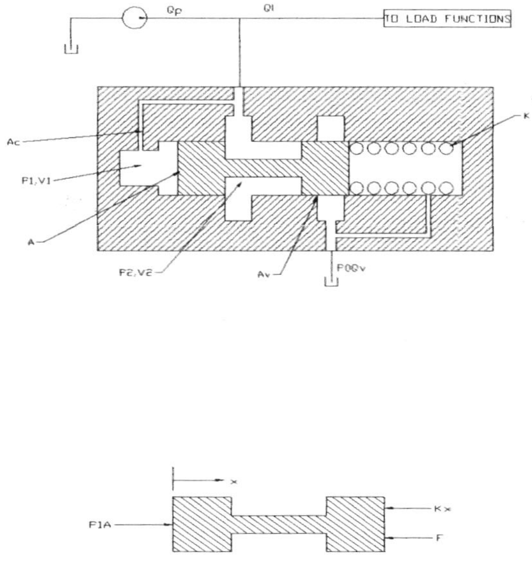

(A)

(B)

Figure 2:(A) Diagram of pressure regulating valve (B) Free body diagram of valve

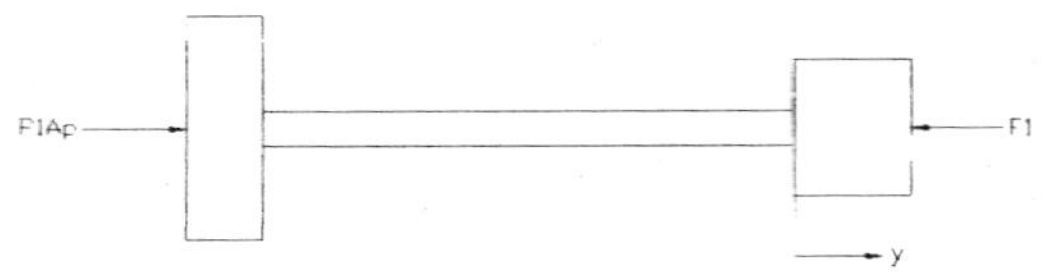

Figure 3: Forces on cylinder piston

In order to model the system, all the components must be defined mathematically. The first step is to model the pressure regulating valve shown in figure 2. The symbols used are defined in Table 1. By summing the forces and using Newton's second law, the valve spool motion can be defined as [7]

$$\ddot{x} = \frac{(P_1 A - F - Kx)}{m_v} \tag{1}$$

The continuity equations can then be used to calculate P_1 and P_2

$$\dot{P}_1 = \frac{\beta}{V_1}(Q_c - A\dot{x}) \tag{2}$$

and

$$\dot{P}_2 = \frac{\beta}{V_2}(Q_p - Q_l - Q_v) \tag{3}$$

The orifice equation can then be used to calculate the value for flow, Q_v

$$Q_v = C_d A_v \sqrt{\frac{2}{\rho}}\sqrt{P_2 - P_0} \tag{4}$$

Table 1. Description of variables

Symbol	Description	Units
A	Valve end area	m^2
A_c	Orifice area	m^2
A_v	Relief valve area	m^2
A_l	Load valve area	m^2
A_p	Piston area	m^2
ρ	Oil density	kg/m^3
β	Oil bulk modulus	N/m^3
C_d	Flow coefficient	
F	Spring preload	N
K	Spring rate	N/m
m_v	Valve mass	kg
P_0	Return pressure	N/m^2
P_1	Control pressure	N/m^2
P_2	Regulated pressure	N/m^2
P_l	Load pressure	N/m^2
Q_p	Pump flow	m^3/s
Q_c	Control flow	m^3/s
Q_v	Valve return flow	m^3/s
Q_l	Load flow	m^3/s
V_1	Cylinder volume	m^3
V_2	Valve volume	m^3
V_l	Cylinder volume	m^3
F_l	Cylinder load	N
m_p	Piston mass	kg
X	Valve spool motion	m/s
Y	Piston motion	m/s

The regulating valve flow area, A_v, is equal to

$$A_v = 5.97*10^{-2}x - 3.03*10^{-5} \quad (5)$$

The valve spool has an overlap of 0.508 mm.

Pressure, P_1, in the load cylinder can be calculated with the equation

$$\dot{P}_l = \frac{\beta}{V_l}(Q_l - A_p\dot{y}) \quad (6)$$

Cylinder volume, V_l, is made up of an initial component and a variable component.

$$V_l = 1.64*10^{-5} + A_p y \quad (7)$$

Flow, Q_l, through the load valve is

$$Q_l = C_d A_l \sqrt{\frac{2}{\rho}}\sqrt{P_2 - P_1} \quad (8)$$

Cylinder piston motion may be described by using Newton's second law of motion

$$\ddot{y} = (A_p P_l - F_l)/m_p \quad (9)$$

RESULTS AND DISCUSSION

This system was then modeled on simulation software in order to see how changing certain parameters would change the response of the system. This idea was used to change the parameters around in the model.

The results shown in figure 4 use the values shown in Table 2.

Table 2. Values applied to the solution

Symbol	Values	Units
A	$2.58*10^{-4}$	m^2
A_c	$4.56*10^{-7}$	m^2
A_p	$4.56*10^{-3}$	m^2
ρ	847.5	kg/m^3
β	$1.222*10^9$	N/m^3
C_d	.6	
F	5226	N
K	210141	N/m
m_v	.114	kg
P_0	0	N/m^2
A_l	$7.74*10^{-5}$	m
V_1	$4.92*10^{-5}$	m^3
V_2	$1.64*10^{-3}$	m^3
m_p	17.5	kg

The purpose of this model was to show how velocity for a hydraulic piston can be controlled with a feedback loop. This system was modeled and the parameters were adjusted until the desired system response was obtained.

The velocity was to be limited to .254 m/s. The pump flow, Q_p, was given an initial value of .00205 m^3/s. The feedback loop was adjusted to obtain that velocity. The feedback loop, as shown in figure 1, has an amplifier, k_a, with the value of 1000. The value of K_a varied in order to maintain the .254m/s. As the velocity of the piston dipped below .254, the value of K_a decreased. This decreased the feedback allowing the velocity to increase as the velocity increased above .254, k_a increased to lower it.

The pump flow was then linearly increased to .00369 m^3/s, between the time, t, of .1 and .3 seconds. This simulated a jump in the flow to see if the system would compensate. This increase in flow is shown on the plots in figure 4 as a slight jump in the piston velocity. However, the system does its job and levels the system out to the desired velocity.

Since the system is modeled on the computer, it allows the user to change certain parameters to see the effect it has on the system and the system's response to jumps.

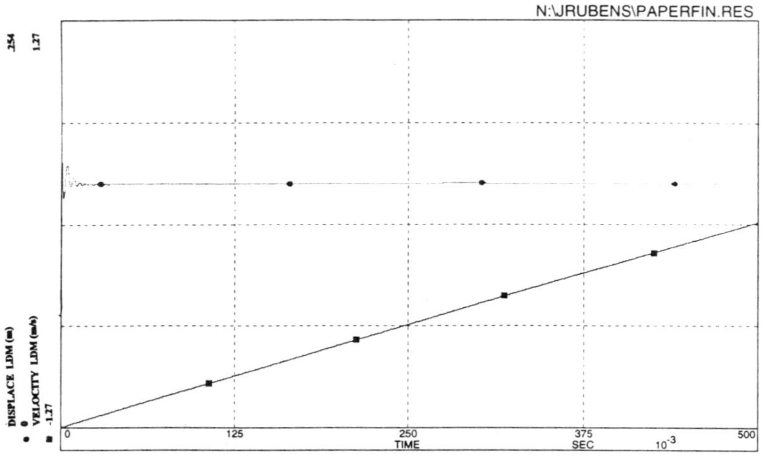

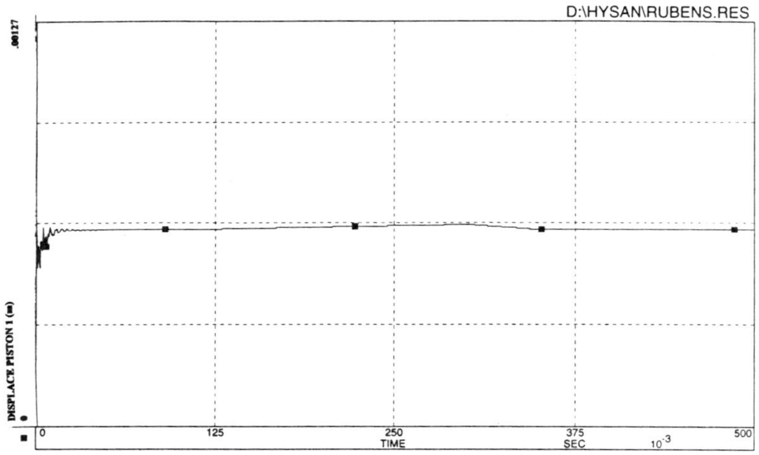

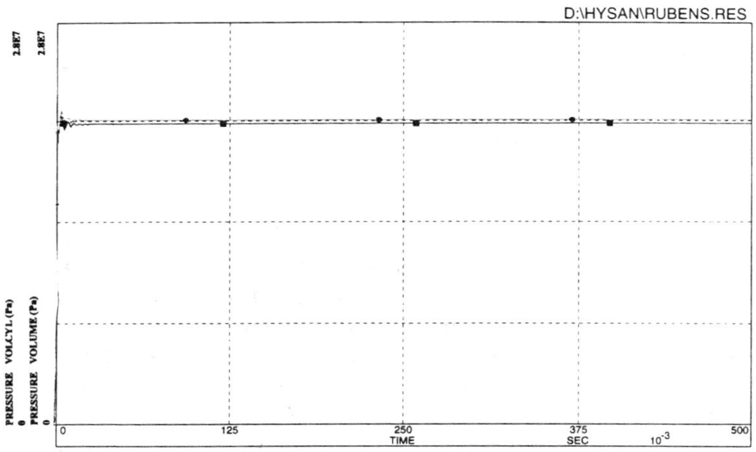

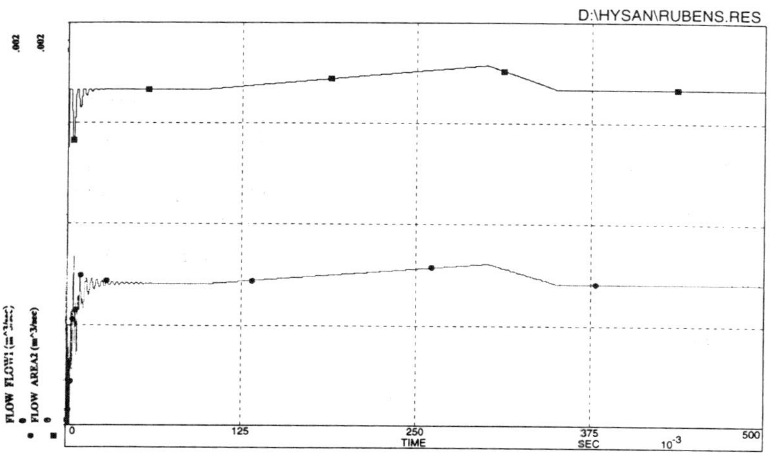

The plots in Figure 4 shows the cylinder piston motion and velocity, cylinder and regulating valve pressures, pump flow and regulating valve flow, and the regulating valve motion. The solution of this model was created using computer software as noted in the ACKNOWLEDGEMENT section. The Software used is of the graphical type[4].

CONCLUSIONS

The model shown in this paper is useful in determining operating parameters for a desired system response. It is also helpful in determining the effect each parameter has on the system response. Due to the nature of the problem, there are many other combinations that will give the same results.

ACKNOLEDGMENTS

The project described was used in the fall 1997 course in ME414 Hydraulics Systems and Controls at Iowa State University.

The authors are also grateful to David L. Persson of the HYDRASOFT Corporation. Mr. Persson provided considerable help in the use of the graphical software used to solve the simulation model[4].

REFERENCES

[1] Bolton, W.: Mechatronics, Addison Wesley Longman Limited, Harlow, England, 1996.

[2] Electrohydraulic controls for proportional flow valve, Off-Highway Engineering, Jan/Feb, pp.43-45, 1997.

[3] Gassman, M.P.:Mathematical model of a fluid flow control valve, SAE Paper 971579, Peoria, IL, 1997.

[4] "Hysan Software for Hydraulic System Analysis," Hydrasoft Corporation, Kankakee, IL, 1996

[5] Johnson, J. E.: Hydraulics for engineering technology, Prentice Hall, Englewood Cliffs, NJ, 1996

[6] Mechatronics, Machine Design, September, pp. 45-46, 1997

[7] Morris, H. C.: Hydraulic System Dynamic Simulation on a Personal Computer, SAE Paper 851954, Long Beach, CA, 1985.

[8] Shetty, K. & Kolk, R. A.: Mechatronics System Design, PWS Publishing Company, Boston, 1997.

960430

Electronic Transmission Control - From Stand Alone Components to Mechatronic Systems

Kurt Neuffer, Kurt Engelsdorf, and Werner Brehm
Robert Bosch GmbH

ABSTRACT

Electronic control systems for automatic transmissions have been gaining acceptance in recent years. Besides the various types of sensors already in use [for speed, temperature, pressure etc.], electro-hydraulic actuators and electronic control units [ECU] are also major constituents of modern control systems.

Former designs utilized discrete components which were wired together. In the interim, actuator modules have been developed for use in new transmission designs.

As well as presenting actuator concepts, this paper describes a highly integrated and advanced control module. This module consists of hydraulic actuators, pilot-solenoid values and an electronic control unit which has been realized in multi-layer hybrid technology.

INTRODUCTION

Various types of solenoid valves are employed in electronically-controlled transmissions [1], [2]. Figure 1 shows a schematic overview of the different actuators in use.

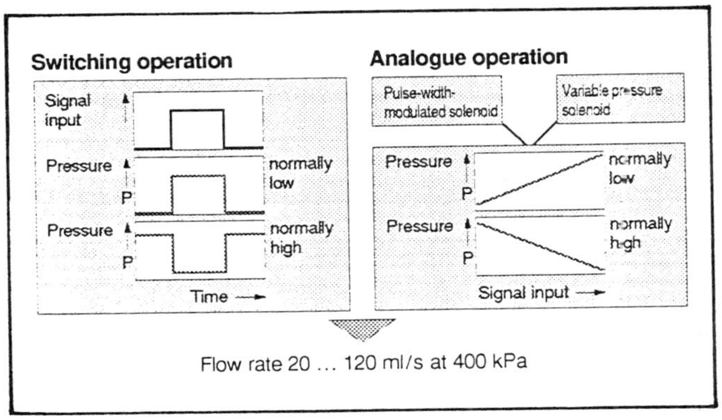

Figure 1: Schematic overview of solenoids for automatic transmissions

The following requirements should thereby be taken into consideration:

1. Size and weight as small as possible
2. Reliability in heavily contaminated automatic transmission fluid [ATF]
3. Operating temperature range from -40°C through +150°C
4. Pressure range up to 5,000 Kpa
5. Maximum vibration acceleration 300 m/s²
6. High number of switching cycles, up to 10^9
7. Low power consumption
8. Low hydraulic resistance of the valve elements for high flow rates
9. Short response times
10. No hydraulic oscillations
11. Low inherent leakage

Extensive durability testing in fixtures as well as in vehicles is required in order to achieve these targets [3], [4]. It should also be noted that the specifications for these components and systems can be adapted over a wide range to meet customer requirements.

ON/OFF SOLENOIDS

2-port solenoids could prove a good choice on account of their simple design, small size and low weight. It is however an inherent disadvantage for oil to be flowing through one or both of the two possible control positions. A 3-port solenoid does not exhibit this power loss in steady state operations. Figure 2 shows a plastic-molded on/off solenoid.

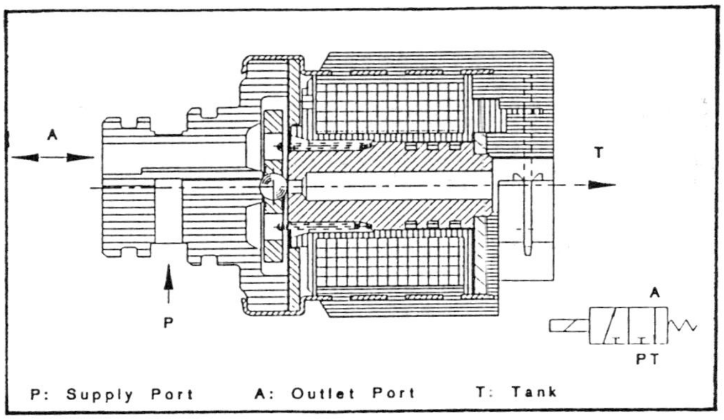

Figure 2: On/off-solenoid, 3 port, normally closed

In addition, there is no danger that entrapped particles of contamination from machining can cause failures.

PULSE-WIDTH MODULATED SOLENOIDS

The pulse-width modulated solenoid [PWM] is used to supply hydraulic pressures which are proportional to the duty cycle of a pulse-width modulated driver signal [9]. In order to prevent the occurrence of oscillations due to natural frequencies in the system, as well as to smooth the pressure pulsation's, the frequency of operation shall be in the range 30 Hz through 100 Hz, depending on the overall system design.

The valve seats must be capable of withstanding up to 10^9 switching operations on account of this frequency. A good solution which fulfills these requirements is given by a pressure-balanced valve incorporating a flat valve seat design [figure 3].

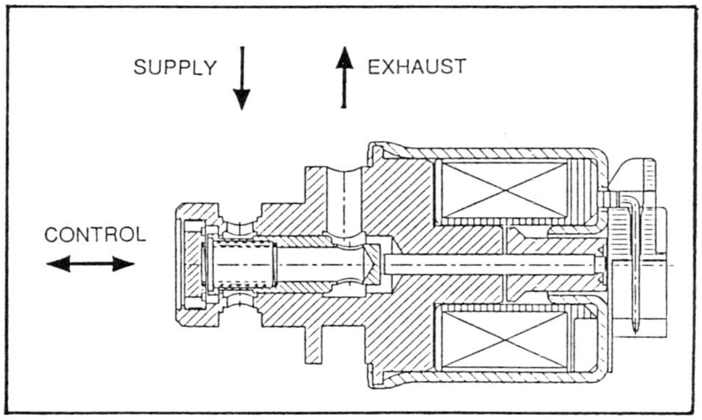

Figure 3: 3-Port PWM with high flow-rate

This pressure-balanced design enables high flow rates and a short armature stroke. Mechanical stress is also reduced on account of the flat valve seat. This type of solenoid can also be used in direct electronic shift control [DESC].

The major advantages of PWM solenoids can be summarized as follows:

1. Minimum hysteresis in the characteristic curve
2. Good hydraulic stability
3. Low sensitivity to contamination
4. Low driver costs compared to analogous pressure regulators

PRESSURE REGULATOR SOLENOIDS [VARIABLE FORCE SOLENOIDS]

Variable force solenoids [VFS] are used to regulate the system pressure. The pressure is a function of the current in accordance with the characteristic curve shown in figure 4.

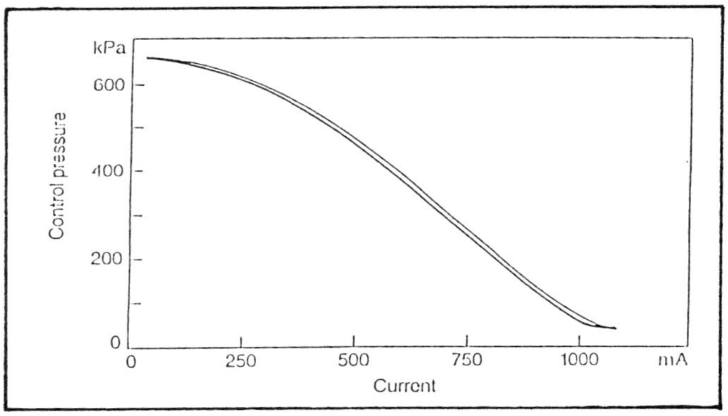

Figure 4: VFS bleed type characteristic curve

These actuators can operate both as a pressure relief valve [bleed type] and as a 3-way pressure regulator [spool type].

There is no need for an additional pressure sensor on account of the high reproducibility of the characteristic curve. A prerequisite for this reproducibility is however the high precision of the mechanical components as well as for the d.c. magnetic system.

Materials of low coercive force must be utilized to reduce the hysteresis as well as to minimize the friction. A special feature of the design in order to achieve this is illustrated in figure 5.

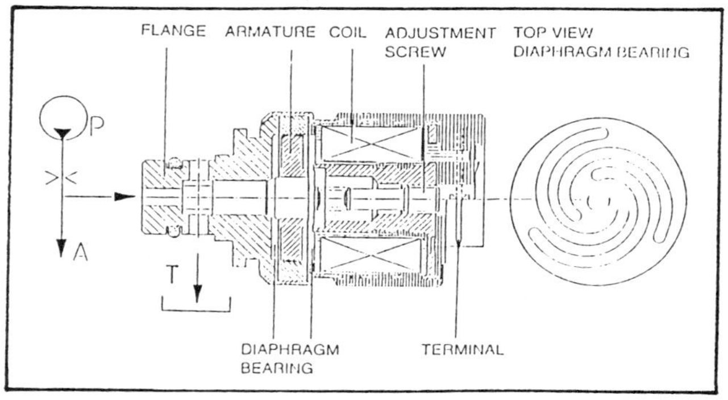

Figure 5: VFS bleed type, 2-port

double diaphragm bearing design

The armature is mounted here between two diaphragm bearings as an example of a bleed-type VFS. This design can be also applied for a spool-type VFS. The thereby resulting characteristic curve is free of frictional hysteresis as shown in figure 6.

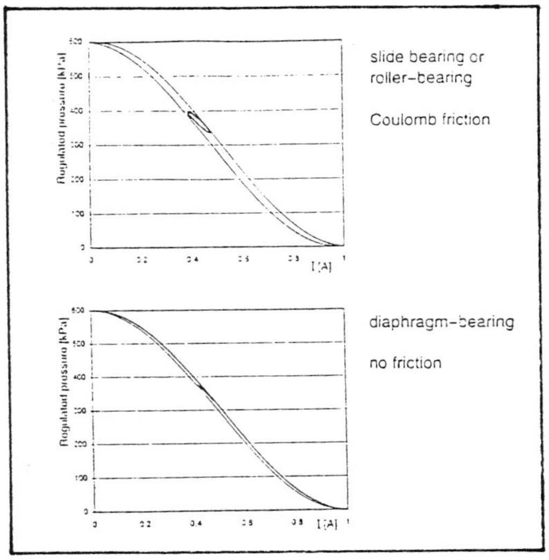

Figure 6: VFS bleed type
comparison friction

This performance is assured throughout the entire lifetime. The diaphragm bearing is insensitive to contamination by particles because of the lack of friction. It is possible to eliminate any material-related hysteresis by the use of application-specific software [5]. This results in an overall hysteresis which is extremely low. A cost-optimized design is depicted in figure 7.

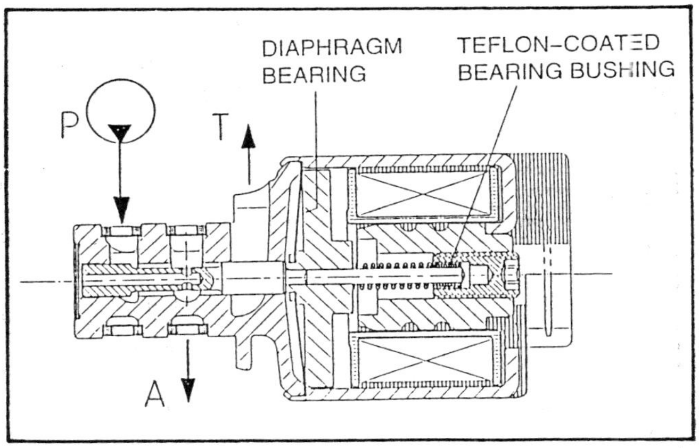

Figure 7: VFS spool type, 3-port

A diaphragm spring on one end is used as the armature bearing. A Teflon-coated bushing is inserted at the other end.

Finite element analysis [FEA] is employed in the design for an efficient geometry of the winding window as well as for the dimensions and position of the working air gap in order to optimize the magnetic circuit [6]. Furthermore, the computer aided design ensures that there are no unwanted high field strengths in the magnetic circuit. Figure 8 shows the magnetic flux path in a VFS resulting from CAD methods.

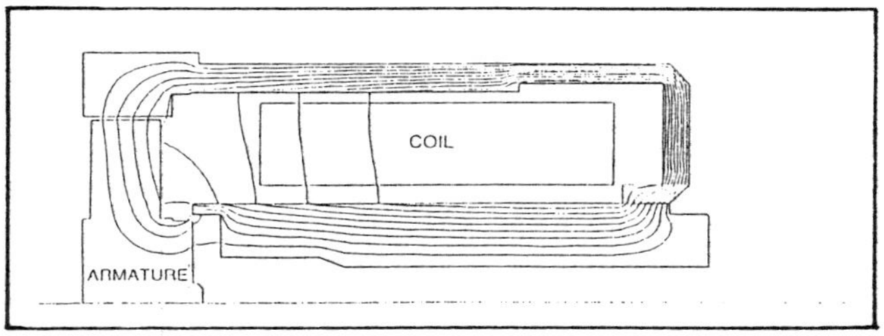

Figure 8: Magnetic flux path of a VFS bleed type

Other simulation techniques are employed for calculations of the dynamic behavior of the solenoids [7][8].

A new bleed-type solenoid of reduced size in comparison to conventional actuators has been designed using the methods described above. The solenoid results are shown in figure 9.

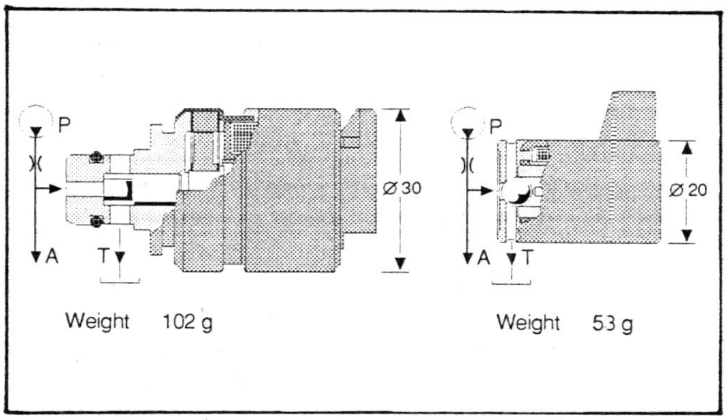

Figure 9: VFS bleed type
Comparison of conventional and new design

This new design contributes to the overall reduction in size for the complete valve body.

SOLENOID MODULES

Solenoids in current applications are mounted in the valve body as single components. Each valve and sensor requires a discrete connector assembly. This results in considerable expenditure in terms of connectors, wiring etc., and damage to solenoid valves and other components during the assembly processes of these cannot be excluded.

From this point of view, it proves more advantageous to mount all actuators and auxiliary components such as switches and speed sensors onto one carrier, or module [figure 10].

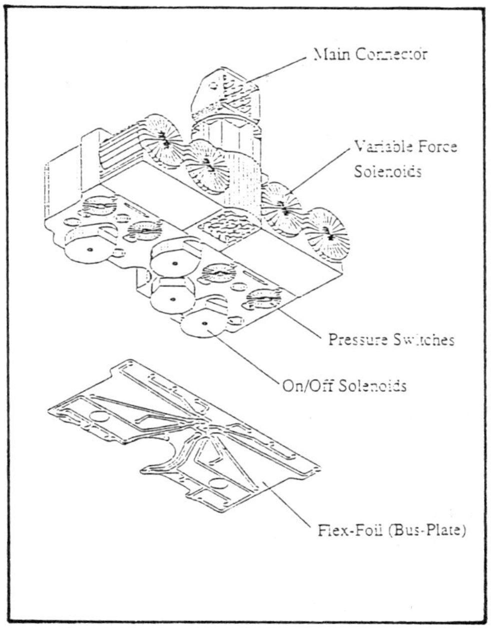

Figure 10: Solenoid module

A modular construction permits the pre-mounting of all sensitive components in a clean environment which results in an improved quality because all elements are assembled and supplied as a complete unit. The reliability of a modular design is increased by the avoidance of electrical failures due to the wiring and connectors being replaced by a "punched grid" in a plastic frame, or by flex foil as the only connection to the main connector. A further benefit is the opportunity of prior electrical and hydraulic testing. This ensures that infant mortality in electromechanical parts can be detected prior to final testing of the unit. Other savings in costs are also achieved due to the reduced amount of wiring required, and in some cases it is also possible to integrate solenoid valve parts into the body of the module.

An example is shown in figure 11.

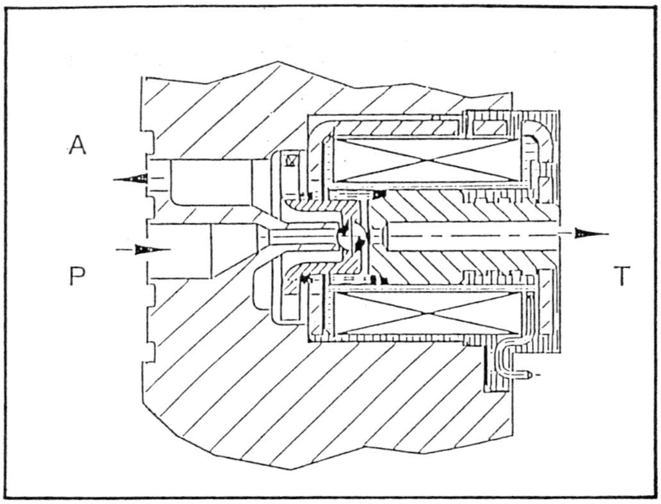

Figure 11: On/off-solenoid, integrated in module body

The solenoid valve requires no fixing clamps or mounting screws. Mounting to the module is realized by ultra-sonic techniques which results in a leak-free sealing of the solenoid.

ELECTRO-HYDRAULIC CONTROL MODULE

The electronic components available in the interim which can be operated at elevated temperatures permit the integration of electronic hardware into the housings of hydraulic actuators. This means that not only can intelligent controls be included but also that the differing physical hardware for signal processing and power operating can be combined into a mechatronic system. What are the reasons for this?

The main points are the technical improvements as mentioned above as well as the thereby associated savings in costs. As can be seen from figure 12, considerable savings in the overall system costs can be made for the customer together with an increasing degree of system integration.

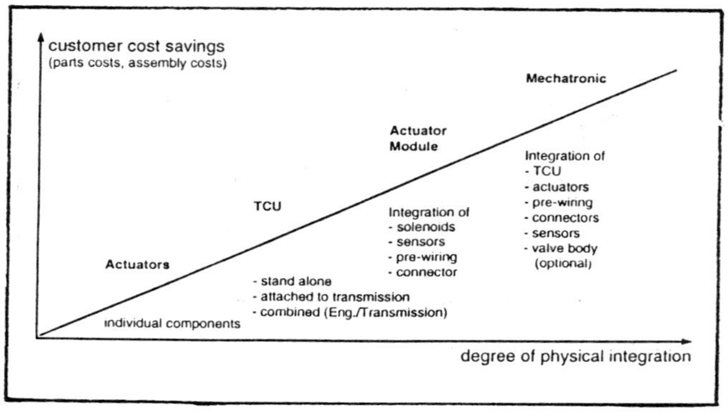

Figure 12: Cost savings by integration

This is possible not only because of the electrical components but also on account of the improvements in assembly processes which can be made, whereby the complexity of interfacing and connecting components can be reduced in particular. This has a positive influence in reducing the failures. Hence there is not only a cost reduction factor but also significant advantages from improved functionality and reliability, as can be seen in figure 13.

Advantages

- cost savings by reduced number of components
- separately pre-tested module (sub-system)
  - reduced test expenditure
  - reduced expenditure of logistics, wiring and mounting
- coupling of electronics and mechanics/hydraulics with transmission enables system optimization by software:
  - optimized adaption of functional parameters
  - compensation of component and production tolerances

Figure 13: Adavantages of Mechatronic concept

The mechatronic approach allows a complete testing of the control system prior to assembly to the transmission as well as the complete application of control unit and transmission prior to assembly to the vehicle. It is therefore possible to compensate for any tolerances in transmission components by software control.

The car manufacturer can thereby increase his productivity by employing simpler handling techniques for a complete transmission which has been pre-tested and calibrated beforehand.

Figure 14 depicts the concept of the mechatronic approach as the design goal for the next generation of transmission control modules.

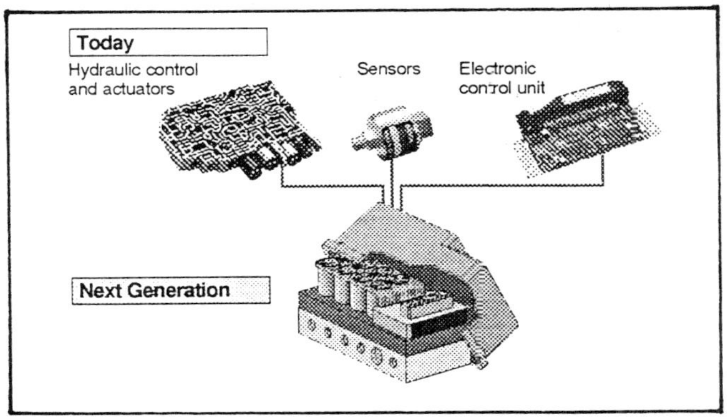

Figure 14: The Mechatronic approach

The present concept of conventional automatic transmissions utilizes stand-alone components which are located throughout the whole vehicle - actuators and sensors in the AT, TCU in the passenger or engine compartments. An alternative is to combine all components necessary for control and including the sensors into one unit for assembly.

A system of standardized sub-components for complete control modules also proves useful for cost effectiveness and flexibility. Depending on the module application, the various sub-module functions such as valve blocks, TCU, connector and sensor blocks could be standardized. This modular system of construction could be particularly useful for those transmission manufacturers intending to offer differing products on the open market to different customers, where a flexibility not only in the software but also in hardware configurations is required. At the same time, it is thereby also possible to reduce cost by the economy of scale.

Figure 15 provides an example of a modular design for the mechatronic control system of a CVT.

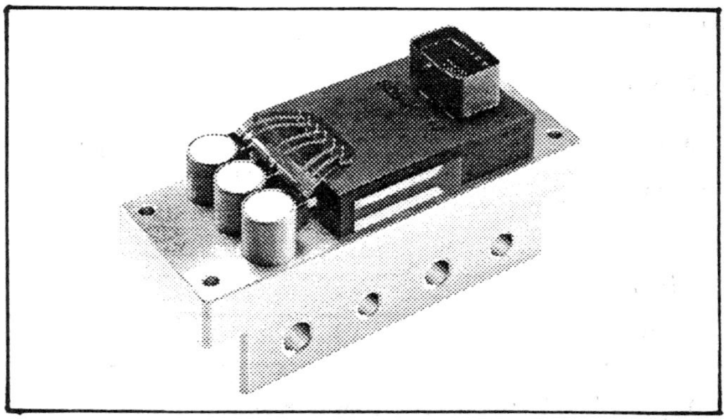

Figure 15: Mechatronic as modular system

All the necessary sub-components are included to give the flexibility required for adaptation of the module to the differing transmissions. Standardized elementary blocks with pilot valves, TCU and connector are assembled on an intermediate plate which serves as the interface to the transmission and a separate valve housing [10].

Integrated electronics incorporate certain special features which have been developed in recent years and enable very compact and reliable designs for ECU's. The fundamental advances made have been the developments in hybrid and micro-hybrid technologies [figure 16].

- Reduced size and weight
- High resistance to
  - temperature
  - vibration
- High reliability due to
  - reduced number of connections (no packaging of components)
  - advanced thermal management by direct mounting
  - stress-free packaging of ICs by soft silicon gel
  - fitted expansion coefficients of ceramic substrate

Figure 16: Advantages of Hybrid ECU technology

Hybrid ECU's are produced by the millions for a variety of vehicle control systems. Micro-hybrid electronics have recently commenced series production for ABS, and applications for other systems will follow. Production dates are given in figure 17 [11].

ECU for:	SOP
- ABS	in production since 1995
- Diesel: pump control	1996
- Engine control	1996
- Airbag	1997
- Diesel: EDC	1998

Figure 17: Products in μ-hybrid technique

Such new technologies open the door for integrating the electronics into control systems on account of the associated benefits. There are however certain difficulties to be resolved on account of the physical environmental conditions encountered and in particular those concerning the high temperature of the surrounding ATF oil in vehicle transmissions.

The design of a mechatronic module shall therefore ensure that:

1. The oil shall not come into contact with the micro-hybrid electronics,
2. There is adequate heat transfer for the electronics even under severe operating conditions.

An insulation against contact with the transmission fluid is assured by encapsulation of the micro-hybrid circuit in a welded steel housing which includes glass feedthroughs for the connection pins [figure 18].

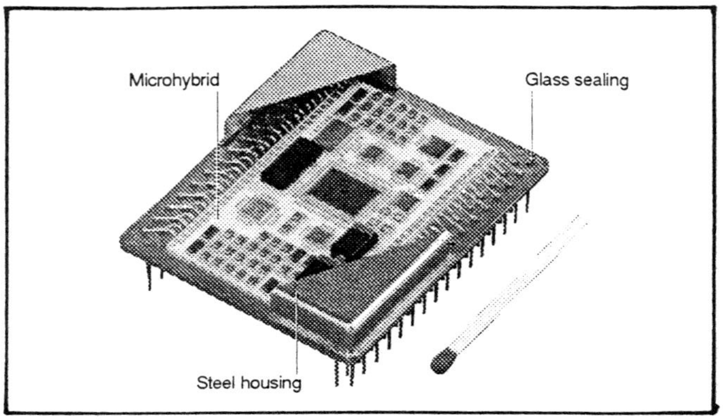

Figure 18: Welded ECU housing with glass sealed eyelets

Figure 19 provides a preliminary analysis of the hybrid design on the basis of realistic corner values and certain physical assumptions.

Parameters	
junction temperature	145 °C
power dissipation (chip)	0.5 W
thermal resistance (chip to base plate)	10 K/W
power dissipation (TCU)	10 W
thermal resistance (base plate to heat sink)	1.5 K/W

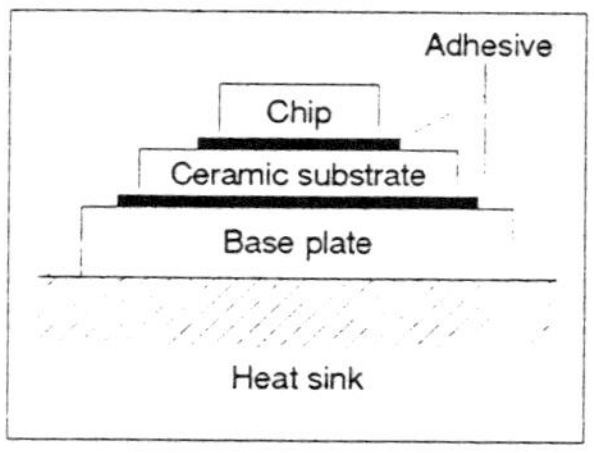

Limit values:

base plate temperature	140 °C
heat sink temperature	125 °C

Figure 19: Thermal analysis for critical μ-hybrid components

An estimation of the thermal limits for such critical components as the microprocessor leads to an upper temperature limit of 125°C for the heat sink which may not be exceeded.

Figure 20 shows a design concept for the realization of the above criteria by utilizing the reflow oil from the oil cooler as a heat sink for an integrated TCU.

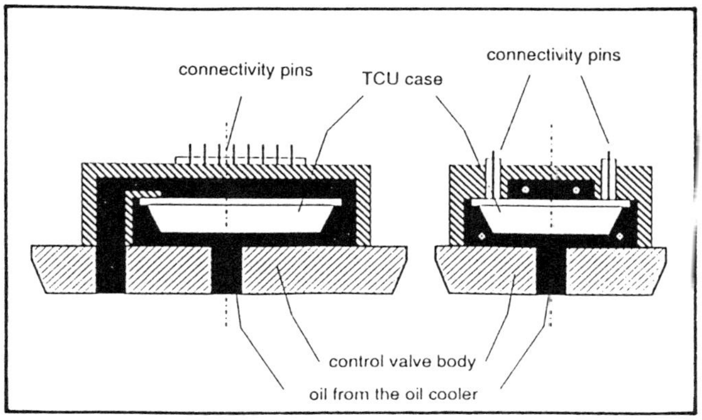

Figure 20: Thermal design for TCU cooling

Temperature measurements from a prototype CVT from Van Doorne's Transmissie [VDT] installed in a VW Golf 2.8 l engine test car indicate that the temperature limits defined for the electronics are not exceeded [figure 21].

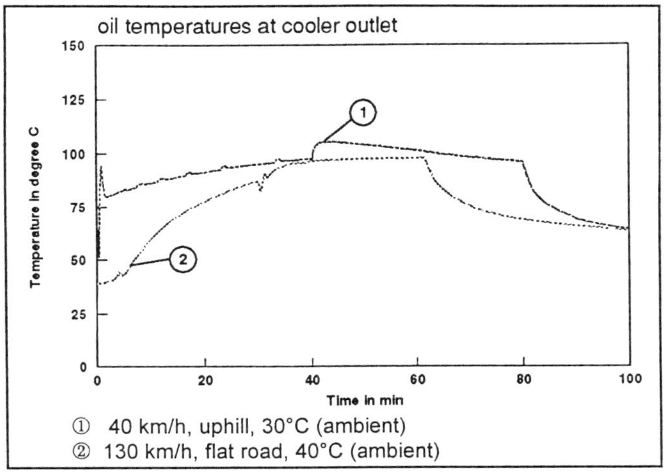

Figure 21: Measured temperatures in the test car CVT

We therefore anticipate the chances as being very good for realizing the integration of control units using this design and the constructional approach given above.

It may be argued that such a solution is inconsistent with the trend of combining the controllers of different vehicle systems into one central electronic unit. The availability of high-speed data bus systems like CAN, can however ensure a realization of both objectives; i.e. the information exchange between different control systems [figure 22] and the cost advantages from integrated electro-hydraulic modules.

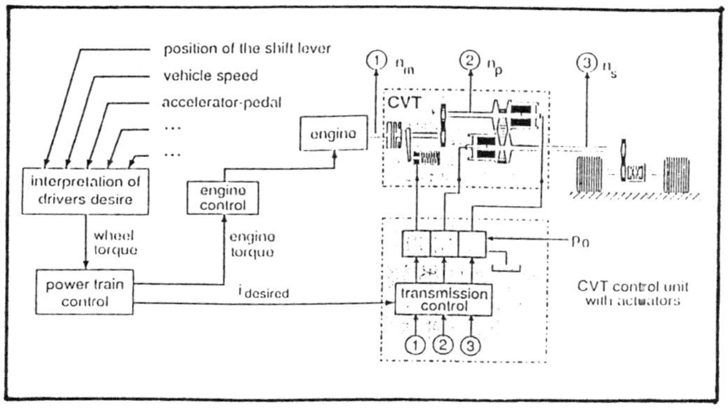

Figure 22: Power train management with CVT

CONCLUSIONS

Actuators which can be reduced in size, together with those electronic components now available which can be operated at temperatures up to 125°C make the combination of hydraulics and electronic hardware a viable proposition.

Such a complete module comprises all the major components of a conventional design, i.e. the individual solenoids and the electronic units and thereby also improves the overall quality of the system.

In conclusion, the handling costs of assembly processes will also be significantly reduced as compared to hitherto conventional designs.

REFERENCES

[1] K. Neuffer
Recent development of AT control: adaptive functions and actuators, Symposium No. 9313, Advanced Technologies in Automotive Propulsion Systems, Society of Automotive Engineers of Japan Inc., 1993, pp 42 - 49.

[2] K. Neuffer, W. Bullmer, W. Brehm:
Transmission Control in: R. Jurgen [editor]: Automotive Electronics Handbook, Chapter 13, McGraw-Hill, Inc. New York, 1995

[3] J.G. Eleftherakis, A. Khalil:
Development of a laboratory test contaminant for t ransmissions, SAE Paper 90 0561
Society of Automotive Engineers, Warrendale, Pa.

[4] A practical approach to automatic transmission reliability, SAE Paper 91 0640,
Society of Automotive Engineers, Warrendale, PA.

[5] Method and apparatus to Convert an Electrical Value into a Mechanical Position by Using an Electromagnetic Element Subject to Hysteresis
U.S. Patent 4,577,143 March 18, 1986

[6] P. Silvester, R. Ferrari:
Finite Elements for electrical engineers, Cambridge, University Press 1990

[7] SABER, Simulation for Circuits Through Systems
Analogy Incorporated, 1993

[8] Henry, P. James, Dennis, S. David:
Predicting Solenoid Transient Performance, SAE-Paper 87 0473

[9] W. Brehm, K. Neuffer:
Fast switching PWM-solenoid for automatic transmissions, Proc. seventh International Conference on Automotive Electronics, London,
Oct. 9-12, 1989, C 391/046 I Mech E, 1989

[10] K. Engelsdorf, M. Mainka, K.-H. Senger
Ein Konzept für integrierte Getriebesteuerungen am Beispiel CVT, VDI-Berichte Nr. 1175, 1995

[11] R. Scleupen, W. Reichert, P. Teuber, G. Walter:
Electronic Control Systems in Microhybrid Technology, SAE 1995, Detroit.

961762

Mechatronic Designs for a Computer Controlled All-Terrain Vehicle

Sanjay I. Mistry
John Deere Product Engrg. Ctr.

David L. Buster
3M Company

Jeff M. Thate, Mark Eubanks, and Satish S. Nair
University of Missouri-Columbia

ABSTRACT

Novel mechatronic designs for the automation and control of an all-terrain vehicle for operation by a C-5/6 quadriplegic are described. As part of a project at the University of Missouri, a Kawasaki Mule 2010 all-terrain vehicle was purchased and fully automated and interfaced with a computer for control, using novel automation designs. This computer controlled all-terrain vehicle (CCATV) is primarily intended to be used for recreational off-road locomotion. A 'drive-by-wire' computer control system has been designed, tested, and installed on a utility vehicle and delivered to the sponsor. A very user-friendly and high-tech control panel was developed with 2 joysticks, 15 push buttons and 3 LCD displays for interfacing with the CCATV systems.

The subsystems that were added or automated include the computer subsystem; sensors; control panel; power subsystem; steering subsystem; throttle/brake subsystem; gear shifting subsystem; parking brake subsystem; the three vehicle settings high/low speed, differential lock/unlock and 2/4 wheel drive; interface subsystem and accessories. The major design issues included the selection of the vehicle features to be automated and their design; actuation strategies for the subsystems; control techniques; computer selection, and power system selection. Issues in ergonomics, ruggedness, reliability and safety were also of considerable significance in the design.

A unique feature of the automation design is the provision for manual mode of operation. That is, each subsystem has been automated such that the automation design does not interfere with the original vehicle subsystems. An auto/manual toggle switch is provided so that whenever needed, an able-bodied person can drive the CCATV in its original, off-the-factory mode. This feature considerably complicated the overall design since reliability could not be compromised. The mechatronic designs reported have potential applications to several areas in addition to that for the disabled operator reported here.

1. INTRODUCTION

Try to think for a minute about all the routines one has to go through to drive a car across town. One would open the door, step inside and sit down, fasten the seat belt, insert the key in the ignition, start the car, put one foot on the brake, shift into gear, let off the brake, and turn the steering wheel, depress the accelerator pedal, while carefully coordinating all the tasks. Several other tasks are also performed concurrently such as activating the turn signals, turning on the lights, adjusting the heat or air conditioning, and the tasks go on and on. Driving a vehicle, clearly, is a very complex task. Now imagine trying to perform all these tasks with the use of only arms and hands, or may be even only arms. Mechatronics (a combination of mechanical and electronics) opens up interesting possibilities in such situations.

More and more people with disabilities are driving vehicles. Advances in servo motor control and computer processing power coupled with a growing public awareness and increased governmental support have allowed many more of the disabled to lead more independent lives through driving their own vehicles. As the need for increased equality through mobility for the disabled rises, so will the need for advances in vehicle automation. Each and every function which can be controlled on a vehicle must be able to be controlled by a disabled person. Different degrees of automation are required for different people. While some would need only hand lever extensions from the gas and brake pedals, others such as the high level quadriplegics, would require a sip-and-puff (a straw-like mechanism used to provide command signals through a series of sips and puffs on the straw) mechanism coupled to a computer to control power provided to different actuators. Through vehicle automation, disabled persons can enjoy some of the independence others have allowing them to lead somewhat normal life styles. One thing is certain that with the ever increasing electronics technology, more disabled persons will be driving vehicles of various types.

There are many more applications of vehicle automation than just mobility for the disabled. In the area of mobile construction equipment, employers and manufacturers are striving to reduce physical strain on the operator incurred by working long hours on construction machinery. The hydraulic systems which currently exists on the machines are being supplemented by electro-hydraulic servovalves and comfortable, low force electronic joysticks. Many of the new electric actuators designed for mobile construction equipment have been used in the project described herein. Another area of application is in farm equipment. Like the construction industry, farm equipment manufacturers are trying to reduce the force required to operate heavy machinery. However, farm equipment manufacturers have an additional group of people to design around, farm workers over the age of 55. Not only does operation effort have to be reduced, but accessibility to tractors and combines must be improved. An all-terrain vehicle which is suited to operation by a disabled person could have many uses for a farmer. The reported project thus has a direct application to increasing accessibility for farmers.

The purpose of this project was to design and install a computer controlled driving system for operation of an all-terrain utility vehicle by a C-5/6 quadriplegic which is a high level of disability. The driving system included all functions that were needed to completely control the CCATV. The functions which were automated include the steering, throttle/brake, parking brake, gear shifting (reverse-neutral-drive), two-wheel drive/four-wheel drive (2WD/4WD) selection, high/low gear selection, differential lock/unlock selection, choke, ignition, starting and lights. Through actuators which are controlled by a computer, the operator is able to fully maneuver the CCATV with minimal amount of hand movements. This is required since the spinal cord is functionally severed between the fifth and the sixth vertebrae of the neck for a C-5/6 quadriplegic. This injury prevents all body movements below the chest area and limits arm movements. The operator moves his arms by using the shoulder muscle, biceps, and some of the forearm muscles. Chest muscles and triceps are not functional, and no finger movement is possible. However, the operator has good coordination and control without shaking or sudden jerks of hands. So the operator can operate joysticks and can apply some force accurately on a half-inch square button. Upper body safety straps on the CCATV are necessary for the operator to remain upright at all times. This paper describes details of the automation carried out on an all-terrain vehicle to be driven by such a disabled person.

Section 2 provides overview of the automated subsystems. Details about the 'brain of the CCATV', the coordination and control computer, are described in section 3. Discussion about the different sensors and the control panel is provided in sections 4 and 5, respectively. A detailed discussion of the automated subsystems is provided in section 6. Insight into some of the design criteria used is also provided. Section 7 briefly describes the design evaluation of the automated systems on the CCATV.

A unique feature of the automation design is the provision for manual mode operation. That is, each subsystem has been automated such that it does not interfere with the original vehicle subsystems. An auto/manual toggle switch is provided so that whenever the need arises, an able-bodied person can drive the CCATV in its originally designed, off-the-factory mode. This considerably complicated the overall design and made it challenging since reliability could not be compromised.

2. AUTOMATION OVERVIEW

Some of the factors considered in designing for automation include hand movements of the operator, amount of force the operator could apply to actuate a particular subsystem, ruggedness and reliability achievable from the automated system, and safety. Based on a careful analysis of these factors, the following factory built subsystems were considered for automation:

- Steering by servo control
- Throttle/brake by servo control
- Parking brake by on-off control
- Gear shift by multi-position control
- High/low speed by on-off control
- Differential lock/unlock by on-off control
- 2WD/4WD by on-off control
- Headlights by switch
- Choke by switch

Automation of these subsystems necessitates adding the following subsystems or components:

- Computer for coordination and control
- Sensors for feedback
- Operator Control panel with 15 push buttons, 3 LCDs, and 2 joysticks
- Power subsystem to power all the added components
- Auto/Manual switch
- Emergency switch

Figure 1 shows a partial overall schematic including several of the automated subsystems. The steering, throttle/brake, parking brake, and gear shift functions require independent actuation units. The speed of response required, actuation power levels, compactness, and reliability are some of the major factors considered. A detailed evaluation of both electric and hydraulic actuation designs was performed. Based on these considerations, the electric option with DC motor was selected for actuating the steering and the throttle/brake subsystems. This design requires a permanent magnet DC motor with a servo amplifier and position sensor; with additional linkages to transfer actuating signal from DC motor to the input to the steering or the throttle/brake subsystems. Two independent single-axis joysticks are used for operator input to the steering and throttle/brake subsystems. For parking brake automation, a commercially available unit has been modified to suit the requirements of the CCATV. For the gear shifting mechanism, a standard commercially available unit with gear shift/park functions is used. This is a compact and reliable design suitable for rugged environments.

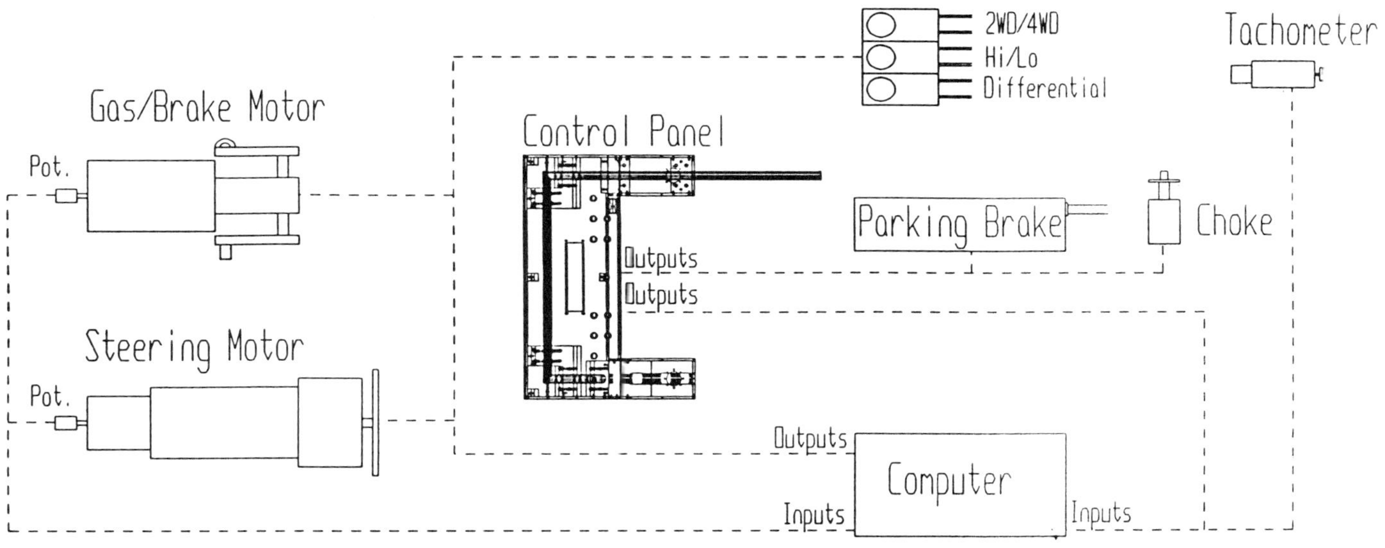

Figure 1: Partial overall schematic of the automated subsystems

The high/low speed, differential lock/unlock, 2WD/4WD subsystems are actuated using three independent pneumatic actuators which are driven using signals from the computer, through relays. For headlights, the existing switch and relay were used. Choke is activated using a separate solenoid driven by a relay which is activated in turn by a signal from the computer. It is designed to be actuated in manual mode also without involving the computer in the loop.

The power switch provides power to all subsystems on the CCATV which also includes the engine and the computer subsystem. The emergency switch can be actuated in case of malfunction of a subsystem or any of its components. This switch disables the engine from the computer subsystem.

3. COORDINATION AND CONTROL COMPUTER

Personal computers are not suited for the application because of its vulnerability against shock, vibration, and harsh environments. After evaluating various options including industrial grade programmable logic controllers (PLCs), micro-controllers and special purpose computers, a 286-CMOS CPU board with solid state disks such as Erasable Programmable Read Only Memory (EPROM) and other data acquisition cards was selected as the control computer for the CCATV. The model can withstand shock and vibration as well as extreme temperatures. The configuration of this computer is as follows:

- AT CMOS CPU Board
- 12 bit CMOS Analog I/O Board
- CMOS Multi-function Adapter Board
- 4 x 4 Keypad
- 4 x 40 LCD Display
- Power Module
- 4-slot Card Cage

The AT CMOS system uses small computer cards which are fully compatible with personal computers (PCs). Its small size, wide temperature range (-40 to + 85 degree C) and rugged CMOS construction, made it ideal for embedded control applications as well as for on-board vehicle mounting. It has COM1 and COM2 serial ports, LPT1 parallel port, a keyboard and speaker port, calendar, clock, watchdog timer, up to 4 MB Dynamic Random Access Memory (DRAM), DOS in Read Only Memory (ROM), coprocessor socket, and three solid state disks. The user has the option to program the card either through the serial port from a PC, or by adding a keyboard monitor and a floppy drive. An external 3.6 V AT clock battery plugs into the card, which powers the clock during power-down, thus providing battery-backup and reliability over long periods of time. The computer stores its setup information in a nonvolatile Electrically Erasable Programmable Read Only Memory (EEPROM), and so is immune to battery or power failure.

The analog I/O expansion card has 16 single-ended analog input channels, two 12 bit analog output channels, eight general purpose digital I/O lines, and interface for a standard module rack. This card converts analog voltages to digital values (A/D) and digital values to analog voltages (D/A or DAC). Analog input voltages are converted with 12 bit plus sign resolution [from -FFFH (-4095 decimal) to +FFFH (+4095 decimal)] with a conversion time of 15 μs. The 16 single-ended analog input lines are multiplexed under software control. It has auto-calibrate, auto-zero, and selectable amplifier gain features which are also under software control. Auto-calibration adjusts positive linearity and full scale errors. Auto-zero corrects the zero error offset during every conversion. The programmable gain amplifier allows reading from low output devices such as thermocouples and transducers and other low voltage sources with a gain of 10 or 100. These analog inputs are protected from overload up to 12 V even when the power is turned off. The A/D converter can convert over 50,000 samples per second. The eight digital I/O lines could be used for reading switches, connecting a keypad, and other general purpose control functions. These lines can be configured as inputs or outputs.

The multi-function I/O card is designed to be used with any PC control card. This provides a modular system which is easy to setup, modify and use. This card integrates two serial ports, one parallel port and 24 digital I/O lines. It is compatible with all Micro PC control cards and micro controllers. It is also compatible with standard size PC CPU cards. Typically it requires 170 mA @ 5 V. The two serial ports may be configured as COM2 and COM3, or COM3 and COM4. COM3 is switchable as either RS-232 or RS-485. The 24 line digital I/O port can be used to interface logic devices, switch inputs, and LEDs. The input and output levels are 0-5 V logic compatible. Keypad and display ports are also included on the card.

The 4 x 40 LCD display consumes very little power (less than 15 mA @ 5 V) and offers high contrast visual feedback. The display has switchable LED back-lighting. When on, this back-lighting allows a user to see display contents during low light conditions. This display is used to provide diagnostic and vehicle status related information continuously to the operator.

The power module provides isolated power to the system for a variety of applications. This is a high efficiency switcher, with low noise output and cool operation. This model takes 24 V DC power as input and provides 5 V DC output to the computer system. The input unit for the module has automatic polarity switching for easy connections. It also has input pulse noise filters.

The 4-slot card cage was designed to provide rugged housing for cards; with table mounting configuration and rubber feet. Mounting holes are also located on the right hand side of the card cage for mounting a power module. Additional holes are also located on the left side and the top to mount other accessories. Figure 2 shows the computer box housing the CMOS computer with data acquisition cards, power module, interface boards and two pulse width modulated (PWM) servo amplifiers.

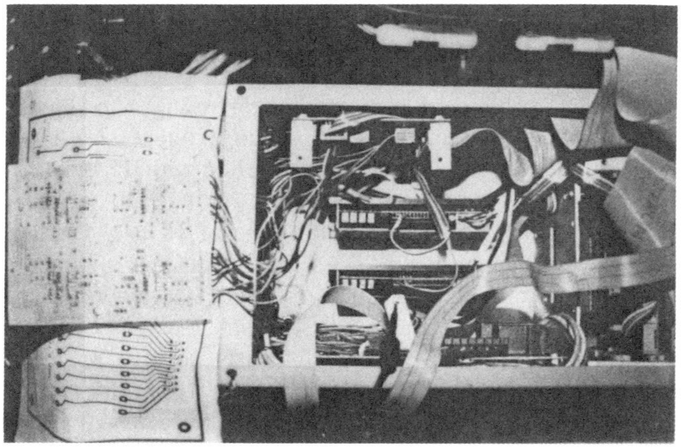

Figure 2: The computer box

4. SENSORS

The automation was carefully performed relying on rugged mechanical designs with minimal sensing to enhance reliability. The sensors used in the automation process include two potentiometers for the steering and throttle/brake subsystems, two joysticks for command inputs to these two subsystems, and a tachometer for speed measurement.

Several types of position sensors such as mechanical helical springs, electrical potentiometers, and optical encoders were considered for this applications. Based on considerations such as ruggedness, reliability, ease of maintenance, mean life, ability to withstand shock and vibration, and compatibility with computer readable output signal, multi-turn potentiometers with reducers are selected to measure the position of the steering wheel and the throttle/brake DC motors. These angular measurements, related to the front wheel position and the throttle/brake pedal position, are used for closed-loop position control of the steering and throttle/brake subsystems.

Two independent single-axis joysticks are used for operator input to the steering and throttle/brake subsystems. Induction type joysticks were selected for the application. These have fewer number of moving parts as compared to the others and made the subsystem more reliable. The joysticks are powered by an input voltage which comes from a regulated 12 V source which is turned on when the CCATV power is turned on. The regulator supply is used to ensure constant voltage supply to joysticks so that the output voltage from the joystick is without fluctuations and noise in spite of varying load conditions on the power system. The joysticks can be operated safely from -15 to +180 °F, thus providing consistent output even under extreme weather conditions.

The joystick has a range of +/- 27.5 degrees (3 to 9 V output) with infinite resolution. The A/D card is able to discern voltage changes of 2.44 mV which implies that the computer is able to discern joystick movements of up to 0.01342 degrees which is a resolution higher than what a hand is capable of.

Design constraints for selection of the speed sensors include:

- output compatibility with computer
- the speed resolution needed at speed as low as 1 mph
- the device must be rugged and dependable for operation in harsh environment
- the speed range for the CCATV is 0 to 22 mph, and the maximum front wheel propeller shaft speed is 8448 rpm

Based on these design constraints, the need for using an angular measurement type transducer is evident. Applicability of various transducers like mechanical including direct counters and centrifugal speed indicator; optical including stroboscope and photocells; electrical including tachometer and reluctance type proximity pickups, was investigated for the application. Considering the capabilities of these transducers and the design constraints, a DC tachometer was selected for sensing vehicle speed. This information is used in the closed-loop steering position control algorithm. For instance, the joystick gain is reduced at high speeds to improve safety during turns. The CCATV speed is also displayed on the control panel.

5. CONTROL PANEL AND ERGONOMICS

The control panel with its 15 buttons, 3 LCD displays, and 2 joysticks provides a complete interface between the CCATV operator and the computer subsystems. All operator inputs in the computer (auto) mode are through this panel. The 15 buttons correspond to the following functions: power, start, ignition, choke, drive, neutral, reverse, high/low speed, differential lock/unlock, 2/4 wheel drive, parking brake, emergency, headlights, diagnostics, and battery voltage. The 3 LCDs are for displaying the status of the CCATV modes (4 line display with 40 characters per line), the CCATV speed, and the 12/24 V battery voltage levels. An important part of the CCATV design considerations is making the vehicle as user friendly and easy to control as possible. Since all of the operating switches are mounted on the control panel located directly in front of the operator, the design of the control panel and switches must be made to accommodate the operator, i.e. ergonomics and easy accessibility had to be kept in mind. Some of the other design considerations used in the control panel design are rigidity, ability to withstand harsh environments such as direct sunlight and rain, vibration damping of the control panel when the CCATV is being operated, and accessibility for repair and maintenance.

The control panel was constructed from 1/4" thick gray PVC sheets. PVC was selected due to its ability to withstand harsh operating environment such as exposure to direct sunlight and rain. The control panel was assembled with 1/8" PVC dowel rods and glued together. However, it has four removable panels: two panels on which the joysticks are mounted, one panel on the bottom, and one under the right hand throttle/brake joystick. These panels allow access to the components mounted inside the control panel for service and maintenance. The control panel is mounted to the arm rest, and since the arm rest is free to rotate, it allows the control panel to swing out of the way for the operator to enter or exit the CCATV. A moveable sleeve with latch type fastener was used to secure the control panel to the arm rest. The control panel was connected to the arm rest by a 7/8" OD, 1/16" thick hollow aluminum tube. This tube encircles the perimeter of the control panel (giving it rigidity) and exits under the right hand throttle/brake joystick to mount to the seat. The aluminum tube is held in place by PVC clamping blocks which are glued to the control panel. These clamps can be disassembled and the tube reinforcement can be removed.

When changing from the auto mode to the manual mode of operation of the CCATV, the control panel can be easily removed from the sleeves under the arm rest and placed in the cargo bed. The steering wheel can be attached to the steering column and the CCATV can be operated in its originally designed manual mode. Figure 3 shows the control panel without the gear shifter and Figure 4 shows the same control panel mounted on the arm rest. The latch type fastener is also visible below the left hand joystick.

6 AUTOMATION DETAILS

This section provides details about the various automated subsystems. These subsystems have been designed to ensure minimal control complexity, i.e. to let the design itself do the work.

6.1 POWER SUBSYSTEM - Although the all-terrain vehicle came equipped with a 12 V electrical system, the electrical requirements of the subsystems and components added exceeded its capabilities. The existing alternator was essentially built around the crank shaft and so increasing its

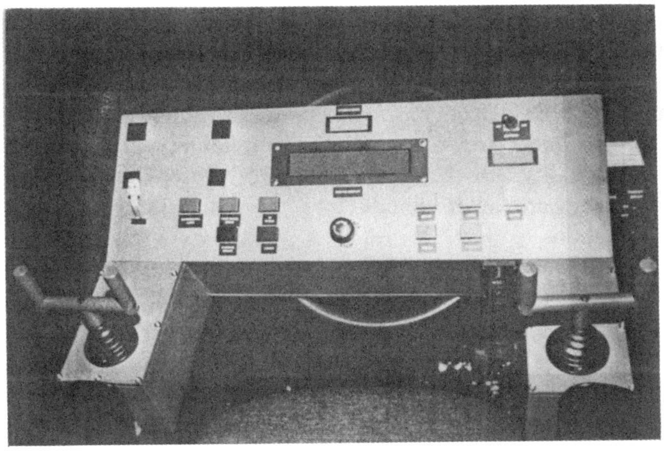

Figure 3: The control panel

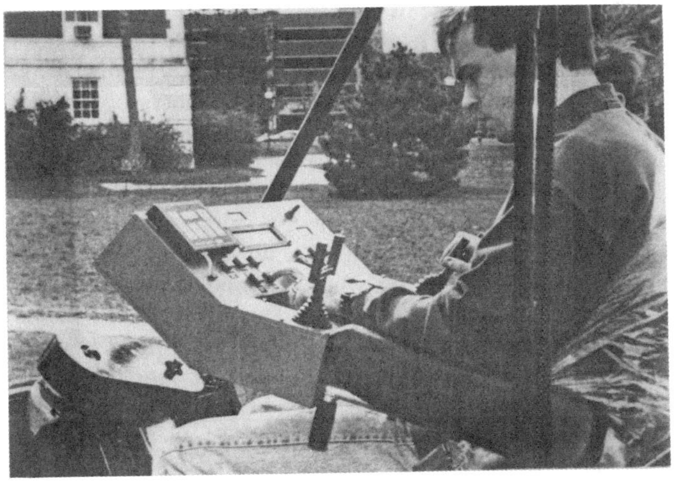

Figure 4: The control panel mounted on the arm rest

output or replacing it was not an option. For this reason, an additional alternator and additional batteries had to be added. The additional electrical system is a 12 and 24 V system. The steering and throttle/brake motors are powered by 24 V while the other automated systems are powered by a 12 V supply. Thus the power subsystem for the CCATV consists of two 12 V batteries with a separate alternator for charging. Connections for directly charging this additional battery-pack using the 110 V AC mains have also been provided on the CCATV.

6.2 STEERING SUBSYSTEM - The steering subsystem is of major importance to the operation of the CCATV. In keeping with the general concepts for vehicle modifications, it is desirable to keep as much of the factory installed steering assembly intact as possible. To minimize modifications to the existing assembly, the computer controlled actuation should be rotational and should simply replace the function of the steering wheel. The computer controlled actuator should be coupled to the steering shaft. The characteristics of the steering assembly are very important when considering the type of actuation to use. Some of the important steering characteristics are peak torque, continuous torque, rotational acceleration, rotational velocity, and the range of rotation. Of these parameters, peak torque and rotational velocity are the most important because they define how much power is required for the actuator. Characteristics of the actuator itself must also be considered. Some of the important actuator characteristics are response time, reliability, ruggedness, controllability, cost, and safety. The following actuation options were considered for the steering subsystem automation:

- Electrical
- Hydraulic
- Electro-Hydraulic (2 designs)

The electrical steering actuation design offers the advantage of good velocity and acceleration control and fast response. It is also relatively easy to control. The fact that no energy is expended in the system until a steering command is given means that the system is inherently stable. The disadvantage of electrical actuation is its high cost and difficulty in finding a suitable high power, low voltage motor.

The main advantage of the hydraulic option is the inherent characteristic of high torque, low speeds and fast response. Although this option seemed well suited for actuation of the steering subsystem, some of the disadvantages included expensive components, being prone to leakage, and additional space required for mounting pump and power pack unit. The electric-hydraulic hybrid designs required modifications to the original steering assembly and so were eliminated.

The final design consisted of a wheel chair electric DC motor as the actuator. This wheelchair motor operates on 24 V and has 20:1 gear reducer molded into the motor housing, thus it is an ideal choice for the application. Typically this motor is directly mounted to the wheel of the wheelchair, which means that it is designed for high torque, high shock loads and varying load conditions, and to be operated in harsh environment. The motor was mounted to the frame using brackets and sliding bars for height adjustments. The motor was coupled to the steering shaft using a roller chain. A position transducer (potentiometer) was attached to the end of the motor shaft through a speed reducer to provide position feedback used to control wheel position. Figure 5 shows the fixture designed for holding the steering motor. The roller chain drive used to transmit rotational motion to the steering shaft and the position sensing potentiometer mounted on the motor shaft are also visible in the figure. The steering command is provided using the left joystick on the control panel. This command is transmitted to the computer and through software control, it is transmitted to the motor via a PWM servo amplifier. The dynamic relationship between applied voltage and the motor position can be mathematically expressed as follows:

$$L_m \frac{di_m(t)}{dt} + R_m i_m(t) + K_b \dot{\theta}_m(t) = E_a(t) \qquad (1)$$

$$(J_m + J_L)\frac{d^2\theta_m(t)}{dt^2} + B_m \dot{\theta}_m(t)$$
$$T_{cm} \operatorname{sign}\left(\dot{\theta}_m(t)\right) = K_T i_m(t) \qquad (2)$$

where L_m - motor inductance; i_m - current through motor; R_m - motor resistance; K_b - back emf constant; θ_m - angular position of motor; E_a - applied voltage; J_m+J_L - motor and load moment of inertia, respectively; B_m - viscous friction; T_{cm} - coulomb friction torque on motor; and K_T - torque constant of motor. The controlled variable for each motor is $i_m(t)$ which is given by the relation

$$i_m(t) = K_p * e = K_p * (\theta_d - \theta) \qquad (3)$$

where K_P = position gain; e = position error; θ_d = desired position; θ = actual position. Figure 6 shows the block diagram of proportional control scheme implemented for closed-loop position control of the steering subsystem.

Figure 5: The steering motor holding fixture

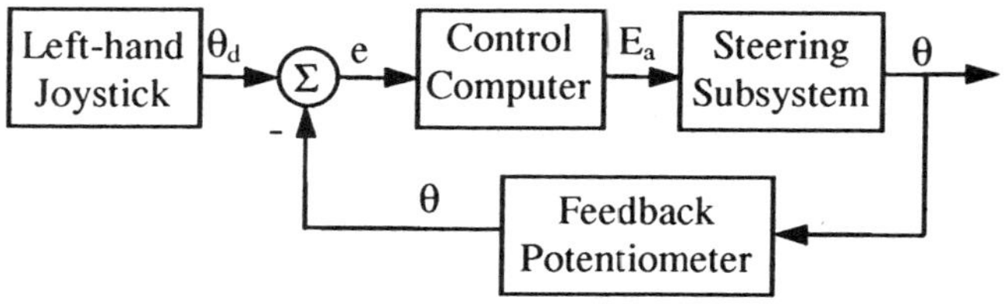

Figure 6: Closed-loop control of the steering subsystem

6.3 THROTTLE/BRAKE SUBSYSTEM - As with the steering system, the throttle/brake system automation had to be carried out in such a way as to minimize modifications to the factory design. From the inception of the throttle and brake subsystem automation design, the need for coordinating the two was apparent. The nature of both subsystems are similar. The throttle provides an acceleration that is positive, while the brake provides negative acceleration. Since the throttle and brake do not need to be used at the same time, only one actuator is necessary to control both functions. As both operations are to be controlled with one joystick, it also makes perfect sense to use one actuator for both purposes. Using a single actuator allows for the same feedback device, a high precision ten-turn potentiometer, to be used for both actuation tasks. Two alternate designs were evaluated for this subsystem automation. One with a linear actuator and the other with a combination of rotary actuator and cam mechanism. Some of the criteria used for evaluation of this design include the amount of force needed with required displacement of the throttle or brake pedal, the speed of actuation, the size of the actuator, and above all, the requirement of fail-safe operation and safety. The linear actuation system was found to be smaller in size and rugged, but its speed of response was not sufficient to provide the required full travel of the pedal quickly.

The alternative design involved a rotary actuator with a cam and cable mechanism. The rotary actuation system was selected over the linear actuation system because of its advantages like faster speed of response by meeting the criteria of full travel within 2 sec, and safety. Also, one of the main advantages of this option was the ability to use manual throttle and brake subsystems without any major modifications to the factory design. The rotary actuated throttle/brake automation involves a similar type of DC motor, as used in the steering system, with a carefully designed cam and cable mechanism. This was one of the nice designs demonstrating the rugged mechanical systems doing the work with minimal reliance on fancy control circuitry. An inductive joystick provides the input and a position sensor provides the feedback signal. In this case the right hand joystick on the control panel provides the command input with the actual position sensed in a manner similar to the steering subsystem. When the motor is turned one way, the throttle system is actuated and when the motor is turned the other way, the brake system is actuated. When the joystick is in the centered position, neither the throttle nor the brake is activated. Moving the joystick in the forward direction actuates the throttle system and in the reverse direction it activates the brake system. Similar to the steering system, the throttle/brake system is also a closed-loop position control system. In order to achieve proper timing on the throttle and brake functions, a special cam and follower system was designed, a cited. Figure 7 shows the throttle/brake motor with cam and cable mechanism. The design, as mentioned, allows for easy operation in the manual mode and does not impede leg movements in any way.

6.4 PARKING BRAKE AND GEAR SHIFTING SUBSYSTEM - The parking brake is a separate unit from the entire vehicle. It is used as a last resort to stop the CCATV in emergency situations. The parking brake function was automated by modifying a commercially available unit and it attaches to the existing brake cable by three clamps. It is activated by a momentary switch mounted on the control panel, as well as by one on the dash board. Thus it can be activated in the computer controlled (auto) mode as well as in the manual mode.

The gear shifting subsystem is required on the CCATV to permit the operator to easily shift from neutral to reverse or to drive gears. An off-the-shelf system designed to be used with automatic transmissions was modified for automating this function. Although the all-terrain vehicle has a torque

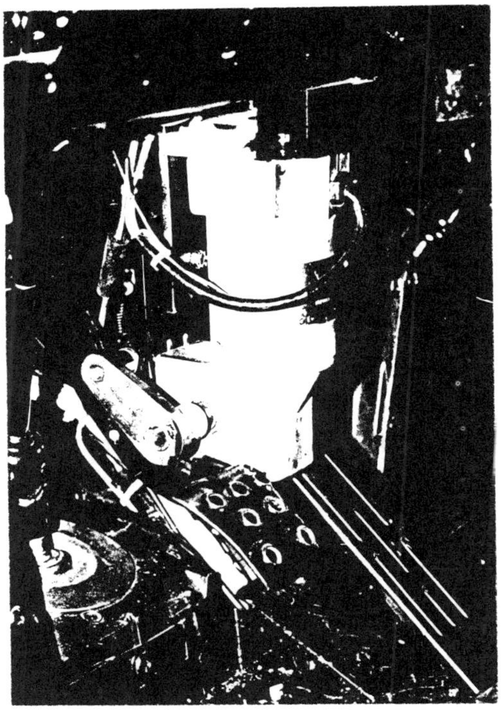

Figure 7: The throttle/brake motor

converter and not an automatic transmission, the operation of the shifter was suitable for use on this vehicle. This gear shifter consists of three components: a digital touchpad to select the gear to shift into, a small electric linear actuator which pushes or pulls the shifting cable, and a microprocessor which interprets the signal from the touchpad and sends the signal to the actuator. The unit is simple and comes in a compact package. The parking brake and gear shifting subsystems are the only ones designed by modifying off-the-shelf units; the other automation subsystems were designed in-house entirely.

6.5 VEHICLE SETTINGS - Initially a high powered single acting pull-in type DC solenoid system was used for automating the following settings: high/low speed selection, differential lock/unlock selection, and 2/4 wheel drive. These solenoids are typically used in construction machinery to set throttle position to a higher idle speed when hydraulic power is needed. They operate on 12 V supply and require 80 A for the 0.5 sec 'pull-in' time. The 'pull-in' exerted a 20 lb. force. After 0.5 sec, the solenoid was automatically switched to a holding coil which drew 1 A current and exerted only a 10 lb. force. Since these solenoids operate in only one direction, a spring had to be added to operate in the opposite direction to disengage the system. The force generated by solenoids was high enough when the gears to be shifted were aligned perfectly, but in many cases the shift required higher forces. In such cases, the solenoid could not achieve the desired position within 0.5 sec and the shift would not be completed. This necessitates investigation of alternative designs. It should be noted that under normal circumstances, an operator compensates for this problem by exerting additional force or moving the shifter until the gear locks in.

Based on the experience with the solenoid system, it was realized that the two position shifting was best suited to a pneumatic double-acting cylinder. Pneumatic cylinders can apply a constant force, independent of the position of the piston. The design, however, requires a compressed air source. Pneumatic cylinders are widely used in industry where large compressors provide compressed air supply to an entire plant, but they are used much less in mobile applications. An adequate compressed air supply requires a compressor which delivers small amounts of air at a large pressure, and an accumulator or tank which stores the compressed air such that larger quantities can be used instantaneously. The accumulator can be sized such that the compressor does not have to turn on every time air is used from the accumulator. Such a pneumatic system design was adopted which consisted of a pneumatic compressor, a pneumatic accumulator, regulator, filter and three pneumatic solenoid valves and three pneumatic cylinders for automating the three CCATV settings cited. Figure 8 shows the solenoid valve bank used for actuating the settings. The compressor is powered by the 24 V power subsystem while the pneumatic cylinders require only 12 V. Figure 9 shows high/low speed selection pneumatic cylinder used for system automation.

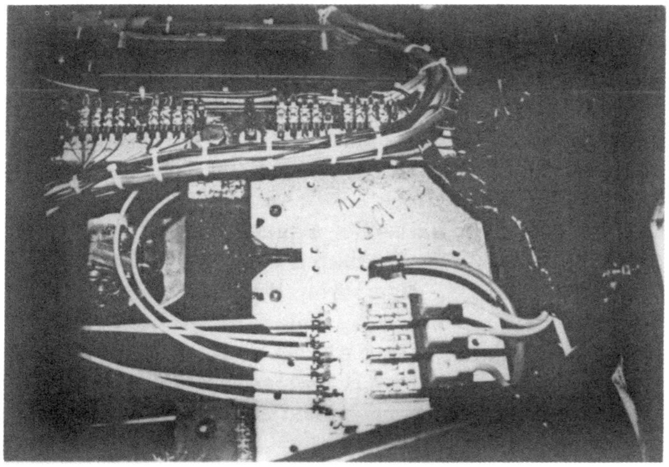

Figure 8: The solenoid valve bank

6.6 INTERFACE SUBSYSTEM - Two DC motor PWM servo amplifiers and a variety of signal conditioning and routing boards were all housed inside the computer box which was attached to the CCATV using three vibration isolation mounts. The interface design outside the computer box consists of regulated 12 and 24 V supplies and a variety of relays, military specification amphenol connectors, and carefully routed bundles of wires. Twelve amphenol connectors were used for this purpose. Three interface boards were designed to provide signal conditioning and wiring connections between the twelve connectors and the components inside the computer box. The control panel wiring and other circuitry was also connected to the computer box using military specification amphenol connectors. Two more amphenol connectors were also mounted on the control panel for wiring connections. Figure 10 shows the computer box, fabricated from aluminum, with amphenol connectors mounted on it.

An additional signal conditioning board with adjustable gain was designed to condition analog signals coming from

Figure 9: The high/low speed selection cylinder

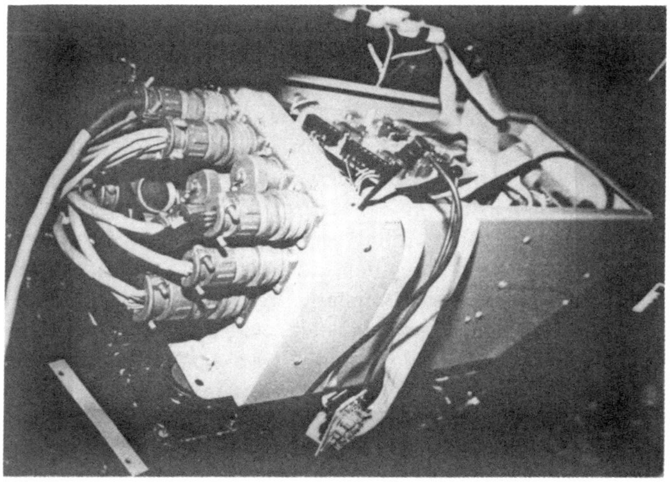

Figure 10: The computer box with amphenol connectors

the steering and throttle/brake subsystem joysticks and potentiometers. Using this board the steering and throttle/brake joystick and potentiometer input signals (which are in the range of 3 to 9 V) are converted to -5 to +5 V range for interface to the control computer. Figure 11 shows this signal conditioning board. A similar conversion is performed for the tachometer input signal which is in the 0 to 15 V range.

The control computer can be interfaced with a desktop PC using a null modem cable/adapter provided in the computer box and standard communication software. Once the communication is established between a desktop PC and the on-board control computer, a user can download and upload data. This communication link is also used to change and upgrade the control software. The virtual drive provides temporary storage for the transferred files until saving them to a battery-backed static RAM or EPROM. The virtual drive also allows a user to test and debug control software files before permanently saving them. Once the control software has been debugged, it can be autoexecuted by modifying the setup of the control computer.

The control software has been developed in-house using C language. Based on the command input and feedback from

Figure 11: The signal conditioning board

sensors, the control software provides actuating signal to actuators and switches to perform a required task. At start up, it goes through a brief diagnostic routine to check functionality of parking brake, steering and throttle/brake system. During this step, it also initializes the data acquisition system and checks the status of various switches provided on the control panel. The control software has capability for executing full diagnostics which can be actuated by a switch provided on the control panel. The computer-software design permits the operator to view and modify the control logic which is useful, if carefully performed, to 'customize' the interface.

7 DESIGN EVALUATION AND SUMMARY

The paper reports the novel mechatronic automation designs developed for an all-terrain vehicle. The designs took approximately two years to complete, and posed several challenges. During the design and implementation stages, all the automated subsystems were extensively tested for proper and reliable operation. The control software was tested and evaluated using a desktop PC and in-house developed simulator, before implementing it on the on-board CMOS control computer. The entire system was tested on the all-terrain vehicle in different types of road and weather conditions. The CCATV was tested for approximately 100 hours before it was handed over to the sponsor. All safety aspects, both hardware and software, were thoroughly checked out. Minor problems such as insufficient current rating due to long cable lengths, and tuning of the CCATV idle speed to solve a shifting problem (inherent to the model) were corrected in the process. The overall mechatronic design was relatively easy for an operator to learn. It took only about 15 minutes for the quadriplegic operator to learn and drive the CCATV. The CCATV has been in field operation for more than two years and has, to date, performed in a consistent and reliable manner.

The CCATV represents a very interesting 'design for control' project wherein careful mechanical design, with minimal control complexity and careful system integration resulted in high reliability and ruggedness. The flexibility of the design permits for its adaptation to operator specific requirements. Although only a specific application related to the disabled has been reported here, the complex mechatronics design, with its mechanical, electrical and computer aspects, has considerable potential for applications to several other areas also.

8 BIBLIOGRAPHY

1 Mistry, S. I., "Automation, Computer Interfacing and Control Software Development for an All-Terrain Vehicle", Technical Report No. 94111, Computer Controlled Systems Laboratory, Department of Mechanical and Aerospace Engineering, University of Missouri - Columbia, Columbia, MO 65211, 1994.

2 Buster, D. L., E. Bramantya, K. Otipoby, J. Thate, and M. Tzang, "Computer Controlled All-Terrain Vehicle", Honors Theses on different aspects of the designs, Computer Controlled Systems Laboratory, Department of Mechanical and Aerospace Engineering, University of Missouri - Columbia, Columbia, MO 65211, 1994.

3 Nair, S. S., "Computer Interfacing and Control", unpublished course notes, Computer Controlled Systems Laboratory, Department of Mechanical and Aerospace Engineering, University of Missouri - Columbia, Columbia, MO 65211.

4 Shigley, J. E. and C. R. Mischke, Mechanical Engineering Design, McGraw Hill Inc., New York, NY, 1989.

5 Franklin, G. F., D. J. Powell and A. Emami-Naeini, Feedback Control of Dynamic Systems, Addison-Wesley Publishing Company, Inc., Reading, MA, 1994.

6 DeSilva, C. W., Control Sensors and Actuators, Prentice Hall, Englewood Cliffs, NJ, 1989.

ACKNOWLEDGEMENT

The authors would like to thank their sponsor Ronald Joe Buster for his enthusiasm and support. This project was performed under grants C-5-34108 and C-5-34189 to the University of Missouri-Columbia. The authors would also like to thank the Mechanical Engineering Workshop and the Electronics Workshop personnel at the University of Missouri-Columbia for their guidance and support through out the duration of this project.

BIOGRAPHY

Sanjay I. Mistry received the B. S. degree in Mechanical Engineering from the University of Bombay, India, in 1987, the M. S. degree in Mechanical Engineering from the Wayne State University, Detroit, MI, in 1989, and the Ph. D. degree in Mechanical and Aerospace Engineering from the University of Missouri-Columbia in 1994. He received the Superior Graduate Achievement Award for his doctoral work from the Department of Mechanical and Aerospace Engineering.

During 1994-1995 he was a Research Associate with the Computer Controlled Systems Laboratory and NSF-Capsule Pipeline Research Center (CPRC) at the University of Missouri-Columbia. He is currently a Senior Engineer in the Electro-Hydraulic Control Systems at the John Deere Product Engineering Center, Waterloo, IA. His research interests are in the areas of modeling and control of dynamic systems, adaptive control, neural networks and fuzzy logic, computer interfacing and hardware implementation, and control software development. He has authored more than 15 journal and conference publications and is currently co-authoring a book on intelligent control. He has co-authored a chapter on "Automation and Control of Coal Log Pipeline Systems" for the "Pipeline System - Manual of Practice."

Please send all correspondence to Satish S. Nair, Associate Professor, Computer Controlled Systems Laboratory, Mechanical and Aerospace Engineering, University of Missouri-Columbia, Columbia, MO 65211.

980201

NAGREMA – A simulation environment for accessory drive optimization

M. Schmidt, R. Isermann
Darmstadt University of Technology

ABSTRACT

Automotive engineering requires dynamic system simulation software. To meet future legislated emission and consumption standards, a vehicle has to be considered as a group of interacting systems. NAGREMA is an automotive simulation software with the focal point on the accessory drive. Variations of the standard V-belt configuration can be compared with decentralized approaches using electric or electro-hydraulic drives. NAGREMA is implemented using MATLAB/Simulink, provides a graphical user interface and a set of vehicle, engine and accessory templates. It reduces model complexity by dividing the vehicle, the engine, the accessory drive and the control system into hierarchically organized subsystems.

Introduction

In automotive engineering, a trend towards electronics and mechatronics becomes obvious [1, 2, 3]. This leads to an increasing electric demand [4, 5]. All electric and electronic devices have to be supplied by the generator, whose efficiency is usually only sufficient at low speed. In combination with increasing customer demand concerning comfort package and safety, this leads to increasing consumption caused by the accessory drive. Consumption and exhaust-emission standards require to fulfill the increasing customer demands with a decreasing consumption. Therefore, the accessory drive is an actual focal point in automotive engineering.

Optimization can only be achieved by regarding the vehicle as a group of interacting systems. Car manufacturers and suppliers have to cooperate very closely. To strengthen this cooperation and to develop the required tools, the "Forschungsvereinigung Verbrennungskraftmaschinen" (FVV) has introduced a council and initialized a research project. At Darmstadt University of Technology, the Institute of Automatic Control and the Laboratory for Combustion Engines are investigating systematically the possibilities and the requirements for a management of the accessory drive.

Starting point was the actual configuration in a mid-size car. The influences on consumption and emissions of the individual accessory were investigated in consumption measuring runs. A simulation program called NAGREMA was developed which has the focal point on the accessory drive and its interactions with the engine, the vehicle body, the passenger compartment and the environment. It is used to investigate variations of the accessory drive configuration.

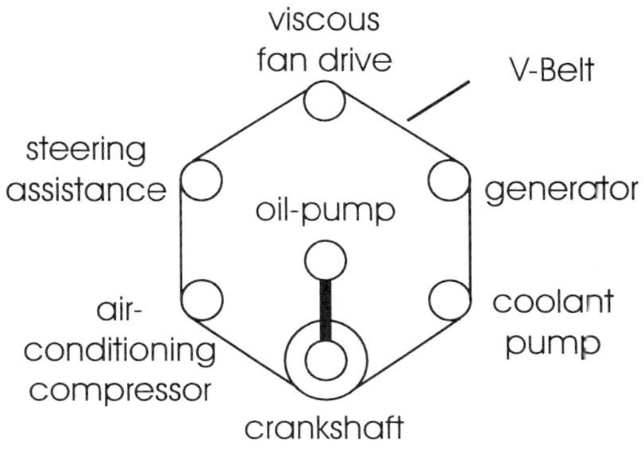

Figure 1: V-belt accessory drive

This paper deals with first results of this research project.

Accessory Drive – State of the Art and Future Variations

Most passenger cars (PC) and commercial vehicles (CV) use a V-belt for the accessory drive, fig. 1.

Table 1 gives a survey of possible accessory drives. A distinction between *centralized mechanical drives* from the crankshaft and *decentralized electro-mechanical drives* is useful. Using a centralized drive, the speed of the accessory depends on the engine speed. For a belt drive, the speed ratio is fixed. Transmission dependant accessory drives can be realized by using gears or viscous drives.

Electric motors enable decentralized drive configurations where the accessory speed can be controlled independently from the engine speed. A switch simply can turn the motor on or off whereas the use of power electronics allows a continuously variable speed.

Fig. 2 shows example for a partially central-

Table 1: Survey of accessory drives

accessory	accessory drive						
	centralized (V-belt)				decentralized (electric engine)		
	without gear		with gear		const. speed		var. speed
	ON	ON/OFF	stepped	vario	ON	ON/OFF	
oil-pump	X	—	—	—	X	—	X
coolant pump	X	X	X	X	X	X	X
cooling fan	X	X	X	X	X	X	X
steering assistance	X	X	X	X	X	X	X
generator	X	X	X	X	—	—	—
air-cond. compressor	X	X	X	X	X	X	X
air compressor (CV)	X	X	X	X	X	X	X

legend	X	possible alternative
	—	no alternative

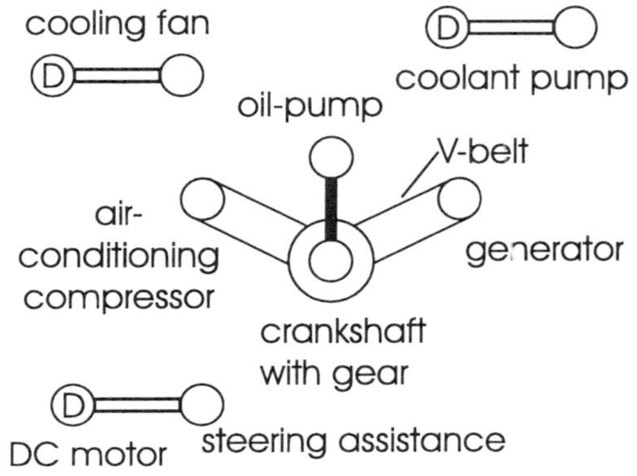

Figure 2: partially decentralized accessory drive

Figure 3: measurement of power demand at mounting position

ized and decentralized accessory drive configuration.

To minimize the accessory's load on the engine, the point of operation should depend on the actual demand. This means to control the speed of each accessory depending on its characteristic curve, driving conditions and environmental factors.

The large number of variations becomes obvious in table 1. The aims of the investigations are

- to figure out promising accessory drive configurations
- to specify the required components (future accessories and drives) and
- to propose a management of the complete accessory drive.

Modeling Engine, Vehicle, Accessories and their Interactions

These aims require simulations on the one hand to reduce development time and on the other hand to determine the potential of improved, not yet existing accessories and drives.

Starting point was the analysis of an state-of-the-art mid-size car. The power demand of each *individual accessory* was measured at its mounting position, fig. 3. The measurements, performed by the Laboratory for Combustion Engines, were divided into steady-state and dynamical experiments [6]. The *characteristic curves* obtained by a steady-state measurement describe the *efficiency* in various operational points and the *average power demand*. Dimensioning the accessory and comparing components of different suppliers might be based on the characteristic curves. Introducing a management of the accessory drive requires the knowledge of the *actual power demand*. Dynamic effects caused by the inertia superpose the steady state effects and increase the demand on accelerations.

$$P = P_S + P_J \tag{1}$$

where P is the actual power demand, P_S the steady-state and P_J the demand caused by the inertia.

As a simplification, first order dynamics is assumed for the input-end torque and power characteristics of the individual accessory.

$$P = \omega \cdot (T_S(\omega, \underline{x}) + \hat{J}\dot{\omega}) \qquad (2)$$

where ω is the rotational speed, $\dot{\omega}$ the rotational acceleration, $\underline{x}$ a vector of additional input values and $\hat{J}$ the overall inertia. $\hat{J}$ not only contains the inertia of the accessory itself but also dynamic effects on the output end. The coolant is an example for additional dynamic effects. During acceleration, the flow rate must increase to reach the new steady-state. This increases the actual torque demand of the coolant pump compared to the steady-state, fig. 4.

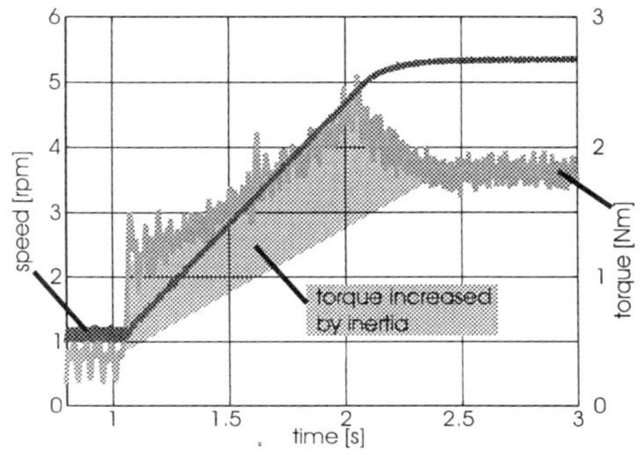

Figure 4: acceleration of the coolant pump

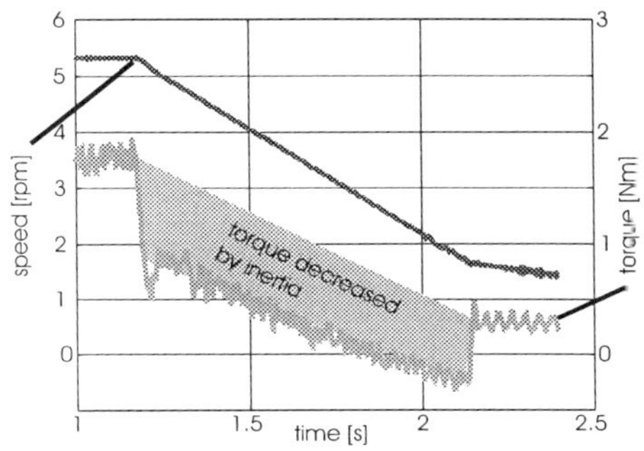

Figure 5: deceleration of the coolant pump

To model the asymmetric behavior on acceleration and deceleration, fig. 5, the total inertia $\hat{J}$ depends on the sign of the rotational acceleration:

$$\hat{J} = \begin{cases} \hat{J}_{acc} & \dot{\omega} \geq 0 \\ \hat{J}_{dec} & \dot{\omega} < 0 \end{cases} \qquad (3)$$

The vector $\underline{x}$ in equation 2 contains the additional input values for the individual accessory.

For the generator (fig. 6), $\underline{x}$ contains the generator current I_{gen} and the battery tension U_{b12}. $\underline{x}$ is calculated in an adapted model of the vehicle electric system including the state-of-charge and the charging characteristics of the battery as well as the actual electric demand of engine and vehicle. The vector $\underline{x}$, determined by the current and the battery model [7, 8] fig. 7, is the second

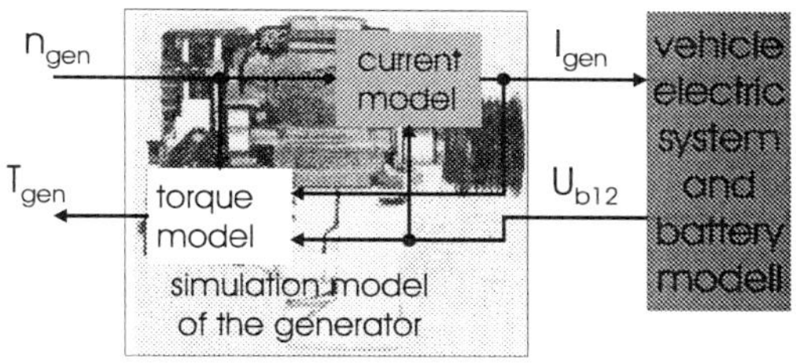

Figure 6: generator model

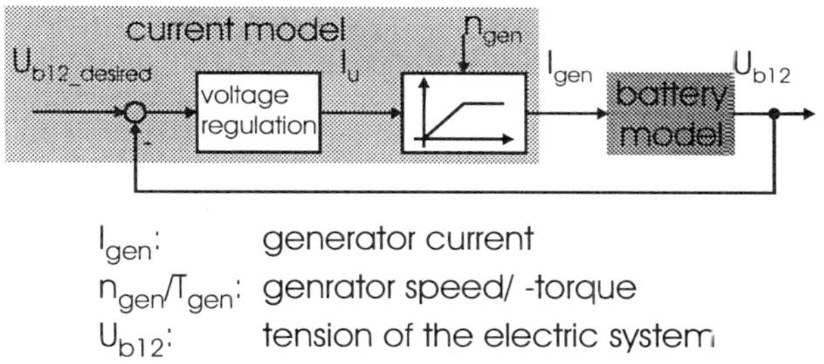

Figure 7: current model

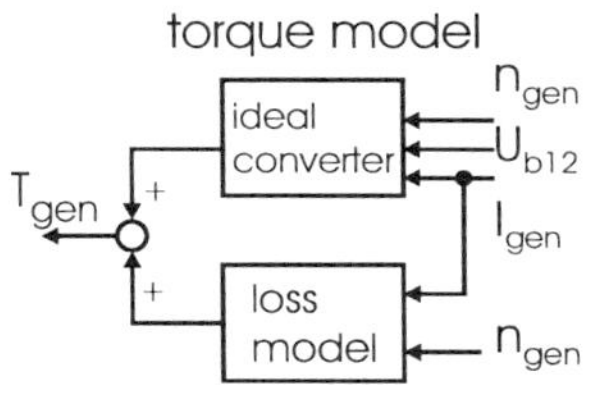

Figure 8: torque model

input of the torque model fig. 8. This model considers an ideal converter and the losses due to the efficiency of the generator fig. 9.

The power demand of the air-conditioning compressor strongly depends on the outside-air temperature, the solar radiation, the interior temperature and the desired interior temperature selected by the driver. Fig. 10 shows a thermal model for the passenger compartment used for the simulation of the interior temperature. Environmental factors are considered to be constant during one simulation run to decrease the number of free parameters.

For the engine model, characteristic curves are used. An overview over existing engine models for control applications is given in [9]. The block diagram of a complete vehicle model shows fig. 11. The drive chain model is derived from [10] whereas further information about the accessory models are contributed in [6]. The representation of characteristic curves is discussed in [11].

NAGREMA – a Tool for Car Manufacturer and Supplier

Based on the accessory models, an automotive simulation tool has been developed. From the beginning, NAGREMA

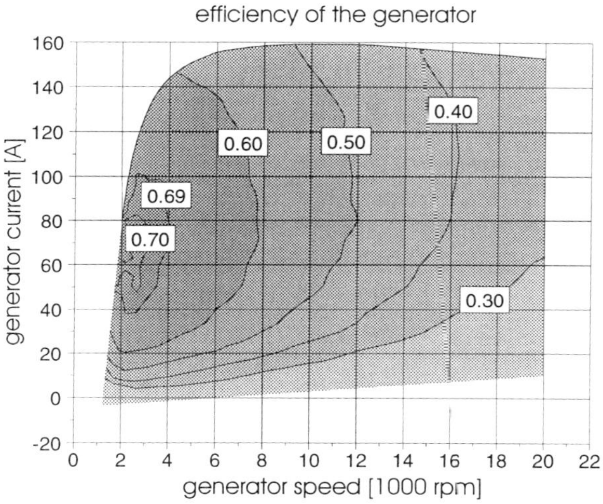

Figure 9: efficiency of the generator

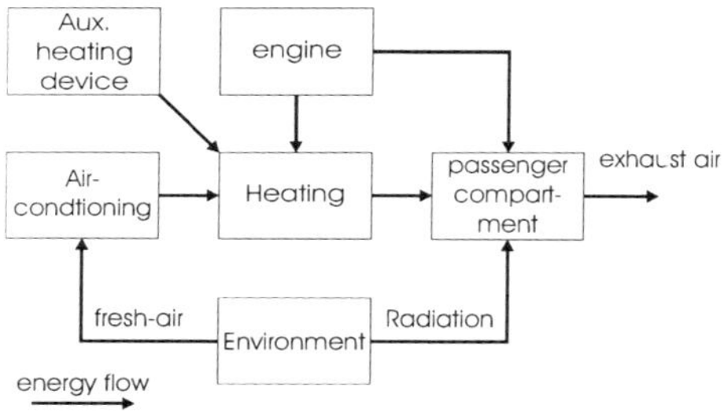

Figure 10: air-conditioning model: energy flows

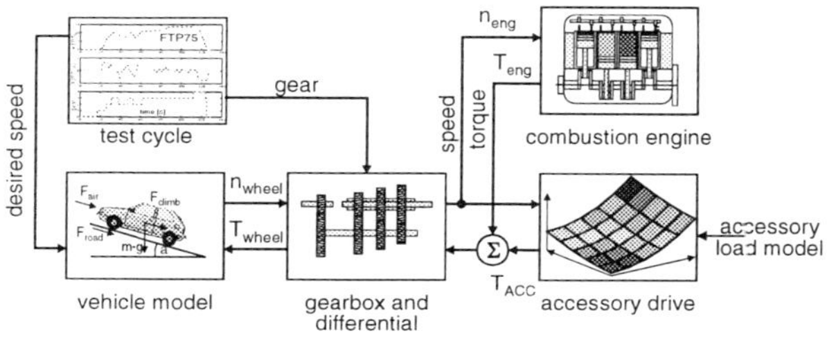

Figure 11: block diagram of a vehicle model

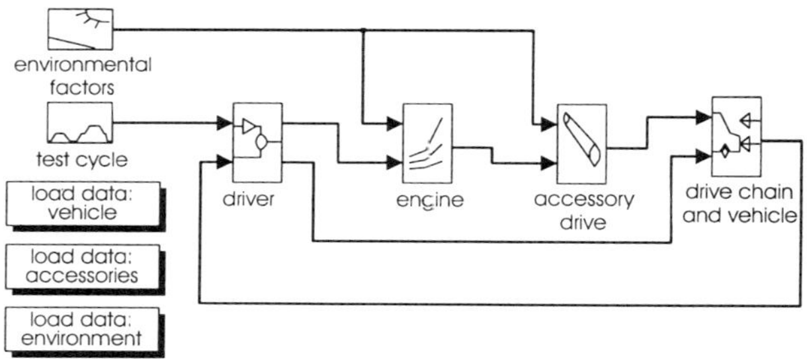

Figure 12: root level vehicle model

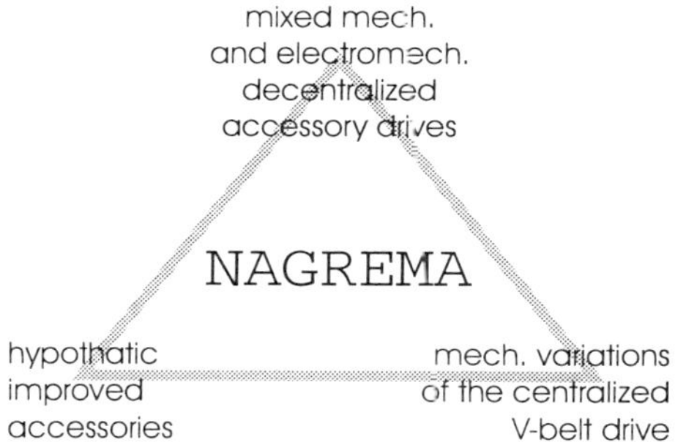

Figure 13: applications for NAGREMA

was intended to be a tool for the car manufacturer and the supplier. To reach this aim, an open software structure is required. MATLAB/Simulink was chosen as the basis for the simulation tool because

- it is platform independent (PC, workstation)
- it is widely spread [12]
- it offers a higher level description language for dynamical systems including a
- graphical user interface
- simulink has a block orientated structure and
- it supports the development of controllers and logic devices.

A block library containing the modules for a vehicle simulation is the starting point. Templates for standard engines, accessory drives, accessories and vehicle bodies are provided. As the required component model libraries are developed, new dynamical systems are easily simulated by reusing previously built subsystems. Furthermore, the integration of existing simulation software like the "Gesamtprozessanalyse" (GPA) [13], the overall vehicle model based on FVV research, was an important point of the software specifications.

From the model of a basic vehicle fig. 12, variations can be derived using simulink as the graphical user interface for model construction. Variations can touch the characteristic curves of the accessories as well as the structure of the model fig. 13. A variation is for example

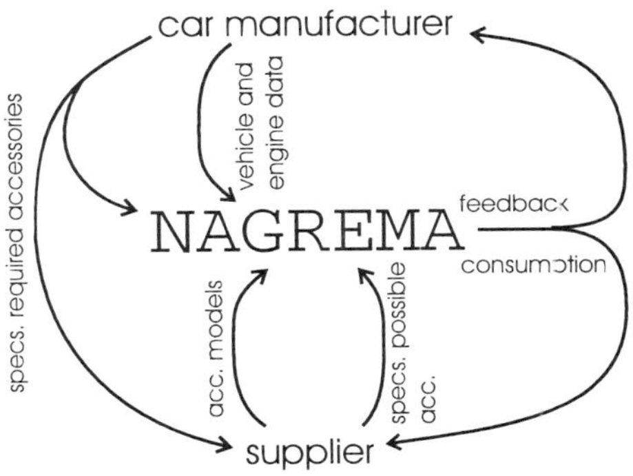

Figure 14: NAGREMA as link between car manufacturer and supplier

replacing the simple V-belt by a V-belt with viscous coupling for individual accessories. Future developments can be specified and their potentials for fuel savings are estimated. As a link between supplier and car manufacturer, NAGREMA can be adapted by the user to introduce existing know-how, measurement data and models fig. 14.

The open structure allows to refine the accuracy of individual models depending on the focal point of the investigations. As a simulation example, the potential of *improved generators* is presented. It is assumed, that the overall efficiency η_{std} can be improved by 5%, 10% or reach the ideal $\eta_{ideal} = 100\%$. 100% is the *ideal* generator, which points out the maximum fuel savings for a given electric demand and continuously working generator. The change of the consumption is given in x l/100km$\hat{=}235.21/x$ mile/gal in tab. 2 and fig. 15. Improving the generators overall efficiency can be obtained by improvements on the mechanical or the electrical side. NAGREMA shows the potential for this improvement and helps to decide if design modifications will pay.

Summary and Conclusions

A trend in automotive engineering towards electronics and mechatronics becomes obvious. This trend includes the accessory drive and the vehicle electric system. Their power demand can't be neglected and the overall efficiency has to be improved. Optimization of the accessory drive is a challenge for both, car manufacturer and supplier. It requires the view of a vehicle as a group of interacting systems which can be reached using dynamical simulation tools. NAGREMA is intended to develop such a management by using a simulation structure open for car manufacturer and supplier. It helps to understand the interactions, to specify required developments and to investigate alternative accessory drive configurations.

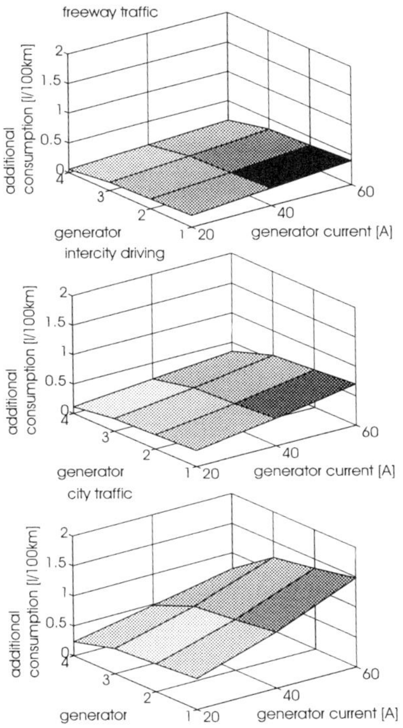

Figure 15: consumption caused by various generators for freeway traffic, intercity driving, city traffic

REFERENCES

[1] R. Isermann. Mechatronic systems - a challenge for control engineering -. In *American Control Conference*, Albuquerque, NM, USA, 1997. IEEE.

[2] R. Isermann. Mechatronic systems for machines and automobiles. In *IEEE/ASME International Conference on Advanced Intelligent Mechatronics*, Tokyo, Japan, 1997. IEEE.

[3] Deußen, N. and Müller, P. Engine temperature management (in german). In *17. internationales Wiener Motorensymposium*, number 267 in Reihe 12: Verkehrstechnik / Fahrzeugtechnik, pages 111–139. VDI, 1996.

[4] Kassakian, G., Wolf, H.-C., Miller, J. and Hurton, C. Automotive electrical systems circa 2005. *IEEE Spectrum*, pages 22–27, 8 1996.

[5] Pfalzgraf, B., Otto, E., Wirth, A., Küper, P.F., Held, W., and Donnerstag, A. *The system Development of Electrically Heated Catalyst (EHC) for the LEV and EU-II Legislation.* Number 951072 in SAE Technical Paper Series. 1995.

[6] Isermann, R., Hohenberg, G., Schmidt, M., and Lenzen, B. Accessory drive management for consumption and emission reduction (in german). In *6. Aachen Colloquium Automobile and Engine Technology*. VDI, 1997.

[7] Reasbeck, P. and Smith, J. G. *Batteries for Automotive Use.* Research Studies Press Ltd., Taunton, Somerset, England, 1997.

Table 2: Consumption [l/100km] caused by various generators

	improved generators $\longrightarrow$			
	Standard η_{std}	$\eta_{std} + 5\%$	$\eta_{std} + 10\%$	$\eta_{ideal} = 100\%$
generator no.	1	2	3	4
electric demand [A]	city traffic [l/100km]			
20	0.52	0.47	0.42	0.24
40	0.94	0.85	0.77	0.49
60	1.48	1.34	1.22	0.74
	intercity driving [l/100km]			
20	0.27	0.24	0.22	0.11
40	0.47	0.42	0.38	0.23
60	0.70	0.63	0.58	0.34
	freeway traffic [l/100km]			
20	0.19	0.16	0.14	0.06
40	0.29	0.26	0.23	0.11
60	0.40	0.36	0.32	0.17

[8] H.P. Schöner. Electrical behaviour of lead/acid batteries during charge, overcharge and open circuit. In *EVS88*, Toronto, Canada, 1988. Electric Vehicle Association of Canada.

[9] Hrovat, D. and Sun, J. Models and control methodologies for ic engine idle speed control design. *Control Engineering Practice*, 5(8):1093–1100, 1997.

[10] Leonhardt, S., Schmidt, Ch., Voigt, K. and Isermann, R. Real-time simulation of drive chaines for use in dynamical engine test stands. In *American Control Conference*, Chicago, USA, 1992. IEEE.

[11] Holzmann, H., Halfmann, C., and Isermann, R. Representation of 3-d mappings for automotive control applications using neural networks and fuzzy logic. In *6. Conference on Control Applications*, Hartford, Connecticut, USA, 1997.

[12] Weeks, R. W. and Moskwa, J. J. *Automotive Engine Modeling for Real-Time Control Using Matlab/Simulink*. Number 950417 in SAE Technical Paper Series. 1995.

[13] W. Schley. *Gesamtprozeßanalyse-Rahmenprogramm (in german)*. Number 434-1 in Forschungsberichte Hefte. Forschungsvereinigung Verbrennungskraftmaschinen, Frankfurt, 1989.

About the author

Martin Schmidt works as a research assistant at the Institute of Automatic Control of Darmstadt University of Technology. His field of research is the optimization of the accessory drive.

The authors are indebted to the Forschungsvereinigung Verbrennungskraftmaschinen (FVV), Frankfurt/Germany for the initialization and the financial support that made this research project possible. The investigations have been accompanied by a council under the chairmanship of Dr.-Ing. K. Kurr. Their support is also acknowledged with thanks.

A LOOK TO THE FUTURE

ACTUATING: THE FUTURE

Randy Frank
Transportation Systems Group
Motorola Semiconductor Products Sector

INTRODUCTION

The amount of computing power on top-of-the-line cars is frequently presented as a measure of the sophistication level of modern vehicles. For example, it is impressive to say that there may be as many as 50 to 80 microcontrollers (MCUs) on a modern vehicle. However, the number of MCUs on these vehicles is easily matched by the number of motors and solenoids that turn the decisions of the MCUs into actions. In the typical vehicle, actuators outnumber MCUs. Without these actuators, all of the MIPs and megabytes of the MCUs could do nothing to effect the supply of fuel from the injector drivers, the spark provided by the ignition coil, the activation of the air bag or antilock brakes, or any other control system on the vehicle. The actuators are required for the system to work—and for the system to perform work.

For every new system that is added to the vehicle, or every change that is made from mechanical to electronic control, new actuators are required. The power electronics and the semiconductors that control these actuators are improving, as are the methodologies used to design the electronic-to-actuator portion of the system. At the same time, the energy required to provide the actuation is increasing at such a rate that future vehicles will, at some point, require a higher-voltage system. The higher system voltage will allow systems and their associated actuators to continue to be added for increasing control, reliability, safety, comfort, and/or convenience. However, a higher system voltage will also dictate new design methodology for future vehicles. In the discussion that follows, these interrelated issues will be put in perspective.

ACTUATORS AND SYSTEMS

There are many types of actuators used in today's vehicles. These include brush motors, brushless dc motors, switched reluctance motors, stepper motors, solenoids, and relays. Displays—and even lamps—may be included here, too, as they play a role in the output side of the system ignition coils.

The type of actuator used in a particular application can vary depending on the system and the amount of work the actuator must perform. Table 1 shows a rather complete list of motors and the vehicle systems in which they are used. In this table, the third column (# Mtrs.) shows the minimum number of motors that may be used, and the fourth column (Upside) shows additional numbers that may be used. In all, 67 motors are listed here. There are, however, almost a dozen solenoids that may be added to this, used for applications such as fuel injectors, bringing the number of electromechanical actuators identified to about 80.

Actuators do the work to raise or lower windows; adjust the antenna, mirrors or seats; turn fans; pump oil, water or fuel; and open and close valves. Although less than 25 percent of these motors are controlled electronically today, that number will increase in future vehicles. The current and power requirements for six different levels of actuators are shown in Figure 1. Switched reluctance (SR) motors are receiving renewed interest as a result of modern digital control and power electronics. Traction control and electric power steering are key applications that may use SR motors.

Table 1 - Systems and motors in today's vehicles.

System	Application	# Mtrs	Upside
Body Electronics			
	Headlight retractors	2	
	Headlight wipers	2	
	Mirror motors	4	
	Washer pump	1	
	HVAC compressor/heatpump	1	
	Speedometer	1	
	Rear wiper motor	1	
	Rear window motor	1	
	Antenna	1	
	HVAC blower	1	
	Wiper motor	2	
	Odometer	1	
	Window lifts	2	2
	Sun roof motor	1	
	Power seat	8	
	Seat back supports	2	
	Power door locks	2	2
	Trunk closing actuator	1	
	Headrest motor	2	
Powertrain			
	Engine cooling fans	2	
	Shift-by-wire	1	
	Cruise control	1	
	Fuel pump	1	
	Idle/air control	1	
	Throttle control	1	
	Drive motor (EV or hybrid)	1	1
	Lubrication pump	1	
Safety and Chassis			
	Seat belt retractors	2	
	Electric power steering	1	
	Suspension motor	4	
	ABS motor and pump	8	
	4WD actuator	1	
ITS/Entertainment			
	Entertainment cooling fan	1	
TOTAL		62	5

Does not include motors in CD or cassette players, navigation, computer peripherals.

ELECTRONIC CONTROL AND POWER ELECTRONICS

With increased control capability and lower energy consumption, the use of power electronics for actuators is increasing. Power semiconductors, including smart power ICs, serve as the semiconductor interface between the actuator and the MCUs in the system. For lower-voltage (<150 V) control in today's vehicle, the power semiconductor of choice is the power MOSFET, a voltage-controlled, cellular structure that can achieve lower on-resistance by simply increasing the number of cells in the design.

Higher vehicle loads are being addressed by power modules that combine IC control with multiple discrete power die, using packaging to isolate the die and to improve thermal dissipation. Insulated gate bipolar transistors (IGBTs) are a design/process variation used for higher-voltage MOSFET-like cellular structure power devices to control higher vehicle loads. IGBTs have been used extensively in the 1990s to switch current to the ignition coil(s). More recent applications include motor control in electric vehicles.

An almost tenfold efficiency improvement has occurred since the first power MOSFETs were used on vehicles (early 1980s). Also, as shown in Figure 2, a 30+ percent reduction in on-resistance has occurred as recently as the 1993 to 1997 period, and further reductions will be possible with the next generation products (around 2000). At the same time, the cost of these power MOSFETs has decreased dramatically, allowing applications that have been possible for many years to break through cost barriers and finally become practical. The change from electromechanical control to solid state control is occurring as a result of the need for improved reliability, more precise control, and increased efficiency. An example of this change is the replacement of relays with smart power ICs or control ICs and power MOSFETs.

Low-voltage power MOSFETs are used in vehicles for switching solenoids in anti-lock brake systems or controlling pumps and motors. Frequently, several power devices are in the same module requiring either a large amount of heatsinking or highly efficient switches. However, the automotive power supply is not a simple 12-V system. A fault condition called "load dump" generates high-energy transients that can easily exceed 100 V when the alternator of the vehicle is disconnected from the battery in an unprotected system. To survive the load dump fault, power devices are rated at 50 to 60 V, and transients are limited by surge suppression devices such as avalanche rectifiers.

Other application objectives have been achieved through the continuous improvement of power MOSFET processing. For example, since the initial use of power MOSFETs on automobiles, the temperature rating has increased from a maximum of 150°C to 175°C operation. Manufacturers of MOSFETs for automobile applications may expect continued on-resistance reduction. However, even more significant is the on-resistance reduction that will occur if central transient suppression is used in vehicles, the supply voltage is limited to a maximum of 30 to 40 V, and MOSFETs are specified with a lower voltage.

Smart Power ICs

One of the roles of the powertrain control module is to control the amount of fuel that a vehicle consumes. Because several functions are included in the module, space is at a premium, and power dissipation is a challenge. Higher levels of integration in a power IC or smart power device have produced sophisticated products with several MOSFET outputs in one monolithic IC capable of driving all of the injectors for four-, six-, or eight-cylinder vehicles. At the same time, complex diagnostic and protection circuitry have been incorporated. Figure 3 shows the circuitry that can be integrated with multiple outputs in a SMARTMOS™ driver*. The four output devices on the right-hand side are isolated from one another and control circuitry on the left-hand side of the figure. The circuitry allows the independent thermal shutdown of each output device. The circuitry can also include serial interfaces for communication with the host microcontroller and more complex diagnostics that can determine open load and the movement of a solenoid as the result of the device being activated.

The more integrated power IC may be used for more applications when the current draw of injectors is reduced through high-impedance design. Furthermore, automakers are being led to the sophisticated diagnostic capability available from a power IC design as a result of legislated diagnostic requirements from the California Air Resources Board (CARB) or the Federal government.

IGBTs

The need for precise control of the spark timing, direct firing of multiple ignition coils, and efficient high-voltage switching has led to the use of IGBTs for this high-voltage application in vehicles. Automotive-specific IGBTs have been designed that incorporate precision output and protective gate clamps. These IGBTs are driven by a gate voltage of 5 V, allowing easy interface to the MCU. Figure 4 shows the circuitry that is integrated in an ignition IGBT.

The applications of IGBTs to controlling the ignition coil are expected to expand, especially as coil-on-plug approach to ignition is implemented. In addition, some improvements forecast for future systems include more protection circuitry for the individual IGBT and the packaging of multiple IGBTS in a single unit.

Electric vehicles (EVs) or hybrid electric vehicles (HEVs) that combine an internal combustion engine with an electric traction motor require higher-voltage power semiconductors. IGBTs have been chosen to drive the electric motors on these modern electric vehicles. The application for the traction motor IGBT is considerably different than that for the ignition IGBT. Efficient, high-frequency, and fault-tolerance switching for the traction motor IGBTs has involved tradeoffs in Vce(sat) and short-circuit withstand capability. Hybrid power module packaging has played a major role in achieving performance and meeting the long-term reliability requirements.

Drive-by-wire systems—including the electronic control of throttle, brakes, steering, and other vehicle functions formerly controlled mechanically or hydraulically—are ideal candidates for a design approach that has been dubbed "mechatronics."

Mechatronics

Mechatronics is defined in McGraw-Hill's *Electronics Dictionary* as "a method for designing subsystems of electromechanical products by the simultaneous simulation of electronic, mechanical, and hydraulic elements to ensure optimum performance. Computer-simulated components are modeled as graphical representations for the creation of schematics, which become simulator input. The schematics are used by the simulator to determine which models are needed and how they are to be interconnected."

In many cases, mechatronics combines the electronics that formerly were located remotely in a stand-alone electronic module, with the electromechanical actuator. This means that an electronically controlled motor may contain the electronics as an integral part, thereby reducing wiring and connections. As available space continues to decrease, and the number of actuators continues to increase, mechatronics is increasingly being viewed as a design methodology for vehicle systems.

Figure 5 shows how digital tools (for the microcontroller) and analog tools (for the drivers of the actuators) are used to model a number of mechatronic system parameters and components. The modeling includes mechanical and thermal analysis. Verifying the proper functionality through simulation can improve reliability and reduce component design time.

HIGHER VOLTAGE FOR FUTURE VEHICLES

With the increasing use of electronic actuators, an increase in power consumption occurs. Today's high-end vehicles may require an alternator that provides 2 kW to maintain the state of charge of the battery. The practical limit of the present 14-V Lundell machine has been projected to be around 3 kW. The power requirement for future vehicles will easily be in the 4 to 5 kW range to power the actuators in systems such as active ride control, electromechanical engine valve actuators, electric water pumps, electric power steering, and electric brakes.

Power-Actuator Interrelation

The new power source for electric and hybrid vehicles in which electricity is part of the energy source is a voltage level in excess of 200 V. Vehicles with fuel cells, high-voltage batteries, and other higher voltage capability will thus comprise a significantly increasing portion of the population of future vehicles. Today's 14-V components are used because they are available and because costly design and tooling changes are avoided. As these systems gain acceptance and demonstrate their benefits/value, new components and system designs will take advantage of higher system supply voltage. This voltage will be greater than 14 V, by some multiple, and less than 200 V.

42-V Power Supply

Global activity currently is underway to ensure that the next voltage used for vehicles is established as a standard before *ad hoc* variations emerge that would cause costly component and systems variations. The MIT (Massachusetts Institute of Technology) Consortium on Advanced Automotive Electrical and Electronic Components and Systems and Forum Bordnetz in Europe consist of automotive industry and supplier representatives actively pursuing standard voltage specifications for future vehicles. An architecture that provides a transition for today's 14-V system to a supply that is three times greater, at 42 V, is gaining acceptance. The proposed architecture for a dual 42 V/14 V vehicle power supply system is shown in Figure 6. Restricting the transients on both the 42 V and the 14 V supplies is essential to achieve lower cost in power electronics.

Once the higher voltage is established, actuators and systems that consume high power can be designed to utilize the higher voltage. New system approaches, such as an integrated alternator, starter, and flywheel, are an indication of how the design of products for future vehicles can be radically different from present methodologies.

SUMMARY

The actuators in today's vehicle transform the performance, passenger compartment environment, and handling of the vehicle to meet legislated requirements and driver expectations. Many actuators are an integral part of electronic controls systems or are electromechanical replacements for earlier mechanical or hydraulic components. To control automotive actuators, several diverse semiconductor processes, products, and packages have been developed. It is projected that these application-specific solutions for power will continue until well into the next century to control the ever-increasing and diversified applications of actuators. Furthermore, the change in the power-supply system of the vehicle to a higher voltage will necessitate continuation of the trend of adding actuators for vehicle control.

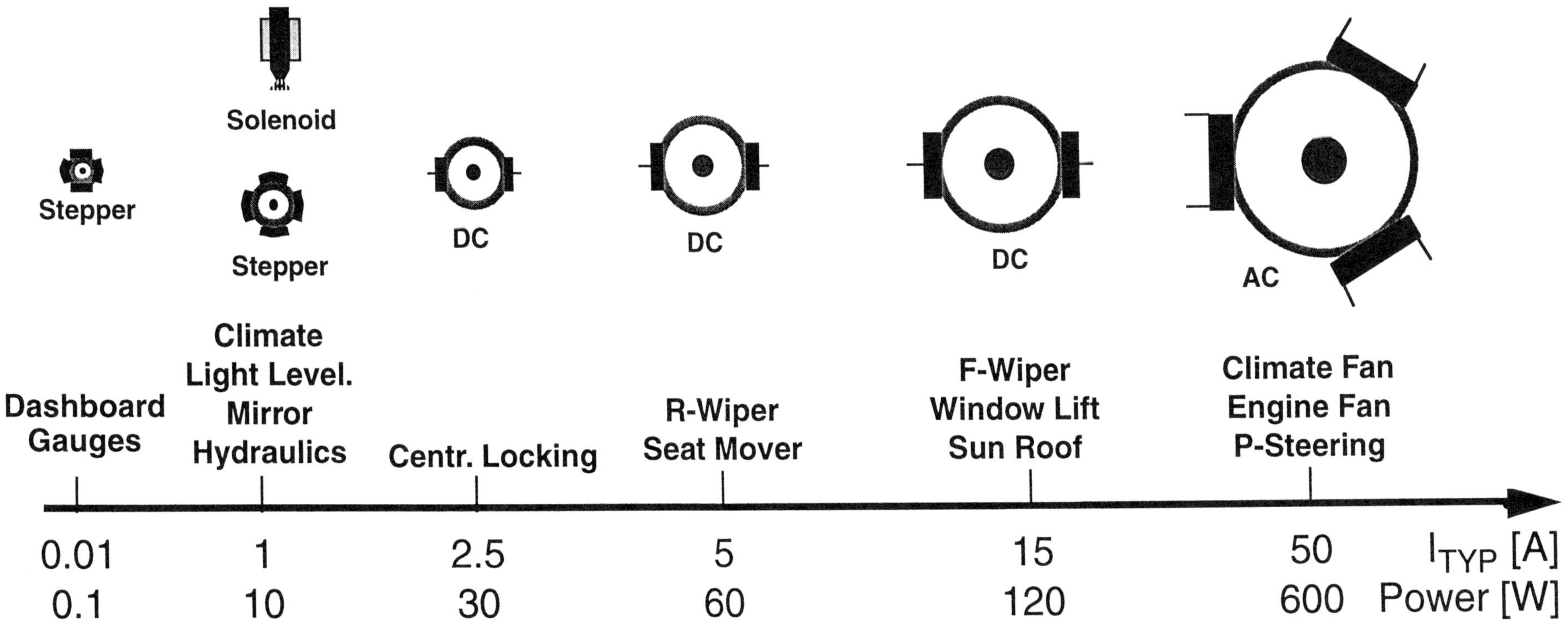

FIGURE 1. Actuator application spectrum.

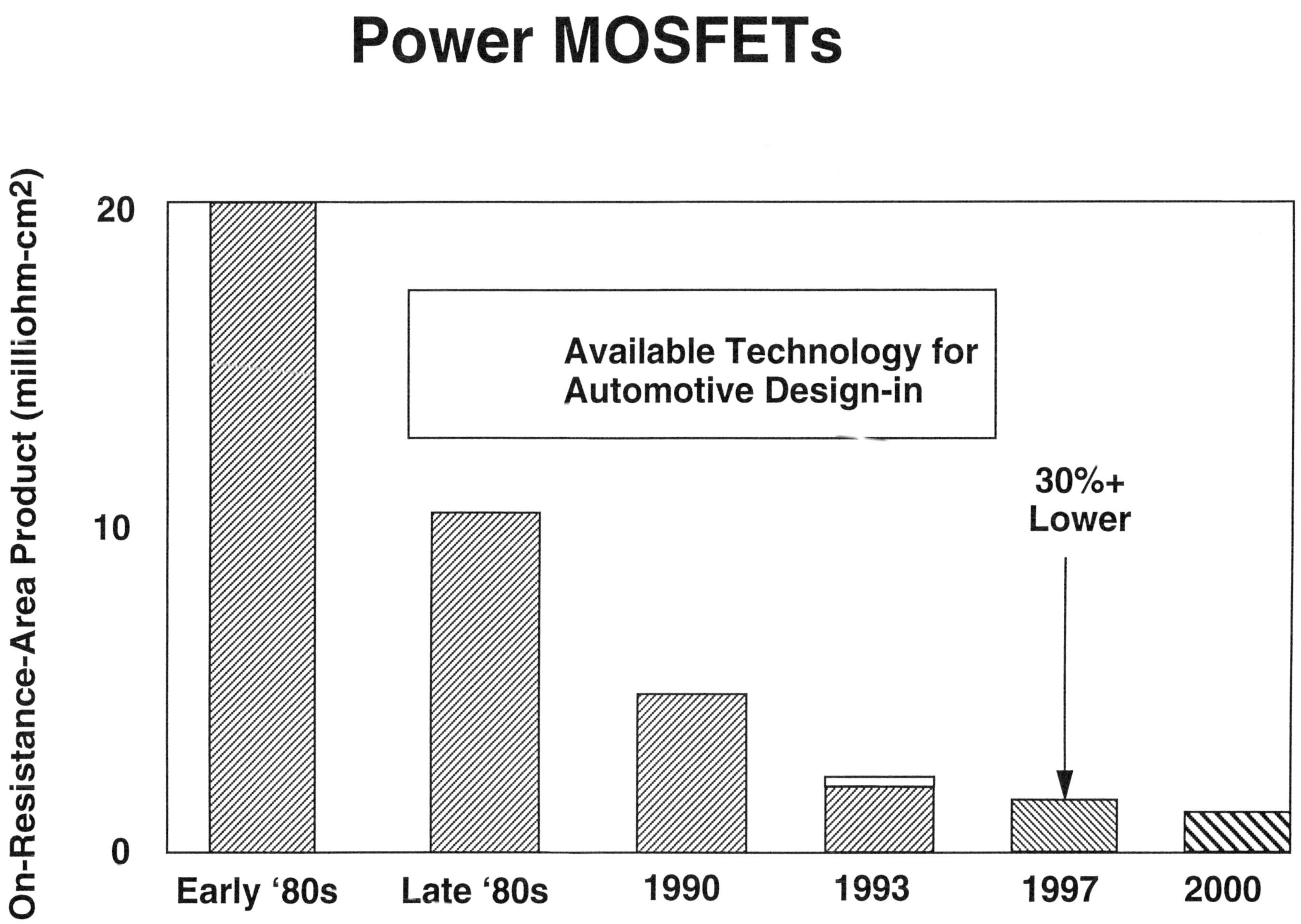

FIGURE 2. Available power MOSFET technology for new designs.

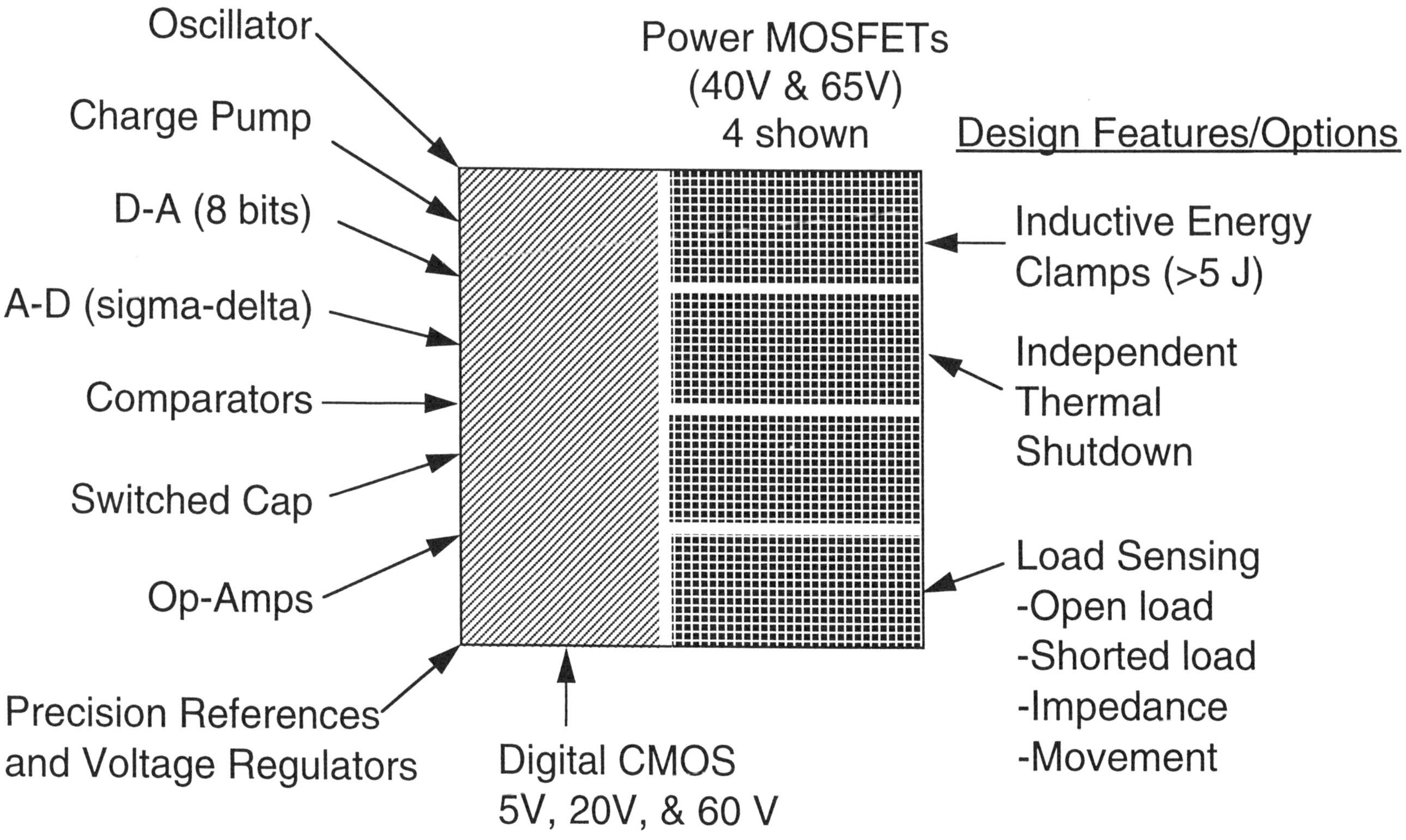

FIGURE 3. Design elements of a SMARTMOS™ power IC.

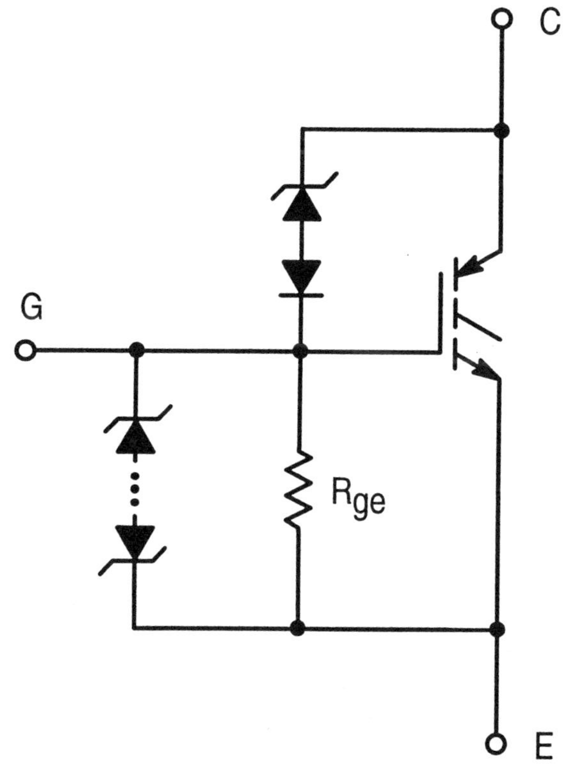

FIGURE 4. IGBT with additional integration for ignition coil driver.

FIGURE 5. Mechatronics tools.

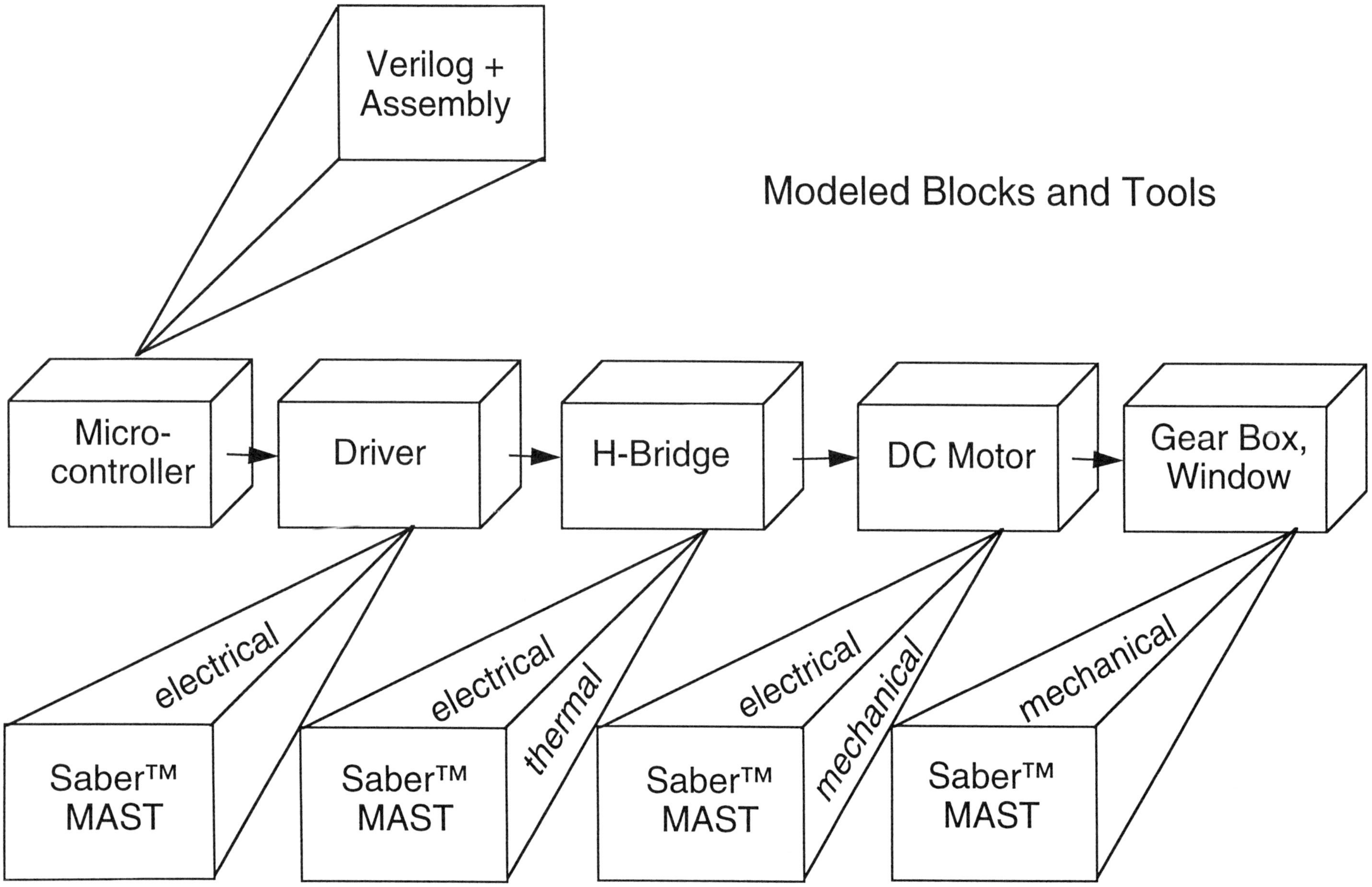

FIGURE 6. Proposed 42 V/14 V electric system.

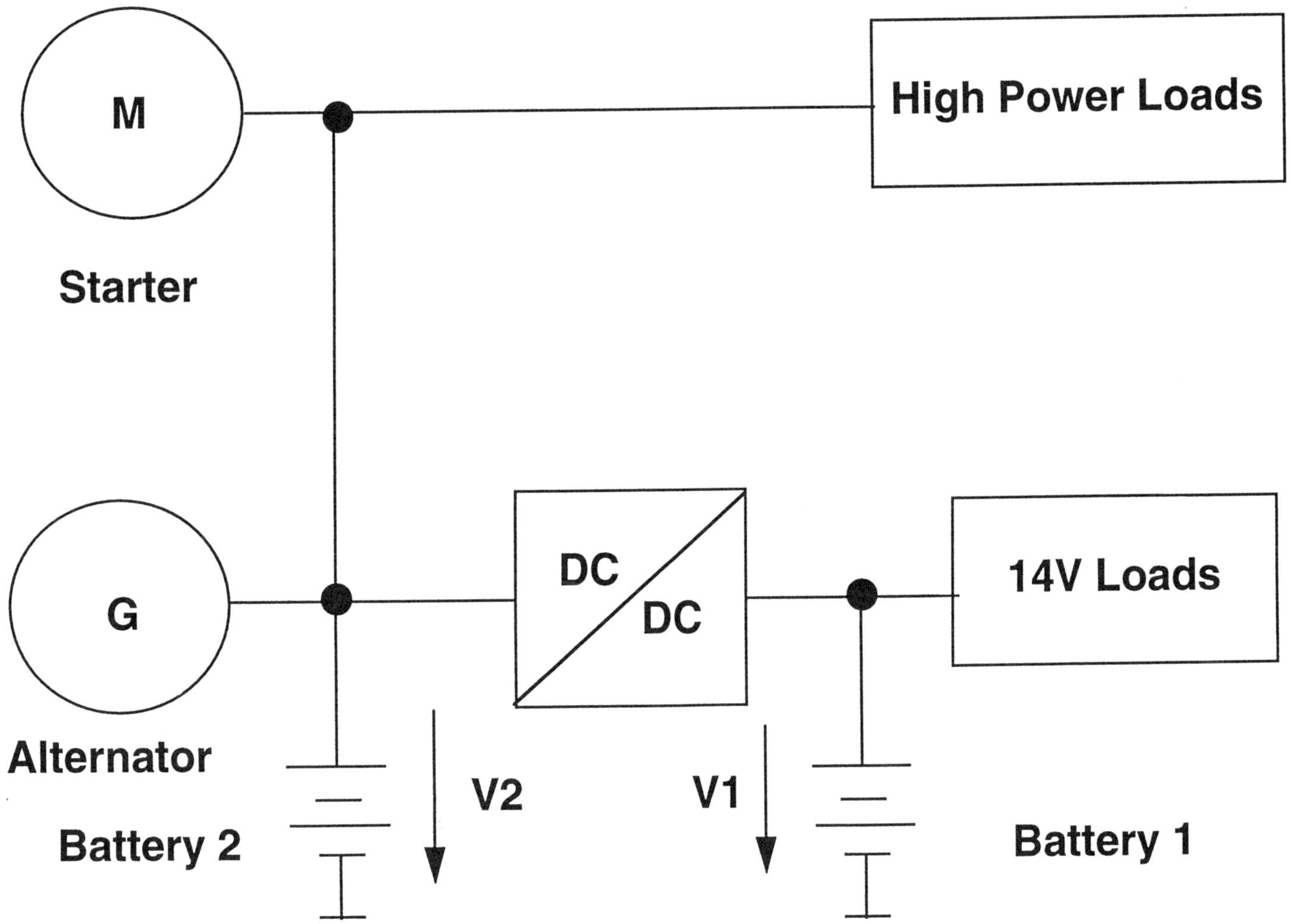